Legal Aspects of Sustainable Development

Volker Mauerhofer
Editor

Legal Aspects of Sustainable Development

Horizontal and Sectorial Policy Issues

Springer

Editor
Volker Mauerhofer
University of Vienna
Vienna
Austria

ISBN 978-3-319-26019-8 ISBN 978-3-319-26021-1 (eBook)
DOI 10.1007/978-3-319-26021-1

Library of Congress Control Number: 2015953273

Springer Cham Heidelberg New York Dordrecht London

Printed on acid-free paper

Springer International Publishing AG Switzerland is part of Springer Science+Business Media (www.springer.com)

Preface

This volume would not have been possible without the work of many. First, to mention, all the helpful support by Springer Publishers, in particular to highlight Fritz Schmuhl, open and supportive to the book idea from the very first proposed structure in August 2014 onwards as well as Devi Ignasy providing quick and precise instructions and responses on all the (more or less) little issues popping up during the implementation of the task. Additionally, two anonymous book reviewers provided several helpful proposals which are hereby especially acknowledged.

However, the main part of the work was done by the contributors who managed to submit their drafts within the—often short—deadlines and to appropriately respond to (perhaps sometimes annoying) requests made by a (perhaps often too) keen editor. All those original contributions constitute the core part of this edited volume, and without any of them, it would not be (hopefully) as helpful and insightful as it is.

The editor would also like to express his gratefulness to those contributors for their dedication as well as efforts who originally committed to submit a chapter but were not able to provide a first draft or—in particular—who were not able to finish a final draft due to time restrictions or other reasons.

The editor would also like to thank all the organizations who provided the setting by means of conferences and other types of meetings which made it possible to bring together in this edited volume such a variety of authors with such a plurality of topics. Alphabetically in the order of their abbreviations, these are the

- the annual European Environmental Law Forum (ELFF) (http://www.elff.info),
- the biannual legal conference track of the European Society for Ecological Economics—ESEE (http://www.euroecolecon.org),
- the annual legal conference tracks of the International Sustainable Development Research Society—ISDRS (http://www.isdrs.org),
- the biannual legal conference track of the International Society for Ecological Economics—ISEE (http://www.ecoeco.org/),

- the Annual Colloquium of the International Union for the Conservation of Nature—IUCN Academy of Environmental Law (http://www.iucnael.org) and
- diverse meetings of the World Commission on Environmental Law (WCL) of the IUCN (http://www.iucn.org/about/union/commissions/cel/).

Finally, the editor takes the freedom to thank, firstly, in the name of all contributors—additionally to acknowledgements at the end of several chapters—the responsible persons within their respective organizational settings who allowed working time to write as well as provided technical, financial and organizational support. Without these essential surroundings, many of the contributions would not have been possible at all or not in this quality and quantity. Secondly, the same freedom is taken to express thankfulness to all the family and other personal surrounding which (at least) respected or even actively supported the dedication of all the contributors to this edited volume. In particular, latter embedding can hardly be overestimated.

Volker Mauerhofer

Contents

Introduction

Volker Mauerhofer

Abstract This chapter provides an overview on the content of this edited volume.

Keywords Sustainable development · Sustainability · Law · Legal · Rule · Legislation · Administration · Court · Enforcement

1 Main Objectives

The edited volume aims to discuss sustainable development and law in a multi-faceted way. It combines a variety of authors, coming from both, the public and the private sector and geographically spreading over five continents. This selection guarantees a broad view that enshrines the more theoretical arguments from the academic side as well as stronger practical applicable perspectives. Thus, this edited volume provides original insights of not only one author but of a geographically and thematically broad team of contributors. Additionally it does—besides horizontal issues—covers several sectorial themes more in detail and does not only focus mainly on one of them. Furthermore, the authors involve very senior as well as younger colleagues. In this way, established scholars holding already professorships as well as early-career academic fellows representing the next scientific generation are included among the authors as well as experienced practitioners and junior professionals. Thus, it belongs also to the aim of the book to provide space

V. Mauerhofer (✉)
Faculty of Life Sciences, University of Vienna, Vienna, Austria
e-mail: volker.mauerhofer@univie.ac.at

V. Mauerhofer
UNU IAS—United Nations University Institute for the Advanced Study of Sustainability, Tokyo, Japan

V. Mauerhofer
PRIMAFF—Policy Research Institute of the Ministry of Agriculture, Forestry and Fishery, Tokyo, Japan

V. Mauerhofer (ed.), *Legal Aspects of Sustainable Development*,
DOI 10.1007/978-3-319-26021-1_1

for thoughtful expansions of established theories as well as the hopeful emergence of innovative ideas. Moreover, the combination of two to four contributions into the ten sub-parts respectively aims toward a compression of likeminded thoughts. This should lead to an intensification of exchange of viewpoints from different angles on a similar theme. Readers have therefore also the opportunity to concentrate on single chapters, but receive comprised knowledge and a variety of thoughts for new ideas on a particular theme.

This book will be launched parallel to the early implementation of the new Sustainable Development Goals adopted by the United Nations General Assembly in September 2015 (UN 2015) and therefore hopefully provide useful input into the discussion and helpful direction for the whole implementation period.

2 Main Arguments

The main arguments within the book are addressing the interrelations and interdependences between sustainable development and law. Law is shown to be an essential precondition to approach solutions towards sustainable development. Furthermore, interrelations of law with other approaches such as the ones purely based on voluntary private activities are discussed. Different situations are analysed towards whether the use of each of these two approaches separately or both approaches together are of advantage. Additionally, it is argued that law is an indispensable basis for the horizontal as well as the sectorial policies respectively. Furthermore, it is advocated that in many situations the horizontal and sectorial policies are necessarily interlinked. Several examples will point out this crucial relationship.

3 Main Themes

The book's main themes are law and sustainable development. Both themes are covered in a very broad sense respectively. Apart of an introduction and conclusions, the book contains twenty-nine contributions. All of them address sustainable development and law together. Thereby, each contribution takes a different geographic and thematic angle. Nevertheless, all the contributions are allocated within one of the two larger parts on horizontal and sectorial policies respectively. "Policies" mainly mean here all kinds of enforceable interactions between the public and the private sector stipulated in and based on law. "Horizontal" means in this connection issues that are not related to a particular part of the natural assets. Within these two larger parts, thematically similar contributions are combined to sub-parts. These sub-parts will respectively enter deeper into the respective theme. Furthermore, they partly indicate cross-connections with other themes of one or both larger parts of the book.

4 Structure

This book is—according to its subtitle—mainly structured into two large parts, a horizontal and a sectorial one. The distinction is artificial and a supporting construct to overcome the shear plurality of subjects, topics and issues covered. Even within each of these two categories, some overlaps are unpreventable such as will be mentioned below.

The part with the horizontal policies has in total five sub-parts (I–V) consisting of fourteen chapters of this edited volume. The second part on sectorial policies contains also five sub-parts (VI–X) with another fifteen chapters.

The following five horizontal policies are covered: (1) general aspects; (2) human and intellectual property rights; (3) communication and social enterprise governance; (4) public participation and (5) assessment tools. While "sectorial" policies mainly address approaches subject to particular environmental assets. Five policies will be also covered in this second larger part, namely: (1) forest and water management; (2) renewable energy; (3) cities, waste and material management; (4) biodiversity, nature conservation, oceans and spatial planning and (5) agriculture and rural policy.

5 Issues Assessed Within the Twenty-Nine Chapters

5.1 Horizontal Policy Issues

The first main part of this edited volume has sub-parts mainly dealing with General Issues (I), Human and Intelectual Property Rights (II), Communication and Social Enterprise Governance (III), Public Participation (IV) and Assessment Tools (V).

In the part with the horizontal policies the first part about general aspects (I) deals in three chapters mainly with the legal circulation history, goals and ways of implementation as well as a regional integration approach of legal sustainability.

Ivano Alogna (Chapter "The Circulation of the Model of Sustainable Development: Tracing the Path in a Comparative Law Perspective") provides an historical and international overview on different types of circulation of the Model of Sustainable Development from a comparative law perspective. He describes the transfer of a rule from a legal system to another or from one person to another as a common practice and distinguishes besides the vertical and horizontal circulation a third type of circulation of legal models, the "oblique" circulation.

Volker Mauerhofer (Chapter "3-D Sustainability and Its Contribution to Governance Assessment in Legal Terms: Examples and Perspectives") then discusses two main objectives of sustainable development law and perspectives *de lege ferenda* (about how law should be newly developed) and *de lege lata* (how existing law should be differently implemented). He provides through '3-D Sustainability' a decision-making concept for the objectives of precautionary trade-offs between the three sustainability dimensions and for maintaining carrying capacities as well as offers proposals for acting *de lege ferenda* and *de lege lata*.

Ferdinand Kerschner and Erika Wagner (Chapter "Sustainability—A Long, Hard Road") afterwards show the legal development of the term ‚Sustainability' during the last decades and some future directions in particular within a situation of regional integration. They discuss the term as one of the most central concepts laid down political strategies as well as in primary environmental law as a basic objective of the European Union and further indicate the implications therefrom on the national level on the example of the legislation of a Member State of the EU.

Human and intellectual property rights such as discussed in the horizontal issues' second sub-part (II) provide an interesting extension of the perspectives on sustainable development law (see also e.g. for an overview Alfredsson 2010, and the recent contributions with Boer 2015).

Rasyikah Md Khalid, Faridah Jalil and Mazlin Bin Mokhtar (Chapter "Environmental Sustainability as a Human Right") take therein a first broad view and discuss Environmental Sustainability as a Human Right. Their discussion looks at sustainability as the capability to ensure that rapid development takes place in harmony with nature in the sense that natural resources such as forest and water resources can be sustained for future users, and focuses on jurisprudence as well as on how environmental sustainability can be regarded as a human right or a right to life under a country's constitution.

The contribution of Vasilka Sancin and Maša Kovič Dine (Chapter "Ensuring Access to Safe Drinking Water as an Imperative of Sustainable Development") within the same sub-part of human rights then focuses more on one important environmental asset when discussing the assurance of access to safe drinking water as an imperative of sustainable development, They analyse based upon different supranational documents whether States are imposed to diligently pursue their efforts of ensuring accessibility to safe drinking water, not only to specifically affected groups of people, but systemically, to the entirety of their populations.

Natalie P. Stoianoff (Chapter "Ensuring Access to Safe Drinking Water as an Imperative of Sustainable Development") deals with a particular type of human rights, namely the rights of indigenous peoples. In her chapter she describes the procedure of developing a legal framework that acknowledges and respects the customary laws and rules of the Indigenous ecological knowledge holders and provides appropriate benefits back to those knowledge holders and that encapsulates principles covered by the *Convention on Biological Diversity*, the *Nagoya Protocol* and the *UN Declaration on the Rights of Indigenous Peoples*.

Communication and Social Enterprise Governance (sub-part III) are other horizontal issues covered in this edited volume.

In this sub-part, Jordi Prades and Aitana de la Varga (Chapter "Framing New Environmental Cultures for Sustainability. Communication and Sensemaking in Three Intractable Multiparty Conflicts in the EbreBiosfera, Spain") assess how communication and sensemaking can contribute to the framing of new environmental cultures for sustainability on the example of three intractable multiparty conflicts in the EbreBiosfera in Spain. They analyse in general the role of communicative processes in environmental conflicts as an engine driving social change to sustainable development and in particular the UNESCO recognition of the Terres de l'Ebre as a Biosphere Reserve (EbreBiosfera) as an alternative proactive, cohesive and consensual frame.

The second Chapter ("Framing New Environmental Cultures for Sustainability. Communication and Sensemaking in Three Intractable Multiparty Conflicts in the EbreBiosfera, Spain") within this sub-part is authored by Aikaterini Argyrou, Tineke Elisabeth Lambooy, Robert Jan Blomme, Henk Kievit, Guus Nieuwenhuijzen Kruseman and Duco Hora Siccama. These authors empirically investigate supportive legal frameworks for social enterprises in Belgium based on a cross-sectoral comparison of case studies for social enterprises from the social housing, finance and energy sector, Their article focuses on the examination of the legal factors of governance as well as the decision-making power of stakeholders in one particular type of social enterprises, the so-called company with a social purpose, 'Vennootschap met Social Oogmerk' (VSO).

The jump from employee participation within social enterprises towards Public Participation in Environmental Matters is not far where latter can be considered to consist—based upon the Rio Declaration from 1992 and the so-called Aarhus Convention of the United Nations Economic Commission for Europe (UNECE) from 1998—of the three pillars access to information, participation in administrative procedures and access to justice (Mauerhofer 2015).

The sub-part IV of this edited volume substantially deals with this horizontal policy issue by means of three chapters.

First, Okubo (Chapter "Judicial Control Over Acts of Administrative Omission: Environmental Rule of Law and Recent Case Law in Japan") introduces the meaning of Public Participation in Environmental Matters and analyses judicial control over acts of administrative omission on the example of the environmental rule of law and recent case law in Japan. She in particular assesses State liability cases and mandamus cases with typical measures for challenging omissions of public authorities and analyses their direct and indirect effects in the field of environmental law where the executive branch is usually given considerable discretion as to how and when to exercise its regulatory power.

Afterwards, Weena Gera (Chapter "Judicial Control Over Acts of Administrative Omission: Environmental Rule of Law and Recent Case Law in Japan") examines the resilience of Public Participation structures for sustainable mining in the Philippines. She in particular analyses—based on the different legislative acts ruling mining in the Philippines—the interplay of institutional-legal architecture, politico-structural conditions and civil society configurations and how it influences the resilience of public participation structures towards sustainable mining in the Philippines.

Finally in this sub-part about Public Participation, Taiwo Odumosu (Chapter "Public Participation and Constitutional Impediments to Sustainable Development in Nigeria") discusses Public Participation and Constitutional Impediments to Sustainable Development in Nigeria. In particular he assesses the Constitution of Nigeria 1999 in terms of its contribution to sustainable society and development. As the imperfections of this constitutions had been the source of a Constitution Review Conference held in 2014, Odumosu analysis whether the amendments proposed there lead to a more sustainable society.

The first main part of this edited volume on horizontal policy ends with a sub-part on "Assessment tools" (sub-part V).

Volker Mauerhofer (Chapter "A Global Conceptual Framework for Categorizing Environmental Change Based on Property Rights and Compensation") starts therein with the presentation of a global conceptual framework for categorizing environmental change based on property rights and compensation. Slightly linked to the previous sub-parts on rights (II) and Public Participation (IV), he in particular focuses on offering a practically applicable system for the assessment of participation of public and private stakeholders in situations of such envisaged changes by distinguishing between governance systems based on command and control as wells as on negotiation.

Next in this sub-part, Sophie Riley (Chapter "Prioritising the Environment in Sustainable Development: Lessons from Australian Environmental Impact Assessment") enters a more concrete assessment tool for projects potentially harmful to the environment. She discusses prioritising the environment in Sustainable Development on the example of lessons from Australian Environmental Impact Assessment (EIA); a tool that—at least in theory—presents decision-makers with a vast array of competing criteria compelling to prioritize and make trade-offs in order to a functionally operative and effective concept of sustainable development.

Finally in the first part of this edited volume and in its last sub-part, Yao-Ming Hsu (Chapter "Reframing Sustainability in Taiwan: Legal Challenges and Opportunities") looks at the different tools and the associated legal challenges and opportunities when reframing sustainability in Taiwan. The policies in Taiwanese programs and action plans, the existing legal frameworks, the approaching establishment of Ministry of Environment and Resources as well as Taiwan's future prospective in international participation are intensively discussed.

5.2 *Sectorial Policy Issues*

The second main part of this edited volume has sub-parts mainly dealing with Forest and Water Management (VI), Renewable Energy (VII), Cities, Waste and Material Management (VIII), Biodiversity, Nature Conservation, Oceans and Spatial Planning (IX) and Agriculture and Rural policy (X).

In the sub-part on Forest and Water Management, Pablo Peña (Chapter "Like a Flap of a Butterfly: Exploring the Effects of Legal Conditions in Costa Rica's Payment for Ecosystem Services Program") explores the effects of legal conditions in Costa Rica's Payment for Ecosystem Services (PES) Program. In doing so, he focuses on laws and policies outside the boundaries of the PES' regulations that shape the way it evolved and functions, such as the forestry regency system, public funds laws, the administrative simplification process across the Costa Rican government as well as foreign policies, in particular by the World Bank.

Next, Henrik Josefsson (Chapter "From River Basins to Landscapes—Holistic Legal Constructs and Their Differentiation") discusses holistic legal constructs and their differentiation on the examples of river basins to landscapes. In particular he explores themes related to the differentiation of spaces of regulation, with regard to the river basin/body of water constructs in the EU Water Framework Directive (WFD) and the landscape construct in the European Landscape Convention. The two spaces of regulation are described and followed by an analysis informed by the sustainability perspective of Elinor Ostrom.

Last but not least in the sub-part on Forest and Water Management, Jonida Abazaj ("Ambitious Goals and Ambiguous Issues: Legal and Policy Challenges for the European Hydropower Sector") discusses, slightly different (to Josefsson in the previous chapter but somehow complementary) legal and policy challenges for the European hydropower sector. Based on secondary data she investigates the conflict occurring between climate change mitigation through hydropower and the protection of good water quality during the implementation of two EU directives: the Renewable Energy Directive (RES) and the WFD.

The next sub-part on renewable energy is partly related to the previous one as water power and wood from forests also constitute potential sources of renewable energy.

Marco Citelli (Chapter "Generating Renewable Energy for the Material Realization of Sustainable Development: What Do We Need from Multilateral Cooperation, the Climate Change and the International Trade Regimes?") investigates certain aspects of multilateral cooperation in the field of renewable energy as well as the role that renewable energy occupies within the climate change regime. Furthermore he addresses the issues that both the generation of energy from renewable sources and the use of renewable energy related technologies are growingly posing to the agents of global trade, with the premise that production of energy from renewable sources is the key driver for the material realization of sustainable development.

Thomas Dromgool and Daniel Ybarra Enguix (Chapter "The Fair and Equitable Treatment Standard and the Revocation of Feed in Tariffs—Foreign Renewable Energy Investments in Crisis-struck Spain") explore a highly topical issue in international investment law, namely the protection of foreign investors' legitimate expectations through the Fair and Equitable Treatment standard (FET) in case of a repeal of renewable energy support schemes. They analyse based of the Spanish case of disruptive cuts, particularly regarding Feed in Tariff regulation supporting photovoltaic energy since 2008, a possible violation of legitimate expectations.

David Grinlinton (Chapter "Horizontal and Vertical Integration of Sustainability into Policymaking, Planning and Implementation of Renewable Energy Projects—The New Zealand Model") addresses the vertical and horizontal integration of sustainability into policy-making, planning and decision-making with a particular focus on renewable energy developments. He thereby focuses on the unique New Zealand approach of incorporating the principle of sustainability as an enforceable concept in domestic legislation which implements a hierarchical model with an environmental sustainability objective at the apex.

The final Chapter "Tackling Climate Change through the Elimination of Trade Barriers for Low-Carbon Goods: Multilateral, Plurilateral and Regional Approaches" to this long sectorial sub-part on Renewable Energy is provided by Christopher Frey. He discusses multilateral, plurilateral and regional approaches for tackling Climate Change through the elimination of trade barriers for low-carbon goods. Frey focuses on efforts to liberalize trade in such goods and on existing trade barriers and the efforts to eliminate these barriers for renewable energy goods and energy efficient products.

The sectorial sub-part on Cities, Waste and Material Management consists of three chapters.

The first Chapter "Environmental Integration in China's Eco-City Development—From An Institutional Perspective" is provided by Ying Yin and deals with the environmental integration in China's Eco-City development taking an institutional perspective. She analyses, from a legal perspective, how environmental integration in sustainable urban planning is achieved or hindered in China's eco-city practice. Her chapter is based on an overview of China's national institutional condition for sustainable urban development and a case study of two Chinese eco-city examples.

The authors of the next Chapter "Identifying the Interaction Between Landfill Taxes and NIMBY. A Simulation for Flanders (Belgium) Using a Dynamic Optimization Model" dealing with waste landfilling and its implications also for tax law are Rob Hoogmartens, Maarten Dubois and Steven Van Passel. In order to inform legislation they are taking primary an economic approach. A dynamic optimization model is constructed to assess the evolution of landfill volumes and landfill prices in time. Carrying out a simulation for Flanders (Belgium), landfill paths and price paths were constructed for two different scenarios, one with legally implied taxes and one without.

The final chapter in this sectorial sub-part on Cities, Waste and Material Management is contributed by Thomas J. de Römph ("Pressing Forward—Developments in the Transition Towards Sustainable Materials Management in EU Environmental Law") who addresses several broader developments in the transition towards Sustainable Materials Management (SMM) in EU environmental law. He briefly explains the legal framework of Sustainable Materials Management in the EU and then sheds light on some important developments in EU law-making and law in the SMM transition, whereupon related opportunities are put forward and illustrated by examples to show the opportunities from a wider perspective.

Within this second main part on sectorial policy issues the next sub-part IX deals with the policy issues related to Biodiversity, Nature Conservation, Oceans and Spatial Planning.

Frederik H. Kistenkas (Chapter "Sustainable Development: New Thoughts, New Policy, New Law?") focuses his assessment on EU nature and planning legislation and aims to fill the gap he identifies between new policy thoughts brought forward such as sustainable growth and Ecosystem Services and their factual implementation by this legislation related to sustainable development. In this way he tries to put some new light on how this could happen regarding the future jurisprudence based on such adapted legislation and draws connections with ongoing EU conservation policy evaluation processes.

Glen Wright, Julien Rochette and Thomas Greiber analyse in the next Chapter "Sustainable Development of the Oceans: Closing the Gaps in the International Legal Framework" how to close the gaps in the international legal framework related to a Sustainable Development of the Oceans. They cover a wide range of international instruments and institutions that aim to regulate maritime activities. Their focus lays in particular on the United Nations Convention on the Law of the Seas (UNCLOS) and the specific situation of the Areas Beyond National Jurisdiction (ABNJ).

Judith Preston (Chapter "Valuing Indigenous Knowledge in Sustainable Resource Management in Australia") looks similar to Stoianoff (Chapter "Ensuring a Sustainable Future through Recognizing and Protecting Indigenous Ecological Knowledge") on Indigenous Knowledge in Sustainable Resource Management in Australia, but more from the angle of nature conservation, biodiversity and protected areas. Her contribution intends to demonstrate how Indigenous Knowledge (IK) can be more effectively incorporated into legal and policy frameworks in Australia for sustainable development. Indigenous Protected Areas (IPAs) are used as case studies to demonstrate how this conservation model can effectively apply IK for sustainable development objectives.

The last sub-part of the main part on sectorial policy issues deals with Agriculture and Rural Policy.

Therein, Marcia Fajardo Cavalcanti de Albuquerque (Chapter "The Sustainable Use of Biodiversity and Its Implications in Agriculture: The Agroforestry Case in the Brazilian Legal Framework") assesses the sustainable use of biodiversity and its implications in agriculture based on the agroforestry case in the Brazilian legal framework. She in particular aims to analyse whether the Brazilian legal framework is well enough structured to implement this system of agro-forestry in all its complexity. More widely, she is seeking solutions for the conflict between traditional farming methods and the biodiversity conservation imperative in Brazil.

Thus, this chapter shows clear connections to other sectorial sub-parts of this edited volume, mainly the previous sub-part (IX) also mainly dedicated to biodiversity as well as the sub-part that includes forest management (VI).

Paul Martin and Jacqueline Williams (Chapter "Next Generation Rural natural Resource Governance: A Careful Diagnosis") provide an intensive diagnosis of the status of rural natural resource governance in Australia and show perspectives for a next generation governance model. They start from the viewpoint that achieving "sustainable development" requires exploitation without diminution of financial capital, manufactured capital, intellectual capital, human capital, social and relationship capital, and (particularly) natural capital.

The fourth and final part of this edited volume on conclusions does not intend to make linkages among all of the chapters but instead provides short summaries of the findings of each of them, shows certain connections among some of them, and makes—where appropriate—brief comparisons within the sub-parts and draw a few own conclusions.

References

Alfredsson G (2010) Chapter 6: human rights and the environment. In: Leary D, Pisupati B (eds) The future of International environmental law. United Nations University Press, Tokyo/New York/Paris

Boer B (ed) (2015) Environmental law dimensions of human rights. Oxford University Press, Oxford

Mauerhofer V (2015). Public participation in environmental matters: compendium, challenges and chances globally, Land Use Policy (in press)

UN (2015) Transforming our World: the 2030 agenda for sustainable development. Finalised text for adoption (1 August). Retrieved August 14, 2015, from https://sustainabledevelopment.un.org/content/documents/7891TRANSFORMING%20OUR%20WORLD.pdf

Part I
Horizontal Policies: General Aspects

The Circulation of the Model of Sustainable Development: Tracing the Path in a Comparative Law Perspective

Ivano Alogna

Abstract The analysis of the diffusion of Sustainable Development at the global level can provide an interesting starting point to see how even one of the most important and universally recognized concepts can give rise to different interpretations and applications. This diffusion is observed here through the mechanism of the circulation of legal models, the cornerstone of comparative legal studies. The circulation of legal models, made famous by Alan Watson with the metaphor of "legal transplant", provides a dynamic approach to the study of comparative law. According to this theory, a transfer of a rule from a legal system to another or from one people to another, not only is not an exception, but also proves to be a common practice since the most ancient of history. Sustainable Development, as a new paradigm adopted at the international level, has shown its dynamics through the vertical and horizontal circulation of its models. Moreover, this contribution will be an opportunity to propose a third type of circulation of legal models: the "oblique" circulation. Thus, the model of Sustainable Development becomes the starting point for the development of regulations based on its principles, but those have different characteristics depending on the context where they are implemented. Therefore, this contribution is an attempt at tracing the path made by Sustainable Development through different stages of its evolution and through various legal systems, trying to shed light on the dynamics of this journey without losing sight of the typical goals of Sustainable Development.

Keywords Circulation of legal models · Diffusion of sustainable development · Horizontal circulation · Vertical circulation · Oblique circulation

---

I. Alogna (✉)
Faculty of Law, University of Paris Panthéon-Assas, Paris, France
e-mail: ivano.alogna@unimi.it

I. Alogna
Doctoral School in Legal Studies, University of Milan, Milan, Italy

I. Alogna
Doctoral School of Private Law, University of Paris Panthéon-Assas, Paris, France

V. Mauerhofer (ed.), *Legal Aspects of Sustainable Development*,
DOI 10.1007/978-3-319-26021-1_2

1 Introduction

"Law is not static. It changes incessantly" (Sacco 1991, p. 390). According to the words of one of the most prominent comparative lawyers, man has always entertained the illusion that he can find a criterion, a legal truth, or principle "for choosing among rules and institutions that is invariable, omni comprehensive and valid everywhere. Reality has so far refuted such illusions, even though this very noble aspiration to find eternal general rules is a powerful stimulus to the improvement of positive law, purging it of irrationality and spurring it on toward higher and higher values" (Sacco 1991, p. 390).

Thus, even in the still young history of environmental law have occurred multiple attempts to provide solutions to the many problems related to the environment that could be forever and universally valid. However, the complexity of the environmental field has quickly made it clear to the various decision-makers that, in front of the utopia derived from the creation of "one-size-fits-all" and eternal principles and rules, stood the economic, social, and political differences of each legal system, in addition to the advancing scientific and technological knowledge and new hazards to the environment.

The analysis of the diffusion of Sustainable Development (*hereinafter* SD) at the global level can provide an interesting starting point to see how even one of the most important and universally recognized concepts can give rise to different interpretations and applications. This diffusion is observed here through the mechanism of the circulation of legal models, the cornerstone of comparative legal studies. According to Alan Watson, one of the most famous scholars of comparative history of law, the circulation of legal models would not only be the object of comparative investigation, but also the orientation criterion of this investigation and its goal. It is therefore considered as the foundation of comparative law (Watson 1977).

The circulation of legal models, made famous by Watson with the metaphor of "legal transplant", provides a dynamic approach to the study of comparative law. According to this theory, a transfer of a rule from a legal system to another or from one people to another, not only is not an exception, but also proves to be a common practice since the most ancient of history (Watson 1974). Indeed, Watson considers borrowing as the most fruitful source of legal change (Watson 1996). On the other hand, Edward M. Wise considers the term "circulation" as 'a more apt metaphor for the phenomenon in question than the term "transplant". The point involves more than terminology: it bears on the perceptions of the kinds of questions it is relevant to ask' (Wise 1990, p. 1). 'It seems less apt to talk in terms of "transplants"; that makes a process almost as natural as breathing sound like major surgery' (p. 12).

In recent years, the attention of scholars has focused in particular on the role that the circulation of environmental law models and concepts has had and continues to have in the development of environmental protection at the global level (Wiener 2001; Yang and Percival 2009; Ruiz Fabri and Gradoni 2009; Morand-Deviller and Bénichot 2010; Alogna 2014).

SD, as a new paradigm adopted at the international level, has shown its dynamics through the vertical and horizontal circulation of its models. Moreover, this contribution will be an opportunity to propose a third type of circulation of legal models: the "oblique" circulation. Thus, the model of SD becomes the starting point for the development of regulations based on its principles, but those have different characteristics depending on the context where they are implemented.

Therefore, this chapter is an attempt at tracing the path made by SD through different stages of its evolution and through various legal systems, trying to shed light on the dynamics of this journey without losing sight of the typical goals of SD. To that end, will be analysed the "model" or "pattern" of SD since its inception (Sect. 2), the types and the reasons for its circulation (Sect. 3), and the dynamics of that circulation around the world (Sect. 4). In the last section, the conclusions will draw up the results of this analysis, allowing the understanding of the goals achieved and the steps still to be done in order to refine this model towards the evolution and betterment of environmental law.

2 The Model of Sustainable Development: History and Conceptualization

When we speak of "model", we refer to the concept accepted within the comparative legal studies of "legal model", that is any "legal object" that could be an example to be copied, an ideal to follow, imitate, or as some commentators would say to borrow or to transplant (from which "legal borrowing" and "legal transplant"). This model may be the object of imitation in the form of a concept, a rule, an institution, a law, or a judiciary decision; though, in the past were even witnessed the imitation of codes and the "reception" of whole areas of law (Sacco 1991).

Even SD, because of its particular origins, its universal character, and its main objectives, takes the form of a legal model, providing an example and finding application in jurisdictions and legal systems other than the one in which it had its origin. Moreover, the very concept of "legal model", thanks to its open and variable boundaries, can provide the right terminology to analyse SD, its unique figure in the history of law that's able to defy all categorization and at the same time to adapt itself to any situation and context.

Accepted everywhere and criticized in several respects, SD is configured as a model formed due to the stratification and the combination of environmental concrete reasons and issues (i.e.: environmental degradation), philosophical and idealistic values (i.e.: the emergence of environmentalism), social (i.e.: the gap of wealth and well-being between the North and South countries), economic (i.e.: limits to growth), and demographic needs (i.e.: overpopulation of the world).

This model is essentially based on the idea that environmental protection cannot be separated from economic and social development problems, depending also on the state of technical and scientific knowledge. Therefore, it is proposed as a "dynamic" model fostering environmental protection through law and finding a balance between opposing and extreme tendencies as, on the one hand, fundamentalist environmentalism and, on the other hand, anthropocentrism, considered until today the dominant paradigm of the relationship between man and nature (Cordini 2007, p. 492).

While the ancestral origins of the model of sustainable development cannot be precisely dated, several authors have traced it back to ancient times, such as Judge Weeramantry who noted that "the concept of reconciling the needs of development with the protection of the environment is...not new. Millennia ago these concerns were noted and their twin demands well reconciled in a manner so meaningful as to carry a message to our age" (Voigt 2009, p. 12). Moreover, already more than 2300 years ago in his *Critias*, the Greek philosopher Plato interrelated the acceleration of the decline of ancient Greece to the deforestation of which he was a witness (Plato IV century B.C. p. 110 d).

However, it would seem that this concept is a particular product of the twentieth century and of its political and economic history—although Philippe Sands dates the idea of *sustainability* at least to 1893, "when the United States asserted a right to ensure the legitimate and proper use of seals and to protect them, for the benefit of mankind, from wanton destruction" (*Pacific Fur Seal* Arbitration, Chap. 10, pp. 415–19; Sands 1995, p. 198). The ideology of "development" as economic growth has been considered a dominant geopolitical imperative since the end of World War II and at the same time the product of the Cold War. In fact, decolonization in Asia in the fifties and in Africa in the sixties posed the "necessity" to provide financial and technical assistance to the new nations that had not yet experienced the industrial revolution, also in order to attract them into the two opposed orbits of influence, by the US and the USSR. However, at the end of the sixties that ideology started to be gradually replaced by a new vision of the world, again by the countries of the North, which relativizes the economic development and stressed the need to respect the limited and not renewable resources of the Planet, considered as one interdependent system (Brunel 2012).

The first steps towards the construction of the model for SD took place just in a particularly iconoclastic year such as 1968, with the questioning of the economic development model *en vogue* in the industrial societies, based on unlimited growth. In fact, in that year UNESCO organised in Paris the first International Conference on the Biosphere. The participants of this conference warned about the irresponsible exploitation of natural ecosystems and, trying to counteract the classic trade-off between environment and development, advanced the idea of an "ecologically sustainable development." Already on this occasion it was discussed the problem of national structures necessary to achieve the objectives of the conference, recognising that a unique formula, which corresponded to the realities existing in each country, could not be recommended, in particular because of the different stages of development. However, the Final Report of the Conference underlined the need to have national laws based on scientific data and invited the developed countries to make available their legislations to inform developing countries (UNESCO 1970, p. 254).

Four years later, the United Nations Conference on the Human Environment (UNCHE), held in Stockholm, laid the basis for the formulation of SD, considering the protection and improvement of the environment as "an imperative goal for mankind" to be followed along with the economic and social development. And for the first time it was declared internationally the fundamental human right "to freedom, equality and adequate conditions of life, in an environment of a quality that permits a life of dignity and well-being" and its "solemn responsibility to protect and improve the environment for present and future generations" (Principle 1). On this occasion appeared the idea of "eco-development", to describe the process of "environmentally sound development" in the sense of a rational development, from an ecological point of view, accompanied by a judicious management of the environment (Prieur 2011, p. 52). Eco-development, indeed, could be considered the predecessor of SD (Ashford and Hall 2011, p. 126).

However, the first official use of the SD expression is contained in the "World Conservation Strategy. Living Resource Conservation for Sustainable Development", a document prepared by two NGOs, the International Union for Conservation of Nature and Natural Resources (IUCN) and the World Wildlife Fund (WWF), together with the United Nations Environmental Programme (UNEP). According to the definition given by the "World Conservation Strategy", SD "must take account of social and ecological factors, as well as economic ones; of the living and non-living resource base; and of the long term as well as the short term advantages and disadvantages of alternative actions." Moreover, this document was a fundamental policy change for the international conservation movement, marking a shift from the traditional focus on a cure rather than prevention and confirming a growing belief that the assimilation of the aims of both conservation and development was the key to a sustainable society (McCormick 1986, p. 178).

In 1982, the "World Charter for Nature", adopted by the United Nation General Assembly, proclaimed five "principles of conservation", of which the fourth proposed that all ecosystems and organisms of the planet "be managed to achieve and maintain optimum sustainable productivity." Unfortunately the latter document, like the Stockholm Declaration ten years before, only succeeded in making the concept of SD "a proposed world 'ethic' that urge[d] nations to simultaneously pursue the perceived competing moral principles of economic/social justice and environmental responsibility" (Hoda 1995, p. 80).

Finally, in 1987 the World Commission on Environment and Development (WCED)—created by the United Nations in 1983—issued the report "Our Common Future", the so called Brundtland report (WCED 1987) where SD was defined as a "development that meets the needs of the present without compromising future generations to meet their own needs. It contains within it two key concepts: the concept of 'needs', in particular the essential needs of the world's poor, to which overriding priority should be given; and the idea of limitations imposed by the state of technology and social organization on the environment's ability to meet present and future needs." This document improved the "ethical idea" of SD derived from past formulations, becoming a "conceptual framework for policy analysis" (Hoda 1995, p. 81), "a broad policy objective, or at least an aspirational goal" (Voigt 2009, p. 15). The Brundtland report was a real breakthrough: for the first time an international commission declared that human activities threatened the world. "The time has come to break out of past patterns. Attempts to maintain social and ecological stability through old approaches to development and environmental protection will increase instability. Security must be sought through change" (WCED 1987 Chap. 12, § 4).

What the Brundtland Commission proposed was a true "paradigm shift": breaking out with past models, it asked to all governmental agencies, international organizations and major private-sector institutions to balance economic growth with environmental protection, making SD an integral part of their mandates. "These [ones] must be made responsible and accountable for ensuring that their policies, programmes, and budgets encourage and support activities that are economically and ecologically sustainable both in the short and longer terms. They must be given a mandate to pursue their traditional goals in such a way that those goals are reinforced by a steady enhancement of the environmental resource base of their own national community and of the small planet we all share" (WCED 1987 Chap. 12, §17).

During the Conference on Ecologically Sustainable Development at Copenhagen in 1991, the United Nations Industrial Development Organization (UNIDO) discussed the priority issue of the relations between industrialized and developing countries, with regard to the application of an environmentally sustainable industrial development. This was really an important topic considering that, from the point of view of industrialized countries, the developing ones represented a threefold risk in the reproduction of the "Western model" of economic growth: an economic competition (for the formers), a growing withdrawal from the resources of the planet (that until then were mostly consumed from the Western world), and an environmental impact exacerbated by rapid population growth in these countries (Brunel 2012, p. 18). The creation of the SD model appeared, therefore, as a necessity and urgent.

On this basis, the United Nations Conference on Environment and Development (UNCED), held in Rio de Janeiro in 1992, officially consecrated SD as the foundation of international cooperation and as "an important global policy that could no longer be ignored" (Voigt 2009, p. 17). The long-awaited "paradigm shift" was taking place thanks to the Rio Declaration (1992) (with its 27 guiding principles), which introduced SD as a "new approach and philosophy" to international relations, and to the Agenda 21 (Agenda for the 21st century), according to which countries should "ensure socially responsible economic development while protecting the resource base and environment for the benefit of future generations" (Agenda 21 1992 § 8.7).

According to Philippe Sands (1995, p. 198), SD is "a general principle" according to which "states should ensure the development and use of their natural resources in a manner which is sustainable". Today, most of the scholarship, various treaties, and judicial decisions have recognized SD as an "emerging principle of customary law" (*ex multis* Voigt 2009; Sands 2003; Hunter et al. 2001; Kiss and Shelton 1994); while, some authors consider it just as a policy objective of international law (Cordonier Segger 2008, p. 117; Lowe 1999, p. 27). This and other disagreements have provided the basis for numerous discussions on the legal status of SD, related both to its vague conceptual boundaries and to its practical applications.

Although SD is widely considered as an elusive concept, its four main components are considered clear: the need to preserve natural resources for the benefit of future generations (intergenerational equity); the 'equitable' use of natural resources, which implies that the use by one state must take into account the needs of the others (intra-generational equity); the aim of exploiting natural resources in a manner which is 'sustainable', or 'prudent', or 'wise', or 'appropriate' (sustainable use); and the need to integrate economic, social and environmental policies (integration) (Sands 1995, p. 199). All these components of the "legal model" of SD give rise to a concept whose objective is even broader than environmental law itself. In fact, to reach its full application it has to be supported in its three typical dimensions or "three E's": Environment, Economy, and Equity.

Some scholars see the development of this new model as an evolution—or a revolution: SD, according to Stéphane Doumbé-Billé, makes the previous law seem old, pushing towards new rules that are more compliant to the evolution in progress. "*Il s'agit là d'une véritable révolution juridique*" (Doumbé-Billé 2007, p. 92)—of environmental law: from a law centred on the protection of the environment to a polycentric SD law, where the centre is to be found in the balance and conciliation between the three pillars of Sustainability ("three E's") and to be reinvented in every public policy or decision, both public or private. The result of such a balance and conciliation is the creation of a new legal field based on a comprehensive and interactive approach concerning the actual complex environmental problems. Thus, this new legal field can actually generate suitable solutions for addressing global and current issues (Meynier 2014, p. 128).

3 Typologies and Reasons for the Circulation of the SD Model

Since the consecration of SD as a model at UNCED in Rio de Janeiro, it has been more than twenty years and meanwhile the circulation of the concept and its components worldwide have occurred. This circulation, as expected, was mainly "vertical", resulting in national applications of the principles of SD in the various policies and regulations, and within the judicial decisions. Even from a chronological point of view this "vertical" type has been the main one, precisely because of the formation of such a model at the international level, as explained above.

However, there has also been a "horizontal" circulation—between national systems, that is the "classic" type of circulation of the legal models or "legal transplants", widely discussed in the literature of comparative law, especially in the field of private law (Watson 1974; Sacco 1991; Ajani 2007; Mattei 2008; Siems

2014)—in which the "importer" country imitated the application of this model in another legal system, more rapid in its adoption and implementation.

Finally, we propose here a third type of circulation, a *tertium genus* between the horizontal and the vertical one: the "oblique circulation". This is the circulation of a legal model derived from the "conditional" imposition by international actors such as the World Bank, the International Monetary Fund (*hereinafter* IMF), or other regional and international organizations.

The main difference between these three types of circulation can be found in their dynamics: in the horizontal one the legal systems, between which the imitation or borrowing of the model occurs, are hierarchically on the same level (according to international law) as two sovereign countries; in the vertical one the model circulates between a higher-level legal system, as international law or European Community law, and a system subordinate to it, as a national legal system, or vice versa (also called "trans-echelon" borrowing, see Wiener 2001, p. 1301); finally, will be called oblique the circulation of a model between an international organization and a national system, without any hierarchical relationship between the former and the latter that can legitimate the imposition of a rule of law, like in the relation between the World Bank and several transitions or developing countries.

It is necessary, however, to point out that on this last particular profile a rich literature already exists in the comparative legal field, which however did not consider this kind of experience as a type of circulation but simply as a "cause" of imitation, halfway between the "prestige" and the military or colonial imposition (Mattei 2008, p. 180). These are considered by most of the comparative law doctrines as the two fundamental causes of imitation. The prestige is considered the most common cause, such as a desire to appropriate the solutions of others because they are considered full of such quality that the doctrine has never been able to clearly define. Sacco believes that "[t]he analysis of this term is, if anything, the province of other disciplines" (Sacco 1991, p. 398). On the other hand, the military conquests and colonization of submitted peoples constitute the second fundamental cause of circulation of legal models: thanks to military force, the diffusion of the law of the most powerful nations takes place. However, as pointed out by the same author, "[r]eceptions due to pure force (…) are reversible and end when the force is removed" (p. 398).

As a matter of fact, the so-called "conditionality" or "conditionality clause", mechanism behind the oblique circulation in the form of numerous legal reforms in the developing countries, has been analysed extensively by the comparative doctrine and not without criticisms. According to some scholars, this paradigm hides, behind the formalistic concept of "conditionality", interests of political and economic opportunity (Ajani 1995, p. 115). Furthermore, this mechanism aims to spread, instead of legal models functional to the development of an economy controlled by the public sector, models to promote the free market. In fact, some political analysts express radical criticism towards institutions such as the IMF and the World Bank, which in their opinion would be "the neo-colonial continuation of Western tutelage" and "thus responsible for the lack of congruence, legitimacy and functionality of modern transplants in developing countries" (De Jong and Stoter 2009, p. 317).

This mechanism provides a system of benefits such as loans, debt relief or bilateral aid by international financial institutions like the World Bank and the IMF, which grant such benefits "on condition" that the recipient country changes certain aspects of its laws or legal institutions in accordance with Western legal models, in order to enhance aid effectiveness. His "obliquity" is due, therefore, both to the structural element of the relation between the international organization and the importing country (neither vertical nor horizontal in its nature), and to the source of the circulating model, legislation or institution, often coming from the United States, which is "transplanted" with a "distorted" *top-down* modality. The reasons of this "americanization" are mainly of two types: first, the importance of the development of the US environmental standards, as well stressed by Sands: "In many respects the United States is rightly considered to have the most highly developed rules of environmental protection of any nation, and is widely recognized as having played the primary role in establishing and developing that branch of international law now known as international environmental law" (Sands 1994, p. 323); secondly, the US role and great influence within these international institutions, such as the World Bank and the IMF, whose voting power is based on a quota system, linked to financial contributions from member governments (US has a percentage of the total number of votes equal to 16.28 % for the World Bank and to 16.75 % for the IMF).

Precisely such a "top-down" approach seems to cause most of the criticisms, being seen as an "undue interference in the national sovereignty and democratic accountability of countries in the developing world" (Siems 2014, p. 277). Several authors have pointed out the bond of this type of circulation to the evergreen movement called "law and development". For instance, the words of John Merryman explain effectively the concept behind the movement born in the United States after World War II: "Development is a euphemism for Progress, and the work of law and development is to lead the way to Progress through law reform" (Merryman 1977, p. 463). Thus, oblique circulation appears as an attempt to impose Western or global standards on developing countries, pursuing instances of post-colonialist and neo-imperialist type whose end is nothing more than the creation of an ideology functional to the exploitation of the resources of developing countries (Mattei and Nader 2008).

Certainly, it is an indicative fact that in recent years this oblique circulation of legal models, through the means of, or directed by supranational institutions have largely replaced the one between individual states (horizontal), which formed the common practice until the first half of the last century. These institutions now support intergovernmental agreements aimed at encouraging reforms and introducing soft law rules, proposals, and recommendations for legal changes. Today, the "prestige" is, therefore, an increasingly insufficient reason to cause a circulation of legal models.

However, apart from these general reasons behind the circulation of legal models, there are additional ones particularly related to the concept of SD. For example, the Rio Declaration provided for cooperation between States with the objective of strengthening "endogenous capacity-building for sustainable development" (Rio Declaration 1992 § 9). The exchange and therefore the circulation of solutions seem to be essential in all areas of science and technology, precisely in order to improve the understanding of environmental issues. Furthermore, the development, but also the adaptation, diffusion, and transfer of technologies are the basis of SD, as a goal to achieve nationally and locally, yet always in a global collaboration, as required by Principle 27 "...in the fulfilment of the principles embodied in this Declaration and in the further development of international law in the field of sustainable development."

Agenda 21 also recognizes the need for the circulation of the model of SD, through what it calls "a new global partnership", which "commits all States to engage in a continuous and constructive dialogue inspired by the need to achieve a more efficient and equitable world economy, keeping in view the increasing interdependence of the community of nations and that sustainable development should become a priority item on the agenda of the international community" (1992 § 2.1). In addition, even the oblique and horizontal dimension of such circulation are recognized, with the identification of its main actors: "The responsibility for bringing about changes lies with Governments (…) in collaboration with national, regional and international organizations, including in particular UNEP, UNDP and the World Bank. Exchange of experience between countries can also be significant. National plans, goals and objectives, national rules, regulations and law, and the specific situation in which different countries are placed are the overall framework in which such integration takes place" (§ 8.2).

Moreover, in order to ensure the effective follow-up of the Conference, Agenda 21 created the Commission on Sustainable Development (replaced in September 2013 by the United Nations High-level Political Forum on Sustainable Development), which represents a fertile ground for the circulation of the model of SD, through its mechanism of reviewing the different countries' progress in achieving the specific goals of Agenda 21. In fact, understanding how other countries have implemented the model of SD may provide an occasion to imitate foreign experiences (Dernbach 1998, p. 281).

However, the Rio Declaration warned against a non-judicious use of the mechanism of legal transplant. In fact, Principle 11 argued the importance of an effective environmental legislation, able to take into account the developmental and environmental context in which it is applied. Indeed, not considering the characteristics of the system in which a particular model of SD has to be implemented, could result in "inappropriate and of unwarranted economic and social cost" to the importing country, particularly in the case of developing countries. The environmental "compatibility" of the model of SD, with respect to the context of application or reception thereof, becomes, therefore, a prerequisite for its effective circulation. This compatibility is the basis of what some comparatist scholars identified as "transplant effect", that is the weak effectiveness of the legal transplant when is not adapted to local conditions, or it is imposed by top-down or conditional mechanisms, or when the population was not familiar with the law (Berkowitz et al.

2003, p. 168). Moreover, according to this scholarship, "only if demand for law is high, will there be high voluntary compliance and will a society invest in the legal institutions necessary for upholding the legal order".

This is one of the most important problems encountered in the circulation of the model of SD, which has two dimensions: on the one hand, the transfer of inappropriate and unsuitable approaches with regard to the local conditions of the importing countries; and on the other, the desire to universalise the solutions to implement SD (Le Prestre 2005, p. 271). In fact, such a model, despite its claimed neutrality and universality in achieving the objectives of environmental protection, can be experienced by developing countries as an interference with their right to development, with the reduction of the available resources in their territory for the sake of a global environmental protection.

In this perspective, the conditionality clause—or in this case the "eco-conditionality" clause.[1] Even "Our Common Future" emphasized that ≪[t]he World Bank has taken a significant lead in reorienting its lending programmes to a much higher sensitivity to environmental concerns and to support for sustainable development≫ and how much ≪[i]t is therefore essential that the IMF, too, incorporate sustainable development objectives and criteria into its policies and programmes≫ (WCED 1987 § 6.2.1. 103–104)—being a kind of diplomatic pressure and in contradiction with the customary principle of non-intervention regulated by the Convention of 26 December 1933 on the Rights and Duties of States, faces very often a social rejection (Borràs Pentinat 2006, p. 400). In fact, the basis for this refusal is the understanding by the loan-applicant countries that such eco-conditionality, just as the financial and more traditional one, is set indirectly by the hegemonic states, which control—economically and in proportion to their participation—the institutions that manage the operations. For this reason there can be a strong social rejection in the importing countries towards the model in question.

To complete the analysis of the reasons for the circulation of such a legal model, we can refer to the classification made by an experienced doctrine in the matters of constitutional transplants and borrowings. The reasons behind this type of circulations, in fact, can be appropriate for the case of the model of SD: functionalist ones (so-called "cost-saving" imitation, to avoid reinventing the wheel); reputational ones (or "legitimacy generating" effects; i.e.: to signal to the world community the breaking with a non-sustainable past); normative universalist reasons (i.e.: the recognition of SD as a universal set of principles); sociological ones (driven by an economic or political elite to promote their own political interest); and "chance" circulations: those that lack any of the precedent reasons, purely determined by chance (Perju 2012, p. 1318).

Finally, it should be emphasized that all these reasons are acting in connection with a particular feature of the model of SD, namely the vagueness in its contours

[1]This means that the projects of the countries applying for funding are subjected to a rigorous analysis to ensure that they don't deteriorate the environment.

and in its application. Then, this feature brings SD within the category of "vague formulas", the importance of which was underscored by the comparative literature with regard to the process of legal transplantation. As a matter of fact, terms such as "due process", "governance", "reasonableness", "rule of law", "transparency", and "accountability" are fundamental "picklocks" to direct the attention of the importer system on a concept that has already demonstrated its experience as "constitutional standard" in other hegemonic jurisdictions. Thereafter, it is just necessary to carry or transplant the operating rules for the adoption of this already accepted concept (Ajani 2007, p. 5).

4 The Dynamics of the Circulation of the SD Model

The circulation of the model of SD is realised through its three dimensions or types: vertical, horizontal, and oblique. However, it must be emphasized that the three dimensions, although they may be considered independently of one another, are strongly interconnected. In fact, the diffusion of the model in question emerges in the global legal landscape as influenced jointly by the circulation of the declarations and international agreements, supported by the example of the countries' most sustainably efficient (potential exporters of the model) and conditioned, in certain contexts, by the pressure of the international and regional financial organizations. The importance of "bottom-up" approaches in applying that model can also not be forgotten: the role of NGOs and associations in defence of the environment and social rights; national and multi-national businesses with their adoption of virtuous behaviour; and the public as green consumers and citizens taking responsibility for everyday actions that can make SD a reality at the local level.

The dynamics of the circulation of the model of sustainable development could be described through a number of examples in order to show their breadth in the global diffusion. However, in this section we will focus on one type of comprehensive and multi-dimensional instrument, fundamental in the circulation of the model under consideration: the national sustainable development strategies and policy plans.

In the case of the vertical circulation of the SD model, the influence of international law, along with regional supranational law, such as the Community law,[2] are main factors of this "circulatory" type. In particular Agenda 21, in its Article 8.13, provided the boost needed to allow the circulation of this model, calling "on governments to adopt and implement law and policies that successfully guide both private and governmental decisions for sustainable development, and to regularly assess and modify them when appropriate to improve their effectiveness"

[2]It should be remembered that the Maastricht Treaty in its Article 2 has assigned to the European Community the task of promoting a harmonious and balanced development of economic activities, through a sustainable growth that respects the environment.

(Dernbach 1998, p. 29). However, the vertical circulation of the SD model is also working through the case law of the International Court of Justice (ICJ), as observed in the case concerning the Gabcikovo-Nagymaros Project, Hungary/Slovakia (ICJ 1997 140; See also Akhtarkhavari 2010, p. 132). Moreover, the WTO Appellate Body (1998) in the Shrimp Turtle Case II also argued SD in the context of article XX (g) of the 1947 General Agreement on Tariffs and Trade (GATT).

As it generally happens as a result of international conventions or regional forums, the first countries able to reproduce and implement policies, standards, or the principles of law laid down therein, are considered exemplary actors of the international community. The chronological criterion in the vertical circulation is usually an interesting index to find out who are the leaders and who are the laggards in a particular field or sector, and this is also true for SD.

We can take into consideration the paradigmatic example of the National Sustainable Development Strategy (NSDS), which in the aftermath of the Rio Conference 1992 was the basis for the application of the SD model globally. Article 37.4 of Agenda 21 provided that "[e]ach country should aim to complete, as soon as practicable, if possible by 1994, a review of capacity- and capability-building requirements for devising national sustainable development strategies, including those for generating and implementing its own Agenda 21 action programme." NSDS constituted the first attempt to achieve better coordination and integration of the SD model at the national level through four (sub-) dimensions: "horizontally (across policy sectors), vertically (across political-administrative levels as well as territorially), temporally (across time), and across societal sectors (public, private, academia, civil society)" (Pisano et al. 2013, p. 6). Thus, in the nineties several European countries emerged as leaders in the application of their NSDSs: Sweden and the United Kingdom in 1994; Switzerland in 1997; Finland in 1998 (Pisano et al. 2013, p. 9).

However, the spread of such a single document that incorporated the economic, social, and environmental dimensions of SD took place, quickly as well, even in systems such as China (1993), Philippines (1996), and South Korea (1996) (Swanson and Pintér 2004, p. 9), where the environmental and social protection were coordinating with economic growth in an initially only theoretical framework of SD. Actually, the socio-economic reality of some developing countries could not permit fully realizing this model unless over a period of time longer than that of the European countries, and with different modalities. Indeed, some commentators argued that many countries in the developing world, while acknowledging verbatim the notion of SD in their respective constitutional and statutory texts, in practice showed a dichotomy between legal rules and effectiveness (Cordini 2007, p. 498).

Moreover, these differences in the applications of the model of sustainable development emerged clearly during the United Nations Conference on Sustainable Development (Rio de Janeiro, from 20 to 22 June 2012), which ended with the adoption of Resolution 66/288 by the General Assembly of the United Nations. Indeed, the final outcome of the Rio+20 Conference took into consideration an important fact: "there are different approaches, visions, models and tools available to each country, in accordance with its national circumstances and priorities, to achieve sustainable development" (UN 2012 § 56).

It should be emphasized that environmental standards, although technical and seemingly free from the resistance found in the transplant of socio-cultural rules, such as those of family law (linked more deeply to their legal-cultural and traditional substratum), also need a political and social acceptance. Indeed, an eminent comparative lawyer has highlighted that the imitation of a rule depends on the circulation of the related political idea (Sacco 1990, p. 151). Thus, the prestige of the origin system is not enough, especially in the case that the legal model expresses immediately a political choice, values, and ideals (Pegoraro and Rinella 2007, p. 97).

This is clearly the case of the horizontal circulation of SD, which as a constitutional model is subject to the phenomenon of imitation-reception between different national legal systems, like what usually happens for the private law models. This circulation generally requires a comparative analysis on the subject (a single rule, a principle, a legal instrument, or even an entire code) by the institutions of the borrower country. In fact, one of the fundamental functions of comparative law is considered to provide materials to aid in the preparation of legal texts. In the process of drafting a new constitution or in its review, usually the specific national organs perform comparisons between the solutions tested elsewhere, and between them and their own frame of reference in terms of values and fundamental political choices. Moreover, we can differentiate between "legal imitations", when the legislator imitates the model produced by another legislator; "scholarly imitations", which operate on a theoretical level; "judicial imitations", the so-called "dialogue between judges" (Pegoraro and Rinella 2007, p. 91). However, a "constitutional model" can affect the ordering of other legal systems, yet not being exactly reproduced as under the effect of a cloning.

An example of such circulation is made by India, which in its Constitution contemplates the two fundamental and conflicting aspects of SD: the right to development and the right to a clean and healthy environment, both considered as necessary and complementary parts of the "right to life" under Article 21 (Sahasranaman 2012, p. 25). In 2006, in a legislative effort directed toward the implementation of SD, India has enacted the National Environmental Policy (NEP) as a guide to regulatory reform, programmes, and projects for environmental conservation. Moreover, the NEP reviews the enactment of legislation by agencies of the central, state, and local government and ensures that the principal objectives correspond with the main elements of SD: conservation of critical environmental resources and efficiency in their use, intragenerational and intergenerational equity, and integration of environmental concerns in economic and social development.

In this case, the model circulated was a German one: the 2002 National Sustainability Strategy (NSS), considered one of the most successful in Europe (Pisano et al. 2013, p. 10), fundamental for drafting the Indian NEP (Alogna 2014, p. 63). As a matter of fact, also the German NSS stressed that "Germany is inseparably linked to the world. It follows from this that there can no longer be local or national island of prosperity and security in the long term. (…) On the other hand, it is precisely the industrial nations that can prove with a strategy for sustainable development that it is also possible to link this with successful economic development. This also offers perspectives for developing countries. A national strategy bringing together economic, ecological and social dimensions in an integrated vision, and succeeding in practice, would also exercise great appeal internationally" (German Government 2002, p. 3).

Another model, even precedent to the Rio Conference in 1992, is of Dutch origin. As pointed out by part of the scholarship, "influenced by domestic environmental pressures as well as the UN-backed concept of sustainable development", the Dutch National Environmental Policy Plan (NEPP) of 1989 received "considerable attention outside the Netherlands" and, therefore, it became an important model of SD (Jörgens 2003, p. 15). The NEPP represented a technocratic vision of SD, with the main objective to reduce the environmental impact, rather than promote social change. Among the main followers of this model there were the European Commission, with the European Union's Fifth Environmental Action Programme of 1992 entitled "Towards an Environmentally Sustainable Development" (p. 16), and some European countries such as Portugal and Latvia who adopted national environmental policy plans, both in 1995 (p. 17).

Even part of the literature on "policy diffusion" took into account the phenomenon of the circulation of models, focusing on the global convergence of environmental policies (Tews 2011; Busch and Jörgens 2012). This convergence would be explained precisely by the international diffusion of ideas, approaches, institutions, and instruments in the field of environmental protection. Although the concept of SD has become successfully institutionalized at the international level, especially thanks to the international conferences organized by the United Nations, its effective implementation at the level of the nation state remains the final goal. This is why the vertical circulation should be considered along with the horizontal one: on the occasion of certain points in time, like in 1972 (UNCHE, Stockholm) or 1992 (UNCED, Rio de Janeiro), corresponding to a high level of international (vertical) communication on environmental issues, the speed of (horizontal) diffusion of models was higher than in other periods, thanks also to a direct dissemination of information about these models (Tews 2011, p. 231).

Then, the information among states, with the creation of highly specialised communication networks or transnational communication channels, can play a key role in the horizontal circulation of the SD model. Indeed, some authors point out an interesting example related to the successful circulation of sustainable development strategies in the early nineties, such as "the important role of the International Network of Green Planners, an issue-specific network that was created with the explicit aim of disseminating the idea of green planning" (Busch and Jörgens 2012, p. 238). Moreover, it was also observed that governments generally orient their choices regarding environmental models toward those that have already been put into practice in other countries (Tews et al. 2002, p. 8), considered that "states are more willing to comply with international rules if they can be sure that other states do the same and that free-riding is discouraged" (Busch and Jörgens 2012, p. 241).

Regarding the oblique dimension of the circulation of the SD model, as seen in the previous section, Agenda 21 (1992 § 8.2) considered the international financial organizations such as the World Bank and IMF among the main actors of the necessary change to convey such circulation. Many scholars regard this type of circulation as a form of imposition—for example, an expert scholarship in environmental policy diffusion denominates this type of circulation "coercive policy transfer" or "domination" (Tews 2011, p. 229)—for implementing SD at the national level in developing and transitional countries.

In fact, the eco-conditionality behind this circulation and under the auspices of international organizations consisted in the pressure toward developing countries to prepare and implement SD strategies. And this, despite the Multilateral Financial Institutions (MFIs), namely the World Bank, IMF, and other international funding organizations, were seemingly prohibited to make provisions for non-economic factors, such as environmental protection. Indeed, in the Articles of Agreement of the World Bank, for example, is written: "The Bank and its officers shall not interfere in the political affairs of any member; nor shall they be influenced in their decisions by the political character of the members concerned. Only economic considerations shall be relevant to their decisions, and these considerations shall be weighed impartially to achieve the purpose [of the World Bank]" (World Bank 1989 Art. IV 10).

However, according to some scholars, SD "represents a bundle of interlocking concepts of very broad environmental, socioeconomic, legal and institutional implication." Organizations such as the World Bank and the IMF "find themselves thrust into the center of a process in which the formal requirement of abstention from "political decision making" might seem to put them at odds with the normative implications of 'sustainable development'" (Handl 1998, p. 644).

Contrary to the horizontal circulation, based on a voluntary mechanism of imitation, the oblique one depends primarily on asymmetric power relationships, from which it takes its connotation of "imposition" (Jörgens 2003, p. 21). Moreover, the experience with the oblique spread of SD strategies in Eastern Europe and in the developing countries shows that "national capacities are a decisive constraint for the domestic implementation" (Jörgens 2003, p. 26) of the SD model. Furthermore, such imposition is a major restriction of the options available to the developing countries to implement such a model. This prevents a more virtuous circulation of legal models, which could be more effective or more problem-adequate than those "imposed" (Jörgens 2003, p. 26).

The most important actor in this process of oblique circulation has been the World Bank, that already in 1987 started to support National Environmental Action Plans (NEAPs) in Madagascar, Lesotho, Mauritius, and the Seychelles (Jörgens 2003, p. 22). But NEAPs did not remain confined within the developing countries: even in Central and Eastern Europe and in the New Independent States (NIS) which emerged from the former Soviet Union, the international organizations (besides the World Bank, also the UN Economic Commission for Europe and the OECD) conveyed such plans. Thus, during the nineties, 16 out of 18 Central and Eastern

European countries adopted a NAEP, followed by the New Independent States, "mainly due to World Bank support" (Jörgens 2003, p. 23).

Finally, it is legitimate to wonder whether the model conveyed by such international organizations is appropriate for the characteristics of the "conditioned" countries. On the one hand, Agenda 21 clearly foresaw that "[l]aws and regulations suited to country-specific conditions are among the most important instruments for transforming environment and development policies into action" (§ 8.13); on the other hand, as a scholar points out, the World Bank's "internal working definition of sustainable development appears to be modulated to suit, first and foremost, the economic proprieties of the Bank as a lending institution, without satisfactorily integrating environmental and human rights concerns" (Civic 1998, p. 241). In fact, the World Bank takes into consideration four categories of capital as interchangeable elements in a sustainability policy: human-made capital (machines, factories, buildings, and infrastructure), natural capital (natural resources), human capital (investment in education, health, and nutrition of individuals), and social capital (the institutional and cultural bases for a society to function). This interchangeability between the different capitals can endanger the established concept and therefore the legal model of SD, which can be transplanted in different developing countries as "a simple trade-off of environmental destruction for the sake of economic development" (Civic 1998, p. 242; see also World Bank 2006; Markandya and Pedroso-Galinato 2009).

5 Conclusions

The analysis of the path traced by the model of SD in its formation and then in its application through the case of national sustainable development strategies has highlighted the role played by the circulation of legal models.

Such circulation, which is the main modality of legal innovation in the various legal systems, showed the positive effect of globalization on environmental protection: a gradual spread to the four corners of the earth of a fundamental legal model with constitutional relevance, such as SD. The merit of this spread is certainly due to the reflections and the efforts made at the international level, which led to the creation of such a model. This model has certainly been an indefinite legal and political creature yet it has shown its effectiveness in producing the paradigm shift from a mere theoretical development to a real one that takes into account the needs of everybody, globally and in a synchronic and diachronic perspective.

Moreover, this contribution wanted to show the strong interconnection and interconnectivity of the different global legal systems, either domestic or international. The fact that countries communicate with each other and share their perceptions and solutions on environmental issues, both in free form as actors of a horizontal circulation between equals, both within appropriate institutions such as the Commission on Sustainable Development and The High-level Political Forum on sustainable development [created at the United Nations Conference on

Sustainable Development (Rio+20) to replace the former], represents a factor of evolution for the model of SD, in particular, and for the protection of the environment and human rights, in general.

An accurate doctrine points out that the different national governments in Europe look at each other because they want to avoid the impression of "falling behind the others or because they seek to draw lessons from successful policies developed elsewhere" (Holzinger 2008, p. 230). However, such behaviour shows a widespread trend, not limited to Europe, as seen in the example of the circulation of the national strategy for the sustainable development model from Germany to India. Thus, it represents a positive trend for which environmental leaders are able to pull along the laggards. And this is also true with regard to environmental standard-setting through international harmonisation, even in the absence of legally binding agreements, like in the case of SD. Therefore, even in the vertical circulation the example of the leaders is able "to set the pace in international environmental harmonisation" (Holzinger 2008, p. 230), towards what can be called "a sustainable race to the top".

Finally, the proposition to introduce a third type of circulation of legal models with the denomination of "oblique circulation", is supported by the "comparative environmental policy" scholarship, adjacent and complementary—and in some respects coincident in research objectives—to that of "comparative environmental law", to which this contribution takes part.

In fact, this neighbour scholarship offers a tripartite division similar to the one proposed here for the circulation of legal models: harmonization (which corresponds in its legal sense to the vertical circulation); diffusion (horizontal circulation); and imposition (oblique circulation).

Despite the many criticisms already expressed towards this oblique mechanism in action, due to its lack of legitimacy, the difficult coordination of foreign models with the particular contexts of the developing legal systems, and the ambiguous nature of the SD model proposed by the international financial organizations, it seems to be the most fruitful type of circulation in conveying the SD model. Future research may address to what extent and by which modality such circulation could be able to achieve the goals of the SD model in transitional and developing countries.

In conclusion, SD as a legal model has resulted in a global diffusion, but there still remains the question of what it represents: is it a principle, a policy objective, a key ideology of modern environmentalism? Is it an action plan or a component of public policies and private actions? Is it a pleonasm, because any development must be sustainable? Or, on the contrary, is it an oxymoron, given that the development by its nature cannot be sustainable? Or, perhaps, it represents a *tertium genus*, a model adaptable to different contexts and situations, a rich mine yet to be explored? The answer can only arrive from the experience, and probably, as well-expressed by an expert in the field, "[t]he comparative law analysis of Environmental Law can significantly contribute to an understanding of how law can further sustainable development" (Robinson 1998, p. 249).

References

Agenda 21 (1992). Programme of action for sustainable development adopted at the United Nations conference on environment and development in Rio de Janeiro. Retrieved February 15, 2015, from https://sustainabledevelopment.un.org/content/documents/Agenda21.pdf

Ajani G (1995) By chance and prestige: legal transplants in Russia and Eastern Europe. Am J Comp Law 43(1):93–117

Ajani G (2007) Transplants, legal borrowings and reception. In: Clark D (ed) Encyclopedia of law and society. Sage Publications, New York (Vol. 3)

Akhtarkhavari A (2010) Global governance of the environment. Environmental principles and change in International law and politics. Edward Elgar, Cheltenham and Northampton

Alogna I (2014) The circulation of legal models: towards the evolution of environmental law. In: Sancin V, Kovič Dine M (eds) International environmental law: contemporary concerns and challenges in 2014. GV Publishing, Ljubljana

Ashford NA, Hall RP (2011) Technology, globalization and sustainable development: transforming the industrial state. Yale University Press, New Haven

Berkowitz D, Pistor K, Richard J-F (2003) The transplant effect. Am J Comp Law 51(1):163–203

Borràs Pentinat S (2006) L'impatto ambientale della clausola di condizionalità: globalizzazione sostenibile? Rivista Giuridica dell'Ambiente 3–4:391–417

Brunel S (2012) Le développement durable. Presses Universitaires de France, Paris

Busch P-O, Jörgens H (2012) Governance by diffusion: exploring a new mechanism of international policy coordination. In: Meadowcroft J, Langhelle O, Ruud A (eds) Governance, democracy and sustainable development. Edward Elgar, Cheltenham and Northampton

Civic MA (1998) Prospects for the respect and promotion of Internationally recognized sustainable development practices: a case study of the World bank environmental guidelines and procedures. Fordham Environ Law J 9:231–259

Cordini G (2007) Lo sviluppo sostenibile per l'ambiente alla luce del diritto pubblico comparato. In: Fois P (ed) Il principio dello sviluppo sostenibile nel diritto internazionale ed europeo dell'ambiente. Editoriale Scientifica, Napoli

Cordonier Segger M-C (2008) Sustainable development in International law. In: Bugge HC, Voigt C (eds) Sustainable development in International and National law. Europa Law Publishing, Groningen

De Jong M, Stoter S (2009) Institutional transplantation and the rule of law: how this interdisciplinary method can enhance the legitimacy of International organisations. Erasmus Law Rev 2(3):311–330

Dernbach JC (1998) Reflexions on comparative law, environmental law, and sustainability. Widener Law Symp J 3:279–286

Doumbé-Billé S (2007) Le droit international de l'environnement et de l'adaptation aux changements planétaires. In: Mélanges en l'honneur de Michel Prieur: Pour un droit commun de l'environnement. Dalloz, Paris

German Government (2002) Perspective for Germany. Our strategy for sustainable development. Retrieved February 15, 2015 from http://www.bundesregierung.de

Handl G (1998) The legal mandate of multilateral development banks as agents for change toward sustainable development. Am J Int Law 92:642–665

Hoda D (1995) The climate change convention and evolving legal models of sustainable development. Pace Environ Law Rev 13(1):75–96

Holzinger K, Knill C, Arts B (eds) (2008) Environmental policy convergence in Europe. The impact of International institutions and trade. Cambridge University Press, Cambridge

Hunter D, Salzman J, Zaelke D (2001) International environmental law and policy. Foundation Press, New York

International Court of Justice (1997) Case concerning the Gabcikovo-Nagymaros Project (Hungary/Slovakia). Judgement of 25 September 1997. Retrieved February 15, 2015 from http://www.icj-cij.org/docket/files/92/7375.pdf

Jörgens H (2003) Governance by diffusion. Implementing global norms through cross-national imitation and learning. Environmental Policy Research Centre. FFU-report 07-2003. Retrieved February 15, 2015 from http://userpage.fu-berlin.de/ffu/download/rep-2003-07.pdf

Kiss AC, Shelton D (1994) International environmental law. Transnational Publishers, New York

Le Prestre P (2005) Protection de l'environnement et relations internationales. Les défis de l'écopolitique mondiale. Dalloz, Paris

Lowe V (1999) Sustainable development and unsustainable arguments. In: Boyle A, Freeston D (eds) International law and sustainable development: past achievements and future challenges. Oxford University Press, Oxford

Markandya A, Pedroso-Galinato S (2009) Economic modeling of income, different types of capital and natural disasters. The World Bank, sustainable development network global facility for disaster reduction and recovery unit. Policy research working paper 4875. Retrieved February 15, 2015 from https://openknowledge.worldbank.org/bitstream/handle/10986/4070/WPS4875.pdf?sequence=1

Mattei U (2008) Circolazione dei modelli giuridici. In Enciclopedia del diritto, Annali I. Giuffré, Milano

Mattei U, Nader L (2008) Plunder: when the rule of law is illegal. Wiley-Blackwell, Oxford

McCormick J (1986) The origins of the World conservation strategy. Environ Rev 10(3):177–187

Merryman J (1977) Comparative law and social change: on the origins, style, decline and revival of the law and development movement. Am J Comp Law 25:457–491

Meynier A (2014) Le rôle des concepts dans la fondamentalisation du droit de l'environnement. In: Milon P, Simon D (eds) Révolution juridique, révolution scientifique. Vers une fondamentalisation du droit de l'environnement? Presses Universitaires d'Aix-Marseille, Aix-en-Provence

Morand-Deviller J, Bénichot J-C (eds) (2010) Mondialisation et globalization des concepts juridiques: l'exemple du droit de l'environnement. IRJS, Paris

Pegoraro L, Rinella A (2007) Diritto pubblico comparato. Profili metodologici. CEDAM, Padova

Perju V (2012) Constitutional transplants, borrowing and migrations. In: Rosenfeld M, Sajó A (eds) Oxford handbook of comparative constitutional law. Oxford University Press, Oxford

Pisano U, Lepuschitz K, Berger G (2013) National sustainable development strategies in Europe 2013. Taking stock and exploring new developments. ESDN Quarterly Report, 29

Prieur M (2011) Droit de l'environnement. Dalloz, Paris

Rio Declaration (1992) Agreement on environment and development adopted at the United Nations Conference on environment and development in Rio de Janeiro. Retrieved February 15, 2015 from http://www.unep.org/Documents.Multilingual/Default.asp?documentid=78&articleid=1163

Robinson NA (1998) Comparative environmental law perspectives on legal regimes for sustainable development. Widener Law Symp J 3:247–278

Ruiz Fabri H, Gradoni L (eds) (2009) La circulation des concepts juridiques: le droit international de l'environnement entre mondialisation et fragmentation. Société de Législation Comparée, Paris

Sacco R (1990) Introduzione al diritto comparato. Giappichelli Editore, Torino

Sacco R (1991) Legal formants: a dynamic approach to comparative law (installment II of II). Am J Comp Law 39(2):343–401

Sahasranaman PB (2012) Oxford handbook of environmental law. Oxford University Press, New Delhi

Sands P (1994) The "greening" of International law: emerging principles and rules. Indiana J Glob Legal Stud 1(2):293–323

Sands P (1995) Principles of international environmental law, vol I. Manchester University Press, Manchester and New York

Sands P (2003) Principles of International environmental law. Cambridge University Press, Cambridge

Siems M (2014) Comparative law. Cambridge University Press, Cambridge

Swanson D, Pintér L (2004) National strategies for sustainable development: challenges, approaches and innovation in strategic and co-ordinated action. International Institute for Sustainable Development, Winnipeg, Manitoba

Tews K (2011) The diffusion of environmental policy innovations. In: Winter G (ed) Multilevel governance of global environmental change: perspectives from science, sociology and the law. Cambridge University Press, New York

Tews K, Busch P-O, Jörgens H (2002) The diffusion of new environmental policy instruments. Environmental Policy Research Centre. FFU-report 01-2002. Retrieved February 15, 2015 from http://userpage.fu-berlin.de/ffu/download/rep_02-01.pdf

UN (2012) Resolution 66/288 "the future we want" adopted by the general assembly of United Nations on 27 July 2012. Retrieved February 15, 2015 from http://daccess-dds-ny.un.org/doc/UNDOC/GEN/N11/476/10/PDF/N1147610.pdf?OpenElement

UNESCO (1970) Utilisation et conservation de la biosphère. Actes de la Conférence intergouvernementale d'experts sur les bases scientifiques de l'utilisation rationnelle et de la conservation des ressources de la biosphère. Paris, 4–13 septembre 1968. UNESCO, Paris

Voigt C (2009) Sustainable development as a principle of International law. Resolving conflicts between climate measures and WTO law. Martinus Nijhoff Publishers, Leiden-Boston

Watson A (1974) Legal transplants: an approach to comparative law. Scottish Academic Press, Edinburgh

Watson A (1977) Society and legal change. Scottish Academic Press, Edinburgh

Watson A (1996) Aspects of reception of law. Scholarly Works. Paper 666. Retrieved February 15, 2015, from http://digitalcommons.law.uga.edu/fac_artchop/666

WCED (1987) Report named "our common future" of the World commission on environment and development headed by Gro Harlem Brundtland. Oxford University Press, Oxford and New York

Wiener JB (2001) Something borrowed for something blue: legal tranplants and the evolution of global environmental law. Ecol Law Q 27:1295–1372

Wise EM (1990) The transplant of legal patterns. Am J Comp Law 38: 1–22 (Supplement U.S. Law in an Era of Democratization)

World Bank (1989) International bank for reconstruction and development (IBRD) articles of agreement as amended effective February 16, 1989. The Bank, Washington DC. Retrieved February 15, 2015 from http://siteresources.worldbank.org/EXTABOUTUS/Resources/ibrd-articlesofagreement.pdf

World Bank (2006) Where is the wealth of Nations: measuring capital for the 21st century. World Bank, Washington DC. Retrieved February 15, 2015 from http://siteresources.worldbank.org/INTEEI/214578-1110886258964/20748034/All.pdf

WTO Appellate Body (1998) Report: United States-import prohibition of certain shrimp and shrimp products, AB-1998-4. Retrieved February 15, 2015 from https://www.wto.org/english/tratop_e/dispu_e/58abr.pdf

Yang T, Percival RV (2009) The emergence of global environmental law. Ecol Law Q 36:615–664

3-D Sustainability and Its Contribution to Governance Assessment in Legal Terms: Examples and Perspectives

Volker Mauerhofer

Abstract Environmental, social and economic capitals, capacities and carrying capacities provide the theoretical construct of the three dimensions of a sustainable development. Based thereon, this chapter aims firstly to provide a conceptual overview on two main objectives of multilevel rule of law systems that should be addressed when adapting these systems towards a more sustainable direction. This first aim is addressed based on '3-D Sustainability', a concept offering six flexibly applicable decision-making criteria for priority setting between these sustainability dimensions based on the burden of proof in the sense of the precautionary principle. The theoretical application of these criteria on several real-world examples of legislative acts indicates the concept's usefulness in practice. The two main objectives identified within this first aim are to stay through international environmental policy within the environmentally sustainable scale and to politically define flexible legal trade-off mechanisms, which more sustainably deal with conflicts among these sustainability dimensions. Secondly, the chapter strives to identify ways to strengthen the application of the existing international environmental legislation. Thus, several innovative mechanisms are identified that overcome current implementation and enforcement deadlocks, without changing existing laws, but also increasing its direct effect. In summary, the chapter innovatively offers—based on ongoing research—several solution proposals for addressing in a sustainable manner geopolitical and organizational scales as well as trade-offs when it comes to re-writing existing environmental legal institutions (de lege ferenda). It further provides proposals for the innovative implementation of existing normative regimes without modifying legal text (de lege lata).

V. Mauerhofer (✉)
Faculty of Life Sciences, University of Vienna, Vienna, Austria
e-mail: volker.mauerhofer@univie.ac.at

V. Mauerhofer
UNU IAS—United Nations University Institute for the Advanced Study of Sustainability, Tokyo, Japan

V. Mauerhofer
PRIMAFF—Policy Research Institute of the Ministry of Agriculture, Forestry and Fishery, Tokyo, Japan

V. Mauerhofer (ed.), *Legal Aspects of Sustainable Development*,
DOI 10.1007/978-3-319-26021-1_3

Keywords Three pillars · Trade-off criterion · Multilateral environmental agreement (MEA) law · Practical politics decision support · Sufficiency policy

1 Introduction

Sustainable Development became internationally during the last four decades a leading concept in terms of safe-guarding the earth (Dobson 1996; De Mooij and van den Bergh 2002; Cordonier Segger 2004; Farley and Smith 2014). Milestones of this endeavour can be seen in the 'World Conservation Strategy' of 1980 (IUCN/UNEP/WWF 1980; Pinto 1995). and especially in the widespread definition within the so-called Brundtland-Report named 'Our Common Future' of the World Commission on Environment and Development (1987, p. 8) which states: 'Humanity has the ability to make development sustainable—to ensure that it meets the needs of the present without compromising the ability of future to meet their own needs'.

Thereon based sustainable development became a main topic of the World Summit of Rio 1992 and got intensively included into the central outcome documents adopted there. This is in particular valid for the Rio Declaration and its additional implementation supplement for the local level, the Local Agenda 21 (Malanczuk 1995). In 2002, sustainable development then provided the term as well as the main content of discussion for the World Summit in Johannesburg. Moreover, 'ensuring environmental Sustainability' has been included into the Millennium Assessment as the seventh out of eight Millennium Development Goals (MDGs) to be achieved by 2015 (MDGs 2015). Subnational, national and supranational (international, regional) law didn't kept away from this trend (Carlman 2008; Mauerhofer 2008a; Ross 2009), but intensively took part which is shown through a constant increase of the use of 'sustainable development' or similar terminology (Mauerhofer 2012). Just some years ago at the global Rio +20 conference all members of the United Nations reassured in the outcome document 'The Future We Want' their dedication to Sustainable Development and its three dimensions, the environmental, social and economic ones (TFWW 2012). However, the interrelation between the three different dimensions of Sustainable Development and the ways of making trade-offs among them and setting priorities, regularly stays quite imprecise in political documents; vague words attributes like, 'consistent', 'coherent', 'inclusive', 'balanced'," do not contribute much to the solution of concrete and obvious conflicts of interest but rather eschew the debate (see also UNEP 2014).

This chapter especially addresses ways towards solutions among these often contradicting positions from the legal viewpoint as law has always to do with decision making. The first part of the chapter looks at these problems from the viewpoint of 'de lege ferenda', that means how the law should look like in the optimal sense. Therein, based on a concept called 3-D Sustainability the chapter strives to offer some new approaches to deal with these trade-off situations and introduces a structured but flexible decision support tool. Here, the chapter focuses on the role of the law for two interrelated main preconditions of environmental sustainability, namely (1) for fixing the environmentally sustainable scale and (2) for setting flexible trade-offs mechanisms among the environmental, economic and social dimensions of sustainable development. The second, shorter part of this chapter takes the viewpoint 'de lege lata', that means it looks at the existing law. Therefrom, some innovative approaches are explained how international (environmental) law can be better implemented and enforced without the necessity to change the wording of the law.

2 The 3-D Sustainability Approach and the Legal Framework

The initial stage for the upcoming part of this chapter is the approach named '3-D Sustainability' which especially concentrates on the viewpoint 'de lege ferenda', thus how to modify legal text towards a sustainable development (Mauerhofer 2008a).

2.1 The Overall Picture Regarding the Environmentally Sustainable Scale

A huge number of attempts exist to show the connection between the environmental, social and economic dimensions of sustainable development within a single picture. This appears, in general, an appropriate way to make the sustainable development more comprehensible and accessible for a larger number of people. However, four main shortcomings of formerly widely used figures of Sustainable Development have already been pointed out which are (1) misinterpretation of embeddings, (2) misjudgement of equity between these three dimensions of sustainability, (3) a lack of expression of limitations and (4) a lack of adequate decision support (with more details Mauerhofer 2008a). The 3-D Sustainabilityapproach described hereinafter intends to circumvent all the four shortcomings previously listed (Mauerhofer 2008a).

The concept of 3-D Sustainability is described by a cone with the following parts (Fig. 1):

1. The base is horizontal with environmental, social and economic capital each represented by concentric rings. The economic capital is embedded in social capital and both are embedded in environmental capital.
2. The cone contains three vertical columns indicating environmental capacity, social capacity (in the sense of capability; Mauerhofer 2013a) and economic capacity, respectively. These columns are:

   (a) connected bottom–up with the different forms of capital (economic, social and natural);

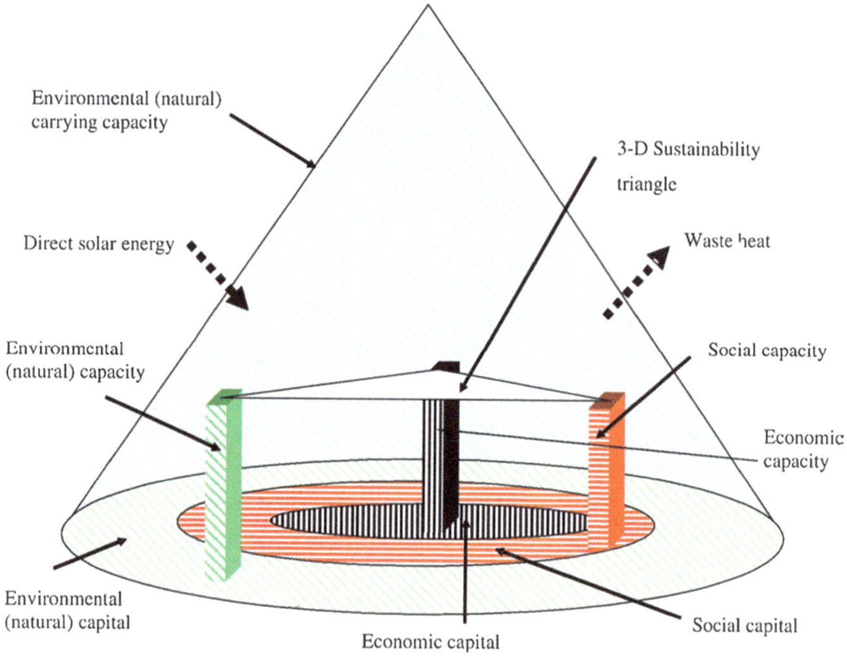

Fig. 1 3-D Sustainability (adapted from Mauerhofer 2008a, p. 498)

   (b) interrelated top–down with the different factors in the 3-D Sustainability triangle (explained in Fig. 2); and
   (c) balancing and re-influencing the 3-D Sustainability triangle.

3. The cone's slopes indicate the boundaries of the environmental system (i.e., the environmental carrying capacity), and the column with the basis on the outer circle (representing the environmental capacity) is able to reach and cross these boundaries (adapted from Mauerhofer 2008a, pp. 497–498).

The 3-D Sustainability triangle balanced on the columns of Fig. 1 is shown more in detail with its individual features in Fig. 2 is widely associated with the triangle of corporate sustainability of Dyllick and Hockerts (2002, p. 135; on slight differences see Mauerhofer 2008a).

2.2 *Definitions and Objectives for the Six Criteria of 3-D Sustainability*

In the following, the six criteria of 3-D Sustainability are more in-depth described in order to increase the understanding for their legal application. An overview is

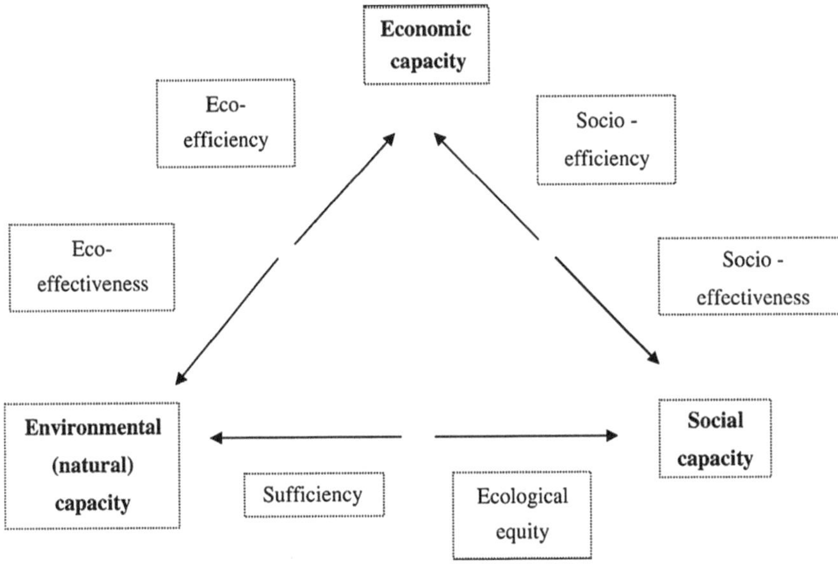

Fig. 2 3-D Sustainability triangle (adapted from Mauerhofer 2008a, p. 498)

presented for each of them including a brief definition respectively and characteristic features of their objectives (Table 1).

Table 1 Shortened definition of quantitative characteristics of the objectives of the six criteria of 3-D Sustainability (adapted from Mauerhofer 2009, 2014)

Criterion	Definition	Characteristic features of its objectives
Sufficiency	Individual's voluntary decision to consume/produce no unit (more) or even unit's less	Absolute, quantitative and/or qualitative, but sustainability depends on current consumption/production level
Eco-effectiveness	Capacity of an action to achieve its expressed ecological objectives	Absolute, at least >50 % improvement, no rebound effects
Ecological equity	Intra- and intergenerationally acting under the 'veil of ignorance' (Rawls 1971)	
Socio-effectiveness	Capacity of an action to achieve its expressed societal objectives	
Eco-efficiency	Input and output ratio (per time) of energy/waste for product/service	Relative, 0–99 % improvement per unit (product/service/person), rebound effects in longer run
Socio-efficiency	Input and output ratio (per time) of an action concerning persons	

Table 1 focuses on quantitative features of the objectives to make them measurable and to legally define them. The decision of an individual towards sufficiency does not automatically indicate sustainability as the decision to produce/consume less could be made at an already unsustainable level and then could still be unsustainable even after such a decrease. The decision to produce or consume no additional unit or even less units than during the previous period should lead to a steady or reduced throughput of energy and matter. Otherwise, also a higher number of units produced can outweigh previous reductions of the use of energy and matter within a single unit what is usually known as the rebound effect (Binswanger 2001). With eco-efficiency and socio-efficiency a rebound effect respectively takes place sooner or later under the premise of a constant growth of the number of units produced. Also with sufficiency, rebound effects can occur (Alcott 2008), but this is not necessarily a following effect when deciding for sufficiency as others may take the same decision or the decision not to consume the additional units of energy and matter available through the sufficiency decision. Sufficiency is—due to its voluntary character—not mainly determined by prescriptive norms such as legal institutions. But it has insofar a regulatory context as actions encouraging people towards sufficiency such as public incentives (such as payments for ecosystem services; see Mauerhofer et al. 2013) or publically sponsored campaigns do often have their basis in the rule of law (see e.g., Schneidewind and Zarnt 2014).

3 Fixing the Environmentally Sustainable Scale and the Law

This part of the chapter will deal with substantial preconditions and issue occurring when addressing the environmentally sustainable scale by means of international environmental law. While how to establish flexible trade-offs mechanisms through legal norms will be subject to the next part of this chapter.

When addressing the environmentally sustainable scale (the ‘carrying capacity’) of exploiting environmental assets (the ‘capital’), the main endeavour should be to arrange this in absolute terms on a safe time scale and within the relevant geographic scale. Issues of scale have been broadly analysed in institutional policy analysis (Young 1994; Adger et al. 2005; Termeer et al. 2010). In this chapter, Gibson et al.’s (2000, p. 218) broadly applicable definitions of scale and level are applied where ‘scale’ is defined as ‘the spatial, temporal, quantitative, or analytical dimensions used to measure and study any phenomenon’, and ‘levels’ are defined as ‘the units of analysis that are located at different positions on a scale’. Cash et al. (2006, p. 8) mention seven different types of scale as of importance for policy studies, namely—additionally to the commonly used spatial and temporal scales—the jurisdictional, institutional, networks, management, and knowledge scales. Regarding environmental carrying capacity, Del Monte-Luna et al. (2004, p. 485) defined it as 'the limit of growth or development of each and all hierarchical levels of biological integration, beginning with the population, and shaped by processes and interdependent relationships between finite resources and the consumers of those resources'. This definition is useful for each abiotic or biotic attribute with a potential for human use as a sink or a source (such as air, water, nitrogen, a wild animal species, a mineral).

The working steps proposed in Table 2 make the spatial aspects of the term 'scale' more explicit such as pointed out by Jordan and Fortin (2002; see also Dovers 1995) when compared with Daly's (1992) original more throughput oriented definition. The correct classification of this scale level is a precondition for the decision about the right action to be taken in order not to exceed the carrying capacity. This applies to the environmental, but also to the social carrying capacity (Mauerhofer 2008a, 2013a).

Among the general barriers that hamper already from the start the maintenance of an environmentally sustainable scale in legal terms, are cases where the geopolitically competent authority is unable to rule the case-related benchmarks for the whole geographic scale of the case. Such a clear mismatch between the relevant geographical scale and the institutional scale (Young 1994; Hein et al. 2006; Satake et al. 2008) occurs for instance in Austria regarding the competence for the conservation of biodiversity. There, this competence is distributed among the nine provinces of this nation without a minimal coordination duties regarding internal cross-border cases, while the supranational level of the environmental integrative legislation of the European Union is mitigating this problem a bit (Mauerhofer 2008a, 2013b).

One case where such a mismatch has been obviated constitutes on the level of the EU the creation of the protected area network NATURA 2000 on the basis of biogeographical regions—following an ecosystems approach—as initial point for the criteria-based selection of relevant sites ensuring the conservation of particular

Table 2 Reduced qualitative methodology of the Legislation-Check based on 3-D Sustainability (adapted from Mauerhofer 2012, 2014)

Overall objective(s)	Interrelationship(s)	Specific questions for geo-political legislative level

Environmentally sustainable scale of critical natural capital (environmental carrying capacity) achieved	Dependent on geographical level
	Dependent factor(s) influencing environmental carrying capacity
	Dependent on time
		Is the scale known?
		• Can the known scale be geo-politically addressed?
		  • Can the known scale be geo-politically addressed by competent authority/-ies?
		    • Can the known scale be geo-politically addressed in a proper way?
		Is the scale unknown/uncertain?
		• Can the unknown/uncertain scale be addressed in a precautionary way?

habitat and species types (European Communities 2005; Mauerhofer 2008b). With this kind of selection procedure as an origin, then the environmental carrying capacity in the sense of the maintenance or restoration of the legally defined favourable conservation status is regulated (Mauerhofer 2008a, p. 504). Another example from the EU can be found within the so called Water Framework Directive (WFD) (European Parliament and Council 2000) which positions its actions on river basins touching the territory of sometimes more than one Member State and being managed by one competent authority that had to be unanimously appointed by all Member States concerned (Grimeaud 2004; Mauerhofer 2012, p. 656).

A case of failing scale designation happens when the geopolitical competent authority is able to cover the scale in geographical terms, but where the designation of an environmentally sustainable scale is hindered by other reasons, for instance e.g. by a lack of financial resources.

Other cases of incorrect scale designation constitute the ones where the precise benchmark is scientifically unknown, but not regulated in a precautionary way (see Table 2).

Among the barriers and obstacles with ruling an environmentally sustainable scale are also general exemptions that are enclosed from the very start into the law. These include formulations such as that the laws is not applied on particular actions at all or declares actions of less impact to be generally permitted because they are not thought to be harmful to the respective scale (without—despite their cumulative effects—laying down an overall limit and/or how many of these actions are allowed). These exemptions can also already be understood as results of trade-offs, however in any way, they effect from the very beginning the designation of an environmentally sustainable scale.

Another initial problem takes place if the scale is not regulated from the start in the correct qualitative and quantitative extent. This happens e.g. when a benchmark for the scale is numerically or by description laid down by law and allows an increased negative effect on the environmental asset while a reduction would be scientifically necessary. Similar is valid if a reduction is legally prescribed but to an insufficient extent.

In comparison to these cases, the trade-offs presented in the next part of this chapter are mainly effecting a scale already legally regulated.

4 Laying Down Flexible Trade-Offs Mechanisms and the Law

This part of the chapter especially presents how flexible trade-off mechanisms for legal norms envisaging a sustainable development can be gained from 3-D Sustainability and also practically implemented. These findings of course also support the aim of the previous part of this paper, namely to ensure that the environmentally sustainable scale is not crossed as a carrying capacity.

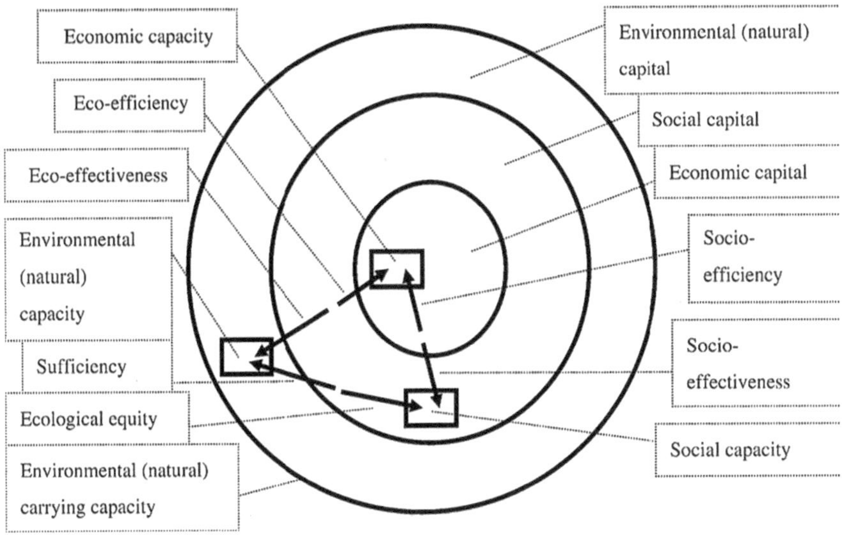

Fig. 3 Simplified view on 3-D Sustainability from above (adapted from Mauerhofer 2008a, p. 501)

4.1 The Flexible Hierarchy Between 3-D Sustainability Criteria Supporting Legal Trade-off Decisions

The assessment of priority setting by 3-D Sustainability takes it initial point from a bird's perspective on Fig. 1 whereby the following picture can be observed (Fig. 3).

From dependencies and the probability of touching or even crossing the social as well the environmental carrying capacities within Fig. 3, a general, but not fixed, hierarchy within the six criteria effecting the triangle has been deducted (Mauerhofer 2008a) to find out the solution(s) more sustainable (Fig. 4). In this way, 3-D Sustainability evidently shows which actions should provisionally be given priority in terms of a holistic view on sustainability. This preliminary, but not

Fig. 4 Hierarchy of priorities (not fixed) within criteria for sustainable development measures (adapted from Mauerhofer 2008a, p. 501)

Sufficiency

Eco-effectiveness

Ecological equity

Socio-effectiveness

Eco-efficiency

Socio-efficiency

fixed hierarchy is therefore conform with the idea of 'prudent decision making' such as defined by Kallio et al. (2007, p. 47), according to which both, the ends and mean(s) of an action are: (a) acceptable, (b) in line with each other and (c) not in conflict with any other end that is considered as more or equally important. The provisional priority given to sufficiency in Fig. 4 is also in accordance with other authors (Lamberton 2005, p. 61; Alcott 2008), and it has been rightly highlighted that '(P)ersonal behaviour change is furthermore not a substitute for international political efforts' (Alcott 2008, p. 770). This hierarchy shown in Fig. 4 can then be implemented on the example of practical legal cases in order to primarily find out the status quo and secondary, to recommend—if appropriate—changes. Changes in this hierarchy can be executed through a shift of a burden of proof. On other occasion, examples especially for the supranational legal level had been geographically distinguished and presented where 3-D Sustainability was respectively applied on real-life situations in order to show the usefulness of the approach (adopted from Mauerhofer 2014 and more in detail Mauerhofer 2008a):

(A) Global examples

1. UNFCCC (1994) and the Kyoto Protocol: the envisaged greenhouse gas reduction level in the Kyoto Protocol as an example for an—in general—eco-effective approach of an absolute reduction of exploiting the environmental capacity, but by applying insufficient quantitative goals (and insufficient implementation mechanisms)
2. Mean's and burden of proof under WTO rules: former negotiations based on the 2001 Doha Ministerial Declaration (para 31i) on the relationship between Multilateral Environmental Agreements (MEAs) and the WTO within the Committee for Trade and Environment (CTE) wherein EU (unsuccessfully) proposed a reversal of the burden of proof, so that a country (party or non-party to the MEA) challenging a measure laid down in a MEA would have to prove that the measures imposed by the other WTO parties do not meet the conditions of WTO rules (see Prévost 2004, p. 52 with hint to WT/CTE/170, esp. point 15);
3. WTO negotiations—the principle of 'less than full reciprocity' (LTFR): this principle promised to developing countries in WTO negotiations on the Doha Development Agenda (Oxfam 2006a, p. 13) means for example on tariffs that developing countries have to cut their tariffs less than developed countries (Oxfam 2006b, p. 17f, Mauerhofer 2008a, pp. 502–503) and therefore lays (in terms of 3-D Sustainability) on the borderline between socio-effectiveness and socio-efficiency.
4. WTO ruling concerning the European Community-Hormones Case: WTO Appellate Body's interpretation of Article 5.7 of the Agreement on the Application of Sanitary and Phytosanitary Measures (the SPS Agreement) in the Hormones Case shifted away burden of proof from the user of a new technology (in this case users in the appealing US) (an opposite

distribution of the burden of proof would be indicated by 3-D Sustainability, taking into account that the measures reflect eco-effectiveness, sufficiency and socio-effectiveness in the case of ignorance) (for more details see Mauerhofer 2008a, p. 503).

5. Fishery—rules concerning the right scale of catches: concern in relation to 3-D Sustainability the environmental capacity of the biotic factor 'fish'. Measures in relation to sufficiency and eco-effectiveness are primarily appropriate to ensure that the carrying capacity is not exceeded and regarding defining the right scale the burden of proof, and resources required to satisfy that burden, should not be the solely responsibility of conservationists (Agardy 2000, p. 762f; Essington 2001, p. 122; Heazle 2006; Freestone 1998).

(B) Regional examples

6. European Union—decoupling of economic growth from resource use and environmental degradation: this goal of the EU (European Union 2002; European Union Council Environment 2002, p. 11.) can be differentiated into an absolute and relative decoupling (OECD 2002, p. 11) Whereas relative decoupling, or the efficiency approach, does not guarantee improvement in absolute terms, at least not in the long run (EEA 2002, p. 7 and pp. 15–16; Azar et al. 2002) due to the rebound effect and absolute decoupling does not appear to be possible due to the entropy laws (Faber et al. 1996; Mauerhofer 2013c). Thus, approaches towards sufficiency and eco-effectiveness are essential.
7. European Court of Justice—biodiversity conservation and double-reversing of the burden of proof: The European Court of Justice—ECJ (2004) has applied a double reversal of the burden of proof in connection with the precautionary principle concerning the environment e.g. in a case related to biodiversity; there it shifted the burden of proof twice to the proponents of the activity in relation to doubts regarding whether to implement a preliminary assessment of a project as well as regarding doubts about the activity's significant effects on a protected site. From the point of view of 3-D Sustainability, the carrying capacity of the biotic factor 'biodiversity', at the regional European level, was at the stake with regards to the occurrence of birds in a particular site (see on this case further Mauerhofer 2008a, p. 504; Opdam et al. 2009; Mauerhofer 2012, pp. 656–657).

These examples related to a supranational scale, in particular a global or a regional one respectively, with various sustainability topics covered. In each example the shift of the burden of proof is shown to be an essential feature when the preliminary hierarchy within the six criteria mentioned above should be modified (Mauerhofer 2008a, p. 502). All these examples arise from the legal side. However, also private individual decision-making between the three dimensions of Sustainable Development can apply 3-D Sustainability including its six criteria. Additionally, a modification distinguishing between different stages in the status quo of "development"—hierarchy following the principle of common, but differentiated responsibility—should be considered when applying the hierarchy.

4.2 Modifying the Hierarchy and the Burden of Proof

Such as already mentioned, the hierarchy of priorities displayed in Fig. 4 is not carved in stone, but subject to modification. To justify a modification in this hierarchy, it is recommended to put the burden of proof on the shoulders of the proponent who is interested in the modification (see also e.g. Princen 1997). That means, if a lower ranked criterion should get priority, the burden of proof is turned around concerning whether the environmental and/or social carrying capacities are threatened. Therefrom follows that the person who intends the application of a criterion of a lower level—although a criterion of a higher level should be applied—has to bear the burden of proof that there are no remaining scientific doubts against the application of the lower criterion (Mauerhofer 2008a). Such an approach supports a precautionary view (Paterson 2007) in order to deal with conflicts of interest between the three sustainability dimensions based on sound science and rationality. Nevertheless, the term 'burden of proof' can be interpreted in various ways. Kokott (1998) pointed out different meanings of the 'burden of proof' in two human rights systems when comparing an inquisitorial system (such as applied by German constitutional law) with an adversarial system (such as applied in the USA). In an inquisitorial system (with a public prosecutor as one party) the term 'burden of proof' covers solely the burden of persuasion (objective burden of proof); whereas in an adversarial system (where fundamentally equally positioned parties bring forward their evidence) the 'burden of proof' contains both, the burden of submitting evidence (i.e. the burden of production) as well as the burden of persuasion (see also e.g., in general for international law Kazazi 1996 and for environmental international law Applegate 2000).

Zandvoort (2005) suggested in this connection a flexible system wherein standards of proofing the absence of risk should be inversely correlated to standards of liability for actors. That would concretely mean that if there is a reduced burden of proof for a prospective actor, a stricter liability should take place, and vice versa. The usefulness of this proposal in practice is questionable. The realization of many environmental risks lead to damages that cannot be financially insured against or compensated for. Often this realization is not appraisable in an economic sense. Additionally, such risks are often embedded in enterprise structures with limited liability (Mauerhofer 2008a, p. 502).

4.3 Flexible Trade-off Mechanisms and Organizational Conflicts of Interest Within the Law

From the beginning of sustainable development, proposals have been brought forward on how to analyse and foster institutions in the sense of norms as the rules of the game (see e.g., Opschoor and van der Straaten 1993; de Graaf et al. 1996; Iovanna and Newbold 2007; Frame and Brown 2008). However, also criticism has been articulated that insufficient influence has been developed on improving decision-making processes in the subject of Sustainable Development (Shi 2004; Ehrlich 2008). Thus, in the following, two simple examples will be shown for the application of 3-D Sustainability on the assessment of administrative organizations (and the norms they are grounded in) based on a 'Governance Check' (Mauerhofer 2013b). Both cases show conflicts of interest within the law and are assessed by flexible trade-off mechanisms based on 3-D Sustainability. While the first example refers to a conflict of interest within one public authority, the second one deals with conflicts of interest between different public authorities.

The first case occurs for instance if one minister is responsible for both, the environmental capital and its capacity. Mauerhofer (2013b, p. 332) has elaborated already more in detail such an example by a practical situation of a political change related to an Austrian Ministry in 2000 (Fig. 5).

Compared with the situation in the year 1999, one and the same Ministry is since about 2000 in charge for the environmental capital as a source and sink (and its carrying capacity as a limit) as well as the use of that capital in the sense of 'capacitating' it (Mauerhofer 2013b, p. 332). This situation could be seen on the one hand as opportunity for improved, because more intense, coordination among Environment on the one side and Agriculture and Forestry, Environment and Water Management; while on the other hand—in practice—the mix between capital

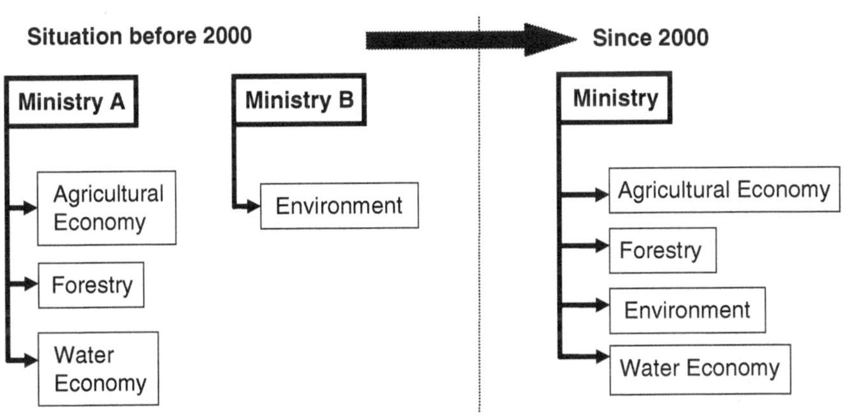

Fig. 5 Time shift of the sub-constitutional distribution of federal competences for agricultural economy, forestry, environment and water economy in Austria (adapted from Mauerhofer 2013b, p. 332)

(including carrying capacity) responsibilities and capacity responsibilities does not support transparent decision making and, thus, prevents public participation when it comes to solving conflicts of interests within this new Ministry (e.g. with regard to the choice of measures concerning afforestation, greenhouse gas reduction, renewable energy sources, and agri-environment subsidies) (Mauerhofer 2013b, p. 332).

The second case, the one of conflicts of interest between different public authorities has been also already presented by the case of a single law and the distribution of rights therein when coordinating between different *capital authorities* and *capacity authorities* (Table 3) (Mauerhofer 2013b, pp. 332–333).

Table 3 indicates on the example of the Austrian Water Act that the federal ministry legally in charge for natural capital 'water' (including its carrying capacity) has for several activities to gain in advance the consent of other ministries mainly in charge for the use ('capacitating') of water or for other forms of capital. Backed by these legally fixed veto rights, these other ministries can—contrary to the principles of 3-D Sustainability—hold back their consent without having any burden of proof about whether their interests will not damage the natural capital 'water' (Mauerhofer 2013b, pp. 332–333).

The examples above focused on the national level and on regulatory conflicts. Doubtless, an extension of the application of the approach to public organizations of subnational or supranational levels could be an interesting subject of future research.

Table 3 Selection of the coordination cases of the Austrian Water Law Act ('Wasserrechts-Gesetz'), assessed according to 3-D Sustainability (adapted from Mauerhofer 2013b, p. 334)

Minister responsible for	Authority	Water act (norm and content)
A. Environmental capital	Minister for environment	Prepares policy and legal acts
B. Environmental carrying capacity	Same than A	
C. Costs of general permission of facility types (economic capital)	Minister for finance	Veto right (Art. 12 Water Law Act)
D. Technical staff + procedure of general permission of facility types (economic capacity)	Minister for commerce	Veto right (Article 12)
E. Data for water contamination register (economic capacity)	Minister for commerce	Veto right (Article 59/2)
F. Effect on authority responsible for waterways (economic capacity)	Minister for transport	Veto right (Article 59/2)
G. Release of water framework ordinances (capital and capacity)	All provinces	Right to be contacted (Art. 54/1)
H. Measure against imminent danger (capital and capacity)	Respective province	Right to be contacted (Article 54/4)

5 Perspectives for International Environmental Law and Its Improved Implementation and Enforcement

This part of the chapter now briefly takes the view of "de lege lata": how to reach improvements without necessarily change the wording of the law.

The insufficient implementation and enforcement of Multilateral Environmental Agreements and related transnational initiatives are widely seen as a serious barrier for endeavours to stop environmental deterioration or even improving situations (Mauerhofer 2011; Visseren-Hamakers et al. 2012).

Thus, to develop and apply for existing MEAs and soft law initiatives more effective implementation and enforcement mechanisms without changing their wording seems to be of utmost importance.

5.1 *Voluntary Approaches*

Most underestimated potential of improved implementation and enforcement of MEAs can be identified in voluntary actions through interested stakeholders who are in the position to provide an added value.

One such example is the execution of a Convention-Check through a protected area authority holding own side-specific implementation pouvoir and budget (Mauerhofer 2011). This check was created upon a bottom–up inquiry by such an authority to offer guidance for an additional positive contribution to the implementation of MEAs. It constitutes a site-specific partnership approach of in total ten working steps, assesses the already existing contribution to the implementation of MEAs, commonly develops recommendations for additional actions and evaluates later the extent of improved implementation regarding relevant MEAs (Table 4) (Mauerhofer 2011).

Table 4 Overview on the methodology and results of a 'Convention-Check' (adapted from Mauerhofer 2014)

Overview on the ten working steps of a 'convention check' (with results from a National Park case study)	Working periods for the case study
Implementation • Six steps incl. table of compliance, workshops in order to gain status quo + recommendations (155 norms of five MEA's assessed, for >50 % activities already set, 82 recommendations developed)	Jan 2007–Sept 2007
Evaluation • Four steps in order to gain ex-post results on the rate of implementation of these recommendations (of those ones relevant for NP-management [n = 44] about 1/3 implemented)	Oct 2009–Dec 2009

This voluntary approach is an example of how to unleash local potential for improved implementation of MEAs without changing the wording of the legislative basis or trying to put pressure on anybody.

5.2 *Increased Capacity Building About Direct Effect of MEAs*

Other examples which also support a "de lege lata approach" are intensified capacity building measures on behalf of judges or civil servants especially on the direct effect of MEAs. Direct effect relates to the opportunity of immediate application of norms of a ratified MEA without having them first been transposed into the national legislation. These measures could be implemented by means such as personal training, direct office support and published guidelines. One example for one of these approaches are the written guidelines outlining the legal norms of the Alpine Convention with potential direct effect as well as the related databases with binding decisions already taken based thereon, both developed by the Austrian Federal Ministry for Agriculture, Forestry, Environment and Water Economy (2007, 2014). An example for the mentioned approach of personal training constitutes the capacity building during the regular ASEAN Chief Justices Roundtable on the Environment projects on behalf of ASEAN judges co-organized by the Asian Development Bank (2013) and co-organized by the International Union for the Conservation of Nature's World Commission of Environmental Law (IUCN WCEL 2013).

5.3 *Extending the Application Scope of MEAs Through 'Accession of Regions'*

A precondition for the direct effect of MEAs and more widely for the intensified implementation of MEAs in nations is the applicability of a particular MEA in a country. This can occur through an individual accession of that nation to MEAs. But this procedure is often hampered by national politics. Another manner for making an MEA applicable to a nation is the accession to that MEA by the accession of an association of regional integration where the nation is party to, such as the EU. In case the EU becomes a party of a particular MEA, even those EU-Member States that are currently not party become obliged to implement and enforce that MEA (see e.g., Verwey 2004). However, there are participatory rights for each Member State in the EU-bodies in charge for the decision whether to access to a MEA.

5.4 Re-Interpretation of Existing MEAs Towards More Scientifically Sound Content

Another strategy that might lead to intensified implementation of MEAs without rewriting them is the (re-)interpretation of them and of the legal norms transposing them.

Propositions for the modified meanings of existing terms can be brought forward from everybody, but—of course—they have to be implemented by civil servants or judged in that proposed sense instead or additionally to the former meaning.

One example was brought forward into the discussion by Mauerhofer and Kim (2014) for the re-interpretation of the term 'Ecological Integrity' that can be found in numerous MEAS and that could be reinterpreted according to these authors in the sense of the nine quantified planetary boundaries (Rockström et al. 2009; Steffen et al. 2015). Another idea would be a more intensified implementation of equity-related norms in MEAs by interpreting and implementing them in the sense of the more quantified scientific Ecological Footprint concept (Wackernagel and Rees 1996).

A precondition for a more scientifically advanced interpretation and harmonized use of terms of MEAs is an idea about where the terms are currently used within official MEAs. Here, the Information Portal for Multilateral Environmental Agreements (InforMEA) of the UN should be mentioned which is implemented within a project of the MEA Information and Knowledge Management (IKM) Initiative with support by the United Nations Environment Programme (UNEP 2015). InforMEA collects in particular news, COP decisions, meetings, and national focal points reports from MEAs, and structures this information by terms from MEA COP agendas (UNEP 2015).

6 Conclusion

3-D Sustainability constitutes an innovative concept for decision support with its separation between capital and capacity as well as with its six criteria. The assessment above indicates that the most relevant contributions of international as well as regional, national and subnational law in collaboration with other instruments (such as economic ones and public awareness) concerning a sustainable development can be identified based on 3-D Sustainability such as following, namely that law

1. regulates the environmentally sustainable scale of capacitating environmental capital in absolute terms on the safe time scale
2. lays down flexible trade-offs mechanisms by the rule of law to support 1.
3. implements and enforces the two points aforementioned.

In addition, voluntary bottom–up approaches for the intensified implementation incl. enforcement of MEAs should be more applied and other strategies that work "de lege lata" and therefore do not include a change of the wording of existing environmental law.

However, many efforts still have to be invested in work on the objectives, definitions and in particular on the indicators for the six criteria of 3-D Sustainability. At the end, it should not be overlooked that 3-D Sustainability itself is a simplified concept, with all advantages and disadvantages related. However, its pictorial expression overcomes the shortcomings of some of the most commonly used figures depicting sustainable development. In summary, 3-D Sustainability offers a flexible instrument for assessment and decision support, which can be applied on various issues and used on all geographic levels. The further perspectives of its legal applications show promises for a more effective sustainable development law.

References

Adger WN, Brown K, Tompkins EL (2005) The political economy of cross-scale networks in resource co-management. Ecol Soc 10(2):9

Agardy T (2000) Effects of fisheries on marine ecosystems: a conservationist's perspective. ICES J Mar Sci 57:761–765

Alcott B (2008) The sufficiency strategy: would rich-world frugality lower environmental impact? Ecol Econ 64:770–786

Applegate JS (2000) The precautionary preference: an american perspective on the precautionary principle. Hum Ecol Risk Assess 6:413–443

Asian Development Bank (2013) Asean chief-justices explore legal solutions. Retrieved 26 Mar 2014. http://www.adb.org/news/asean-chief-justices-explore-legal-solutions-environmental-threats

Austrian Federal Ministry for Agriculture, Forestry, Environment and Water Economy (2007) Handbook alpine convention. Retrieved 26 Mar 2014. http://www.lebensministerium.at/publikationen/umwelt/archiv/die_alpenkonvention-handbuch_fuer_ihre_umsetzung.html (in German)

Austrian Federal Ministry for Agriculture, Forestry, Environment and Water Economy (2014) Database alpine convention. Retrieved 26 Mar 2014. www5.umweltbundesamt.at/alpenkonvention (in German)

Azar C, Holmberg J, Karlsson S (2002) Decoupling—past trend and prospects for the future. A report of the Department of Physical Resource Theory of the Chalmers University of Technology and Götheburg University with contributions by Tobias Person, Robert Ayres, Thomas Sterner and Jonas Nässèn to the Environmental Advisory Council of the Swedish Ministry of the Environment. Retrieved 31 Mar 2014. http://www.mvb.gov.se

Binswanger M (2001) Technological progress and sustainable development: what about the rebound effect? Ecol Econ 36:119–132

Carlman I (2008) Control system for sustainable development. In: Dubois DM (ed) Proceedings of the computing anticipatory systems: CASYS = 07 C eighth international conference, 8. American Institute of Physics, vol 1051, pp 187–194

Cash DW, Adger WN, Berkes F, Garden P, Lebel L, Olsson P, Pritchard L, Young O (2006) Scale and cross-scale dynamics: governance and information in a multilevel world. Ecol Soc 11(2):8

Cordonier Segger M-C (2004) Significant developments in sustainable development law and governance: a proposal. Nat Resour Forum 28:61–74

Daly HE (1992) Allocation, distribution and scale: towards an economics that is efficient, just and sustainable. Ecol Econ 6:185–193

de Graaf HJ, Musters CJM, ter Keurs WJ (1996) Sustainable development: looking for new strategies. Ecol Econ 16:205–216
De Mooij RA, van Den Bergh JCJM (2002) Growth and the environment in Europe: a guide to the debate. Empirica 29:79–91
Del Monte-Luna P, Brook BW, Zetina-Rejón MJ, Cruz-Escalona VH (2004) The carrying capacity of ecosystems. Glob Ecol Biogeogr 13:485–495
Dobson A (1996) Environment sustainabilities: an analysis and a typology. Environ Politics 5:401–428
Dovers SR (1995) A framework for scaling and framing policy problems in sustainability. Ecol Econ 12:93–106
Dyllick T, Hockerts K (2002) Beyond the business case for corporate sustainability. Bus Strategy Environ 11:130–141
EEA (2002) Environmental signals. Benchmarking the millennium. European environmental agency regular indicator report Luxemburg: office for official publications of the European communities
Ehrlich PR (2008) Key issues for attention from ecological economists. Environ Dev Econ 13:1–20
Essington TE (2001) The precautionary approach in fisheries management: the devil is in the details. Trends Ecol Evol 16:121–122
European Communities (2005) Natura 2000—conservation in partnership. Luxembourg: office for official publications of the European communities. Retrieved 31 Mar 2014. http://ec.europa.eu/environment/nature/info/pubs/docs/nat2000/conservation_in_partnership.pdf
European Court of Justice—ECJ (2004) Case C-127/02 Waddenvereniging and Vogelbeschermingsvereniging [2004] ECR I-7405
European Parliament and Council (2000) Directive 2000/60/EC of the European Parliament and of the Council of 23 October 2000 establishing a framework for Community action in the field of water policy. Official Journal L 327 of 22.12.2000. Retrieved 27 July 2015. http://eur-lex.europa.eu/legal-content/EN/TXT/?uri=celex:32000L0060
European Union Council Environment (2002) 2457th meeting of the EU Council Environment at 17th December 2002 in Luxemburg
European Union (2002) Decision No. 1600/2002/EC of the European Parliament and of the Council of 22 July 2002 laying down the Sixth Community Environment Action Programme. Official Journal L, vol 242, p 1. 10 Sept 2002. Office for Official Publications of the European Communities, Luxemburg
Faber M, Proops J, Baumgärtner S (1996) The use of the entropy concept in ecological economics. In: Faber M, Proops J (eds) Ecological economics: concepts and methods. Edward Elgar, Cheltenham, pp 115–135
Farley HM, Smith ZA (2014) Sustainability: if it's everything, is it nothing? Routledge, London, 176 pp
Frame B, Brown J (2008) Developing post-normal technologies for sustainability. Ecol Econ 65:225–241
Freestone D (1998) The burden of proof in natural resources legislation. In: FAO Legislative Study, no 63, Rome
Gibson CC, Ostrom E, Ahn TK (2000) The concept of scale and the human dimensions of global change: a survey. Ecol Econ 32:217–239
Grimeaud D (2004) The EC water framework directive—an instrument for integrating water policy. RECIEL 13(1):27–39
Heazle M (2006) Lessons in precaution: the international Whaling commission experience with precautionary management. Mar Policy 30(5):496–509
Hein L, van Koppen K, de Groot RS, van Ierland EC (2006) Spatial scales, stakeholders and the valuation of ecosystem services. Ecol Econ 57:209–228
Iovanna R, Newbold SC (2007) Ecological sustainability in policy assessments: a wide-angle view and a close watch. Ecol Econ 63:639–648

IUCN/UNEP/WWF (1980) World conservation strategy, international union for conservation of nature and natural resources (IUCN)/united nations environment programme (UNEP)/world wide fund for nature (WWF). IUCN/UNEP/WWF, Gland

IUCN WCEL (2013) Final program judicial colloquium biodiversity. Retrieved 26 Mar 2014. http://ajne2.itkeep.com/wp-content/uploads/2013/12/FINAL-PROGRAM-judicial_colloquium_biodiversity.pdf

Jordan GJ, Fortin M-J (2002) Scale and topology in the ecological economics sustainability paradigm. Ecol Econ 41:361–366

Kallio TJ, Nordberg P, Ahonen A (2007) Rationalizing sustainable development'—a critical treatise. Sustain Dev 15:41–51

Kazazi M (1996) Burden of proof and related issues: a study on evidence before international tribunals. Kluwer Law International, Hague

Kokott J (1998) The burden of proof in comparative and international human rights law: civil and common law approaches with special reference to the American and German legal systems. Kluwer Law International, Hague

Lamberton G (2005) Sustainable sufficiency—an internally consistent version of sustainability. Sustain Dev 13(1):53–68

Malanczuk P (1995) Sustainable development: some critical thoughts in the light of the Rio Conference. In: Ginther K, Denters E, de Waart PJIM (eds) Sustainable development and good governance. Martinus Nijhoff Publishers, Dordrecht, pp 23–52

Mauerhofer V (2008a) 3-D sustainability: an approach for priority setting in situation of conflicting interests towards a sustainable development. Ecol Econ 63:496–506

Mauerhofer V (2008b) Conservation of wildlife in the European union with a focus on Austria. In: Panjwani P (ed) Wildlife law: a global perspective. American Bar Association (ABA) Publishing, Cleveland, pp 1–55

Mauerhofer V (2009) Legislation check based on '3-D sustainability'—an approach for global precautionary governance change'. In: Proceedings of the 15th ISDRC 2009, Utrecht/The Netherlands, 5th–8th July 2009. Retrieved 8 July 2009. http://globalchallenge2009.geo.uu.nl

Mauerhofer V (2011) A bottom-up 'convention-check' to improve top-down global protected area governance. Land Use Policy 28:877–886

Mauerhofer V (2012) A legislation check based on '3-D sustainability'—addressing global precautionary land governance change. Land Use Policy 29:652–660

Mauerhofer V (2013a) Social capital, social capacity and social carrying capacity: perspectives for the social basics within environmental sustainability. Futures 53:63–73

Mauerhofer V (2013b) The 'governance-check': assessing the sustainability of public spatial decision-making structures. Land Use Policy 30:328–336

Mauerhofer V (2013c) Lose less instead of win more: the failure of decoupling and perspectives for competition in a degrowth economy. Environ Values 22:43–57

Mauerhofer V, Hubacek K, Coleby A (2013) From polluter pays to provider gets: distribution of rights and costs under payments for ecosystem services. Ecol Soc 18(4):41

Mauerhofer V (2014) Challenging the labyrinth: sustainable scales, trade-offs and perspectives through environmental law. In: Sancin V, Kovič-Dine M (eds) International environmental law: contemporary concerns and challenges in 2014. Ius Software, d.o.o, GV Založba, Ljublijana, pp 25–42

Mauerhofer V, Kim RE (2014) Interpretation of ecological integrity in the sense of planetary boundaries: a new way forward for implementing multilateral environmental agreements. In: ISEE (ed) Book of abstracts of the biannual International Society for Ecological Economics (ISEE) 2014 conference, Reykjavik, Iceland, 13th–15th August, p. 14. Retrieved at 13 Aug 2014. http://isee2014.yourhost.is

MDGs (2015) Millenium development goals. Retrieved at 26 July 2015. http://www.un.org/millenniumgoals/

OECD (2002) Indicators to measure decoupling of environment pressures from economic growth. SG/SD. Organization for Economic Co-operation and Development, Paris

Opdam PFM, Broekmeyer MEA, Kistenkas FH (2009) Identifying uncertainties in judging the significance of human impacts on natura 2000 sites. Environ Sci Policy 12:912–921
Opschoor H, van der Straaten J (1993) Sustainable development: an institutional approach. Ecol Econ 7:203–222
Oxfam (2006a) What happened in Hong Kong? Briefing Paper No. 85. Retrieved 27 July 2015. http://www.oxfam.de/files/20051221_whathappenedhongkong_161kb.pdf
Oxfam (2006b) A recipe for disaster. Will the Doha round fail to deliver for development? Briefing Paper No. 87. Retrieved 27 July 2015. http://www.oxfam.de/files/20060427_arecipefordisaster_282kb.pdf
Paterson J (2007) Sustainable development, sustainable decisions and the precautionary principle. Nat Hazards 42:515–528
Pinto MCW (1995) Reflections on the term sustainable development and its institutional implications. In: Ginther K, Denters E, de Waart PJIM (eds) Sustainable development and good governance. Martinus Nijhoff Publishers, Dordrecht, pp 72–99
Prévost D (2004). Selected international developments regarding health and environmental regulation of relevance to the European Union. Eur Environ Law Rev 13:38–60
Princen T (1997) The shading and distancing of commerce: when internalization is not enough. Ecol Econ 20:235–253
Rawls J (1971) A theory of justice. Harvard University Press, Cambridge
Rockström J, Steffen W, Noone K, Persson Å, Chapin FS III, Lambin EF, Lenton TM, Scheffer M, Folke C, Schellnhuber HJ, Nykvist B, de Wit CA, Hughes T, van der Leeuw S, Rodhe H, Sörlin S, Snyder PK, Costanza R, Svedin U, Falkenmark M, Karlberg L, Corell RW, Fabry VJ, Hansen J, Walker B, Liverman D, Richardson K, Crutzen P, Foley JA (2009) A safe operating space for humanity. Nature 461:472–475
Ross A (2009) Modern interpretations of sustainable development. J Law Soc 36:33–54
Satake A, Rudel TK, Onuma A (2008) Scale mismatches and their ecological and economic effects on landscapes: a spatially explicit model. Glob Environ Change 18(4):768–775
Schneidewind U, Zarnt A (2014) The politics of sufficiency. Oekom Publisher, Munich
Shi T (2004) Ecological economics as a policy science: rhetoric or commitment towards an improved decision-making process on sustainability. Ecol Econ 48:23–36
Steffen W, Richardson K, Rockström J, Cornell SE, Fetzer I, Bennett EM, Biggs R, Carpenter SR, de Vries W, de Wit CA, Folke C, Gerten D, Heinke J, Mace GM, Persson LM, Ramanathan V, Reyers B, Sörlin S (2015) Planetary boundaries: guiding human development on a changing planet. Science 347:1259855
Termeer CJAM, Dewulf A, van Lieshout M (2010) Disentangling scale approaches in governance research: comparing monocentric, multilevel, and adaptive governance. Ecol Soc 15(4):29
TFWW (2012) The future we want (TFWW). Outcome document adopted at the United Nations Conference on Sustainable Development, in Rio de Janeiro, Brazil in June 2012. Retrieved from 31 Mar 2014 from http://www.uncsd2012.org/content/documents/727The%20Future%20We%20Want%2019%20June%201230pm.pdf
UNEP (2014). Post 2015 Note 1# 'Integrating the three dimensions of sustainable development'. Retrieved from 31 Mar 2014 from http://unep.org/unea/docs/UNEP%20Post%202015%20Note%201%20final.pdf
UNEP (2015) Informea initiative. Retrieved 26 July 2015. http://www.informea.org/
UNFCCC (1994) United nations framework convention on climate change, 1771 UNTS 107, UN Doc.A/AC.237/18 (Part II)/Add.1
Verwey DR (2004) The European Community, the European Union and the international law of treaties - a comparative legal analysis of the Community and Union's external treaty-making practice. TMC Asser Press, The Hague
Visseren-Hamakers I, Leroy P, Glasbergen P (2012) Conservation partnerships and biodiversity governance: fullfilling governance functions through interactions. Sustain Dev 20:264–275
Wackernagel M, Rees WE (1996) Our ecological footprint: reducing human impact on the earth. New Society Publications, Gabriola Island/BC

World Commission on Environment and Development (1987) Our common future. Oxford University Press, Oxford

Young OR (1994) The problem of scale in human-environment relations. J Theor Politics 6:429–447

Zandvoort H (2005) Globalisation, environmental costs, and progress, the role of consensus and liability 52(6) Water Science and Technology 43–50, as cited in: J. P. van der Sluijs, Uncertainty and precaution in environmental management: Insights from the UPEM conference, Environmental Modelling and Software, In Press, 2006, Corrected Proof

Sustainability—A Long, Hard Road

Ferdinand Kerschner and Erika Wagner

Abstract Sustainability—a term compromising a group of themes and being of immense importance for today's and all future generations. It has developed in last decades to one of the most central concepts in environmental law and poses an important component of political and juridical aims. The Treaty of Amsterdam implemented sustainability in primary law as a basic objective of the European Union. As a consequence, the EU established the Sustainable Development Strategy (EU SDS) with the aim to improve the living situation of all generations. Also on a national level, many measures have been taken to enhance the "sustainable development" approach. But one has to ask the question: does the concept of sustainability work? Do the measures taken, inter alia, in the fields of transport, energy, climate change and waste management lead to the desired success or is it just a sublime goal, a program without (sufficient) implementation? Or is it even just a tokenism?

Keywords Sustainable development · Environmental law · Climate change law · Traffic law · Industrial plant law · Energy law

1 Historical Way to Sustainability

1.1 International Law

In his opus "silvicultura oeconomica" the German Hans Carl von Carlowitz formulated the concept of sustainability for the first time for the forestry sector (von Carlowitz 1713, pp. 105–106). His approach of "Nachhaltigkeit" (German

F. Kerschner (✉) · E. Wagner
Johannes Kepler University, Linz, Austria
e-mail: Ferdinand.Kerschner@jku.at

E. Wagner
e-mail: Erika.Wagner@jku.at

V. Mauerhofer (ed.), *Legal Aspects of Sustainable Development*,
DOI 10.1007/978-3-319-26021-1_4

term for "sustainability") influenced the Austrian Forestry Law. On an international level, the thought of sustainable development was defined 1987 for the first time in the so-called Brundtland Report issued by the World Commission on Environment and Development (World Commission on Environment and Development 1987; Epiney and Scheyli 1999). "Sustainable development" was defined therein as "development that meets the needs of the present without compromising the ability of future generations to meet their own needs" (World Commission on Environment and Development 1987, p. 8; see also e.g. Wagner 2000; Mauerhofer 2004a; Arbter 2005). In the course of the United Nations Conference on Environment and Development (UNCED) of 1992, the Rio Declaration on Environment and Development and the Agenda 21 were accepted. Although "soft law" instruments (and therefore non-binding), both acts are seen as important achievements in international environmental law (Wagner 2000). In order to review the progress in the implementation of the Rio Declaration and the Agenda 21, the Commission on Sustainable Development (CSD) was established. Disadvantages resulting from the non-enforceability and possible solutions for that were discussed at the World Summit on Sustainable Development which took place in Johannesburg in 2002—with only moderate success. Also the following World Summit in Rio de Janeiro in 2012 (Lengauer 2011; Madner 2011) stayed—to sum up—behind the expectations and initiated the liquidation of the CSD (Kanning 2013; Gareis and Varwick 2014).

1.2 European Union

In EU law, the concept of sustainability was recognized for the first time in the Single European Act—SEA concluded 1986 (Article 130r(1) SEA footnote 9). The term "sustainable development" was, however, not explicitly mentioned, but the legislature referred to prudent and rational utilization of natural resources. With the Treaty of Amsterdam entered into force 1996, the principle of sustainability has been anchored in the Union's primary law (Article 2 Treaty establishing the European Economic Community—TEEC "harmonic, balanced and sustainable development" and Article 6 TEEC, the horizontal clause relating to environmental protection). Article 6 TEEC stated that the horizontal character should support the achievement of sustainable development. Therefore the horizontal clause obliges to consider the need to promote sustainability when planning or taking measures.

Article 37 Charter of Fundamental Rights of the EU (CFR) was introduced by the Treaty of Nice concluded in 2001 and entered into force in 2003. Its novelty is questionable given that it does not grant a "real" fundamental right to single persons but rather states a general duty of the EU and the Member States to consider environment protection in their policies (Raschauer 2008). Expectations one might have had about a "procedural environmental fundamental right" were disappointed; Article 37 CFR turned out as a purely objective-juridical repetition of Article 6 and 174 TEEC and seems to be misplaced in the Charter of Fundamental Rights due to its character as a non-enforceable objective for the states (Calliess 2011a, b).

2 Sustainability—A Non-Comprehensible Term?

An exact and generally valid definition for "sustainable development" is nearly impossible. As a result of its broad scope, the shimmering and (maybe already too) much used construct of sustainability contains certain ambiguities. One reason for that is the linguistic imprecision caused by the authentic but varying linguistic versions of the acts. The German version uses the concept "nachhaltig", the English one "sustainable" and the French one "durable" (permanently). Another reason for the lack of clarity originates from the multidisciplinarity; countless areas are concerned in many different ways. It can clearly be noted that the basic concept of sustainable development is the integration of the interests of future generations. Human behaviour should not endanger livelihood of future generations. Future generations have a right for a clean environment in which the satisfaction of their needs is ensured (Epiney and Scheyli 1999).

The following postulates can be derived from the principles that were set by the Rio Declaration:

(a) Anthropocentric Approach

Humans are central for the implementation of sustainable development. Protection of nature without considering the needs of people, who live in it, is not intended. The protection of the environment should ensure a healthy and productive life of people under natural living conditions.

(b) Economic Growth

States are allowed to exploit their natural resources according to their own environmental and economic development policy. When doing so, the states have to ensure that the environment of other countries or areas outside their sovereign territory do not take any damage. One could argue that this does not encompass an obligation to protect the domestic environment, and consequently, that a state could, in principle, completely destroy its environment as long as the environmental media of other states are not harmed. This is, in the authors' opinion, not a legitimate interpretation of the concept of sustainable development. According to its clear wording ("development"), the principle refers to progress as well as economic growth. However, an eternal economic growth is not possible. Facing this limitation, it should in this context, in the authors' opinion, be rather talked about "environmental economy". Progress should not be disrupted, but it should be made with due consideration of the impacts on domestic and foreign environment.

(c) Intergenerational Justice

Future generations should not already be deprived today of all opportunities and resources; thus, the present generation should take into account the future needs when taking actions. For the purpose of intergenerational equity, living conditions between members of different generations should not continue to deteriorate.

(d) Ecological Approach

In order to ensure sustainable development, environmental protection must be seen as an integral part—at least equal to economic and social aspects—of the development process. Anything else bears the risk to foil the achievement of sustainable development.

(e) Intragenerational Justice

Sustainable development does not only aim at the protection of future generations, also the improvement of living conditions of the present generation is an important factor. Poverty reduction is a mandatory requirement for the success of the implementation of sustainability because current problems can displace future ones from the field of view and take away all relevance from them.

(f) Access to Information

Raising people's awareness for the need of sustainable development is one of the most crucial factors for the concept's success. Therefore, a first step is environmental education and ecological consciousness-raising of individual citizens (Kerschner et al. 2009). Access to information is necessary to ensure a public counterbalance when tackling environmental problems. Without participation of the general public, the struggle for sustainability is similar to a fight against windmills.

Today, the term "sustainability" is often interpreted as social development, which encompasses economic, environmental and social aspects (Lewis 2011; Bauer and Seebacher 2012). As regards content, the concept of sustainability can be expressed by using a three-pillar model: Environmental dimension (intergenerational equity), social dimension (intragenerational equity) and economic dimension (Grunwald and Kopfmüller 2012). These three dimensions have to be considered in an equal way in when shaping policies or preparing legal acts that are intended to contribute to sustainable development or have an effect on it. It has to be noted that within the three pillars the balance between environmental, economic and social issues was and is not given at all. Since the economic crisis in 2007 economic dimensions have prevailed significantly. The sustainability assessment mentioned in the Amsterdam Treaty (Arbter 2005), which is about a systematic assessment process of legislative acts according to the dimensions of sustainability, cannot change this impression; sustainability is emphasized merely in non-binding documents and seems to be an overused term. The ambitious climate goals are pursued only half-heartedly in all Member States. Acts adopted before 2007, especially environmental impact assessment (EIA), strategic environmental assessment (SEA) and nature impact assessment (NIA), which make a significant and effective contribution of European environmental protection, are not able to overcome the challenges of the 20th century by their own. In addition, they are increasingly hollowed out and stripped of their effectiveness. CO_2 certificate trading has proved to be a "climate poker" with a lot of winners, but barely any losers. The real losers seem to be the future generations, to whom the concept of sustainability is dedicated.

One way to ensure a concrete implementation of programs and targets would be a strict "sustainability assessment", i.e. an integrated assessment of the potential environmental, social and economic impacts of (legislative) proposals. At a European level, the Commission introduced with the "Impact Assessment" a procedure that serves as kind of test for sustainability of European legislation and policies (EC 2002a). Considerable time pressure when conducting these assessments may affect the quality of the results in a negative way (Arbter 2005). Other points of criticism are the insufficient involvement of Member States and the (by tendency) stronger involvement of interest groups representing the industrial or business sector compared to the involvement of stakeholders representing environmental or social interests (Leith and Speth 2003; Arbter 2005).

SEA, EIA and NIA provide for sustainability insofar as they are designed to safeguard that the absorption capacity of environment is not exceeded and an overuse of resources is prevented. The potential impacts of projects in industry, agriculture, transport, regional planning etc. must be known in advance.

Such upstream sustainability assessments are unavoidable for a comprehensive implementation of the principle of sustainability. It requires, in essence, an orientation on the long-term carrying capacity of the environment and should follow the principle that pollution does not exceed the environment's absorption capacity and that the consumption of resources is not greater than their regeneration rate (Reese 2010). Consume of fossil fuel should be avoided at all. In order to act sustainable, the actual impact must already be known in advance.

In the following, it will be shown that the effective European development regarding sustainability is at odds with the primary EU law.

3 Legal Framework at European Level

3.1 Primary Law

At primary law level the principle of sustainable development was expressly stipulated in several norms such as shown in the following subchapters.

(a) Preamble of the Treaty of the European Union (TEU)

The preamble of the TEU inter alia states: "DETERMINED to promote economic and social progress for their peoples, taking into account the principle of sustainable development and within the context of the accomplishment of the internal market and of reinforced cohesion and environmental protection, [...]". In contrast to the objectives of Article 3 TEU described in the following, the preamble is not legally binding but is of expressive importance and used to interpret the treaty.

(b) Article 3(3) TEU

Article 3(3) TEU normalizes inter alia that "The Union shall establish an internal market. It shall work for the sustainable development of Europe based on balanced economic growth and price stability, a highly competitive social market economy, aiming at full employment and social progress, and a high level of protection and improvement of the quality of the environment [...]". Article 3(3) TEU enumerates all areas of sustainable development. Thus, the question which dimension in the context of sustainability predominates (ecological, economic or social) (Wagner 2000), should be off the table (Windoffer 2011). However, it is debatable whether Article 3(3) TEU obliges EU institutions to take action in order to ensure the pursuit of sustainability. The prevailing opinion derives from the aforementioned objectives, which are mentioned in Article 3 TEU and the main provisions of the environmental chapter (Article 191 ff Treaty on the Functioning of the European Union [TFEU]) an obligation to elaborate and implement a general environmental policy (Raschauer 2008; Nettesheim 2014a). But a duty to act in specific cases cannot be deduced (Wagner 2000). However, the provision has to be interpreted in a way that the various aspects of sustainability must be incorporated into the activities of the Union.

A broad discretion for the (ordinary) EU legislature on whether to consider sustainability or promote exclusively economic aspects would infringe primary law. Every promotion of economic aspects must be contemplated in context of the environmental objectives. In the authors' opinion, focusing on economic aspects only is inadmissible. This result derives a fortiori from Article 11 TFEU which is described later.

(c) Article 3(5) TEU

Article 3(5) TEU states that the European Union "... shall contribute to peace, security, the sustainable development of the Earth, solidarity and mutual respect among peoples, [...]".

Another new feature that was introduced by the amendment of the Treaty of Lisbon is Article 3(5) TEU; the provision expressly states that the Union shall contribute to global sustainable development.

Article 21(2) lit d and f TEU is closely connected referring thereto by stipulating:

"The Union shall define and pursue common policies and actions, and shall work for a high degree of cooperation in all fields of international relations, in order to:

(d) foster the sustainable economic, social and environmental development of developing countries, with the primary aim of eradicating poverty;
(f) help develop international measures to preserve and improve the quality of the environment and the sustainable management of global natural resources, in order to ensure sustainable development;"

Here, Article 3(3) and (5) as well as Article 21(2) lit d and f TEU are made concrete and the objectives of the Union to promote sustainable development (also globally) are emphasized.

(d) Article 11 Treaty of the Functioning of the European Union (TFEU)

Article 11 TFEU (ex Article 6 TEEC) reads as following "Environmental protection requirements must be integrated into the definition and implementation of the Union's policies and activities, in particular with a view to promoting sustainable development". It contains a horizontal clause concerning the environment that states an obligation to include environmental considerations when formulating and implementing Union policies and activities. When referring to the "environmental protection requirements", Article 11 TFEU adds the principles set out in Article 191 TFEU—namely high level of protection, the principles of prevention and precaution, the origin principle and the polluter-pays principle—to the principle of sustainability (Wagner 2000; Nettesheim 2014a, b).

The focus of EU environmental policy should be placed on the ecological capability and the obligation to consider not only economic objectives but also the interests of future generations when taking decisions that may impact environmental quality and resource supply. Bearing the goal of the horizontal clause in mind, this explains all the more that any legislative proposal should undergo a sustainability assessment (Epiney and Scheyli 1999; Calliess 2011a, b).

The integration clause has also unmistakably procedural aspects since the obligation to consider and integrate environmental issues can only be met if it is ensured by appropriate institutional arrangements that the effects of the proposed measure are properly identified and registered (Nettesheim 2014b) and taken into account at the factual implementation. The Sustainable Development Strategy (EU SDS), for instance, can be considered as one of the (indeed not sufficient) attempts of a procedural implementation of Article 11 TFEU (EC 2001).

3.2 *The European Sustainable Development Strategy*

In 2001, the European Council of Gothenburg adopted the previously mentioned "Strategy for Sustainable Development" (EC 2001; Calliess 2011a, b). This so-called Gothenburg Strategy was intended to ensure and support the concept of sustainability in environmental, economic and social policy issues on a long-term basis by setting the framework conditions.

In 2005, a new "EU Sustainable Development Strategy—A platform for action" was adopted by the European Council (EC 2010a, b).

The guiding principles of this strategy are:

- Support and protection of fundamental rights;
- Solidarity within and between generations;
- Guarantee of an open and democratic society;
- Involvement of citizens, businesses and social partners;
- Policy coherence and integration;
- Use of the best available knowledge;
- Precautionary principle and the principle of holding the polluter liable.

The latest strategy was adopted 2010 under the title "Strategy Europe 2020" (EC 2010b).

The Commission identified seven main problematic areas, which urgently call for solutions in order to achieve sustainability. These areas are climate change, the impact of transport, production and consumption, sustainable management of natural resources, public health, social exclusion and poverty and the ageing of society (Council 2009).

Member States are required to develop sustainability strategies and report on them and the progress they make on a regular basis.

Every two years the implementation of the strategy is evaluated by the Commission on the basis of sustainable development indicators (EC 2002b) wherein the 'open list' of environment-related headline indicators is analysed, and was subsequently evaluated by the European Council (Council 2009). These indicators consist of the seven most relevant environmental indicators, which have been selected to assess the sustainability strategy.

The European Council is concerned with the implementation of the sustainability strategy in Member States, but obviously without putting sanctions. The strategies are existent on paper and in files. One cannot escape the impression that goal of many of these strategies, programs, principles etc. is to appease. Even its incorporation in primary law seems to be a fig leaf hiding the deficiencies regarding its implementation and enforceability (Calliess 2011a, b).

3.3 The Charter of Fundamental Rights

Article 37 CFR has the heading "Environmental Protection" and the following wording:

"A high level of environmental protection and the improvement of the quality of the environment must be integrated into the policies of the Union and ensured in accordance with the principle of sustainable development".

According to the prevailing interpretation, Article 37 CFR does not grant a right to a high level of environmental protection (Calliess 2003). Such an interpretation would mean that Article 37 CFR is simply a repetition of the content of the TFEU/TEU rules. But it begs the question, why the provision was placed precisely in the Charter of Fundamental Rights if there is no intention to improve the legal position of individuals? But even if it is just a programmatic objective at EU level it has—if taken seriously—normative effect (Kerschner 1996).

3.4 The Article 191 Treaty on the Functioning of the European Union

Article 191(1) TFEU determines, in an exhaustive manner, the goals of EU environmental policy: protection of environment and human health, ensuring efficient use of resources and promoting international action to deal with regional or worldwide environmental problems, with special emphasis on the climate. All these elements relate to the principle of sustainability in one way or another; they are of highest importance for the establishment of a sustainable framework of action. Prerequisite is, however, that they are taken seriously.

In summary, the principle of sustainable development on primary law level has to be understood as an objective and interpretation principle of the Union (Epiney and Scheyli 1999); furthermore, it ensures that promotion of development and growth does not take place without considering environmental aspects.

4 Selected Areas of Law with Sustainability Dimension

4.1 Climate Protection

Climate protection is deemed to be one of the main challenges in sustainable development.

The Kyoto Protocol gives the possibility for the implementation of emissions trading schemes by which compliance with the obligations to reduce emissions should be facilitated. For the implementation of the Kyoto Protocol, the Directive 2003/87/EC establishing a scheme for greenhouse gas emission allowance trading was issued.[1] Currently, the Union aims also to conclude an international agreement to reduce greenhouse gas emissions from aviation, in order to achieve a globally coordinated action towards the air transport industry.[2]

According to the prevailing view, the greenhouse gas allowance trading in this form has unfortunately proven to be ineffective (Fütterer and Pichl 2010). More than ever the question occurs, whether it is possible to solve environmental problems on the free market. Moreover, companies can even make handsome profit from allowance trading. National expenditure for the free allowances issued in the

[1]Directive 2003/87/EC of the European Parliament and of the Council of 13 October 2003 establishing a scheme for greenhouse gas emission allowance trading within the Community and amending Council Directive 96/61/EC, OJ L 275, 25.10.2003, pp. 32–46.

[2]Regulation (EU) No 421/2014 of the European Parliament and of the Council of 16 April 2014 amending Directive 2003/87/EC establishing a scheme for greenhouse gas emission allowance trading within the Community, in view of the implementation by 2020 of an international agreement applying a single global market-based measure to international aviation emissions, OJ L 129, 30.4.2014, pp. 1–4.

first trading period was immediately transferred to the consumers, which had an enormous increase in electricity prices for consumers as consequence—windfall profits (Fütterer and Pichl 2010). To ensure an effective application of this market-based instrument, the start-up difficulties have to be overcome; essential is, in the authors' opinion, that the "carbon leakage" list is eliminated.

To address the problems in climate change policy, the Commission has published several documents. In 2009, the White Paper "Adapting to climate change: Towards a European framework for action" was released (EC 2009); it deals with strategies to diminish the impacts of climate change. In 2013, a Green Paper providing a 2030 framework for climate and energy policies was issued (EC 2013a), which shows, on the one hand, the achievements already made and, on the other hand, future improvements that need to be achieved (Madner and Hartlieb 2013). On 22th January 2014, the Commission established the framework for the future European energy and climate policy for the period from 2020 to 2030 (EC 2014a). By 2030, the greenhouse gas emissions shall be reduced by 40 % compared to 1990 levels. For this purpose the Directive on the revision of the EU Emissions Trading Scheme was adopted, which aimed on an improvement and standardization of the scheme as well as on a massive change in the allocation of allowances (Madner and Isepp 2014). Also renewable energies should be expanded and energy efficiency should support the reaching of the aims, which are defined in the framework for the future European energy and climate policy (EC 2014a).

In Austria the Climate-Protection Act, which entered into force on 22nd November 2011, was adopted on a national level.[3] Even the negotiations between representatives of government and opposition parties were a difficult task and took three and a half years to be completed. The result was immensely criticized. The Climate-Protection Act is not an ordinary environmental protection act and the novelty of its content is questionable. It does not oblige any third parties, provide funding or shift competences; it does not define specific protection targets: the status quo is maintained (Schwarzer 2012). Therefore, the Austrian Climate-Protection Act is a pure planning tool. Reduction targets for each Member State (that are to be achieved by 2020) are set by the so-called Effort-Sharing Decision.[4] Sectors affected by this decision are waste management, energy and industry (if they are not included in emissions trading), fluorinated gases, building, agriculture and transport. These are sectors that have not already been subjected to the greenhouse gas emission trading.

The reduction commitments for the period 2013–2020 are allocated between the federal state and the nine states of Austria according to an agreement that is concluded by the federal state and the state governments pursuant to Article 15a of the

[3]The "Klimaschutzgesetz" published in the federal law gazette ("BGBl") I number 106/2011, in the currently valid version.

[4]Decision No 406/2009/EC of the European Parliament and of the Council of 23 April 2009 on the effort of Member States to reduce their greenhouse gas emissions to meet the Community's greenhouse gas emission reduction commitments up to 2020, OJ L 140, 5.6.2009, pp. 136–148.

Federal Constitutional Law[5] (the Bundes-Verfassungsgesetz, B-VG). A national climate change committee has been established as a coordination hub between federal and state governments (Granner and Raschauer 2012). For each area of climate protection suitable measures shall be negotiated by representatives from the Ministry of Agroeconomy, Forestry, Environment and Water Management, the Ministry of Finance, the Federal Chancellery, the Ministry of Justice, Ministry of Science, Research and Economy, Federal Ministry of Family and Youth, Ministry of Labour, Social Affairs and Consumer Protection, and for particular organization such as the Chamber of Commerce and the Chamber for Labour.

4.2 Energy

In 2006, the Commission issued the "Green Paper—a European Strategy for Sustainable, Competitive and Secure Energy" (EC 2006). With this Green Paper, the Union wanted to take the first step towards a sustainable, competitive and secure European energy policy. When implementing the three dimensions of sustainability, the EU faces great challenges like the increasing dependency on imports, climate change and the increasing energy demand in Europe. In order to overcome these obstacles, the Commission asked the Member States to adjust their action according to three main objectives:

- Sustainability
- Competitiveness
- Security of supply

In 2013, the "Green Paper, A 2030 framework for climate and energy policies" followed (EC 2013b).

In general, there are two control mechanisms for sustainable and climate friendly development (Luger 2012):

- Reduced use of resources (resource policy) and
- Energy Policy (efficient energy use)

Following this approach, a number of secondary legislation has been enacted. Due to the high number of legislative acts, just a few examples of recent years shall be highlighted. Directive 2009/28/EC[6] establishes a common framework for the promotion of energy from renewable energy sources and sets binding national targets for the share of renewable energy (for an overview see e.g. Rabl and Brenner

[5]The "Bundes-Verfassungsgesetz" (B-VG) published in the federal law gazette ("BGBl") I number 1/1930, in the currently valid version.

[6]Directive 2009/28/EC of the European Parliament and of the Council of 23 April 2009 on the promotion of the use of energy from renewable sources and amending and subsequently repealing Directives 2001/77/EC and 2003/30/EC, OJ L 140, 5.6.2009, pp. 16–62.

2009). Referring to the requirements of the Directive, Austria must achieve a share of 34 % of renewable energy sources in the energy mix. Furthermore, each Member State has to adopt an action plan for the promotion of renewable energy (Urbantschitsch and Hofer 2009). The aim is to reduce greenhouse gas emissions (further details see e.g. Piska 2010).

In 2014, the Commission released a Communication on energy efficiency in which an evaluation of developments in the field of energy efficiency is carried out (EC 2014a), and where it is assessed whether the energy efficiency targets for 2020 (increasing energy efficiency by 20 %) can be achieved. The Communication also includes a proposal for a new energy savings target for 2030 (increasing energy efficiency by 30 %). Thereby the Commission completed the climate and energy policy framework for 2020–2030 presented in January 2014 (Falke 2014a, b, c). The Council has already approved the binding aim of the European Union by the end of the year 2013.

Existing legislation that should ensure the achievement of the energy savings target in 2020 consist of (Falke 2014a, b, c):

- Energy Efficiency Directive (Directive 2012/27/EU)[7];
- Directive on the Energy Performance of Buildings (Directive 2010/31/EU)[8];
- Product regulations on the energy efficiency of buildings;
- Product rules establishing the minimum requirements for energy efficiency and the requirement to inform on labels on Energy Efficiency (Directive 2009/125/EC and Directive 2010/30/EU[9], and its implementing rules);
- CO_2 emissions norms for cars and light commercial vehicles;
- An increased funding by the EU Structural and Investment Fund (Horizon 2020, ELENA and the European Energy Efficiency Fund);
- Electricity Market Directive (Directive 2009/72/EC)[10];
- TEU Emission Trading Scheme.

In addition to the Building-Efficiency Directive mentioned second in the list above, the Commission presented a Communication on resource efficiency opportunities in the building sector in July 2014 (EC 2014b). The aim is to reduce

---

[7]Directive 2012/27/EU of the European Parliament and of the Council of 25 October 2012 on energy efficiency, amending Directives 2009/125/EC and 2010/30/EU and repealing Directives 2004/8/EC and 2006/32/EC Text with EEA relevance, OJ L 315, 14.11.2012, pp. 1–56.

[8]Directive 2010/31/EU of the European Parliament and of the Council of 19 May 2010 on the energy performance of buildings, OJ L 153, 18.6.2010, pp. 13–35.

[9]Directive 2009/125/EC of the European Parliament and of the Council of 21 October 2009 establishing a framework for the setting of ecodesign requirements for energy-related products, OJ L 285, 31.10.2009, pp. 10–35 and Directive 2010/30/EU of the European Parliament and of the Council of 19 May 2010 on the indication by labelling and standard product information of the consumption of energy and other resources by energy-related products, OJ L 153, 18.6.2010, pp. 1–12.

[10]Directive 2009/72/EC of the European Parliament and of the Council of 13 July 2009 concerning common rules for the internal market in electricity and repealing Directive 2003/54/EC, OJ L 211, 14.8.2009, pp. 55–93.

environmental impact of new and refurbished buildings by increasing resource efficiency and improving the information available on the environmental performance of buildings. In addition, the Commission intends to create a regulatory framework with a limited number of core indicators to assess the environmental impact of buildings (Falke 2014a, b, c).

In Austria, the Green Electricity Act 2012[11] is an important pillar of the national climate, energy efficiency and environmental activities. The Act of 2012 amended the old—due to numerous amendments hard to read—Green Electricity Act; all provisions concerning the promotion of electricity from renewable energy sources were transferred into a separate law (Rihs 2012). From the synopsis of binding and non-binding EU measures in energy policy and from the fact that fossil fuels will be on one point scarce it can be deducted that a move to renewable energy source is indispensable to ensure economic development. Closely connected to this is the goal to maintain ecological capacity (among others to combat climate change) and idea of generational justice. A shift to renewable energy sources comes with many difficulties (funding policy, no cost transparency, misled subsidization strategies etc.). The transition to a renewable energy system is essential for the European Sustainability Strategy and its dimensions. But it is doubtful whether the actions taken in this regard have enough dynamics. There is still some strong opposition from those who gain profit by exploiting fossil energy (Christian et al. 2015).

Not only that the transition to renewable energy sources is counteracted thereby, in fact several debatable methods are used to exploit fossil resources (such as high-volume hydraulic fracturing to exploit shale gas or shale oil) or to "hide" the CO_2 in geological formations (Carbon Capture and Storage technology, CCS). The "Energiewende" (energy turnaround) is, in any event, a logical requirement of the objective of sustainability.

Thus, the failure of the EU climate change policy is a failure of the sustainability strategy; the climatic deterioration shows that the above-mentioned objectives of sustainability were not implemented adequately.

4.3 *Waste Management*

The field of waste management is one of the oldest and most regulated fields of European environmental law. Currently, the Commission is paying increased attention to plastic in a Green Paper dealing with a European strategy on plastic waste in the environment (EC 2013b). The European waste law contains no specific rules for this kind of waste. The Green Paper (EC 2013b) intended to start a public debate on the sustainable design of plastic products and the use of plastic waste. In addition, the Commission developed in the same year an interim report on the implementation

[11]The "Ökostromgesetz" published in the federal law gazette ("BGBl") I number 75/2011, in the currently valid version.

of the European waste management legislation adopted in the years 2007 till 2009. In conclusion, it can be said that the transposition into national law largely went without a hitch; however there are significant problems in the implementation and enforcement of some Directives. The Commission is especially concerned about the Waste Framework Directive[12] and the Landfill Directive.[13] In the Landfill Directive, there are a number of gaps as regards the fulfilment of requirements and in the Waste Framework Directive there are considerable difficulties in the implementation of the requirements for certain landfills (Falke 2013a, b).

The Austrian Waste Management Act (Abfallwirtschaftsgesetz, AWG) 2002[14] regulates the waste prevention; preparation for re-use of waste; recycling, duties of persons who are concerned with waste management; and requirements for waste treatment plants. The AWG explicitly refers to the precautionary principle and the principle of sustainability. Waste prevention should be an important guiding principle. § 1(3) cipher 3 AWG states that material has to be collected, stored, transported and treated as "waste" if the sustainable use of water or land may be affected. § 9 AWG defines the goals of sustainable waste prevention. The demonstratively mentioned measures deemed to achieve these goals are put into perspective by reference to the criterion of economic viability.

It can be concluded that as long as the discussion centres around refraining from plastic bags, there will be no real improvements for plastic-contaminated oceans, mountains of garbage etc. The principle of waste prevention is largely on paper.

4.4 *Industry*

Economic activities often collide with environmental interests. Therefore, the EU has adopted various acts to prevent environmental damage and ensure sustainable development. They are intended to set limits and standards for the industry and thereby mitigate the exploitation and continuous destruction of the environment.

The Directive on Industrial Emissions requires an approval for certain industrial and agricultural activities with a high pollution potential. The authorization is linked to certain environmental requirements. The IPPC Directive intends to implement the preventive and precautionary principle as well as the polluter-pays principle in the industrial sector.

[12]Directive 2008/98/EC of the European Parliament and of the Council of 19 November 2008 on waste and repealing certain Directives, OJ L 312, 22.11.2008, pp. 3–30.

[13]Council Directive 1999/31/EC of 26 April 1999 on the landfill of waste OJ L 182, 16.7.1999, pp. 1–19.

[14]The "Bundesgesetz über eine nachhaltige Abfallwirtschaft" published in the federal law gazette ("BGBl") I number 102/2001, in the currently valid version.

In 2012, the revised version of the "Seveso II-Directive"[15], named according a place of an major industrial accident, entered into force. The Seveso III Directive[16] incorporates new developments in international and European chemical classification law. The amendments concern, inter alia, easy access to information about the dangers of nearby industrial plants and rules of conduct in case of an accident, the improvement of the participation of the public concerned in planning projects, access to justice for the citizens to whom no such right has previously been granted, and stricter standards for the planning of new establishments (Falke 2012).

In 2014, the Commission published a "Green Action Plan for SMEs" (EC 2014c) in order to support small and medium-sized enterprises and to show them—in application of the Smart Business Act (EC 2008)—how to make use of the business opportunities a fight against environmental problems offers them. An increased use of circuit models and green technology could be highly profitable for the companies. The SME Action Plan aims to increase resource efficiency, to promote environmentally friendly business practices, to open new business opportunities and to facilitate market access for eco-friendly SMEs (Falke 2014a, b, c).

4.5 *Traffic*

The traffic sector is still a stepchild in the implementation of the sustainability debate (Kerschner 2001). It is a much-debated topic, but finally hardly anybody is willing to give up the convenience of motorized transportation. Moreover, the public transport systems are usually not covering all areas; especially in rural areas people are often dependent on individual transport. There is definitely no lack of awareness regarding this problem; nevertheless, there are barely any sustainable solutions in this field (Tichler et al. 2011); it seems that the car is the "holy cow" of Europe.

At an international level, the Convention for the Protection of the Alps (short the "Alpine Convention") was concluded among several European entities (including Austria and the predecessor of the EU at November 07, 1991) and entered into force in March 06, 1995.[17] The Alpine Convention is a framework agreement that aims to facilitate sustainable development in the Alpine region. On this basis, the Traffic-Protocol of the Alpine Convention was adopted. In the interest of

---

[15]Council Directive 96/82/EC of 9 December 1996 on the control of major-accident hazards involving dangerous substances, OJ L 10, 14.1.1997, pp. 13–33.

[16]Directive 2012/18/EU of the European Parliament and of the Council of 4 July 2012 on the control of major-accident hazards involving dangerous substances, amending and subsequently repealing Council Directive 96/82/EC, OJ L 197, 24.7.2012, pp. 1–37.

[17]Retrieved May 22, 2015 from http://www.alpconv.org/en/convention/framework/default.html and see also the "Alpenkonvention", published in the federal law gazette ("BGBl") III number 477/1995, in the currently valid version.

sustainability, the Parties are obliged to implement an efficient and safe flow of traffic in a cross-border traffic network, which has to comply with certain requirements (Mauerhofer 2004b).

Also at a European level, there are efforts to curtail the problem areas of traffic by various measures. The Directive on the promotion of clean and energy-efficient road transport vehicles[18] (Directive 2009/33/EC) was enacted in order to improve energy efficiency and to reduce pollutant emissions caused by road traffic. Regulation No. 510/2011[19] of the EU tries to tackle to problem at the source, at the point of formation of the emission; collective emission norms according to new light commercial vehicles are defined in order to reduce their CO_2 emissions. In 2011, the Commission summarized the most urgent tasks, objectives and measures referring to traffic problems in the White Paper called "Roadmap to a Single European Transport Area—Towards a competitive and resource efficient transport system" (EC 2011).

In particular the use of alternative fuels is considered to be a possible way to reduce the transport sector's pressure on the climate. The EU program "Horizon 2020", which was established by Regulation No. 1291/2013[20], intended, among other things, to promote research on alternative fuels for vehicles (such as electricity, hydrogen or natural gas). Minimum requirements for the construction of infrastructure necessary to provide the vehicles with the alternative fuels are given by Directive 2014/94/EU.[21] In fall of 2012, the Commission submitted a proposal with the aim to limit the use of land for the purpose of biofuels production to an acceptable degree (EC 2012a). Moreover, the impact of European biofuels on the climate should be improved. This was in reaction to the vehement criticism on the production of biofuels, i.e. the use of food crops for biofuel production and thereby caused rise in price of basic food (Falke 2013a, b).

In May 2014, a strategy to reduce CO_2 emissions of trucks and buses was developed by the Commission (EC 2014d) with the purpose to supplement the legislation on passenger cars and light commercial vehicles. Anyhow, approximately a quarter of the emissions caused by road traffic are produced by heavy commercial vehicles. The current strategy still does not establish effective measures to avoid or reduce emissions. A first step to change this is the implementation of certification, notification and monitoring of pollution causing heavy commercial vehicles (Falke 2014a, b, c).

---

[18]Directive 2009/33/EC of the European Parliament and of the Council of 23 April 2009 on the promotion of clean and energy-efficient road transport vehicles, OJ L 120, 15.5.2009, pp. 5–12.

[19]Regulation (EU) No 510/2011 of the European Parliament and of the Council of 11 May 2011 setting emission performance standards for new light commercial vehicles as part of the Union's integrated approach to reduce CO_2 emissions from light-duty vehicles, OJ L 145, 31.5.2011, pp. 1–18.

[20]Regulation (EU) No 1291/2013 of the European Parliament and of the Council of 11 December 2013 establishing Horizon 2020—the Framework Programme for Research and Innovation (2014–2020) and repealing Decision No 1982/2006/EC, OJ L 347, 20.12.2013, pp. 104–173.

[21]Directive 2014/94/EU of the European Parliament and of the Council of 22 October 2014 on the deployment of alternative fuels infrastructure, OJ L 307, 28.10.2014, pp. 1–20.

Considering its current development, European transport policy will only comply with the aim of sustainability if, on the one hand, a change to alternative driving systems is implemented and, on the other hand, the mobility behaviour of Europeans changes drastically. The tendency to travel bigger distances within shorter time periods—in particular with airplane—needs to experience a trend reversal. The development of a European high-speed rail network, which has already been improved in the last years, seems to be a step in the right direction. Moreover, the EU traffic policy acknowledges that rural areas need to be better connected in many Member States.

The state of air quality in Europe is closely related to the topic of traffic. In order to achieve the objectives for 2020 as set out in the relevant legislation, the Commission adopted a new strategy "Clean Air for Europe" (EC 2013c). Its goal is to ensure compliance with the legislation and that the full potential of existing measures is used; furthermore, health effects of air pollution shall be reduced by a third. The following measures should contribute to these purposes (Falke 2014a, b, c):

- Proposal for a revision of Directive 2003/35/EC on National Emission maximum amount (EC 2013d);
- Proposal for a reduction of emissions of certain air pollutants by medium-sized combustion plants (EC 2013e);
- Ratification of Gothenburg Protocol 2012 and the
- Reduction of ammonia-emissions from agriculture.

Air pollutants have a negative effect on human health. This is indeed essential for the concept of sustainable development: Damage to health caused by environmental toxins raises the macroeconomic costs and can consequently impair economic development. Making environmental standards—in particular those related to human health—enforceable for individuals, seems to be the most suitable way to realize the current sustainability strategy. It certainly requires a widespread application in Europe. Similar considerations arise with respect to toxic food constituents and hazardous substances in products. In this regard, more and more is revealed. Many of the toxins contained in products are suspected to involve a residual risk of impairing the reproductive capacity or the progeny.

4.6 *Nature Protection and Biodiversity*

In 2010, the Commission confirmed in its report on the implementation of the Action Plan for the conservation of biological diversity that the targets were missed and the loss of biodiversity could not be stopped (EC 2010b). Pollution, invasive species and climate change still threaten biodiversity. A new target regarding halting the loss of biodiversity was set by the Commission for the year 2020 (EC 2010c).

In October 2014, the Regulation of the European Parliament and of the Council on the prevention and management of the introduction and spread of invasive alien species[22] was enacted. Invasive alien species have a massive negative impact on biodiversity; their spread causes adverse changes in the domestic ecosystem and affects native species seriously. Furthermore, it is to be feared that they pose a risk to human health as well. The Regulation aims to prevent, control and eliminate the spread of invasive alien species. In order to do so, it obliges to release a list by beginning of 2016 latest that contains all the dangerous species that may not be imported into the EU, acquired, used, released or sold.

Protection of various ecosystems is still a concern of the Union. Specifically, the EU became active now regarding the habitat "forest". Forests serve economic, social and environmental aspects; so their population has an impact on all dimensions of sustainability. Healthy and functioning forests reduce CO_2 in the atmosphere and contribute in this way to climate protection. Moreover, they serve as habitat for many animal and plant species. In 2013, the Commission has developed a new EU Forestry Strategy in order to implement sustainable forest management. On this basis, Member States have to compile national action plans and supporting programs (EC 2013f).

The Environmental Impact Assessment Directive[23] states an obligation to assess the environmental impacts of certain projects before the competent authorities may approve them. If the result of the impact assessment is unsatisfactory, the approval will be refused or, if possible, only granted under certain conditions. In 2013, the Commission has issued detailed guidelines on the application of the EIA.

In April 2014, the EIA Directive was amended in some areas by the Directive 2014/52/EU[24] (see also EC 2012b) in order to contribute to achieve the goal of a "resource-efficient Europe". In summary the most important changes are (Falke 2014a, b, c):

- Obligation for Member States to simplify the EIA process;
- Introduction of deadlines for the various phases of environmental assessments with the obligation to ensure a timely handling of the case;
- Simplification of the screening procedure;
- Quality and content of the EIA reports should be improved;
- Reports shall be written in a comprehensible way;

---

[22]Regulation (EU) No 1143/2014 of the European Parliament and of the Council of 22 October 2014 on the prevention and management of the introduction and spread of invasive alien species, OJ L 317, 4.11.2014, pp. 35–55.

[23]Directive 2011/92/EU of the European Parliament and of the Council of 13 December 2011 on the assessment of the effects of certain public and private projects on the environment, OJ L 26, 28.1.2012, pp. 1–21.

[24]Directive 2014/52/EU of the European Parliament and of the Council of 16 April 2014 amending Directive 2011/92/EU on the assessment of the effects of certain public and private projects on the environment Text with EEA relevance, OJ L 124, 25.4.2014, pp. 1–18.

- Authorities shall demonstrate their objectivity;
- Reasons for a decision of approval shall be disclosed in a transparent and clear way; and
- Projects with significant adverse effects shall be observed as part of the mitigation, prevention and reduction activities (Falke 2014a, b, c).

The Environmental Liability Directive (ELD)[25] on environmental liability with regard to the prevention and remedying of environmental damage establishes a framework for a uniform system of liability for environmental damages. In application of the polluter-pays principle, the person who causes a risk of damage or damage should bear the costs of preventive and remediation measures, respectively. If the perpetrator cannot be held liable for the damages on environmental media or the threat thereof, Member States and the general public can bear the costs in order to avoid that the damage is not remedied at all and an impaired environment is left over (Hinteregger and Kerschner 2011).

5 Conclusion

Sustainability cannot be solely achieved by strategies, programs, plans, long-term goals etc. The legislature has to choose more effective ways in order to ensure that actions are actually implemented in reality. The path to a sustainable economy will be a steep and stony one. It requires common efforts of the whole society. At the moment, we are at the beginning of the path; there is still some temptation to decelerate or even turn back. We have to be aware that any revision of the idea of sustainability would be to the disadvantage of future generations. Are we really that selfish?

References

Arbter K (2005) Policies and instruments that contribute to sustainable development. Which roads lead there? Juridikum, 35- (in German)

Bauer M, Seebacher U (2012) Sustainable action in business and everyday life. SozSi, 161 (in German)

Calliess C (2003). The environmental competences of the EU under the treaty of nice—the framework of European environmental legislation. Zeitschrift für Umweltrecht, 129 (in German)

Calliess C (2011a) Article 11 TFEU. In: Calliess C, Ruffert M (eds) TEU/TFEU (para. 5). C.H.Beck, Munich (in German)

[25]Directive 2004/35/CE of the European Parliament and of the Council of 21 April 2004 on environmental liability with regard to the prevention and remedying of environmental damage, OJ L 143, 30.4.2004, pp. 56–75.

Calliess C (2011b) Article 37 CFR. In: Calliess C, Ruffert M (eds) TEU/TFEU (para. 1). C.H.Beck, Munich (in German)
Christian R, Kerschner F, Wagner E (eds) (2015) Law of energy turn (in German)
Council (2009) Council document 10917/06 of 09th June 2006. Retrieved 22 May 2015, from http://register.consilium.europa.eu/doc/srv?l=EN&f=ST%2010917%202006%20INIT
EC (2001) Communication from the commission, a sustainable Europe for a better world: a European Union strategy for sustainable development, COM (2001) 264 final
EC (2002a) Communication from the commission on impact assessment, COM (2002) 276 final
EC (2002b) Report from the commission to the council, analysis of the 'open list' of environment-related headline indicators, COM (2002) 524 final
EC (2006) Green paper, a European strategy for sustainable, competitive and secure energy, COM (2006) 105 final
EC (2008) Communication from the commission to the European parliament, the council, the European economic and social committee and the committee of the regions "think small first" a "small business act" for Europe, COM (2008) 394 final
EC (2009) White paper, Adapting to climate change: towards a European framework for action, COM (2009) 147 final
EC (2010a) Communication from the commission, Europe 2020 a strategy for smart, sustainable and inclusive growth, COM (2010) 2020 final
EC (2010b) Report from the commission to the council and the European parliament. The 2010 assessment of implementing the EU action plan, COM (2010) 548 final
EC (2010c) Communication from the commission to the European parliament, the council, the European Economic and social committee and the committee of the regions, options for a EU vision and target for biodiversity beyond 2010, COM (2010) 4 final
EC (2011) White paper, roadmap to a single European transport area—towards a competitive and resource efficient transport system, COM (2011) 144 final
EC (2012a) Proposal for a directive of the European parliament and of the council amending directive 98/70/EC relating to the quality of petrol and diesel fuels and amending directive 2009/28/EC on the promotion of the use of energy from renewable sources, COM (2012) 595 final
EC (2012b) Proposal for a directive of the European parliament and of the council amending directive 2011/92/EU on the assessment of the effects of certain public and private projects on the environment, COM (2012) 628 final
EC (2013a) Green paper, a 2030 framework for climate and energy policies, COM (2013) 169 final
EC (2013b) Green paper, on a European strategy on plastic waste in the environment, COM (2013) 123 final
EC (2013c) Communication from the commission to the European parliament, the council, the European economic and social committee and the committee of the regions, a clean air program for Europe, COM (2013) 918 final
EC (2013d) Proposal for a directive of the European parliament and of the council on the reduction of national emissions of certain atmospheric pollutants and amending directive 2003/35/EC, COM (2013) 920 final
EC (2013e) Proposal for a directive of the European parliament and of the council on the limitation of emissions of certain pollutants into the air from medium combustion plants, COM (2013) 919 final
EC (2013f) Communication from the commission to the European parliament, the council, the European economic and social committee and the committee of the regions, a new EU forest strategy: for forests and the forest-based sector, COM (2013) 659 final
EC (2014a) Communication from the commission to the European parliament, the council, the European economic and social committee and the committee of the regions, a policy framework for climate and energy in the period from 2020 to 2030, COM (2014) 15 final

EC (2014b) Communication from the commission to the European parliament, the council, the European economic and social committee and the committee of the regions on resource efficiency opportunities in the building sector, COM (2014) 445 final

EC (2014c) Communication from the commission to the European parliament, the council, the European economic and social committee and the committee of the regions, green action plan for SMEs, COM (2014) 440 final

EC (2014d) Communication from the commission to the council and the European parliament, strategy for reducing heavy-duty vehicles' fuel consumption and CO_2 emissions, COM (2014) 285 final

Epiney A, Scheyli (1999) The concept of sustainable development in the European union. In: Lang W, Hohmann H, Epiney A (eds) The concept of sustainable development. Nations—and European legal aspects. Stäpfli, Zurich, p 43 (in German)

Falke J (2012) New developments in European environmental law. Zeitschrift für Umweltrecht, 639 (in German)

Falke J (2013a) New developments in European environmental law. Zeitschrift für Umweltrecht, 51 (in German)

Falke J (2013b) New developments in European environmental law, Zeitschrift für Umweltrecht, 310 (in German)

Falke J (2014a) New developments in European environmental law, Zeitschrift für Umweltrecht, 249 (in German)

Falke J (2014b) New developments in European environmental law, Zeitschrift für Umweltrecht, 501 (in German)

Falke J (2014c) New developments in European environmental law, Zeitschrift für Umweltrecht, 635 (in German)

Fütterer M, Pichl M (2010) Climate Protection with market and deals: neo-liberal and post-neo-liberal shifts in environmental law and policy on climate protection. Juridikum, 500 (in German)

Gareis SB, Varwick J (2014) The united nations. Tasks, instruments and reforms. Verlag Barbara Budrich, Opladen (in German)

Granner G, Raschauer N (2012) Climate Protection. Recht der Umwelt, 10 (in German)

Grunwald A, Kopfmüller J (2012) Sustainability. Campus Verlag, Frankfurt-Main (in German)

Hinteregger M, Kerschner F (eds) (2011) Commentary on the Bundes-Umwelthaftungsgesetz (Federal environmental liability act). Manz, Vienna (in German)

Kanning H (2013) Sustainable development—the social challenge for the 21st century. In: Baumast A, Pape J (eds) Corporate sustainability management. Ulmer, Stuttgart (in German)

Kerschner F (ed) (1996) National objective environmental protection. Manz, Vienna (in German)

Kerschner F (ed) (2001) Austrian and European traffic law—on the road to sustainability. Manz, Vienna (in German)

Kerschner F, Funk B, Priewasser R (eds) (2009) Education for Sustainable development. Manz, Vienna (in German)

Leif T, Speth R (eds) (2003) The silent power. Lobbyism in Germany. Westdeutscher Verlag, Wiesbaden (in German)

Lengauer A (2011) Union up-to-date. Zeitschrift für Rechtsvergleichung 27:219 (in German)

Lewis P (2011) Multinational enterprises and environment: regulatory framework and voluntary incentives. Recht der Umwelt, 86 (in German)

Luger M (2012) European political visions and strategies for sustainable resource and energy industry. In: Yearbook energy industry, 174 (in German)

Madner V (2011) Communication from the European commission according to the EU position at the UN-conference Rio + 20: towards the green economy and better governance in june 2012 in Rio de Janeiro 20th June 2011, COM (2011) 363 final, Recht der Umwelt, 130 (in German)

Madner V, Hartlieb J (2013) Green paper of the European commission: a 2030 framework for climate and energy policies 27th March 2013, COM (2013) 169 final. Recht der Umwelt, 61 (in German)

Madner V, Isepp K (2014) Communication from the commission of the European parliament and the council: a policy framework for climate and energy in the period from 2020 to 2030, COM (2014) 15 final. Recht der Umwelt, 34 (in German)

Mauerhofer V (2004a) On the legal implementation of sustainability in Austria (part I). Recht der Umwelt 46 (in German)

Mauerhofer V (2004b) On the legal implementation of sustainability in Austria (part II). Recht der Umwelt 74 (in German)

Nettesheim M (2014a) Article 191. In: Grabitz E, Hilf M, Nettesheim M (eds) The law of the European Union: TEU/TFEU, 35th supplemental set (para. 14). Verlag C.H. Beck, Munich (in German)

Nettesheim M (2014b) Article 11. In: Grabitz E, Hilf M, Nettesheim M (eds) The law of the European Union: TEU/TFEU, 35th supplemental set (para. 24). Verlag C.H. Beck, Munich (in German)

Piska C (2010) Environment and waste. In: (eds) Yearbook European law, p 375 (in German)

Rabl T, Brenner W (2009) Promotion of renewable energies: the new directive 2009/28/EC, ecolex, 910 (in German)

Raschauer N (2008) The impact of the Lisbon treaty reform on the European and Austrian Environmental law, HLR 45 (in German)

Reese M (2010) Guiding principles of environmental law. To the future viability of leading protection concepts. Zeitschrift für Umweltrecht 339 (in German)

Rihs G (2012) Green electricity act 2012: flashback and highlights. Recht der Umwelt-Umwelt- und Technikrecht 24:71 (in German)

Schwarzer S (2012) Target agreements between political actors as a management tool in the new climate protection law. Recht der Umwelt 23:49 (in German)

Tichler R, Steinmüller H, Schneider F (2011) Thoughts about an optimal sustainable energy system. In: Steinmüller H, Hauer A, Schneider F (eds) Yearbook energy industry 2011. Neuer Wissenschaftlicher Verlag (NWV), Vienna (in German)

Urbantschitsch W, Hofer E (2009) Environmental Law, ecolex, 908 (in German)

von Carlowitz HC (1713) Silvicultura oeconomica. Johann Friedrich Braun Verlag (in German)

Wagner E (2000) European environmental protection in the light of the Amsterdam treaty. Recht der Umwelt, 43 (in German)

Windoffer A (2011) The impact assessment process as a tool for legal safeguards for sustainability. Mohr Siebeck, Tübingen (in German)

World Commission on Environment and Development (1987) Our common future. Oxford University Press, Oxford

Part II
Horizontal Policies: Human and Intellectual Property Rights

Environmental Sustainability as a Human Right

Rasyikah Md Khalid, Faridah Jalil and Mazlin Bin Mokhtar

Abstract Sustainability means capable of being sustained. The term sustainable development is introduced under this concept to ensure that rapid development takes place in harmony with nature; and natural resources such as forest and water resources can be sustained for future users. Within the legal fraternity, the word 'sustainability' has seeped into the legal discourses as to whether it can be regarded as a legal principle which can be used and argued upon in the court of law. This chapter highlights on prospect and challenges of establishing environmental sustainability as a valid legal principle. It focuses on the concept of environmental jurisprudence and how environmental sustainability can be regarded as a human right or a right to life under a country's constitution. The paper concludes that this can be done through judicial activism and by relaxing the doctrine of legal standing.

Keywords Constitution · Environment · Jurisprudence · Human right · Legal standing · Sustainability

R. Md Khalid (✉)
Faculty of Law, Universiti Teknologi MARA (UiTM), Shah Alam 40450, Selangor, Malaysia
e-mail: kmr572000@yahoo.com

R. Md Khalid · F. Jalil
Faculty of Law, Universiti Kebangsaan Malaysia (UKM), Bangi 43600, Selangor, Malaysia

M. Bin Mokhtar
Institute of Environment and Development, Universiti Kebangsaan Malaysia (UKM), Bangi 43600, Selangor, Malaysia

V. Mauerhofer (ed.), *Legal Aspects of Sustainable Development*,
DOI 10.1007/978-3-319-26021-1_5

1 Theoretical Framework

Jurisprudence refers to the theory or philosophy of law; hence environmental or earth jurisprudence is a philosophy of law that seeks to understand the relationship between law and living things based on different disciplines like ecology, sociology and economy. It concerns with identifying the 'rights' of living and non-living things within the environment that is often left out in the development process that concern with economic and social prosperity (Baber and Bartlett 2009). Pollution, loss of biodiversity and environmental degradation caused by development process will affect each living things that depend on a good environmental flow. Nevertheless, if pollution is to be regarded as a crime, it requires different understanding of law because it is not concerned with the rights and liabilities between individual, but rather between man and the entire environment (Banerjee 2009). Environmental jurisprudence can assist by explaining the interconnectedness of one living thing to another and the rights and duties of a man towards the environment.

From the naturalist point of view, environmental jurisprudence seeks to uphold the notion that natural law is the law of nature and is the fundamental law behind freedom or rights as a human (Kimbrell 2011). Equally, this right should extend to the environment through ecological principles that give due rights to the ecosystem and their services. Nevertheless, industrialisation and modernisation have been motivated and bounded by the efficiency principle which makes efficient use of natural resources to the detriment of the entire ecosystem. In this respect Kimbrell (2011) argues that efficiency must be balanced with objectivity and natural law will treat nature in its own right. However since nature cannot bring the case against man, it is man who will act as its guardian against environmental destruction.

It appears that the naturalists see close relationship between man and nature. Nevertheless, as Bosselmann (2008, p. 13) puts it, the dominance of positive law has led to ecological ignorance and do not regards sustainability of natural resources as a right that must be protected. Judge Christopher Weeramantry in one of his judgement has also argued that the dominance of positivism increases the separation between morality and law as it diminished the 'long tradition of living harmony in nature' and deprived the 'need of human activity to respect the requisites for its maintenance and continuance' [Gabcikovo-Nagymaros Project (Hungary/Slovakia) Case 1997]. Bosselmann (2008, p. 13) further argues that modern environmental law completes this separation by saying:

> The very existence of environmental law as a distinct subject area is proof of the fact that sustainability values have not found their way into the legal system. Both domestic and international environmental law are characterized by the absence of integration and ecological sustainability. Moreover, with their anthropocentric, resources oriented and non-integrative approach, they tend to foster industrialization rather than changing it.

As a result, environmental sustainability becomes a loose concept and does not have contents and directions as a correlative right and duties. This is ironic as present generation owes various duties towards future generation and these duties can be the legal standard by which sustainability must be observed. If Dworkin (1977) describes a legal principle as a standard that is to be observed as a requirement of justice, fairness or morality, then environmental jurisprudence provide a solid foundation by which environmental sustainability should succeed as a legal principle.

By legalizing environmental sustainability, future development may continue without threatening the survival of other species that depend on the same resources. The question of duty that man owes towards another and towards nature generates the idea that such duty arises as he borrows the nature from future generation. Brown-Weiss (1992, p. 385) coined this as intergenerational equity when she says:

> We, the human species, hold the natural environment of our planet in common with all members of our species: past generations, the present generation, and future generations. As members of the present generation, we hold the Earth in trust for future generations. At the same time, we are beneficiaries entitled to use and benefit from it.

It appears that intergenerational equity aims to ensure that the present generation should not abuse non-renewable resources so as to deprive the future generation of its benefit. It concerns not only with relationship between generations but also environmental conservation. Such generational rights corresponds with religious approaches such as Islam that regards human rights not as an individual right but as rights of the community of believers as a whole (Khadduri 2001). Under international law, this principle has found its way in Principle 3 of the Rio Declaration which states that "the right to development must be fulfilled so as to equitably meet developmental and environmental needs of present and future generations" (UN 1992).

In all, it is apparent that there is an urgent need to reinforce environmental sustainability as a legal principle. This can be achieved through a specific legislation or by recognizing environmental sustainability as a guaranteed right under a country's constitution. This must be supported by the judicial system that strictly adheres to environmental jurisprudence. However, as will be discussed below, strict adherence on the rule of legal standing may impede judicial activism in this area.

2 Sustainability as a Constitutional Right

It is interesting to note that several countries have incorporated the concept of environmental sustainability into their constitutions. The Federal Constitution of the Swiss Confederation 1999 (Swiss Parliament 1999) provides the best illustration on this matter where a large number of provisions have been promulgated to incorporate the concept of sustainability and environmental protection. Article 73 for instance provides that the Confederation and the Cantons shall try to achieve "a balanced and sustainable relationship between nature and its capacity to renew itself". Article 74(1) further requires the Confederation to legislate on the protection of the population and its natural environment against damage or nuisance while Article 74(3) makes the Cantons responsible for the implementation of such regulations.

The Swiss Confederation has also shown great interest in sustainability of water resources as it provides a specific provision on the matter. Under Article 76 of the constitution, the Confederation shall ensure the economic use and the protection of water resources as well as protection against harmful effects of water. The Confederation shall lay down principles on the conservation and exploitation of water resource and other measures affecting the water-cycle as well as "legislate on water protection, on ensuring appropriate residual flow, on hydraulic engineering and the safety of dams, and on measures that influence precipitation". In return, the Cantons will manage their water resources and impose charges for the use of water, even on the Confederation for using water for its transport operations. It is the duty of the Confederation to take accounts the concerns of the Cantons where the water originates in fulfilling its duties.

Within the Association of South East Asian Nations (ASEAN), Philippine leads the notion of environmental sustainability as both human and constitutional right when judicial notice was given to the issue of intergenerational equity towards environmental sustainability. Philippine is also at the forefront amongst other ASEAN member states to incorporate sustainability of a healthful environment in its 1987 Constitution (Philippines Parliament 1987). In this respect, section 15 of Article II preserves the right to health while and section 16 provides that "the State shall protect and advance the right of the people to a balanced and healthful ecology in accord with the rhythm and harmony of nature". Read together, states are required to protect and promote the right to a healthful environment.

Following this spirit, the court in the landmark case of *Oposa v. Factoran* (Supreme Court of the Philippines 1993) held that a minor had the right to sue on behalf of his and succeeding generations for a balanced and healthful ecology under Article 16 of the Constitution. The cause of action arose after the Department of Environment and Natural Resources was held to have a clear duty to protect and advance that right and should cease from accepting or approving new timber license agreement. The court held that the right of the people of Philippines to a balanced ecology implied "judicious management and conservation of the country's forest" and elaborates:

> These basic rights need not even be written in the Constitution for they are assumed to exist from the inception of humankind. If they are now explicitly mentioned in the fundamental charter, it is because of the well-founded fear of its framers that unless the rights to a balanced and healthful ecology and to health are mandated as State policies by the Constitution itself, thereby highlighting their continuing importance and imposing upon the State a solemn obligation to preserve the first and protect and advance the second, the day

> would not be too far when all else would be lost not only for the present generation, but also for those to come- generations which stand to inherit nothing but parched earth incapable of sustaining life.

The case becomes the leading case that uses the concept of intergenerational equity which concerns with the moral and philosophical standing to allow the rights and interests of future generations to be taken into account in today's decisions making process. It enforces the present generation's duty not to abuse the environment because it is borrowed from the future generation.

Extensive provisions on environmental sustainability are rarely found in the constitution of Commonwealth countries like India, Malaysia and Pakistan. In fact, the development of sustainability as a legal principle will depend on judicial activism in construing constitutional rights. In India and Pakistan, the judges have used the existing rights under their constitution to include the right to clean environment. This has been done through the provision of "the right to life". In this respect, Article 21 of the Indian Constitution (Indian Parliament 1949) provides that 'no person shall be deprived of his life or personal liberty except according to procedures established by law.' As the law is prescribed in a negative right, the court in *Subhash Kumar v. State of Bihar* (Supreme Court of India 1991) interpreted the right to life to encompass the right to a wholesome environment which include pollution-free water or air.

Similarly in Pakistan, the right to life is provided under Article 9 of the Constitution of the Islamic Republic of Pakistan (National Assembly of Pakistan 2010). As such, the court in *Shehla Zia v. Water and Power Development Authority (WAPDA)* (Supreme Court of Pakistan 1994) held that Article 9 includes 'all such amenities and facilities which a person born in a free country is entitled to enjoy with dignity, legally and constitutionally'. As this case concerns with electro-magnetic field hazards exposure from installation of grid station and the likes, Justice Salem Akhtar commented;

> Under our Constitution, Article 14 provides that the dignity of man and subject to law, the privacy of shall be inviolable. The fundamental right to preserve and protect the dignity of man and right to 'life' are guaranteed under Article 9. If both are read together, question will arise whether a person can be said to have dignity of man if his right to life is below bare necessity line without proper food, clothing, shelter, education, health care, clean atmosphere and unpolluted environment.

It is interesting to note that the Indian government has amended the Indian constitution in 1976 to incorporate important environmental protection provisions (Indian Parliament 1976). They are Article 48A which provides that 'the State shall endeavour to protect and improve the environment and safeguard the forests and wildlife of the country' and Article 51A(g) that requires individuals 'to protect and improve the natural environment including forests, lakes, rivers and wildlife, and to have compassion for living creatures'. These provisions direct the judiciary to be more focussed towards achieving environmental sustainability in the fast developing India. Article 51A(g) for instance has provided a valid ground for the court in the case of *MC Mehta v Union of India* (Supreme Court of India 1987) to order the

Kanpur Municipal Corporation to stop the practice of throwing burnt and semi-burnt corpses into the Ganges River.

In contrast to judicial activism in interpreting human right in India and Pakistan, other common-law country like Malaysia has not incorporated words like environment, let alone the word sustainability, in the country's constitution. As such, it is difficult for the judiciary to determine the status of environmental sustainability within the Malaysian legislative framework. Although there are provisions for land, water and forest as the main natural resources for the state government, there is no mentioning of the word environment or the duty to protect it. Thus, water pollution and landslide caused by poor maintenance of drains and water courses are usually brought before to the court on the basis of tortious acts of negligence as seen in the case of *Majlis Perbandaran Ampang Jaya v Steven Phoa Cheng Loon & Ors* (Federal Court of Malaysia 2006) and Lim Teck Kong v. Dr. Abdul Hamid Abdul Rashid & Anor (Court of Appeal Malaysia 2006), as a nuisance in *Renal Link v. Dato' Dr Harnam Singh* (Court of Appeal Malaysia 1997b), a trespass in *Sin Heap Lee-Marubeni v. Yip Shou Shan* (Court of Appeal Malaysia 2005), and through the Ryland v Fletcher's strict liability rule in *Woon Tan Kan v. Asian Rare Earth Sdn Bhd* (High Court Malaya 1992).

Malaysian judiciary have attempted, albeit obliquely, to infer that the right to clean environment may be part of the right of life under Article 5(1) the Federal Constitution. This was made by Justice Gopal Sri Ram in an unfair dismissal case of *Tan Teck Seng v Suruhanjaya Perkhidmatan Pendidikan* (Court of Appeal Malaysia 1996) when he said:

> …the expression 'life' appearing in Article 5 does not refer to mere existence. It incorporates all those facets that are an integral part of life itself and those matters which go to form the quality of life. Of these are right to seek and be engaged in lawful and gainful employment and to receive those benefits that our society has to offer to its member. It includes the right to live in a reasonable healthy and pollution free environment.

Unfortunately, Justice Gopal Sri Ram did not expand this interpretation in an environmental appeal case of *Ketua Pengarah Alam Sekitar & Anor v Kajing Tubek & Ors* (Court of Appeal Malaysia 1997a). The issue raised in this case was whether the controversial Bakun Dam project was subjected to the Environmental Impact Assessment (EIA) under section 34A of the Environmental Quality Act (EQA) 1974 or the state of Sarawak's Natural Resources Ordinance 1949. The court held that the Sarawak's EIA (Amendment) Order 1995 had retrospectively excluded the operation of section 34A of the EQA 1974 which imposes a duty upon person who carries out any of the prescribed activities to submit an EIA report.

It is submitted that there is still lack of judicial activism by the Malaysian judiciary for environmental protection and conservation. Although a liberal interpretation was given to Article 5 of the Federal Constitution to include the right to life and healthy environment in Tan Teck Seng's case, it has been abandoned in the *Kajing Tubek* case merely because the issue was not raised on that particular appeal and was overwhelmed by the issue of legal standing. As aforesaid, lack of inextricably link between the right to sue and the right to clean environment will impede development of environmental jurisprudence in Malaysia. It is thus submitted that the word environment or sustainability of natural resources to be incorporated into the Malaysian Federal Constitution to provide it with the legal basis and to be upheld in the court of law.

3 The Rule on Locus Standi and Environmental Degradation

One of the difficulties faced by the common law judges in pursuing the development of environmental jurisprudence is the requirement of legal standing or locus standi. This is an entrenched principle in the common law system which means that a plaintiff must have a sufficient or personal stake in the outcome of a case, traceable to the defendant, to distinguish the individual from other persons or the public at large. In Malaysia and Singapore, the rule on standing was confirmed as it stands in the case of *United Engineers Malaysia (UEM) Berhad v. Lim Kit Siang* (Supreme Court of Malaysia 1988). In this case, the respondent asked for a declaration that the letter of intent issued by the government to UEM to construct the North-South Highway was invalid. The Supreme Court allowed the UEM appeal as the respondent could not show that his private rights were infringed or he had suffered damages. In his judgment, Justice Abdul Hamid Omar elaborated:

> The law of standing to sue has two fundamental rules. First, apart from certain cases in which standing to sue is in the discretion of the court, the plaintiff must possess an interest in the issues raised in the proceedings. Second, where the private plaintiff relies on an interest in the enforcement of a public right and not of a private right, standing will be denied unless the Attorney-General consents to a relator action, or the plaintiff can demonstrate some special interest beyond that possessed by the public generally.

Concurring with the above ruling, Justice Gopal Sri Ram in the *Kajing Tubek* case (Court of Appeal Malaysia 1997a) enlightened on the difference between threshold locus standi and the substantive locus standi when he said:

> Threshold locus standi refers to the right of a litigant to approach the court in relation to the facts which form the substratum of his complaint. Although a litigant may have threshold locus standi, he may, for substantive reason, be disentitled to declaratory relief. This, then, is substantive locus standi.

The judge however believed that it is the court's interpretative jurisdiction that will determine the question of locus standi and that may depend on issues like national security or public interest. It is only very recently that the Federal Court has relaxed the interpretation of locus standi to include those with at least a genuine interest in an issue. This was the ruling in the case of *Malayan Trade Unions Congress (MTUC) & Ors v Minister of Energy, Water and Communications & Government of Malaysia* (Court of Appeal Malaysia 2014). In this case, the MTUC pursued against the Government for access to an audit report and the concessionaire agreement with Syarikat Bekalan Air Selangor Sdn Bhd (SYABAS). The court satisfied that the MTUC had a genuine interest in the document and it is not necessary for the applicant to establish infringement of a private right or the suffering of special damage as required in the UEM case. The Court however denied access to both documents on the ground that the audit report was an official secret document and that there was no proof that the Minister's decision to deny access to the concession agreement was illegal or flawed for procedural impropriety. This decision clarified the law on locus standi and allows the general public who has genuine interest in clean environment to bring an action if the environment is destructed.

On a different note, attempts have been made to develop legal standing or the right to sue onto natural resources likes tress and rivers. Stone (2010) argued that if rivers were to be polluted, the owner of land where the river flows will initiate a civil claim against the polluter and the river will be protected as a result; but if the rivers were to have a standing or its own rights, new rules are needed to empower the guardian or trustees of the tree to bring a suit against the factory in the name of the river. This idea was written down in an article called "Should Trees Have Standing?" and has amazingly founds its way in an appeal case of *Sierra Club v Morton* (Supreme Court of USA 1972). In this case, the club brought a suit for an injunction against Walt Disney to construct a recreational park in Sierra Nevada Mountains. Although majority of the judges conquered that the club has no standing and was not adversely affected within the meaning of standing, a dissenting judge paved way for the new rule by saying:

> The idea of standing would be simplified if we allowed environmental issues to be litigated in the name of the inanimate object to be despoiled, defaced or invaded. Contemporary public concern for protecting nature's ecological equilibrium should lead to the conferral of standing upon environmental objects to sue for their own preservation. See Should Trees Have Standing?

This dissenting judgment has brought about mix reaction within the legal fraternity, but the nature-rights movements had used this argument in their plights whilst lawyers had begun to file suits in the name of non-legal person such as, amongst others, a river in *Bryam River v Village of Port Chester* (District Court Connecticut 1974) and an endangered Hawaiian bird called the Palilla in *Palilla v. Hawaiian Department of Land and Natural Resources* (District Court Hawaii 1979).

The Sierra Club judgement marked new development of environmental jurisprudence which becomes highly relevant within the context of climate change. The release of greenhouse gases into the atmosphere, mass scale deforestation and urbanization for instance, has led to changes in the climate temperature which intensify the hydrological cycle. However, development continues as it is driven by competition and efficiency so much so that it disrespects the needs of the ecology and disrupts the entire environmental flow. These concerns have been discussed in *Massachusetts v Environmental Protection Agency (EPA)* (Supreme Court of USA 2007) where the court upheld a petition initiated by a group of private organization, later joined by intervener Massachusetts and other state and local government, for the

EPA to begin regulating the emission of greenhouse gases under section 202(a)(1) of the Clean Air Act. The court held that Massachusetts is a sovereign state and has a prerogative to force the Federal Government to reduce greenhouse gases emissions. The judges had gone beyond the normal causation argument to include 'injury' to the environment as a valid ground of standing.

It is submitted that wavering the rule of standing is much easier in a civil law system like Philippines as was done in the already mentioned *Oposa v. Factoran* case (Supreme Court of the Philippines 1993) than in common law country like Malaysia. At the 2nd Roundtable Discussion for ASEAN Chief Justices on Environmental Enforcement 2012 held in Malacca, Malaysia, the majority of the ASEAN judges felt that wavering of the traditional standing rules as rather challenging. With the exception to the court in the Philippines that allows legitimate parties to have standing for the preservation of environment and some readiness on the part of the Thailand judiciary to do the same, common law judges of Malaysia and Singapore are still adhering religiously to the legal standing requirement. Such attitude has and will delay this interesting and significant improvement of environmental jurisprudence in the thriving and developing South East Asian region.

4 Sustainability of Water Resources as a Human Right

Apart from making environmental sustainability as constitutional right, efforts have been made at the international level provide every person with access to clean water and sanitation. The Universal Declaration of Human Rights 1948 concluded in Paris can be regarded as an earliest attempt to incorporate the right to water as part of human rights. This was followed by the United Nations Conference on the Human Environment in 1972 and the Ministerial Conference on Drinking Water Supply and Environmental Sanitation in 1994. During the 1977 United Nations Water Conference in Mar del Plata, commitments were made to improve water development for agriculture and to provide safe drinking water and sanitation through international cooperation in training program and facilities (UN 1977). In 2004, the Bonn Charter for Safe Drinking Water was promulgated to lay down key principles for better management for reliable provision of safe drinking water which include management of the whole water supply chain, close co-operation between stakeholders and redefining the roles and responsibilities of institution in delivering safe drinking water (IWA 2004). The right to safe drinking water has also become a subset of "the right of everyone to an adequate standard of living" in the International Covenant on Economic, Social and Cultural Rights 1966 (UN 1966).

In addition to these initiatives, the historical plight towards establishing access to water and sanitation as a human right started since the declaration of several General Assembly Resolutions namely:

- Resolution 54/175 of 17 December 1999 on the Right to Development (UN 1999),
- Resolution 55/196 of 20 December 2000 which proclaimed 2003 as the International Year of Freshwater (UN 2000),
- Resolution 58/217 of 23 December 2003 that proclaimed 2005–2015 as the International Decade for Action (UN 2003),
- Resolution 61/192 of 20 December 2006 that proclaimed 2008 as the International Year of Sanitation (UN 2006), and
- Resolution 64/198 of 21 December 2009 for the midterm comprehensive review of the implementation of the International Decade for Action on "Water for Life" (UN 2009).

On July 28, 2010, the General Assembly Resolution 64/292 adopted the "Human Right to Water and Sanitation" which "recognizes the right to safe and clean drinking water and sanitation as a human right that is essential for the full enjoyment of life and all human rights' and "calls upon States and international organizations to provide financial resources, capacity-building and technology transfer, through international assistance and cooperation, in particular to developing countries, in order to scale up efforts to provide safe, clean, accessible and affordable drinking water and sanitation for all" (UN 2010). The passing of this Resolution is seen as highly opportune and complements international commitment to halve the proportion of people who cannot afford safe drinking water by 2015 as prescribed by Goal 7 of the Millennium Development Goals (UN 2013).

Issues on lack of access to basic water supply is also deemed as a substantive issue arising from the implementation of Articles 11 and 12 of the International Covenant on Economic, Social and Cultural Rights (UN 1966). The Covenant specifies a number of rights for securing an adequate standard of living and water becomes an essential element in fulfilling those rights. Although the Covenant does not specifically prescribe the right to water, water is fundamental for human survival and the right to water falls within the purview of other rights such as the right to adequate food, right to health, the right to gain a living by work and the right to take part in cultural life. As the right to food can only be achieved with a sustainable supply of water, the nexus between water and food security becomes inextricably linked. Hence, the duty for states to ensure that the right to water is fulfilled will encompass the right to adequate food and the right to gain a living.

It appears that although the right to water is still new, it has been an important aspect of other recognised rights under international law. In addition, the international human rights law entails specific obligations related to access to safe drinking water. In this respect, Resolution 64/292 requires member states to ensure adequate access to "a sufficient amount of safe drinking water for personal and domestic uses" that includes water for drinking, sanitation, washing, food preparation and hygiene. It also includes the right to maintain access to existing water supplies and the right to be free from interference such as arbitrary disconnection or contamination (UN 2010).

Be that as it may, it becomes rather unfortunate when government decides to develop more dams to meet the majority citizen's right to water. Dam development that requires deforestation and extensive land clearance may disrupt the right to self-determination and livelihood of the locals and the indigenous people in the affected area. This right is provided under Article 3 of the United Nations Declaration on the Rights of Indigenous Peoples (UN 2007) which allows them to freely determine their political status and freely pursue their economic, social and cultural development. Article 4 of the UNDRIP further provides that in exercising the right, indigenous peoples "have the right to autonomy or self-government in matters relating to their internal and local affairs, as well as ways and means for financing their autonomous functions".

Such possible violation of the minority's right to self-determination has started to emerge in the case of indigenous people in Malaysia whose land were taken to give way for an inter-state raw water transfer project from the state of Pahang to the state of Selangor. The federal government of Malaysia embarked on the project after a national study projected future water crisis in the most industrialised state of Selangor which also locates the capital city of Federal Territory of Kuala Lumpur. Although the project aims to meet the right to water of the majority, hundreds of indigenous people had to be relocated as a result and that raised the issue whether their right to self-determination has been infringed (Khalid and Hassan 2013).

It is submitted that in ensuring the right to water is fulfilled, governments must strike the correct balance between these conflicting rights. This can be done by shifting the traditional water supply management approach that requires large water infrastructure projects to meet the increasing demand, towards a more sustainable water demand management that promotes water conservation, recycling or rain-water harvesting. Governments can also introduce incentives or tax rebates for water saving or increase water rate through payment for ecosystem services approach. With more education and awareness campaign, the general public will value water and save more. The overall reduction in water demand will ensure that water resources can be sustained into the future without sacrificing the entire environment.

5 Conclusion

It can be seen that although the notion of environmental sustainability struggles with the positive theory of law, it has found its way into the discourses on environmental jurisprudence as well as intergenerational equity. Be that as it may, none has provided a significant impact on decision making process both at the national and international level that seriously considers and represents the welfare of future generation. It is important that policy makers appreciate that development process can continue with less environmental degradation when environmental sustainability is ensured in decision making process and can be enforced in the court. As such, the status of environmental sustainability which is a vital need for human survival has to start with international agreements related to sustainable development or sustainability of natural resources like water resources. This must be adopted by countries around the world through national laws that uphold environmental sustainability as a person's right under the constitution as seen in the Swiss and Philippine Constitution. This allows the judiciary to punish any person whose activities interfere with the environment or inconsistent with the notion of environmental sustainability. In addition, judges in the common law system must also be bold enough to relax or waive the doctrine of legal 'standing' so that the trees will remain 'standing'.

References

Baber WF, Bartlett RV (2009) Global democracy and sustainable jurisprudence: deliberative environmental law. MIT Press, London

Banerjee D (2009) Environmental jurisprudence in India with reference to initiatives of supreme court for enviro-social justice. J Environ Res Dev 3(4):992–997

Bosselmann K (2008) The principle of sustainability: transforming law and governance. Ashgate Publishing, Surrey

Brown-Weiss E (1992) Intergenerational equity: a legal framework for global environmental change. In: Brown-Weiss E (ed) Environmental change and international law: new challenges and dimensions. United Nations University Press, Tokyo

Court of Appeal Malaysia (1996) Tan Teck Seng v Suruhanjaya Perkhidmatan Pendidikan [1996] 1 MLJ 261

Court of Appeal Malaysia (1997a) Ketua Pengarah Alam Sekitar & Anor v Kajing Tubek & Ors [1997] 3 MLJ 23. Retrieved 22 May 2015. http://www.ecolex.org/server2.php/libcat/docs/COU/Full/En/COU-157029.pdf

Court of Appeal Malaysia (1997b) Renal Link v. Dato' Dr Harnam Singh (1997) 2 MLJ 373

Court of Appeal Malaysia (2005) Sin Heap Lee-Marubeni v. Yip Shou Shan (2005) 1 MLJ 515

Court of Appeal Malaysia (2006) Lim Teck Kong v. Dr. Abdul Hamid Abdul Rashid & Anor (2006) 3 MLJ 213. Retrieved 22 May 2015. http://www.hba.org.my/laws/CourtCases/L/lim_teck_kong_v.htm

Court of Appeal Malaysia (2014) Malayan Trade Unions Congress (MTUC) & Ors v Minister of Energy, Water and Communications & Government of Malaysia (Civil Appeal No. W-01-424-2010). Retrieved 22 May 2015. http://www.kehakiman.gov.my/directory/judgment/file/w-01-424-2010.pdf

District Court Connecticut (1974) Bryam River v Village of Port Chester (1974) 5 E.L.R. 20440. Retrieved 22 May 2015. https://elr.info/sites/default/files/litigation/5.20440.htm

District Court Hawaii (1979) Palilla v. Hawaiian Department of Land and Natural Resources (1979) 471 F. Supp. 985

Dworkin R (1977) The philosophy of law, oxford readings in philosophy. Oxford University Press, New York

Federal Court of Malaysia (2006) Majlis Perbandaran Ampang Jaya (MPAJ) v Steven Phoa Cheng Loon & Ors [2006] 2 MLJ 389. Retrieved 22 May 2015. http://www.hba.org.my/laws/CourtCases/M/majlis.htm

High Court Malaya (1992) Woon Tan Kan v. Asian Rare Earth Sdn Bhd [1992] 3 CLJ 786. Retrieved 22 May 2015. https://archive.org/stream/463359-woon-tan-kan/463359-woon-tan-kan_djvu.txt

Indian Parliament (1949) The Constitution of India. Retrieved 22 May 2015. http://lawmin.nic.in/olwing/coi/coi-english/coi-english.htm

Indian Parliament (1976) The Constitution (Forty-second Amendment) Act. Retrieved 10 Oct 2015. http://india.gov.in/my-government/constitution-india/amendments/constitution-india-forty-second-amendment-act-1976

International Water Association (IWA) The bonn charter for safe drinking water (September 2004). Retrieved 22 May 2015. http://www.who.int/wsportal/bonn/en/

Khadduri M (2001) The Islamic conception of justice. Johns Hopkins University Press, Maryland

Khalid RM, Hassan MS (2013) Sustaining water supply in Malaysia: an overview of possible conflict of rights under international law. Paper presented at the 2nd international conference on law and society, International Islamic University Malaysia

Kimbrell A (2011) Nature's law: reinventing environmental jurisprudence. Retrieved 22 May. http://www.gaiafoundation.org/galleries/podcasts/nature%E2%80%99s-law-reinventing-environmental-jurisprudence-andrew-kimbrell

Malaysian Parliament (1957) Federal Constitution of Malaysia. Retrieved 22 May 2015. http://www.agc.gov.my/images/Personalisation/Buss/pdf/Federal%20Consti%20(BI%20text).pdf

National Assembly of Pakistan (2010) The Constitution of the Islamic Republic of Pakistan. Retrieved 22 May 2015. http://www.na.gov.pk/publications/constitution.pdf

Philippines Parliament (1987) The 1987 Constitution of the Republic of the Philippines. Retrieved 22 May 2015. http://www.gov.ph/constitutions/the-1987-constitution-of-the-republic-of-the-philippines/

Stone CD (2010) Should tress have standing and other essays on law, morals and the environment. Oxford University Press, New York

Supreme Court of India (1987) MC Mehta v Union of India 1987 AIR 965. Retrieved 22 May 2015. http://indiankanoon.org/doc/1285836/

Supreme Court of India (1991) Subhash Kumar v. State of Bihar 1991 AIR 420. Retrieved 22 May 2015. http://indiankanoon.org/doc/1646284/

Supreme Court of Malaysia (1988) United Engineers Malaysia (UEM) Berhad v. Lim Kit Siang [1988] 2 MLJ 12

Supreme Court of Pakistan (1994) Shehla Zia v. Water and Power Development Authority (WAPDA) (PLD 1994 SC 693). Retrieved 22 May 2015. https://cmsdata.iucn.org/downloads/shehla_zia_v__wapda__hr_no__15_k_of_1992___pld_1994_sc_693.pdf

Supreme Court of the Philippines (1993). Oposa v. Factoran (G.R. No: 101083). Retrieved 22 May 2015. http://www1.umn.edu/humanrts/research/Philippines/Oposa%20v%20Factoran,%20GR%20No.%20101083,%20July%2030,%201993,%20on%20the%20State's%20Responsibility%20To%20Protect%20the%20Right%20To%20Live%20in%20a%20Healthy%20Environment.pdf

Supreme Court of USA (2007). Massachusetts v Environmental Protection Agency (EPA) 549 U.S. 497. Retrieved 22 May 2015. http://www.supremecourt.gov/opinions/06pdf/05-1120.pdf

Supreme Court of USA (1972) Sierra Club v Morton 405 U.S. 727 (1972). Retrieved 22 May 2015. https://supreme.justia.com/cases/federal/us/405/727/case.html

Swiss Parliament (1999) Federal Constitution of the Swiss Confederation. Retrieved 22 May 2015. https://www.admin.ch/opc/en/classified-compilation/19995395/index.html

UN (1966) United Nations International Covenant on Economic, Social and Cultural Rights. Retrieved 22 May 2015. http://www.ohchr.org/Documents/ProfessionalInterest/cescr.pdf

UN (1977) United Nations Report of the United Nations Water Conference. Retrieved 22 May 2015. http://www.internationalwaterlaw.org/bibliography/UN/Mar_del_Plata_Report.pdf

UN (1992) Rio Declaration on Environment and Development. Retrieved 22 May 2015. http://www.unep.org/Documents.Multilingual/Default.asp?documentid=78&articleid=1163

UN (1999) United Nations General Assembly Resolution 54/175. Retrieved 22 May 2015. http://www.un.org/en/ga/search/view_doc.asp?symbol=A/RES/54/175

UN (2000) United Nations General Assembly Resolution 55/196. Retrieved 22 May 2015. http://www.un.org/en/ga/search/view_doc.asp?symbol=A/RES/55/196

UN (2003) United Nations General Assembly Resolution 58/217. Retrieved 22 May 2015. http://www.un.org/en/ga/search/view_doc.asp?symbol=A/RES/58/217

UN (2006) United Nations General Assembly Resolution 61/192. Retrieved 22 May 2015. http://www.un.org/en/ga/search/view_doc.asp?symbol=A/RES/61/192&Lang=E

UN (2007) United Nations Declaration on the Rights of Indigenous Peoples. Retrieved 22 May 2015. http://www.un.org/esa/socdev/unpfii/documents/DRIPS_en.pdf

UN (2009) United Nations General Assembly Resolution 64/198. Retrieved 22 May 2015. http://www.un.org/en/ga/search/view_doc.asp?symbol=A/RES/64/198

UN (2010) United Nations General Assembly Resolution 64/292. Retrieved 22 May 2015. http://www.un.org/en/ga/search/view_doc.asp?symbol=A/RES/64/292

UN (2013) United Nations Millennium Development Goals. Retrieved 22 May 2015. http://www.un.org/millenniumgoals/pdf/Goal_7_fs.pdf

Ensuring Access to Safe Drinking Water as an Imperative of Sustainable Development

Vasilka Sancin and Maša Kovič Dine

Abstract The principle of sustainable development has become a constant phrase of international documents adopted in the last few decades leading to the development of an international legal obligation of ensuring sustainable development. This chapter argues, that ensuring access to safe drinking water is one of the imperative due diligence obligations derived from sustainable development, requiring from a State to take all necessary actions that can be reasonable expected from it to take to ensure sustainable development. This obligation of ensuring access to safe drinking water is an obligation of conduct and as illustrated in the chapter stems from the two intertwined pillars of sustainable development: social development and environmental protection. Such an understanding of the obligation to ensure sustainable development opens up a possibility of additional legal reasoning, alongside the human rights arguments, when advocating for respect of an international legal obligation imposed on States to diligently pursue their efforts of ensuring accessibility to safe drinking water, not only to specifically affected groups of people, but systemically, to the entirety of their populations.

Keywords Access to safe drinking water · Due diligence · Obligation of conduct · Human right · Rational utilization

1 Introduction

> Water is critical for sustainable development, including environmental integrity and the eradication of poverty and hunger, and is indispensable for human health and well-being.

V. Sancin (✉) · M. Kovič Dine
Faculty of Law, University of Ljubljana, Ljubljana, Slovenia
e-mail: vasilka.sancin@pf.uni-lj.si

M. Kovič Dine
e-mail: masa.kovic-dine@pf.uni-lj.si

V. Mauerhofer (ed.), *Legal Aspects of Sustainable Development*,
DOI 10.1007/978-3-319-26021-1_6

(General Assembly Resolution 58/217, International Decade for Action, "Water for Life", 2005–2015)

Water is at the core of sustainable development.

(General Assembly Resolution 66/277, The future we want, Rio +20)

Without ensuring access to safe drinking water sustainable development becomes an anachronistic notion. While today the term sustainable development has become a part of our daily vocabulary, it seems that its real meaning is often misunderstood to the advantage of environmental protection. This likely stems from the initial definition of the term sustainable development adopted by the World Commission on Environment and Development in 1987 in its report entitled Our Common Future (WCEL 1987):

> development that meets the needs of the present without compromising the ability of future generations to meet their own needs.

The term "needs" in this definition has been mostly understood in relation to natural resources, which has influenced the strong environmental perception of the term sustainable development. However, the Rio Declaration on Environment and Development (hereafter Rio Declaration 1992), extended the meaning of sustainable development beyond its environmental focus. The three components of sustainable development—economic development, social development and environmental protection—are today considered as independent and mutually reinforcing pillars and access to safe drinking water is a necessary precondition to ensure at least the social development, while it significantly contributes also the achievement of economic development and environmental protection. Water is the necessary element for the survival of the humankind and as such satisfies the social and human element of the sustainable development. Water is also the necessary element for the survival of the fauna and flora on the planet Earth and for all the other chemical processes on the planet ensuring a liveable ecosystem. Finally, water figures as a commercial good necessary for the functioning of agriculture and numerous other industries.

So far scholarly research has mostly focused on the question whether access to safe drinking water can be qualified as a human right (Winkler 2012), prompting some scholars to argue that human rights are not an instrument to solve issues of resource scarcity (Sunstein 1995, p. 730; Scheuring 2009, pp. 155–156). However, this chapter attempts to demonstrate that States need to ensure access to safe drinking water not only as their obligation deriving from human rights law, but separately and additionally also in order to fulfil their obligations under international law on sustainable development. While it is true that the human rights dimension brings with it also wider possibilities of invocation of these obligations, including the possibility of bringing claims against States by individuals either through UN mechanisms or regional human rights courts, pursuing sustainable development's substantiated arguments seems so far an underexplored avenue for addressing States' failure of ensuring access to safe drinking water. Moreover, the human rights dimension of the access to safe drinking water at the present stage of development can at most be regarded as an individual and not a collective human right,[1] while sustainable development intrinsically calls for a collective approach requiring systemic and comprehensive solutions benefiting the entire State's population with potential positive cross-border implications.

Hence, this chapter argues that States are under the sustainable development legal framework under an obligation of conduct of ensuring access to safe drinking water on the entirety of their populated territories in particular under the two strongly intertwined pillars—social development and environmental protection. In ensuring the obligation of sustainable development, the States thus have an underlying due diligence obligation to ensure access to safe drinking water stemming from various elements of these two pillars as illustrated in the following sections of the chapter.

2 Three Pillars of Sustainable Development

The Rio Conference on Environment and Development, was convened with the intention to address the issue of sustainable development in its entirety. The Rio Declaration (1992) consequently recognizes that sustainable development can only be reached once environmental protection becomes a fundamental element of the development process and not a separate concept (Principle 4) (Kasimbazi 1995, p. 89), while clearly indicating the human dimension of sustainable development (Principle 1). Human beings are thus put at the center of concern of sustainable development and are entitled to a healthy and productive life in harmony with nature. Scholars emphasize that such a healthy and productive life cannot be achieved without fulfilling the fundamental needs of the people (Schadendorf 2015, p. 172; Sunstein 1995, p. 730; Scheuring 2009, pp. 155–156).

The international community clearly confirmed that social development is one of the three main pillars of sustainable development in Johannesburg Plan of Implementation (UN 2002) adopted in 2002 at the World Summit on Sustainable Development. The Johannesburg Plan of Implementation (UN 2002) provides a framework for activities of States to ensure the implementation of the Rio principles and unequivocally extends the application of sustainable development from environmental to economic and social development (Cordonier Segger and Khalfan 2004, pp. 26–28). It confirms the need to promote the "integration of the three components of sustainable development—economic development, social development and environmental protection—as independent and mutually reinforcing pillars" (UN 2002, p. 2). The Johannesburg Plan of Implementation contains

[1] Although human rights obligations are *erga omnes* obligations, meaning they can be invoked not only by affected individuals, but also other States, inter-state cases in front of regional human rights courts are extremely rare.

numerous actions by States towards reduction of poverty, unsustainable patterns of consumption, conservation and management of natural resources and the improvement of the general health of the populations worldwide (UN 2002). It recognizes that sustainable development is the main tool for reducing these problems and improving the well-being of the world population. All of these actions are also undeniably connected to access to safe drinking water as none of them can be achieved without it.

2.1 International Legal Obligation of Ensuring Sustainable Development

The Johannesburg Plan of Action not only confirmed the existence of the three pillars of sustainable development, but also the existence of a legal obligation of ensuring sustainable development (UN 2002). Moreover, the International Court of Justice in the *Gabčikovo-Nagymaroš Case* (ICJ 1997) recognized the trend of adoption of new legal norms and standards that require the reconciliation of economic development with the protection of the environment, as expressed in the concept of sustainable development (ICJ 1997, p. 78). Judge Weeramantry in his separate opinion clarified that sustainable development is "more than a mere concept, but a principle with normative value." (Weeramantry 1997, p. 89). Several studies have been prepared trying to define the particular obligations of States arising out of the obligation sustainable development (UNSCD 1996; IUCN 2010; New Delhi Declaration 1992). Among these The New Delhi Declaration of Principles of International Law Related to Sustainable Development (hereafter New Delhi Declaration 1992) prepared by the Committee on Legal Aspects of Sustainable Development of the International Law Association deserve particular attention as it is the most comprehensive document on the legal obligations of sustainable development (French 2011, p. 536). The New Delhi Declaration (1992) defines seven core obligations that are derived from the obligation to ensure sustainable development. These are: (1) the duty of States to ensure sustainable use of natural resources, (2) equity and the eradication of poverty, (3) common but differentiated responsibilities, (4) the precautionary approach to human health, natural resources and ecosystems, (5) public participation and access to information and justice, (6) good governance, (7) integration and interrelationship, in particular in relation to human rights and social, economic and environmental objectives. These obligations can be divided in two groups: obligations specific for sustainable development that take into account environmental protection and social rights while steering economic development and the general principles of international environmental law (Attapatu 2002, p. 273).

For the purpose of this chapter, special attention is given to the duty of States to ensure sustainable use of natural resources. This obligation recognizes the sovereign right of States to exploit its natural resources, while respecting the obligation not to cause harm to the environment of other States and areas beyond the jurisdiction of any States and to exploit the natural resources in a manner, that contributes to the development of its populations while also respecting the needs of future generations (New Delhi Declaration 1992, Principle 1; French 2011, p. 537; Cordonier Segger and Khalfan 2004, p. 110; Saunders 1991, pp. 883–884; Palassis 2011, pp. 62–63.). The goal of this duty is to ensure the conservation of the natural resources for the present and future generations.

Water as a natural resource is of outmost importance for the survival of humans. Hence ensuring that the human population has access to safe drinking water is essential for its development. Therefore, ensuring exploitation of this natural resource in a manner that grants access to safe drinking water is a specific obligation that ensures the fulfilment of the obligation of sustainable development. Fulfilling the obligation of ensuring access to safe drinking water is thus a precondition for the fulfilment of the obligation of sustainable development. When international documents require States to ensure sustainable development, they directly transfer the obligation to ensure availability of natural resources on these States (Maggio 1996–1997, pp. 216–217). Hence, the obligation of sustainable development directly transfers the obligation of access to safe drinking water on the States.

3 Access to Safe Drinking Water as an Element of Social Development

Access to safe drinking water is a necessary requirement for social development. Article 11 of the International Covenant on Economic, Social and Cultural Rights adopted in 1966 (ICESCR 1966) provides:

> The States Parties to the present Covenant recognize the right of everyone to an adequate standard of living for himself and his family, including adequate food, clothing and housing, and to the continuous improvement of living conditions.

While article 11 does not explicitly address the right to water, the Committee on Economic, Social and Cultural Rights confirmed in 2002 with the acceptance of the General Comment No. 15, that the right to adequate standard of living includes also the right to water (CESCR 2002). Moreover, the majority of States, that have ratified ICESCR have confirmed in their political declarations that the right of adequate standard of living necessarily includes the right to safe drinking water (Right to Water 2015). Adoption of the general comment has an important function in the recognition of access to safe drinking water as a social right. Namely, before its adoption access to water has been strongly considered only as an economic commodity (Scheuring 2009, pp. 147–148), mainly because it was considered that management of water as an economic good would provide the background for its best utilization and ensure equitable use (Scheuring 2009, p. 148). Unfortunately, this has had the opposite effect leading to privatization of water resources and water prices too high for the impoverished and marginalized communities to afford (Scheuring 2009, p. 148). This consequence was the main drive for the adoption of the general comment. This is confirmed in its paragraph 11 where it is specifically stated that water should be treated as a social good and not primarily as an economic one, calling for the recognition that the right to water must be realized in a sustainable manner, available for current and future generations (CESCR 2002, para. 11).

The General Comment has put forward three main criteria for satisfying the right to access to safe drinking water (CESCR 2002, para. 12). These are availability of water for sufficient and continuous use for personal and domestic needs of each person, quality of water that is "free from micro-organisms, chemical substances and radiological hazards that constitute a threat to a person's health" (CESCR 2002, para. 12b) and the need to ensure that water is physically and economically accessible to every person without discrimination. These criteria further confirm that access to water is treated generally as an element of social development of any society and as the General Comment also outlines, it should never be used as an instrument of political and economic pressure (CESCR 2002, para. 32).

The General Comment No. 15 claims further that the right to safe drinking water can be derived also from article 12 of the ICESCR recognizing "the right of everyone to the enjoyment of the highest attainable standard of physical and mental health." (CESCR 2002, para. 3).

Following the abovementioned interpretations, States Parties to the ICESCR have the obligation to ensure at least the minimal standards of access to safe drinking water for their population. Nine core obligations of States in this regard are to (CESCR 2002, para. 37): (a) ensure access to the minimum essential amount of water, (b) ensure the right and water facilities on a non-discriminatory basis, (c) ensure physical access to water facilities or services with sufficient amount of safe and drinking water, (d) ensure that personal security is not threatened when having physical access to water, (e) ensure equitable distribution, (f) adopt and implement a national water strategy and plan of action for the whole population with periodical reviews, (g) monitor the realization of the right to water, (h) adopt low-cost water programs targeted to protect the vulnerable and marginalized groups, and (i) take measures to prevent, treat and control diseases linked to water. Each of these core obligations bears an element of sustainability, directly linking the access to safe drinking water to the obligation of ensuring sustainable development.

On 28 July 2010 the United Nations General Assembly (UNGA) in its Resolution 64/292 (UNGA 2010) declared that "the right to safe and clean drinking water and sanitation as a human right is essential for the full enjoyment of life and all human rights." (UNGA 2010, para. 1). This resolution confirms what many have argued with regard to the ICESCR (Salman 2012, p. 45) that access to water is central for the fulfilment of all the other ICESCR rights and also rights enshrined in the International Covenant on Civil and Political Rights (ICCPR 1966). Hence, confirming that water is at the centrality of States' obligations under Article 1(2) of ICESCR (1966), which states that people shall not be "deprived of their own means of subsistence", confirms that access to safe drinking water is a basic need of humanity (Salman 2012, p. 45).

Just a few months later in September 2010 the United Nations Human Rights Council (HRC) adopted the Resolution 18/1, which recognizes that access to safe drinking water is a right under international law included in several international human rights documents under the right to the adequate standard of living (HRC 2010, para. 7a). The HRC resolution also calls upon the States to monitor and analyse the realization of the right to access to safe drinking water on the basis of the criteria set up in the General Comment No. 15 i.e. "availability, quality, acceptability and affordability". As will be presented below these criteria represent also the principle of rational utilization of water, as the water can be utilized only according to the scope of the availability and the quality of the water resource itself.

The report of the former UN Independent Expert on the Human Right of Access to Safe Drinking Water and Sanitation, Catarina de Albuquerque, noted that the right to safe drinking water is a human right of equal value to all the other human rights (de Albuquerque 2010).

While the ICESCR (1966) might be the only general treaty recognizing the right to water in an "indirect" way, there are several other human rights documents that address the right to water in an implicit way. It has been argued by several authors (Winkler 2012, p. 49; Ziganshina 2008, p. 114), that article 6 of the ICCPR on the right to life can be considered to cover also the right to water, as water is indispensable for life, which has also been confirmed by the majority of States through the adoption of the UNGA Res. 64/292. Additionally, water is essential for the health of each individual. Lack of water and especially safe and drinking water leads to diseases among population that can cause death. The survival of the human population is dependent on sufficient access to safe drinking water. Water is hence one of those elementary natural resources that essentially needs to be managed in the most sustainable way to ensure its availability also for the future generations.

Access to safe drinking water is generally the most limited for marginalized groups, like children, women, disabled and the indigenous peoples. Hence all the international documents addressing the needs of these marginalized groups recognize their right to access to safe drinking water. The Convention on the Rights of the Child from 1989 (UN 1989) recognizes the right to access to safe drinking water in articles 24 and 27. Article 24 grants each child the right to achieve the highest possible health level, which cannot be attained without access to adequate amount of food and water of satisfactory standards for health. Thus, this right dependent on the realization of the right to a sufficient standard of living entailed in article 27 of the Convention.

The call for improvement of the standard of living including to provide access to safe and drinking water is also included in the Convention on Elimination of all Forms of Racial Discrimination Against Women from 1979 (UN 1980). Here the right to water is directly expressed as part of the right to non-discrimination of women in article 14, which also expressly recognizes the developmental dimension of the right. States have the obligation to ensure women participation in the development processes by ensuring the right to a sufficient standard of living, especially through ensuring boarding, sufficient hygienic conditions, access to electrical energy, transportation routes and connections and access to safe drinking water. The wording of article 14 is one of the most direct expressions of the connection between access to water and sustainable development in an international document.

Finally, also the International Convention on the Rights of Persons with Disabilities in Article 28 recognizes that disabled persons and their families have the right to the adequate standard of living, which includes adequate nutrition, clothing and shelter. The States are also bound to ensure access to safe drinking water (UN 2007).

As already mentioned above, states have continuously confirmed that access to safe drinking water is a human right and several States have included such a right in their constitutions, e.g. South Africa, Kenya, Uruguay (Langford 2006, p. 443; Ziganshina 2008, p. 118), Uganda and Ecuador (Ziganshina 2008, p. 115). In Argentina, Belgium, Germany and India the judiciary has derived this right as its own right from other human rights (Langford 2006, pp. 443–444).

The obligation of ensuring access to safe drinking water has been included not only in human rights documents but also in numerous environmental law documents. Already in 1977 the United Nations Water Conference in its outcome document recognized that all peoples regardless of their social and economic status have the right to access to safe drinking water of the quality and quantity necessary to fulfil their basic needs (UN 1977). Similarly, the Council of Europe (CoE) at the 2001 meeting of the Committee of Ministers to Member States on the European Charter of Water Resources, confirmed that everyone has the right to a sufficient quantity of water that meet their basic needs (CoE 2001). This confirmation was based on the recognition of the fundamental right to safe drinking water by international human rights documents (CoE 2001). It also needs to be pointed out that the preamble of the Recommendation of the Ministers of the Council of Europe recognizes already a year before at the Johannesburg World Summit on Sustainable Development that "water is an ecological, economic and social asset that is a prerequisite for sustainable development." (CoE 2001 recital 5 of the preamble).

At the 2002 Johannesburg World Summit on Sustainable Development governments not only confirmed the three abovementioned pillars of sustainable development, but also recognized that access to safe drinking water is a human right (Langford 2006, p. 438), thus falling under the social and human rights pillar. This has been reiterated also in later documents. The United Nations Millennium Declaration (UN 2000) called for "halving, by the year 2015, the proportion of people who are unable to reach or to afford safe drinking water." (UN 2000, para 19). The General Assembly Resolution 66/288 "The Future we want" endorsing the Rio +20 United Nations Conference on Sustainable Development in 2012 (hereafter Rio +20 Declaration 2012) reaffirmed the existence of a human right of access to drinking water (Rio +20 Declaration 2012, para. 121).

4 Environmental Protection and Access to Safe Drinking Water

Water cannot sustain the human population, if it is not safe for drinking. The General Comment No. 15 alone requires that water has to be of a certain quality. Hence the human rights documents already set the basic standard for water quality and protection. These are then further defined in numerous water protection documents that recognize the environmental element of access to safe drinking water. However, the environmental protection aspect of sustainable development does not only cover the issues of water quality, but also questions of water management and equitable use of water resources. There is a slight difference between the latter two notions. Equitable use is a principle of general water management law and has evolved and is still exclusively referred to for international watercourses, while the issue of appropriate and rational water utilization and management is in the interest of all States and thus refers to all water resources, not only transboundary.

The principle of equitable utilization is the most important common denominator of international documents on international watercourse management and the fundamental principle of international watercourse law. The 1997 Convention on the Law of Non-Navigational Use of International Watercourses, 1992 UNECE Convention on the Protection and Use of Transboundary Watercourses and International Lakes and its protocol, as well as the 1966 ILA Helsinki Rules on the Uses of Waters of International Rivers all recognize the obligation of States sharing an international watercourse to use the watercourse in an equitable manner. The principle of equitable utilization calls for equitable and reasonable sharing of the watercourse between riparian States considering several factors, such as the size of the population, geography and availability of other freshwater resources, among others (Versteeg 2007, p. 374). Equitable use of water resources thus requires from all the States that share a common water resource to use it in a way that enables all the other States such a use of the same water resource, as is necessary for the needs of their population. This means that the use of the water resource by one State is limited by the needs of the other States. Equitable utilization is thus as such already a form of expression of the principle of sustainable development, where every State has the right to use the water resources for its population to the limit that enables such use also to the population of other riparian States, as well as for the future generations of all the States sharing the resource.

The principle of equitable utilization hence stands as a barrier for States to prevent overuse of the shared water resource. As it has been developed as a principle of international watercourse law, it is often seen as applicable only to shared resources. However, a closer look at international law documents on water management indicates, that equitable utilization applies to non-shared water sources as well. The phrase generally used in these documents is not equitable utilization but rational utilization, however the idea of preservation of water for all generations both current and future, remains the backbone of this term.

The International Court of Justice in the *Pulp Mills Case* (ICJ 2010, p. 4) confirmed that the principle of reasonable utilization is an obligation under sustainable development legal framework, deciding that the principle of reasonable utilization reflects also "the need to strike a balance between the use of the waters and the protection of the river consistent with the objective of sustainable development." (ICJ 2010, para. 177).

While all documents dealing with water management of non-shared resources have the nature of soft-law, they still indicate the general opinions of the majority of States, as most of them were adopted at international conferences or by international organizations with near universal participation of States. Agenda 21 adopted at the United Nations Conference on Environment and Development in Rio de Janeiro in 1992, is currently the most authoritative international document on water management of non-shared freshwater resources with its Chap. "Coherence Issues Between Climate, Energy and Water in the Case of European Hydropower: Can We Have It All?", addressing the Protection of quality and supply of freshwater resources (Agenda 21 1992). In paragraph 18.2 Agenda 21 confirms that "water is needed in all aspects of life" and further calls on States "that adequate supplies of water of good quality are maintained for the entire population of this planet..." recognizing that access to water is an essential obligation for ensuring sustainable development of the worldwide populations. In paragraph 18.3 Agenda 21 itself confirms that rational utilization of water resources is essential to this achievement.

The United Nations Millennium Declaration finally confirmed that rational utilisation is a necessary element of sustainable development by committing "to stop the unsustainable exploitation of water resources by developing water management strategies at the regional, national and local levels, which promote equitable access and adequate supplies." (UN 2000, para. 23). Furthermore, the Rio +20 Declaration reiterated that access to safe drinking water is necessary to improve the implementation of integrated water resource management (Rio +20 Declaration 2012, para. 120), which cannot be carried out without respecting the principle of rational utilisation.

5 Access to Safe Drinking Water as a Due Diligence Obligation of Sustainable Development

The elementary obligation of sustainable development imposes on States a due diligence obligation of ensuring access to safe drinking water for their present and future populations. The due diligence obligations entails "diligence reasonably expected from, and ordinarily exercised by, a person who seeks to satisfy a legal requirement or to discharge an obligation." (Garner 2007, p. 488). *Mutatis mutandis*, States have to exercise diligence reasonably expected from, and ordinarily exercised by them to satisfy their obligation of ensuring access to safe drinking water in concrete circumstances of each case. The notion of due diligence, as "an obligation to deploy adequate means, to exercise best possible efforts, to do the utmost, to obtain th[e] result" (ITLOS 2011, para. 110) is gaining a momentum in particular in the spheres of environmental law and human rights. Both, the human right to access to safe drinking water and the environmental principle of rational utilization, are essential elements of the obligation to ensure sustainable development. They are of equal importance for its implementation and no preference can be given to anyone of them (Versteeg 2007, p. 386). Even some national legislations like the 1998 South African National Water Act (National Water Act 36 of 1998 cited after Ziganshina 2008, p. 118) provide for rational utilization of water resources, by requiring that water resources should be used to meet the basic human needs and ecological needs and only then can they be used for other purposes (Ziganshina 2008, p. 118).

The recognition of access to safe drinking water as a human right in the ICESCR (1966) and the need for rational utilization of any water resource as derived from the customary law principle of equitable utilization are both obligations of conduct. Article 2(1) of the ICESCR (1966) provides for progressive realization of the right to access to safe drinking water. It obliges States to take all the necessary steps to the maximum of their available resources to achieve progressively the full realization of the rights under ICESCR (Salman 2012, p. 46). ICESCR (1966) acknowledges that some States (mostly developing States) may have difficulties fulfilling certain covenant rights due to the limits of available resources. Access for safe drinking water is specifically one of such rights, as States may lack substantial water resources or access to water resources is prohibitively expensive due to geographical and geological restrictions.

Similarly, ensuring sustainable development requires due diligence in States' behaviour. As can been derived from the judgment in the *Pulp Mills* case, due diligence means that a State "is obliged to use all the means at its disposal" to fulfil an obligation of conduct (ICJ 2010, para. 101). Consequently, a State will not breach its obligation of sustainable development if it can prove that it did everything in its power to ensure access to safe drinking water (ICJ 2010, para. 204). If a State can prove, that it made every effort to its maximum available resources to ensure access to safe drinking water by showing that it acted vigilantly, by preventing any damage (ICJ 2010, para. 204) to the natural resource (McIntyre 2010, pp. 476, 485), managing it rationally, as per the principle of rational utilization, then the failure to reach the end result will not count as a violation.

6 Conclusion

The obligation to ensure sustainable development requires States to act in consideration of social, environmental and economic development. This notion often leaves one wondering what exactly has a State to do to fulfil this obligation deriving from numerous international law documents. The New Delhi Declaration adopted by the International Law Association has determined seven duties and principles that need to be fulfilled every time a state has to implement its obligation of ensuring sustainable development. While these seven core duties significantly aid in the understanding of the substance of the obligation, they represent more a set of guidelines than determining concrete obligations.

As the *Pulp Mills Case* shows, the obligation of ensuring sustainable development calls for due diligence by States in fulfilling certain obligations of conduct. These obligations of conduct have to satisfy the three pillars of sustainable development. As the chapter has illustrated, the obligation of access to safe drinking water is one of such obligations, as it carries the social, environmental and economic elements of development. However, the social and environmental dimensions of access to safe drinking water, have been questioned many times, as water has been primarily seen as an economic good. As was presented in this chapter access to safe drinking water satisfies also the social dimension, as a human right under international human rights law, as well as the environmental dimension requiring rational utilization of the water resources.

Recognition that the fulfilment of the due diligence obligation of ensuring access to safe drinking water is imperative for the implementation of the principle of sustainable development has a further legal implication. It allows for an additional legal reasoning, alongside the human rights arguments, when advocating for respect of an international legal obligation imposed on States to diligently pursue their efforts of ensuring accessibility to safe drinking water, not only to specifically affected groups of people, but systemically, to the entirety of their populations.

References

Agenda 21 (1992) Programme of action for sustainable development, UN GAOR, 46th Sess., Agenda Item 21, UN Doc A/Conf.151/26 (1992)

Attapatu S (2002) Sustainable development, Myth or Reality? A survey of sustainable development under International Law and Sri Lankan Law. Georget Int Environ Law Rev 14:265–300

CESCR (2002) General Comment No. 15 (2002), The Right to Water, UN Doc. E/C.12/2002/11

Cordonier Segger MC, Khalfan A (2004) Sustainable development law, principles, practices and prospects. Oxford University Press, Oxford

de Albuquerque (2010) United Nations Human Rights, Office of the High Commissioner for Human Rights, UN united to make the right to water and sanitation legally binding. Retrieved 15 July 2015 from http://www.ohchr.org/SP/NewsEvents/Pages/DisplayNews.aspx?NewsID=10403&LangID=E

French D (2011) From the depths: rich pickings of principles of sustainable development and general international law on the ocean floor—the seabed disputes chamber's 2011 advisory opinion. Int J Marine Coast Law 26:525–548

Garner BA (2007) Black's law dictionary. St. Paul, Thomson West

HRC (2010) Resolution adopted by the Human Rights Council, 18/1 The Human Right to Safe Drinking Water and Sanitation, UN Doc. A/HRC/RES/18/1

ICJ (1997) Gabčikovo–Nagymaroš project (Hungary/Slovakia). Judgment, ICJ Reports 1997, 25 Sept 1997, pp 7–87

ICJ (2010) Pulp Mills on the River Uruguay (Argentina v. Uruguay). Judgment, ICJ Reports 2010, 20 April 2010, pp 14–107

ICESCR (1966) International Covenant on Economic, Social and Cultural Rights (ICESCR), GA Res. 2200 A (XXI) (1966), 21 UN GAOR Supp. (No. 16) at 49, UN Doc. A/6316 (1966)
ITLOS (2011) Responsibilities and obligations of States with respect to activities in the Area, Advisory Opinion, ITLOS Reports 2011, 1 Feb 2011, pp 10–78
IUCN (2010) Draft International Covenant on Environment and Development, Environmental Law Programme of the International Union for the Conservation of Nature and Natural Resources in cooperation with The International Council of Environmental Law. Retrieved 24 July 2015 from http://data.iucn.org/dbtw-wpd/edocs/eplp-031-rev3.pdf
Kasimbazi E (1995) An international legal framework for forest management and sustainable development. Annu Surv Int Comp Law 2:67–97
Langford M (2006) Ambition that overleaps itself? A response to Stephen Tully's critique of the general comment on the right to water. Neth Q Hum Rights 24:433–445
Maggio G (1996–1997) Inter/intra-generational equity: current applications under international law for promoting the sustainable development of natural resources. Buffalo Environ Law J 4:161–223
McIntyre O (2010) The proceduralisation and growing maturity of international water law: case Concerning Pulp Mills on the River Uruguay (Argentina v Uruguay). J Environ Law 22 (3):476, 485
New Delhi Declaration (1992) New Delhi declaration of principles of international law related to sustainable development, Committee on Legal Aspects of Sustainable Development of the International Law Association, UN Doc. A/CONF.199/8, 9 Aug 2002
Palassis S (2011) Beyond the global summits: reflecting on the environmental principles of sustainable development. Colo J Int Environ Law Policy 22:41–77
Right to Water (2015) The right to water and sanitation, International Timeline. Retrieved 10 May 2015 from http://www.righttowater.info/progress-so-far/international-timeline/#1948
Rio Declaration on Environment and Development (1992) UN Doc. A/CONF.151/26 (vol. I), 31 ILM 874 (1992)
Rio +20 Declaration (2012) Resolution adopted by the General Assembly on 27 July 2012, 66/288 The future we want, UN Doc. A/RES/66/288
Salman MA (2012) The human right to water—challenges of implementation. Am Soc Int Law Proc 106:44–46
Saunders M (1991) Valuation and international regulation of forest ecosystems: prospects for a global forest agreement. Wash Law Rev 66:871–892
Schadendorf S (2015) Investor-state arbitrations and the human rights of the host state's population: an empirical approach to the impact of Amicus Curiae submissions. In: Weiss N, Throuvenin JM (eds) The influence of human rights on international law. Springer, Heidelberg, pp 167–188
Scheuring S (2009) Is there a right to water in international law? UCL Jurisprud Rev 15:147–171
Sunstein CR (1995) Rights and their critics. Notre Dame Law Rev 70:727–757
UN (1977) Report of the United Nations water conference, Mar Del Plata, 14–25 Mar 1977, UN. Doc. E/CONF.70/29. Retrieved 24 July 2015 from http://www.ielrc.org/content/e7701.pdf
UN (1980) Convention on elimination of all forms of racial discrimination against women, GA res. 34/180, 34 UN GAOR Supp. (No. 46) at 193, UN Doc. A/34/46; 1249 UNTS 13; 19 ILM 33 (1980)
UN (1989) Convention on the Rights of the Child, GA res. 44/25, annex, 44 UN GAOR Supp. (No. 49) at 167, UN Doc. A/44/49 (1989); 1577 UNTS 3; 28 ILM 1456 (1989)
UN (2000) Resolution adopted by the General Assembly, 55/2 United Nations Millennium Declaration, UN Doc. A/RES/55/2
UN (2002) Johannesburg Plan of Implementation, Report of the World Summit on Sustainable Development, UN Doc. A/CONF.199/20 (2002), 4 Sept 2002
UN (2007) Convention on the Rights of Persons with Disabilities GA res. 61/106, UN Doc. A/RES/61/106, Annex 1
UNGA (2010) Resolution adopted by the General Assembly on 28 July 2010, 64/292, The human right to water and sanitation, UN Doc. A/RES/64/292. Retrieved 24 July 2015 from http://www.un.org/es/comun/docs/?symbol=A/RES/64/292&lang=E

UNSCD (1996) Report of the Expert Group Meeting on Identification of Principles of International Law for Sustainable Development, United Nations Commission on Sustainable Development (26–28 Sept 1995). Retrieved 24 July 2015 from http://www.un.org/documents/ecosoc/cn17/1996/background/ecn171996-bp3.htm

Versteeg M (2007) Equitable utilisation or the right to water? Legal responses to global water scarcity. Tilburg Foreign Law Rev 16:368–392

WCEL (1987) Report of the World Commission on Environment and Development: Our Common Future, UN GA Res., A/42/427 (1987), 4 Aug 1987

Weeramantry (1997) Separated opinion. In: ICJ. Gabčikovo–Nagymaroš project (Hungary/Slovakia), Judgment, ICJ Reports (1997), 25 Sept 1997, pp 88–119. ICJ, The Hague

Winkler IT (2012) The human right to water: significance, legal status and implications for water allocation. Hart Publishing, Oxford and Portland

Ziganshina D (2008) Rethinking the concept of the human right to water. Santa Clara J Int Law 6:113–128

Ensuring a Sustainable Future Through Recognizing and Protecting Indigenous Ecological Knowledge

Natalie P. Stoianoff

Abstract This paper sets out the way in which Indigenous ecological knowledge has received increasing recognition as a holistic mechanism through which Australia's natural resources can be sustainably managed. This increased recognition and consequent utilization needs to take place within a legal framework that acknowledges and respects the customary laws and rules of the Indigenous ecological knowledge holders and provides appropriate benefits back to those knowledge holders. This paper considers the nature of such a legal framework and reports on the research conducted by the author and her research team, through the use of action research and Indigenous research paradigm methodologies, in developing such a legal regime that encapsulates the principles established in the *Convention on Biological Diversity 1992*, expanded in the *Nagoya Protocol* to the Convention, and reinforced in the *United Nations Declaration on the Rights of Indigenous Peoples 2007*. The result was a White Paper espousing a sui generis legal framework of recognition and protection of Indigenous knowledge associated with natural resource management focussed on the Aboriginal Communities of the state of New South Wales in Australia and accordingly reflects the concerns and interests of those communities while incorporating the international law principles described above. This was achieved through an initial comparative analysis of regimes already in existence in other nations, the establishment of a highly skilled and multidisciplinary Working Party representing both Indigenous and non-Indigenous individuals and stakeholders, and finally through Aboriginal Community consultation.

Keywords Indigenous ecological knowledge · Aboriginal peoples · Natural resource management · Free prior informed consent · Mutually agreed terms

---

N.P. Stoianoff (✉)
Faculty of Law, University of Technology Sydney, Sydney, Australia
e-mail: Natalie.Stoianoff@uts.edu.au

V. Mauerhofer (ed.), *Legal Aspects of Sustainable Development*,
DOI 10.1007/978-3-319-26021-1_7

1 Introduction

It has been recognized that for Australia to sustain the wellbeing of its population over the long term, it is necessary to find ways of supporting economic growth without degradation of the environment but instead implement wise stewardship of natural resources (National Sustainability Council 2013). Indigenous ecological knowledge has become increasingly recognised as a more effective means of managing the Australian landscape particularly since that knowledge has a holistic approach of understanding the seasons, biodiversity, land and water (Commonwealth of Australia 2013). An example is the managed burning practices of Indigenous peoples. Such knowledge and associated practices were ignored in Australia during the most part of the last century to the detriment of the land resulting in extinctions of biodiversity due to wildfires that probably would not have occurred had the traditional land management practices been allowed to continue (Australian Government National Parks 2015). Such attitudes have been changing and the knowledge of Indigenous elders are being implemented more and more across Australia (EMR Projects 2012). This has been occurring through the Australian Government's Environmental Stewardship Program which 'assist[ed] private land managers to maintain and improve the condition and extent of environmental assets of high public value on, or impacted by activities on, their land' (Commonwealth of Australia 2009, p. i).

Accordingly, this chapter sets out the way in which Indigenous ecological knowledge has received increasing recognition as a holistic mechanism through which Australia's natural resources can be sustainably managed. For example, through the Australian Government's National Landcare Program funding is provided to projects that encourage partnerships with Indigenous people and communities enabling participation in 'land and sea management, drawing on their significant and unique knowledge, skills and responsibilities' (Australian Government National Landcare Program 2015a, p. 1). This chapter argues that increased recognition and consequent utilization of Indigenous ecological knowledge needs to take place within a legal framework that acknowledges and respects the customary laws and rules of the Indigenous ecological knowledge holders and provides appropriate benefits back to those knowledge holders. This chapter considers the nature of such a legal framework.

Further, the theoretical framework for this paper is drawn from the principles established in the Convention on Biological Diversity (CBD 1992), expanded in the Nagoya Protocol (2010) on Access to Genetic Resources and the Fair and Equitable Sharing of Benefits Arising from their Utilization to the Convention on Biological Diversity, and also reinforced in the United Nations Declaration on the Rights of Indigenous Peoples (2007). These principles provide Indigenous knowledge holders with a right to share in the benefits obtained from the use of their knowledge. In doing so there is emphasis on the need for free prior informed consent to be given by such knowledge holders before others access that knowledge. In that context, access and use of that knowledge needs to be on mutually agreed terms. This paper reports on the research conducted by the author and her research team in developing such a legal regime that encapsulates these principles through the use of action research and the Indigenous research paradigm methodologies.

The result of the research undertaken is the development of a legal framework that recognizes and protects Indigenous ecological knowledge associated with natural resource management. The focus was on the Aboriginal Communities of the state of New South Wales in Australia and accordingly reflects the concerns and interests of those communities while incorporating the international law principles described above. This was achieved through an initial comparative analysis of regimes already in existence in other nations, the establishment of a highly skilled and multidisciplinary Working Party representing both Indigenous and non-Indigenous individuals and stakeholders, and finally through Aboriginal Community consultation.

How does this ensure a sustainable future? Sustainable development focuses on social and human, natural and economic factors. By recognising and protecting Indigenous ecological knowledge each of these factors are addressed. Australia's natural environment is improved through a holistic approach to natural resource management. The wrongs endured by Indigenous Australians are being addressed through such a legal framework of recognition and protection since, for Aboriginal peoples, land and knowledge are inherently connected and access to traditional lands is an important aspect of cultural expression and well-being. And finally, through benefit sharing economic self-sufficiency can be achieved.

2 A Sustainable Australia

The Australian Federal Government has recognised that to build a sustainable Australia 'improved information is needed about the economy, environment and society, and the linkages between them, to better inform decisions and policy making' (Australia Government Measuring Sustainability Program 2015, p. 1). It has also been recognised that an intergenerational perspective needs to be taken to consider the future impact of actions and decisions made today (Australian Government Measuring Sustainability Program 2015). To help achieve these objectives, a set of sustainability indicators have been devised under Australia's Measuring Sustainability Program (Australian Government Measuring Sustainability Program 2015). These indicators are divided into three areas: social and human capital, natural capital and economic capital. Together they form a matrix of sustainability indicators divided as follows:

- headline indicators, divided into themes, to provide information on key sustainability issues
- supplementary indicators for each theme, to provide additional information and support a more detailed understanding of the issues represented in the headline indicators
- cross-cutting, contextual indicators covering key demographic information (Australian Government Sustainability Indicators for Australia 2015).

Themes under social and human capital include: skills and education; health; employment; security; institutions, governance and community engagement. Themes under natural capital include: climate and atmosphere; land, ecosystems and biodiversity; natural resources; water; and waste. While themes under economic capital include: wealth and income; housing; transport and infrastructure; productivity and innovation (Australian Government Sustainability Indicators for Australia 2015; Australia Government Measuring Sustainability Program 2015). In this matrix of indicators, Australia's Indigenous population is recognized as a contextual indicator under the topic of cultural diversity (Australian Government Sustainability Indicators for Australia 2015).

In carrying out natural resource management activities and conserving biodiversity, the Australian Government claims a long history of working with Indigenous people and recognizing and respecting their knowledge in such management. Specifically, the Department of Environment states that

> 'Indigenous Australians are key partners with us in managing Australia's environment and cultural heritage' (Australian Government Indigenous Australians Caring for Country 2015).

The Department goes on to identify several initiatives aimed at engagement of Indigenous Australians. One such initiative is the Indigenous Carbon Farming Fund where three projects, each receiving $300,000, regarding the use of fire have been funded to date (Australian Government Indigenous Carbon Farming Fund 2015), while five feasibility and assessment projects, each receiving $50,000, have been funded to date (Australian Government Indigenous Carbon Farming Fund 2015).

Other initiatives engaging Indigenous Australians include: Caring for our Country; Indigenous Heritage Program; Indigenous Land Management Facilitators; and Indigenous Advisory Committee (IAC) under the *Environment Protection and Biodiversity Conservation Act 1999* (Cth). The Caring for our Country initiative includes several Indigenous-specific programs, namely, Working on Country, Indigenous Protected Areas, Reef Rescue Indigenous Land and Sea Country Partnerships and the Indigenous Emissions Trading commitment (Australian Government Indigenous Australians Caring for Country 2015).

The Working on Country Indigenous ranger program is an extensive ongoing program designed to deliver environmental outcomes and address Indigenous disadvantage. This is important when one realizes that Indigenous Australians' own an estimated 20 % of the Australian continent, upon which lies some of [Australia's] most environmentally precious natural assets and … is rich in cultural and spiritual meaning', and have cultural and traditional responsibilities to protect and manage their Country both land and water (Australian Government Working on Country 2015). Until 2013, the program received over $244 million, and then a further $320 million has been made available from 1 July 2013 for five years (Australian Government Working on Country 2015). It is responsible for employing 'more than 680 Indigenous rangers … in around 95 ranger teams across Australia' with the expectation that the number of rangers trained and employed through Working on Country will reach 730 by June 2015 (Australian Government Working on Country 2015). The program has improved the 'management of fire, feral animals, invasive weeds, threatened species and coastal and marine systems' across regional and remote Australia as the Indigenous rangers utilize and build upon their traditional knowledge to protect and manage their Country (Australian Government Indigenous Australians Caring for Country 2015).

Success under the Caring for Country program is exemplified when an Indigenous community decides to declare an Indigenous Protected Area on their land (AIATSIS 2011). An Indigenous Protected Area is an area of Indigenous-owned land or sea where traditional owners have entered into an agreement with the Australian Government to promote biodiversity and cultural resource conservation (Australian Government About Indigenous Protected Areas 2015). Indigenous Protected Areas protect Australia's biodiversity while providing training and employment for Aboriginal people on their own country (Australian Government About Indigenous Protected Areas 2015). There are 60 declared Indigenous Protected Areas covering just over 48 million hectares across Australia making a significant contribution to Australian biodiversity conservation as it accounts for around 36 % of our National Reserve System (Australian Government About Indigenous Protected Areas 2015). This program demonstrates a clear link between the recognition given to Indigenous Australians for the tens of thousands of years of management of their Country and the resulting movement toward closing the gap of Indigenous disadvantage, with communities engaged in managing Indigenous Protected Areas reporting better health, social cohesion and higher school attendance (Australian Government Cultural and Social Benefits 2015). In addition to the goals of supporting Indigenous landowners/interests to manage the Indigenous Protected Area and develop cooperative management arrangements with Government agencies, the Indigenous Protected Area Program is aimed at supporting 'the integration of Indigenous ecological and cultural knowledge with contemporary protected area management practices' (Australian Government About Indigenous Protected Areas). While a positive step toward recognition of the value of Indigenous ecological knowledge, the problem with such 'integration' is whether it is carried out in accordance with the customary laws of the community whose knowledge it is. Further, is due compensation given for the use of that knowledge? How would it be calculated? Who benefits and how is it distributed?

Another successful initiative is the Indigenous Heritage Program supporting the identification and conservation of Indigenous heritage, and where appropriate, its promotion (Australian Government Indigenous Heritage Programme 2015). This has been a long running program with more than 500 projects being funded across Australia since 2004 amounting to over $30 million in support (Australian Government Indigenous Heritage Programme 2015). This program has demonstrated the significance of heritage to the wellbeing and sustainable development of Indigenous communities. Projects requiring access to traditional lands in order to carry out cultural heritage activities have 'delivered improved health outcomes and healthier lifestyles for participants' (Australian Government Indigenous Heritage Programme 2015). While providing short-term employment, the program has enabled the development of a variety of marketable skills including database management, site survey and recording techniques, and fencing (Australian Government Indigenous Heritage Programme). Such greater economic participation has even seen the creation of tourism businesses around cultural knowledge and heritage (Australian Government Indigenous Heritage Programme 2015). In so doing, this has enhanced both pride in culture and respect for the cultural knowledge of Elders thus leading to greater community cohesion, safety and support.

However, despite the above claimed successes of the various programs described (Productivity Commission 2011), there are still issues to be resolved. For example, Indigenous representation should be improved in mainstream natural resource management bodies not just those dedicated to Indigenous programs. What happens when the project or program finishes? This issue requires engagement with sustainable development to enable communities to develop opportunities and negotiate on their own behalf, improve access to Country and cultural resources, improve understanding and relationships between community organisations and Traditional Owners or Custodians which may require assistance to resolve disputes where they arise in relation to Country. This brings us back to the question of how Indigenous ecological knowledge is to be accessed, recognized and valued. The Australian Government speaks of a 'two-way transfer of knowledge' when engaging with Indigenous ecological knowledge indicating that 'it is important that Indigenous peoples have access to scientific knowledge and best practice for natural resource management' (Australian Government National Landcare Programme 2015b, p. 1) but this fails to recognize that Indigenous ecological knowledge in Australia is the relevant best practice—however, it must be knowledge from that Country. The utilization of such knowledge by the National Landcare Program needs to make arrangements to protect such knowledge and ensure benefit sharing takes place with the relevant Indigenous communities holding such knowledge. Accordingly, there is a need for a legal framework that acknowledges and respects the customary laws and rules of the Indigenous ecological knowledge holders and provides appropriate benefits back to those knowledge holders and their communities where such knowledge is utilized outside of those communities and for the natural resource management of the nation.

3 Principles for a Legal Framework of Recognition and Protection

Australia is a party to a number of international treaties and declarations which recognise the significance of traditional and Indigenous knowledge and cultural expressions, and emphasise the need to respect, preserve and maintain knowledge, innovations and practices of Indigenous and local communities. For example, the Convention on Biological Diversity (CBD 1992) provides member nations with the opportunity to establish regimes that would regulate foreign and domestic access to valuable genetic resources and traditional and Indigenous knowledge while establishing benefit-sharing mechanisms relating to that access.

The CBD has also led to significant international debate on the interrelationship with intellectual property rights, particularly patent rights and plant breeders' rights, which are often the end goal of the desire to access such genetic resources. The role of Indigenous knowledge in this context is significant and brings into the equation the broader cultural property of Indigenous and local communities. This has been reinforced by the United Nations Declaration on the Rights of Indigenous Peoples (2007).

Specifically, Article 11 of the Declaration recognises the right of Indigenous people 'to practise and revitalize their cultural traditions and customs'. This extends to 'the right to maintain, protect and develop the past, present and future manifestations of their cultures, such as archaeological and historical sites, artefacts, designs, ceremonies, technologies and visual and performing arts and literature'.

The expectation is that the State will develop with Indigenous peoples effective compensation mechanisms 'with respect to their cultural, intellectual, religious and spiritual property taken without their free, prior and informed consent or in violation of their laws, traditions and customs' (Article 11 United Nations Declaration on the Rights of Indigenous Peoples 2007).

There are many nations and regions around the world that have already adopted legislative regimes to accommodate such rights. Nations utilising *sui generis* (or stand alone) legislation (Intergovernmental Committee on Intellectual Property and Genetic Resources, Traditional Knowledge and Folklore 2012) recognising and protecting traditional or Indigenous knowledge include Brazil, Peru, Panama, and the Philippines, requiring the establishment of registers or databases and a representative authority of some sort.

Some are based on the *WIPO-UNESCO Model Provisions* (WIPO-UNESCO 1982), which contain intellectual property type provisions. More recently the Pacific Regional Framework for the Protection of Traditional Knowledge and Expressions of Culture (2002), provides the *Model Law For The Protection Of Traditional Knowledge And Expressions Of Culture 2002* establishing Traditional Cultural Rights and Moral Rights over such knowledge and expressions, the need for prior informed consent as discussed above, a regime for applications for use and identifying the traditional owners, authorised user agreements, an enforcement regime covering both civil and criminal actions and providing defences and finally, establishing a Cultural Authority to oversee the entire regime. Other regional solutions are found in Africa, the Andean Community of Nations (formerly, Andean Pact nations) and ASEAN.

What needs to be kept in mind is that Australia's obligations under the CBD provide alternative mechanisms for respecting, preserving and maintaining Indigenous or traditional knowledge. However, what began as an attempt to establish a national approach became a 'nationally consistent' approach, which is yet to be achieved. Some Australian jurisdictions have their own approach to dealing with Indigenous or traditional knowledge but not all. Of those that do have an approach, voluntary protocols rather than mandatory obligations are utilised to varying degrees of success. The Australian Government is of the opinion that its domestic measures are consistent with the *Nagoya Protocol on Access to Genetic*

Resources and the Fair and Equitable Sharing of Benefits Arising from their Utilization to the Convention on Biological Diversity 2010, to which Australia became a signatory in January 2012 (Australian Government The Nagoya Protocol in Australia 2015). That may well be in the case of protecting Australia's genetic resources, however, in respect of the Indigenous knowledge associated with such resources, such an opinion is questionable. The Protocol requires that where '… Indigenous traditional knowledge [is utilized] countries have to make sure that the knowledge was acquired in accordance with the rules of the country where those Indigenous people live' (Australian Government The Nagoya Protocol in Australia 2015). This requires two key elements:

- the prior informed consent of the Indigenous community providing the knowledge must be obtained prior to access, and
- that access must be on mutually agreed terms.

Meanwhile, as a result of collaboration between the *World Intellectual Property Organization* (WIPO) and the *United Nations Environment Programme* (UNEP) responsible for the introduction of the CBD, in 2000 the WIPO General Assembly established the *Intergovernmental Committee on Intellectual Property and Genetic Resources, Traditional Knowledge and Folklore* (IGC or Intergovernmental Committee). This forum was responsible for negotiating international instruments for the protection of traditional or Indigenous knowledge and culture from an intellectual property perspective. The result thus far has be three draft international instruments with one dedicated to traditional or Indigenous knowledge. Adjei and Stoianoff (2013) point out that there are 8 key elements to a framework of protection for such knowledge:

(1) The need to satisfy the meaning of traditional knowledge and its scope.
(2) The identification of the beneficiaries.
(3) The scope of protection encompassing elements of confidentiality and moral rights in the protection against misappropriation and misuse.
(4) The nature of sanctions and remedies not too dissimilar to those used in intellectual property law.
(5) The need for disclosure in the patent and plant variety rights regimes.
(6) The establishment of an administrative body or competent authority to manage the data, or the rights conferred or the enforcement, dispute resolution and national treatment.
(7) The creation of databases.
(8) Accommodating trans-boundary co-operations where knowledge and biodiversity extend across artificial borders (Adjei and Stoianoff 2013).

WIPO is basically exploring a *sui generis* model for the protection of Indigenous or traditional knowledge. This represents a recognition that protecting Indigenous knowledge does not fit into the existing intellectual property paradigm for a number of reasons such as the need for protection in perpetuity in accordance with cultural norms, the difficulty of identifying the 'author' or 'creator' of the knowledge but most importantly because intellectual property does not recognize the communal rights over that knowledge. Consequently, the IGC at WIPO embarked upon a process of developing a model of protection that would accommodate the peculiarities of Indigenous knowledge. The White Paper for the Office of Environment and Heritage NSW, *Recognising and Protecting Aboriginal Knowledge Associated with Natural Resource Management*, prepared by the author and her research team under the Indigenous Knowledge Forum and the North West Local Land Services (formerly Namoi Catchment Management Authority) in 2014 (UTS Indigenous Knowledge Forum and North West Local Land Services 2014), followed suit and developed such a *sui generis* model law for implementation at state level but with the potential for nationwide application.

The arguments for and against *sui generis* law were acknowledged (UTS Indigenous Knowledge Forum and North West Local Land Services 2014). International support was evident from WIPO, UNEP and the Conference of the Parties for the CBD (UTS Indigenous Knowledge Forum and North West Local Land Services 2014). Customary laws could be incorporated into such a regime taking 'into account needs and expectations of Indigenous and local Communities, [enabling] protect[ion of the] integrity of traditional knowledge and [punishing] use that offends Indigenous and local Communities while encouraging acceptable use by third parties' (UTS Indigenous Knowledge Forum and North West Local Land Services 2014, p. 15). Accordingly, the impetus for such an approach stemmed from the fact that if Indigenous knowledge is to be recognized as part of a living culture that requires access to Country for it be preserved, maintained respected and developed in accordance with customary laws and crossing the thresholds of intellectual property type rights as well as environmental responsibilities, then the legal framework that must be created will be inevitably unique (Oguaman 2006, pp. 217–219).

4 Method and Methodology

The research undertaken by the author and her research team under the umbrella of the Indigenous Knowledge Forum was designed to establish a comparative framework facilitating the development of a regime for recognition and protection of Indigenous ecological knowledge engaging the Namoi Catchment Aboriginal Community of north west New South Wales in developing a standard-setting model of involvement in natural resource management and access to Country. This project design came out of the inaugural Indigenous Knowledge Forum held at UTS in August 2012 (Indigenous Knowledge Forum Report 2012). Indigenous and non-Indigenous Australians participated in that forum and with the guidance of the Advisory Board comprising significant Indigenous representatives the project was developed. An application was made to the Aboriginal Communities Funding Scheme of the Namoi Catchment Management Authority [now North West Local Land Services (NWLLS)] and with the advice of the Aboriginal Officer and the Aboriginal Advisory Committee of the NWLLS the project was chosen to be funded as it represented project partnership research that was of great interest to the communities and satisfied their 2010–2020 Action Plan. The project was carried out in three stages and consultation with the Aboriginal Officer, among others, took place at each stage.

The three stages of the research project were: development of a comparative framework; the drafting of the *sui generis* regime; and the Aboriginal community consultation to refine the regime. First this project carried out a doctrinal comparative study, collecting and analysing legislative and policy regimes already in existence in other parts of the world. Key criteria in each regime were identified and then compared to international obligations. This provided the comparative framework upon which a standard-setting model of involvement in natural resource management and access to country could be developed to ensure the recognition and protection of Indigenous ecological knowledge as part of a living culture. To this end a Comparative Study Report marked the end of stage one.

In stage two a working party was formed to assist in developing a draft *sui generis* regime. The working party comprised of both Indigenous and non-Indigenous members drawn from the UTS Indigenous Knowledge Forum committees, interested participants from the August 2012 Indigenous Knowledge Forum, key personnel from the Namoi Catchment Management Authority (CMA) and Namoi Aboriginal Advisory Committee (NAAC) meeting at least three times during this stage to guide the development of a draft regime. The working party utilised the Comparative Study Report prepared in stage one to provide a basic framework for discussion of how such a regime might operate to ensure the elements of supporting a living Aboriginal culture with improved access to country through recognition and protection of Indigenous ecological knowledge about country are met. A Discussion Paper, comprising the Comparative Study Report and Draft Regime, was then prepared.

Stage three saw the distribution of the Discussion Paper through the Namoi CMA to all members of the Namoi Catchment Aboriginal Community and other interested parties as determined by the Namoi CMA and consultation sessions were held in key locations in the Namoi Catchment region. These consultation sessions, akin to focus groups, were subject to university human research ethics approval and were organized by the Aboriginal Officer of the Namoi CMA in accordance with appropriate cultural norms and protocols of each community. The consultations formed an important aspect of engagement with the Indigenous communities this proposed legislation was designed to benefit. With the recommendation of the Aboriginal Officer, the research team included an Aboriginal Advisor being a member of the Indigenous Knowledge Forum Advisory Board and member of the working party assisting with the drafting of the proposed model legislation. The Aboriginal Advisor together with the Aboriginal Officer assisted in ensuring that the focus groups take place in accordance with cultural norms. And in order to ensure the views and recommendations of the focus group, participants were able to be incorporated in the eventual refining of the legal regime and drafting of the White Paper (UTS Indigenous Knowledge Forum and North West Local Land Services 2014), information sheets were distributed at each consultation meeting and signed consent forms obtained from each participant.

The purpose of the consultations was to test the draft legal framework against Aboriginal community concerns and expectations thereby enabling the draft regime to be refined into a model capable of application not only to the north west region of New South Wales but also to other regions, preferably through legislative implementation. To this end a White Paper (UTS Indigenous Knowledge Forum and North West Local Land Services 2014) was prepared and delivered to the Office of Environment and Heritage.

Examples of regimes from other nations provided some direction to the research project as did the key criteria from existing international conventions, agreements, protocols and declarations. However, it is important to note that this project sought to address the need for recognition and protection of Indigenous ecological knowledge by starting at the local, grassroots level, employing an action research methodology coupled with an Indigenous research paradigm at both stages two and three of the project, albeit to varying degrees. The project gave Indigenous Australians an opportunity to actively participate in the process of formulating legislation relating to access to and benefit sharing arising from utilisation of their knowledge. The action research methodology emphasizes cooperative or collaborative inquiry (Heron 1996) whereby all active participants, Indigenous and non-Indigenous, are fully involved in research decisions as co-researchers (Reason and Bradbury 2007). Through the internet, the project provided all interested parties, not just Indigenous Australians, with access to analysis of current models for and outcomes of implementing similar legislation in other countries to assist in the process of identifying how best to accommodate unique aspects of Indigenous ecological knowledge and culture as it relates to Aboriginal and Torres Strait Islander peoples of Australia into formulation of access and benefit sharing legislation.

Participation in turn assisted in generating ownership of the outcomes, understanding of any resulting legislation and its intent and an opportunity to deliver robust legislation that not only meets Australia's international obligations but also effectively protects the interest of an important sector of the Australian community. During stage two the Indigenous research paradigm played an important role by engaging all participants in the collection of research data through the method of storytelling by Indigenous Elders in the group, exploring meaning and working through issues together to ensure accurate interpretation of language (Czaykowska-Higgins 2009). This process was also adopted during the focus group meetings on Country being mindful of the culture of place and the privilege of sharing in the flow of cultural knowledge.

5 Results and Conclusions

The methodological approach emphasized the key outcome of this project, namely the development of a legislative regime that facilitates the recognition and preservation of Indigenous knowledge about Country and the protection of that knowledge enabling its custodians to share in the benefits of its use. To this end the resulting draft legislation, explained in this chapter of the White Paper (UTS Indigenous Knowledge Forum and North West Local Land Services 2014), sets out key principles rather than detailed prescriptive provisions—these are left to regulations that would need to be implemented, once again with the participation of those to be protected under the legislation.

The scene is set with a preamble recognising the impact of European arrival on the Knowledge and connection to Country of Aboriginal Peoples in New South Wales, as well as the aims of the legislation. Section 1 goes on to establish the rights of Aboriginal communities over their knowledge. Key terms are then defined under Sect. 2 and the beneficiaries under the legislation are identified at Sect. 3. The process of access to the knowledge is described under Sect. 4 and the guidelines for benefit sharing are provided in Sect. 5. Sanctions and remedies for breaches of the legislation are provided under Sect. 6 ranging from remedies similar to those available for infringement of intellectual property rights to penalties ranging from fines to imprisonment. Meanwhile, Sect. 7 requires the establishment of a Competent Authority to administer such a regime including managing the necessary databases to enable the access process to operate. Section 8 provides for dispute resolution in the event there are multiple communities claiming 'ownership' of the same or similar knowledge. As with every regime there needs to be a set of express exceptions and Sect. 9 provides for that eventuality. The nature of the three types of database Registers and the obligations regarding disclosure are dealt with in Sect. 10. The remaining three provisions are more general in nature dealing with interaction with existing laws (Sect. 11), acknowledging mutual recognition of rights and compliance (Sect. 12) and transitional provisions (Sect. 13).

To demonstrate the operation of this draft regime the White Paper (UTS Indigenous Knowledge Forum and North West Local Land Services 2014) provides at Chap. 8 a fictitious case study example. Even so, the White Paper does not claim to provide a complete solution for the recognition and protection of Aboriginal knowledge associated with natural resource management. Shortcomings include: the need to clarify the form and nature of the Competent Authority and the governance processes to be implemented; the way the databases are to be formed, funded and managed from community level to state and ultimately national levels; and the administration processes for access and benefit-sharing including guidance on mutually beneficial terms, model agreements and processes for negotiation.

While this may seem to be all too hard, one must recall the first stage of the research which gathered together a database of already existing regimes from around the world that address many of the elements of the regime espoused by the White Paper. And if there is still doubt as to the benefit of introducing such a regime one need only consider the flow-on effects which include:

A. Recognition that Indigenous ecological knowledge is part of a living culture that requires access to Country for it to be preserved, maintained respected and developed;

B. Providing a mechanism for the documenting, recording and recovering of Indigenous knowledge, to make it available for future generations of Aboriginal and Torres Strait Islander community members;
C. Dealing with intergenerational loss of knowledge about Country (land and water) by encouraging the younger generation to spend time on Country with their Elders to regain their traditional language in which the culture and the knowledge is maintained through oral tradition;
D. Improving natural resource management by facilitating access to Country, recalling the various Australian Government programs described above aimed at both Indigenous engagement and the sustainable use of Indigenous ecological knowledge; and perhaps most importantly;
E. Recognition that Indigenous knowledge is to be valued and utilised in accordance with Indigenous protocols that govern use and dissemination of this knowledge including the need for prior informed consent and the establishment of an appropriate benefit sharing arrangement in accordance with mutually agreed terms.

This brings us back to the sustainability indicators that the Australian Government has established as a means of measuring 'our stocks of social and human, natural and economic 'capital' and ensuring that the resources inherited by future generations allow for the same (or greater) levels of wellbeing as enjoyed by Australians today' (Australian Government Sustainability Indicators for Australia 2015). If as the Sustainability Report (2013) (National Sustainability Council 2013) suggests 'the wellbeing of individuals, communities and society has been widely accepted as an appropriate objective of governments' then providing a mechanism of empowerment to Australia's Indigenous communities such as through the proposed *sui generis* legislation for recognising and protecting their Indigenous ecological knowledge goes some way to addressing all three forms of capital (social, natural and economic) as they relate to improving the wellbeing of Australia's Indigenous communities.

References

Adjei P, Stoianoff NP (2013) The World Intellectual Property Organisation (WIPO) and the Intergovernmental Committee: developments on traditional knowledge and cultural expressions, intellectual property forum issue 92, Mar 2013, pp 37–48

AIATSIS (2011) The benefits associated with caring for country. Literature Rev June 2011, p 14

Australian Government About Indigenous Protected Areas (2015) Retrieved 20 July 2015 http://www.environment.gov.au/indigenous/ipa/index.html

Australian Government Cultural and Social Benefits (2015) Retrieved 20 July 2015 http://www.environment.gov.au/indigenous/ipa/cultural.html

Australian Government Indigenous Australians Caring for Country (2015) Retrieved 19 July 2015, from http://www.environment.gov.au/indigenous/

Australian Government Indigenous Carbon Farming Fund—Successful Applicants (2015) Retrieved 19 July 2015 from http://www.environment.gov.au/cleanenergyfuture/icff/pubs/icff-category2-projects.pdf

Australian Government Indigenous Heritage Programme (2015) Retrieved 20 July 2015 http://www.environment.gov.au/indigenous/ihp/index.html

Australian Government Measuring Sustainability Program (2015) Retrieved 19 July 2015 from http://www.environment.gov.au/topics/sustainable-communities/measuring-sustainability

Australian Government National Landcare Programme (2015a) Indigenous Natural Resource Management. Retrieved 19 July 2015 from http://www.nrm.gov.au/indigenous-nrm

Australian Government National Landcare Programme (2015b) Indigenous Natural Resource Management, Utilising Knowledge Retrieved 19 July 2015 from http://www.nrm.gov.au/indigenous-nrm/knowledge

Australian Government National Parks (2015) Conserving Uluru-Kata Tjuta National Park. Retrieved 22 July 2015 from http://www.environment.gov.au/parks/uluru/management/programs/fire.html

Australian Government Sustainability Indicators for Australia (2015) Retrieved 19 July 2015 from http://www.environment.gov.au/topics/sustainable-communities/measuring-sustainability/sustainability-indicators

Australian Government The Nagoya Protocol in Australia (2015) Retrieved 20 July 2015 http://www.environment.gov.au/system/files/pages/9fc06ac0-f5af-4b47-a80f-d9378088d743/files/nagoya-factsheet_1.pdf

Australian Government Working on Country (2015) Retrieved 20 July 2015 http://www.environment.gov.au/indigenous/workingoncountry/index.html

CBD (1992) Convention on biological diversity, opened for signature 5 June 1992, 31 ILM 818 (entered into force 29 Dec 1993)

Commonwealth of Australia (2009) Department of the environment water heritage and the arts. Environmental stewardship box gum grassy woodland project: information booklet, (i). Retrieved 20 July 2015 from http://nrmonline.nrm.gov.au/downloads/mql:2397/content

Commonwealth of Australia (2013) Caring for Our Country—Outcomes 2008–2013, pp 21, 31, 36–37, 39

Czaykowska-Higgins E (2009) Research models, community engagement, and linguistic fieldwork: reflections on working within Canadian indigenous communities, language documentation and conservation, 3(1), June 2009

EMR Projects (2012) Three action research projects: (i) Traditional Knowledge Revival Pathways Fire Program, (ii) Kuku Thaypan Fire Management Research Project and (iii) The Importance of Campfires to Effective Conservation—Cape York Peninsula Australia, EMR project summaries. J Ecol Manag Restor, January 2012. Retrieved 20 July 2015 from http://site.emrprojectsummaries.org/2012/01/26/three-action-research-projects-itraditional-knowledge-revival-pathways-fire-program-iikuku-thaypan-fire-management-research-project-and-iiithe-importance-of-campfires-to-effective-conservation-2/

Heron J (1996) Co-operative inquiry: research into the human condition. Sage, London

Indigenous Knowledge Forum Report (2012) Retrieved 22 July 2015 from http://indigenousknowledgeforum.org/images/ikf-report.pdf

Intergovernmental Committee on Intellectual Property and Genetic Resources, Traditional Knowledge and Folklore (2012) Glossary of key terms related to intellectual property and genetic resources, traditional knowledge and traditional cultural expressions, twenty-second session, Geneva, 9–13 July 2012

Nagoya Protocol (2010) Protocol on access to genetic resources and the fair and equitable sharing of benefits arising from their utilization (2010), adopted by the conference of the parties to the convention on biological diversity at its tenth meeting on 29 Oct 2010 in Nagoya

National Sustainability Council (2013) Sustainable Australia report 2013, conversations with the future. DSEWPaC, Canberra, 2013. Retrieved 20 July 2015 from http://www.environment.gov.au/system/files/resources/e55f5f00-b5ed-4a77-b977-da3764da72e3/files/sustainable-report-full.pdf

Oguaman C (2006) International law and indigenous knowledge: intellectual property, plant biodiversity and traditional medicine. University of Toronto Press, Toronto

Pacific Regional Framework for the Protection of Traditional Knowledge and Expressions of Culture (2002) Model law for the protection of traditional knowledge and expressions of culture 2002, Suva, Fiji

Productivity Commission (2011) Chapter 8, overcoming indigenous disadvantage: key indicators 2011. Retrieved 20 July 2015 http://www.pc.gov.au/research/recurring/overcoming-indigenous-disadvantage/key-indicators-2011

Reason P, Bradbury H (2007) Handbook of action research, 2nd edn. Sage, London

United Nations Declaration on the Rights of Indigenous People GA Res 61/295, UN GAOR, 61st Sess., Supp No 49, UN Doc A/RES/61/295 (2 Oct 2007)

UTS Indigenous Knowledge Forum and North West Local Land Services (2014) Recognising and protecting aboriginal knowledge associated with natural resource management, white paper for the office of environment and heritage NSW. Retrieved 20 July 2015 http://indigenousknowledgeforum.org/components/com_content/models/forms/white_paper.pdf

WIPO-UNESCO (1982) WIPO-UNESCO model provisions for national laws on the protection of expressions of folklore against illicit exploitation and other forms of prejudicial action, 1982, WIPO Lex No.: UNESCO001

Part III
Horizontal policies: Communication and Social Enterprise Governance

Framing New Environmental Cultures for Sustainability. Communication and Sensemaking in Three Intractable Multiparty Conflicts in the EbreBiosfera, Spain

Jordi Prades and Aitana De la Varga

Abstract Since communication defines conflict and intervenes in the discursive construction of sustainable development, it is more than just a mere tool in mediation processes aimed at resolving environmental intractable conflicts, but is itself a constitutive component of the conflict. Through communication, the discursive and organizational practices and logics of institutions like the law, government and social movements frame and make sense of conflicts regarding the environment. Our objective is to analyse communicative processes in environmental conflicts as an engine driving social change to sustainable development. From an interactional approach to framing we analyse three environmental disputes related to water and energy in the Terres de l'Ebre (Southern Catalonia). We observe how, as an alternative to a conflictive frame, the UNESCO recognition of the Terres de l'Ebre as a Biosphere Reserve (EbreBiosfera) is configured in a proactive, cohesive and consensual frame. In all three conflicts new meanings for sustainable development associated with environmental and social justice and democracy have resulted in new environmentally sustainable cultures, specifically, new water and energy cultures that produce local results of global application. In terms of implementation of these cultures, communicative legislation or "soft law", understood as a horizontal interactive two-way dialogue, is more effective and offers more satisfactory long-term results than a traditional top-down approach.

Keywords Intractable environmental conflict · Organizational communication · Interactional framing · Sensemaking · Sustainability culture

---

J. Prades (✉)
ASTERISC Communication Research Group, Universitat Rovira i Virgili (URV), Tarragona, Spain
e-mail: jordi.prades@urv.cat

A. De la Varga
CEDAT Centre for Environmental Law, Universitat Rovira i Virgili (URV), Tarragona, Spain
e-mail: aitana.delavarga@urv.cat

V. Mauerhofer (ed.), *Legal Aspects of Sustainable Development*,
DOI 10.1007/978-3-319-26021-1_8

1 Communication, Conflict and Law as Communication Processes

A conflict inevitably includes at least one element in dispute, two opposing players and communicative interaction. Since every environmental conflict is socially constructed, it involves communication, which itself is constitutive of the conflict.

Intractable conflicts are conflicts that are persistent and resolution-resistant but also have some degree of tractability, they can be transformed in order to become more tractable and reach agreement (Campbell 2003; Lewicki et al. 2003; Gray et al. 2007; Asah et al. 2012) through communication processes. Thus, rather than consider conflicts as irresolvable, institutions and individuals could, in their discourses and practices, adopt a transformative approach, as Campbell (2003, p. 365) proposes: "In transformative mediation, disputes and conflict are not seen as necessarily destructive events but rather as opportunities for growth and transformation in the disputing parties".

There has been a great deal of research into how the media participate in the construction and social perception of the environment and into how they affect the social production of environmental awareness and risk (Weingart et al. 2000; Cottle 2006; Hansen 2010).

But going beyond a media-centred perspective, conflict is primarily built, denied, resolved or prolonged in time through interactions between the actors in dispute. Institutions such as the law, governments and social movements create meanings regarding the environment and sustainable development through discursive and organizational rather than media practices.

Thus, legislation, planning, evaluation, authorization, inspection and other organizational processes constitute a particular notion of the environment with their own means for communicating outside of the media, for instance, through the United Nations Environment Programme, international conferences and summits, EU climate and energy objectives, EU treaties and secondary law, Agenda 21, ministries of the environment, environmental projects promoted by corporations as part of their corporate social responsibility and the use of ecological footprint information on labels.

When we talk about institutions we refer to formal in contrast with informal institutions, in the sense of North (1990). Institutions could be also classified as "internal" or "external": "If rule breaking is sanctioned by the state, institutions are called 'external', if rule breaking is sanctioned by members of society, institutions are called 'internal'" (Voigt 2013, p. 8). Internal institutions may be unorganized actors or organized actors like social movements organised through associations or platforms to face conflicts.

We understand conflict as a form of communication that feeds on action-reaction dynamics between parties in opposition, in such a way that conflict becomes the subject rather than the object of communication, as Castelló (2012, p. 19) suggests when he says that "any content can then be used, since the centrality of the communicative act is not to transmit information or knowledge about the object but to keep the conflict going".[1]

This perspective, which challenges the utilitarian view of communication, proposes a reflection on the functions of conflict and dysfunctional communication, rather than on conflict as social malfunctioning that can be remedied with "proper" communication. Communication has often been called for to resolve conflicts—yet it was through communicative processes that those conflicts arise: without communication, parties cannot express their divergent views, whereas too much or inappropriate communication can exacerbate and entrench a conflict.

In the risk society (Beck 1992), environmental concerns have become as much a part of all social structures as political, ethical, economic and legal conflicts. Conflict is in itself a risk in that it introduces changes in power relations as a defining element of the actors and their legitimacy (Beck and Kropp 2008). Power relations determine actor responses to risk and conflict, whether these are addressed or not and how quickly, using what resources, using what approach and applying what rules. Ultimately, according to Hajer (1995, p. 275), "the politics of discourse is not about expressing power-resources in language but is about the actual creation of structures and fields of action by means of story-lines, positioning, and the selective employment of comprehensive discursive systems (such as law or physics), etc.".

The social construction of environmental conflict implies that observable facts (increasing temperatures, atmospheric CO_2 levels, water pollution, etc.) have multiple interpretations regarding a phenomenon, its causes and possible solutions —including perceptions as a risk, problem or conflict. Indeed, "sometimes environmental problems are intractable because we focus more on solving the problem than on carefully identifying and defining the problem" (Asah et al. 2012, p. 117). Undoubtedly, rethinking environment in terms of conflict draws new crossroads from a sustainability perspective, especially in fields such as water and energy. In the case of water:

> Water, our unifying theme, is simultaneously a prerequisite for all life forms on earth, a crucial agent of change in human affairs, and a fulcrum of diverse social interests. It is, at times, the focal point of intense contestation. At other times, it is the hub of cooperative resource management. Water is unique in the ways it supports, links, and is embedded within complex networks of physical and ecological processes, and in the ways it is configured and reconfigured through political, discursive, economic, and technological processes. It is arguably the most salient element of connectivity in the local to planetary continuum of ecological and social systems. Yet despite the universal characteristics of water, the precise path of its transformation and use by human societies is highly contingent on historical and geographical context (Sneddon et al. 2002, p. 664).

As for nuclear power, the titles of OECD reports on radioactive waste reveal a shift from confrontation to cooperation, from information and consultation to scaled approaches, long-term decision making and citizen empowerment (OECD 2003,

[1]All translations from Spanish or Catalan are by the authors unless already available in English in published form (see References). Page numbers, when provided, refer to original works.

2004, 2010b, 2013). Regulatory, socioeconomic and cultural variables require approaches that pool different meanings and build bridges between parties.

The ambiguity and complexity regarding wind farms is yet another illustrative example:

> Wind farm development draws on knowledge as varied as that required by aerodynamics, management, engineering, ethology and law. It also involves the alignment, in some stable way, of birds, folklore associations, green certificates or grid connections. Moreover, it is virtually impossible to establish borderlines to the application process, regarding for instance the actors who are and those who are not part of it, what topics or contexts are relevant to it, or whether or not there is a specific time schedule. Faced with such ambiguous conditions, wind farm developers put their projects into contexts, describe the technical features of the installations, and answer potentially damaging criticism. Their aim is to manage the indecisiveness of, and surrounding, their project by way of the formal permission process, all with a view to gaining legal acceptance for their plans to build a wind farm (Corvellec and Risberg 2007, pp. 308–309).

Although explaining opposition to water transfer from a river, tendering to be the location for a nuclear waste site and objecting to wind power deployment may appear to represent contradictions, all three cases do, in fact, share common features of intractability, beyond the not-in-my-backyard (NIMBY) effect.

To manage intractability in environmental conflicts, the law intervenes as a framework for action by public authorities and individuals, by normatively establishing the environment as a legal object and right and by shaping the legal concept of the environment and its protection. Environmental law enacts legislation that provides guidelines for dealing with environmental conflicts—whether general laws, sectoral laws or regulations—and so determines what level of risk is deemed acceptable or unacceptable (Esteve Pardo 2006). Each state, through its legislation and implementation of these laws, thus establishes the importance it attaches to the environment and its protection.

In the Spanish state, the concept of environmental protection linked to sustainable development derives from the Constitution, specifically from Articles 40, 128 and 130 and also Article 45,[2] which lays down guiding principles such as the individual right to enjoyment and duty to preserve an environment suitable for

---

[2]Artículo 45. 1. Todos tienen el derecho a disfrutar de un medio ambiente adecuado para el desarrollo de la persona, así como el deber de conservarlo. 2. Los poderes públicos velarán por la utilización racional de todos los recursos naturales, con el fin de proteger y mejorar la calidad de la vida y defender y restaurar el medio ambiente, apoyándose en la indispensable solidaridad colectiva. 3. Para quienes violen lo dispuesto en el apartado anterior, en los términos que la ley fije se establecerán sanciones penales o, en su caso, administrativas, así como la obligación de reparar el daño causado.

(Article 45.1. Everyone has the right to enjoy an environment suitable for personal development, as well as the duty to preserve it; 2. The public authorities shall safeguard rational use of all natural resources with a view to protecting and improving the quality of life and preserving and restoring the environment, by relying on essential collective solidarity; 3. Criminal or, where applicable, administrative sanctions, as well as the obligation to make good the damage, shall be imposed, under the terms established by the law, against those who violate the provisions contained in the previous clause).

personal development and the public obligation to ensure rational use of natural resources, with the aim of protecting and improving quality of life and protecting and restoring the environment based on collective solidarity. Viewing the provisions referred to in these articles in conjunction, it can be concluded that state responsibilities include protecting the environment, as well as promoting progress, modernization and socioeconomic development and equitably distributing regional and personal incomes.

The state, charged both with ensuring rational use of all natural resources in its territory in the general public interest and with guaranteeing the right to enjoy the environment, must ensure that the environment as a right must be balanced with other rights, which may, a priori, seem contradictory; hence, all these rights need to be weighted when in conflict. The state both protects the environment and fosters its exploitation so as to promote socioeconomic development and the wellbeing of its citizens. Nonetheless, some governments perceive development purely in terms of the creation of material wealth, confusing general interest with the accumulation of capital in the hands of a few.

This is the contradiction faced by sustainable policies that, in short, try to marry the need for socioeconomic development with environmental protection and social justice. Through sustainable development policies, governments assume the environmental dimension of all public actions; thus, all policies must be sustainable to not only ensure the protection of the environment and socioeconomic development but also social justice and quality of life for its citizens.

Such contradictions in the state remit should be viewed less as limitations and more as opportunities for innovative interpretations, for reconsideration of the dominant discourse regarding sustainable growth and ecological modernization as strategies to accommodate radical environmentalism and for seeking out legitimate ways to conceptualize and analyse environmental conflict (Hajer 1995).

The discursive system of law legitimizes the legal framework governing conflict between the three dimensions of sustainable development, namely, economic, conservationist and social. The Spanish legal system has introduced, at least formally, instruments for communicative legislation. The application of the rules of the game can respond either to a model based on order, punishment and top-down one-way communication from the authorities to the citizens, or, alternatively, to a model based on communicative legislation, that is, horizontal interactive two-way communication between authorities and citizens. In this case, authorities solicit dialogue but do not directly intervene; rather, as Witteveen and van Klink (2011) point out, the law establishes fundamental principles and values aimed at promoting a gradual change in attitudes and behaviours. The same authors further affirm that communicative legislation is often characterized as "soft law" or even as "symbolic law" and, in their interactionist interpretation of the concept of "law", they also consider communicative legislation as a form of responsive law.

In this sense, environmental legislation permits enhanced interaction between all actors promoting responsive law, although sometimes merely in a formal way; for instance, it facilitates access and provides environmental data and information for decision making and policy development purposes.

Our approach involves interactions between institutions and sensemaking, with social actors making sense of controversies by developing coherent accounts of reality in the form of narratives and constructing meanings that, in turn, draw on institutionalized principles to justify the vision of a situation. Individuals and organizations invoke and actively mobilize institutions through sensemaking practices.

1.1 Sensemaking and the Framing of Environmental Conflicts

Sensemaking must be understood as an ongoing communication process by which situations, actions, circumstances and events such as conflicts are "talked into existence" (Weick et al. 2005). This process involves the perceptions, interpretations and actions of organizations and also their discourse, with their relationships making more sense in a constitutive way than is usually attributed (Taylor and Robichaud 2004; Weber and Glynn 2006). Institutions and social movements shape their sensemaking regarding the environment and sustainable development through their discursive practices and frames. The forms in which they communicate offer a productive field to be analysed in terms of insights to constitutive organizational communications.

This organizational communication perspective on conflict—which presents communication as a process whose outcome "is not the linear transmission of information but the nonlinear production of interpretations" (Aula and Siira 2010, p. 131)—is useful to identify sensemaking and framing processes in complex interactions such as occur in environmental conflicts. As the same authors stated: "We maintain that a social complexity perspective on conflict enables an interpretative view of organizational communication. In other words, whereas the mechanistic model treats nonlinearity, disorder, chaos and emergence as system defects, social complexity embraces these qualities as natural parts of human, and thus organizational, communication" (Aula and Siira 2010, p. 131).

In the broadest sense, to frame a problem is to selectively define it by choosing a perspective that highlights one aspect and conceals other aspects so that the issue is viewed only from one perspective (Goffman 1974; Gamson and Modigliani 1989; Entman 1993). According to the constructionist and interactional paradigm in framing research (Van Gorp 2007; Dewulf et al. 2009), framing is a dynamic process in which multiple actors and meta-communicational aspects of discourse intervene.

From a sensemaking perspective, framing refers "to a pattern of highlighting similar aspects of experience to give a coherent account of what is going on that is continuously shaped and reshaped in interactions" (Brummans et al. 2008, p. 28).

Through this discursive interaction, actors construct shared visions of reality, align themselves with existing coalitions or constitute others and attribute sense to an environmental conflict and to the role played by each actor in regard to causes, consequences and possible solutions (Lewicki et al. 2003). Conflicting interests, values and expectations generate divergent frames anchored in both historical facts of collective memory and in personal and world experiences lived in a particular setting.

When this setting is the environment, complementary views appear that match the three dimensions or frames corresponding to sustainable development: (1) the economic dimension, which perceives the environment as an asset or good to be exploited and focuses on managing supply; (2) the conservationist dimension, which views the environment as a resource to protect and focuses on managing demand; and finally, (3) the social dimension, which focuses on environmental justice and democracy (public participation and access to information and justice) and takes into account cultural values.

In this last dimension, the environment is perceived as "an immersive space that provides the grounding experiences and material" for social "relations-in-place". Here, where the focus is on "environmental discourses rooted in culture and grounded in history, identity and community", it is useful to study discourse as a practice relating people and the environment as Milstein et al. (2011, pp. 487–489) propose.

Identifying the frames in play and understanding how they are built and evolve helps manage conflict. Even if consolidated over time, frames can still be changed: "Linked as they are to information processing, message patterns, linguistic cues, and socially constructed meanings, framing and reframing are vital to the communications underlying negotiations" (Shmueli et al. 2006, p. 210). Understanding frames in action enables intervention in communication processes in order to bring parties together and reach agreement. Below we describe a case study in which we observe and depict processes of sensemaking and reframing.

2 The EbreBiosfera as a Case Study

The Catalan Government Delegate to the Terres de l'Ebre,[3] on concluding his term in office, described government action in the region in the period 2007–2010 as follows:

> We have inherited counties with severe deficits in infrastructure and services in virtually all areas, deficits resulting from a historic neglect by the authorities of the Terres de l'Ebre. This was where they located nuclear power plants or where they obtained water … .Today … the model advocated by the government is to bring an end to the concentration of

[3]EbreBiosfera was the first brand and the promotional name of the UNESCO-recognized Biosphere Reserve of the Terres de l'Ebre. This southwestern region of Catalonia gets its name from the Ebre river which rises in northern Spain. The Ebre Delta is an internationally protected important Mediterranean wetland area.

dumped waste from elsewhere in the Ebre area: no more nuclear waste, no more landfills. A model of territory which also defines its candidacy as a Biosphere Reserve …..

A Biosphere Reserve (BR) is a UNESCO-recognized area within the Man and Biosphere (MaB) Programme. Worldwide there are currently[4] 651 BRs in 120 countries (UNESCO 2015).[5] Spain and the USA are first in the world in terms of number of BRs, with 47 each one.[6]

Since the concept was devised in 1974, socioeconomic aspects of sustainable development have gradually been added to the ecological dimension of BRs, which have become "exceptional areas for research, long-term monitoring, training, education and public awareness" where local communities participate in the management of resources (UNESCO 1995). Through conservation, development and logistical support actions conducted in core, buffer and transition areas, BRs become "learning sites for sustainable development" (UNESCO 2008) and experimentation and testing laboratories that implement sustainable development principles through new approaches to the management of environmental conflict and the integration of interests.

Although it is not essential to integrate BRs in legislation to achieve good results in resource management and overall planning (Elbakidze et al. 2013), legal recognition and an institutional structure legitimize and reinforce BRs. Such legal recognition is the outcome of political will. In Spanish legislation[7] governing natural heritage and biodiversity (L42/2007 of 13 December), BRs are recognized as "another means for protecting areas" and as "areas protected by international instruments".

Specifically, Article 3.31 of L42/2007 defines BRs as "territories declared as such [BRs] within the UNESCO MaB Programme, subscribed to by the Kingdom of Spain, whose heritage and natural resources are managed in an integrated, participatory and sustainable way". The same law further refers to BRs in Chapter I of Title IV (Sustainable use of natural heritage and biodiversity), defining BRs, their purpose and characteristics within the Spanish BR network and the MaB Programme. There is also explicit legal provision that the Natural Heritage and Biodiversity Fund should promote ecological production and finance specific measures to prevent erosion and desertification in BRs.

The EbreBiosfera, declared a BR by UNESCO in 2013, is described as follows on the website of the MaB Programme:

---

[4]20 July 2015.

[5]http://www.unesco.org/new/en/natural-sciences/environment/ecological-sciences/biosphere-reserves/ (retrieved 20 July, 2015).

[6]Two new Spanish reserves have been included this year in the MaB Network: The Spanish-Portuguese Iberian Transfrontier Meseta and the Anaga Massif in Tenerife (the Canary Islands).

[7]Spanish legislation abbreviations as used in this manuscript are as follows: Law, L; Decree, D; Royal Decree, RD; Royal Decree Law, RDL.

> This biosphere reserve covers 367,729 hectares and has a population of 190,000. Located in the Catalunya [sic] region, it includes the delta and watershed of the Ebro [sic] River, which is Spain [sic] largest river in terms of volume. It has a large number of different ecosystems ranging from inland to coastal areas. The main part of the land is used for cattle. Alternative energy sources such as hydrological, solar and windpower, are being developed in the region, respecting biological conservation and landscape values.[8]

UNESCO recognition represents a paradigm shift in the management of an environmental conflict in which communication is the engine behind sustainable development. The defining moments for the EbreBiosfera, in fact, were two periods that illustrate how communicative dynamics (planning, organization, participation and information) were both constitutive and interlinked from the outset: pre-candidacy in the three-year run-up to 2010 (participative consultation, exploratory analyses and strategic information) and candidacy thereafter (organization, briefings, communication plan).

These two periods are quite explicitly described in the first proposal submitted for application, namely the document titled "La Reserva de Biosfera de Terres de l'Ebre" (Aragonés and Miró 2012[9]). "The strategic pre-candidacy information process is perceived … not so much as a participatory moment but more as a communicative moment, integrated in a natural way in an existing participatory process" (Aragonés and Miró 2012, p. 290).

The EbreBiosfera is an appropriate context for researching the communicative dynamics generated by recent environmental conflicts in the Terres de l'Ebre as a consequence of three projects: Ebre river water transfer, radioactive waste management and windpower generation.

In addition to these three conflicts, other environmental controversies have contributed to the Terres de l'Ebre being perceived as Catalonia's "backyard". They include the failed Enron project to build a combined-cycle power plant—which united all protest in 2001 under the slogan "Stop the Onslaught on our Lands"—and the more recent Escal UGS Castor offshore underwater gas storage facility, not to mention other offshore operations (Repsol oil wells and the Zèfir wind project), the management of reservoirs for electricity production purposes, the regulation of environmental flows, flood control, the proliferation of macrophytes and the black fly (Simuliidae family), the illegal introduction of invasive species (*Silurus glanis, Dreissena polymorpha and Pomacea bridgesii*), the role played by companies and authorities in relation to chemical spills in the Flix reservoir, urban waste landfill projects, nuclear power plant management and high-voltage power lines.

Our decision to focus on the water transfer, radioactive waste management and windpower generation projects was motivated by factors that mark them as intractable environmental conflicts (Campbell 2003; Lewicki et al. 2003; Gray et al. 2007; Asah et al. 2012). These factors include their capacity for social mobilization, their persistence over time, their proximity in both space and time, the numerous

---

[8]http://www.unesco.org/new/en/media-services/multimedia/photos/mab-2013/spain-terres-de-lebre/ (retrieved 29 April, 2015).

[9]This public document is not yet published or available online.

actors involved, the distribution of powers between state, regional and local levels, media attention, politicization and judicialization and, finally, risk perceptions.

As the main actors in conflict we focus especially on public authorities supporting or authorizing the projects and the civil platforms in opposition. On one hand, we examine the language they used to define the projects; on the other hand, we focus on their discursive and organizational logics. According to Thornton et al. (2012, p. 2), institutional logics may be considered as "frames of reference that condition actors' choices for sensemaking, the vocabulary they use to motivate action and their sense of self and identity".

We accordingly formulated the following research questions:

- How have three intractable multiparty conflicts helped shape the EbreBiosfera?
- What (re)framing and sensemaking processes regarding sustainable development and communicative legislation can be identified in the three conflicts affecting the EbreBiosfera?

We describe the three conflicts in terms of four key moments in the sections below, described and numbered as follows: (1) initiation (with relevant legislation as contextual information); (2) reaction; (3) development; and (4) the current situation.

2.1 Water Transfer Planning

1. In 2000 the conservative Popular Party (PP) government presented a National Hydrology Plan (NHP) that provided for the transfer of water from the Ebre River to Barcelona and Southern Spain. In response, an Ebre Defence Platform (PDE) was launched that mobilized thousands of people in favour of a "new water culture" (NWC) in demonstrations in Barcelona, Madrid and Brussels. Social pressures led to the appointment of a Catalan Government Delegate to the Terres de l'Ebre. Despite protests against the NHP, it was approved as L10/2001. Mass demonstrations against the NHP again took place in Barcelona in 2002 and 2003.
2. With the victory of the socialist party (PSOE) in the 2004 elections, L10/2001 was repealed (RDL2/2004) and a new NHP was enacted as L11/2005. The PDE, however, also protested against this law.
3. A lengthy drought in 2008 led both the PSOE government in Spain and the three-party coalition government in Catalonia to propose a transfer of water from the Ebre river to Barcelona (RDL3/2008). Another mass demonstration against the water transfer brought thousands of people to Amposta, a town in the Ebre Delta. Southern Catalonia again rebelled against decisions made in Barcelona and Madrid. However, the law was never implemented as rains intervened to bring an end to the emergency. There were further demonstrations

in 2010, this time before the passing of legislation regarding water plan for Catalonia (D188/2010)[10] and again in 2014 in Deltebre.

4. In 2014, the PP formed a majority government and adopted the Ebre Hydrology Plan (RD129/2014), with a five year delay and two years after Spain was convicted of breach of the EU Water Framework Directive (WFD) in a judgment of the Court of Justice of the EU of 4 October 2012 (Case C-403/11). The Ebre Hydrology Plan has been appealed by the Catalan government before Spanish and EU authorities, with disagreement focusing on the environmental flows proposed by the Spanish government, considered insufficient by the Sustainability Commission of the Terres de l'Ebre (CSTE).

2.2 *Radioactive Waste Management*

1. In 2004 a community waste management programme for democratic management of radioactive waste was launched, based on a proposal of the Association of Municipalities in Nuclear Areas (AMAC). Also in 2006, the PSOE government approved the 6th General Radioactive Waste Plan (6th PGRR), which provided for the construction of an ATC (*almacén temporal centralizado*, literally, a "centralized temporary store") as an interim facility for storing spent nuclear fuel and high-level waste. Between 2005 and 2006 seven roundtable meetings were held to discuss the development of nuclear energy in Spain. In 2006 the Catalan Anti-Nuclear Graveyard Coordinator (CANC) was created to oppose the project.
2. An Interministerial Commission (IMC) was created (RD775/2006), which, in 2008, approved the ATC and the conditions of the public tender for interested municipalities to compete for the site and its associated technology centre.
3. In 2010, the Council of Ascó (plenary of 26 January 2010) approved the town's candidacy and Ascó was pre-selected along with three further sites. There were demonstrations for and against the Ascó's candidacy.
4. In late 2011 Ascó was definitively ruled out. In 2012 the ATC was awarded to Villar de Cañas, located in Cuenca (that is outside Catalonia). Also in 2012, Ascó and the six municipalities nearest to this town were excluded from the EbreBiosfera, given that due to the existence of two nuclear power reactors in Ascó, BR compatibility with nuclear facilities was restricted after the Fukushima accident of 2011.

[10]The Spanish state has exclusive powers regarding rivers which, like the Ebre, cross several autonomous regions (interregional basins), whereas each region, including Catalonia, has exclusive powers regarding their own river basins.

2.3 Windpower Generation

1. In 2000, the Catalan conservative Convergència i Unió (CiU) government submitted a wind energy proposal for public approval that included a map and a draft decree regulating the implementation of windpower in Catalonia. The plan was opposed by various platforms, given the concentration of these projects in the Terres de l'Ebre. In 2001 the Catalan parliament passed a motion urging the government to withdraw the plan and to draw up an alternative map.
2. In 2002 a sectoral territorial plan was approved to implement windpower in Catalonia (D174/2002). The corresponding windpower map failed to satisfy the platforms, which continued to mobilize as windpower was deployed in the territory. Windpower companies also organized, creating the EolicCat association in 2006.
3. In 2009, a new progressive three-party coalition government in Catalonia approved a decree (D147/2009) that, partially repealing the previous law, established priority development areas (ZDPs) defined according to wind conditions, capacity to draw off the energy produced and environmental, landscape and land planning viability. In 2010, the presentation of a proposal for the first offshore windpower project (the Zèfir test station) also raised suspicions.
4. Currently, all windpower projects have been halted due to the economic crisis and to PP government changes to renewable energy legislation in Spain (RDL1/2012 and L24/2013, which came into force in 2014).

Table 1 summarizes key events in the EbreBiosfera conflicts.

We addressed the research questions using a qualitative and interpretive approach to the sample includes plans and projects referring to the three conflicts, legislation, minutes of meetings, press releases, institutional publications, platform manifestos, slogans on pamphlets, posters and banners, etc.

Organizational logics were examined by analysing legislative, public participation, popular mobilization, symbolization, staging and mediation processes. The time period and the number and variety of sources covered by the sample reflected the spatial-temporal dimensions, the contexts and the relationships between actors and stakeholders.

To detect indicators of frames in the conflicts we focused on how parties negotiated the meanings of issues and how they worked out definitions of their identities and relationships. We also considered the communication itself and how parties made sense of the interaction (Dewulf et al. 2009).

3 Review of Language and Practice

According to Rull (2010), language analysis shows how three Catalan or Spanish political language mechanisms were used to distort or reshape the original concept, as follows:

Table 1 Timeline of water, nuclear and windpower conflicts and contexts in relation to the EbreBiosfera project

Year	Socio-political context: Government – Spain	Socio-political context: Government – Catalonia	Socio-political context: Social movements	Environmental conflict: Water	Environmental conflict: Nuclear	Environmental conflict: Wind	Biosphere project
2000	PP 7th legislature 5/4/2000-20/1/2004	CiU 6th legislature 5/11/1999-23/9/2003; Catalan Govt. Delegation in Terres de l'Ebre created	PDE and windpower platform created	NHP and Ebre transfer proposed		First windpower map proposed	
2001			Joint rally by platforms and NWC rallies	NHP and Ebre transfer L10/2001		Withdrawal of first proposal	
2002			Barcelona PDE rally			D174/2002	
2003		Three-party PSC-ERC-ICV 7th legislature 5/11/2003-8/9/2006	Barcelona PDE rally			Windpower deployed	
2004	PSOE 8th legislature 2/4/2004-15/1/2008			Transfer repealed RDL 2/2004	Community waste management programme launched		
2005				New NHP L11/2005	Roundtable talks		
2006			Anti-nuclear graveyard platform created		Roundtable talks 6th PGRR	EolicCat created	
2007		Three-party PSC-ERC-ICV 8th legislature 17/11/2006-5/10/2010		Drought decree approved	AMAC visit to Habog	Windpower deployed	Approach to UNESCO
2008	PSOE 9th legislature 1/4/2008-27/9/2011		Ebre Delta PDE rally (Amposta)	Drought law RDL3/2008	IMC RD775/2006	Horta plebiscite[a]	BR model selected
2009				Drought law repealed		ZDPs D147/2009	Project created
2010		CiU 9th legislature 16/12/2010-2/10/2012	Barcelona PDE rally; Rallies for and against the ATC	Catalan Hydrology Plan D188/2010	Plenary of the Council of Ascó; Vinebre plebiscite[b]	First Zèfir offshore pilot project	Organigram
2011					Ascó ruled out	Deployment halted by crisis	Application
2012	PP 10th legislature 27/12/2011-2015	CiU 10th legislature 17/12/2012-2015		EU Court of Justice ruling (C-403/11)	ATC project awarded to Villar de Cañas (outside Catalonia)	RDL1/2012 Suspension (funding withdrawn)	Proposal rejected
2013						L24/2013	Proposal approved
2014			Rally on the new bridge over the Ebre Delta (Deltebre)	Ebre Hydrology Plan RD129/2014		L24/2013 (entry into force)	First steps as a BR
2015	Elections	Elections 27/09/2015					New logo/managing body

[a]On 16 March 2008, the inhabitants of Horta voted regarding a windfarm project, with 630 votes against and 160 votes for. The projected windfarm was not built but not as a result of the non-binding plebiscite but as a result of a court ruling

[b]Vinebre (a small town located near Ascó) held a non-binding plebiscite regarding the ATC on 28 March 2010. Of the 478 inhabitants, 109 voted against and 12 for. The Catalan parliament and many other local bodies also approved motions against the ATC

(a) "An alternative synonymous expression. A well-formed descriptive term is coined to designate the reality, perfectly synonymous with the term to be avoided and also avoiding negative connotations. One example is that of the *almacén temporal centralizado* (ATC) mentioned above, in which storage is of nuclear waste, although this is not made explicit; another is *captación temporal de agua* (temporary water collection) rather than *trasvase* (water transfer) ... Note that they tend to be lengthy, descriptive expressions, which, in itself, possibly indicates that such terms conceal negative connotations." (Rull 2010, p. 120).

(b) "A new term that shifts the focus or reference frame, that is, use of a term that applies to a slightly different reality. Examples are *desaceleración* (deceleration) to refer to an economic crisis, *nódulo* (nodule) to refer to a tumour, *nouvingut* (newly arrived) to refer to an immigrant, etc." (Rull 2010, p. 120).

(c) "A negative association (which works in the opposite way to the other two mechanisms). Examples are *cementerio nuclear* (nuclear graveyard), *gobierno nacionalsocialista catalán* (Catalan national-socialist government), *chapapote* (oil spill residue, especially used in the aftermath of the *Prestige* oil spill off the coast of western Spain in November 2002) etc. The fact that a negative connotation already exists creates an association of ideas" (Rull 2010, p. 121).

Our analysis goes beyond this classification, as, in line with our approach to framing as sensemaking, we understand factors other than language to operate in these mechanisms. We also identify the use of metaphors as yet another device in the semantic distortion/manipulation toolbox.

3.1 More than Words

Some examples of application in our sample of the mechanisms mentioned above are as follows. Referring to an "inter-basin transfer" from "wet Spain to dry Spain" to "prevent Ebre water from flowing wasted into the sea" and to ensure "water for all" or proposing a "temporary supply" to respond to a "national drought emergency" is not the same as referring to a "water transfer" or to positions reflected by slogans such as "Not a Drop More", "*The River is Life*: No to Transfer", "For a New Water Culture", "Transfers are Not the Solution", "The Ebre Without Water is Death for the Delta" or "Respect EU Environmental Directives".

According to Jiménez and Martínez-Gil (2005, p. 10) "the need for justifying an old list of dams and out-of-basin diversions has generated a 'false official speech' (Martínez-Gil 1997), based upon propagandistic concepts such as 'deficitary' and 'excedentary' basins, the handy 'general interest' (a concept that should be reviewed to fit sustainability), the already traditional 'hydrologic unbalanced condition of Spain' or the very much appreciated by politicians 'hydrologic solidarity'".

Powerful water metaphors have also been used to refer to events such as the "human flood" or "anti-transfer wave" of Ebre protesters that "inundated Madrid", reproducing the idea of "the country coming to town" in a "blue march".

As for radioactive waste, the OECD (2010a) referred to interpretations, values and perceptions that define this waste as negative when referring to control, risks, or mortgaging future generations, and as positive when referring to economic investment and research. The Dutch Habog facility, taken as a model for the Spanish ATC, is an example of the symbolic dimension of radioactive waste. As Rull clearly puts it:

> Look at the words used to describe this facility. The Spanish government refers to *centralized temporary storage*, not saying what is being stored. And that is no accident, since what is being stored is nuclear waste. If what is stored is not named, it is as if what is there is being concealed … in short, this definition seems innocuous. Opponents, in contrast, refer to a *nuclear graveyard*. The word *graveyard* has clear negative connotations associated with death. Therefore, they also play at using the word or expression that suits their interests because it best represents what they want. (Rull 2010, p. 113; Rull's italics).

Frequently terms that are not synonyms are used to refer to the same concept as a device to introduce positive associations in the debate. Examples are "spent fuel" to refer to "nuclear waste" (suggesting possibilities for reusing), "storage" (from "store", implying "to keep, set aside, or accumulate for future use") instead of "disposal" ("the act or means of getting rid of something") and "incentives" instead of "compensation" (an additional payment and a motivating influence rather than reparation for loss).

In regard to windpower also, the use of metaphors associating it with industry and light pollution has been documented in settings where the natural landscape is represented almost as a spiritual force (Woods 2003). In the Terres de l'Ebre, two values in addition to the river define the landscape in local artistic and historical but also universal terms: Cubist paintings by Pablo Picasso, who spent some time in Horta, and the scenes of the Battle of the Ebre in the closing days of the Spanish Civil War, which have been converted into a historical tourism resource to promote the local Movement for a Culture of Peace. Here, to refer to the "deployment of wind farms" and "wind mills" is not the same as referring to the "massification of wind farms" and "wind turbines".

The EbreBiosfera itself has not been immune to semantic discussions regarding the term "reserve" due to the negative connotations. In fact, the MaB Strategy Group suggests using the term "reserve" only for the core and referring to the entire area covered by the MaB programme as a Biosphere Region—with positive connotations as it includes the regional identity of BRs. Another quite neutral synonymous term for BR is "park", used in all the Austrian BR designations.

Regarding the EbreBiosfera, this term was coined during the candidacy process to avoid using "reserve" (it also had the advantage of being sufficiently succinct to fit in media headlines). However, the current designation—Terres de l'Ebre Reserva de la Biosfera—includes the word "reserve".

As we have seen, in the three cases the actors in conflict use language in a similar way to guide discourse and create story-lines to frame the projects throughout the conflict timeline. But it is not just a matter of language but also of practices.

3.2 Framing Action

Supporters of the all three platforms have repeatedly mobilized jointly to share causes, strategies and scenarios, thus forming a significant critical and compact mass of leaders and activists. Social movements are capable of contesting final decisions, providing there is a margin of discretion in decision making. Against the platforms, various groups, defending their own interests in a setting marked by intense pressures and great socio-political tension, have accused the platforms of promoting a culture of "no". These groups mainly include licensed irrigation water users, residents in Ascó and nearby municipalities, windpower companies operating in Catalonia, not to mention a number of mayors and local and regional politicians. Regarding risk perceptions and acceptance, institutions decide what risk can be assumed based on objective regulations and criteria and applying, to a greater or lesser extent, precautionary and preventive principles in decision making. Climate change is the thread unifying all three conflicts: (1) rising sea levels due to global warming threaten the Ebre Delta, which is already undergoing subsidence and salinization; (2) nuclear power is stigmatized and perceived as dangerous, although for years the sector tries to improve its image with the argument that nuclear power can help fight climate change (OECD 2000); and (3) although windpower risks are local compared to the overall benefits, opponents try to preserve their most emblematic landscapes by preventing local damage from a global cause (to reduce greenhouse gas emissions).

The analysis of practices shows how some frames—mainly identity and institutional dimensions—reflect the features of intractable multiparty conflicts.

- **The identity dimension**
  In referring to the Terres de l'Ebre in terms of "we" and the rest of the country as "they", a historical identity is defined, with the river as the axis of cohesion that socially structures the territory and landscape as the place where the local people live everyday experiences that eventually personal reminiscences and historical memory and that evoke feelings related to wellbeing (tranquillity, spiritual connection with nature, outdoor leisure, gastronomy, etc.). Due to spatial-temporal concurrence, the conflicts generated by different projects are all perceived as the same attack on territory and identity.
  As for the geographical identity, the tensions between general and local interests surface in all three conflicts, with all three platforms of the unanimous opinion that the territory has already shown sufficient "solidarity" with the rest of the country, for several reasons. First, water has been transferred from the Ebre river since 1989 (L18/1981). Second, two of the three Catalan nuclear power plants are located in Terres de l'Ebre (in Ascó), with political and institutional aspects

of the water-energy nexus (Scott et al. 2011) arising in terms of water transfers, hydroelectric dams and cooling plants for the nuclear stations. Thirdly, the fact that most wind farms in Catalonia are located in the Terres de l'Ebre could be considered an environmental injustice linked to the historical marginalization of this region, and threatening "life projects" described by Zografos and Martínez-Alier (2009, p. 1736).

- **Institutional dynamics**

  Conflict "becomes institutionalized over time as the disputants' behaviours reflect repetitive, habitual patterns of actions that are reinforced by social consensus"; furthermore, "from a regulative perspective, deinstitutionalization involves relaxing the coercive forces that prevent exploring alternative behaviours" (Gray et al. 2007, pp. 1420–1425).

  The legislation, planning, environmental and other regulatory intervention practices are very distinct from the logics of the street (demonstrations, protest rallies and symbolic stagings), although platforms also use the institutions—by means of allegations, claims, appeals and complaints to administrative, political and judicial bodies—to achieve their goals. In all three conflicts, platforms, environmental groups and local authorities have resorted to the courts to halt conflictive projects (judicialization).

  Electoral interests and changes in government mark milestones in environmental policy. Relationships between parties that alternate in state and regional governments and power relationships between the state and Catalonia politicize the conflict (politicization).

  The platforms have been particularly adept at adopting the "media logic"—a term coined by Altheide and Snow (1979)—to capture the attention of the media, acquire legitimacy and create an impact, furnishing news makers with powerful symbolic images such as the "anti transfer knot" (a knot of water pipes against a blue background), a radioactive skeleton and reproductions of Cubist and Spanish Civil War landscapes. These icons are reproduced regularly in the media, which, according Putman and Shoemaker (2007), use the conflicts to frame news production (media attention).

Just as in the language analysis, it can be observed that the three conflicts share several frames in relation to identity, characterization of opponents, power, risks versus benefits, conflict management and the role played by frames in this management, etc. (Shmueli et al. 2006). Note that we are not establishing a typology of frames, as, in fact, selecting frames is itself a way of framing. Furthermore, "a key difficulty with typologies is that conflicts often occur in unique and often complex contexts" (Young et al. 2010, p. 3979) and these categories often overlap.

Even so, classifications like those by Young et al. (2010)—who define six categories of conflicts, namely beliefs and values, interests, decision-making processes, available information, questions of legal, social and economic structure and personal differences between individuals and groups—are useful to explore the many dimensions of a conflict, communicative interactions between actors and sensemaking processes.

In our discussion, therefore, going beyond the limitations regarding previously established frames, we revisit the interactional approach to framing and propose new labels for the conflicts unfolding in the Terres de l'Ebre. Thus, instead of dissent and protest, we consider consensus and a new culture of environmental sustainability regarding the EbreBiosfera.

4 Discussion

The EbreBiosfera illustrates the switch from a negative discourse reflecting victimization and resignation to a purposeful discourse reflecting action, enthusiasm and hope that points to something more than just agreeing to disagree. Rather, there is an ongoing effort by disputants to configure a frame of consensus to the conflict-as-frame:

> … There is both social commitment to environmental conservation and consensus on the need to diversify and strengthen the local economy. In the 21st century, presenting conservation and active heritage management as compatible with socioeconomic development is an achievable challenge ….

Therefore, the reserve of the future has to be articulated as an ongoing process of harmonizing interests and objectives of local and regional socioeconomic actors; As a meeting point or better, as a road to travel together, obtaining consensus in order to undertake concrete actions and projects that can improve quality of life of citizens while successfully managing the territory. (Aragonés and Miró 2012, pp. 296–297).

The EbreBiosfera shapes environmental conflict as a form of proactive communication and as the engine of a social shift towards sustainability. A unifying thread in the candidacy proposal was sustainable development expressed as three pillars summarized in a definitive slogan: "Local territorial values help build economies and societal values." Emerged from the three conflicts are new interpretations of sustainable development related to environmental and social justice, democracy, identity and culture.

Undoubtedly one of the greatest contributions of the anti-transfer grassroots movement has been a recuperation of territorial identity linked with the Ebre river. Since the river is not just a natural accident, but reflects traditions and heritage, its management goes beyond technical issues and is, in fact, a matter of resource "control" with moral components (Jiménez and Martínez-Gil 2005; Mollinga 2008).

Mass mobilizations against the NHP of the early 2000s fostered a social reconstruction of the Ebre basin that reformulated ancient symbols and created new icons reflecting a retrospective vision of the river and theoretical premises regarding the environment. As a result, for the people of the Terres de l'Ebre, the river is viewed as an asset or good that must be preserved as a legacy for future generations (Boquera Margalef 2009).

In addition to contributing to this changed vision at a local level, the PDE has developed and disseminated the principles of the NWC—that is, managing water as a resource and conserving it as heritage—implemented in the EbreBiosfera and in Catalonia and of universal application.

4.1 The New Water Culture in Practice

The NWC, defined in opposition to the former regenerationist concept of water as an inexhaustible resource, anticipated the change of paradigm reflected in the WFD by prioritizing "river happiness" over "river power" and "river business" (Martínez-Gil 1997). The NWC "is a social and scientific movement that, assuming the ecosystemic perception of the EWFD, also considers the relationships of water with human spiritual facets, therefore proposing a certain water ethics. The NCW positions itself in a "holistic" perception of water." (Jiménez and Martínez-Gil 2005, p. 1).

This new frame, initially theorized by academics, has been disseminated by social movements and has been partially assumed by public authorities in the face of public pressures and scientific evidence. Consequently, as a tangible result of the application of the principles of sustainable development advocated by the NWC, methods of water efficiency, savings, purification, reuse and recovery have been implemented in the Terres de l'Ebre and more widely in Catalonia. Although new irrigation plans threatens achieved goals.

4.2 Towards a New Energy Culture

The ATC and windpower conflicts pose energy model questions that challenge the status quo in terms of the role of different energy sources in a mixed supply model, centralized generation away from consumption points, self-production restrictions and the cost of electricity, among others. At the Spanish level, noteworthy is the New Energy Model Platform and its #OligopolyOFF campaign.

We are in transition towards a "new energy culture" that replicates the NWC as a frame with promising outcomes for sustainability. As one example, despite obstacles posed by legislation that favours large facilities, new types of social organizations calling for energy sovereignty—like Som Energia—are taking shape as "prosumer" cooperatives that generate and sell green energy.

Public and private sectors organizations are also taking this new energy culture frame on board, as demonstrated by actions in favour of savings and efficiency, a recent agreement regarding energy poverty in Catalonia and the creation of new partnerships such as roundtables and clubs of mayors in favour of sustainable energies (Mesa d'Alcaldes de l'Energia de Catalunya, Club del Pacte d'Alcaldes de Catalunya), local authority windpower associations (Associació de Municipis Eòlics), etc.

5 Conclusions

Research into communication reveals new ways to move towards sustainability in the face of problems that require technically and economically feasible responses but that are also socially acceptable in complex risk communication environments such as those represented by environmental conflicts. The EbreBiosfera needs not only to manage environmental communication, but also to communicate environmental conflict management.

As a new frame and complex adaptive system for sustainability and governance, EbreBiosfera represents unity of action among the actors in conflict, although this fact does not necessarily means that the gap between them has been closed.

Social movements supported by scientific evidence that legitimize opposition to conflict are more productive for sustainability than one-way politicized or instrumental legislation and planning: they bring about better long-term outcomes in terms of changes in public environmental policies and in individual consumption habits and, eventually, a communicative style of legislation.

Communicative legislation integrates platform interests in political and decision-making processes. The EbreBiosfera was created, therefore, by politicized legislation; that is, the decision was made, ultimately, under pressure from social movements.

A communicative perspective on the law suggests new ways of responsible law making that enhances the possibilities of harmonizing interests. It means real interaction and simultaneously integrates law, policies and society.

Parties in conflict and policy makers in the Terres de l'Ebre can benefit from adaptive and participatory approaches at a time when the EbreBiosfera is still being consolidated.

Legal and administrative powers frequently fail to effectively implement environmental legislation and the precautionary and preventive principles in decision making. Authorities, when acting within a discretionary margin and balancing interests, need to always ensure adequate protection of the environment, since to do otherwise has negative repercussions for sustainable development. It would be desirable for decisions to consider the "3-D sustainability" approach (Mauerhofer 2008) as a way of setting priorities in conflict situations.

Acknowledgements This research was conducted within the framework of: (1) the Martí Franquès Research Fellowship Programme of the URV (2013PMF-PIPF-68), funded by the Endesa Foundation and the Ascó-Vandellòs Nuclear Association and (2) the research project titled "From sustainable development to environmental justice: towards a conceptual matrix for global governance", financed by the Ministry of Economy and Finance for the period 2014-2016 (DER2013-44009-P).

References

Altheide D, Snow P (1979) Media logic. SAGE, Beverly Hills

Aragonés J, Miró N (2012) EbreBiosfera. Propuesta de Reserva de Biosfera de Terres de l'Ebre. CODE/URV, Amposta

Asah S, Bengston D, Wendt K, Nelson K (2012) Diagnostic reframing of intractable environmental problems: case of a contested multiparty public land-use conflict. J Environ Manage 108:108–119

Aula P, Siira K (2010) Organizational communication and conflict management systems. A social complexity approach. Nordicom Rev 31(1):125–141

Beck U (1992) Risk society. Towards a new modernity. SAGE, London

Beck U, Kropp C (2008) Environmental risks and public perceptions. In: Pretty J et al (eds) The SAGE handbook of environment and society. SAGE, London

Boquera Margalef M (2009) Primer la sang que l'aigua. Els pilars d'una nova identitat ebrenca. Benicarló, Onada

Brummans B, Putnam L, Gray B, Hanke R, Lewicki R, Wiethoff C (2008) Making sense of intractable multiparty conflict: a study of framing in four environmental disputes. Commun Monogr 75(1):25–51

Campbell M (2003) Intractability in environmental disputes: Exploring a complex construct. J Plann Lit 17(3):360–371

Castelló E (2012) Conflicto político y medios: marcos, narrativas y discursos. In Castelló E (ed) La mediatización del conflicto político. Discursos y narrativas en el contexto español. Laertes, Barcelona, pp 9–38

Corvellec H, Risberg A (2007) Sensegiving as mise-en-sense—the case of wind power development. Scand J Manag 23:306–326

Cottle S (2006) Mediatized conflict. Open University Press, New York

Dewulf A, Gray B, Putnam L, Lewicki R, Aarts N, Bouwen R, van Woerkum C (2009) Disentangling approaches to framing in conflict and negotiation research: a meta-paradigmatic perspective. Hum Relat 62(2):155–193

Elbakidze M, Hahn T, Mauerhofer V, Angelstam P, Axelsson R (2013) Legal framework for biosphere reserves as learning sites for sustainable development: a comparative analysis on Ukraine and Sweden. Ambio 42:174–187

Entman R (1993) Framing: towards clarification of a fractured paradigm. J Commun 43(4):51–58

Esteve Pardo J (2006) El derecho del medio ambiente como derecho de decisión y gestión de riesgos. Dissertation, Universidad de la Rioja

Gamson W, Modigliani A (1989) Media discourse and public opinion on nuclear power: a constructionist approach. Chic J 95(1):1–37

Goffman E (1974) Frame analysis: an essay on the organization of experience. Northeastern University Press, Boston

Gray B, Coleman P, Putnam L (2007) Introduction. Intractable conflict: new perspectives on the causes and conditions for change. Am Behav Sci 50(11):1415–1429

Hajer MA (1995) The politics of environmental discourse. ecological modernization and the policy process. OUP, Oxford

Hansen A (2010) Environment, media and communication. Routledge, Abingdon

Jiménez N, Martínez-Gil J (2005) The new culture of water in Spain: a philosophy towards a sustainable development. E-WAter, 7, 1–20. Retrieved 17 Feb 2015, from: http://www.ewa-online.eu/tl_files/_media/content/documents_pdf/Publications/E-WAter/documents/58_2005_07.pdf

Lewicki R, Gray B, Elliott M (eds) (2003) Making sense of intractable environmental conflicts: concepts and cases. Island Press, Washington DC

Martínez-Gil J (1997) La nueva cultura del agua en España. Bilbao: Bakeaz/FNCA, Zaragoza

Mauerhofer V (2008) 3-D sustainability. an approach for priority setting in situation of conflicting interests towards a sustainable development. Ecol Econ 64:496–506

Milstein T, Anguiano C, Sandoval J, Chen Y, Dickinson E (2011) Communicating a "new" environmental vernacular: a sense of relations-in-place. Commun Monogr 78(4):486–510

Mollinga P (2008) Water, politics and development: framing a political sociology of water resources management. Water Altern 1(1):7–23

North D (1990) Institutions, institutional change and economic performance. Cambridge University Press, Cambridge

OECD (2000) Nuclear energy in a sustainable development perspective. OECD, Paris

OECD (2003) Public information, consultation and involvement in radioactive waste management. OECD, Paris

OECD (2004) Stepwise approach to decision making for long-term radioactive waste management. Experience, issues and guiding principles. OECD, Paris

OECD (2010a) More than just concrete realities, The symbolic dimension of radioactive waste management. OECD, Paris

OECD (2010b) From information and consultation to influence and power. 10-year evolution in public involvement in radioactive waste management. OECD, París

OECD (2013) Stakeholder confidence in radioactive waste management. an annotated glossary of key terms. OECD, Paris

Putnam L, Shoemaker M (2007) Changes in conflict framing in the news coverage of an environmental conflict. J Disput Resolut 1(10):167–175

Rull X (2010) Política i manipulació del llenguatge. Comunicació: Revista de Recerca i d'Anàlisi 27:105–123

Scott C, Pierce S, Pasqualetti M, Jones A, Montz B, Hoover J (2011) Policy and institutional dimensions of the water-energy nexus. Energy Policy 39:6622–6630

Shmueli D, Elliot M, Kaufman S (2006) Frame changes and the management of intractable conflicts. Confl Resolut Q 24(2):207–218

Sneddon C, Harris L, Dimitrov R, Özesmi U (2002) Contested Waters: conflict, scale, and sustainability in aquatic socioecological systems. Soc Nat Resour 15:663–675

Taylor J, Robichaud D (2004) Finding the organization in the communication: discourse as action and sensemaking. Org Artic 11(3):395–413

Thornton P, Ocasio W, Lounsbury M (2012) The institutional logics perspective: a new approach to culture, structure, and process. OUP, Oxford

UNESCO (1995) Biosphere reserves. The seville strategy and the statutory framework of the World Network. UNESCO, Paris

UNESCO (2008) Madrid action plan for biosphere reserves (2008–2013). UNESCO, Paris

Van Gorp B (2007) The constructionist approach to framing: bringing culture back in. J Commun 57:60–78

Voigt S (2013) How (not) to measure institutions. J Inst Econ 9(1):1–26

Weber K, Glynn M (2006) Making sense with institutions: context, thought and action in Karl Weick's theory. Org Stud 27(11):1639–1660

Weick K, Sutcliffe K, Obstfeld D (2005) Organizing and the process of sensemaking. Organ Sci 16 (4):409–421

Weingart P, Engels A, Pansegrau P (2000) Risks of communication: discourses on climate change in science, politics, and the mass media. Public Underst Sci 9:261–283

Witteveen W, van Klink B (2011) Why is soft law really law? a communicative approach to legislation. In: Luo H, Bi H (eds) The challenges of soft law. The Commercial Press, Beijing, pp 332–361

Woods M (2003) Conflicting environmental visions of the rural: windfarm development in Mid Wales. Sociol Rural 43(3):271–288

Young JC, Marzano M, White RM, McCracken DI, Redpath SM, Carss DN, Quine CP, Watt AD (2010) The emergence of biodiversity conflicts from biodiversity impacts: characteristics and management strategies. Biodivers Conserv 19:3973–3990

Zografos C, Martínez-Alier J (2009) The politics of landscape value: a case study of wind farm conflict in rural Catalonia. Environ Plann A 41:1726–1744

An Empirical Investigation of Supportive Legal Frameworks for Social Enterprises in Belgium: A Cross-Sectoral Comparison of Case Studies Concerning Social Enterprises in the Social Housing, Finance and Energy Sector

Aikaterini Argyrou, Tineke Elisabeth Lambooy, Robert Jan Blomme, Henk Kievit, Guus Nieuwenhuijzen Kruseman and Duco Hora Siccama

Abstract This study aims to investigate how the existing legal framework for social enterprises in Belgium affects the activity of social enterprises in the social housing, finance and energy sector. The focus is thereby on the legal factor of governance and the decision-making power of stakeholders. These matters are examined in respect of one particular type of social enterprises, the so-called company with a social purpose, 'Vennootschap met Sociaal Oogmerk' (VSO). The authors conducted three case studies in Belgium. They examined in which way the VSO law has been implemented in three social enterprises which are active in different sectors, i.e. the energy, finance and housing sector and compared the

A. Argyrou (✉) · G.N. Kruseman · D.H. Siccama
Utrecht University, Utrecht, Netherlands
e-mail: A.V.Argyrou@gmail.com

A. Argyrou
Nyenrode Business University, Breukelen, Netherlands

T.E. Lambooy
Corporate Law, Nyenrode Business University, Breukelen, Netherlands
e-mail: T.Lambooy@nyenrode.nl

R.J. Blomme
Organisational Behaviour, Nyenrode Business University, Breukelen, Netherlands
e-mail: R.Blomme@nyenrode.nl

H. Kievit
Center for Entrepreneurship, Governance and Stewardship, Nyenrode Business University, Breukelen, Netherlands
e-mail: H.Kievit@nyenrode.nl

V. Mauerhofer (ed.), *Legal Aspects of Sustainable Development*,
DOI 10.1007/978-3-319-26021-1_9

results. By comparing the case studies, this article aims to generate (i) a cross-sectoral theoretical analysis regarding the practical application of the legal factor of governance in the three Belgian social enterprises and, (ii) a comprehensive understanding of the involvement of different stakeholders in the social enterprises' governance in these sectors. Useful conclusions were drawn for the improvement of the legal framework for social enterprises in Belgium as well as for the improvement of the social enteprises functioning.

Keywords Social enterprises · Cooperatives · Governance · Stakeholders · Belgium

1 Introduction

An alternative type of entrepreneurial activities has emerged in Europe, the so-called social enterprises. The growth and the activities of social enterprises have contributed substantially to various national economies in Europe. Primarily, social enterprises have provided solutions to the problem of unemployment in a post-crisis environment by employing a significant amount of individuals (over 11 million since 2011). They have also contributed to the sustainable growth and development of the various European Union (EU) member states by adopting and achieving long term financial objectives and social goals that are aimed at finding solutions for major social challenges, such as poverty, social exclusion and stigma. Furthermore, social enterprises take into consideration their environmental impact on society by showcasing a high level of environmental responsibility (EC 2011).

On 25 October 2011, the European Commission (Commission) launched a Communication regarding the Social Business Initiative (SBI), i.e. the SBI Communication. The SBI Communication acknowledges the significant contribution of social enterprises to the national economies of various EU member states. It also endorses the development of an enabling environment with favourable policies and supportive legislation for social enterprises in Europe. The creation of a favourable and enabling legislative environment would entail either the design of appropriate legal forms for social enterprises by the national legislators or the improvement of existing legislative environments for social enterprises in order to alleviate existing obstacles. In the SBI Communication, the Commission introduced uniform criteria for the identification of social enterprises among the different EU member states (EC 2011, pp. 2–3).

In the context of the SBI, the Commission launched a mapping study to explore the ecosystems of social enterprises in 28 member states and Switzerland (EC 2014a). Another objective of the mapping study was the exploration of the legal forms and business models of social enterprises in a national context.

Among many important conclusions, the mapping report showcased that all 29 examined countries (28 EU member states and Switzerland) have organisations that fulfil the uniform criteria for social enterprises that the Commission introduced. Additionally, the report concluded that up to 19 EU member states have introduced legal frameworks for social enterprises. Some of the identified legal frameworks have been introduced as tailor-made legislation for social enterprises, or as legislation offering a specific legal form to social enterprises. Among the 19 EU member states, there are countries with: (i) legislation which introduces the cooperative legal form for social enterprises (mostly in Southern-European countries, e.g. Greece, Italy, Spain, and Portugal); (ii) legislation which introduces the corporate legal form for social enterprises, e.g. the Community Interest Company (CIC), a limited liability company form (in the United Kingdom); and (iii) legislation which introduces a legal label for social enterprises, e.g. the 'Vennootschap met Sociaal Oogmerk' (in Belgium; hereafter: VSO). This label can be adopted by various legal forms (Lambooy and Argyrou 2014, p. 73). Furthermore, the mapping report illustrated that there are also member states without special legislation, i.e. the Netherlands, Austria, Germany, and member states that are in the process of developing legislation for social enterprises, e.g. Latvia.

The emergence and the development of social enterprises in various EU member states have also attracted the attention of researchers and academics who aim to comprehend the concept of social enterprises. Several studies have been published elaborating on a conceptual definition for social enterprises based on the creation of uniform characteristics or taxonomies (Dees 1998; Dees and Anderson 2003; Kerlin 2006; Alter 2007). Other studies focus on specific aspects of social enterprises such as the governance models of social enterprises (Spear 2004; Low 2006; Campi et al. 2006; Mason et al. 2007; Spear et al. 2009, 2014), the success factors for social enterprises (Sawhill and Williamson 2001; Alvord et al. 2004; Sharir and Lerner 2006; Elkington and Hartigan 2008; Cox and Schmuecker 2010), and the performance of social enterprises (Kaplan 2001; Paton 2003; Austin et al. 2006; Bagnoli and Megali 2011; Lee and Nowell 2014). However, the existing academic literature which elaborates on the concept of social enterprises tends to neglect the discussion regarding the influence of the institutional environment and the influence of legislation on social enterprises (Haugh 2006; Peattie and Morley 2008). Haugh (Haugh 2006, p. 6) points out the necessity for conventional research to examine and compare diverse existing legal forms of social enterprises which differ in terms of ownership and financing. Other scholars have identified more pragmatic reasons which necessitate research on the legal forms of social enterprises. Spear et al. (2009, pp. 261–262) point out that the variety of existing legal forms generates frustration and complexity for social enterprises. Melmoth (2005) notes the disadvantages of choosing an inappropriate legal structure for social enterprises which can be time-consuming, complex and expensive. Cox and Schmuecker (2010, p. 12) showcase with empirical evidence that social enterprises can be hampered by the selection of an inappropriate legal and organisational structure. The lack of information on legal and organisational structures for social enterprises creates confusion amongst entrepreneurs as to whether their organisation is a social enterprise.

In a previous study (Argyrou et al. 2015), we theoretically elaborated on and we empirically examined tailor-made legislation for social enterprises in Greece, which country introduced the *cooperative legal form* for social enterprises. The current study aims to investigate tailor-made legislation for social enterprises in Belgium, which has introduced the *legal label* for social enterprises. Particularly, we aim to examine how the legal provisions regarding governance encourage social enterprises to include stakeholders and employees in the decision-making processes. In a future study, we aim to investigate participatory and inclusive governance schemes in social enterprises in the UK. We will examine to what extent the UK tailor-made legislation concerning the *corporate legal form* for social enterprises (the CIC), stimulates the participation of employees and stakeholders in the decision-making processes.

2 Governance of Social Enterprises

The international academic society has agreed that social entrepreneurship applies to a broad range of diverse organisations which do not share one, but rather various organisational and legal characteristics. From a legal point of view, social enterprises are institutionalised economic entities with a social purpose. They use existing legislation to promote their commercial activities and transactions aiming at achieving social goals as opposed to ordinary commercial enterprises which aim to fulfil solely financial objectives. However, social enterprises constitute a 'facet' of social entrepreneurship (Galera and Borzaga 2009, p. 216). They use existing or tailor-made legal forms to promote their mission-based entrepreneurial activities and commercial transactions to achieve a social impact (Defourny and Nyssens 2010, p. 44). Social enterprises do not share unique entrepreneurial or legal characteristics but they are rather hybrid entities using various legal forms that combine for-profit with non-for-profit characteristics (Haugh 2006). The organisational and legal differences between the various types of social enterprises make 'governance' in social enterprises a complicated concept. Furthermore, due to the diversity of the organisational and legal forms of social enterprises, research regarding governance in social enterprises has been segmented. It focuses either on the examination of governance in particular legal forms, for instance concerning the cooperative legal form or the UK corporate legal form (CIC) (Spear 2004; Spear 2009; Ebrahim and Rangan 2014), or on the organisational forms of social enterprises, such as Work Integration Social Enterprises (WISE) (Campi et al. 2006). Moreover, various governance theories apply to organisations that belong to different sectors including the for-profit sector and the non-profit sector (Low 2006). Finally, in some studies a specific national context is examined, e.g. the UK national context of social enterprises (Mason et al. 2007). However, Low (2006, p. 337)–elaborating on the governance of social enterprises–emphasises that we need to conduct research on the aspects of governance which at least transcends the discrepancies of social enterprises that belong to different sectors (i.e. for-profit, non-profit, social economy sector).

The issue of corporate and organisation governance has been significantly elaborated in academic literature on social enterprises. Pestoff (2013, p. 56) emphasises the importance of researching the governance of social enterprises in Europe and the underlying values of the social economy sector, which tend to be characterised by a "quest" for democracy. Ebrahim and Rangan (2014) note the importance of research on the challenges of governance that social enterprises encounter when striving to achieve a trade-off between their commercial activities and the fulfilment of their social mission. Spear et al. (2014, p. 138), aiming at generating theoretically grounded foundations for the discussion on governance of social enterprises, suggest a typology of governance structures, including: (1) self-selecting governance models for social enterprises in which the board of directors is the most powerful organ and its members have no ownership rights; (2) governance models in which the decision-making organs are selected by the members following the principle of democracy; and (3) hybrid governance models in which self-selecting governance and membership governance schemes are combined.

Governance of social enterprises is also characterised by the 'internalisation' of stakeholders in the decision-making processes and their active participation as internal components of the social enterprise, which ultimately lead to more open and democratic decision-making processes (Campi et al. 2006, p. 35; Mason et al. 2007, pp. 288–289). Stakeholders can either participate in the organisational decision-making processes as formal members and co-owners of the social enterprises, or they can influence informal processes of decision-making (Campi et al. 2006, p. 36).

The stakeholder theory elaborates on the influence of various stakeholders on the decision-making processes of a social enterprise. Stakeholders have been widely defined by Freeman and Reed (1983, p. 91), to contain any group or individual who can affect or is affected by the activities of an organisation to achieve its purpose. Di Domenico et al. (2010, pp. 695-696) emphasise the active participation of stakeholders in the social enterprises' governance and management as a prerequisite for social enterprises to generate social benefits such as social capital. Spear et al. (2009, p. 256) stress that it is an advantage for multi-stakeholder organisations to bring together and balance different perspectives and interests, however, there are also disadvantages towards achieving a clear purpose and coming to an agreement. Campi et al. (2006, pp. 40–42) finally, elaborate on the advantages of multi-stakeholder governance such as ensuring organisational stability, providing better access to resources including legitimacy and reducing external constraints.

However, the use of the stakeholder theory has been criticised to have direct implications in the governance of social enterprises. The underlying foundation of the stakeholder theory is the responsibility of the decision-making organs to give attention to the legitimate interest of stakeholders and to reconcile the conflicts of interest that occur between the organisation's interests and the interests of the various stakeholder groups (Di Domenico et al. 2010, p. 682; Donaldson and Preston 1995; Phillips et al. 2003). Thus, directors and managers become accountable to a large variety of stakeholders with diverse interests that need to prioritise (Ebrahim and Rangan 2014). However, Mason et al. (2007, p. 289) argue that the stakeholder theory is incompatible with corporate governance of social enterprises because the theory envisages accountability of directors towards multiple stakeholder groups which might not directly be the case in respect of social enterprises. Furthermore, Mason et al. (2007, p. 289) point out that the inclusion and participation of stakeholders in the decision-making processes of social enterprises can be doubtful if it is not transparent, well-established and subject to external scrutiny. In this respect, the Lambooy and Argyrou (2014) study, which elaborates on tailor-made national legislation for social enterprises in Belgium, the UK and Greece, showcases that legal concepts such as governance, transparency and accountability have been addressed and regulated differently from legal regimes available for ordinary companies. Tailor-made legislation for social enterprises in various national contexts, i.e. in the UK and Belgium, imposes legal duties on the directors of social enterprises concerning the fulfilment of the enterprises' social purpose. Particularly, the Belgian legislation for social enterprises provides legal rights to stakeholders to that end. They can request the dissolution of a social enterprise by court decision if the social enterprise's Articles of Association (AoA) do not comply with the applicable legal obligations that protect the social purpose of the enterprise. Another example is the tailor-made legislation in the UK for social enterprises which imposes obligations on CIC directors to adopt formal stakeholder consultation processes. Annually, they must repot on this in detail in a social report. The social enterprise is obliged to submit the social report to an external institution. Thus, the examination of tailor-made legislation for social enterprises, and the examination of the legal rights and duties that such legislation confers to the governing organs of social enterprises, can contribute to the academic discussion, as well as to the effective and efficient implementation of stakeholder governance in social enterprises.

2.1 The Legal Factor of Governance in Tailor-Made Legislation for Social Enterprises and the Development of Legal Theory

Lambooy and Argyrou (2014, p. 74), elaborating on the legal factors that characterise tailor-made and enabling legislation for social enterprises, identified that the legal factor of governance is a crucial one. They revealed that governance has a particular meaning within the existing tailor-made legislation. Primarily, governance reflects the decision-making power of participants in the function of the social enterprise which is not *per se* based on capital ownership (Lambooy and Argyrou 2014; Argyrou et al. 2015). Most importantly, governance entails the role and the legal rights of the various categories of stakeholders in the decision-making processes of the tailor-made legal forms for social enterprises (Lambooy and

Argyrou 2014; Argyrou et al. 2015; Borzaga and Defourny 2001; Campi et al. 2006).

Legislation provides particular rights and obligations to decision-making bodies in social enterprises. We note, however, that there are various tailor-made legal forms for social enterprises in the different EU member states. Thus, the rights and obligations for the governance organs of social enterprises vary in the different national contexts. The emergence of tailor-made legislation for social enterprises is a milestone in the development of social enterprises. However, legal theory regarding the governance of the newly introduced tailor-made legal forms for social enterprises is still underdeveloped. In its definition for social enterprises in the SBI Communication, the European Commission mentions a uniform criterion that applies to the governance of social enterprises. According to this criterion, social enterprises are "managed in an open and responsible manner and, in particular, involve employees, consumers and stakeholders affected by its commercial activities" (EC 2011, pp. 2–3). However, according to the mapping report on the ecosystems of social enterprises "important differences remain, especially with respect to the interpretation and relevance of the 'governance dimension'", even in countries with tailor-made legislation for social enterprises (EC 2014a, p. vi). Thus, we hypothesise that the rights and obligations that tailor-made legislation for social enterprises provides to the governance organs of social enterprises primarily vary on a national level, and secondarily differ from ordinary governance schemes for commercial enterprises in the sense that they safeguard open, transparent and participatory decision-making processes. The various rights and obligations that national tailor-made legislation confers to the decision-making organs of social enterprises, as well as to the different types of stakeholders, need to be theoretically elaborated and empirically examined. Theoretical elaboration will generate the foundations for the development of legal theory relevant to the newly introduced tailor-made legislation for social enterprises whereas empirical examination will contribute to the existing theoretical discussion regarding the governance of social enterprises and the participation of stakeholders.

2.2 *The Necessity for Empirical Research in the Governance of Social Enterprises*

Mason et al. (2007, p. 297) point out the need for more empirical research concerning the governance of social enterprises. Campi et al. (2006) while looking at the functioning of stakeholders in the governance of Belgian WISEs, examine the importance of stakeholders as members of the organisation thereby taking into consideration the legislation regarding stakeholder participation and the influence of stakeholders within the decision-making formal processes. In the Campi et al. (2006, p. 43) study, it is also emphasised that deeper analysis is needed of the real influence of stakeholders in decision-making processes, especially regarding informal processes such as informal coalition building and the access to information.

Campi et al. (2006, pp. 38–39) also conclude that the legal status constitutes a weak indicator of whether a WISE has a single or multi-stakeholder nature. They state that the choice of a multi-stakeholder status *"tends to depend on the autonomous decision of its founders (more generally, members/owners) within legal frameworks that often permit -explicitly or implicitly- but do not require the involvement of more than one category of stakeholders"*. We note that this conclusion may however vary depending on the specific rules provided in the tailor-made legislation for social enterprises in the various countries. For example, in the UK, the CIC Regulation 2005 *requires* the involvement of more than one category of stakeholders in the decision-making processes of the CIC. The CIC Regulation 2005 considers stakeholders to be all the 'persons affected by the company's activities'(Article 14 (b), CIC Regulation 2005); a definition which entails the involvement of various types of stakeholders.

In the following sections, we examine by means of the case study method (see Sect. 5 below) how the Belgian VSO legal provisions are implemented in practice. We assess how specific provisions stimulate cooperatives with a social purpose in different sectors to include, primarily, employees, and subsequently, other stakeholders in the decision-making processes. The legal form of the cooperative was chosen because it is the dominant legal form adopted by Belgian VSO social enterprises. Subsequently, we provide a cross-sectoral explanation of the involvement of employees and other stakeholders in the decision-making processes of three Belgian cooperatives with a social purpose in different sectors. We compare the incentives for employees to acquire membership rights, which provide formal access to decision-making processes as it is prescribed by law, with the incentives for other categories of stakeholders, which can only participate indirectly in decision-making processes using informal processes and means of communication. If we take into consideration the Campi et al. (2006) conclusion, that the legal provisions applicable to VSOs permit but do not require the participation of employees in the decision-making processes, we can hypothesise that the concept of employee participation will differ in the various VSO social enterprises in the different sectors, depending on the autonomous decisions of their founders. Furthermore, by examining the Belgian VSO legislation which contains only enabling provisions for the participation of one category of stakeholders in the decision-making processes of the governing bodies, i.e. only for employees (See Sect. 3 below), we hypothesise that other types of stakeholders are less incentivised to participate in the decision-making processes of the social enterprises. By comparing the incentives for employees with the incentives for other categories of stakeholders, we also aim to contribute to the discussion on whether the VSO legislation should be revised and improved.

Furthermore, the empirical examination of the practical implementation of the Belgian tailor-made legislation by social enterprises will enrich the theoretical framework of governance for these types of social enterprises. The discussion will ultimately lead to conclusions that will improve the understanding of social enterprises' organisational function and corporate activity (Argyrou et al. 2015).

The following section elaborates on the existing tailor-made legislation for VSO social enterprises in Belgium, which allows the participation of employees in governance and in decision-making processes. The position of other types of stakeholders in the VSO decision-making processes will be also discussed later in the article. The sections that follow contain our findings derived from the empirical examination of the implementation of the Belgian legislation on the issue of participation and influence of employees and other types of stakeholders in the governance of these social enterprises.

3 The Belgian Vennootschap Met Sociaal Oogmerk (VSO) Legal Label for Social Enterprises

In Belgium, in 1995, the legal label for social enterprises, the so-called 'VSO' label, was included in the Belgian Companies Code 1999. The VSO legal label was introduced to increase the marketability and recognition of existing social enterprises such as cooperatives and mutual societies. Furthermore, it provided an alternative entrepreneurial type to commercial enterprises with for-profit objectives that wish to shift to social purpose seeking activities (Coates and Van Opstal 2009; Cafaggi and Iamiceli 2009). However, in Belgium, the VSO legislation has not been widely used (EC 2014a, p. i). Defourny and Nyssens (2008) reported that the VSO label has only had little success over the years. Cafaggi and Iamiceli (2009)–based on the research results of Defourny and Nyssens (2001, p. 47)–also argued that the evolution of the company with a social purpose was unsuccessful "due to the burden of the requirements imposed by the law or the lack of substantial tax incentives." (Cafaggi and Iamiceli 2009, p. 42). Additionally, in a country-specific empirical research project implemented by the Belgian King Baudouin Foundation in 2013, it was pointed out that "the dedicated legal form 'with a social purpose' (met sociaal oogmerk/avec finalité sociale) has so far not widely spread across the sector of social enterprises. This finding raises some doubts on the adequacy and added value of this legal form for social enterprises." (King Baudouin Foundation 2013, p. 15). It has been reported that up to 700 organisations have adopted the VSO legal label so far (EC 2014b, p. 3).

In an earlier legal study, Lambooy and Argyrou (2014, p. 75) provided an overview of the legal regime for the VSO label. It appeared that the VSO label can be adopted by all types of business organisations with legal personality, including companies, regulated by the Belgian Companies Code 1999. Article 661 in conjunction with Article 2 §2 of the Belgian Companies Code 1999 provide the rules on the VSO label. Article 661 contains a list of cumulative requirements that all forms of business organisations with a share capital, and established under Belgian law, can adopt if they wish to acquire the VSO label. These are the private limited liability company, the limited liability cooperative and the unlimited liability cooperative, the public limited liability company, and the European economic interest groups. The Article 661 requirements have to be included in the business organisation's AoA either prior to incorporation or by amending the existing AoA. The most characteristic requirements include: (i) explicit reference in the AoA that the shareholders only seek limited profit or no profit; (ii) a description of the social purpose; (iii) a policy for the distribution of profits according to the social purpose; and (iv) a voting cap which cannot exceed one tenth of the votes deriving from all shares represented in the general assembly (i.e. the imposition of a 10 % voting cap to each shareholder). Additionally, legal entities with the VSO label are required to annually issue a special report which explains the means that the business organisation has undertaken to implement the statutory social purpose. The special report must be in the form of an overview that showcases how investments, operating expenses, rewards and remuneration have been allocated towards the fulfilment of the social purpose.

Article 661 does not contain provisions that create a special corporate governance regime for social enterprises with the VSO legal label. Thus, the corporate governance is subject to the pertinent rules provided by the Belgian Companies Code 1999 for the specific legal type of such VSO social enteprise. However, Article 661 (viii) of the Belgian Companies Code 1999 introduces the obligation for VSO social enterprises to include provisions in the AoA which permit employees to acquire membership rights/shares after the completion of one working year and to lose this capacity a year after the employment relationship has been terminated (Cafaggi and Iamiceli 2009, p. 43; Coates and Van Opstal 2009, p. 38; EC 2014a, pp. i–ii; Breesch and Coekelberg 1995 cited in Coates and Van Opstal 2009, p. 38). The purchase of shares, either existing or new, and the entrance into the organisation as members/shareholders, is also subject to the approval of the competent governing body of the VSO; either the general assembly or the board of directors.

Furthermore, various rights are conferred to employees who acquire membership rights/shares in an organisation that has adopted the VSO label. These include: (i) administrative rights; (ii) voting rights; (iii) rights to profits and dividends; and (iv) rights to acquire information. Regardless of the number of shares which an employee is allowed to purchase, Article 661 requires that each VSO include in its AoA a provision with respect to the exercise of voting rights attached to such membership rights/shares. According to this provision, everyone who takes part in a vote at the general assembly, can exercise only a certain number of votes which cannot exceed the maximum of one tenth of the votes deriving from all the shares represented. The percentage is reduced to one twentieth if employees are members/shareholders. Legislation imposes only this maximum rate of voting rights that can be exercised by the members/shareholders. However, VSO social enterprises are also allowed to stipulate in their AoA more stringent restrictions to further reduce the voting rights of the members/shareholders. This can ultimately result in the application of the democratic 'one man, one vote' rule, which is usually applicable to cooperatives (Coates and Van Opstal 2009, p. 38).

As regards the rights of members/shareholders to profits and dividends, Article 661 (v) allows members/shareholders to participate in the enterprises' profits, by receiving dividends only to the limited extent prescribed by law. Currently, the regulations provide for a cap of 6 %.

Finally, the right to information entails the rights of members/shareholders to particular information concerning the company's operations, financial situation, access to books and other documentation. With respect to these rights, different rules may apply depending on the type of legal entity that has adopted the VSO label.

4 Method

The research that has been employed is partly theoretical and partly empirical. We did desk research regarding the application, interpretation and implementation of the Belgian legal requirements to the governance structure of VSOs. This research was complemented by the conduct of three case studies concerning Belgian social enterprises that have adopted the VSO legal requirements in their AoA. According to the report published by the Belgian King Baudouin Foundation, the majority of social enterprises in Belgium have activities that belong to more than one industrial sector (King Baudouin Foundation 2013, p. 16). We selected VSO social enterprises from three random sectors, i.e. the energy sector, the financial sector, and the social housing sector.

Based on the results of the three case studies and the subsequent comparison, we aim to formulate cross-sectoral theoretical statements (Eisenhardt 1989) regarding the practical application of the legal factor of governance in social enterprises. In addition, we aim to develop a comprehensive baseline theory–generated inductively- regarding the involvement of various stakeholders in the organisational structure of social enterprises in different sectors (Eisenhardt and Graebner 2007). Finally, the three case studies aim to highlight the various ways in which Belgian VSO provisions can be implemented in practice and how this affects the activity of the social enterprise to enable the involvement of stakeholders in the decision-making processes. The development of the case studies and their comparison will also contribute to the discussion regarding the legal forms of social enterprises in Belgium. Moreover, the comparison of the case studies will generate cross-sectoral suggestions for improving the organisational functioning of social enterprises in different sectors in Belgium.

4.1 Semi-structured Interviews, Interview Topics and the Approach Used

The development and execution of the research project was conducted between May 2014 and July 2015. For the collection of data, various methods were employed. Primarily, we examined the Belgian legislation regarding VSO and we collected legal data concerning the three social enterprises such as their AoAs, annual reports and other documents produced and published by them. We also collected empirical data using the method of semi-structured interviews and developing subsequent transcriptions of the interviews. In total, we conducted 14 semi-structured interviews with three social enterprises in Belgium that have adopted the VSO label, i.e. CORE, Microstart, and Volkshuisvesting. A minimum of four interviews were conducted with each social enterprise. From every social enterprise in our sample, we interviewed at least four persons who represented four corresponding organisational layers. These layers were: (a) shareholders; (b) external stakeholders, i.e. client or beneficiary; (c) directors; and (d) employees. Different questionnaires were developed for the respondents per organisational layer. With CORE we conducted two additional interviews, i.e. with a stakeholder and an employee, to enrich our data. In that way we achieved collecting responses from at least one respondent per organisational layer from every social enterprise while collecting at least three interviews per organisational layer in total. By means of selecting respondents from different and various organisational layers, we achieved plurality in the responses and variety in the representation of various angles and perceptions. We also achieved to validate and cross-examine the existing data.

The interviews were of a broad nature and had a semi-structured character. Draft tables and memos were used to systemise the methodological, observational and theoretical input during the analytical process, whereas during the interview stage interview reports were also used and coded. Interview transcriptions were supplemented with observations and information that was retrieved from the relevant legal documents. All interviews were transcribed by professional transcribers in the original language in which the interview was conducted, i.e. English or Dutch. All transcriptions were translated by professionals in the English language.

4.2 Procedure

The validity of the collected data and the analysis were tested with the technique of qualitative data triangulation. Triangulation entails the use of a plurality of methods to validate data which describe various facets of the same social phenomenon (Yin 2013, p. 119). Initially, some data were retrieved from applicable regulation and each social enterprise's legal documents, whereas empirical in-depth data were collected with the conduct of semi-structured interviews. Data accuracy and validation were achieved primarily with follow-up questions and through the thorough revision of the interview transcriptions and the interview reports by the respondents. Subsequently to data collection, the constant comparison method (coding) was used for the data analysis, according to which codes were created to constantly compare and contrast the text ideas collected in the interviews. By using codes, all data were sorted, grouped, subdivided and matched to categorised themes. Even though some pre-selected codes were used deductively, which emerged from the legal analysis of the topic, i.e. governance and stakeholder participation, the majority of the codes were extracted from the existing data inductively. We used an *a priori* framework with a list of the definitions of the selected concepts and contrasted these definitions with codes emerging from the data. The codes emerged from the content of the interview transcription. The process consisted of creating codes using open in-vivo coding with the use of software for qualitative analysis, i.e. ATLAS t.i. The emerging codes from the interview transcriptions were: influence, board of directors, decisions of directors, stakeholder participation, stakeholders meeting, informal meetings, shareholder types, trust, voting process, general assembly process, social report, voting rights, consensus, services, stakeholder information, and values. Following the completion of the interviews, we processed the data further by refining the already existing codes to directly relate them to the research question of this article; we created the following themes, i.e. decision-making processes, employee participation, membership rights, incentives for membership, incentives for employment, profit distribution, formal participation, informal participation, voting cap, and incorporation. Template analysis was the main analytical method (King 2004; Crabtree and Miller 1999). By applying template analysis we contrasted pre-existing concepts with emergent concepts from the data and we integrated them into patterns with respect to stakeholder and employee participation in the governance of social enterprises (Corbin and Strauss 2014). In the following section, we will discuss the results of the empirical investigation.

5 Results: Cooperatives with a Social Purpose (VSO Label) in the Renewable Energy, Financial and Housing Sector

In Europe, special types of cooperatives with new organisational and operational aspects have emerged, i.e. social cooperatives, WISE (Huybrechts and Mertens 2014, p. 195; Yildiz et al. 2015, p. 62), and multi-stakeholder cooperatives (Spear 2004). Cooperatives by definition are organisations which are owned by their members rather than by investors (Huybrechts and Mertens 2014). They are characterised by democratic ownership and (equal) representation in the decision-making processes. Cooperatives also tend to involve a variety of stakeholders in decision-making by assigning ownership and membership rights to stakeholders (Spear 2004). The governance of cooperatives involves different types of stakeholders, such as producers, consumers, investors, etc. In contrast, the various interests of the different groups of stakeholders in traditional commercial for-profit companies will not always be taken into account in the decision-making processes by the governing bodies. For example, decisions can favour particular members/shareholders over other stakeholders. In cooperatives, equality, democratic governance and the involvement or participation of stakeholders in decision-making converge the interests of stakeholders towards the achievement of a common mission. Huybrechts and Mertens (2014) elaborate on pragmatic legitimacy that cooperatives enjoy from stakeholders when they offer them options for ownership and membership. Additionally, they address normative legitimacy that cooperatives enjoy as an alternative organisational form, which differs substantially from for-profit organisational entities. Campi et al. (2006), examined stakeholder participation of various categories of stakeholders in WISE and other social enterprises in Belgium. His study indicates that the dominant type of stakeholders, which participates in decision-making, is employees (24 %). However, the study concludes that (i) none of the different categories of stakeholders has an overwhelming influence at boards of directors, and that (ii) the participation and the existence of stakeholders is not only reflected on the board level but it means actual sharing of voting power (Campi et al. 2006).

According to the Belgian legislation, cooperatives and other business organisations, i.e. public and/or private limited liability companies can adopt the VSO label. A cooperative can become "a cooperative with a social purpose" by reforming its AoA to include all the cumulative requirements that have been introduced in Article 661 of the Belgian Companies Code (1999). Spear (2004, p.105), and subsequently Coates and Van Opstal (2009), stress that the majority of VSOs in Belgium have the legal form of a cooperative. However, the introduction of the VSO legal requirements into the cooperatives' AoA will not suffice. The legal requirements need to be implemented in practice too. Belgian legislation contains sanction mechanisms in Article 667 applicable to VSO social enterprises that do not implement the legal requirements. The lack of implementation can ultimately lead to the termination of the VSO-status by court decision. Members/shareholders, interested third parties and the public prosecutor can assert a legal claim that an organisation declares itself as a VSO, while not having its AoA in line with the VSO legal requirements. They can also claim that although a VSO meets the legal requirements in its AoA, it does not act in practice accordingly. In both cases, the court has the competence to judge whether a termination of the VSO-status is reasonable in relation to a particular breach. In the following part, we will examine how the three social enterprises of our case studies have implemented the VSO legal requirements in practice. We will start with introducing the cooperatives which are the subjects of the three case studies.

5.1 Cooperatives with a Social Purpose in the Renewable Energy Sector—First Case: Cooperative Enterprise in Rational Energy (CORE)

The number of social enterprises in the Belgian renewable energy sector has recently increased (EC 2014b, p. 24). Huybrechts and Mertens (2014) justify the emergence of cooperatives and social enterprises in the Belgian renewable energy sector by exhibiting the emergent motivation of consumers to better control the production and the supply of energy with positive effects to the environment and the community. Belgian energy social enterprises have been developed to raise awareness concerning the necessity to reduce energy consumption, climate change issues, and other current environmental challenges.

The Cooperative Enterprise in Rational Energy (CORE) is a cooperative with a social purpose, located in Leuven, Belgium. CORE was incorporated as a cooperative with a social purpose to bridge social entrepreneurship and education with the promotion of technical projects and social awareness for rational energy consumption in society. In its educational programmes, CORE also introduces the values of a cooperative, social entrepreneurship, and of sustainable development to students.

CORE has been operating and executing technical projects since 2012. The projects of CORE are either designed or commissioned by members/shareholders or they are developed by CORE to supply other cooperatives and social enterprises. The managing director of CORE mentions "If they [clients] ask us to do projects, we first ask them 'can you become a shareholder' […]. We invite them, we don't force them and it's also that they might say no […]. If they become a shareholder, the price for the services will be lower." [Interview with SJ, 16 February 2015]. Hence, CORE has promoted the creation and development of technology solutions for rational and sustainable use of energy which can be used by other cooperatives and social enterprises, and by the members/shareholders of CORE [Interview with SJ, 16 February 2015]. In that way, CORE, promotes to its clients and potential stakeholders the idea of social and cooperative entrepreneurship.

CORE is not operated by employees with a typical employment relationship with the organisation. The development of CORE's projects as well as the operationalisation of the daily business is performed by students-volunteers who attend the 'Postgraduate Innovative Entrepreneurship Programme' for engineers at the Katholieke Universiteit in Leuven (KU Leuven). Students-volunteers are mainly involved in the execution of the projects in exchange of student credits for the provision of part-time services. Subsequently, CORE's deliverables are communicated by the involved students-volunteers to the academic world in the form of classes, courses, academic outcome (i.e. student theses), and in student events focused on raising awareness regarding rational and sustainable energy consumption through social entrepreneurship and cooperative membership [Interview with YG&G, 13 November 2014].

CORE is a limited liability cooperative with a social purpose which has adopted three types of shares that correspond to three different types of members (AoA, Article 7). The categories include: (i) type A shares for CORE's founders and structural partners, either legal or natural persons; (ii) type B shares for legal or natural persons. Type B shares are offered to CORE's clients and other stakeholders, for instance community partners; and (iii) type C shares for students. Students are only allowed to purchase one share each. Type C shares are offered to: (1) students-volunteers who participate in the operationalisation of CORE; (2) any individual who is a student enrolled in an accredited public institution of higher education in Belgium and wishes to participate in CORE's projects; and (3) student alumni or former CORE volunteers who maintain their share; they constitute a very important part of CORE type C members/shareholders with the responsibility to promote and communicate the idea of CORE in the business society [Interview with YG&G, 13 November 2014]. All the shares of CORE have the same nominal value of € 100.

CORE is governed by a board of directors which is composed of 14 members and a general assembly which meets yearly. The board of directors has been appointed to be the organ with the most extensive powers at CORE towards the achievement of the social purpose of the organisation and the management of the daily business. In the board of directors, each type of member/shareholder ought to be represented as follows: (i) a minimum of three directors from the members/shareholders with type A shares, i.e. founder/structural partners. CORE has appointed four directors to represent type A members/shareholders; (ii) a maximum of three directors from the members/shareholders with type B shares, i.e. clients and community. CORE has appointed three directors to represent type B members/shareholders; and (iii) a maximum of three directors from the members/shareholders with type C shares, i.e. students-volunteers. CORE has appointed three directors to represent type C members/shareholders. A total of 10 out of 14 directors are also CORE members/shareholders. Additionally, the board comprises external directors who are not members/shareholders of CORE. This is the case for the managing director, the director of legal support, the director of financial support, and the secretary who is a student with the responsibility to check and control the organisation [Interview with SJ, 16 February 2015]. The board of directors is the competent organ to decide on the acceptance, the resignation, and the exclusion/dismissal of members/shareholders, and on the provision to employees of membership/shareholdership rights [Interview with SJ, 16 February 2015]. Each member/shareholder must act in compliance with CORE's AoA and with the decisions that are taken by the general assembly and the board of directors.

CORE's general assembly is annually convened with the responsibility to authorize the board of directors to continue activities for the following year. In compliance with the VSO legal requirement, every member/shareholder participates in the voting. The voting power per member/shareholder is however limited to a maximum of 10 % of the total voting power of all the shares represented in the general assembly [Interview with SJ, 16 February 2015]. The 10 % voting cap differs from the democratic rule of 'one man one vote' but it results in a better balance and democracy at CORE according to the managing director. He explains "one of the reasons why we don't go for one share-one vote, is because type A and B shareholders are companies, they can only be represented by one person in the general assembly while the students-shareholders in five years or ten years time can grow to a group of 200 shareholders. Then, you will have a different balance." [Interview with SJ, 16 February 2015]. However, as of today, decisions in the general assembly of CORE have been taken unanimously [Interview with YG, 13 November 2014]. The voting power of members/shareholders can be diluted further down to 5 % on the occasion of employees-members/shareholders participating in the voting process. However, there are no persons with a traditional employment relationship at CORE.

5.2 *Cooperatives with a Social Purpose in the Financial Sector-Second Case: Microstart*

Microfinance has been defined as the provision of direct or intermediate financial services to marginalised individuals, entrepreneurs and communities that have no access to commercial financial banking services (Morduch 1999; Périlleux 2015; Périlleux et al. 2011). Microfinance services are provided by a variety of microfinance institutions in Belgium among others, cooperatives and non-profit organisations, with double bottom line objectives which aim at creating not only financial but also social returns (Périlleux 2015). Microfinance institutions in Belgium are considered social enterprises (Périlleux 2015; Périlleux et al. 2011). Many microfinance social enterprises have been incorporated as cooperatives and are accredited by the Belgian National Cooperative Council (CNC) that they fully comply with the cooperative principles (Périlleux 2015). Other cooperatives have adopted the VSO social purpose label by introducing the legal requirements in their AoA (Münkner 2004).

Microstart is a microfinance social enterprise which is located in Brussels, Belgium. Microstart was incorporated in 2011 by the French non-profit organisation 'Association pour le droit à l'initiative économique'(Adie) and the bank 'BNP Paribas Fortis' with the financial endorsement of the European Investment Fund (EIF). Adie is the founder of Microstart, and one of the largest microfinance institutions in Europe. It provides microcredit and microfinance to French micro-entrepreneurs. With the establishment of Microstart, Adie aimed at the extension of its successful microfinance practice to Belgium [Interview with LH, 16 June 2015].

Microstart's organisational structure combines the operating and financing aspects of two legal entities and corresponding organisations, i.e. Microstart, a limited liability cooperative with a social purpose, and Microstart Support, a non-profit organisation. Microstart has been assigned with the task to provide microcredit and microfinance to clients who are excluded from the traditional Belgian banking system, such as unemployed people, jobseekers, recipients of welfare support and self-employed persons. Microstart Support is a non-profit organisation with the task to provide coaching and business development services to clients who receive microcredit and microfinance.

Microstart is a cooperative that has adopted the social purpose and accordingly it should comply with the VSO legal requirements. The members/shareholders of Microstart are not allowed to strive for any pecuniary gain (profit) (AoA, Article 1). The cooperative is required to act both financially and commercially, directly or indirectly, in accordance with the social purpose mentioned in its AoA. Microstart is a for-profit legal entity. However, according to Microstart's AoA, the distribution of dividends to its members/shareholders is not allowed (AoA, Article 29). On the contrary, profits are used for the development of new agencies, the design of new projects, the elimination of costs, and rewards to employees (AoA, Article 29).

Adie and BNP Paribas Fortis have purchased the majority of Microstart shares. The AoA of Microstart provide for three types of shares: (i) type A shares that are reserved for Adie; (ii) type B shares that are reserved for BNP Paribas Fortis; and (iii) type C shares that can be purchased by other legal and natural persons. In Microstart, all types of shareholders ought to be represented in the board of directors. Microstart Support is also represented with one member in the board of directors. The general assembly has the competence to appoint additional directors if the capital of the cooperative exceeds a certain amount (€ 1,210,000) (AoA, Article 17). The president of the board is an independent director elected and appointed by the general assembly.

Microstart is governed by a board of directors. The board is appointed as the competent organ to exercise the broadest powers and activities necessary for the cooperative to achieve the social purpose. The board of directors has been elected directly by the general assembly. At Microstart the general assembly is convened annually to decide on issues of major importance for the continuation of Microstart's activity, i.e. to review and accept the annual accounts, to decide on the use of profits, and to discharge the liability of directors (AoA, Article 21). Decision-making processes at a general assembly level are subject to the VSO legal requirement according to which no one is allowed to participate in the voting procedure with more than 10 % of the total number of votes. However, as of today, all Microstart decisions in the general assembly have been taken unanimously [Interview with LH, 16 June 2015]. Finally, even though Microstart has adopted the majority of the legal VSO requirements, its AoA exclude the provision of membership rights/shares to employees and other stakeholders. A member of the board of directors mentioned that "we don't give membership rights to employees and so I would be surprised if that particular aspect would be in the statutes." [Interview with LH, 16 June 2015].

5.3 *Cooperatives with a Social Purpose in the Social Housing Sector—Third Case: Volkshuisvesting*

Social housing is a regional competence in Belgium for the various different regional governments, i.e. the Flemish region, the Walloon region, and the Brussels-capital region. Each regional government has the competence to generate and implement its own housing and social housing policy. In the Flemish region, the Flemish Housing Code 1997 has been introduced to regulate the Flemish social housing policy (Decree for the Flemish Housing Code 1997–Decreet houdende de

Vlaamse Wooncode 1997). The Flemish Housing Code 1997 constitutes the primary legislative source for the housing policy in the Flemish region. Furthermore, the Flemish government has the competence to recognise organisations with a social purpose as social housing corporations which can implement the objectives that the Flemish government aims to fulfil with respect to the social housing policy. In the Flemish region, social housing corporations are not public organisations but independent legal entities. They are obliged to execute the tasks prescribed by the Flemish Housing Code 1997 and its supporting regulations. Social housing organisations are operated in the Flemish region under the supervision and monitoring of the thereto established supervision and monitoring organisations (Flemish Housing Code 1997, Article 40 §1). Social housing corporations are also allowed to benefit from favourable loans and subsidies.

However, due to their public task, social housing corporations are highly regulated. According to the Flemish Housing Code 1997, social housing corporations are obliged to take the legal form of either a cooperative or a public limited company and to adopt the VSO legal label. Therefore, social housing corporations are in principle subject to the provisions of the Belgian Companies Code 1999. However, the provisions of the Belgian Companies Code 1999 only apply in so far as the Flemish Housing Code 1997, or the AoA designed by the Flemish government for social housing corporations, do not deviate from the Belgian Companies Code 1999 (Flemish Housing Code 1997, Article 40 §2).

One deviation, which emanates from the Flemish Housing Code 1997 is the following: social housing VSO organisations cannot provide membership rights/shares to employees (Flemish Housing Code 1997, Article 40 §2). Thus, the 10 % voting cap on the exercise of the voting rights does also not apply to the public members/shareholders in social housing corporations (Flemish Housing Code 1997, Article 40 §3).

'Volkshuisvesting' is a social housing cooperative in the Flemish region of Belgium. It provides renting and social housing facilitation services to people requiring social housing according to income or (eligibility to dispose of) property. The enterprise was incorporated in the early 1920s as a private limited liability company whereas in 2009 it converted into a cooperative with a VSO label.

Volkshuisvesting has adopted a tripartite social objective to fulfil the regional housing policy (AoA, Article 4). Primarily, it aims to improve the living conditions of families and individuals in need of social housing, such as low-income families and individuals, the elderly, and people with disabilities. Secondly, Volkshuisvesting aims to realise social housing neighbourhoods by purchasing property. Finally, Volkshuisvesting aims to contribute to society by renovating or otherwise improving the quality of existing old houses.

In Volkshuisvesting's AoA, it is explicitly mentioned that the members/shareholders are allowed to seek pecuniary (capital) gain only to a limited extent (AoA, Article 5). In principle, profits are not distributed to the members of Volkshuisvesting. Members can only receive dividends subject to the regulated cap, i.e. currently 6 % (AoA, Article 21).

Volkshuisvesting has different types of members/shareholders: 23 from the public domain and 49 from the private domain [Interview with MP, 18 February 2015]. However, the classification of members/shareholders is not explicitly mentioned in Volkshuisvesting's AoA. According to Volkhuisvesting's AoA (Article 11), public institutions that are allowed to acquire the shares of Volkshuisvesting are the Flemish government, any province, any community, and any public institutions for social welfare. Private institutions that own Volkshuisvesting shares are banks and insurance companies (Interview with MP, 18 February 2015).

The social cooperative is managed by a management committee of seven members. According to Volkshuisvesting's AoA, the board of directors can entrust the day-to-day management to a management committee assigned for that purpose (AoA, Article 14). The management committee takes the decisions regarding the daily management and operations. However, these decisions can be subject to the approval of the board of directors.

Volkshuisvesting is governed by the board of directors, which is the competent organ to decide on any matter that concerns the cooperative including all matters that are not reserved to the competence of the general assembly (AoA, Article 14). It decides with a unanimous majority but if unanimity cannot be reached, simple majority applies. The board of directors is also the competent organ to decide on the issuance of new shares (AoA, Article 8). In such case, only the members/shareholders of the cooperative are in principle allowed to purchase newly-issued shares. A person or entity can however be nominated by two existing members/shareholders, after which the board of directors decides in a confidential meeting on the allowance of such new member/shareholder. The regular transfer of shares between members/shareholders is also subject to the approval of the board of directors (AoA, Article 9).

The board of directors is required to comprise of at least 11 members in order to represent both the private and the public members/shareholders, i.e. five directors from the public domain and five directors representing the private members/shareholders (AoA, Article 11). Volkshuisvesting has a board of directors, which consists of 13 members. At Volkshuisvesting, the board of directors prepares and publishes an annual social report in which it is showcased how the cooperative's activities pursue the social purpose. The social report particularly indicates in which way the expenditures on investment, operating costs and salaries have contributed to achieving the social purpose of the cooperative. The report is integrated in the annual report of Volkshuisvesting (AoA, Article 13, Interview with MP, 18 February 2015).

At the top of the governance hierarchy, Volkshuisvesting is also governed by the general assembly of the members/shareholders. The general assembly is assigned to convene annually to decide on any proposals regarding amendment of the AoA and/or on proposals formulated by the board of directors (AoA, Article 26). However, Volkshuisvesting, and any other recognised social housing corporation in the Flemish region, is not allowed to amend the AoA without the permission of the Flemish minister responsible for housing (Decision of the Flemish Government

2010, Article 8 §1). This provision does not apply to several designated minor changes (Decision of the Flemish Government 2010, Article 8 §2).

The general assembly is entitled to appoint and dismiss the members of the board of directors (AoA, Article 12). According to Volkshuisvesting's AoA, members/shareholders from the public domain, i.e. Flemish government, any province, any community, and any public institution for social welfare, can exercise their voting rights with as many votes as are attached to the membership rights/shares which they own. The remaining members/shareholders from the private domain are subject to the VSO legal requirement, according to which no one is allowed to participate in the voting procedure with more than 10 % of the total number of votes (AoA, Article 24).

However, in practice, according to the managing director of Volkshuisvesting, "the cooperative strives for a unanimous majority in the general assembly, which has worked out fine so far." [Interview with MP, 18 February 2015].

Furthermore, employees are not allowed to become a member/shareholder of Volkshuisvesting. Article 40 §2 Flemish Housing Code 1997 explicitly excludes the application of Article 661 (vii) and (viii) of the Belgian Companies Code 1999, which regulates the legal right of employees to acquire membership/shareholdership rights. Volkshuisvesting's AoA mention in this context that the employees cannot become members/shareholders (AoA, Article 9). The same rule applies vice versa to Volkshuisvesting's members/shareholders. They can only become an employee of Volkshuisvesting provided that they dispose of their shares. Due to the exclusion of membership rights/shares, the employees of Volkshuisvesting do not have a direct access to the decision-making processes of the cooperative.

In the section that follows, we will elaborate on the cross-sectoral comparison and synthesis of our findings.

6 Discussion

6.1 *Governance*

CORE, Microstart and Volkshuisvesting are cooperatives with a social purpose, which operate in three different industrial sectors, respectively, in the energy sector, the financial sector and the social housing sector. All of the examined cooperatives with a social purpose have adopted the VSO legal label by incorporating the VSO legal requirements into their AoA. The VSO legal label can be adopted by social enterprises operating in different sectors. However, in particular industrial sectors, e.g. the housing sector, the VSO legal label has been prescribed for organisations that implement the regional public policy regarding social housing.

The governance of cooperatives is characterised by a standard correlation between membership/shareholdership and decision-making. The general assembly of the members/shareholders is the competent organ to elect the board members. Therefore, we contend that the traditional self-selecting governance models, which areemployee participation often applied in for-profit organisations and which models are characterised by a powerful board of directors comprised mainly of directors who are not also members/shareholders, is not the prevailing model in cooperatives with a social purpose. In the governance models of cooperatives with a social purpose, the decision-making organs are closely associated with the rights that membership/shareholdership confers. However, they are not characterised by the principle of democracy *per se* as it is expressed by the concept of 'one man, one vote'. Contributing to the taxonomy of Spear et al. (2014), we argue that cooperatives with a social purpose in Belgium, showcase hybrid governance models in which self-selecting governance and membership governance schemes are combined (Spear et al. 2014, p. 138). Thus, we identify the following types of cooperatives with a social purpose that apply hybrid governance schemes:

(1) Cooperatives with a social purpose that apply a regulated 10 % cap on the exercise of voting rights of the cooperative members/shareholders according to the VSO legal requirements. The reduction of the voting rights of the members/shareholders in the form of a regulated cap is required by the VSO legislation. Even though, the voting cap differs from the principle of democratic participation embodied in the 'one man, one vote' rule, it introduces a hybrid decision-making model which allows members/shareholders to achieve balance and democracy in the decision-making processes of the general assembly. At CORE, the voting cap incorporated in the AoA reduces the ordinary voting rights that are attached to the cooperative shares to a certain extent, i.e. one tenth. This 10 % voting cap ensures a balance between the voting power of type A and B shareholders-which are legal persons that can be only represented by one person at the general assembly-with the voting power of type C shareholders, i.e. the growing number of students-volunteers who represent themselves personally at the general assembly. At Volkshuisvesting, the 10 % voting cap only applies to members/shareholders of the private domain, whereas the members/shareholders of the public domain may always exercise their voting rights in full. Resuming, with respect to the implementation of the legal VSO requirement which imposes a 10 % voting cap to the exercise of the voting rights, we identify: (i) cooperatives with a social purpose which apply a 10 % cap to the exercise of voting rights but which cap is reduced to 5 % for employees-members/shareholders (CORE); (ii) cooperatives with a social purpose, which do not allow employee participation, but which do apply a 10 % voting cap to the exercise of voting rights of members/shareholders (Microstart); and (iii) cooperatives with a social purpose, which do not allow employee participation, and where the 10 % cap only applies to members/shareholders belonging to the private domain (Volkshuisvesting).

(2) Cooperatives with a social purpose that strive for unanimity. Even though the implementation of the 10 % voting cap has been introduced in the AoA of the

examined cooperatives with a social purpose, it follows from the interviews that at the board level and general assembly level, the board members and the members/shareholders strive for unanimity and consensus. At CORE, a director mentions "We try to vote as few times as possible, I think we almost never really vote […]. We vote that we all agree, we discuss the problem and we go for reaching consensus. Until now we didn't have situations in which we had to really vote." [Interview with YG, 13 November 2014]. Both Microstart and Volkshuisvesting also strive for a unanimous majority in the general assembly [Interview with MP, 18 February 2015; Interview with LH, 16 June 2015].

(3) Cooperatives with a social purpose that appoint directors who represent various types of members/shareholders. In addition, the general assembly can appoint independent directors to either (i) supervise and control the board of directors; (ii) provide expertise with respect to legal, financial and management issues, or with respect to the fulfilment of the social purpose; or (iii) represent any supportive organisations to the operation of the cooperative with a social purpose.

6.2 *Employee Participation in the Decision-Making Processes of Cooperatives with a Social Purpose*

Employee participation in the decision-making processes of cooperatives with a social purpose is required by the Belgian VSO legislation. The legal requirement imposes on the VSO social enterprise to provide a right to the employees to become a member/shareholder and to participate in the decision-making processes in the general assembly. As noted above, according to Article 667 of the Belgian Companies Code 1999, the lack of implementation of the cumulative VSO legal requirements in the AoA can lead to sanctions. Claims submitted by a member/shareholder, interested stakeholders, or the public prosecutor could cause the termination of the VSO label. However, in practice, there is no great pressure on a VSO social enterprise to implement the employee co-ownership legal requirement. Apparently, there is no monitoring public institution/body with the competence to validate the compliance of the VSO social enterprise with the legal requirements of the VSO regime.

Furthermore, the implementation of the legal requirement regarding employee membership/shareholdership may be also barred by later and more specific legislation. The latter can derogate the earlier and less specific legislation (*lex specialis derogat legi generali and lex posterior derogate legi priori*). We noted that this is the case with respect to social cooperatives in the housing sector.

Subsequently, as regards employee participation in the decision-making processes, based on our interview data, we can distinguish various attitudes in social cooperatives:

(1) *Cooperatives with a social purpose which are keen to allow employee participation and to provide membership rights/shares to employees.* At CORE, even though the AoA allow the provision of membership rights/shares to employees, there are no 'regular' employees to purchase such shares. In this social cooperative, students-volunteers design, operate and execute CORE's projects for as long as they are students. In return they receive study points and they acquire knowledge in the field of social cooperative entrepreneurship and sustainable energy innovation, working experience, and a big network in industrial engineering. Working at CORE is also part of their education. Student-volunteers are considered the "real employees" at CORE and that is why they are invited to become members/shareholders by purchasing one type C share [Interview with SJ, 16 February 2015; Interviews with YG&G, 13 November 2014; Interview with PO, 16 February 2015]. As members/shareholders, they participate in the decision-making processes in the general assembly, they exercise voting rights, they elect and appoint representatives at the board of directors, and they influence the daily decisions that are taken at CORE. Students-volunteers who graduate are allowed to maintain their membership rights/shares but they are not allowed to provide services at CORE anymore [Interview with YG&G, 13 November 2014]. They however still constitute a very important sub-category of type C members/shareholders as they promote the values and the idea of CORE to the market, attract new projects and new members/shareholders. They also share with the new students-volunteers their experiences by coaching and mentoring [Interview with YG&G, 13 November 2014].
(2) *Cooperatives with a social purpose which are not keen to allow employee participation nor do they provide membership rights/shares to employees.* At Microstart, employees are not allowed to purchase shares. The VSO legal requirements to that end have not been introduced in Microstart's AoA. The involvement of employees in the decision-making processes and the provision of membership rights/shares is perceived to be related with risks in governance due to the fact that the organisation is in the start-up phase. [Interview with LH, 16 June 2015] An employee at Microstart mentions "we are still at the first steps, so we first have to make the company very stable and then maybe we will open the shares to old clients and maybe to employees and some volunteers." [Interview with CO, 16 June 2015].
(3) *Cooperatives with a social purpose which are excluded from offering membership rights/shares to employees.* Volkshuisvesting, is a cooperative with a social purpose in the housing sector and is subject to the regional social housing policy of the Flemish region. The Flemish Housing Code 1997, forbids the provision of membership rights/shares to employees. Hence, Volkshuisvesting cannot provide membership rights/shares to employees. Furthermore, members/shareholders of this social cooperative can only become employees after they have disposed of their shares.

6.3 Stakeholder Participation

In the context of social enterprises, a 'stakeholder' can entail any group or individual who can affect or is affected by the achievement of a social enterprise's purpose. The emphasis is put on the external stakeholders of a social enterprise, such as supporters, clients and beneficiaries (Freeman and Reed 1983). According to the VSO legal regime, cooperatives with a social purpose, are not required to include in their AoA provisions regarding stakeholders (other than employees); e.g., which allow stakeholder participation or provide membership rights/shares to various kinds of stakeholders. Based on our interview data, we can identify cooperatives with a social purpose with the following attitudes as regards stakeholder participation in the decision-making processes.

(1) *Cooperatives with a social purpose, which are keen to allow stakeholder participation in the decision-making processes by implementing provisions regarding membership rights/shares in the AoA as well as by allowing stakeholder participation through informal processes.* It has been already explained how student-volunteers have access to decision-making processes at CORE. In addition, to student-volunteers, clients and customers are invited to purchase type B shares [Interview with SJ, 16 February 2015]. For CORE, it is important that the projects that are executed either stem from clients, or are directed to clients, which are or aim to become a member/shareholder [Interview with SJ, 16 February 2015]. Upon a request for a new project, CORE invites the new client to join the cooperative as a member/shareholder. Membership/shareholdership confers to such client formal power to participate in the decision-making processes because it will have the right to vote in the general assembly and to elect and appoint representatives for the board of directors.

Furthermore, both students-volunteers and clients, besides being allowed to participate in formal decision-making processes, can also participate in informal meetings. Students-volunteers and alumni at CORE can participate in monthly type C member/shareholder meetings to discuss ideas, problems, strategies or claims with their representatives in the board [Interview with YG&G, 13 November 2015]. Existing clients or prospective clients at CORE can participate in monthly thematic events organised by CORE's students-volunteers. During the thematic-events the progress of CORE's projects is explained with reference to the specific social mission that CORE's projects address, e.g. rational energy consumption, mobility or sustainable energy in housing. Hence, knowledge is shared between the existing and potential clients and the cooperative's members/shareholders. They all are interested in adopting a multi-stakeholder approach and support the social mission of CORE [Interview with JW, 13 November 2013]. At CORE, the application of democratic procedures between shareholders, directors and employees/volunteers is a crucial issue. The principle of democracy is not only reflected in the application of the 10 % voting cap but also in the open and participatory decision-making processes, in which employees and stakeholders can equally contribute. Furthermore, the communication between the members/shareholders is open and informal [Interview with GY&G, 13 November 2014; Interview with SJ, 16 February 2015]. A student-volunteer at CORE was allowed to attend the meeting of the board of directors as a visitor in order to finally decide on the advantages of becoming a member/shareholder [Interview with G, 13 November 2014]. Stakeholders perceive that they influence decision-making both directly and indirectly by sending emails, by reporting on projects, by voting, and by communicating directly with directors. At CORE, there is no decision taken either by the board of directors or the general assembly that remains secret from employees or stakeholders [Interview with SJ, 16 February 2015]. The managing director of CORE mentions "when a meeting is finished, I don't mind to tell everything. Before a meeting, I'm always trying to ask everybody what their opinion is." [Interview with SJ, 16 February 2015; Interview with YG&G, 13 November 2015]. Between the members/shareholders, this is a feeling of trust and respect because communication is open.

(2) *Cooperatives with a social purpose which are keen to allow informal stakeholder participation or self-selecting representation of stakeholders in the decision-making processes but which are not keen to allow in the AoA for the provision of membership rights/shares to any type of stakeholders*. Those cooperatives with a social purpose, have stakeholders who cannot influence the decision-making formally, e.g. by the exercise of voting rights. However, they can do so in an indirect way through informal means of communication, e.g. by the involvement of intermediaries, such as managers or self-selected representatives/directors who are not members/shareholders. We found an example thereof at Microstart, where employees and volunteers have never attended physically any meeting of the board of directors. However, they communicate their interests to intermediaries and trust their representatives at the board level [Interview with CO & Interview with LH, 16 June 2015]. They also feel that they actually influence decision-making. For instance, Microstart employees contributed substantially in the decision-making process regarding the modification of Microstart's business plan. Initially, Microstart's business plan was perceived by Microstart employees to be ambitious including very high objectives in terms of numbers of microcredit and returns [Interview with CO, 16 June 2015]. The business plan was transplanted from the French practice and was not really adapted to the Belgian context. Microstart employees managed to communicate with the board of directors that it was not feasible to achieve the objectives in the due time. Furthermore, employees and volunteers at Microstart indicate that they trust the members of the board of directors because governance decisions comply with the social objectives of the operational part of the cooperative [Interview with CO, 16 June 2015; Interview with E, 13 November 2014]. The board of directors serves its role to equally guarantee the social and the financial objectives of the cooperative:

(i) the financial objectives are safeguarded by directors who represent the main financiers of Microstart, i.e. BNP Paribas Fortis and the EIF; and (ii) the social goals of Microstart are safeguarded by directors who are representatives of the non-profit organisation Adie. Even though physical contact between the board of directors and employees or volunteers is not regular, meetings and informal discussion take place between volunteers, employees, managers and directors [Interview with CO, 16 June 2015; Interview with LM, 13 November 2014]. One Microstart employee mentioned that although employees, volunteers and managers are invited to meet and discuss with the board members, there is however still the perception that in respect of certain information and decisions, the board members are not allowed to disclose any information. But generally, communication at Microstart between the board members, managers and employees is relatively open. Information in the form of a newsletter is distributed weekly to employees and managers regarding the operational activities of Microstart [Interview with CO, 16 June 2015]. Annually employees are provided with access to financial information. Other information is disseminated to employees and other managers, e.g. decisions that have been taken by the board of directors and/or the general assembly. One Microstart employee mentions "if there is a decision which involves everybody, everybody is going to receive the information." [Interview with CO, 16 June 2015]. The Microstart website also contains information accessible for stakeholders, including newsletters, annuals reports, financial statements and organograms. Besides employees and volunteers, also customers can influence indirectly the decision-making processes at Microstart. Annually, Microstart organises a client satisfaction survey. Questions are submitted to clients with respect to (i) their level of satisfaction with the services that Microstart provides in terms of microcredit and support; (ii) the level of satisfaction with the interest rate that Microstart applies to its microcredits; and finally (iii) the level of satisfaction concerning the contact with Microstart's advisors and consultants. Additionally, Microstart organises informal events where clients are invited to evaluate and discuss Microstart's services and activities. For example, a Microstart employee mentions that meetings are organised with clients in the same sector. It is also common practice at Microstart to involve clients in the operational decisions [Interview with CO, 16 June 2015].

At Volkshuisvesting, both the employees and the tenants of the social housing units are not allowed to acquire membership rights/shares nor to participate in the decision-making processes. The reason hereof lies in the provisions of the Flemish Housing Code 1997, which bar the issuance of employee and stakeholder membership rights/shares [Interview with MP&S, 18 February 2015; Interview with IGT, 19 February 2015]. The Flemish government, any province, any community, and any public institutions for social welfare are considered the only stakeholders that are allowed to participate in the decision-making processes through the exercise of membership rights/shares. Even though specific types of stakeholders, i.e. employees and tenants are not allowed to participate formally and directly in the decision-making processes, they can still participate in informal meetings [Interview with MP&S, 18 February 2015]. At Volkshuisvesting, employees can participate in staff meetings. The outcome there of is communicated informally with the managing director of Volkshuisvesting and the President of Volkshuisvesting [Interview with MP&S, 18 February 2015; Interview with IGT, 19 February 2015]. Furthermore, Volkshuisvesting also organises meetings with the tenants of the social housing units–directly or via the tenants committee. The tenants and their representatives are allowed to propose ideas for the improvement of the housing units. They can also submit claims and complaints. The tenants' committee comprises at least four members. A delegate from Volkshuisvesting participates as a member of the committee. The tenant committee is allowed to make proposals and submit requests. The managing director of Volkhuisvesting has the responsibility to forward these to the Volkshuisvesting governing committee, which will deal with such proposal, claims and complaints.

6.4 Incentives for Formal Versus Informal Participation of Employees and Stakeholders

In this section we will compare (i) the incentives that stimulate employees and stakeholders to acquire membership rights/shares, which provide them formal access to decision-making processes, with (ii) the incentives that stimulate other categories of stakeholders to participate indirectly to decision-making using informal processes and means of communication. With this part we aim to contribute to the theoretical discussion that was initiated by Campi et al. (2006; see Sect. 2.2.)

6.4.1 Employees and Stakeholders with Membership Rights/Shares

Employees and stakeholders can formally participate in the decision-making processes of social enterprises by exercising their voting rights at the general assembly and by electing representatives at the board of directors. The provision of membership rights/shares to employees is a legal requirement under the VSO legal regime that applies to cooperatives with a social purpose. It can be implemented by including this right in the AoA. A variation is to introduce a special type of shares for employees. The VSO legal regime does not require the provision of membership rights/shares to other type of stakeholders like clients, supporters and beneficiaries. Although not legally required, a VSO can provide membership rights/shares to other type of stakeholders. Some have formalised this by creating a special category of shares for stakeholders, e.g. CORE. The existence of provisions regarding membership rights/shares to both employees and stakeholders in the AoA generates an enabling environment for employee and stakeholders' participation. At CORE, all students-volunteers who operate the cooperative have purchased type C shares [Interview with YG&G, 13 November 2014]. They perceive the possession of membership rights/shares as a means to acquire insights as to how the organisation functions and how the legal framework can be used to safeguard their interests (which are of a non-financial character) in decision-making [Interview with YG&G, 13 November 2014]. In respect of the purchase of one share of 100 euro, a student-volunteer mentions "I immediately filled in the information form to become a shareholder type C […]." [Interview with G, 13 November 2014]. The underlying idea behind CORE's success and development is that participation, membership/shareholdership, and the communication of the social mission, will on a continuous basis generate new projects with a social purpose and new clients (who will also become a member/shareholder) [Interview with SJ, 16 February 2015]. CORE is a for-profit cooperative with a social purpose, which aims to bridge entrepreneurship with education. Profits cannot be distributed to the cooperatives' members/shareholders. They are rather reinvested in the cooperatives' social purpose and growth in the form of new projects. A part of CORE's mission is the promotion of social and cooperative entrepreneurship for the fulfilment of social objectives, which ultimately leads to the execution of projects and to decision-making processes, which is influenced substantially by values of equality, democracy and participation. Thus, students-volunteers are incentivised to participate and become members/shareholders. Simultaneously, they are educated and they acquire professional experience in industrial engineering aimed at promoting rational energy use, sustainable mobility and social entrepreneurship. They also build up a strong network, which allows them to communicate and promote further the idea of sustainable energy use in all aspects of daily life and social entrepreneurship [Interview with YG&G, 13 November 2014].

Clients and stakeholders also purchase CORE shares to support the cooperative financially and to have access to the decision-making processes by voting. A type B member/shareholder, who is an academic scientist, mentions "I'm not interested in getting that money back, so for me that money is gone, but it is well-spent." [Interview with JW, 13 November 2014]. He believes that investing in human capital, i.e. students-volunteers who work together with stakeholders and members/shareholders on projects concerning rational energy use in mobility, is very valuable, especially because in the current business landscape, the major business players have not yet developed solutions in this area [Interview with JW, 13 November 2014]. Another example is the Belgian company Pantarein, also a type B member/shareholder of CORE. Pantarein designs the construction and operation of factory plants. In collaboration with CORE students-volunteers, they advise on energy optimisation for the consumption of water in factory plants such as textile factories [Interview with FM&PO, 16 February 2015]. Even though the representatives of Pantarein are invited to participate in all the decision-making processes of CORE, they have only been engaged in a few of them. They feel that they contribute to the fulfilment of the social mission by influencing the decision-making processes, directly or indirectly, by sending emails and by reporting on existing and new projects to the board of directors [Interview with FM&PO, 16 February 2015].

6.4.2 Employees and Stakeholders Who Informally Participate in the Decision-Making

If a cooperative with a social purpose does not offer to employees and/or to other stakeholders the chance to buy membership rights/shares, they thus have no formal power to participate in the decision-making processes. In that situation, the only way of exercising influence is through informal participation. Interests, suggestions and ideas communicated to managers, directors and to self-selected representatives. The contribution by employees and stakeholders to the fulfilment of the social purpose is in that situation restricted to the operational side of the cooperative while decisions and information are communicated indirectly, and only to those parties that need to be involved. At Microstart, employees are sceptical about whether they would purchase shares or acquire membership rights/shares even if the opportunity of acquiring membership rights/shares was offered to them. [Interview with CO, 16 June 2015]. Personal interests and motivations generally have not yet been aligned with the fulfilment of the social purpose that the cooperative serves. An employee mentions "[…] it is very important where I put my money and I like my independence, so in my opinion I would probably put my money somewhere else […]. Maybe for symbolic purposes I would put some money [in Microstart shares] to be able to have more power in the decision-making. Just for personal interest, I would be very interested to see how the decisions are taken in the board." [Interview with CO, 16 June 2015]. Even though the employees trust the decisions that are taken by the board of directors and the general assembly, they feel the urge to participate in a more direct way in the decision-making processes.

It is perceived by Microstart's employees that if Microstart would provide membership rights/shares to employees and clients, the mix of such different types of members/shareholders could improve the decision-making processes because the operational side of the social enterprise would then be formally represented. Furthermore, our interview data suggest that informal communication between the board, management, employees and stakeholders has generated the impression that there is asymmetry of information between the Microstart branches in Brussels and outside Brussels.

Finally, our interview data indicate that at Volkshuisvesting, employees and stakeholders have no urge to be involved in the cooperatives' affairs nor are they actually aware of the opportunities that exist to informally participate in the decision-making processes [Interview with MP&S, 18 February 2015].

7 Conclusions

Social enterprises have contributed substantially to the promotion of sustainable development and inclusive growth in Europe. This chapter elaborated on tailor-made legislation regarding social enterprises. We particularly examined how the VSO tailor-made legislation for social enterprises in Belgium influences the participation of employees and other types of stakeholders in the decision-making processes. We tested this question by conducting three case studies relating to three cooperatives with a social purpose, each in another sector; respectively, the renewable energy sector, the microfinance sector, and the social housing sector. Main achievements, from an academic perspective, are (i) the contribution to the emerging legal theory with respect to the VSO tailor-made legislation and (ii) the generation of empirical findings which build on and contribute to existing theoretical foundations regarding the participation of stakeholders in the governance of social enterprises.

Primarily, our findings in the three case studies confirmed that even though the Belgian tailor-made legal framework regarding social enterprises is conducive to *employee participation* in the decision-making processes, the concept of employee participation differs in the examined cooperatives with social purpose in the three sectors. In some cases, the direct and formal participation of employees in the decision-making processes of the social enterprise came with the acquisition of a legal right, i.e. membership right/share. The acquisition of membership rights/shares allows employees to participate in a direct way in the decision-making processes of VSO social cooperatives, by (i) exercising voting rights in the general assembly, subject to a regulated cap, i.e. 10 %, and indirectly by (ii) electing the members of the board of directors. Additionally, we found indirect participation of employees via informal and unregulated participation settings in all three cases in the different sectors. In one sector, employee participation was excluded by special sectoral legislation.

From a practical point of view, this chapter also contributes to a social entrepreneur's understanding of the substance and the implementation of the VSO legislation in Belgium. The enrichment of the discussion with empirical findings of current best practices will also induce Belgian social entrepreneurs to consider creating more inclusive and participatory models in governance and decision-making.

In this chapter we limited our research and focused mainly on the examination of the formal participation of employees and stakeholders in the decision-making processes of social enterprises and the effects of tailor-made legislation for social enterprises. We provided only limited references to informal means of employee and stakeholder participation as a more elaborated examination is required to test the informal and noninstitutionalised dynamics that take effect in the absence of enabling legislation.

Furthermore, even though the participation of other types of stakeholders is not supported by the VSO legislation, cooperatives with a social purpose have in practice independently developed both formal and informal means to include stakeholders in the decision-making processes. One example of stimulating formal stakeholder participation is the creation of a specific category of shares for certain types of stakeholders. The opportunity to purchase membership rights/shares enhances the incentives for stakeholders to participate in the decision-making processes, either directly by (i) exercising voting rights in the general assembly, often subjected to a regulated cap, i.e. 10 %, and indirectly by (ii) electing members-representatives in the board of directors. We also found examples of informal stakeholder participation in the decision-making processes in the examined case studies. Taking into consideration the existing legal framework, the theoretical framework regarding governance, and the criticism regarding the application of stakeholder theory to the governance of social enterprises, there are additional issues that future research can address. What needs to be further examined is how formal and informal participation of stakeholders can improve (i) the accountability of the decision-making organs within social enterprises and (ii) the transparency of the decisions towards stakeholders and the society. The empirical findings of this chapter indicate that employees and stakeholders have better and direct access to information in the cooperatives with a social purpose which are keen to encourage formal participation of stakeholders and employees in the decision-making.

References

Alter K (2007) Social enterprise typology. Virtue Ventures LLC 12:1–124

Alvord SH, Brown LD, Letts CW (2004) Social Entrepreneurship and societal transformation. J Appl Behav Sci 40(3):260–282

Argyrou A, Lambooy T, Blomme RJ, Kievit H (2015) An understanding how social enterprises can benefit from supportive legal frameworks: a case study report on social entrepreneurial models in Greece. Int J Bus Glob (in press)

Austin JE, Gutierrez R, Ogliastri E, Reficco E (2006) Effective management of social enterprises: lessons from businesses and civil society organizations in Iberoamerica. Harvard University Press, Cambridge, MA

Bagnoli L, Megali C (2011) Measuring performance in social enterprises. Nonprofit Volunt Sect Q 40(1):149–165

Belgian Companies Code 1999 Official translation Wetboek van vennootschappen available in French and Dutch, May 7, 1999. Retrieved 23 July 2015, from http://www.ejustice.just.fgov.be/cgi_loi/loi_a.pl?N=&=&sql=(text+contains+(%27%27))&rech=1&language=nl&tri=dd+AS+RANK&numero=1&table_name=wet&cn=1999050769&caller=image_a1&fromtab=wet&la=N&pdf_page=27&pdf_file=http://www.ejustice.just.fgov.be/mopdf/1999/08/06_2.pdf

Borzaga C, Defourny J (eds) (2001) The emergence of social enterprise. Routledge, London

Breesch T, Coeckelbergh D (1995) De VSO. Non Profit, recht & management-reeks. Gent: Mys & Breesch, 110

Cafaggi F, Iamiceli P (2009) New frontiers in the legal structure and legislation of social enterprises in europe: a comparative analysis. In: Noya A (ed) The changing boundaries of social enterprises. OECD Publishing, Paris, pp 25–86

Campi S, Defourny J, Gregoire O (2006) Work integration social enterprises: are they multi-stakeholder and multiple goal organizations? In: Nyssens M (ed) Social enterprise: at the crossroads of market. Routledge, London, pp 29–49

Coates A, Van Opstal W (2009) The joys and burdens of multiple legal frameworks for social entrepreneurship: lessons from the Belgian case. EMES conferences selected papers series, ECSP-T09-01. Retrieved 19 July 2015 from http://papers.ssrn.com/sol3/papers.cfm?abstract_id=1432427

Corbin J, Strauss A (2014) Basics of qualitative research: techniques and procedures for developing grounded theory. SAGE Publications, London

Cox E, Schmuecker K (2010) Growing the big society: encouraging success in social and community enterprise in deprived communities. IPPR, Newcastle

Crabtree B, Miller W (eds) (1999) Doing qualitative research, 2nd edn. Sage, London

Decision of the Flemish Government of October 22, 2010, Laying down the additional conditions and the procedure for recognition as social housing and establishing the procedure for assessing the performance of social housing. Official translation Besluit van de Vlaamse Regering van 22 Oktober 2010. Retrieved 24 July 2015 from http://www.ejustice.just.fgov.be/cgi_loi/loi_a.pl

Dees JG (1998) The meaning of social entrepreneurship. Draft Report for the Kauffman Center for Entrepreneurial Leadership, Stanford University, pp 1–5

Dees JG, Anderson BB (2003) For-profit social ventures. In: Kourilsky ML, Walstad WB, Osborne AE (eds) Social entrepreneurship. Senate Hall Academic Publishing, Birmingham

Defourny J, Nyssens M (2001) Belgium: social enterprises in the community services sector. In: Borzaga C, Defourny J (eds) The emergence of social enterprise. Routledge, London, pp 47–65

Defourny J, Nyssens M (2008) Social enterprise in Europe: recent trends and developments. Soc Enterp J 4(3):202–228

Defourny J, Nyssens M (2010) Conceptions of social enterprise and social entrepreneurship in europe and the United States: convergences and divergences. J Soc Entrepr 1(1):32–53

Di Domenico H, Haugh H, Tracey P (2010) Social bricolage: theorizing social value creation in social enterprise. Entrepr Theor Pract 34(4):681–703

Donaldson T, Preston LE (1995) The stakeholder theory of the corporation: Concepts, evidence, and implications. Acad Manag Rev 20(1):65–91

Ebrahim A, Rangan VK (2014) What impact? a framework for measuring the scale and scope of social performance. Univ Calif Berkeley 56(3):118–141

EC (2011) Communication from the commission to the european parliament, the council, the european economic and social committee and the committee of the regions social business initiative: Creating a favourable climate for social enterprises, key stakeholders in the social economy and innovation. COM 682 final, Brussels, 25 Oct 2011. Retrieved 23 July 2015 from http://ec.europa.eu/internal_market/social_business/docs/COM2011_682_en.pdf

EC (2014a) European Commission, A map of social enterprises and their ecosystems in Europe: Synthesis report, 14 Oct 2014. Retrieved 19 July 2015 from http://ec.europa.eu/social/main.jsp?langId=en&catId=89&newsId=2149

EC (2014b) European Commission, A map of social enterprises and their ecosystems in Europe-Country Report: Belgium, 14 Oct 2014. Retrieved 19 July 2015 from http://ec.europa.eu/social/keyDocuments.jsp?advSearchKey=socentcntryrepts&mode=advancedSubmit&langId=en&policyArea=&type=0&country=0&year=0&orderBy=docOrder

Eisenhardt KM (1989) Building theories from case study research. Acad Manag Rev 14(4):532–550

Eisenhardt KM, Graebner ME (2007) Theory building from cases: opportunities and challenges. Acad Manag J 50(1):25–32

Elkington J, Hartigan P (2008) The power of unreasonable people: how social entrepreneurs create markets that change the world. Harvard Business Press, Cambridge

Freeman RE, Reed DL (1983) Stockholders and Stakeholders: A New Perspective on Corporate Governance. California Management Review XXV (3): 88–106
Flemish Housing Code 1997 Official translation Decreet houdende de Vlaamse Wooncode, 15 July 1997. Retrieved 23 July 2015, from http://www.ejustice.just.fgov.be/cgi_loi/change_lg.pl?language=nl&la=N&cn=1997071539&table_name=wet
Galera G, Borzaga C (2009) Social enterprise: an international overview of its conceptual evolution and legal implementation. Soc Enterp J 5(3):210–228
Haugh H (2006) A research agenda for social entrepreneurship? Soc Enterpr J 1(1):1–12
Huybrechts B, Mertens S (2014) The relevance of the cooperative model in the field of renewable energy. Ann Public Cooperative Econ 85(2):193–212
Kaplan RS (2001) Strategic performance measurement and management in nonprofit organizations. Nonprofit Manag Leadersh 11(3):371–386
Kerlin AJ (2006) Social enterprises in the United States and Europe: understanding and learning from the differences. Voluntas 17:162–179
King N (2004) Using templates in the thematic analysis of text. In: Cassell C, Symon G (eds) Essential guide to qualitative methods in organizational research. Sage Publications, London
King Baudouin Foundation (2013) Mapping social enterprises in Belgium, October 2013. Retrieved July 20, 2015 from http://www.oksigenlab.eu/sites/default/files/selusi_belgium_-full_report.pdf
Lambooy TE, Argyrou A (2014) Improving the legal environment for social entrepreneurship in Europe. Eur Co Law 11(2):71–76
Lee C, Nowell B (2014) A framework for assessing the performance of nonprofit organizations. Am J Evaluation 1–21
Low C (2006) A framework for the governance of social enterprises. Int J Soc Econ 33(5/6):376–385
Mason C, Kirkbride J, Bryde D (2007) From stakeholders to institutions: the changing face of social enterprise governance theory. Manag Decis 45(2):284–301
Melmoth G (2005) Simply legal: an introduction to legal structures in the social economy, 'foreword' in cooperatives UK. Cooperatives UK, Manchester
Münkner H (2004) Multi-stakeholder co-operatives and their legal framework. In: Borzaga C and Spear R (eds) Trends and challenges for co-operatives and social enterprises in developed and transition countries. (pp.49–84).Trento: Edizioni 31
Morduch J (1999) The microfinance promise. J Econ Lit XXXVII:1569–1614
Paton R (2003) Managing and measuring social enterprises. SAGE, London
Peattie K, Morley A (2008) Eight paradoxes of the social enterprise research agenda. Soc Enterp J 4(2):91–107
Périlleux A (2015) When social enterprises engage in finance: agents of change in lending relationships, a Belgian typology. Strateg Change 24(3):285–300
Périlleux A, Hudon M, Bloy E (2011) Surplus distribution in microfinance: Differences among cooperative, nonprofit, and shareholder forms of ownership. Nonprofit Voluntary Sect Q DOI: 0899764011406287
Pestoff VA (2013) The role of participatory governance in the EMES approach to social enterprise. J Entrep Org Divers 2(2):48–60
Phillips R, Freeman ER, Wicks AC (2003) What stakeholder theory is not. Bus Ethics Q 13 (4):479–502
Sawhill JC, Williamson D (2001) Mission impossible? measuring success in nonprofit organizations. Nonprofit Manag Leadersh 11(3):371–386
Sharir M, Lerner M (2006) Gauging the success of social ventures initiated by individual social entrepreneurs. J World Bus 41(1):6–20
Spear R (2004) Governance in democratic member-based organisations. Ann Public Cooperative Econ 75(1):33–59

Spear R, Cornforth C, Aiken M (2009) The governance challenges of social enterprises: evidence from a UK empirical study. Ann Public Cooperative Econ 80(2):247–273

Spear R, Cornforth C, Aiken M (2014) Major perspectives on governance of social enterprise. In: Defourny J, Hulgård L, Pestoff V (eds) Social Enterprise and the Third Sector: Changing European landscapes in a comparative perspective. Routledge, London and New York, pp 133–157

Yildiz Ö, Rommel J, Debor S, Holstenkamp L, Mey F, Müller JR, Radtke J, Rognli J (2015) Renewable energy cooperatives as gatekeepers or facilitators? Recent developments in Germany and a multidisciplinary research agenda. Energy Res Soc Sci 6:59–73

Yin RK (2013) Case study research: design and methods, 5th edn. SAGE Publications, London

Part IV
Horizontal Policies: Public Participation

Judicial Control Over Acts of Administrative Omission: Environmental Rule of Law and Recent Case Law in Japan

Noriko Okubo

Abstract Judicial control of administrative acts of omission is indispensable for ensuring sustainable development. Despite advancements of environmental legislation, deficient implementation is a serious problem in many countries. This issue grows in importance, especially because in the field of environmental law the executive branch is usually given considerable discretion as to how and when to exercise its regulatory power. This article analyses recent developments in case law on judicial control of administrative acts of omission in Japan. State liability cases and mandamus cases are typical measures for challenging omissions of public authorities. Recently, State liability lawsuits, such as Minamata and asbestos cases, have proven to be an effective measure to challenge the non-use of the regulatory power. Although such kind of lawsuit is a direct measure to compensate the victims, it also plays an important role as an indirect measure to push the government to exercise its regulatory power. The mandamus action is more a direct measure to control illegal failures. It is a new type of administrative litigation expected to demonstrate more effectiveness in the future.

Keywords Access to justice · Principle 10 · Environmental rule of law · State liability · Mandamus action

1 Introduction

'Environmental rule of law', a concept first promoted by the United Nations Environment Programme (UNEP 2015), is essential to meeting the goal of sustainable development, according to Decision 27/9, adopted by the UNEP Governing Council in 2013 (UNEP 2013). The text of the Decision states that 'the violation of environmental law has the potential to undermine sustainable development and the

N. Okubo (✉)
Graduate School of Law and Politics, Osaka University, Osaka, Japan
e-mail: okku@law.osaka-u.ac.jp

V. Mauerhofer (ed.), *Legal Aspects of Sustainable Development*,
DOI 10.1007/978-3-319-26021-1_10

implementation of agreed environmental goals and objectives at all levels and the rule of law and good governance play an essential role in reducing such violations'. It requests the Executive Director 'to lead the United Nations system and support national Governments upon their request in the development and implementation of environmental rule of law with attention at all levels to mutually supporting governance features', thus emphasizing the importance of rule of law in the field of the environment.

The UN-Secretary-General defines the rule of law as 'a principle of governance in which all persons, institutions and entities, public and private, including the State itself, are accountable to laws that are publicly promulgated, equally enforced and independently adjudicated, and which are consistent with international human rights norms and standards. It requires, as well, measures to ensure adherence to the principles of supremacy of law, equality before the law, accountability to the law, fairness in the application of the law, separation of powers, participation in decision-making, legal certainty, avoidance of arbitrariness and procedural and legal transparency'. (UN 2004, p. 4).

In Asian countries, there have been remarkable developments in environmental law following the United Nations Conference on Environment and Development of 1992 (the Rio Earth Summit), such as the introduction of detailed provisions concerning the environment in the 2007 Constitution of Thailand, and enactment of the Basic Environment Law of 1993 in Japan (Okubo 2005, 2015a). Despite such advancements, a problem of deficient implementation of these and related laws persists in many Asian countries, for various reasons, including the lack of detailed regulation, insufficient financial resources, and corruption among officials (Sakumoto 2011).

In such cases, environmental litigation often contributes to ensuring environmental rule of law; in addition, various kinds of judicial reforms have been carried out in some Asian countries to improve the effectiveness of judicial control in environmental matters (Okubo 2015b). There has been a clear tendency to introduce legislation enabling public interest litigation. In the late 2000s, citizen access to justice was greatly strengthened, not only through the expansion of legal standing (*locus standi*), but also through the introduction of various new types of litigation and the improvement of provisional remedies. In addition, recently, some countries have established an environmental court or specialized environmental division within the judicial system (Pring and Pring 2009; see also Osaka University 2015).

Compared with other Asian countries, environmental law has been relatively well implemented in Japan. However, non-use of regulatory power by the competent administrative authority has sometimes led to serious health damage or environmental pollution. In 2004, the Administrative Case Litigation Act (ACLA)[1] was revised, introducing the 'mandamus action' to push administrative agencies to

[1]Act No. 139 of 16 May 1962, latest revision: Act No. 59 of 17 July 2015. The text of the 2007 law is available in English at http://www.env.go.jp/en/laws/policy/basic/index.html (last accessed on 15 June 2015).

exercise their competence appropriately. In addition, several courts recognised State liability for failure to develop regulations or for inaction to enforce regulations. This article analyses recent developments in case law on judicial control of such administrative acts of omission in Japan.

2 Types of Environmental Litigation

The Japanese judicial system is composed of three levels of courts: the Supreme Court, the high courts, and the district courts. Article 76 of the Japanese Constitution[2] prohibits the establishment of extraordinary tribunals. This means that there is no specialized court, such as an administrative court or an environmental court; there are only specialized divisions within the ordinary courts.

Similarly, there is no specific environmental legal action. Litigation involving the environment could be of a civil, administrative, or criminal nature, depending on its legal basis. In addition to judicial procedure, methods of Alternative Dispute Resolution (ADR)[3] also play an important role in environmental litigation.

Main types of environmental civil litigation comprise claims for compensation of damages in torts and claims for injunctive relief against polluters based on the Code of Civil Procedure.[4] According to Article 709 of the Civil Code,[5] a person who has intentionally or negligently infringed any right of others, or any legally protected interest of others, shall be liable for any consequential damages.

Civil injunctive relief is a very important measure for preventing pollution and environmental damage. The legal basis for an injunction is mainly ‘personal right’, which consists of a physical personal right concerning life and health and mental status.

State liability cases are handled as civil litigation in Japan. The State Redress Act[6] regulates State liability concerning the exercise of governmental power by governmental officials (Article 1) and defects in the establishment and management of public facilities, such as airports, roads, etc. (Article 2).

---

[2]Constitution of 3 November 1946. The text is available in English at http://www.env.go.jp/en/laws/policy/basic/index.html (last accessed on 15 June 2015).

[3]Act on the Settlement of Environmental Pollution Disputes, Act No. 108 of 1 June 1970, latest revision: Act No. 69 of 13 June 2014. The text of the 2007 law is available in English at http://www.env.go.jp/en/laws/policy/basic/index.html (last accessed on 15 June 2015).

[4]Act No. 109 of 26 June 1996, latest revision: Act No. 30 of 8 May 2012. The text of the 2011 law is available in English at http://www.env.go.jp/en/laws/policy/basic/index.html (last accessed on 15 June 2015).

[5]Act No. 89 of 27 April 1896, latest revision: Act No. 94 of 11 December 2013. The text of the 2006 law is available in English at http://www.env.go.jp/en/laws/policy/basic/index.html (last accessed on 15 June 2015).

[6]Act No. 125 of 27 October 1947. The text is available in English at http://www.env.go.jp/en/laws/policy/basic/index.html (last accessed on 15 June 2015).

Administrative case litigation involves actions for the judicial review of administrative dispositions ('Kokoku Sosho'), public law-related actions, citizen actions (*actio popularis*), and interagency actions, according to the Administrative Case Litigation Act. The primary types of environmental administrative litigation are actions for the judicial review of administrative dispositions and 'suits by residents', or citizen actions. 'Judicial review of administrative dispositions' refers to appeals against the exercise of public authority by an administrative agency. These actions are classified into six groups: (a) 'action for the revocation of the original administrative disposition'; (b) 'action for the revocation of an administrative disposition on appeal'; (c) 'action for the declaration of nullity, etc.'; (d) 'action for the declaration of illegality of inaction'; (e) 'mandamus action'; and (f) 'action for an injunctive order' (Article 3, ACLA).

According to Article 42 of the ACLA, citizen actions may be filed only by persons specified, and in cases provided for, in the particular Act. Section 242-2 of the Japanese Local Autonomy Law[7] provides for such citizen action, or 'suits by residents', and states that any inhabitant of an ordinary local government may resort to judicial remedies for illegal financial acts of its own local authority. Such acts may be related to expenditures, the acquisition, management, or disposal of property, or decisions by relevant officers. This law was based on the American 'taxpayer suit', a type of financial litigation. In some cases, local residents have tried to use this type of litigation for nature conservation. The residents filed complaint in these suits against public works agencies for damaging the natural environment and claiming that the expenditures of the local authority for public works was unlawful because its action to permit the public works was also illegal.[8]

Among these types of legal actions, State liability cases and mandamus cases are typical measures for challenging omissions of public authorities.

3 Liability for Non-use of Regulatory Power

3.1 State Liability for Non-use of Regulatory Power

Civil litigation against polluters has been relatively effective in Japan in public health cases, especially with claims for damages. Recently, State liability lawsuits have proven to be an effective means of challenging the non-use of the regulatory power.

Article 1, Paragraph (1) of the State Redress Act sets forth provisions on State liability for public officers exercising the authority of the State:

[7]Act No. 67 of 17 April 1947, latest revision: Act No. 50 of 26 June 2015.

[8]For example, Awase Wetland case. See Naha Branch, Fukuoka High Court 15 October 2009, Hanrei Jiho No. 2066, 3.

When a public officer who exercises the public authority of the State or of a public entity has, in the course of his/her duties, unlawfully inflicted damage on another person intentionally or negligently, the State or public entity shall assume the responsibility to compensate therefor.

The 'exercise of authority' also includes the non-exercise or inaction by the executive branch, i.e. the non-use of its regulatory power. The number of actions based on liability for non-use of regulatory power has increased in the fields of environmental protection, consumer protection (such as lawsuits for adverse drugs), and workers protection.

According to Japanese case law, the failure of public officers to enforce regulations is unlawful if such failure is deemed to be extremely unreasonable in light of three factors: (a) the purport and purpose of the laws and ordinances that are the basis of the authority; (b) the nature of the authority; and (c) the specific circumstances of the case.[9]

Under this strict condition defined by case law, there had been no decision by the Supreme Court that has ruled in favour of State liability in a long time.[10] On 27 April 2004, the Supreme Court held for the first time that an administrative omission was illegal, and the State was liable under the State Redress Act.[11] The case was related to the failure of the Minister of International Trade and Industry to exercise his authority to enforce safety regulations under the Mine Safety Law[12] to prevent the outbreak of pneumoconiosis in coal mines.

In the field of environmental law, the executive branch is usually given considerable discretion as to how and when to exercise its regulatory power. Therefore, it may often be difficult to determine in what specific circumstances the non-exercise of regulatory power constitutes fault or negligence on the part of governmental authorities.

3.2 The Kansai Minamata Disease Case

The Kansai Minamata disease case produced a landmark decision related to the State's liability for not using its regulatory power in environmental matters. Discovered in Minamata city, Kumamoto prefecture, in 1956, Minamata disease is one of the most serious infirmities caused by water pollution during a high growth period after World War II in Japan (Ui 1992). The disease was attributed to the methyl mercury that had accumulated in fishes, and those who ate them had been

[9]Supreme Court 24 November 1989, Minshu Vol. 43, No. 10, 1169, Supreme Court 23 June 1995, Minshu Vol. 49, No. 6, 1600.

[10]Ibid.

[11]Supreme Court 27 April 2004, Minshu Vol.58, No. 4, 1032. The judgement is available in English at http://www.courts.go.jp/app/hanrei_en/detail?id=696 (last accessed on 15 June 2015).

[12]Act No. 70 of 16 May 1949, latest revision: Act No. 69 of 13 June 2014.

poisoned with it. The pollution caused also a congenital version of Minamata disease. Even now, there are still many victims.

The many lawsuits that resulted were basically divided into two groups. On one hand, some of the victims since the 1970s have won cases seeking compensation from the polluter, the company Chisso. This type of litigation is one of the typical and classical environmental cases in Japan. For example, on 20 March 1973, Kumamoto District Court held that when there is a risk of damage to people's life or health, the industries concerned should take all possible measures to avoid the consequences of pollution, regardless of the cost and other economic implications of such measures. When there are doubts as to the safety of its activities, the company has the duty to suspend its operations.[13]

On the other hand, some victims litigated against the State and the Kumamoto prefecture on the grounds of non-use of their regulatory power against Chisso. The Kansai Minamata disease case is also such a case, filed in 2001 by the Minamata victims who lived in the Kansai area, in Western Japan, at the time of litigation.

On 15 October 2004, the Supreme Court rejected a demand for revision by the appellants, the State and Kumamoto prefecture, and ruled that each of the appellants must pay the appellees up to 2,500,000 yen respectively, according to the damage.[14] As for the liability of State, it held that 'authority should be exercised in a timely and appropriate manner, aiming at protecting the life and health of people living in areas that might suffer water quality deterioration in the designated water area' (III4(2)A).

The Court pointed out that the national government, as of the end of November 1959, recognised that a large number of people had contracted Minamata disease, and a considerable number of them had died of the disease. Furthermore, the government also could recognise, with the high probability, that the causative substance of Minamata disease was a kind of organic mercury compound discharged from the company Chisso's plant, and was capable of quantitatively analysing a small amount of mercury contained in effluent discharged from the plant. Finally, the Court also pointed out that the subsequent spread of Minamata disease could have been prevented, if authority had been exercised at that time, by identifying the area as a designated water area, establishing water quality standards, and ordering Chisso to improve the method of treating effluent, temporarily stops using the facility, etc.

As a result, the Court concluded that the failure of the national government, by the end of December 1959, to exercise the authority to make regulations and to enforce them under the then Water Quality Laws in order to prevent the spread of health hazards that caused the Minamata disease, was an unlawful inaction.

---

[13]Kumamoto District Court 20 March 1973, Hanrei Jiho No. 696, 15.

[14]Supreme Court 15 October 2004, Minshu Vol. 58, No. 7, 1802. The judgement is available in English at http://www.courts.go.jp/app/hanrei_en/detail?id=1260 (last accessed on 15 June 2015).

3.3 Asbestos Cases

After this decision, several courts also recognised State liability in other public health cases. Most recently, on 9 October 2014, the Supreme Court held in the Sennan Asbestos case that the State was liable for failing to promptly issue the ministerial ordinances to regulate asbestos based on the Labor Standards Act[15] and the Industrial Safety and Health Act.[16]

Asbestos is a hazardous substance that induces diseases such as mesothelial tumour and asbestosis attributable to exposure to asbestos dust. Until now, its production and use have been gradually and totally prohibited in Japan. However, asbestos had been used since the end of 19th century. In 2005, Kubota, which is one of the biggest manufacturing companies in Japan, announced the existence of many occupational victims of asbestos, as well as victims of asbestos dust from environmental exposure around the Kanzaki plant in Amagasaki City (Kubota Corporation 2006). After that, it was revealed that the number of victims was increasing. It led not only to strict regulation of asbestos, but also to enactment of the Act on Asbestos Health Damage Relief.[17]

The Sennan area in Osaka prefecture is where many small asbestos textile factories have located from 1907 to 2005. Asbestos yarn, cloth, and other insulation materials that were used for ships, cars, and construction, etc. were manufactured in this area. There were only several big factories, and most of the factories were household industries typically operated by families and a few employees. Therefore, workers, their families, and owners of small factories, as well as neighbourhoods, have suffered from disease caused by asbestos.

Twenty-six asbestos victims brought action in court against the Japanese Government, seeking compensation in 2006 (First Sennan case). Subsequently, 33 victims followed suit, and filed the second case in 2009 (Second Sennan case) (Osaka Lawyers Advocacy for the Victims of Asbestos Disease 2009). These parties argued that the government knew the hazards of asbestos, was able to warn people about the toxicity of asbestos and to control asbestos, but did not exercise its competence appropriately and timely.

Osaka District Court recognised State liability in the first case in 2010.[18] However, Osaka High Court overruled the original decision and denied State

---

[15]Act No. 49 of 7 April 1947, latest revision: Act No. 31 of 29 May 2015. The text of the 2012 law is available in English at http://www.env.go.jp/en/laws/policy/basic/index.html (last accessed on 15 June 2015).

[16]Act No. 57 of 8 June 1972, latest revision: Act No. 82 of 25 June 2014. The text of the 2006 law is available in English at http://www.env.go.jp/en/laws/policy/basic/index.html (last accessed on 15 June 2015).

[17]Act No. 4 of 10 February 2006, latest revision: Act No. 69 of 13 June 2014. The text of the 2006 law is available in English at http://www.env.go.jp/en/laws/policy/basic/index.html (last accessed on 15 June 2015).

[18]Osaka District Court 19 May 2010, Hanrei Jiho No. 2093, 3.

liability in 2011.[19] In contrast, Osaka District Court in 2012,[20] and Osaka High Court in 2013,[21] both recognised State liability in the second Sennan case. Finally, the Supreme Court affirmed State liability and allowed the claim by the ex-workers.

The Supreme Court dismissed the final appeal in the second case by the appellant, the State, and ruled that the State must pay the appellees 330 million yen in total.[22] As for the first case, the Court overruled the original decision and ordered the case sent back to the lower court.[23] The government settled with the victims of the first case in December 2014.

The Supreme Court pointed out that, as of 1958: (1) the medical findings on asbestosis had been established and the State was also aware that asbestos dust caused serious damage; (2) there were practical technical findings as necessary for making it obligatory to install local exhaust ventilation systems, which was the most effective measure to prevent asbestos dust at the abovementioned factories, etc.; and (3) despite the administrative guidance provided previously, the progress in installation of local exhaust ventilation systems for operations relating to asbestos remained unsatisfactory. The Court also stated that the Minister of Labour failed to exercise its authority to enact ministerial ordinances based on the Labour Standards Act (prior to the revision by Act No. 57 of 1972), and failed to make it obligatory under penal provisions to install local exhaust ventilation systems at said factories, etc., on or after 26 May 1958, and that such failure is illegal.

After these judgments, the Minister of Labour apologized, announced that the government was willing to settle with the plaintiffs of other similar actions, and pay compensation under the same conditions as the Sennan cases.

4 The Mandamus Action

4.1 Introduction of the Mandamus Action in Administrative Litigation

As mentioned earlier, the Administrative Case Litigation Act establishes four types of actions. The primary type is the Kokoku Appeal, i.e. the judicial review of administrative dispositions. There are several types of Kokoku Appeals to control omissions by public authorities.

For a long time, the ACLA only provided an 'action for the declaration of illegality of inaction' as the means to control omissions by public authorities.

---

[19]Osaka High Court 25 August 2011, Hanrei Jiho No. 2135, 60.

[20]Osaka District Court 28 March 2012, Hanrei Times No. 1386, 117.

[21]Osaka High Court 25 December 2013, Minshu Vol. 68, No. 8, 900.

[22]Supreme Court 9 October 2014, Minshu Vol. 68, No. 8, 799. The judgement is available in English at http://www.courts.go.jp/app/hanrei_en/detail?id=1296 (last accessed on 15 June 2015).

[23]Supreme Court 9 October 2014, Hanrei Jiho No. 2241, 13.

This refers to actions seeking the declaration of illegality of an administrative agency's failure to make an administrative disposition that it should have made within a reasonable period of time in response to an application filed under laws and regulations [Article 3, Para. (5)].

Action for the declaration of illegality of inaction has not been effective as a control measure. There are two reasons: first, only the applicant for an administrative disposition has standing for this type of litigation. However, existing environmental law in Japan does not grant the right for local residents or NGOs to require a certain control measure against illegal polluters in environmental matters. Second, even if the claim was allowed, it is still uncertain whether the administrative agency would have accepted or rejected the application. The administrative agency certainly is obligated to make some disposition. However, if the application is rejected, the applicant has to file suit again, to seek revocation of the rejection.

In 2004, the ACLA was revised, and the 'mandamus action' was introduced [Article 3, Para. (6)]. This is an action through which the plaintiff is able to seek a certain act of an administrative agency (e.g. an order to suspend a factory's operation). The mandamus action is available in the following two cases:

(a) where the administrative agency has not made a certain original administrative disposition which it should make (excluding the case in the following item);
(b) where an application has been filed under laws and regulations to request that the administrative agency make a certain administrative disposition, but the administrative agency has not made it.

In the second case, only the applicant, the addressee of the disposition, has standing. This is the same requirement as for actions for the declaration of illegality of inaction. However, here, the applicant is able to request a certain and concrete disposition, which could result in more direct and effective redress for the applicant.

The first case provides a new possibility for controlling omissions for the third party who is not the addressee of the disposition, such as local residents who live near a factory or other facilities that could cause adverse effects on health and the environment.

4.2 *Mandamus Action for the Minamata Disease Cases*

In Japan, there is a unique administrative redress program for victims by the environmental pollution (ERCA n.d.). In 1973, serious health damage caused by environmental pollution, such as Minamata disease, forced the government to enact the Pollution-Related Health Damage Compensation Law.[24] The purpose of this law was to provide quick relief and compensation to victims. Under this law, compensation payments, such as medical care expenses and handicap compensation, were

[24]Act No. 111 of 5 October 1973, latest revision: Act No. 69 of 13 June 2014.

made to victims who satisfied three requirements of: (a) designated illness, (b) designated region, and (c) exposure. They also had to be certified as patients through a screening by the governor of the prefecture in which the designated region was located. According to the polluter-pays principle, the industry covers the main costs for the patient.

However, it is not always easy to diagnose Minamata disease, because in many cases its typical symptoms only partially appear. Although the purpose of this program was to provide quick relief for the victims, it has taken very long time in grey zone cases of Minamata disease to decide about the application for certification as a patient. In addition, in the case of dismissal of application, applicants had to go to court to seek a revocation of such administrative decision.

After the introduction of the mandamus, some victims filed mandamus suits seeking the disposition to certify as Minamata disease patients. On 27 February 2012, the Fukuoka High Court ordered Kumamoto prefecture to certify the applicant as a Minamata disease patient.[25] Finally, on 16 April 2013, the Supreme Court dismissed the final appeal by the appellant, Kumamoto prefecture, and the plaintiff was certified as a Minamata disease patient.[26]

This case shows that the mandamus could also be an effective measure for other relief programs, such as compensation for workers' accidents based on the Industrial Accident Compensation Insurance Act, in which certification by the government is also required.

4.3 Disposal Site Case

The mandamus litigation has also been proving effective on environmental pollution cases, especially those related to illegal dumping. Before the introduction of the mandamus action, civil injunctive relief against the polluter (not the government) was the typical measure to prevent pollution and damage by illegal dumping.

In some disposal site cases, the lower courts have recognised that the right of neighbourhoods to live in a healthful and peaceful environment needs to be protected as a personal right,[27] and issued injunctions when there was a high risk of pollution or contamination of surface and groundwater arising from waste disposal sites, for example.

Local residents have been using the mandamus action since it was adopted in 2004. On 7 February 2011, the Fukuoka High Court ordered the Governor of Fukuoka prefecture to take action against illegal waste traders who were damaging

---

[25] Fukuoka High Court 27 February 2012, Shomu Geppo Vol. 59, No. 2, 209.

[26] Supreme Court 16 April 2013, Hanrei Jiho No. 2188, 42. See also another Judgement in the Osaka Minamata mandamus case on the same day, Supreme Court 16 April 2013, Minshu Vol. 67, No. 4, 1115. The latter judgement is available in English at http://www.courts.go.jp/app/hanrei_en/detail?id=1197 (last accessed on 15 June 2015).

[27] For example, Sendai District Court 28 February 1992, Hanrei Jiho No. 1429, 109.

the living environment.[28] On 7 March 2012, such decision became definitive when the Supreme Court rejected the final appeal by the local government.[29] The plaintiff also sought administrative subrogation. Although this case was dismissed by the Fukuoka High Court, such claim could be effective when the polluter does not have enough financial resources for restoration.

5 Administrative Action for Injunctive Order Concerning Public Facilities

Health damage and environmental pollution have been caused not only by private facilities, but also by public facilities, such as airports, roads, etc. Article 2, Paragraph (1) of the State Redress Act sets forth provisions on State liability concerning defects in the establishment and management of public facilities.

When damage to another person is caused because of a defect in the placement or administration of a road, river, or other public structure, the State or public entity shall assume the responsibility to compensate therefor.

The State has strict liability for defects in the establishment or management of public facilities. Such liability is not limited to physical and external defect of equipment or imperfection for users. It should include risk of damage to persons other than users, such as neighbourhoods, even when public facilities are used in accordance with their intended purpose.[30]

The Supreme Court has been awarding compensation for damages caused by noise from airports and air bases in several cases. The first important one was the Osaka International Airport case, which involved noise pollution. On 16 December 1981, the Supreme Court ruled affirmatively for payment of damages.[31]

However, civil litigation seeking injunction has been dismissed in airport cases. The Supreme Court stated that civil litigation is not available because the management of such facility is related to the exercise of public authority. It has been unclear what kind of administrative litigation is available in these cases.

The actions seeking compensation based on the State Redress Act have been effective to some extent for the improvement of pollution, e.g. air and noise pollution caused by roads, for which civil injunction relief has been allowed in some cases.[32] However, this has not been the case for the air bases; though the Supreme Court permitted compensation,[33] the problem has not been solved.

---

[28]Fukuoka High Court 7 February 2011, Hanrei Jiho No. 2122, 45.

[29]Supreme Court 3 July 2012, LEX/DB25482345.

[30]Supreme Court 16 December 1981, Minshu Vol. 35, No. 10, 1369. The judgement is available in English at http://www.courts.go.jp/app/hanrei_en/detail?id=66 (last accessed on 15 June 2015).

[31]Ibid.

[32]For example, Kobe District Court 31 January 2000, Hanrei Jiho No. 1726, 20.

[33]For example, Supreme Court 25 February 1993, Shomu Geppo Vol. 40, No. 3, 452.

Consequently, local residents have filed administrative litigation seeking injunctive orders. Atsugi Air Base is one of such recent cases. This Base, located in Kanagawa prefecture, has been used not only by the Self-Defense Force, but also by the U.S. Navy. On 21 May 2014, Yokohama District Court ordered the Self-Defense Forces to cease flights between 10 p.m. and 6 a.m. to avoid noise pollution.[34] This is the first court decision that resulted in an injunction for the Air Base.

However, the Yokohama District Court rejected the call for an injunctive order against U.S. military flights, on the grounds that they are not matters under the control of the Japanese government. It also stated that there are currently no legal measures available for the improvement of noise pollution caused by U.S. military aircraft.

6 Future Perspectives

There are various types of litigation in Japan to control omissions by administrative agencies, through civil litigation and administrative litigation. Such measures have been strengthened in this century by case law, as well as by legislation.

Civil litigation, in the form of a State liability lawsuit, is basically the measure for relief for the victims. However, it also plays an important role as an indirect measure to push the government to take action. Recently, State liability lawsuits have proven to be an effective measure to challenge the non-use of the regulatory power. State liability litigation concerning health damages caused by asbestos is expected to increase, because old asbestos producers do not exist anymore, so current victims cannot not call them to account. In combination with this litigation, the plaintiffs have requested for the better relief program for asbestos victims and stronger measures to prevent further health damage. The production and use of asbestos is already prohibited, but it is important to prevent further health damage, such as through incidental exposure to asbestos during building demolition (Ministry of Environment 2011). Recently, preventative measures were strengthened by the revision of the Air Pollution Control Act. Asbestos litigation, including the lawsuits by ex-workers of building companies, seems to have contributed to promoting this process.

The mandamus action, a new type of administrative litigation, is expected to demonstrate more effectiveness in the future to control illegal failures by public authorities, especially since it is available for people affected by the non-use of administrative power.

However, a narrow interpretation of standing still has been a serious barrier for the access to justice in administrative cases. According to Article 9, paragraph (1), of the Administrative Case Litigation Act, a person who has a legally protected

[34] Yokohama District Court 21 May 2014, LEX/DB 25446437.

interest has standing. However, according to case law, a legally protected interest must be concrete and individualistic, to be protected by legal provision that gives a basis for administrative disposition. Therefore, standing has not always been permitted for local residents affected by the non-use of administrative power.

In 2014, the Administrative Procedure Act[35] was revised, such that anyone is able to request an administrative agency for an administrative act or necessary guidance to rectify an illegal act. In this case, the administrative agency is obligated to investigate the case and to take appropriate measures, if necessary. This new administrative appeal system could also contribute to strengthen the control of the non-use of regulatory power.

In the field of environmental law, the executive branch is usually given considerable discretion as to how and when to exercise its regulatory power, and the particular administrative agencies have often been reluctant to do it. If only the addressee of the administrative disposition has the possibility of challenging the administrative disposition, the administrative agency would take the economic interest of the entrepreneur more into account. Therefore, judicial and administrative control of administrative acts of omission is indispensable for ensuring environmental rule of law, and such measures should continue to be improved.

References

ERCA—Environmental Restoration and Conservation Agency of Japan (n.d.). Operation of the pollution-related health damage compensation system. Retrieved 15 June 2015, from http://www.erca.go.jp/erca/english/activities/ac_01.html

Kubota Corporation (2006) Kubota's response to the issue of Asbestos. Retrieved 15 June 2015, from http://www.kubota-global.net/csr/report_past/pdf/2006/p34.pdf

Ministry of Environment (2011) Summary of countermeasures against Asbestos in Japan. Retrieved 15 June 2015, from https://www.env.go.jp/en/air/asbestos/summary_caa.pdf

Okubo N (2005) Environmental policy instrument diversity. In: JEC—Japan Environmental Council (ed) The state of the environment in Asia 2005/2006. Springer, Tokyo, pp 337–348

Okubo N (2015a) The development of the Japanese legal system for public participation in land use and environmental matters, Land Use Policy. Accepted 29 Dec 2014

Okubo N (2015b) Principle 10 and developments in Asia. In: Ramos GE, Spitzkatz M (eds) Rule of law for good environmental governance. University of Cebu and Konrad-Adenauer-Stiftung, pp 154–168

Osaka Lawyers Advocacy for the Victims of Asbestos Disease (2009) Serious health hazard of Asbestos in Sennan Area. Retrieved 15 June 2015, from http://www.asbestos-osaka1.sakura.ne.jp/english.html

Osaka University (2015) Discussion papers and materials of Osaka University's international workshop on 'participation principle indicators under the environmental law: towards establishing international collaboration in pursuit of environmental justice', held 9–10 March 2015. Retrieved 27 July 2015, from http://greenaccess.law.osaka-u.ac.jp/english/resources/enworkshop2015march

---

[35]Act No. 88 of 12 November 1993, latest revision: Act No. 69 of 13 June 2014. The text of the 2005 law is available in English at http://www.env.go.jp/en/laws/policy/basic/index.html (last accessed on 15 June 2015).

Pring G, Pring C (2009) Greening justice: creating and improving environmental courts and tribunals. The access initiative. Retrieved 15 June 2015, from http://www.accessinitiative.org/sites/default/files/Greening%20Justice%20FInal_31399_WRI_0.pdf

Sakumoto N (2011) Development of Asian Environmental Law and the enforcement. Horitsu Ronso The Meiji Law Rev 83(2–3):205–223 (in Japanese)

Ui J (1992) Minamata disease. In: Ui J (ed) Industrial pollution in Japan. United Nations University Press, Tokyo, pp 103–132

UN (2004) The rule of law and transitional justice in conflict and post-conflict societies. Report of the Secretary-General (S/2004/616). Retrieved 15 June 2015, from http://www.ipu.org/splz-e/unga07/law.pdf

UNEP (2013) Decision 27/1: new rules of procedure of the Governing Council Decision 27/9. In: Decisions adopted by the Governing Council/Global Ministerial Environment Forum at its first universal session. Retrieved 15 June 2015, from http://www.unep.org/gc/gc27/docs/Decisisions_adopted_by_the_first_universal_session_(advance).pdf

UNEP (2015) Environmental rule of law: critical to sustainable development. Retrieved 15 June 2015, from http://www.unep.org/delc/Portals/24151/Documents/issue-brief-environmental-justice-sdgs.pdf

UNEP (n.d.) Environmental rule of law. Retrieved 15 June 2015, from http://www.unep.org/delc/worldcongress/Home/tabid/55710/Default.aspx

Examining the Resilience of Public Participation Structures for Sustainable Mining in the Philippines

Weena Gera

Abstract This paper analyses the interplay of institutional-legal architecture, politico-structural conditions and civil society configurations and how it influences the resilience of public participation structures towards sustainable mining in the Philippines. It illustrates how intersecting forces of predominant political economic interests and emergent civil society networks navigate through legal frameworks to influence dependencies and the shifting boundaries of public engagement in mining governance. It argues that the country's mining sector is in a state of impasse with a government struggling to restructure its mining policies to accommodate a growing public clamour for environmental and social protection while continuing to provide incentives to large-scale mining corporations within a neoliberal framework. This has expanded and created new spaces, including legal and juridical paths, for civil society to pursue accountability mechanisms and challenge entrenched constraints. However, this attempt by the government to reconcile incompatible interests in the mining sector, translates to an ambivalent framework of governance that could not identify its priority. This results to arbitrary policy compromises that exacerbate prevailing tensions amid power imbalances in the sector, leaving civil society in perpetual square-off against mining corporations.

Keywords Public participation · Mining governance · Sustainable mining · Governance for sustainable development · Resilience of governance structures

1 Introduction

Amid an expanding discourse on resilience as an integral framework for sustainable development, a gap in the analytical trajectory is an examination of the resilience of specific governance structures in the promotion of sustainable development.

W. Gera (✉)
University of the Philippines Cebu, Cebu, Philippines
e-mail: wsgera@up.edu.ph

V. Mauerhofer (ed.), *Legal Aspects of Sustainable Development*,
DOI 10.1007/978-3-319-26021-1_11

A fundamental element of governance is public participation including the relations between state and people, and the mechanisms through which collaboration can be generated across sectors. It is broadly acknowledged that public participation is critical in promoting environmental sustainability. Chapter 23 of Agenda 21, the programmatic declaration emerging from the Rio de Janeiro United Nations (UN) Conference on Environment and Development (1992), noted that one of the fundamental prerequisites of the achievement of sustainable development is broad public participation in decision-making. The necessity of the interactions between the public and public authorities in environmental matters has also been enshrined in the UN Economic Commission for Europe, Aarhus Convention on 25 June 1998. Moreover, this is pursued in Principle 10 Guidelines, otherwise known as Guidelines for the Development of National Legislation on Access to Information, Public Participation and Access to Justice in Environmental Matters adopted at the 11th Special Session of UN Environment Programme Governing Council/Global Ministerial Environmental Forum in Bali, Indonesia on 26 February 2010.

Most recently, in its Discussion Paper on 'Governance for Sustainable Development', United Nations Development Programme (UNDP) (2014, p. 2, 8) highlighted that "for development to be sustainable—economically, socially and environmentally—and equitable, a new approach is needed that addresses the political, as well as the technical, aspects of development solutions… one of the major findings of various national and thematic consultations on the post-2015 development agenda is the need for strong participatory monitoring mechanisms by non-state actors to hold government accountable." Kanie et al. (2014, p. 1768) added that "governance is indispensable for the implementation of integrated Sustainable Development Goals (SDGs), and this is the added value of SDGs as compared to the Millennium Development Goals (MDGs)… Such governance requires scale-specific actions… and engage stakeholders." With these international policy frameworks, it becomes critical to examine how participatory policies are institutionally designed and enacted in countries, and how these are enforced on the ground amid the complexities of its practice vis-à-vis specific political economy constraints and inherent pluralities among sectoral interests.

The Philippines is a country noted for its comprehensive legal framework for participatory governance, including in environmental matters and in promoting sustainable development (Lin-Heng 2002; Gera 2015). Public participation is enshrined in the 1987 Constitution with Article II Section 23, Article XII Section 5 and Article XIII Sections 15–16 recognizing the rights of independent people's organizations, including indigenous communities, "to pursue their legitimate and collective interests and aspirations through peaceful and lawful means, at all levels of social, political and economic decision-making." Hosting about 60,000 registered non-government organizations (NGOs), the country has been regarded as potentially having "the largest NGO density in the world" (Holden 2005, p. 227). On environmental matters, national laws as well as local ordinances were legislated to expand access to participation of non-state actors and ordinary citizens in determining environmental policies and monitoring their enforcement.

The country however is also replete with criticisms on the systemic limits of its implementation of public participation (Rood 1998, 2005; Wurfel 2006; Holmes 2011; Rivera 2011; Gera 2015). It has been argued that while the country appears to have a strong institutional design for public participation, there is a decisively low level of institutionalization of actual participatory processes that could not achieve substantive autonomy from power structures. Gera (2015, p. 19) posited that a critical factor to this is the "weakness in public deliberations and interactions among civil society groups in the country that constrain legitimacy of representation, coherence and necessary consolidation in engagements with government institutions."

This paper proceeds with a more nuanced sectoral analysis to illustrate specific conditions that both constrain and facilitate the institutionalization and resilience of public participation structures in promoting sustainable development. By examining the context of the mining sector in the Philippines, the study demonstrates how intersecting forces of predominant political economic interests and emergent civil society networks navigate through complicated legal and juridical systems to influence dependencies and the shifting boundaries of public engagement in mining governance. It also analyses how institutional arrangements and stakeholder configurations have been calibrated, adjusted, reconstituted, or transformed to negotiate competing sectoral concerns. The study argues that in its successive policy strategies in negotiating conflicting demands to achieve economically viable and sustainable mining, the Philippine government has expanded spaces and legal paths for public participation to advance environmental and social policies. Nonetheless, it highlights the precariousness of the position of civil society organizations (CSOs) vis-à-vis large mining corporations, as they operate within an ambivalent policy framework that indiscriminately attempts to reconcile the competing agenda of neoliberal mining and sustainable development.

2 Framework

The concept of sustainable development has evolved since the 1972 UN Conference on the Human Environment held in Stockholm until the 2012 UN Conference on Sustainable Development or Rio +20. It however was popularized in *Our Common Future* also known as the Brundtland Report with the classic definition: "development which meets the needs of the present without compromising the ability of future generations to meet their own needs" (World Commission on Environment and Development 1987). A current prominent notion expands the concept of sustainable development as "development that meets the needs of the present whilst safeguarding Earth's life-support system upon which the welfare of current and future generations depends" (Griggs et al. 2013, p. 305). In the specific context of the mining industry, this paper anchors its conceptualization of sustainable mining to the UN Economic Commission for Africa's (UNECA 2004, p. 2) statement that "one of the major challenges of sustainable development in the context of non-renewable resources such as minerals is for the resources and the revenues from their production to be used, while they are available, in such a manner as to contribute to the transformation of the economies of those developing countries from which they are produced and sustain them long after the minerals have been depleted."

The study is particularly concerned with the institutionalization—as an indicator of resilience—of public participation structures for sustainable mining. Such structures include the institutional arrangements, configurations and relationships among stakeholders, mechanisms and boundaries of public engagement in the broad spectrum of mining governance. Gera (2015) posits that institutionalization of public participation, particularly in engaging the state requires the "creation of enduring and non-arbitrary, non-ad hoc systems of collaboration, and establishment of clear standards, parameters, criteria, procedures and mechanisms of engagement and coordination with government involving legitimate and fully accredited CSOs" (Gera 2015, p. 2). It is anchored to North's widely used definition of institutions as the "rules of the game in a society, or more formally, as the humanly devised constraints, that provide a stable structure of human interaction" (1990, p. 3). Process-wise, it means the application of clear criteria of engagement that discourages arbitrary, irregular and clientelist decisions and power play (Batalla 2000). This is parallel with the main attributes identified by Gerring et al. (2005) of an institutionalized polity that includes stability of processes, endurance, regularity, and non-variance. Essentially, institutionalization requires that formal institutions (i.e. laws) prevail over power structures—one key aspect of the rule of law.

However, beyond the legal framework, institutionalization of public participation structures also means that informal institutions (i.e. deliberative processes, relationships and cultures of participation and interest aggregation) are reflexively adjusted and transformed to be able to respond to current urgencies, and withstand inherent political constraints that could undermine public participation. As emphasized by Leuven (2009, p. 50), an important part of a more inclusive consensus is a "result of a deliberative and interactive process and mutual learning…" to calibrate responses and criteria of support or resistance of state programs. This means that civil society maintains not only accountability achieved through clear mandates, but also substantive autonomy from clientelist interests, through strategic consolidation of 'overlapping consensus' (Rawls 1987) and likeminded initiatives, and reconciliation of contradictory frameworks that transforms collaborative processes (Holmes 2011; Rivera 2011).

Public participation specifically toward sustainable mining requires, according to UNECA (2004, p. 1), "at the minimum… that procedures exist by which those that would be affected by the decision as to whether or not a mineral resource is to be developed can make meaningful inputs into the making of that decision." However, in this paper, institutionalization of public participation is understood as not only the capacity to make meaningful inputs in decision-making but also the transparent and accountable enforcement of such stakeholder inputs in inclusive agreements and standards. Analysis looks into the broad spectrum of public participation in mining governance from policy formulation, monitoring and enforcement down to revenue allocation and equity participation. Equity participation includes ownership of equity interest in mining projects by the community as participants in the benefits of both small and large-scale mining.

Finally, public participation is a comprehensive concept within the framework of social inclusion in governance. The bounds dealt with in this paper are however delimited to collaborations and engagement of major organized groups in civil society or CSOs that include, NGOs, community-based associations and sectoral groups, with the government and its agencies. These CSOs also often pursue public interest/good, rather than individual or private interests, and thus excludes business organizations and chambers. The aim is to focus on examining the dynamics of public participatory structures in mining governance involving collaboration, accountability and representation by legally organized entities (i.e. duly registered and has legal identity) in civil society.

3 Philippine Mining Governance and Structures of Public Participation Toward Sustainable Development: The Enduring Paradox

Mining represents one of the dilemmas in the sustainable development initiatives in many countries, particularly in the developing regions. The benefits of mining to the national economy is seen in terms of fiscal revenues, foreign exchange earnings, employment and transfer of skills. However, mining is also widely seen as extractive causing widespread ecological destruction and adverse social impact on mining communities. This dilemma is stark in the case of the emerging Philippine economy, signatory to both the World Trade Organization and the Rio +20 UN Conference on Sustainable Development. This section illustrates the contradictions in the Philippines' mining governance: across successive policy strategies, it has remained solidly anchored within a neoliberal framework that privileges large-scale mining corporations, at the same time maintaining a rich policy narrative of sustainable development and public participation.

3.1 The Philippine Mining Industry: An Overview

The Philippines is rich in mineral resources with 9 million ha or about 30 % of the country's total land area identified to have rich mineral deposits. It is known to be the 5th most mineralized country in the world, 3rd in gold, 4th in copper and 5th in nickel. The industry has generated US$5.1 billion in gross production value of metallic minerals in 2011 and 2012 alone. Philippine export data of minerals from January to September of 2012 to major destinations such as Japan, China, US and Canada amounted to US$1.59 billion, mostly of copper, gold and nickel

(Salamat 2013). The country also has an estimated US$840 billion worth of untapped mineral resources, according to the Mines and Geosciences Bureau of the Department of Environment and Natural Resources (MGB-DENR). The Chamber of Mines of the Philippines also estimated a potential growth of mining contribution to about 5–6 % of the country's GDP by 2018 through growth in investments projected at US$15.2 Billion from 2011 to 2018 with a total annual revenue of US $12.1 Billion by 2018 (Recidoro 2014). Mining is thus vital to the country's development particularly for economic growth, industrialization and modernization.

However, there are arguments that the mining industry's minimal current economic contribution could not offset the ecological and social hazards it generates. Mining contributed no more than 0.91 % of the GDP from 2003 to 2012 on the average (Senate Economic Planning Office 2013, p. 5). It has a minimal 2 % mandated excise tax collections, contributing only 0.16 % to the country's total revenue of 1.2 trillion pesos, and employing only about 0.6–0.7 % of the total workforce in 2012 (*Kalikasan*-PNE 2012; Salamat 2013). Mining in the Philippines caused pollution of upland, agricultural, and aquatic ecosystems due to acid mine drainage, hazardous wastes, laterite and other toxic chemical spills. Provisions exempting mining companies from total log ban also result to a loss of forest cover in critical watershed and biodiversity areas. These translate to vulnerability of communities to disaster risks such as flooding and typhoons, as well as water contamination and diseases. "The main risk posed by typhoon to a mine concerns the stability of its tailings dam because the heavy rains associated with a typhoon can cause tailings impoundment to fail either through excess water pressure or by overtopping" (Holden and Jacobson 2012, p. 79).

A huge case recorded was the Marcopper Mine tailing disaster in Marinduque in March 1996. It was noted as one of the largest mining catastrophes in the world, where the breakage of a drainage plug holding toxic mining waste resulted to the release of more than three million metric tons of mine tailings along the 26 km long Boac River and the coastal areas. The toxic spill resulted in a flash flood, which affected around 4400 residents from five villages who were displaced, lost their fishing livelihoods and whose, water, agricultural lands and livestock were contaminated. Leaching of heavy metals such as lead, zinc, copper, arsenic, among others, into the river were linked to health hazards endured by the residents, with cases of lead poisoning, skin diseases, respiratory problems and other infections reported by the Department of Health in 1996 and 1997 (Coumans 1999; Asuncion 2005; Macdonald and Southall 2005; World Bank 2007).

Another catastrophe was in 2012, when some 20 million metric tons of sediments have flowed into water channels from the Philex Mine tailings pond in Itogon Benguet, when its drainage tunnel was breached according to a report from the MGB-DENR. After months of recurring leakages, the Philex mine spill has become the "biggest mining disaster" in the Philippines in terms of volume surpassing the Marcopper disaster. The affected Balog Creek has been declared biologically dead as a result of the disaster (Dinglasan 2012; Quitasol 2013).

It has also been argued that large-scale mining in the country is predominantly export-oriented and dominated by transnational corporations and local mining elites, and dependent on foreign capital and technologies. The use of advanced technologies has resulted in the assignment of higher paid jobs to expatriates or workers who are not from local communities (Hatcher 2014). These often result to increased poverty incidence due to economic dislocation and/or decreasing productivity and income among mining affected peasants, fisher-folks and small-scale miners. Cases of prostitution among women were also reported due to livelihood losses combined with increased number of men miners, which created the demand (Salamat 2013; Quitasol 2013).

Mining industry is also replete with reported cases of human rights violations. These include land-grabbing, killings and harassment against locals and anti-mining activists via legal suits and red-baiting to discourage public opposition, as well as militarization over lands which are ancestral domains of indigenous tribes in the country (Coumans 1999; Doyle et al. 2007; *Kalikasan*-PNE 2012; Salamat 2013). Nearly 4 % of lands approved by the government for mining (1.14 million ha as of January 2012) were still covered with mining tenements, subject to mandatory relinquishment provided under the law, as noted by the MGB-DENR (Salamat 2013). This means there are still ongoing tensions over these lands approved.

3.1.1 Brief History

The boom of the mining industry in the Philippines started during the 60s and 70s, when the government under Marcos regime embarked on a radical reform process to develop a systematic exploration and exploitation of the Philippines' mineral resource to attract foreign direct investment and bolster the national economy. New mineral policy, legal, regulatory and administrative frameworks were formulated and enforced. Presidential Decree 463 otherwise known as the Mineral Resources Development Decree of 1974 was established. This decree revised the Commonwealth Act No. 137 otherwise known as the Mining Act, with the aim to modernize administration and disposition of mineral lands and to promote and encourage the development and exploitation thereof.

The dwindling foreign investments in the 1980s paved the way for the proliferation of thousands of small-scale miners who have been digging for minerals using crude methods. To regulate allocation of access and control for mining rights, Presidential Decree 1899 which was issued in 1984 by then President Marcos, defined small-scale mining as "artisanal," meaning it does not make use of sophisticated mining equipment, involves minimal investments in infrastructure and processing plants, relies heavily on manual labour, and is owned, managed, or controlled by an individual or entity qualified under existing mining laws, rules, and regulations. This was reinforced by Republic Act (RA) 7076 or Small Scale Mining Act, issued in 1991 by President Corazon Aquino, which defined small-scale mining as an extractive activity "relying heavily on manual labor… and does not use explosives or heavy mining equipment."

Small-scale miners however broadly perceive these laws as restrictive on their growth and unresponsive to the realities of expansion among small-scale industries, which are using heavy equipment such as excavators, backhoes, and dozers, among others, as necessity (i.e. to extract nickel). These restrictions have forced many operations underground using poor infrastructure and hazardous substances, chemical tools and dynamites resulting to fatal accidents. These miners did not apply for permits nor pay taxes because that would expose them to possible criminal prosecution for using explosives and chemicals. Other problems noted in small-scale mining include illegal gold trading, smuggling, and employment of child labourers.

On March 1995, President Fidel V. Ramos signed into law the Philippine Mining Act (RA 7942), which liberalizes foreign investments in the industry allowing 100 % ownership of the claimed mining land area and minerals by multinational mining corporations. This law signalled the government's expanded neoliberal reforms that drive the country's mining sector toward large-scale mining activities. Foreign mining corporations were allowed to mine a maximum area of 81,000 ha for a period of 25–50 years in exchange for an investment of US $50 million in the country's industry (Begonia and Leonen 1996). Companies were allowed to repatriate up to 100 % of their invested capital prior to taxation, and were also given up to 10 years of tax holidays with possible extension (*Kalikasan*-PNE 2012). Since then, corporate mining permits multiplied in the belief that large-scale mining tax revenues would spur economic growth.

Small-scale miners have become increasingly threatened generating antagonisms in the industry, save for a handful of ore-sharing agreements entered into by some large companies with small-scale mining cooperatives. New large mines also often bar local communities from accessing artisanal mining sites, thus depriving them of their livelihood (Hatcher 2014). Prevailing plans to ban all small-scale mining in nickel production illustrates the tenuous legal position of small-scale mining in general.

Small-scale mining is considered dirty, unsafe, environmentally destructive, and fundamentally unsustainable, particularly in the practice of high grading operations, which refers to taking out only the best ores while leaving behind lower grade ores (Fonbuena 2013). Civil society groups however asserted that despite perceived sophisticated systems of rehabilitation among large-scale mining, its operations have worse adverse impact on the sustainability of the environment. It is destructive as it uses the method of open-pit mining which entails clearing thousands of hectares of rainforests and agricultural lands, deep excavations to extract minerals, the use of toxic heavy metals and chemicals to process mineral ores, and the consumption of millions of liters of water (Espiritu 2015; Salamat 2013).

It is also important to consider that small-scale mining is historically an important source of income for poor Filipinos in many parts of the country. The rise of settlements of small-scale miners in remote mountains and valleys has helped boost the economic growth of towns and provinces. It also accounts for a sizeable share of the country's output of metallic minerals. In 2007, small-scale miners dug 32,282 kg of gold worth P33.2 billion, or more than five times the output of the country's big gold producers such as Philex Mining Corp. or Canada's Toronto Ventures Inc. (Fonbuena 2013). The small-scale gold mining alone employs about

200,000 people (Artajo 2012, p. 7). Small-scale mining activities nonetheless continue to be limited to areas declared as "*Minahang Bayan*" in effect also limiting most areas where there are nickel operations. Most nickel miners operate on small-scale permits as they await approval for their permits for medium or large-scale operations (Rappler 2013).

3.2 Legal Structures for Public Participation in Mining Governance

Legal structures for public participation in mining governance are a direct response to the investment risks asserted by CSOs. These created avenues for public concerns over environmental and social impacts of mining to be articulated and legislated. While broadly seen as a critical feature of democratic governance, certain sectors such as the Chamber of Mines of the Philippines as well as the Mines and Geosciences Bureau perceive this to have created a challenging environment for the development of the mining industry in the country (Christian Aid and PIP Links 2004, p. 9, 17).

(a) *Presidential Decree* 1586: *Environmental Impact Statement System*

The Philippines was one of the earliest to initiate Environmental Impact Assessments (EIA) among developing countries. It formally established the Environmental Impact Statement (EIS) System in 1978 through Presidential Decree 1586, which provides the legal and procedural framework for conducting EIAs for projects of the government and private companies likely to have significant environmental impact. The EIS system has provisions for public involvement including public consultation and public hearings. It also provides for alternative dispute or conflict resolution which may be used to reach consensus if there are complex or unresolved issues or constraints to scoping meetings and/or the social acceptability of the Department of Environment and Natural Resources (DENR)'s decisions on Environmental Compliance Certificates (ECCs). The implementation of the 1995 Mining Act requires the DENR to administer an EIA. In 2010, DENR issued Memorandum Circular 2010–14 allowing for greater public participation in the implementation of the Philippine EIS system and for EIA findings to be presented during public consultations.

(b) *Local Government Code of* 1991

The Local Government Code of 1991 (RA 7160), which devolved environmental functions to the local government units (LGUs), elaborated provisions to ensure public participation in local environmental governance. These are expressed in Sections 2c, 26 and 27, highlighting the imperatives for consultation with CSOs on projects and programs involving the maintenance of ecological balance. Sections 35–36 of the Code also highlight the role of people's organizations and NGOs as partners of LGUs in pursuing various projects, including environmental matters. Two critical features of the Code on public participation include the Local Sectoral Representation system (Section 41c) for community stakeholder involvement in legislation, and the Local Development Council (LDC) (Title VI, Sections 107–108) in development planning. The LDC is crucial in mining governance as one of its tasks is to craft the Comprehensive Land Use Plan, which serves as a blueprint in the rational and sustainable use of land resources for future physical development of cities and municipalities to be adopted and enacted by the local legislative council into a Zoning Ordinance.

(c) *Mining Act of* 1995

As stated above, the Philippine Mining Act (RA 7942) was signed into law in 1995 liberalizing foreign investments in the industry. The law has been largely opposed and protested against by the public due to questions of its constitutionality vis-à-vis national sovereignty, particularly its provisions on Financial and Technical Assistance Agreement, which allows foreign ownership. Nonetheless, the same law provides for public consultation and participation, specifically in environmental monitoring. Section 16 of the law states that "no ancestral land shall be opened for mining operations without the prior consent of the indigenous cultural community concerned." It also reiterates in Section 70, which provides for the Environmental Impact Assessment, that people's organizations and NGOs shall be allowed and encouraged to participate in ensuring that contractors observe all the requirements of environmental protection.

(d) *Indigenous Peoples Rights Act of* 1997

The 1997 Indigenous Peoples Rights Act (IPRA) (RA 8371) promotes the participation of communities in the management of natural resources through the Ancestral Domain Claim framework. IPRA requires that indigenous communities be informed and consulted, for their free, prior and informed consent before any mining concessions or other projects are started. Section 6 of the law provides for mandatory representation in all policy-making bodies and in local legislative councils. In consultation with indigenous peoples, the National Commission on Indigenous Peoples, in close coordination with the Department of Interior and Local Government, must come up with appropriate measures to ensure the full participation of indigenous peoples in matters affecting their development.

(e) *2004 Executive Order 270*

In 2004, President Gloria Macapagal-Arroyo issued Executive Order (EO) 270 National Policy Agenda on Revitalizing Mining in the Philippines. It pushes for the critical role of investments in the minerals industry for national development and poverty alleviation. It articulated that in order to enhance the contribution of mining, minerals and metals to promoting sustainable development, there is a need to address the environmental, economic, health and social impacts and benefits and enhance the participation of stakeholders throughout the life cycles of mining operations.

(f) *2010 Supreme Court Rules of Procedure for Environmental Cases*

In 2007, the Supreme Court designated 117 special environmental "green" courts within the judiciary in the interest of pursuing improved environmental adjudication (Ito 2011). In 2010, the Supreme Court issued the Rules of Procedure for Environmental Cases (A.M. No. 09-6-8-SC). These rules include provisions on civil suits including Strategic Lawsuits Against Public Participation (SLAPP) (Rule 6) ensuring pre-emptive defence for those parties engaged in enforcing environmental laws or asserting environmental rights. SLAPPs are civil lawsuits filed against an NGO or individual for speaking out and/or criticize projects or practices that harm both the environment and affected communities. The SLAPP method involves intimidation, silencing or 'chilling' of environmental defenders who had to endure the stress and costly expenses of several years of litigation, as a 'punishment' or 'price' for exercising their political rights. These consequently decrease citizen willingness to participate in policy-making or to act as watchdogs (Gleason 2003; Tapang 2009; *Kalikasan*-PNE 2010). Under the rules, persons or organizations can immediately file a counter case as defence against a SLAPP.

The *Rules* empowers the courts to issue environmental protection orders as an immediate action to protect the environment and the communities affected. Other remedies and orders direct government agencies to protect, preserve or rehabilitate the environment. The *Rules* also enable communities to petition for the suspension or stoppage of destructive, environmental and development activities through the Citizen's Suit provision. Specifically, Rule 7 contains the Writ of *Kalikasan*, (or the writ of nature) "a remedy available to a natural or juridical person, entity authorized by law, people's organization, non-governmental organization, or any public interest group accredited by or registered with any government agency, on behalf of persons whose constitutional right to a balanced and healthful ecology is violated, or threatened with violation by an unlawful act or omission of a public official or employee, or private individual or entity, involving environmental damage of such magnitude as to prejudice life, health, or property of inhabitants in two or more cities or provinces." Moreover, Rule 8 provides for the Writ of Continuing Mandamus which states that, "When any agency or instrumentality of the government or officer thereof unlawfully neglects the performance of an act which the law specifically enjoins as a duty resulting from an office, trust or station in connection with the enforcement or violation of an environmental law rule or regulation or a right therein … the person aggrieved thereby may file a verified petition in the proper court…"

(g) *2012 Executive Order 79*

The most recent policy provision came in the form of EO 79 (2012), titled "Institutionalizing and Implementing Reforms in the Philippine Mining Sector Providing Policies and Guidelines to Ensure Environmental Protection and Responsible Mining in the Utilization of Mineral Resources". This is also in response to the heightening public clamour against the environmental hazards of mining activities, as well as against unjust sharing of profit from mining, toward a repeal of provisions that allow very minimal share for host communities and for the government (Cabacungan 2012). It imposed a moratorium on new mining contracts until a law rationalizing the revenue sharing scheme between the government and mining firms have taken effect. It seeks to implement industry reforms and expanded the "no-go" areas for mining such as (a) tourism destinations, which total around 78 sites identified by the Department of Tourism based on the National Tourism Development Plan, (b) agricultural lands, including an area in Leyte; and (c) island ecosystems identified by the National Integrated Protected Areas System.

This policy order also creates the Mining Industry Coordinating Council to implement reforms and conduct dialogue with stakeholders and review all existing mining-related laws and rules. The council will be co-chaired by the chairpersons of the Climate Change Adaptation and Mitigation and the Economic Development clusters of the Cabinet. Other members will include the Justice Secretary, the chairperson of the National Commission on Indigenous Peoples and the president of the Union of Local Authorities of the Philippines. The Chamber of Mines of the Philippines criticized this policy to be counterproductive to future investments in mining (Chamber of Mines of the Philippines 2013; Recidoro 2014).

3.3 *Civil Society Configurations: Networks, Relationships and Solidarity Actions*

Distinctive in the mining sector in the Philippines is the presence of enduring civil society groups and environmental networks. Since the enactment of the Mining Act in 1995, a growing constituency against large-scale mining was formed in the Philippines to protest its constitutionality, particularly the provision on Financial and Technical Assistance Agreement. Hatcher (2014, p. 61) noted that, "the Philippines is characterized by the existence of strong anti-mining coalitions bridging the concerns of a number of actors such as indigenous peoples, environmental groups, as well as a wide range of other regional and national civil society organizations." Major activities and participatory processes range from posting press releases, conduct of investigative reports, scientific research and analysis on the destructive impact of proposed large-scale mining projects, monitoring and education, legislative lobbying, provincial resolutions, protest movements, mass mobilizations, and solidarity actions through sectoral and national formations, international lobbying, up to adjudication and legal actions against human rights violations and tax evasion cases.

Local solidarity actions are organized among indigenous and tribal community associations. Such is the case of Alliance for Buguey Committed for Development Association collaborating with Aparrianos Movement for the Conservation and Environmental Protection and Concerned Laloeno Against Illegal Mining—in lodging complaints with the MGB-DENR which led to the suspension of black-sand mining in the town of Buguey, Cagayan. Other mining-affected areas from Nueva Viscaya, Romblon, Leyte, Cagayan, Cantilan and Zambales also join solidarity protests of environmental groups by performing local solidarity actions in their respective areas. These actions also mobilize the indigenous and tribal mining communities, as well as cooperate with small-scale miners in protesting against the government's favour of large-scale corporate mining.

Alliances and coalitions such as *Kalikasan*-PNE, *Alyansa Tigil Mina* (Alliance against Mining), Defend Patrimony, Center for Environmental Concerns, Task Force-Justice for Environmental Defenders, Philippine Indigenous Peoples' Link, Philippine Movement for Climate Justice, among others, take leadership in mobilizing advocacy and lobbying campaigns against alleged environmental and human rights violations by mining corporations and their underlings. These coalitions link different organizations, local community associations and forums to bring together a common campaign, as well as collaborate with international environmental organizations and global civil society networks such as War on Want, London Mining Network, Banktrack, among others. Llorito (2006, para. 8) noted that together, these actors represent a 'formidable political force' against miners.

Alliances established among these non-state actors also extend their solidarity and partnerships with local governments and national political parties, specifically in formulating legislative proposals and lobbying alternative mining bills or resolutions to Congress/local legislative councils. In particular, there is a common call for the repeal of the Mining Act of 1995 toward the passage of a new law to stop large-scale mining in the Philippines. Notably, the Catholic Church is solidly entrenched within the anti-mining movement in the country, with the Catholic Bishops Conference of the Philippines petitioning for the repeal of the Mining Act. These coalitions lobbied members of Congress to reorient the industry towards needs-based extraction, sustainable economic development and environmental protection.

Three nearly identical but separate bills were filed in the fifteenth Congress. In July 2010, House Bill (HB) 206 or Alternative Mining Bill was filed by House Deputy Speaker Lorenzo Tanada III. In December 2010, HB 3763 or Philippine Mineral Resources Act of 2010 was filed by *Akbayan* Representatives Kaka Bag-ao and Walden Bello, Ifugao Rep. Teddy Brawner Baguilat, Cagayan de Oro Second District Rep. Rufus Rodriguez, Rep. Maximo Rodriguez, Nueva Vizcaya Rep. Carlos Padilla and Paranaque Rep. Roilo Golez. Then in March 2011, seven Representatives led by *Bayan Muna* Rep. Teddy Casino filed a similar bill, HB 4315 or the People's Mining Act of 2011 in the lower house.

HBs 206 and 3763 similarly propose that mining should be guided by a Minerals Utilization Framework which defines what minerals should be extracted, how much and when these should be extracted. HB 4315 proposes the development of a strategic National Industrialization Program by the government. The three bills also propose to reorient the Mines and Geosciences Bureau (MGB) away from serving as an approving body for permits, a role which has been criticized as contradicting its mandate as a sub-agency of the DENR. Instead, they affirm the MGB's role as a scientific research institution, which shall exclusively and directly be responsible for undertaking mining exploration activities on behalf of the State (Ito 2011). All three bills have been consolidated, along with other mining-related bills, and deliberated in the Natural Resources Committee of the House of Representatives. Moreover, on February 2012, Senate Bill 3126 was filed by Senator Sergio "Serge" Osmena III, providing the senate version of HB 3763.

4 Examining the Resilience of Public Participation Structures in Mining Governance and Implications on Sustainable Development

There is a complex combination of factors that contributes to an evolving capacity of CSOs to make significant inroads, defining the staying capacity of public participation structures in mining governance. Essentially, public participation in mining governance is characterized by a dynamic configuration of green pluralism among environmental, social development and human rights groups whose rights and privileges for public participation are sanctioned by ambivalent legal-institutional frameworks.

4.1 Dynamic Configuration of Green Pluralism

Inherent in pluralist societies, public participation in the mining sector is characterized by variegated modes of participation among fragments of initiatives and frameworks. Moreover, across regime changes, shifting political blocs and changing systems of resource allocation, public participation structures have gone through periodic reconfigurations that also define their shifting alliances, access and capacities in negotiating influence. Civil society configurations constitute a dynamic interaction among home-grown multi-sectoral groups which are often mediated by local and/or national NGOs and environmental advocacy groups to: (1) organize local solidarity actions among indigenous communities whose stakes are their livelihoods and lands; (2) secure the alliance of key sectors which have critical influence in electoral mobilization; (3) build strategic policy support with pro-environment and/or anti-mining local government leaderships and progressive national political parties, and/or capacity to establish their own political arm to penetrate the legislative arena. These local and national public participation structures are also fast integrated into a growing global civil society movement that responds to the transnational dimensions of the mining industry. Notably, local CSOs are often logistically capacitated through funding from their global connections of development networks.

A critical force contributing to the strength of public participation in Philippine mining governance is the involvement of the Catholic church. The Catholic church is known to exert significant influence in electoral mobilization, organizing voter support among a Filipino constituency which is about 80 % Christian (World Fact Book 2015). Their voice is a crucial influence in shaping policy decisions that bolsters the position of the anti-mining camp in civil society. Particularly that mining is a deeply embedded issue in elections, there's a political incentive behind why the government remains compelled to engage non-state actors, particularly the local governments that accommodate demands from alliances among small-scale mining elites, as well as the tribal associations with ancestral domain. This ultimately shifts dependencies and influences the access to, and boundaries of, public engagement.

In an industry with clear and tangible dichotomies of interests—on one camp among corporate and elite mining companies lobbying against economic costs, and on the other camp, indigenous communities, including small-scale miners and environmental groups lobbying against environmental and social costs—legally, it becomes pragmatic for civil society to consolidate forces to overcome a common adversary. The goal is also unified—to repeal the Mining Act, and thus the struggle and advocacy is along a single trajectory, despite divergences in the avenues of protest, as well as in the ideological framing of legal alternatives.

However, certain civil society alliances are captured and steered by political party factions. These predominate the realities of ongoing tensions among politically allied CSO networks in their mobilization and lobbying agenda. This was evident in the fragmented Congressional legislative lobbying for alternative mining laws that dilutes rather than fortify the voice of the public. Particularly in the case of repealing the Anti Mining Act of 1995, one camp headed by *Kalikasan*-PNE and Defend Patrimony lobbied the *Bayan Muna* party list representative in Congress to enact House Bill 4315 or the People's Mining Bill calling for the nationalization of the mining industry. On the other hand, House Bill 3763 and Senate Bill 3126 (re-filed as the Alternative Minerals Management Bill (HB 984/Senate Bill 43 were was spearheaded by *Akbayan* partylist supported by networks not allied with the national democratic movement. These include *Alyansa Tigil Mina*, *Aniban ng Manggagawang sa Agrikultura*, Philippine Misereor Partnership, Inc., Philippine Movement for Climate Justice, Sanlakas, Sentro, Task Force Detainees of the Philippines, SOS-*Yamang Bayan* Network, Philippine Alliance of Human Rights Advocates and *Kongreso ng Pagkakaisa ng mga Maralita ng Lungsod* (*Alyansa Tigil Mina* 2014).

Nonetheless, amid a unified agenda to repeal the Mining Act of 1995, rival political party-list groups such as *Bayan Muna* sponsoring the HB 4315 and *Akbayan* sponsoring HB 3763, along with the ruling Liberal Party sponsoring HB 206, agreed to set aside political differences to fight mining law (Meruenas 2013). Such agreement is seen as a huge advance among anti-mining and environmental movements. Lansang (2011) earlier observed that advocacy coalitions, specifically *Kalikasan*-PNE and Defend Patrimony Alliance and the *Alyansa Tigil Mina*, are crucial in influencing consolidation processes in the advocacy campaigns. This was a significant contribution made by these coalitions amid prevailing fragmentation among CSOs divided along political party factions and alliances.

The continued cooperation among these political blocs however remains fragile, particularly on matters of representation in the sponsorship of the consolidated alternative mining bill. In July 2013, *Akbayan* and its alliance with SOS *Yamang*

Bayan and *Alyansa Tigil Mina*, together with Senator Sergio "Serge" Osmena III, re-filed the Alternative Minerals Management Bill via House Bill 984 and Senate Bill 43 to repeal the Mining Act of 1995 toward a People's Mineral Resources Act (SOS *Yamang Bayan* 2013). This notably excludes the camp of the *Bayan Muna* and its alliances in the bill sponsorship. Deep political tensions notably resurface constraining a sustained consolidation among the different political factions that expectedly extends to their base organizations from the CSOs. As the Mining Act of 1995 marks its 20th year in 2015, these bills still couldn't be enacted into law.

4.2 Thriving Through Contradictory Frameworks and Policy Ambivalence

It is evident that the political economy constraints involved the national government's historical bias towards large-scale mining corporations amid a policy thrust to boost foreign investments and national economic growth. In a highly patrimonial political context, large-scale mining corporations could easily create financial incentives for political officials and bureaucrats to sponsor and support neo-liberal mining policies. A wide array of investigative reports and court cases is available documenting incidence of buy-outs, co-optations, including government-contracted paramilitary services to protect mining corporations.[1] The government nonetheless simultaneously pursues a policy narrative supportive of sustainable development, and sanctions public participation in mining governance as outlined in the previous section.

It can be argued that the government's successive policy strategies in negotiating conflicting demands in relation to neo-liberal-oriented large-scale mining and sustainable mining, created ambivalence in its legal provisions. This has opened new spaces for civil society to pursue tactical legal recourse indicating paths to resilience in challenging entrenched constraints in the mining sector. However, this also translates to a contradictory framework of governance that could not identify its priority nor segregate conflicting mandates of its agencies. This results to arbitrary policy compromises that are easily manipulated, placing civil society including local communities in perpetual square-off against powerful mining corporations.

[1]See Philippine Center for Investigative Journalism Portfolio of Investigative Reports; The Law Phil Project—Philippine Laws and Jurisprudence Databank; Compendium of Annual Country Reports on Human Rights Practices: Philippines by US Department of State; Human Rights Watch Annual World Report: Philippines; International Coalition for Human Rights in the Philippines Reports.

4.3 On the Constitutionality and Overlaps of the 1995 Mining Act

In 1995 upon the enactment of the Mining Act, the La Bugal—B'Laan Tribal Association questioned the constitutionality of the Mining Act and its Implementing Rules and Regulations before the Supreme Court. By January 2004, responding to civil society petitions, the Supreme Court declared parts of the Mining Act as unconstitutional, since full foreign ownership in mining through Financial and/or Technical Assistance Agreements (FTAAs) is contradictory to the 1987 Philippine Constitution's provision on national patrimony (Supreme Court, La Bugal—B'Laan Tribal Association v. Ramos 2004a). Despite the country's slow judicial system (Sereno et al. 2009; Macaraig 2015), this was seen as a significant milestone among civil society, environmental groups and indigenous communities in their long struggle. However, by December of the same year, the Supreme Court reversed its decision declaring the constitutionality of the Mining Act (Supreme Court, La Bugal—B'Laan Tribal Association v. Ramos 2004b). This Supreme Court resolution states that, "full control is not anathematic to day-to-day management, provided that the State retains the power to direct overall strategy." This gave the government the authority to issue FTAAs, which was supportive of Executive Order 270, a National Policy Agenda on Revitalizing Mining in the Philippines, issued by President Macapagal-Arroyo in January 2004—a policy designed to counter the previous Supreme Court decision.

The Mining Act is also widely criticised of its provisions that either overlap or contradict with other national and local laws such as:

- the Indigenous Peoples Rights Act of 1997;
- the National Integrated Protected Areas System Act of 1992;
- the Local Government Code of 1991;
- the Agriculture and Fisheries Modernization Act of 1997;
- the Climate Change Act of 2009; and
- the Environmental Impact Statement System.

This is noted in the draft midterm review of the Philippine Development Plan (2011–2016) (Senate Economic Planning Office 2013). For example, there is a policy inconsistency between the Mining Act and the authority given to LGUs in mining regulation that creates either resistance or co-optation among certain LGUs. Amid a context of several mining prohibitions by the LGUs, the national government reasserts its authority via Executive Order 79.

4.4 Environmental Impact Assessment

As previously discussed, the implementation of the 1995 Mining Act requires the Environmental Management Bureau to administer an EIA, which includes provisions to carry out public consultations over a proposed mining project. The result of these public consultations should subsequently be reflected in the EIS—which is needed to secure an ECC—before a mining company can obtain permission to operate. An important caveat, however, is that the public consultation component of the EIA is not required when a mining company is only securing an exploration permit (Begonia and Leonen 1996). Moreover, a great deal of controversy surrounds a number of mining projects with alleged informal negotiations and political buy-outs/bribery involving mining corporations (IBON 2006). There is also a lack of monitoring and transparency in the different stages of mining resulting to fraudulent EIA and EIS (Godio 2012).

Doyle et al. (2007) noted that a fact-finding team found that in practice, the participation rights including the right to information and participation in decision making required by the EIA is lacking. The authors further took note of the complaints of local communities and NGOs regarding the difficulty communities had in obtaining copies of EIAs and the lack of independent analysis or explanation of their contents and implications, which points to a distinct lack of enforcement of environmental standards. Accessing such mining information has proven to be quite difficult in the Philippines due to the lack of an enabling law. Gera (2015) pointed out that while the Supreme Court argued in a 1987 decision that the guaranteed provisions of the Constitution of the right to information are "self executing" and enforceable even without an implementing legislation, it contradicted itself by ruling that "neither the right to information nor the policy of full public disclosure is absolute, there being matters which, albeit of public concern or public interest, are recognized as privileged in nature" (Supreme Court, Akbayan v. Aquino 2008).

In recent Supreme Court hearings on the petitions by civil society against the Mining Act, Chief Justice Maria Lourdes Sereno, based on a list submitted by the DENR, reportedly found out that most of the 350 registered mining companies in the country do not have or do not indicate an "environmental work program" in their mining concessions with the government (Salamat 2013). Bravante and Holden (2009, p. 523) conclude that the EIA process in mining is a tokenism designed to make it appear that mining projects are being assessed for their environmental impact while actually already approved, allowing for a minimum amount of opportunity for members of the public to participate.

4.5 Dual Role of Environmental Agencies

The weakness of the country's EIA system also stems from the contradictions in the mandate of the government's top environmental agency—the DENR. Hatcher (2014) noted of the intrinsic contradictory nature of DENR's mandate, which is to act both as a promoter and as a regulator for the sector. The DENR is the ultimate department responsible for making decisions pertaining to mining leases. It has been reported however that the DENR regional offices and field personnel were actively and aggressively helping mining corporations in convincing the people to accept the mining project (Christian Aid and PIP links 2004, p. 12). Bello et al. (2004, p. 226) also argued that the MGB-DENR in particular has no effective power to sanction firms that violate regulations while operating along conflicting goals—that of promoting the mining industry and that of protecting the lands. This reflects the ambivalence of the country's policy framework that could not segregate conflicting mandates and thus creates a duality that leave CSOs fighting the very agencies designed to protect them.

4.6 *Free, Prior and Informed Consent of the IPRA Versus Executive Order 270*

The 1997 IPRA specifically mandates for the free, prior and informed consent (FPIC) of indigenous communities before any mining concessions or other projects are started. Local communities are expected to have a say in the mining projects but these indigenous peoples don't have time, sometimes don't have education, to read technical reports. There are also claims that in many instances, the local communities' consent to the mining activities in areas declared as ancestral lands of indigenous peoples by the IPRA is obtained through unfair consultation processes, misinformation, misrepresentation, bribery or promises of financial and material benefits, and intimidation or coercion through military or paramilitary means (Doyle et al. 2007; Longid 2012).

Whitmore (2006) listed different modes of abusing the FPIC: ignoring or misrepresenting joint meetings; creation of bogus community organizations; falsifying documents of community assent; asking communities to sign agreements in languages they don't understand; bribery or intimidation of community leaders. There are reports that industries use the attendance sheets for information meetings to prove the consent of the community for a project. FPIC as a mechanism of voice and representation could only subject community members to violence and intimidation amid powerful economic and political interests. Thus despite the IPRA, it has been reported that roughly 60 % of all approved mining applications in the country covering up to more than 1 million ha, are all in the ancestral lands of indigenous peoples (Salamat 2013). EO 270 easily sanctions these approvals based on the priority to revitalize the mineral industry as a national policy agenda.

4.7 *Justice System for Indigenous Peoples*

Section 15 of the IPRA stipulates the State's recognition of indigenous people's "right to use their own commonly accepted justice systems." Moreover, the Supreme Court has created the green courts that hear environmental cases. The Office of the Ombudsman also established the Environmental Ombudsman to hear environmental violations by public officials. The CSOs have made inroads in keeping government officials in check either by media exposure or by filing legal cases. CEC-Philippines (2012) however argued that these do not substantially change the regular judicial process and its inherent limitations. The green courts for instance were not absolved of their regular cases to handle in addition to those involving environmental laws. The Asian Legal Resource Center (2012) also noted that these institutions are either weak or dysfunctional such that even non-indigenous community members are discouraged to file court actions, let alone indigenous people who are often subjected to intimidation.

This can be gleaned from the case of a group of fishermen who sued Placer Dome Inc. and Barrick Gold Corporation, the company which absorbed the former, concerning the Marcopper mining disaster that took place in 1996 in Marinduque. The petitioners sued the company in July 2004 for the damages caused by 16 years of dumping approximately 200 million tons of mine tailings into the Calancan Bay, resulting to damages in people's health (with 22 children found to have elevated lead and mercury content in their blood) and livelihood. They also contended that Placer Dome was liable for expelling some 2 million m^3 of toxic industrial waste in the area and failing to rehabilitate the waters of Romblon, Batangas, Marinduque and Quezon (Reformina 2011). In June 2011, the Supreme Court granted them a Writ of *Kalikasan* (Supreme Court, Hernandez et al. v. Placer Dome, Inc. 2011). However, in September 2013, the regional trial court in Marinduque threw out the case. The dismissal order, penned by Judge Emmanuel Recalde of the Regional Trial Court Branch 38 in Boac, came as negotiations for a settlement continued between the Marinduque provincial government and Barrick Gold Corporation (Cinco 2013).

There are still many environmental cases dismissed due to failure to exhaust administrative remedies. Other barriers to access justice for ordinary citizens include: high costs, lengthy delays in the court process, and lack of information about the judicial system and how it works (Co et al. 2010). According to Malaluan (2009, p. 4), the judicial remedy of Mandamus is also "inaccessible to the public" based on a Social Weather Station survey whereby only 12.7 % would file a case in court if refused access to a public document. The Task Force for Indigenous People's Rights reported extrajudicial killings of indigenous peoples who lobbied against mining companies encroaching indigenous land. Civil society groups working for indigenous communities were subjected to harassment such as lawsuits and red-baiting. Rood (2005, p. 18) noted that, "It is quite difficult for aboveground organizations to articulate grievances in ways that might resemble the rhetoric of underground organizations. Security organizations of the government, both police and armed forces, may well hamper their activities." Since 2001, there have been 36 recorded cases of anti-mining activists killings, seven of them in 2011 (Kalikasan-PNE 2012).

4.8 *Political Intimidation Through SLAPPs Vs. Rules of Procedure for Environmental Cases*

Rivera (2011, p. 197) argued that, "…civil society organizations have done an invaluable job of monitoring the exercise of power and initiating reform advocacies but these same groups continue to be repressed and harassed by powerful political clans and oligarchic blocs." Public interest lawyers and NGOs in general also actively participate in environmental litigation (Lin-Heng 2002). However, the incidence of SLAPPs in the Philippines poses considerable constraint to sustained resistance against mining corporations. Viewed as harassment suits, many SLAPP cases against environmental defenders were filed by mining firms or individuals associated with them, as a response to community opposition to mining projects in their communities. In July 2007, two board members of the CEC-Philippines found themselves "SLAPPed" with a libel suit worth P10 million (around US$200,000) for publishing a primer discussing the ecological, economic, and social impacts of the project on the island and its residents. The libel suit was filed by the Philippine partners of Lafayette Mining Ltd., an Australian-owned mining firm in Albay (CEC-Philippines 2012, p. 43). In 2010, Bautista, National Coordinator of *Kalikasan*-PNE noted that the group has recorded more than 110 environmental advocates and community leaders facing SLAPP cases in court mostly opposing big mining and logging projects (Gera 2015). Political intimidation, financial burdens, and a court case can distract or pose challenges to campaigners and NGOs. By shifting the arena of dispute away from the community and into the courts, SLAPPs are seen to significantly impact on advocacy campaigns (Tapang 2009).

There has been concrete enforcement of the Rules of Procedure for Environmental Cases issued in 2010 by the Supreme Court. One case is the September 2011 dismissal of a SLAPP suit against nine indigenous Ifugaos of Nueva Vizcaya sued by DENR for allegedly occupying forest lands illegally. The suit was filed after their vocal opposition to a large-scale mining project awarded to Oceana Gold. However, despite the existence of Supreme Court rules, there has yet to be a legislation that provides a comprehensive anti-SLAPP mechanism in the Philippines. House Bill 3593, or the Anti-SLAPP Act of 2010 was filed in Congress by representatives from *Bayan Muna*, *Anakpawis*, Gabriela Women's Party, *Kabataan* and Alliance of Concerned Teachers. It aims to prohibit the filing of harassment suits, and provide measures for its dismissal. It however remains shelved in the same way that alternative mining bills remain deliberated and subjected to political lobbying from influential pro-mining camps. Arguably, the establishment of the *Rules* was a breakthrough in mining governance. Their effective implementation however becomes defeated without an enabling law and substantial awareness and capacity-building among stakeholders, particularly the judiciary.

4.9 Compromises Within Executive Order 79

Seen as another advance by civil society participation was the decision under President Aquino's administration to issue EO 79 in July 2012 seeking Mining Act amendments on revenue sharing. As previously noted, this is designed to impose a moratorium on new mining contracts as the government works to legislate revenue sharing scheme. The Chamber of Mines of the Philippines (2012, para. 12) however sees the policy as a counterproductive provision that "blatantly impinge on the constitutional and legal rights of the mining industry, and would affect future investments in mining." It criticized Section 9 of the Implementing Rules and Regulations (IRR), citing it as patently illegal as it effectively shortens mining contracts to 25 years in violation of Section 32 of the Mining Act. It alleges that such provision was not contained in EO 79 or in the draft IRR during the Minerals Industry Consultative Council consultation with the industry. Mining corporations threatened to pursue legal action against the government on the reduced contract term to 25 years from 50. Moreover, in a motion filed with the Supreme Court, the Chamber of Mines asked the high tribunal's permission to intervene in two petitions again challenging the constitutionality of the Mining Act of 1995.

Following a threat from the mining industry that it would challenge the rules in court, the Mining Industry Coordinating Council—the inter-agency body tasked to implement EO 79—revised the IRR. Section 9 was revised to state that "mining contracts/agreements that *may be renewed shall be subject to existing laws*, rules and regulations at the time of renewal." Under the original version, expiring 25-year tenements may be renewed but subject to new terms and conditions set by laws existing at the time of renewal. The original version also held that expiring mining contracts may be declared as mineral reservations (Despuez 2012).

Sections 3 and 7 were also revised with the final revision of Section 3 defining expired mining tenements as contracts whose 25- or 50-year terms have lapsed. It considers a mining contract as expired *if, after the initial 25-year term lapses, the parties concerned fail to agree on the terms of renewal*. Section 7, which declares that no new mineral agreement will be granted until a new law rationalizing revenue-sharing schemes has been approved, was revised to likewise include the issuance of permits for expansion of existing contract areas "*unless there is an imminent and/or threatened economic disruption*, such as a shortage of critical commodities and raw materials, that could adversely affect priority government projects and/or economic activities" (Despuez 2012).

Initially seen as a progressive policy that gave momentum to civil society campaign for sustainable development, the revisions in the IRR represent the government's continued accommodation, or lack of autonomy from the influence of the rich mining corporations. This ambivalence touted by the government as a "middle ground" (Mines and Geosciences Bureau-Department of Environment and Natural Resources 2014) poses critical implications not just in terms of placing public participation structures in a legal tightrope vis-à-vis certain government officials and the mining corporations, but also ultimately in putting sustainable development policy subject to perpetual legal debates on its merits vis-à-vis the national economic revitalization.

4.10 Mandatory Environmental Insurance Coverage

In 2005, DENR issued Administrative Order 2005–06 requiring Mandatory Environmental Insurance Coverage (MEIC) for all applications of ECCs. The MEIC covers payment of damages to health and property, environmental rehabilitation, remediation, clean-up costs, or facility to recompense for environmental impairment or damage by a project.

Affected industries however resisted its implementation on the ground of redundancy due to the Environmental Guaranty Fund required to set up compliance with the ECC terms and conditions. Concerns were also raised with regards to the premiums payable since insurance companies normally offer all-risk insurance that covers other risks such as natural calamities and terrorism, not just environment and pollution damage. Moreover, the Foreign Chambers of Commerce of the Philippines reported that out of the 41 insurance companies in the country as of 2006, only one attested to cover environmental insurance as underwriter, with the other 40 only acting as agents. Due to these issues and political pressure, the DENR indefinitely suspended the requirement of MEIC in February 2006 (Lerma and Batan 2015). This policy has not been reinstated since then.

5 Conclusion

Public participation in mining governance represents the very limits of the Philippine's democratic governance framework—one that is not only characterized by poor enforcement of legal standards, but of contradictory legal-institutional regimes that open ample rooms for political manoeuvre. The policy ambivalence represents an attempt by the government to reconcile incompatible interests in the mining sector. On one hand it responds to pressures to deal with environmental and social risks of mining investments, and constructs spaces for dissent articulation and policy formulation within the framework of inclusiveness. On the other hand, it continues to formulate mining-friendly policies based on neo-liberal foreign-led large-scale mining. It is a strategy of dualism used by the state as a convenient mechanism to maintain legitimacy while not upsetting specific interests within the mining industry. This, however, exacerbates prevailing tensions amid power imbalances in the mining sector, leaving civil society networks and local communities to contend with multinational mining corporations. This emphasizes the degree to which the sustainable development provisions in mining governance and its framework of public participation are vulnerable to the whims of patrimonial structures, often between elements of the state and large-scale mining corporations.

Nonetheless, one thing is apparent about public participation structures in Philippine mining governance—these are enduring. Anti-mining civil society coalitions in the country have gained significant inroads in compelling large-scale mining corporations as well as the government to address their legacy of socio-environmental hazards, particularly through the courts (Hatcher 2014). Constraints to sustainable development however also endure amid these arbitrary policies that could not regulate unsustainable extractive systems that pay premium to economic viability of mineral deposits, without serious attention to environmental and social costs.

Thus the endurance of civil society could not be translated as resilience in terms of institutionalization of the rule of law that consistently prevails over power structures, or in terms of autonomy or ability of civil society to effectively resist patrimonial political powers exercising clientelist influence. Nor it can be regarded as resilience in terms of regularity and stability of clear procedures, parameters and standard non-ad hoc criteria of engagement with the government. Moreover, public participation in mining governance cannot be regarded as resilient in terms of the ability of certain CSOs to transcend beyond political differences and reconcile contradictory frameworks to engage in deliberative processes, towards strategic consolidation and synergy in challenging legal and political economy constraints.

The urgency is clear for sustainable mining in the Philippines: civil society has to engage the government in crafting a more precise formulation of environmental and social obligations of mining corporations that leaves little or no room for manoeuvre. The urgency for regulating the mining industry by effectively enforcing the ECCs and the MEICs cannot be overemphasized, which can be facilitated through a clear segregation of conflicting mandates among environmental agencies specifically the DENR and its sub-agency the MGB, so they can best play a regulatory role, rather than as advocates of the mining industry vis-à-vis the Chamber of Mines of the Philippines. This imperative for clear mandates is parallel with what UNECA (2004, p. 37) has noted that, "the relationship between newly created institutions with a primarily environmental focus, on one hand, and older existing institutions, which have traditionally regulated mining, can be reasonably arranged in different ways…".

Moreover, public participation could move to a trajectory that builds a strategic partnership with the government in institutionalizing the EIA system toward an integrated assessment of environmental and social impact analysis. A particular social consideration would be the allocation of benefits from mining, where a reasonable share goes to the communities disproportionately affected. Such is critical in the process of legislating a new mining law to replace the 1995 Mining Act, along with a legislation to facilitate access to information.

Ultimately, there is the imperative for a reorientation of the mining corporate culture, if the country is to reconcile the mining industry interest and sustainable development goals. Legal-institutional regimes should aim at giving effect to the new orientation of mining policy that does not undermine its policy narrative of sustainable development. The prevailing dichotomy between corporate mining interests and the sustainable development movement creates a unique imperative for effective consolidation of actions, specifically in the push for alternative legislative proposals. Deliberative processes toward strategic consolidation however remain a huge challenge among the ranks of CSOs in the country, which are well-penetrated by politically-allied factions and pressure groups. New deliberative strategies are thus required of civil society if they are to effectively challenge prevailing unsustainable mining practices and steer the mining industry toward a sustainable development-anchored model of mining operation.

References

Alternative Mining Bill, House Bill No. 206 (July 2010)

Alyansa Tigil Mina (2014) Mendiola march to press Malacanang against mining. Sentro. Retrieved 28 July 2015. http://www.sentro.org/?p=281

Artajo MI (2012) Enhancing decent work outcomes in small-scale gold mining. Discussion paper series, Institute for Labour Studies. Retrieved 25 July 2015. http://ilsdole.gov.ph/wp-content/uploads/2012/12/artajo-with-edits.pdf

Asian Legal Resource Center (2012) Philippines: obstacles concerning access to justice and protection for indigenous people. Statement submitted to Asian Human Rights Commission. Retrieved 25 July 2015. http://www.alrc.net/doc/mainfile.php/hrc21/729/

Asuncion M (2005) Righting the wrongs. Legal Rights and Natural Resources Center, Kasama sa Kalikasan, Friends of the Earth Philippines. Retrieved 25 July 2015. http://www.miningwatch.ca/sites/www.miningwatch.ca/files/Righting_the_Wrongs_0.pdf

Batalla E (2000) De-institutionalizing corruption in the Philippines: identifying strategic requirements for reinventing institutions. Paper presented at the conference on institutionalizing strategies to combat corruption: lessons from East Asia. Konrad Adenauer Stiftung and De La Salle University, Makati. Retrieved 25 July 2015. http://unpan1.un.org/intradoc/groups/public/documents/APCITY/UNPAN013117.pdf

Begonia FG, Leonen MM (1996) Mining: legal notes and materials. Legal Rights and Natural Resources Center Inc, Quezon City

Bello W, Docena H, de Guzman M, Malig M (eds) (2004) The anti-development state: the political economy of permanent crisis in the Philippines. Zed Books, London

Bravante MA, Holden WN (2009) Going through the motions: the environmental impact assessment of nonferrous metals mining projects in the Philippines. Pac Rev 22(4):523–547

Cabacungan GC (2012) 78 no-mining zones in Aquino EO. Inquirer news. Retrieved 28 July 2015. http://newsinfo.inquirer.net/226212/78-no-mining-zones-in-aquino-eo#ixzz3Xa3ZySaX

CEC-Philippines (2012) Confronting the ecological crisis: a situationer on Philippine environmental issues and struggles. Center for Environmental Concerns-Philippines, Quezon City. Retrieved 25 July 2015. http://nafconusa.org/wp/wp-content/uploads/2014/11/State-of-the-philippine-environment-web041312.pdf

Chamber of Mines of the Philippines (2012). Shaping the future of Philippine mining: press release. Chamber of Mines of the Philippines. Retrieved 28 July 2015. http://www.chamberofmines.com.ph/09232012pr.html

Chamber of Mines of the Philippines (2013) Mining Philippines 2013 a resounding success. Mining Philippines News, 3. Retrieved 28 July 2015. http://www.chamberofmines.com.ph/download/mpnews03.pdf

Christian Aid and PIP Links (2004) Breaking promises, making profits: mining in the Philippines. Retrieved 25 July 2015. http://www.piplinks.org/system/files/Philippines+Mining+report+-+corrected.pdf

Cinco M (2013) Marinduque folk lose case vs. mine firm. Inquirer news. Retrieved 28 July 2015. http://newsinfo.inquirer.net/479345/marinduque-folk-lose-case-vs-mine-firm

Co E, Malaluan N, Neame A, Manuel M, Musngi MRV (2010) Philippine democracy assessment: rule of law and access to justice. Action for economic reforms, Philippines. Retrieved 25 July 2015. http://www.idea.int/publications/philippine_democracy_assessment/

Coumans C (1999) Boac tragedy aftermath: Canadian transnational dumps waste, responsibility in Marinduque. Philippine Center for Investigative Journalism. Retrieved 25 July 2015. http://pcij.org/stories/1999/marcopper.html

Despuez O (2012) DENR issues revised implementing rules of EO 79. InterAksyon. Retrieved 28 July 2015. http://www.interaksyon.com/business/45108/denr-issues-revised-implementing-rules-of-eo−79

Dinglasan R (2012) Philex spill 'biggest mining disaster' in PHL, surpassing Marcopper—DENR. GMA News. Retrieved 28 July 2015. http://www.gmanetwork.com/news/story/281988/news/nation/philex-spill-biggest-mining-disaster-in-phl-surpassing-marcopper-denr

Doyle C, Wicks C, Nally F (2007) Mining in the Philippines: concerns and conflict. Society of St. Columban, West Midlands

Espiritu B (2015) The destructive impacts of corporate mining in the Philippines: the Tampacan copper-gold mining project in Mindanao. Global research. Retrieved 28 July 2015. http://www.globalresearch.ca/the-destructive-impacts-of-corporate-mining-in-the-philippines-the-tampacan-copper-gold-mining-project-in-mindanao/5436594?print=1

Establishing an environmental impact statement system including other environmental management related measures and for other purposes, Presidential Decree 1586 (Malacanang 11 June 1978). Retrieved 28 July 2015. http://www.emb.gov.ph/ECA%20Center/PD1586.pdf

Fonbuena C (2013) On shaky ground. Rappler. Retrieved 28 July 2015. http://www.rappler.com/newsbreak/2860-on-shaky-ground

Gera W (2015) Public participation in environmental governance in the Philippines: the challenge of consolidation in engaging the state. Land Use Policy. doi:10.1016/j.landusepol.2014.02.021

Gerring J, Thacker S, Moreno C (2005) A centripetal theory of democratic governance: a global inquiry. Am Polit Sci Rev 99:567–581

Gleason J (2003) Strategic lawsuits against public participation. In Stec S (ed) Handbook on access to justice under the Aarhus convention (pp 59–60). The Regional Environmental Center for Central and Eastern Europe, Szentendre, Hungary. Retrieved 25 July 2015. http://documents.rec.org/publications/accesstojustice_March2003_EN.pdf

Godio MJ (2012) Transforming extractive industries in the Philippines: locating spaces for people's participation in mining policies. UN Research Institute for Social Development. Retrieved 25 July 2015. http://www.unrisd.org/80256B3C005BE6B5/%28httpNews%29/E60DD0711B8264FDC1257A0F004BB3D1?OpenDocument

Griggs D, Stafford-Smith M, Gaffney O, Rockström J, Ohman MC, Shyamsundar P, Steffen W, Glaser G, Kanie N, Noble I (2013) Sustainable development goals for people and planet. Nature 495:305–307

Guidelines for the institution of a system of mandatory environmental insurance coverage, administrative order 2005–06, Department of Environment and Natural Resources (Quezon City 2005). Retrieved 25 July 2015. http://www.denr.gov.ph/policy/2005/dao/dao2005−06.pdf

Hatcher P (2014) Regimes of risk: the World Bank and the transformation of mining in Asia. Palgrave Macmillan, New York

Holden WN (2005) Civil society opposition to nonferrous metals mining in the Philippines. Voluntas Int J Voluntary Non-profit Organ 16(3): 223–249

Holden WN, Jacobson RD (2012) Mining and natural hazard vulnerability in the Philippines: digging to development or digging to disaster?. Anthem Press, London

Holmes R (2011) The curious cases of Philippine civil society and decentralization. In: Miranda F, Ronas M, Rivera T, Holmes R (eds) Chasing the wind: assessing Philippine democracy. Commission on Human Rights, Quezon City, pp 139–181

IBON (2006) The state of the Philippine environment, 3rd edn. IBON Databank and Research Center, Quezon City

Indigenous Peoples Rights Act, Republic Act 8371, Congress of the Republic of the Philippines (Manila 1997). Retrieved 25 July 2015. http://www.gov.ph/1997/10/29/republic-act-no−8371/

Institutionalizing and implementing reforms in the Philippine mining sector providing policies and guidelines to ensure environmental protection and responsible mining in the utilization of mineral resources, Executive Order No. 79 (Malacanang Manila 2012). Retrieved 25 July 2015. http://www.gov.ph/2012/07/06/executive-order-no−79-s−2012/

Ito L (2011) High time for a pro-environment, pro-people mining policy. Center for Environmental Concerns-Philippines, January–March. Retrieved 28 July 2015. http://www.cecphils.org/content/challenging-mining-act-1995

Kalikasan-PNE (2010) Kalikasan—people's network for the environment, Environmental activists laud Supreme Court's issuance of Writ of Kalikasan as an early earth day gift. Mines and Communities. Retrieved 28 July 2015. http://www.minesandcommunities.org/article.php?a=10060

Kalikasan-PNE (2012) Kalikasan—People's Network for the Environment, The Philippines deserves more from mining (LTE). Bulatlat. Retrieved 28 July 2015. http://bulatlat.com/main/2012/03/23/the-philippines-deserves-more-from-mining-pr/

Kanie N, Abe N, Iguchi M, Yang J, Kabiri N, Kitamura Y, Mangagi S, Miyazawa I, Olsen S, Tasaki T, Yamamoto T, Yoshida T, Hayakawa Y (2014) Integration and diffusion in sustainable development goals: learning from the past, looking into the future. Sustainability 6:1761–1775. doi:10.3390/su6041761

Lansang L (2011) NGOs, coalition building and the campaign for a minerals management policy in the Philippines. Philippine Polit Sci J 32(55):127–166. doi:10.1080/01154451.2011.9723534

Lerma B, Batan T (2015) Environment and climate change law 2015—Philippines. International comparative legal guides. Retrieved 25 July 2015. http://www.iclg.co.uk/practice-areas/environment-and-climate-change-law/environment-and-climate-change-law−2015/philippines

Leuven KU (2009) J. Rawls's idea of 'overlapping consensus' and the complexity of 'comprehensive doctrines.' Ethical Perspect 16(1): 21–60

Lin-Heng L (2002) Public participation in the environment: a south-east asian perspective. In: Zillman DM, Lucas A, Pring G (eds) Human rights in natural resource development: public participation in the sustainable development of mining and energy resources. Oxford University Press, UK, pp 651–678

Llorito DL (2006) Philippines' mining down in the dumps. Asia Times. Retrieved 28 July 2015. http://www.atimes.com/atimes/Southeast_Asia/HE26Ae02.html

Local Government Code, Republic Act No. 7160, Congress of the Republic of the Philippines (Manila 1991). Retrieved 25 July 2015. http://ppp.gov.ph/wp-content/uploads/2015/01/The-Local-Government-Code-of-the-Philippines.pdf

Longid B (2012) States hear human rights challenges in the Cordillera due to Philippine policies. Statement delivered at the Universal periodic review pre- session discussion of the Philippines. Unrepresented Nations and People's Organizations, Geneva. Retrieved 25 July 2015. http://www.unpo.org/article/14135

Macaraig A (2015) Sereno: justice must be real-time, high-tech. Rappler. Retrieved 28 July 2015. http://www.rappler.com/nation/99497-sereno-justice-technology-reforms

Macdonald I, Southall K (2005) Mining ombudsman case report: Marinduque island. Oxfam Australia, Australia

Malaluan N (2009) Right to know, right now: a primer on access to information. The Philippine Center for Investigative Journalism. Retrieved 25 July 2015. http://pcij.org/resources/nepo-malaluan-right-to-know-right-now.pdf

Meruenas M (2013) Akbayan, Bayan Muna set aside differences to fight mining law, GMA News Online. Retrieved from http://www.gmanetwork.com/news/story/304067/news/nation/akbayan-bayan-muna-set-aside-differences-to-fight-mining-law

Mines and Geosciences Bureau-Department of Environment and Natural Resources (MGB-DENR) (2014) Creating the policy regime for the mining industry, Department of Environment and Natural Resources. Retrieved 25 July 2015. http://mgb.gov.ph/art.aspx?artid=666

Mining Act, Republic Act No. 7942, Congress of the Republic of the Philippines (Manila 1995). Retrieved 25 July 2015. http://www.mgb.gov.ph/Files/Policies/RA%207942.pdf

National Policy Agenda on Revitalizing Mining in the Philippines, Executive Order No. 270 (Malacanang Manila 2004). Retrieved 25 July 2015. http://www.gov.ph/2004/01/16/executive-order-no−270-s−2004/

North D (1990) Institutions, institutional change and economic performance. Cambridge University Press, Cambridge

People's Mining Act, House Bill No. 4315 (March 2011). Retrieved 25 July 2015. http://www.congress.gov.ph/download/basic_15/HB04315.pdf

Philippine Mineral Resources Act, House Bill No. 3763 (December 2010)/ Senate Bill 3126 (February 2012)/Re-filed as House Bill 984 and Senate Bill 43 (July 2013). Retrieved 25 July 2015. http://alyansatigilmina.net/download/files/HB%203763%20MMB.pdf; http://www.senate.gov.ph/lis/bill_res.aspx?congress=15&q=SBN−3126; http://www.ecoteneo.org/?p=1585

Quitasol K (2013) Critics of large-scale mining bat for 'pro-people' mining. Bulatlat. Retrieved 28 July 2015. http://bulatlat.com/main/2013/04/24/critics-of-large-scale-mining-bat-for-pro-people-mining/#sthash.kZS7g6WX.dpuf

Rappler (2013, December). Timeline: Philippine mining laws and policies. Rappler. Retrieved 25 July 2015. http://www.rappler.com/business/special-report/whymining/infographics/15910-timeline-philippine-mining-laws-and-policies

Rawls J (1987) The idea of an overlapping consensus. Oxford J Leg Stud 7(1):1–25

Recidoro R (2014) EO 79 and beyond: the state and prospects of the Philippine mining industry. Chamber of Mines of the Philippines. Retrieved 25 July 2015. http://www.slideshare.net/ronnierecidoro/mining−101-the-basics-of-mining-and-the-philippine-mining-industry

Reformina I (2011) SC issues writ v. Placer Dome for Marcopper disaster. ABS-CBN News.com. Retrieved 28 July 2015. http://www.abs-cbnnews.com/nation/regions/03/08/11/sc-issues-writ-vs-placer-dome-marcopper-disaster

Rivera T (2011) Rethinking democratization in the Philippines. In: Miranda F, Ronas M, Rivera T, Holmes R (eds) Chasing the wind: assessing Philippine democracy. Commission on Human Rights, Quezon City, pp 182–198

Rood S (1998) Decentralization, democracy, and development. In: Timberman DG (ed) The Philippines: new directions in domestic policy and foreign relations. Asia Society, Makati, pp 1–25

Rood S (2005) Forging sustainable peace in Mindanao: the role of civil society. East-West Center, Washington

Rules of Procedure for Environmental Cases, A. M. No. 09-6-8-SC Supreme Court of the Republic of the Philippines (Manila 29 April 2010). Retrieved 25 July 2015. http://philja.judiciary.gov.ph/assets/files/pdf/learning_materials/A.m.No.09−6−8-SC_Rules_of_Procedure_for_Envi_Cases.pdf

Salamat M (2013) Philippine mining situation at a glance. Bulatlat. Retrieved 28 July 2015. http://bulatlat.com/main/2013/09/12/philippine-mining-situation-at-a-glance/

Senate Economic Planning Office (2013) Realizing the Philippines' mining potential. SEPO Policy Brief. Retrieved 28 July 2015. http://www.senate.gov.ph/publications/PB%202013-12%20-%20Mining_Policy%20Brief_final_revised_010614.pdf

Sereno ML, De Dios E, Capuno J (2009) Justice and the cost of doing business in the Philippines. Philippine Rev Econ 46(1):87–102

SOS Yamang Bayan (2013) Congress champions re-file alternative minerals management bill. Alyansa Tigil Mina. Retrieved 28 July 2015. http://alyansatigilmina.net/2013/07/03/congress-champions-re-file-alternative-minerals-management-bill/

Supreme Court La Bugal—B'Laan Tribal Association v. Ramos (2004a) G.R. No. 127882, J. Carpio Morales, En Banc Decision (Manila 27 January). Retrieved 28 July 2015. http://sc.judiciary.gov.ph/jurisprudence/2004/jan2004/127882.htm

Supreme Court La Bugal—B'Laan Tribal Association v. Ramos (2004b) G.R. No. 127882, J. Panganiban, En Banc Resolution (Manila 1 December). Retrieved 28 July 2015. http://sc.judiciary.gov.ph/jurisprudence/2004/dec2004/127882.htm

Supreme Court Akbayan v. Aquino (2008) G.R. No. 170516 (Manila 16 July). Retrieved 28 July 2015. http://sc.judiciary.gov.ph/jurisprudence/2008/july2008/170516.htm
Supreme Court Hernandez et al. v. Placer Dome Inc (2011) G.R. No. 195482: (Manila 21 June). Retrieved 28 July 2015. http://www.chanrobles.com/scresolutions/2011juneresolutions.php?id=189
Tapang L (2009) SLAPP us not! Engaging harassment suits against Philippine environmental defenders. Center for Environmental Concerns- Philippines, Quezon City. Retrieved 25 July 2015. http://www.cecphils.org/system/files/CEC%20SLAPP%20primer%20without%20cover.pdf
United Nations Conference on Environment and Development (UNCED) (1992) Agenda 21: Programme of Action for Sustainable Development, U.N. GAOR, 46th Sess., Agenda Item 21, UN Doc A/Conf.151/26, Rio Declaration, Forest Principles. United Nations, New York. Retrieved 28 July 2015. https://sustainabledevelopment.un.org/content/documents/Agenda21.pdf
United Nations Development Programme (UNDP) (2014) Discussion paper on 'Governance for sustainable development': Integrating governance in the post-2015 development framework. Retrieved 25 July 2015. http://www.undp.org/content/dam/undp/library/Democratic%20Governance/Discussion-Paper–Governance-for-Sustainable-Development.pdf
UNECA (2004) United Nations Economic Commission for Africa: Improving public participation in the sustainable development of mineral resources in Africa. ECA/SDD/04/02. Retrieved 25 July 2015. http://www.africaminingvision.org/amv_resources/AMV/public_participation.pdf
United Nations Economic Commission for Europe (UNECE) (1998) Convention on access to information, public participation in decision-making and access to justice in environmental matters. Fourth Ministerial Conference in the 'Environment for Europe', 25 June, Aarhus, Denmark. Retrieved 28 July 2015. http://www.unece.org/env/pp/treatytext.html
United Nations Environment Programme (UNEP) (2010) Guidelines for the development of national legislation on access to information, public participation and access to justice in environmental matters. 11th Special Session of UNEP Governing Council/Global Ministerial Environmental Forum, 26 February, Bali, Indonesia. Retrieved 28 July 2015. http://www.unep.org/environmentalgovernance/Portals/8/GuidelinesAccesstoJustice2010.pdf
Whitmore A (2006) The emperors new clothes: sustainable mining? Viewpoint J Clean Prod 14:309–314
World Bank (2007) The Philippines environment monitor 2006. The World Bank, East Asia and Pacific Region, Philippines. Retrieved 25 July 2015. http://siteresources.worldbank.org/INTPHILIPPINES/Resources/FullPEM06bw.pdf
World Commission on Environment and Development (1987) Our common future. Oxford University Press, Oxford
World Fact Book (2015) Religions. Retrieved 25 July 2015. https://www.cia.gov/library/publications/the-world-factbook/fields/2122.html
Wurfel D (2006) Mining and environment in the Philippines: the limits of civil society in a weak state. Philippine Stud 54(1):3–40

Public Participation and Constitutional Impediments to Sustainable Development in Nigeria

Taiwo Odumosu

Abstract A sustainable society is an imperative of good governance and constitutionality in the modern times. It is defined as one that can progress without catastrophic setbacks in the foreseeable future. This view is strengthened by pre-independence Africa when crude oil exploration has not destroyed the flora and fauna. The advent of military dictatorship and reliance on crude oil brought unwholesome development. Military rule is antithetical to constitutionalism. The constitution of nations should be a holistic document that gives palliative to citizens in time of need, most significantly when their fundamental inalienable rights are infringed either by the State, persons in high places or an individual. The citizens must have access to justice. This paper assesses the Constitution of Nigeria 1999 in terms of its contribution to sustainable society and development. Though there is no perfect constitution, yet its imperfections had been the source of a Constitution Review Conference held in 2014. Can the amendments proposed there justify sustainable society? It posits that the provisions of the constitution must be made justiciable for sustainable society and development to be achieved; else the constitution will be burdened for non-functionality and lack of societal sustainability. We must change from a worldview that places human beings at centre of the universe, and see sustenance of the integrity of the whole earth system as overriding concern. Thus, only an acceptable and functional constitution can bring about good governance which Nigerians presently seek.

Keywords Access to justice · Functional leadership · Political principle · Social justice · Popular participation · Political commitment

T. Odumosu (✉)
Faculty of Law, Freedom of Information Unit, Law Clinic, Department of Public Law, Olabisi Onabanjo University, Ago-Iwoye Ogun State, Nigeria
e-mail: taiodu@gmail.com

V. Mauerhofer (ed.), *Legal Aspects of Sustainable Development*,
DOI 10.1007/978-3-319-26021-1_12

1 Introduction

Democracy is charting a new constitutionalism in the world today. The up-rising in places like Libya, Syria, Egypt and other countries is indicative of the peoples' anticipation of a functional leadership which can engender good governance (The Guardian 2011a, b). Beyond this however, is the need for a constitutional framework that captures the yarning of the people as well as encapsulating their aspirations. In recent years, a good constitution has been seen as a *sine qua non* to good government (Nwabueze 2006). It is already historically believed that when there is a good constitution that gives the people free hand and access to justice, there will be development that will capture the essence of the social contract (Rousseau 1762). This in turn will enhance the journey to the destination of sustainable development.

2 Attempts of Definitions

Sustainable development is defined in Our Common Future (Brundtland Report 1987) as:

> … development that meets the needs of the present without compromising the ability of future generations to meet their own needs. It contains two key concepts: (a) the concept of needs, in particular the essential needs of the world's poor, which need overriding priority; and (b) the idea of limitations imposed by the state of technology and social organization on the environment's ability to meet present and future needs.

Definitions of sustainable development see the world as a system that connects space and a system that connects time (IISD 2013). *Development*, on the other hand, is defined as a collective process of change toward improvements in quality of life for human beings and their communities. *Sustainability* refers to need for development to be integrated, socially, economically and environmentally, oriented to the long-term, and hence, able to last (Cordonier Segger and Khalfan 2004). Earth's resources are over-used beyond its carrying capacity, leading to high waste yields and emissions into the atmosphere and water through man's careless exploitations and inefficient use of natural resources. Thus, the only way to improvement (on) this is sustainable development (Nowosielski et al. 2007).

3 History of Sustainable Development

The History of Sustainable Development dates back to 1972 when the United Nations Conference on the Human Environment in Stockholm brought the industrialized and developing nations together to champion the 'rights' of the human family to a healthy and productive environment. Several meetings thereafter addressed the rights of people to adequate food, sound housing, safe water, and access to means of family planning. The recognition to revitalize humanity's connection with Nature, led to the creation of global institutions within the UN system [(United Nations Conference on Sustainable Development (UNCSD 2012)]. By 1980, the International Union for the Conservation of Natural Resources (IUCN) published the World Conservation Strategy (WCS), which was a precursor to the concept of sustainable development (UNCSD 2012). It was observed that conservation of nature cannot be achieved without alleviating poverty and misery of hundreds of millions of people. Unless the fertility and productivity of the planet are safeguarded, the human future is at risk (UNCSD 2012). In 1982, the WCS initiative led to the approval of the World Charter for Nature (UNCSD 2012). The Charter stated that "mankind is a part of nature and life depends on the uninterrupted functioning of natural systems."

By 1983, the World Commission on Environment and Development (WCED) has been established (UNCSD 2012). In 1984, it was constituted as an independent body by the United Nations General Assembly (UNCSD 2012). WCED's objective was to formulate 'A global agenda for change' (UNCSD 2012). In 1987, in its report *Our Common Future*, the WCED advanced the understanding of global interdependence and the relationship between economics and the environment previously introduced by the WCS (Brundtland Report 1987, p. 5). The report wove together social, economic, cultural and environmental issues and global solutions. It reaffirmed that "the environment does not exist as a sphere separate from human actions, ambitions, and needs, and therefore it should not be considered in isolation from human concerns. The environment is where we all live; and development is what we all do in attempting to improve our lot within that abode. The two are inseparable." (Brundtland Report 1987, p. 7)

In June 1992, the first UN Conference on Environment and Development (UNCED) was held in Rio de Janeiro and adopted an agenda for environment and development in the 21st century. This is tagged "Agenda 21: A Programme of Action for Sustainable Development". It contains the Rio Declaration on Environment and Development, which recognizes each nation's right to pursue social and economic progress and assigned to States the responsibility of adopting a model of sustainable development; and, the Statement of Forest Principles (UNCSD 2012). Agreements were also reached on the Convention on Biological Diversity and the Framework Convention on Climate Change. UNCED mobilized the Major Groups and legitimized their participation in the sustainable development process. UNCSD observes rightly that for the first time the lifestyle of the current civilization was addressed in Principle 8 of the Rio Declaration. The urgency of a deep change in consumption and production patterns was expressly and broadly acknowledged by State leaders. This Agenda 21 further reaffirmed that sustainable development was delimited by the integration of the economic, social and environmental pillars (UNCSD 2012).

The spirit of the conference was captured by the expression "Harmony with Nature", encapsulated in the first principle of the Rio Declaration: "Human beings are at the centre of concerns for sustainable development. They are entitled to a healthy and productive life in harmony with nature". In 1993, UNCED instituted the Commission on Sustainable Development (CSD) to follow-up on the implementation of Agenda 21. In June 1997, the General Assembly dedicated its 19th Special Session (UNGASS-19) to design a "Programme for the Further Implementation of Agenda 21" (UNCSD 2012). In 2002, ten years after the Rio Declaration, a follow-up conference, the World Summit on Sustainable Development (WSSD) was convened in Johannesburg to renew the global commitment to sustainable development. The conference agreed on the Johannesburg Plan of Implementation (JPoI) and further tasked the CSD to follow-up on the implementation of sustainable development. On 24 December 2009 the UN General Assembly (UNGA) adopted a Resolution (A/RES/64/236) agreeing to hold the UNCSD in 2012—also referred to as 'Rio+20' or 'Rio 20' (UNGA 2009). The Conference seeks three objectives: (a) securing renewed political commitment to sustainable development, (b) assessing the progress and implementation gaps in meeting already agreed commitments, and (3) addressing new and emerging challenges (UNCSD 2012). The Member States agreed on two themes for the Conference: green economy within the context of sustainable development and poverty eradication, and institutional framework for sustainable development (UNCSD 2012).

The Conference, accordingly, seeks the following three objectives:

(a) securing renewed political commitment to sustainable development,
(b) assessing the progress and implementation gaps in meeting already agreed commitments, and
(c) addressing new and emerging challenges.

The Member States therefore agreed on two themes for the Conference:

- green economy within the context of sustainable development and poverty eradication, and
- institutional framework for sustainable development

Since UNCED, sustainable development as a universal concept has come to stay with us. Social and legal commentators often point to the short comings of the concept and its complexity. For instance, it has been a moving factor for agitation for better life for the human species both at the world's institutions and organizations working in the economic, social and environmental sectors. Notwithstanding, it has been pretty difficult to grant the environmental pillar the same recognition enjoyed by the other two pillars despite the recognition that our earth is precariously vulnerable (UNCSD 2012). The concept of sustainable development has been incorporated in many UN declarations and has been at the forefront of world's institutions and organizations working in the economic, social and environmental sectors. However, they all recognize how difficult it has proven to grant the environmental pillar the same recognition enjoyed by the other two pillars despite the many calls by scientists and civil society signalling the vulnerability and precariousness of the Earth since the 1960 s. According to CISDL, the concept of sustainable development requires accommodation, reconciliation and integration between economic growth, social justice (including human rights) and environmental protection objectives, towards participatory improvement in collective quality of life for the benefit of both present and future generations (CISDL 2005).

The term 'sustainable development law' describes an emerging corpus of international and municipal legal principles and instruments which address the intersections between international economics, environmental and social law (including human rights law), towards development that can last for the benefit of present and future generations (CISDL 2005). According to Leal Filho (2011, p. 1), sustainable development may have many meanings such as: (a) the systematic, long-term use of natural resources—as defined in the Brundtland Report—so that these are available for future generations (here the concept refers to national and local policies); (b) the modality of development that enables countries to progress, economically and socially, without destroying their environmental resources (with reference to country policies); (c) a development which is socially just, ethically acceptable, morally fair and economically sound (referring to the social ramifications of development); (d) a development where environmental indicators are as important as economic indicators (here referring to the close links it bears with economic growth) (Leal Filho 2011, p. 1). He further suggested that there are some assumptions that need to be made, if sustainable development efforts are expected to yield the expected benefits. They are:

1. Sustainable development efforts should involve everyone.
2. It should be life-long.
3. It should be holistic and about connections.
4. It should be practical and action-oriented to convince people it works and makes sense (Leal Filho 2011, p. 3).

Moreover, it should be stated that, in order to yield the expected benefits, efforts towards implementing sustainable development need to provide people with the knowledge, understanding and capacity to influence mainstream society in a way which moves environmental objectives along with other legitimate social and economic objectives (Leal Filho 2011, p. 3).

What is clear from the above is that in places where sustainable development as a concept has been practised, there has been considerable progress in terms of physical and spatial development (Tarasofsky 2007, p. 27). In spite of the support the idea had, Richard Tarasofsky observed that: "despite this strong show of support, twenty years later, no country is fully on a sustainable path". As Jim MacNeill (2006, p. 2) puts it, "… in no case has [progress in implementing the strategic imperatives] been at the pace and scale needed to keep up with the unsustainable trends that we charted in *Our Common Future.*" The International Law Association (ILA) adopted the following principles as benchmark to monitor the compliance of states on sustainable development. The ILA seven principles are:

1. The duty of States to ensure sustainable use of natural resources;
2. The principle of equity and the eradication of poverty;

3. The principle of the precautionary approach to human health, natural resources and ecosystems;
4. The principle of public participation and access to information and justice;
5. The principle of good governance;
6. The principle of common but differentiated obligations;
7. The principle of integration and interrelationship, in particular in relation to human rights and social, economic and environmental objectives (New Delhi Declarations 2002, pp. 1–20).

Having the forgoing at the background, the political development which could have pushed most African countries to the fringes of sustainable development was not to be due to military interventions and wars. According to Albert (2012, p. 10) the history of development in many African countries (including Nigeria) in post-independence era has been tales of thwarted hopes and aborted goals. After attainment of independence in the 1960s, many African countries were plunged into cold war from the Democratic Republic of Congo to Angola, Mozambique and the military intervention in Ghana and later Nigeria. These events in a way might have affected sustainable development in Africa in general as well as in Nigeria in particular.

4 Defining Constitution

Constitution has been defined as the fundamental law, written or unwritten, that establishes the character of a government by defining the basic principles to which a society must conform; by describing the organization of the government and regulation, distribution, and limitations on the functions of different government departments; and by prescribing the extent and manner of the exercise of its sovereign powers (West's Encyclopaedia of American Law 2008).

Consequently, a constitution is a system for government, codified as a written document, which contains fundamental laws and principles. It usually contains fundamental political principles, and establishes the structure, procedures, powers and duties, of a government. Nigerians have been agitating for national constitutional conference because of their tenacious belief that a good constitution can engender sustainable society. A national conference is a formal platform for dialogue by constituent units of the nation convened by the national government of a country to discuss issues or problems that inhibit national progress or challenge national cohesion.

5 The Nigerian Constitution and Sustainable Development

Based on this definition and the ILA principles, it is imperative to examine the provisions of the Constitution of the Federal Republic of Nigeria 1999 (as amended) and other laws and see whether they are capable of (or can be said to be tailored towards) creating a sustainable development for the citizens of Nigeria. The Constitution was enacted by the military on 5th May 1999 but came into operation on May 29, 1999. Chapter II of the Constitution is so germane because it is "the political, economic and social blue print and inbuilt manifesto … to be pursued by the government and people of Nigeria to ensure an ideal nation. Public officials whether exercising legislative, executive, judicial or other functions are to purposefully implement these objectives and principles in order to make Nigeria a just, progressive and happy country; the country of our dream" (Malemi 2006, p. 46).

The Constitution (1999) provides that the Federal Republic of Nigeria shall be a State based on the principles of democracy and social justice (Section 14(1)). It also declares that: sovereignty belongs to the people of Nigeria from whom government through this Constitution derives all its powers and authority (Section 14(2)(a)); the security and welfare of the people shall be the primary purpose of government (Section 14(2)(b)); and the participation by the people in their government shall be ensured in accordance with the provisions of this Constitution (Section 14(2)(c)).

An interesting aspect of Chap. II of the Constitution is Section 20. It provides that:

> The State shall protect and improve the environment and safeguard the water, air and land, forest and wild life of Nigeria.

Nigeria is a fertile ground for mining activities in terms of crude oil, quarry, precious stones, bauxite etc. and of mobile telecommunication mast installation activities. The prevalent destruction of nature through the activities of the crude oil and other prospectors is affecting the livelihoods of the poorest people and is likely to cause increased suffering and conflict in the future. Violence is destroying the social fabric of communities especially the Niger Delta where the nation's oil is procured, as well as the trust needed for peaceful human wellbeing (Fatona et al. 2011, p. 202). Without a quick shift to sustainable practices respectful of universal and inviolable dignity of all humans, the living conditions of future generations will only get worse—especially as our world population is still growing significantly (Goepel 2010a, b).

Nigeria relies on crude oil for its income. According to OPEC report, oil and gas sector accounts for about 35 % of gross domestic product, and petroleum exports revenue represents over 90 % of total exports revenue. (OPEC 2015). The multi-national oil companies often pretend to carry out Environmental Impact Assessments (EIAs). For instance, Namibia is a magnet for uranium prospectors just as Niger Delta spews crude oil in Nigeria. The uranium in Namibia is accessible because of its rare combination of accessible surface deposits and a stable pro-mining government (Namibia Ministry of Mines and Energy 2014). Nigerian government should not wait for the oil firms to carry out EIA for them. Similarly the mobile telecom firms do not carry out EIA reports (Nigeria Communications Week 2015). The masts are built near homes and sometimes in compound of natives without the full implications being disclosed to the property owners whose precarious economic conditions often propel them to take such course (Nigeria Communications Week 2015). There is need for disclosure and consent. It seems the monetary inducement is the over-riding factor here. The Federal Government should commission independent bodies to carry out Strategic Environmental Assessments (SEA) for it. Mines and minerals including oil fields, oil mining, geological surveys and natural gas, are on the Exclusive Legislative List (Item 39 of the Second Schedule Nigerian Constitution 1999 as amended in 2011). Over time, Nigerian government appears to have agreed that the state has no business in some ventures which resulted to deregulation of some aspects of the economy into the hands of the private sector. The National Conference, in its report, made far reaching recommendations which if adopted will guarantee sustainable society. Amongst others, it recommends as follows:

- That as a means of promoting social citizenship, there is a need to make the provisions on socio-economic rights in Chap. II of the Constitution justiciable. This is derived from the fact that the lack of basic amenities and social mobility is at the root of the various communal strife (Constitutional Conference Report 2014. p. 109)
- Ensure Chap. II which contains all citizenship rights and obligations become (justiciable) enforceable (Constitutional Conference Report 2014, p. 110)

The National Conference thus viewed Chap. II of the 1999 Constitution of Nigeria as ideological aspiration which represents the basic law of citizens' rights and duties of the state. It therefore advised that the chapter be given necessary force of law to—for the first time—build national integration and cohesion (Conference Report 2014, p. 110) which, in our estimation, will engender sustainable society.

Consequently, it is submitted that some of the items on the Exclusive Legislative List mentioned above should be placed in the care of the state governments. This will provide enough funds for those state governments to execute their pro-people programmes as well as enhancing the principle of federalism as enshrined in the same constitution thus leading to sustainable development. The SEA, according to a report by IIED (2009), does not aim at finding faults. Rather, it looks for opportunities to solve problems and improve practices (IIED 2009). It checks that SEA process is:

*fully compliant with relevant national and international requirements (including laws, guidelines and commitments),
*fit for purpose and relevant to the needs of decision makers;
*effective in achieving positive environmental benefits and good outcomes in development cooperation (IIED 2009).

That Nigeria has not been doing this is pardonable because such reviews of SEAs are rear until now. There was no established generic methodology for them.

Chapter II of 1999 Constitution frames these basic human rights (Yerima 2009) in this manner because the Constitution recognises the reality that economic and social development can only occur within the limits of what can be sustained ecologically. However worthy our reasons for 'development' may be, if we catch fish or cut down trees faster than they can be replenished we are degrading our world, impoverishing our children, and in the long run, destroying the habitat on which we, and other species, depend. Despite this constitutional right, in many cases we are not only failing to achieve the goal of ecological sustainability, but also accelerating away from it as suggested above. The magnitude of the environmental hazards that faces Nigeria especially in the Niger Delta region because of oil exploration and deforestation going on in the Southwest is difficult to overstate. And justice is slipping away from the grips of the victims (Fatona et al. 2011, p. 204). Human pressures on Earth's life-supporting systems are causing them to deteriorate rapidly and possibly irreversibly. The gap between rich and poor continues to widen, and communities continue to disintegrate. The inescapable conclusion is that if we continue with our present modes of behaviour, our species (and many others) do not have a viable future (Cullinan 2007; Rapport and Maffi 2013, p. 106).

Sustainable livelihoods and eradication of poverty are core areas where sustainable development could easily be achieved. Livelihood is sustainable when it:

> …comprises the capabilities, assets (stores, resources, claims and access) and activities required for a means of living: a livelihood is sustainable which can cope with and recover from stress and shocks, maintain or enhance its capabilities and assets, and provide sustainable livelihood opportunities for the next generation; and which contributes net benefits to other livelihoods at the local and global levels and in the short and long term (Chambers and Conway 1992, p. 7f).

Nigeria is endowed with human and natural resources which include cultural, ecological and economic diversity and yet struggles with developmental issues (Imam-Tamim 2012, p. 105). The country remains poor and lags behind in development. Statistics from Debt Management Office (The Sunday Punch 2015) showed that both the Federal Government and the 36 States of the Federation as well as the Federal Capital Territory had a total debt profile of about N12.06tn ($60.9 billion). The country's public debt had risen to this value as of March 31, 2015 from N10.16tn as of March 31, 2014, translating to the N1.9tn rise (or 18.7 %) within a 12-month period (The Sunday Punch 2015). No country can sustain its livelihood with such debt burden hanging on its neck. Nigeria which produces oil for exportation was crippled by scarcity of Premium Motor Spirit (PMS) for almost three months and epileptic power supply. Fuel pump price which official rate was N87 suddenly in May 2015 rose to N500 in Lagos, the country's commercial capital, with similar experiences in other parts of the country (The Sunday Punch 2015). Power also dropped from 4800 Megawatt (MW) to about 1300 MW. This affected businesses, including banks, radio stations and telecommunication companies which had depended on power generators thereby reducing hours of service (The Sunday Punch 2015).

The constitution in its current version is incapable of bringing about sustainable development because the entire Chap. II is not justiciable. In *Ezeoukwu v. Ezeonu II*, where inhuman and degrading treatment, slavery, servitude, force labour were challenged under Chap. II of the Nigerian Constitution, the Supreme Court of Nigeria (2009) emphatically stated that:

> It is not every right of the Petitioner, which is involved here. It is only the right, which falls within the provisions of Chap. IV. No right outside the provisions of Chap. IV can found an action under the jurisdiction of court… Clear, unambiguous and serious as the infringed right may be, the court cannot raise its status to that of a fundamental right, if, in fact, it cannot be spelt out in Chap. IV (Nigeria Weekly Law Report.

Laws and regulations suited to country-specific conditions are among the most important instruments for transforming environment and development policies into action not only through *command and control* methods but also as a normative framework for economic planning and market instruments (Kurukulasuriya and Robinson 2006, p. 17; UN Economic and Social Development 2010). We recognise the continuous need for improvement of our laws, but Nigeria has been affected by the shortcomings in the laws and regulations. These laws include for example the Harmful Waste Act (Chap. (Cap) H1, Laws of Federation of Nigeria, the Oil in Navigable Waters Act (Cap 06), the Nuclear Safety and Radiation Protection Act (Cap 07) and the Environmental Impact Assessment Act (Cap E12) which beg for effective and efficient implementation and enforcement mechanism (Shehu 2012, p. 194f).

In spite of these laws, the adverse effects of large scale exploitation of natural resources in the Delta Region and other States bordering them has not shown that Nigerian leadership is doing enough to sustain the environment. All affected communities must be properly consulted and companies operating there challenged to the doctrine of greening now being preached all over the world as an aspect of sustainable development; and the Nigerian government must adopt a legislative framework setting forth the obligation to abide by strict environmental standards as well as fair and equitable revenue distribution.

The fact that Chap. II of the Constitution is not justiciable is the pinnacle of political unseriousness. Odinkalu (2012) asserted that the idea that the provisions of Chap. II of Nigeria's 1999 Constitution are outside the ambit of judicial enforcement is ideology and not law. He condemned this ideological posture as 'bad policy, incomprehensible logic, lousy politics, and at best, out-dated law' (Odinkalu 2012, p. 1). It is an abdication of political and social responsibility on the part of the state to regard some aspect of the constitution as in-justiciable. This aspect of the constitution should therefore be made an extension of Chap. II which focuses on the Fundamental Human Rights of the Nigerian people and through which the courts have been making progressive pronouncements to the joy and well-being of the Nigerian citizens and non-indigenes alike.

6 Any Role for the Judiciary?

What can the courts do to make socio-economic rights realisable, justiciable and enforceable? Sustainable development requires mediation between the interests of current generations and those of future generations as well as between competing interests of current generations (Kameri-Mbote and Odote 2009). Not surprisingly, the judiciary has been called upon in the quest for enforcing sustainable development policies owing to its traditional role in dispute resolution and interpretation of laws. As Kaniaru et al. (1998, p. 2) stated:

> The judiciary plays a critical role in the enhancement and interpretation of environmental law and the vindication of the public interest in a healthy and secure environment. Judiciaries have, and will certainly continue to play a pivotal role in the development and implementation of legislative and institutional regimes for sustainable development.

A judiciary, well informed on the contemporary developments in the field of international and national imperatives of environmentally friendly development will be a major force in strengthening national efforts to realise the goals of sustainable development and the rights of individuals accessing the judicial process (Kaniaru et al. 1998). The role of the judiciary is particularly important in developing countries, such as those in Africa, where the bulk of the population is poor and relies on natural resources for livelihood and sustenance, and where the countries' economies have those same resources as the bedrock of the gross domestic product.

At the World Summit on Sustainable Development in Johannesburg in 2002, chief justices and senior judges from around the world presented the Johannesburg Principles on the Role of Law and Sustainable Development. The Principles had been adopted at the Global Judges Symposium on the Role of Law and Sustainable Development. The Principles underscored the critical role that judiciaries around the world can and should play in efforts to promote sustainable development Johannesburg Principles 2002, p. 2). The judges admitted the fact that "an independent Judiciary and judicial process is vital for the implementation, development and enforcement of environmental law, and that members of the Judiciary, as well as those contributing to the judicial process at the national, regional and global levels, are crucial partners for promoting compliance with, and the implementation and enforcement of, international and national environmental law …" (Johannesburg Principles 2002, p. 1). The judges then made a commitment to contribute towards the realization of the goals of sustainable development through the judicial mandate to implement, develop and enforce the law, and to uphold the Rule of Law and the democratic process." (Johannesburg Principles 2002).

Nigerian judges should develop sustainable approach to the ways cases before them are treated. For instance, remedy for victims of oil spillage is through compensation as provided by the Constitution (Section 251(1)) and various oil and gas laws. The State High Courts have been usual avenue for redress. Nevertheless, through decrees, the military government had been able to vest only the Federal High Court with jurisdiction in the State where the spill occurs. Sometime, the State may not have a Federal High Court or it may be sited in faraway state capital and the victim might need to travel far (Fekumo 2004). A Litigant may not be able to afford the money, resources and other logistics to prosecute his case, and this might brood an unnecessary anger which often propel citizens to take arms against their nation.

In *The Shell Petroleum Development Co. of Nigeria Limited v. Otelemaba Maxon* (Federation Weekly Law Report (FWLR) 2001) the court held:

> This court had decided in *C.G.C (Nig) Ltd. v. Asagbara,* following *Barry v. Eric (supra)* that in an action arising from activities or operations of companies engaged in these businesses, the Federal High Court is the competent court to assume jurisdiction. The decision of this court in *Shell v. Isaiah (supra)* has now been consigned to history…

If Nigeria's courts adopt such attitude, then the greening of Nigeria's environment is in trouble. We might not be able to achieve any sustainable development just as, it now appears, we are not meeting up with the rest of the world in the quest for Millennium Development Goals (Chiejina 2015, p. 1).

7 Suggestions

There should be taken a proactive approach towards the implementation of Nigeria's laws. The aspect of the 1999 Constitution (as amended in 2011) which is now in-justiciable should be made justiciable to enhance Nigeria's march to sustainable development. Its justiciability has the potential of gearing the people as well as the government into action to achieving better socio-economic, political development. While the fear of the unknown may affect the government decision to make this aspect of the Constitution justiciable, it can still provide guidelines on the procedure people should adopt while seeking redress on the justiciability or otherwise of an item under the provision of the Constitution. In fact, the usefulness of the Freedom of Information Act 2011 and the positive manner it is being used in the society at present emboldens the view expressed in this chapter in this regard. The frivolity which the government apprehends so much before acceding to appending presidential assent to the Act is now a fear of the past which the government finds to be superfluous anyway.

Second, Nigerians should work more on an efficient and effective monitoring mechanism beyond the present use of ethnic militia in checking activities of vandals of oil pipelines from which environmental degradation ensues. Oil companies should be made to find effective ways of tackling oil spillage in host communities. The judicial officers should also be trained and encouraged to adopt new rules rather than relying on traditional stereotypes in dealing with modern environmental issues which often bring hardship to the indigent in communities where environmental degradation has become live issues. Breaches of the principles of sustainable development should be made actionable per se. In this wise, the old rule of *locus standi* must be relaxed to encourage public interest litigation on sustainable development matters.

8 Conclusion

In conclusion, we need a pragmatic approach to law making to achieve sustainable development. This becomes necessary because a suffering nation is not likely to engender a sustainable nation let alone sustainable development. For instance, the Nigerian Senate hurriedly passed into law about 46 laws at the eve the winding up of the 8th Senate of the National Assembly. This is not too good for laws expected to galvanise a sustainable society (Aborisade 2015, p. 1). While it can be agreed upon that some of the provisions of Chap. II of 1999 Constitution are mere ideals or duties of the State to its citizens, some of them that have to do with the environment need be enshrined and made enforceable for people to achieve sustainable development. Both government and individuals are careful in their treatment of others because if their acts fall under any of the provisions of Chap. IV of the Constitution, the aggrieved can easily get redress in the court. This sort of feeling is necessary in dealing with our environment. The constitution must be justiciable for sustainable development to be achieved. Non-justiciability will foist a burden on the constitution for non-functionality and ecological sustainability. As said above, we must change the outlook that places human beings at the centre of the universe and emphasise the maintenance of the integrity of the whole Earth system. Thus, only a good and functional constitution can bring about good governance, which Nigerians presently seek. The best time to do this is now when the constitution is under review. Expectedly, the National Conference's report had been submitted. Its findings and submissions were all encompassing. For instance, it did an overview of the perennial agitations for national conference in Nigeria, surveyed various national constitutional conferences held in Nigeria up to 2005, examined the contemporary socio-economic and political challenges to national development and cohesion and made resolutions requiring constitutional or legislative amendments and policy implementation. Given the tenacity the government of the All Progressive Congress intends to work to create a sustainable environment for all Nigerians, the implementation of the resolutions of the National Conference Report will create a sustainable society for all Nigerians and it may be dawn yet on creation day for the nation.

References

Aborisade S (2015). Senate passes 46 bills in 10 minutes. Retrieved on 26 June 2015 from www.punchng.com

Albert IO (2012) Peace and sustainable development in Africa in law and sustainable development in Africa. In: Egbewole WO, Etudaiye MA, Olatunji OA (eds) Law and sustainable development in Africa. Al-Fattah Publications Ltd., Ilorin, pp 1–25

Brundtland Report (1987) Our common feature; report; report of the UN World Commission on Environment and Development; NY, USA, 4 Aug 1987; UN Doc A/42/427, New York: UN

Chambers R, Conway G (1992) Sustainable rural livelihoods: practical concepts for the 21st century. IDS Discussion Paper No. 296, Brighton, IDS

Chiejina A (2015) Nigeria in last minute move to meet health-related MDG target. Retrieved 26 June 2015, from www.businessdayonline.com

CISDL (2005) What is sustainable development law? A CISDL Concept Paper1, Montreal. Retrieved 26 June 2015, from http://cisdl.org/public/docs/What%20is%20Sustainable%20Development.pdf

Constitution of the Federal Republic of Nigeria (Promulgation) Decree No. 24 (1999). Retrieved 30 July 2015, from www.nigeria-law.org/ConstitutionOfTheFederalRepublicOfNigeria.htm

Constitution of the Federal Republic of Nigeria (1999) (as amended in 2011). Retrieved 30 July 2015, from www.nassnig.org/document/download/5830

Constitutional Conference Report (2014) Nigeria mational conference. Retrieved 30 July 2015, from www.premiumtimesng.com

Cordonier Segger M-C, Khalfan A (2004) Sustainable development law: principles, practices & prospects. Oxford University Press, Oxford

Cullinan C (2007) Sustainable development and the law. Retrieved 27 June 2015, from http://www.enviropaedia.com/topic/default.php?topic_id=266

Fatona PO, Musah KT, Odumosu T (2011) Environmental injustice and human rights abuse: the burden of Nigeria's Niger Delta. Olabisi Onabanjo University Journal of Public Law 1 (1):195–211

Federal High Court (2001) The Shell Petroleum Development Co. of Nigeria Limited v. Otelemaba Maxon (Federation Weekly Law Report (FWLR) 2001)

Fekumo FJ (2004) The problem of jurisdiction in compensation for environmental pollution and deregulation in Nigeria (Oil and Gas): a fundamental rights enforcement alternative. Paper presented at the Nigerian Bar Association annual delegates conference, Abuja

Freedom of Information Act (2011). Laws of federation of Nigeria, 2011

Goepel M (2010a) Formulating future just policies: applying the Delhi sustainable development law principles (Electronic version). Sustainability 2(6):1694–1718. Retrieved from www.mdpi.com/journal/sustainability

Goepel M (2010b) Applying the Delhi sustainable law principles. Retrieved from www.worldfuturecouncil.org/fileadmin/user.../sustainability-02-01694.pdf

IIED (2009) Annual report, Doorway to a green economy

IISD (2013) What is sustainable development? Retrieved 28 June 28 2015, from http://www.iisd.org/sd/

Imam-Tamim MK (2012) Challenges of sustainable development in Nigeria: legal perspectives in law and sustainable development in Africa. In: Egbewole WO, Etudaiye MA, Olatunji OA (eds) Law and sustainable development in Africa. Al-Fattah Publications Ltd., Ilorin, pp 96–135

Johannesburg Principles (2002) Johannesburg principles on the role of law and sustainable development adopted at the global judges symposium held in Johannesburg, South Africa on 18–20 Aug 2002

Kameri-Mbote P, Odote C (2009) Courts as champions of sustainable development: lessons from East Africa (Electronic version). Sustain Develop Law Policy 10(1):31–84. Retrieved 27 June 2015, from http://digitalcommons.wcl.american.edu/cgi/viewcontent.cgi?article=1008&context=sdlp

Kaniaru D, Kurukulasuriya L, Okidi C (1998) UNEP judicial symposium on the role of the judiciary in promoting sustainable development, (Electronic version). Conf Proc 22:21–23

Kurukulasuriya L, Robinson N (eds) (2006) Training manual on international law. Retrieved 27 June 2015, from https://books.google.com.ng/books?isbn=9280725548

Leal Filho W (2011) About the role of universities and their contribution to sustainable development (Electronic version). Higher Educ Policy 24

MacNeill J (2006) The forgotten imperative of sustainable development. Green Law, J Pace Environ Law Programs 10(1):1–15

Malemi E (2006) The nigerian constitutional law. Princeton Publishing Co., Lagos

Ministry of Mines and Energy (2014) Strategic environmental management plan for the central Namib Uranium Province 2012 Report prepared by Geological Survey of Namibia, pp 1–155

Nigeria Communications Week (2015). Retrieved 25 June 2015, from www.nigeriacommunicationsweek.com
Nowosielski R, Spilka M, Kania A (2007) Strategies of sustainable development in practice. J Achiev Mater Manuf Eng 20(1–2):555–558
Nwabueze B (2006) Constitutional democracy in Africa, vol 1–5. Spectrum Books, Ibadan
Odinkalu C (2012) Lawyering for a cause: the imperative of justiciability of socio-economic rights in Nigeria, Vanguard. Retrieved 27 June 2015, from www.vanguardngr.com
OPEC (2015) Nigeria: facts and figures. Retrieved 27 June 2015. from http://www.opec.org/opec_web/en/about_us/167.htm
Rapport D, Maffi L (2013) The dual erosion of biological and cultural diversity: implications for the health of ecocultural systems. In: Pilgrim S, Pretty J (eds) Nature and culture. Routledge, London, pp 103–119
Rousseau J (1762) The social contract or principles of political right. In Hutchins RM (ed) (1952) The great books of the western world. Encyclopaedia Britannica, Chicago, pp 387–439
Shehu AT (2012) Constitutionalism and sustainable development in Africa: the Nigerian legislative and institutional responses. In: Egbewole WO, Etudaiye MA, Olatunji OA (eds) Law and sustainable development in Africa. Al-Fattah Publications Ltd., Ilorin, pp 166–201
Supreme Court of Nigeria (2009). Ezeoukwu v. Ezeonu II (1991) 6 NWLR (Pt. 200) 708
Tarasofsky R (2007) Canada's progress in addressing the strategic imperatives set out in "Our Common Future". Retrieved 7 June 2015, from https://www.iisd.org/pdf2007/facing_canadaprogress.pdf
The Guardian (2011a) Arab league denounces civilian casualties" Sunday, Oct 30 2011, p 14
The Guardian (2011b) Gaddafi's death renews Syria protests, 20 Killed" Sunday Oct 23 2011, p 12
The Principles of International Law Related to Sustainable Development (2002) New Delhi Declarations, Retrieved 25 June 2015, from www.cisdl.org
The Sunday Punch. 31 May 31 2015, vol 19, p 9–12
UN Economic and Social Development (2010) Core Publications: AGENDA 21 para 8.13. Retrieved 7 June 2015, from http://www.un.org/esa/dsd/agenda21/res_agenda21_08.shtml
UNGA (2009) UN general assembly resolution (A/RES/64/236). Retrieved on 26 June 2015, from http://www.uncsd2012.org/history.html#sthash.LND7PcsZ.dpuf
United Nations Conference on Sustainable Development (UNCSD) (2012) Retrieved 25 June 2015, from http://www.uncsd2012.org/rio20/index.php?menu=22
West's Encyclopedia of American Law (2008) (2nd ed) Retrieved 25 June 2015, from http://legal-dictionary.thefreedictionary.com/constitution
Yerima TF (2009) Appraising the significance of the liberal and utilitarian conceptions of human rights in the 1999 constitution of Nigeria. Ikeja Bar Rev 2(1):210–215

Part V
Horizontal Policies: Assessment Tools

A Global Conceptual Framework for Categorizing Environmental Change Based on Property Rights and Compensation

Volker Mauerhofer

Abstract The paper offers a new conceptual framework based on property rights and financial compensation in different nature conservation situations in order to provide a globally applicable system for the assessment of participation of public and private stakeholders in envisaged changes within those situations. These situations of change represent a modification from conservation toward non-conservation and vice versa. The framework distinguishes further between governance systems based on command and control as wells as on negotiation. Within these main change situations and governance types, the framework allows the distribution of change situations into 8 main sections. These main sections are further separated into 32 sub-sections by means of different property right and compensations situations among public and private stakeholders. The theoretical utility of this new framework is then demonstrated by testing it by means of a random sample of 74 papers (25 %) out of representative 297 papers from the academic literature dealing with property rights. These 74 papers provided practical examples for situations of change in conservation as evidence for most of the 32 sub-section. Several papers provide examples for more than one sub-section. The allocation of papers to these different subsections is described and discussed in detail. This widely possible allocation proofs in general the global applicability and usefulness of the new framework. The framework also proofed to be appropriate for formally (rule of law based) and informally (customary law based) institutionalized situations where rights are given to public and private stakeholders for other practical cases of public participation.

V. Mauerhofer (✉)
Faculty of Life Sciences, University of Vienna, Vienna, Austria
e-mail: volker.mauerhofer@univie.ac.at

V. Mauerhofer
UNU IAS—United Nations University Institute for the Advanced Study of Sustainability, Tokyo, Japan

V. Mauerhofer
PRIMAFF—Policy Research Institute of the Ministry of Agriculture, Forestry and Fishery, Tokyo, Japan

V. Mauerhofer (ed.), *Legal Aspects of Sustainable Development*,
DOI 10.1007/978-3-319-26021-1_13

Keywords Environmental protection · Negotiation framework · Legal position · Ownership · Disaccord

1 Introduction

Property rights are seen nowadays to be paramount in contributing to environmental problems (Schlager and Ostrom 1992; Ostrom and Schlager 1996; Penker 2008; Quinn et al. 2010; Roy et al. 2012). The vast literature on property rights appears to concentrate rather on the description of particular property rights (e.g. Ostrom and Schlager 1996) and provides less focus on the overall legal situation how these rights (and corresponding duties) are distributed among the stakeholders when it comes to situations of an envisaged change. This is even more astonishing as this framework wherein these rights and duties are embedded is of major importance for the outcome of such a situation of an intended change, especially when it comes to disputes among stakeholders about the extent of environmental protection.

Formal institutions provide rights such as property rights to people and they define also the relationship among the right-holders (North 1990). Each right means also duties for other people, for example the duty to respect other rights (Schlager and Ostrom 1992). In case a right is restricted (Gerber et al. 2009), the restriction means also a duty for the right-holder. The restriction may base on the individual right of another right-holder. For example ownership of land may be restricted through the right to collect wood by another right-holder. The restriction may also be based on general prescriptions. This is for instance the case if the ownership of a land is affected by public conservation restrictions on behalf of wildlife, such as a ban not to fish. The restriction sometimes goes along with the formal (institutionalized) duty to offer compensation (Gerber et al. 2009).

The identification of the holders of the rights over the natural resources and the rights to compensation in a negotiation situation on conservation is essential to analyse cost-effectiveness of conservation actions. Because it can help to envisage the likeliness of a compensation solution and the relative extent of the financial effort involved in comparison to the other negotiation situations described. Therefore, the aim of this research is to create as well as theoretically and practically test a more comprehensive framework for assessing the most cost-effective strategies for conservation actions in connection with negotiation situations as well as related bundles of property rights.

The paper is structured such as following. First, two conceptual approaches are briefly discussed and then combined to the new assessment framework. This new framework is afterwards theoretically tested by applying it on a representative sample out of 297 papers from the academic literature dealing with property rights in the sense of Schlager and Ostrom (1992). The papers of the sample were assessed towards whether it can be allocated to the different categories and subcategories in the new framework. Each allocation is in detail qualitatively described and also a quantitative overview is given. In the following, I substantially discuss the scope and the limits of the new framework and explore future application fields.

2 Compensation Situations and Property Rights: Two Conceptual Approaches

In the following, the concept of Schlager and Ostrom (1992) on property rights as well as a new concept based on Mauerhofer and Nyacuru (2013) on compensation situations are described respectively. They provide the intellectual basis for the new framework and are then combined.

Schlager and Ostrom (1992) present an approach to explain resource management in regimes usually categorized as common property, describing different situations where multiple stakeholders have diverse bundle of rights over a natural resource (Table 1).

The explanations for the different rights in Table 1 are such as following (adapted from Schlager and Ostrom 1992, p. 252):

- Access: The right to enter a defined physical property.
- Withdrawal: The right to obtain the products of a resource
- Management: The right to regulate internal use patterns and transform the resource by making improvements.
- Exclusion: The right to determine who will have an access right, and how that right may be transferred.
- Alienation: The right to sell or lease either or both of the above collective-choice rights.

These authors also emphasize that different bundles of property rights, whether de facto or de jure, influence incentives for individuals, the types of actions they take, and the outcomes they achieve.

Beside the property rights regime also the compensation situation is crucial as it reflects the distribution of different interests and positions regarding natural resources in connection with rights. Furthermore it determines the formal power

Table 1 Bundles of rights and positions associated

Bundles of rights associated with positions				
	Owner	Proprietor	Claimant	Authorized user
Access and withdrawal	×	×	×	×
Management	×	×	×	
Exclusion	×	×		
Alienation	×			

Adapted from Schlager and Ostrom (1992, p. 252)

distribution from the beginning that may influence the outcome of negotiations about payments for any environment-related rights. Mauerhofer and Nyacuru (2013) based on Rothgang (1997) distinguish between five negotiation situations that are extended for the means of this paper into eight compensation situations in conservation (Table 2).

Table 2: Binding rules and negotiation opportunities in situations of change affecting right-holder

The situations 1. to 4. involve some sort of formal (or even informal) binding social rule which is of course only as strong as its enforcement. The consequence of its non-compliance may be sanctions (e.g. sentence to prison) or—in the worst (illegal) informal case—killing. While the cases 5. to 8. reflect voluntary situations not underlined by pressure but solely by one or more financial offers. The situations 5. and 6. usually provide substantial space for negotiations and of course also determine the positions of the offeror and the accepter. This is of quite utmost importance in negotiations.

The situations 1., 2., 7. and 8. do not provide any space for negotiations and the two latter ones fall under "rule of inalienability" (Mauerhofer and Nyacuru 2013). Each of these cases can base either on a formalized legal regime based on laws issued according to institutionalized procedures by the competent legislative body, or the cases can have their origin in informal (customary) law regimes. All the eight cases can also reflect a combination of these two different legal regimes.

While a description of compensation situations highlights the relationship among stakeholders, property right systems concentrate more on the position of the individual stakeholder.

Table 2 Binding rules and negotiation opportunities in situations of change affecting right-holders

Right-holder and situation of change (extended based on Mauerhofer and Nyacuru 2013)	Binding rule	Negotiation
1. Conservationist has to accept change without compensation	Yes	No
2. Non-conservationist has to accept change without compensation	Yes	No
3. Conservationists has to accept change with compensation opportunity	Yes	Yes
4. Non-conservationist has to accept change with compensation opportunity	Yes	Yes
5. Conservationist may accept change with compensation opportunity	No	Yes
6. Non-conservationist may accept change with compensation opportunity	No	Yes
7. Conservationist does not accept compensation at all	No	No
8. Non-conservationists does not accept compensation at all	No	No

3 Towards a Systematic Framework

In the following, I present a framework that incorporates bundles of property rights as described by Schlager and Ostrom (1992) into compensation situations as described by Mauerhofer and Nyacuru (2013) to distinguish different situations to better understand negotiation positions when a form of compensation can be negotiated with the holders of rights over natural resources (Table 3).

Table 3 Situation of change for right-holder and their legal position

Right-holder and situation of change (extended based on Mauerhofer and Nyacuru 2013)	Right-holder's legal position (Schlager and Ostrom 1992)
1. Conservationist has to accept change without compensation	a. Owner
	b. Proprietor
	c. Claimant
	d. Authorised user
2. Non-conservationist has to accept change without compensation	a. Owner
	b. Proprietor
	c. Claimant
	d. Authorised user
3. Conservationist has to accept change with compensation opportunity	a. Owner
	b. Proprietor
	c. Claimant
	d. Authorised user
4. Non-conservationist has to accept change with compensation opportunity	a. Owner
	b. Proprietor
	c. Claimant
	d. Authorised user
5. Conservationist may accept change with compensation opportunity	a. Owner
	b. Proprietor
	c. Claimant
	d. Authorised user
6. Non-conservationist may accept change with compensation opportunity	a. Owner
	b. Proprietor
	c. Claimant
	d. Authorised user
7. Conservationist does not accept compensation at all	a. Owner
	b. Proprietor
	c. Claimant
	d. Authorised user
8. Non-conservationist does not accept compensation at all	a. Owner
	b. Proprietor
	c. Claimant
	d. Authorised user

This framework describes in each case a situation where a wide range of stakeholders could be considered conservationists as well as non-conservationists depending on different interests over natural resources, and where the right to receive compensation is disentangled into the bundle of rights of each stakeholder over the resource. It provides a theoretical and conceptual basis to better analyse the costs and assess cost-effectiveness of the conservation actions.

The sole opportunity or the definite duty to accept a compensation for the partial or full restriction of a right can correspond in negotiations in different ways with the sole opportunity or duty to offer such compensation. The whole negotiation situation is also influenced by which type of right is on stake at all. The explicit expression of these relationships often predetermines the costs of implementation, monitoring and enforcement, and decision-making regarding a conservation action.

In the following I show results of the theoretical application of the new framework on a sample of 297 papers from the academic literature dealing with property rights in the sense of Schlager and Ostrom (1992). The sample was gained by a survey in the Web of Knowledge and Scopus (15th December 2013 WoK) where I extracted all papers that cite Schlager and Ostrom (1992) by that date. I show for seven of eight basic compensation situations at least one concrete example which I found within the overall sample of 74 papers (=25 %) out of the 297 papers (Table 4).

Table 4 Situation of change for right-holder, their legal position and literature examples from sample

Right-holder and situation of change (adopted from Mauerhofer and Nyacuru 2013)	Right-holder's legal position (Schlager and Ostrom 1992)	References in the literature from a sample (total n = 74 [25 %] of 297 papers citing Schlager and Ostrom 1992)
1. Conservationist has to accept change without compensation	a. Owner	Glück (2002), Penker (2009), Quinn et al. (2010), Rodgers (2010), Lescuyer et al. (2012), Lescuyer (2013)
	b. Proprietor	Quinn et al. (2010), Dumyahn and Pijanowski (2011), Rodgers (2010)
	c. Claimant	Naidu (2011), Quinn et al. (2010), Dumyahn and Pijanowski (2011), Rodgers (2010), Nagendra and Gokhale (2008)
	d. Authorised user	Naidu (2011), Campbell et al. (2001), Hayes (2007), Behera and Engel (2006), Dumyahn and Pijanowski (2011), Khan and Hague (2010)

(continued)

Table 4 (continued)

Right-holder and situation of change (adopted from Mauerhofer and Nyacuru 2013)	Right-holder's legal position (Schlager and Ostrom 1992)	References in the literature from a sample (total n = 74 [25 %] of 297 papers citing Schlager and Ostrom 1992)
2. Non-conservationist has to accept change without compensation	a. Owner	Persha and Blomley (2009), Glück (2002), Lescuyer et al. (2012), Lescuyer (2013), Behera and Engel (2006)
	b. Proprietor	Lund and Treue (2008), Bogale et al. (2006), Behera and Engel (2007), Barsimantov et al. (2011)
	c. Claimant	Persha and Blomley (2009),
	d. Authorised user	Persha and Blomley (2009), Roy et al. (2012), McCay et al. (1995), Wiber et al. (2010), Caballero-Miguez et al. (2012), Aburto et al. (2013)
3. Conservationist has to accept change with compensation opportunity	a. Owner	
	b. Proprietor	Bottazzi and Dao (2013)
	c. Claimant	
	d. Authorised user	Kusters et al. (2007)
4. Non-conservationist has to accept change with compensation opportunity	a. Owner	Glück (2002), Khan and Khan (2009)
	b. Proprietor	
	c. Claimant	Cinner et al. (2012a,b)
	d. Authorised user	Nagendra and Gokhale (2001)
5. Conservationist may accept change with compensation opportunity	a. Owner	Nielsen and Treue (2012), Ahmed et al. (2008), Naidu (2011), Campbell et al. (2001)
	b. Proprietor	
	c. Claimant	Trawick (2001)
	d. Authorised user	Hayes (2007)
6. Non-conservationist may accept change with compensation opportunity	a. Owner	Persha and Blomley (2009), Mahanty et al. (2013)
	b. Proprietor	
	c. Claimant	Ahmed et al. (2008), Persha and Blomley (2009)
	d. Authorised user	Persha and Blomley (2009)
7. Conservationist does not accept compensation at all	a. Owner	Persha and Blomley (2009)
	b. Proprietor	
	c. Claimant	
	d. Authorised user	Sekhar (2004)
8. Non-conservationist does not accept compensation at all	a. Owner	
	b. Proprietor	
	c. Claimant	
	d. Authorised user	

With regard to the empty fields within the third column next to 6.b., 7.b. and 7.c. it has to be kept in mind that the owner position (or in general the higher position) regularly includes all the rights and positions below (but this has not been implemented in any field in order to keep the table simple).

In the following I will describe more in detail how these cases found are fitting into our framework.

3.1 Conservationist Has to Accept Change Without Compensation

Glück (2002) describes the Austrian state as owner of mountain forests with all rights described by Schlager and Ostrom (1992), but restricted by some de iure and de facto access and withdrawal rights of other users who may even cause detrimental effects to these forests (see also Penker 2008, for all owners). Quinn et al. (2010) show how two legal acts providing access rights to moorlands and forests for tourists as authorized entrants prevail—without compensation—all other property rights held by Conservationists as owners (which also include the rights of proprietors, claimants and authorized users). Similar, Dumyahn and Pijanowski (2011) describe regarding noise mitigation in National Parks that the Park authorities currently have regarding the soundscape only the rights of proprietors (and visitors only the rights of authorized users), but would need based on legal acts ownership rights against flight sound pollution. Rodgers (2010) shows a situation for the United Kingdom where all owners, proprietors or claimants of common land were burdened by the introduction of historically authorized user rights (namely grazing rights) through simply registering them 1965 into a new law, without reassessing these right in terms of their sustainability. Lescuyer et al. (2012) and Lescuyer (2013) illustrate a situation where local communities which are currently in an ownership position have to accept a change of the legal situation which allows the government to issue of logging concessions to third parties without a rule prescribing compensation.

In Indian sacred forests ('kan'), local communities are considered claimants, but they have to accept the State, as holder of ownership rights, to provide leasing of extraction rights to local contractors (Nagendra and Gokhale 2008).

Naidu (2011) describes the situation of communities related to three different forests management systems in India with different bundles of property rights even towards shutting down forests internally (in the sense of "exclusion" and proprietor positions within the systematic of Schlager and Ostrom 1992). However, it is explicitly mentioned that the ownership position remains with the Indian government and that mining contracts are concluded outside these forest management system (leading to conflict situations and the elimination of certain case studies from the scope of Naidu's paper).

A similar situation is described by Campbell et al. (2001) on users in the common property resource systems of social forests of Zimbabwe in relation to outside intruders obtaining permits from state agencies, bypassing the local village structures. Quite the same situation is shown by Hayes (2007) in a comparative analysis of agricultural expansion in the Mosquitia Forest Corridor. There, a Honduran indigenous community with sole common property user rights had to accept negative changes in the forests through new settlers because of insufficient enforcement of ownership rights by the state (while a similar Nicaraguan community holding all property rights could prevent such changes).

In Bangladesh, the state has the legal ownership as well as absolute control over natural resources such as fisheries in waterbodies and enacted 2005 a new auction of lease rights policy that lead to the loss of de facto use rights of small local fishermen communities, to major leasing through outside investors and—following—to profit maximising and degradation of natural resources (Khan and Hague 2010).

3.2 *Non-conservationist Has to Accept Change Without Compensation*

In Austria, mountain-forest owner—who usually have the right to exploit the forest—are obliged by the forest act to manage so-called protection forests (that protect for example infrastructure or human settlements against avalanches) according to special guidelines, as long as the management costs are covered by the revenues from timber sale (Glück 2002). Thus, no compensation for the change is provided. Similar would be valid if an Austrian provincial government responsible for conservation would—without any financial offer—restrict or even prohibit the use of a forest in order to protect biodiversity. Lescuyer et al. (2012) and Lescuyer (2013) also describe a situation where local communities which are currently in an ownership position have to accept a change of the legal situation which allows the government to designate this land a protected area without a rule prescribing compensation for the restrictions related.

Lund and Treue (2008) describe the case of a decentralization of forest use in Tanzania providing to local communities even rights to arrest offenders in the sense of exclusions right and also discuss the possibility of re-decentralization in connection with this and other African decentralization projects if environmental goals are not reached, thus indicating that the proprietor rights can be withdrawn again without compensation.

Bogale et al. (2006) show for Ethiopia an example wherein the disclosure of a watershed protection area restricted the access and where farmers even launched armed attacks against guards to "reclaim the alienated land" (p. 137). Although in this case the right to alienate cannot be further transferred, the importance of access and withdrawal rights is clear as without them, nobody is allowed to harvest anything, and thus the own right of exclusion is restricted,

Behera and Engel (2006) show for Joint Forrest Management (JFM) in India that this only did not include the transfer of alienation rights but allowed the Forest Departments (FDs)—in 12 out of 23 states that implemented JFM—unilateral power to cancel the JFM agreements and in most cases to even dissolve the Forest Protection Committees holding those rights.

Roy et al. (2012) describe the situation of authorized users in the Sunderbans which had to accept the establishment of reserves without compensation (but with use-rights in the also established puffer zones). The situation of users and withdrawers of individual transferable quotas (ITQs) in Canadian and US fisheries such as described by McCay et al. (1995) is rather similar whereas the government retains the right to determine the overall quota and other fishery aspects that influence the sustainable use of fish stock. Again for Canada, Wiber et al. (2010) illustrate a situation wherein a single firm was granted rights of first refusal to depurate all clams harvested from closed (polluted) beaches in a region, resulting even in the de facto privatization of all (polluted and unpolluted) clam beaches by reducing the rights of other users.

Again related to open-access ocean resources Caballero-Miguez et al. (2012) show a situation where users of mussels harvested by means of floating raft cultivation in Galicia are able to obtain access and withdrawal rights in the sense of licenses. Even these rights are legally for example in a narrow sense prescribed regarding equipment, duration (a maximum of 30 years) and geographic space. Furthermore, the law determines the management, who is excluded and under which (narrow) circumstances the rights can be transferred. After the maximum period the change happens and the authorized users lose their rights.

In Chile, a local fisherman community had to accept the legal closure of former open-access fishery regarding the main relevant species and to accept instead a spatially restricted common-pool fishery area concerning that species (Aburto et al. 2013).

3.3 Conservationists Has to Accept Change with Compensation Opportunity

In the Bolivian Pilon Lajas Biosphere Reserve and Indigenous Territory, the State has the ownership rights of the territory but local communities are the proprietors of the land and natural resources (Bottazzi and Dao 2013). In that case local communities are conservationists that prefer to preserve the forest around the reserve but they have to accept compensation—if offered—for new migrant settlements that arise in the buffer zone of the reserve.

In Sumatra/Indonesia an original agroforest land-use system developed by local farmers more than hundred years ago under a customary system was in 1997 first declared to a state forest land by the government (which enables the government e.g. to issue logging concessions to outsiders) and—after protests—the government enacted in 1998 a degree acknowledging the local user rights and allowing the locals to register their traditional land use rights (Kusters et al. 2007). Although nobody registered its rights in order not to acknowledge the government's ownership, the acknowledgement helps to protect the agroforestry region as the source of livelihood of the locals, as a puffer zone of a National Park and against logging ambitions of outsiders (Kusters et al. 2007).

3.4 *Non-conservationist Has to Accept Change with Compensation Opportunity*

Another negotiation situation occurs on the opposite case, when non-conservationists hold the rights over natural resources but have to accept compensation for the execution of conservation activities by other stakeholders.

Glück (2002) describes for Austrian mountain forests an instrument for 'banning' a protective forest under certain conditions by the forest authority. The beneficiary protected (e.g. against avalanches) by the ban of any economic use of this forest has to compensate the mountain-forest owners for the necessary forest but the forest authority always covers the transaction costs and public funds may take over the compensation (Glück 2002).

Khan and Khan (2009) indicated for the Swat valley in Pakistan that the provincial government declared the forests as protected in 1975, converted them into state property and compensated the former de jure owners with a 60 % share in royalty for timber extraction. However, for this example the allocation of the term "non-conservationist" to the de-iure owners could be questioned as the state authorities continued to use the forests, paid royalties—if at all—delayed and were mentioned to be corrupt, all leading to an degradation of the forests together with other factors such as tourism and increasing population.

Local communities living within the Peten in Guatemala that have held concession rights had originally—with the feeling of being offered no benefits in exchange—to accept the establishment of a Biosphere Reserve by the State, but latter received certain user and proprietor rights in terms of forests concessions in and outside the Biosphere reserve (Barsimantov et al. 2011). This can be considered a sort of compensation, although the total extend of the historical rights might have been reduced.

Fishermen of Madagascar have to accept compensation (if offered) for marine preservation, as they have management rights—they are claimants—in marine parks (Cinner et al. 2012a, b).

In Nepal local communities as authorized users are restricted in their withdrawal of forest products in buffer zones of protected areas, as they are limited to those forest products authorized by the fairly strict management guidelines provided to the community while this restriction does not exist in community forests outside of buffer zones (Nagendra and Gokhale 2008). However, they are allowed to certain touristic activities such as to organize Elephant rides in these buffer zones and this can be clearly seen as a sort of compensation in exchange.

In last situations described the stakeholders have to accept the change but (adequate?) compensation has to be offered. While in the following cases, conservationists or non-conservationist may accept the change and, in exchange, may accept compensation, both by voluntary agreements.

3.5 Conservationist May Accept Change with Compensation Opportunity

When the government of Tanzania, as holder of ownership rights over natural resources, voluntarily decides to introduce a community forest based management by transferring ownership rights to local communities in a protected area (Nielsen and Treue 2012), it is voluntarily accepting loss of rights to preserve endangered forest species. This situation shows a case where the holder of property rights could have introduced compensation by law in the change process. Similarly, in Bangladesh, wetlands are owned by the public and the government confers exclusive use rights to individuals or groups who hold the lease, after successful bidding, mainly for fishing (Ahmed et al. 2008). Naidu (2011) and Campbell et al. (2001) describe situations for different countries where the government as owner of the resources respectively grants resource extraction permits to outsiders (overwhelming local user communities' rights but in both cases it is not described that the conservation aims regarding the resources are maintained and if and in how far a compensation was involved.

Trawick (2001) describes the case of a village community which successfully governs an Andean Irrigation System for centuries based on common management and use of water. This community would have the opportunity to sell their rights to third parties such as they bought the rights of a community member family who left to town (p. 16). Other cases where local communities hold claimant rights for example over a forest that they plan to preserve, for example, and they actually can decide if accept compensation from a logging company to exploit that forest, would fit into that situation.

Hayes (2007) indicates in a comparative analysis of agricultural expansion in the Mosquitia Forest Corridor that in a Honduran indigenous community with sole common property user rights some members accept—by selling their rights to new settlers negative—the changes in the forests through them. Without reference to Schlager and Ostrom, Gómez-Márquez et al. (2011) describe the situation in Spain where pastoralist hold authorized user rights over traditional transhumance routes. These authors indicate that, when a private mining operator is licensed to open an exploitation where routes pass through, pastoralists may accept compensation for relocation of water resources, routes, etc. As a result when pastoralists do not accept compensation the private operator would not be allowed to extend the mining exploitation.

3.6 Non-conservationist May Accept Change with Compensation Opportunity

Persha and Blomley (2009) illustrate a situation where non-conservationists—a forestry cooperative—, are the owners of a forest in Tanzania. The cooperative may decide if it accepts compensation, for example from a REDD+program, to conserve the forest. Similar situations are described in seven case studies related to Payments for Ecosystem Services (PES) by Mahanty et al. (2013).

Ahmed, et al. (2008) find that local communities in the Hakaluki Haor wetland in Bangladesh may have willingness to participate in wetland conservation activities. They hold the right to manage the natural resources of the wetland as claimants and they may accept conservation practices under monetary or non-monetary compensation.

Pastoralist in Kenya's Maasailand who are authorized users may accept aggregate their grazing land via a voluntary agreement with other pastoralists to improve the livestock production and sustainability of the traditional practices (Mwangi 2007).

3.7 Conservationist Does not Accept Compensation at All

Sekhar (2004, p. 264) showed for a fishermen community at Chilika lake/India that a High Court—upon claim—"ordered the state to uphold the traditional access rights to lake resources and continue leasing to local fishermen" which in the following could restrict outsiders' entry and make the state to recognize of the community's role in fisheries management. It can be assumed that the fishermen would not have accepted a financial compensation of their rights, although the question of compensation to the fishermen was not directly at stake as outsiders did not pay lease fees to the fishermen but to the state as owner.

3.8 Non-conservationists Does not Accept Compensation at All

It is rather not astonishing that not any case for the eight category has been found. This fits well into the overall picture of the "homo economicus". This type of persons does likely accept money in exchange to depletion of nature. Nevertheless there are situations imaginable which do fit into this 8th category. These could be cases where people based on social, cultural or religious reasons to not accept compensations, no matter how much is offered. These could be called a socio-lexicographic behaviour (Mauerhofer et al. 2013).

4 Discussion

4.1 Discussion of the Method

The method of the selection of the papers assess was—despite random—not optimal as it was observed that the list with 297 papers did not include every ISI-listed paper (e.g., Thanh and Sikor 2006) that cited Schlager and Ostrom (1992) while papers of the same journal published in the same year were included (e.g. Nguen 2006). In any way, this issue was only found related to one journal and one year.

Whereas authors of papers described the practical situation and made an allocation to property rights and positions, this allocation was reordered if the practical situation indicated a different allocation.

For example, Quinn et al. (2010) extensively describe management rights of the owners in the uplands of the United Kingdom but conclude that there are no persons with claimant rights found in their study. Another example constitutes the paper of Caballero-Miguez et al. (2012) who indicate that the legal "system for granting licenses for floating raft cultivation in Galicia guaranteed the access, withdrawal and exclusion rights, whereas the rights of management and alienation were more highly restricted." (p. 89) In fact they describe earlier that the right to exclude is not in the hand of the licensees but with the legislative body.

The new framework is well applicable for both, de jure or de facto rights. Case studies that contain both types could be also analysed, for instance, by inserting into the third column of the Table 4 the persons with de jure and de facto rights separately.

4.2 Discussion of the Theoretical Application

The system is not applicable when alienation rights for individuals are not legally in place. The allocation within the proposed Table is not possible if the right holder is the Community but an individual person "sells" this right. Legally spoken, such a sell is invalid as usually nobody can transfer a right which this person does not have but de facto the transfer happens especially if enforcement is weak or not existent (see e.g. Hayes 2010). The allocation is also not possible as in these situation one person often belongs to both groups, the conservationists and non-conservationist (if a distinction is possible at all).

The differentiation proposed does also not work if the paper lacks the information weather compensation has to be offered or not. Also those cases are difficult to allocate into the scheme where for example the public authority gives away rights and—in exchange—other benefits are derived than direct compensation (e.g. higher employment, revenues from taxes or export customs).

A crucial differentiation is between de facto right positions and de iure right positions. If de iure positions are not appropriately implemented, de facto positions can overwhelm with unintended effects (Schlager and Ostrom 1992; Barsimantov et al. 2011; Khan and Khan 2009). Other studies indicate that occasionally also de facto rights are compensated for (Trawick 2001).

Some studies elaborate for example only the position that people perceive to have with regard to natural resources while the exact legal background, whether there is a clear rule or not, is not in detail provided for (e.g. Ahmed et al. 2008). In any way, withstanding, unsettled de iure and de facto positions are in many cases the origin and ongoing reason of conflicts among stakeholders. This was shown for example by Sekhar (2004) in a well-described fishery case study in Chilika lake/India regarding the rights of management and exclusion.

Even if no property rights are withstanding against a certain project, the projects might be not permitted due to 'public interests' by a public authority. This is for example illustrated by Munton (1995) on the case of an opencast coalmine project that had already obtained all property rights necessary for the extraction of the minerals but was finally not permitted by an public council due to overall environmental conservation aspects initiated by people without the relevant property rights.

This paper only accesses immediate change situation, e.g. the permission about an envisaged project and whether e.g. a conservationist/non-conservationist has to agree or agrees to the change. Papers that only show a certain position but do not refer to a concrete situation of change are not included in the table.

On the other hand, the person who want to bring forward the change has also a certain legal position such as all rights but one obtained and by not obtaining the last right all the other rights already obtained somehow become useless to that person and—it can be argued—also change in their value, and therefor constitute a situation of change on the side of that person. But actually all those other rights have not been useful at all but just would become the first time useful at that moment when the last right missing has been obtained.

Persha and Blomley (2009) show that a variety of right-holding private persons (such as owners, claimants and authorized users) that do not follow conservation interests regarding forests can have the possibility to partly or fully transfer these rights, but can also be fully restricted by a public authority's ban (regardless an ownership position of that authority).

My approach also—due to the focus on Schlager and Ostrom (1992)—does not include those papers that focus on Schlager and Ostrom (1993) or Ostrom and Schlager (1996) where these authors elaborated their framework more towards a clearer distinction between access and withdrawal, leading to five types of users. The focus on the Thompson Reuters Web of Science of course excludes all journals and articles not listed in this data base but cite Schlager and Ostrom (1992).

The new framework described in Tables 3 and 4 has also some advantages in comparison with the concept of Schlager and Ostrom (1992), as it in particular does make clear that every legal position can be—in theory and often in practice—

restricted by an authoritative act based on democratic legitimation due to certain public interest that override private property right positions.

These authoritative acts can restrict every private property right position from its very beginning or be introduced afterwards. The introduction of these acts can be done in a formal-legally correct way, but based on illegal reasons such a bribery or other forms of corruption.

Thus, the quality of each property right is also case-sensitive upon the opportunities for authoritative acts. Hence, the position of an owner could be considered by a holder weaker than the position of a proprietor depending on the respective extent of potential authoritative acts.

Not only authoritative acts that actively weaken the property right-holder position are essential, but also those which defend property right-holder position. Apart of self-defence, the question whether a property right-holder position can be defended against outside intruders with the help of public authorities based on democratic legal acts ("enforcement") is crucial too. The most comprehensive property right position is of only formal value if not protected by effective enforcement mechanism. Enforcement mechanisms can be distinguished for example into legal or illegal, formal or informal and community- or individual based ones.

The application of the scheme presented in the Table 4 has another limitation, namely in the cases where it is unclear whether an activity towards a change is voluntarily or not, such as in cases of resettlements due to conservation reasons.

Situations of decentralization and privatization (Agraval and Ostrom 2001; Ybarra 2011) are also more difficult to allocate to the table as the law that provides the basis for the transfer of property rights from the public forestry administration to local people is usually voluntarily released despite local pressure on the public legislative bodies. Also it is not easy to decide (1) to allocate the state (who initiates the change) still to the conservationists and (2) whether or not anything has in exchange the quality of an compensation (e.g. fees for right transfers, higher income tax revenues, reduced costs of control). These difficulties always occur when the public is both, rule setter and right-holder.

The scheme in the table is limited applicable also to situations of change between two non-conservationists or two conservationists.

The distinction between conservationists and non-conservationist is hardly possible if papers lack to describe the conservation outcome of a situation of change or the outcome is not visible yet.

For example, the legally imposed or facultative transposition of use rights of agricultural land and forests in collective ownership to individuals happened recently (Yin et al. 2013) and therefor does not provide information on the outcome in terms of conservation of these natural resources.

In comparison, Hayes and Persha (2010) analyse five studies of different ownership situations and show that the communally managed reserves in both studies, where local residents held sole decision-making authority to make rules regarding the access, use and management of forest lands, showed more positive forest conservation outcomes than the other three cases where residents held minimal or no forest rulemaking rights.

Quinn et al. (2010) describe legal restrictions of conservation organizations as owner through access rights on behalf of tourists. The same legal restriction is in this study valid for ownership positions of water companies, forestry (state/private), agriculture and owners of moorland dedicated to grouse hunting. With regard to these user groups and their relationship to tourists the allocation of the terms "conservationists" and "non-conservationists" is for example rather difficult.

Changes that happen within the substance of the right, such as the re-interpretation by the right-holder of a de-jure use and withdrawal right towards a de facto exclusion right due to an unclear legal formulation such as described by Aburto et al. (2013) for two fishery rights in Chile, are also difficult to include into the Table developed.

5 Conclusion

The paper describes a new conceptual framework based on property rights and financial compensation in different nature conservation situations in order to provide a globally applicable system for the assessment of participation of public and private stakeholders in envisaged changes within those situations.

These situations of change represent a modification from conservation toward non-conservation and vice versa. The framework distinguishes further between governance systems based on command and control as wells as on negotiation. Within these main change situations and governance types, the framework allows the distribution of change situations into 8 main sections. These main sections are further separated into 32 sub-sections by means of different property right and compensations situations among public and private stakeholders.

The theoretical utility of this new framework is then demonstrated by testing it by means of a random sample of 74 papers (25 %) out of representative 297 papers from the academic literature dealing with property rights. These 74 papers provided practical examples for situations of change in conservation as evidence for most of the 32 sub-section. Several papers provide examples for more than one sub-section. This widely possible allocation proofs in general the global applicability and usefulness of the new framework. The framework also proofed to be appropriate for formally (rule of law based) and informally (customary law based) institutionalized situations where rights are given to public and private stakeholders. Furthermore, the results allow simple assumptions on the regularity of occurrence of situations representing the 32 sub-sections. The discussion reveals also limits of the new framework. Despite, the frameworks' potential future extensions are manifold such as regarding relative predictions on the length of negotiation procedures and on the cost-effective amount of the compensations to be paid by one stakeholder to the other.

The research establishes a general framework to assess the legal preconditions under which the extent of environmental protection in situations of disagreement among stakeholders can be further assessed. It therefore contributes to the assessment of the fundamental question whether further marginal steps towards an additional usage of the environment and on which aspects of the legal systems the solution of these questions mainly depends upon among public and private stakeholders.

References

Aburto J, Gallardo G, Stotz W, Cerda C, Mondaca-Schachermayer C, Vera K (2013) Territorial user rights for artisanal fisheries in Chile—intended and unintended outcomes. Ocean Coast Manage 71:284–295

Agrawal A, Ostrom E (2001) Collective action, property rights, and decentralization in resource use in India and Nepal. Politics Soc 29:485–514

Ahmed I, Deaton BJ, Sarker R, Virani T (2008) Wetland ownership and management in a common property resource setting: a case study of Hakaluki Haor in Bangladesh. Ecol Econ 68:429–436

Barsimantov J, Racelis A, Biedenweg K, DiGiano M (2011) When collective action and tenure allocations collide: Outcomes from community forests in Quintana Roo, Mexico and Petén, Guatemala. Land Use Policy 28:343–352

Behera B, Engel S (2006) Institutional analysis of evolution of joint forest management in India: a new institutional economics approach. Forest Policy Econ 8(2006):350–362

Bogale A, Taeb M, Endo M (2006) Land ownership and conflicts over the use of resources: implication for household vulnerability in eastern Ethiopia. Ecol Econ 58:134–145

Bottazzi P, Dao H (2013) On the road through the Bolivian Amazon: a multi-level land governance analysis of deforestation. Land Use Policy 30:137–146

Caballero-Miguez G, Garza-Gil MD, Varela-Lafuente MM (2012) Legal change, property rights system and institutional stability: the case of the floating raft culture in the Galician mussel sector. Ocean Coast Manage 55:84–93

Campbell B, Mandondo A, Nemarundwe N, Sithole B, De Jong W, Luckert M, Matose F (2001) Challenges to proponents of common property recource systems: despairing voices from the social forests of Zimbabwe. World Dev 29:589–600

Cinner JE, McClanahan TR, MacNeil MA, Graham NAJ, Daw TM, Mukminin A, Feary DA, Rabearisoa AL, Wamukota A, Jiddawi N, Campbell SJ, Baird AH, Januchowski-Hartley FA, Hamed S, Lahari R, Morove T, Kuange J (2012a) Comanagement of coral reef social-ecological systems. Proc Nat Acad Sci USA 109(14):5219–5222

Cinner JE, Daw TM, McClanahan TR, Muthiga N, Abunge C, Hamed S, Mwaka B, Rabearisoa A, Wamukota A, Fisher E, Jiddawi N (2012b) Transitions toward co-management: the process of marine resource management devolution in three east African countries. Global Environ Change-Human Policy Dimensions 22:651–658

Dumyahn SL, Pijanowski BC (2011) Beyond noise mitigation: managing soundscapes as common-pool resources. Landscape Ecol 26:1311–1326

Gerber JD, Knoepfel P, Nahrath S, Varone F (2009) Institutional resource regimes: towards sustainability through the combination of property-rights theory and policy analysis. Ecol Econ 68:798–809

Glück P (2002) Property rights and multipurpose mountain forest management. Forest Policy Econ 4:125–134

Gómez-Márquez I, Alejano LR, Bastante FG (2011) Mining compatibility with other projects in Spain: solutions and benefits. Resour Policy 36(1):22–29

Hayes TM (2007) Does tenure matter? A comparative analysis of agricultural expansion in the Mosquitia Forest Corridor. Human Ecol 35:733–747

Hayes TM (2010) A challenge for environmental governance: institutional change in a traditional common-property forest system. Policy Sci 43:27–48

Khan SR, Khan SR (2009) Assessing poverty-deforestation links: evidence from Swat, Pakistan. Ecol Econ 68:2607–2618

Khan S, Haque CE (2010) Wetland resource management in Bangladesh: implications for marginalization and vulnerability of local harvesters. Environmental Hazards-Human and Policy Dimensions 9:54–73

Kusters K, de Foresta H, Ekadinata A, van Noordwijk M (2007) Towards solutions for state vs. local community conflicts over forestland: the impact of formal recognition of user rights in Krui, Sumatra, Indonesia. Human Ecol 35:427–438

Lescuyer G (2013) Sustainable forest management at the local scale: a comparative analysis of community forests and domestic forests in Cameroon. Small-Scale For 12:51–66

Lescuyer G, Assembe-Mvondo S, Essoungou JN, Toison V, Trébuchon JF, Fauvet N (2012) Logging concessions and local livelihoods in Cameroon: from indifference to alliance? Ecol Soc 17(1):7

Lund JF, Treue T (2008) Are we getting there? Evidence of decentralized forest management from the Tanzanian Miombo Woodlands. World Dev 36:2780–2800

Mahanty S, Suich H, Tacconi L (2013) Access and benefits in payments for environmental services and implications for REDD+: lessons from seven PES schemes. Land Use Policy 31:38–47

Mauerhofer V, Nyacuru F (2013) Distributional dimensions of PES: general issues and critical aspects. In: Proceedings of the 19th annual international sustainable development research conference, 1–3 July 2013, Stellenbosch University, South Africa, Retrieve 1 July 2013, from www.isdrc19.co.za

Mauerhofer V, Hubacek K, Coleby A (2013) From polluter pays to provider gets: distribution of rights and costs under payments for ecosystem services. Ecol Soc 18(4):41

McCay BJ, Creed CF, Finlayson AC, Apostle R, Mikalsen K (1995) Individual transferable quotas (ITQs) in Canadian and US fisheries. Ocean Coast Manage 28:85–115

Munton R (1995) Regulating rural change—property rights, economy and environment—a case study from Cumbria, UK. J Rural Stud 11:269–284

Mwangi E (2007) Subdividing the commons: distributional conflict in the transition from collective to individual property rights in Kenya's Maasailand. World Dev 35(5):815–834

Nagendra H, Gokhale Y (2008) Management regimes, property rights, and forest biodiversity in Nepal and India. Environ Manage 41:719–733

Naidu SC (2011) Access to benefits from forest commons in the Western Himalayas. Ecol Econ 71:202–210

Nielsen MR, Treue T (2012) Hunting for the benefits of Joint Forest Management in the Eastern Afromontane Biodiversity Hotspot: effects on bushmeat hunters and wildlife in the Udzungwa Mountains. World Dev 40(6):1224–1239

Nguyen TQ (2006) Forest devolution in Vietnam: differentiation in benefits from forest among local households. Forest Policy Econ 8:409–420

Ostrom E, Schlager E (1996) The formation of property rights. In: Hanna S, Folke C, Mäler K-G (eds). Rights to nature: ecological, economic, cultural and political principles of institutions for the environment. Island Press, Washington, DC, pp 127–156

Penker M (2009) Landscape governance for or by the local population? A property rights analysis in Austria. Land Use Policy 26:947–953

Persha L, Blomley T (2009) Management decentralization and montane forest conditions in Tanzania. Conserv Biol 23(6):1485–1496

Quinn CH, Fraser EDG, Hubacek K, Reed MS (2010) Property rights in UK uplands and the implications for policy and management. Ecol Econ 69:1355–1363

Rodgers C (2010) Reversing the 'Tragedy' of the commons? Sustainable Management and the Commons Act 2006. Mod Law Rev 73:461–486

Rothgang M (1997) Ökonomische Perspektiven des Naturschutzes. Duncker & Humblot, Berlin, p 315

Roy AKD, Alam K, Gow J (2012) A review of the role of property rights and forest policies in the management of the Sundarbans mangrove forest in Bangladesh. Forest Policy Econ 15:46–53

Schlager E, Ostrom E (1992) Property-rights regimes and natural resources: a conceptual analysis. Land Economics 68(3):249–262

Schlager E, Ostrom E (1993). Property rights regimes and coastal fisheries: an empirical analysis. In: Anderson T, Simmons R (eds) The political economy of customs and culture: informal solutions to the commons problem. Rowman & Littlefieldpp, Lanham, pp 13–41

Sekhar NU (2004) Fisheries in Chilika lake: how community access and control impacts their management. J Environ Manage 73:257–266

Thanh TN, Sikor T (2006) From legal acts to actual powers: devolution and property rights in the Central Highlands of Vietnam. Forest Policy Econ 8:397–408

Trawick PB (2001) Successfully governing the commons: principles of social organization in an Andean irrigation system. Human Ecol 29:1–25

Wiber MG, Rudd MA, Pinkerton E, Charles AT, Bull A (2010) Coastal management challenges from a community perspective: the problem of 'stealth privatization' in a Canadian fishery. Mar Policy 34:598–605

Ybarra M (2011) Slashed and burned: the debate over privatization of Q'eqchi' Lands in Northern Guatemala. Soc Nat Resour 24:1027–1041

Yin RS, Yao SB, Huo XX (2013) China's forest tenure reform and institutional change in the new century: what has been implemented and what remains to be pursued? Land Use Policy 30:825–833

Prioritising the Environment in Sustainable Development: Lessons from Australian Environmental Impact Assessment

Sophie Riley

Abstract It is a truism that to be effective, the concept of sustainable development (SD) needs to be functionally operational. Environmental impact assessment (EIA) provides such an opportunity; yet EIA also presents decision-makers with a vast array of competing criteria compelling decision makers to prioritize and make trade-offs. Moreover, legislation provides little guidance on how to prioritize these criteria and still achieve SD. Using the Australian state of New South Wales as a case study, the discussion evaluates the relationship between policy, legislation and the weight given to SD. The topic is important for policy makers, decision-makers, proponents of development and conservationists. The paper draws on two bodies of work: the literature on prioritising and trade-offs in decision-making by authors such as Brownlie and Retief et al.; and the paradigms and models of science identified by Cashmore. It proffers a means of curtailing the wide discretion available to decision-makers using civic science, which to be effective, needs to be legislatively-based.

Keywords Sustainable development · Environmental impact assessment · Science-based decision-making · Trade-offs

1 Introduction

In its early formulations, sustainable development (SD) focussed on integrating economic and environmental factors (WCED 1987, para. 72) with a third pillar, that of social considerations added in 2002 (United Nations 2002). At the project level, environmental impact assessment (EIA) has developed into one of the most important means of facilitating SD, providing a framework for decision-makers to appraise the consequences of proposed development against the principles of SD

S. Riley (✉)
Faculty of Law, University of Technology Sydney, Sydney, Australia
e-mail: Sophie.Riley@uts.edu.au

V. Mauerhofer (ed.), *Legal Aspects of Sustainable Development*,
DOI 10.1007/978-3-319-26021-1_14

(Makmor and Ismail 2014, p. 21). As decision-makers undertake their assessments they exercise wide discretion. However, where environmental, social and development matters are not sufficiently compatible, achieving integration has proved elusive. Instead, decision-makers weigh and balance relevant criteria in an attempt to accommodate a plethora of often-conflicting stakeholder expectations (Colyvan et al. 2011, p. 1246; Cashmore 2004, p. 404).

The purpose of this paper is to evaluate the relationship between policy and legislation, at the project level, with respect to the weighing and balancing of criteria relevant to SD. The discussion adopts a case study methodology, using developments in policy and planning law in the Australian state of New South Wales (NSW). To provide a frame of reference, the paper commences with a précis of SD in the NSW planning system before moving to an examination of recent case law in the NSW Land and Environment Court. Although the Land and Environment Court has expounded the core elements of SD, the challenges inherent in balancing and weighing criteria in order to achieve SD cannot be resolved solely by judicial intervention.

In particular, judicial review has a limited remit in adjudicating on administrative and executive decision-making (Fisher 2013, p. 48). Primarily, such review safeguards the public interest by ensuring that the executive is held accountable if it does not adhere to lawful decision-making processes. This has limited bearing on how decision-makers weigh and balance criteria, unless they fail to follow legislative guidance, or they make a decision that is manifestly unreasonable. For these reasons, EIA runs the risk of focusing on process at the expense of achieving sustainable outcomes. In practice, decision-makers in EIA generally enjoy large amounts of discretion that can be exercised in political and unsustainable ways (Mauerhofer 2012, pp. 657, 658). Moreover, they rarely need to adhere to guidance on the weighing and balancing of criteria (Stein 2000, pp. 3, 8), except in those instances where government has explicitly elevated socio-economic considerations (SEPP 55 1998; SEPP Mining 2013). In Australia, for example, the economy is heavily reliant on resource extraction, especially with respect to the energy product sector (Australian Bureau of Statistics 2013, Chap. 4) and the executive faces many challenges in balancing short-term economic benefits against environmental protection. This, and analogous dilemmas have unfurled in the context of the NSW planning regime, where governments, keen to promote mining and film production, promote those objectives at the expense of the environment.

This paper argues that managing the wide discretion available to decision-makers is vital to aligning their choices with principles of SD. This is an important issue, because where the executive exercises discretion on matters that cannot be integrated, the decision-maker invariably engages in prioritisation, which leads to explicit and implicit trade-offs. It is rare that such trade-offs foster SD (Morrison-Saunders and Retief 2013). One means of guiding discretion is to make better use of scientific material that can help reign in unsustainable prioritization and resultant trade-offs. EIA practitioners, themselves, are aware that science is not used as effectively as it could be, especially in the decision-making component of EIA (Morrison-Saunders and Bailey 2003, pp. 683–685, 688).

The discussion draws on two bodies of work: first, the literature on prioritising and trade-offs, by authors such as Brownlie et al. (2013) and Retief et al. (2013); and second, the paradigms and models on the role of science in EIA identified by Cashmore (2004). Cashmore recognizes that science operates along a continuum, ranging from applied science to civic science. The paper argues that decision-making located towards the "civic science" part of the spectrum is compatible with the inclusion of societal and community concerns. In addition, the different types of science can be appropriate at different stages of EIA. However, to diminish or manipulate the role of science for political purposes lowers the threshold for good decision-making, which triggers a decline towards unsustainable development.

It should also be noted, that making better use of scientific material is not the only approach to using science to strengthen the "environment" in SD. Mauerhofer (2012, p. 653 and 2008, pp. 501–502) for example, has noted that those who advocate for land use policies that operate at a weak sustainability level, should bear the burden of proof that "there are no doubts (from a scientific point of view) about the appropriateness of these policies." Mauerhofer's viewpoint focusses on the importance of science. Similarly, the argument in this paper acknowledges this importance by calling for strengthened legislative guidance on the use of science in the context of the NSW planning regime in order to counterbalance short-sighted weighing and balancing in EIA.

2 Sustainable Development and the NSW Planning Scheme

Initial references to SD at the international level regarded policy and legislation as essential to supporting the integration of conservation and development, as well as being pivotal to managing the rational use of resources (IUCN 1980, parts 9 and 10). The NSW Planning Scheme pre-dates the emergence of the concept of SD and to some extent this is reflected by the lack of regard for environmental matters in early planning laws. The *Local Government (Town and Country Planning) Amendment Act* (1945) (NSW), for example, empowered local councils to regulate development in their areas. However, by 1979 it had become clear that these schemes were not adequately dealing with environmental protection. This prompted the NSW government to enact the *Environmental Planning and Assessment Act* (1979) (NSW) (the EPAA), establishing the first comprehensive planning scheme in NSW. The first version of the EPAA did not refer to SD, although the objectives of the Act, set out in section 5, included the protection of the environment and the proper management of natural resources for the purpose of promoting social and economic welfare.

Some eight years later, the 1987 *Report of the World Commission on Environment and Development: Our Common Future* (Brundtland Report) embraced one of the most recognisable characterizations of SD: “development that meets the needs of the present without compromising the ability of future generations to meet their own needs” (WCED 1987, part 2.1) While this formulation is short and to the point, the concept’s details remain far from settled (Bond and Morrison-Saunders 2011, p. 2). Indeed, to guide implementation of SD, in 1992 States adopted both the Rio Declaration on Environment and Development as well as Agenda 21. The former sets out 27 guiding principles for attaining SD; while the latter comprises a blueprint of some 40 Chapters for achieving the same objective. The notion of social integration was included in the concept of SD following the World Summit on Sustainable Development in 2002 (United Nations 2002, p. 26).

In Australia, SD was known as ecologically sustainable development and quickly found its way into policy documents such as the *National Strategy for Ecologically Sustainable Development* (1992) as well as objectives in legislation, including section 3 of the *Threatened Species Conservation Act* (1995) (NSW) and an amendment to the EPAA in 1997, that introduced SD as an objective in section 5 (vii). These instruments portray SD by a variety of purposes, objectives and descriptions. Australia’s National Strategy, for example, depicts SD as “using, conserving and enhancing the community’s resources so that ecological processes, on which life depends, are maintained, and the total quality of life, now and in the future, can be increased” (part 1). The EPAA describes the concept by traditional references to core values such as to the precautionary principle, inter-generational equity, conservation of biological diversity, improved valuation, pricing and incentive mechanisms and the polluter pays principle (EPPA, section 4). However, SD must also accommodate socio-economic objectives that range from promoting and coordinating the “orderly and economic use and development of land” (EPAA section 5[a][ii]), to providing affordable housing (EPPA, section 5[a][viii]), and enhancing public participation (EPPA, section 5[c]). These objectives are consistent with the Rio Declaration on Environment and Development and provide a timely reminder that SD should not be confused with environmental conservation (Kidd 2008, p. 102). Rather, SD centres on the integration of a vast range of economic, social and environmental matters (Kidd 2008, p. 85).

For these reasons, documentation generated by EIA should identify the benefits and detriments of proposed development (Schaffer *v Hawkesbury, para.* 2) including highlighting “the likely consequences of environmental change” (Wathern 1988, p. 6). Rather than pinpointing “right” or “wrong” solutions, EIA is designed to provide decision-makers with sufficient information to make informed choices (Demidova and Cherp 2005, p. 413). However, given the variety of principles and potential activities that broach SD it is not surprising that both the meaning, and application of the concept, generate much dissidence (Bond and Morrison-Saunders 2011, p. 1). Broadly speaking, the concept has strong and weak versions, with the former providing a robust eco-centric approach to environmental protection, while the latter tends towards subordinating environmental issues to economic development (Cashmore 2004, p. 417). For practical purposes, the strong version considers *whether* development should go ahead, while the weaker version uses balances and trade-offs in order to determine *how* proposed development should go ahead.

Section 79C of the EPAA provides a list of matters that decision-makers must take into account when considering proposals. These not only include relevant planning instruments, environmental and socio-economic values, but also extend to the public interest. The NSW Land and Environment Court has held that the latter is equated with "the broader public good" (Taralga Landscape Guardians Inc v Minister for Planning and RES Southern Cross Pty Ltd (2007), para. 3), which itself is consistent with objectives of SD (Telstra Corporation Limited v Hornsby Shire Council, para. 123). This does not necessarily mean that consideration of SD is obligatory under the rubric of public interest (Australians for Sustainable Development Inc v Minister for Planning (2011) NSWLEC 33 para. 241-3). Although as Bates points out, SD is a sufficiently wide concept to warrant inclusion of the public interest in almost every instance (Bates 2013, para. 7.40). Notwithstanding the lengthy criteria enumerated in section 79C, the section is silent on matters of weighting and balancing; and this remains one of the thorniest regulatory challenges.

3 Trade-offs, Priorities and Policy

World-wide, objectives of sustainable development are not being met (Brownlie et al. 2013, p. 24). One of the reasons for this lies with the practice of implementing weak sustainability which is entrenched by systemic failures in EIA processes that focus on mitigation of harm as a means of facilitating development, rather than considering whether the project should in fact proceed (Brownlie et al. 2013, p. 24). Essentially, this has occurred as a result of trade-offs where government and decision-makers accept that a certain level of environmental degradation is the cost that society pays for socioeconomic advancement (Gibson 2013, p. 2). Brownlie et al. define trade-offs as activities that

> occur in development when two or more conflicting objectives are being pursued in a situation where resources are limited, and result in a specific negative outcome being exchanged for another positive outcome in time and/or space (2013, p. 25).

Although the strategy set out in the "Brundtland Report" was based on SD, which itself was squarely grounded on integration, (WCED 1987, para. 72) some of the objectives of environmental protection and development may not be sufficiently compatible to permit integration. In these circumstances, decision makers have no choice but to weigh and balance the relevant criteria, which leads to trade-offs (Bates 2013, para. 7.33; McShane et al. 2011, pp. 968–969). Government, for example, may wish to promote a product sector or activity and specifically advance trade-offs; while decision-makers engage in trade-offs, although less explicit ones, when they exercise their discretion in the weighing and balancing of criteria.

Examples of the former are provided by the *Filming Approval Act* (2004) (NSW) and SEPP Mining (2013), which are examined later in this article; while examples of tacit trade-offs stem from decision-making processes, such as those discussed in Australians for Sustainable Development Inc v Minister for Planning, Lend Lease (Millers Point) Pty Ltd and Barangaroo Delivery Authority (2011) NSWLEC 33.

Some trade-offs are procedural and relate to the level of public engagement, or the time and resources spent on EIA. Other trade-offs are substantive and relate to prioritisation that translates into terms and conditions in development approvals; or may involve legislative schemes that provide frameworks for trade-offs, such as occurs with biodiversity offsets (State of NSW and Office of Environment and Heritage 2014). Offsets in particular, allow proponents of development to demonstrate that they have responded appropriately to environmental impacts, even though they involve trading-off a negative environmental outcome in one location, which may not be adequately compensated by protection measures elsewhere (Walsh 2014, p. 397). Whether trade-offs occur by way of process or substance, they invariably weaken the achievement of SD (Morrison-Saunders and Retief 2013). They also reveal what society and/or government consider worthy of protection (Palmer 1997, pp. 5–6, 114–115), an issue that is frequently brought to light, at least in NSW, in environmental litigation and its aftermaths.

A case in point is the decision in Blue Mountains Conservation Society Inc v Director-General, National Parks and Wildlife Service (2004) 133 LGREA 406 (Blue Mountains case). The case involved judicial review of the grant of a licence to use of part of the Grose Wilderness in the Blue Mountains National Park to film scenes for the movie "Stealth". The applicants objected on the basis that the area select for filming was environmentally-sensitive and alternative sites were available (Blue Mountains case, para. 28). The Court noted that the proposed activity was not consistent with either the management objectives for wilderness areas as set out in section 9 of the *Wilderness Act* (1987) (NSW) or with the objectives of the *National Parks and Wildlife Act* (1974) (NSW) in section 2A. Accordingly, the Court found that the licence had been improperly granted. In coming to this decision his Honour Lloyd *Judge* also noted community concerns that the public would have been excluded from the site during filming (Blue Mountains case, para. 20). In effect, the granting of the licence would have traded-off a part of the environmental and social objectives of the legislation in exchange for economic benefits flowing from the filming. The Court was not prepared to sanction this trade-off as it contradicted key purposes of the legislation.

Yet, following this decision, the NSW government passed the *Filming Approval Act* (2004) (NSW) (the Film Act). The second reading speech made it very clear that the purpose of the Film Act was to encourage the development of the film industry in NSW (MacDonald 2004). Section 5(1) of the Act provides that filming activities may be carried out without the need for development consent under the EPAA and may be carried out "even if the development would be prohibited, or would require development consent, in the absence of this section". As a safety measure, the Film Act also stipulates that the Minister (who is the decision-maker) may not grant approval for filming in wilderness areas unless the filming is undertaken for educational or tourist purposes (the Film Act, section 4[4]). Furthermore, the Minister must be satisfied that filming activities minimise adverse environmental impacts and are also carried out in the shortest time feasible (the Film Act, section 4[7]). The notion of "minimising" environmental impacts implicitly concedes that some environmental damage is acceptable, as the price to pay for expansion of the film industry. This observation on its own does not necessarily offend against the principles of SD. However, the tenor of the Film Act arguably offends against the notion of integration, because the Act emphasises economic gain at the expense of environmental and social considerations. In particular, the Film Act did not sufficiently grapple with one of the reasons for the decision in Blue Mountains case, namely that the public would be excluded during filming and such exclusion was contrary to the objectives of the *National Parks and Wildlife Act* (1974) (NSW).

Another, and comparable, illustration stems from the litigation in Australians for Sustainable Development Inc v Minister for Planning, Lend Lease (Millers Point) Pty Ltd and Barangaroo Delivery Authority (2011) NSWLEC 33 (the Barangaroo Case). The Barangaroo case involved the re-development of a 22 hectare site on the shores of Sydney Harbour to the west of the Sydney central business district. The Minister for Planning had given consent for early excavation and the construction of a basement car park. The applicant was seeking orders that the approvals were invalid as well as an injunction to restrain the works.

The site had been used for gas production and other industrial purposes for many decades and it was common knowledge that the land was contaminated. (Barangaroo case, para. 21). The litigation centred on the extent and quality of the remediation deemed necessary and the applicants' concern that the building works would release toxins into Sydney Harbour. For these reasons, the applicants argued that stricter levels of remediation were required than the defendants were prepared to undertake. The appeal clustered on two broad issues: first, whether the decision-maker, who was the Minister for Planning, had given proper regard to SD as a component of the public interest; and, second, the consequences flowing from the failure of the second defendant to have a remedial action plan in place in accordance with clause 17(1)(c) of SEPP 55 (1998) (Barangaroo case, para. 246–247).

At the conclusion of the hearing the Court reserved its judgement, which was expected within two weeks. Yet, before the decision was handed down, the Minister for Planning made an order exempting the Barangaroo site from the operation of SEPP 55. This limited the range of orders open to the Court. Had the Minister not promulgated this exemption, the Court would have found in favour of the applicant on the basis of a breach of SEPP 55. Biscoe *Judge* held that although the Minister did in fact consider the contamination impacts on Sydney Harbour when he granted approval for the building works, this was not sufficient to comply with the very specific requirements of SEPP 55 (Barangaroo case para. 249–50). However, his Honour would not have found in the applicant's favour with respect to the arguments relating to SD (Barangaroo case para. 252, 260 and 298). Biscoe *Judge* noted that the Minister had material before him that squarely addressed issues relevant to SD, including site contamination, climate change and the impacts of the project on Sydney Harbour. Moreover, the Minister had clearly considered these issues (Barangaroo case para. 249–280). The fact that the construction phase of the project breached SEPP 55 did not of itself indicate that the Minister had failed to give due regard to SD (Barangaroo case para. 251). Accordingly, faced with the amendments to SEPP 55, Biscoe *Judge* had no option but to dismiss the appeal (Barangaroo case para. 309).

The decision is instructive for demonstrating how dramatically government policy can shape the implementation of SD. The purpose of SEPP 55 was to ensure that contaminated land was remediated to a level that would mitigate harm to humans and the environment. That type of determination needed scientific evidence that would identify an optimum level of remediation. Yet, government policy overrode the need for a science-based approach in favour of advancing the economic importance of the development. The outcome was that principles of SD became buried in the prioritising of a plethora of conflicting socio-economic considerations. On this point, the Court acknowledged that there were serious environmental impacts that the Minister needed to consider (Barangaroo case para. 245), which meant that the Minister could not discharge his obligation by a perfunctory appraisal (Barangaroo case para. 243). At the same time, the weighting and balancing of the criteria was a matter for the Minister's discretion. Accordingly, the Court rejected arguments pertaining to SD that hinged on the need to investigate how this discretion had been exercised. In reality, the weighting and balancing of criteria comprises a finely-tuned process that enables decision-makers to make whatever evaluative determinations are "rationally open" to them (H and J Standen Pty Ltd v Minister for Planning and Infrastructure 2014, para. 111).

The third example, stems from the aftermath of the decision in Bulga Milbrodale Progress Association Inc v Minister for Planning and Infrastructure and Warkworth Mining Limited (2013) NSWLEC 48 (Warkworth's case). In that case, the Land and Environment Court refused an application for expansion of a mine. The mine had originally started operating in 1981 and over the years the original development consent had been varied. The latest variation, made in May 2003 was subject to a biodiversity offset, which meant that part of the land on which the mine operated could not be disturbed. The litigation arose because Warkworth was seeking to mine in the biodiversity-offset area and also to mine closer to residential areas. The Minister approved the mine's expansion, but imposed a number of conditions intended to mitigate environmental damage and also to deal with the community's objections.

On appeal, the Land and Environment Court held that the conditions were inadequate and refused Warkworth's application. Preston *Chief Justice* noted that there were significant adverse impacts relating to "biological diversity, noise and dust, and social impacts" that were inadequately addressed by the Minister's approval (Warkworth's case, para. 14). Furthermore, his honour noted that the economic evaluations provided by Warkworth did not address these issues appropriately (Warkworth's case, para. 14–20). Warkworth's appeal against this decision was dismissed.

However, following this decision the government amended State Environmental Planning Policy (Mining, Petroleum Production and Extractive Industries 2007 (Mining SEPP)) to elevate the economic significance of mining. In accordance with clause 12AA, the determining factors are the project's economic benefits and consideration whether "other industries or projects are dependent on the development of the resource" (Mining SEPP clause 12AA(4)). At the same time, the Mining SEPP does not alter the fact that pursuant to section 79C of the EPAA, decision-makers need to consider the public interest. Although the ramifications of clause 12AA are not fully clear, the amendments represent an explicit trade-off of monumental proportions. The Mining SEPP elevates the socio-economic significance of proposed mining; and arguably does this to a level that conflicts with those elements of SD, such as the precautionary principle, protection of biodiversity and intergenerational equity (The Law Society of New South Wales 2013, 5). Walsh (2014, p. 403) in particular notes that the clause "could be used to justify conditions which do not fully address biodiversity impacts on the basis that such conditions would be onerous and limit exploitation" of a significant resource.

These three cases illustrate the willingness of government to engage in trade-offs that shift the balance and weighting of decision-making to promote economic objectives at the expense of environmental and social matters. It is an approach that is entirely inconsistent with achieving SD. Retief et al. (2013) posit that acknowledgment of the importance of environmental values is essential for resolution of the trade-off and prioritising dilemma. Core environmental values need to be generally accepted, as well as being made "explicit in public law and policy" (Retief et al. 2013, p. 21). One means of emphasising the importance of environmental values is to make more effective use of science.

4 A Science-Based Approach?

As EIA evolved, it developed multi-faceted criteria to accommodate a range of human and environmental concerns. In a comparable way, the use of science evolved along a spectrum ranging from applied science, as a "rational process of objective enquiry", to civic science, where the emphasis is placed on social values (Cashmore 2004, p. 403, 405, 410; Morrison-Saunders and Bailey 2003, p. 685). This has led to discussion on how stakeholders use science in decision-making and whether science can provide an objective counterbalance to the subjective prioritisation and trade-offs that otherwise occurs.

Cashmore (2004, p. 408) categorizes the application of science and socio-economic values across five models: the Analytical Science Model, where science plays a crucial role and is seen as a methodical means of undertaking an objective inquiry (Cashmore 2004, p. 408); the Environmental Design Model where science identifies a level of damage that the decision-maker deems tolerable

(Cashmore 2004, p. 410); the Information Provision Model that provides sufficient information, including appraisal of alternatives to the proposed development to enhance decision-making (Cashmore 2004, p. 411); the Participation Model that views socio-economic issues as being more central to decision-making (Cashmore 2004, p. 412); and, the Environmental Governance Model, where science has an important role to play, but the emphasis is placed on community participation, transparency and accountability. In the latter case, the goal is to provide "a framework for negotiation and compromise" (Cashmore 2004, p. 413). In practice, these classifications are not discrete. They overlap, and in many instances it will be difficult to categorize decisions as belonging solely to one group or another. However, underpinning Cashmore's approach is the notion that that science does not automatically exclude socio-economic matters, and in EIA different versions of science can be appropriate for different parts of the decision-making process.

The further the role of science shifts towards civic science, the more likely it is that socio-economic matters will find a voice. This can include input from industry, or NGOs and community groups that have a strong eco-centric approach. Incorporating this range of views can bring mixed blessings. While civic science is more likely to give credence to social and community values, it also heightens the potential for controversies that centre on conflicting power-plays. The latter may have less to do with science and more to do with subjective reasons for resistance or politicisation of the process of EIA. In these situations, decision-makers can use science to shape the outcome, prejudicing its role as an objective determinant. The circumstances surrounding the Bald Hills wind farm in the South Gippsl and area of the State of Victoria is a case in point.

The proposed wind farm consisted of 52 wind turbines which would have provided a renewable energy source, reducing greenhouse gas emissions. Yet, the Minister for Planning refused his consent because the wind farm threatened the orange-bellied parrot. The Minister based his decision on a second report he commissioned that was prepared pursuant to modified scoping. The original report had concluded that the wind farm would not adversely affect the conservation status of the orange-bellied parrot. Prest (2007, pp. 236–237) argues that the Minister's actions were politically motivated, not only because they manipulated the science, but also because they appeared to be driven by community opposition to the project, in what was an election year. In the ensuing backlash, the proponent commenced litigation, which was ultimately settled, followed by a re-submission and approval of the project. Prest (2007, pp. 255–256) notes that:

> The Bald Hills incident also demonstrates the broad discretion available to the Environment Minister…The flaw in the Minister's decision-making was that an extremely remote possibility of significant impact to a threatened bird – the orange bellied parrot – was given weight out of all proportion to its true significance. Thus the difficulty is with politicisation of the planning and environmental approval decision-making process.

In Prest's view, the Minister had exploited the science in order to justify an exercise of discretion that was unreasonable. Although science may be regarded as an objective discipline, a great deal depends on how it is used. In particular, science needs to be treated in a balanced way; neither should it be ignored or overruled, nor should it be commandeered for political purposes.

Where government overrules the science, as for all intents and purposes the NSW government did when it amended SEPP 55 and the Mining SEPP, the science becomes virtually ineffectual. One consequence is that economic considerations become the predominant imperative. It is also equally wrong to use science to authenticate refusal of a project for political reasons, as occurred in the Bald Hills case; or to use science to justify policy that abrogates social and community concerns. The latter are crucial to environmental management (Grumbine 1997, p. 47; Ecosystem Approach 2000, p. 103), and must be considered equally along with the scientific evidence. If social and community concerns are given scant regard, it not only shapes the type of science decision-makers favour, but also leads to decisions that impact harshly on communities in a way that is arguably inconsistent with SD. Martin and Williams (2014, p. 315) have examined this issue in the context of water allocation licences in the Murray-Darling system. In those cases, a form of applied science was used to carry out modelling in order to reduce water allocations. Yet, as the authors point out, modelling will rarely capture nuances such as community needs and values

> Through these arrangements the *Water Act* privileges knowledge and values held by selected scientific experts and economists. By implication this subordinates views and interests that are not readily supported by scientific or economic evidence, or values that differ from these interests, potentially hindering access to social justice for marginalised communities affected by these decision-making processes.

For Martin and Williams (2014, p. 315), the problems lie not so much with science per se, but with the type of science that is used, and with the way that the science is used. Given the social consequences of the new licence regime, it would have been fairer for regulators to use a type of "social science". Otherwise, a strict application of applied science runs the risk that regulators distort the science to conform to a pre-determined outcome. This falls short of a genuine application of science, and administrators' preferences can lead to environmental, economic and social costs. Practitioners in the field of EIA consider that science is crucial for empirical work such as gathering and presenting information on data, environmental impacts and mitigation of impacts (Morrison-Saunders and Bailey 2003, pp. 683–685, 688); a point with which, Martin and Williams (2014, p. 320) concur. EIA practitioners also perceive that the use of science is neither consistent nor as high as it could be (Morrison-Saunders and Bailey 2003, p. 683). Its use depends as much on the economic capacity of the developer as it does on legislative and policy requirements (Morrison-Saunders and Bailey 2003, p. 692).

Decision-makers also regard science as less important for the decision-making component of EIA, where they consider it just one of many aids to reaching a decision (Morrison-Saunders and Bailey 2003, pp. 683–685, 688). Yet, the two extremes in decision-making, ignoring the science, or giving science too much prominence, can both lead to unsustainable trade-offs because of the way that criteria are prioritised. If science is to be given a more meaningful role, important issues stem from the type of science that is most appropriate to EIA and whether the different components of EIA warrant different forms of science. Applied science, for example could be more relevant to the initial stages of EIA, including risk prediction, while decision-making would be better served by civic science. This would allow the accommodation of core values of SD, which would also provide a foundation of knowledge to help prevent one criterion from dominating. One issue that does arise is whether this approach merely transfers the problem from trade-offs that are unjustified because they are environmentally-detrimental to trade-offs that are equally unjustified because they favour science-based approaches. McShane et al. (2011, p. 968) note that in environmental regulation trade-offs are inevitable, but their benefit lies in the fact that they bring together a range of stakeholders. Thus, the drawing power of trade-offs makes it more likely that "the right people are…at the negotiating table". Accordingly, the differing roles of science advanced by Cashmore can be seen as one means of pulling together "the right people", a concept that can vary in accordance with the level of decision-making involved and the role of science at that particular level.

5 Implications for a Sustainable Development

Undoubtedly EIA has improved decision-making by comparison to the situation before its introduction; however, it is questionable whether EIA can be viewed as having achieved SD (Cashmore 2004, p. 404). During the time that EIA developed, it needed to accommodate a range of human and environmental concerns and, as already discussed, policy-makers and decision-makers turned to prioritising and trade-offs to achieve desired outcomes. Again, as already discussed, decision-makers in EIA generally have a large amount of discretion that can be exercised in political and unsustainable ways. This means that EIA focuses more on process at the expense of outcomes. One means of slowing this trend is to provide clear legislative guidance on how to deal with competing claims. Rather than prioritizing economic values, decisions should be science-based, and informed by values that underpin environmental protection as an equal consideration to social and economic factors.

The type of science used in EIA, and the way that the science is used influences whether regimes develop strong or weak versions of SD. Where science is used to give the environment equivalent weighting to that of economic and/or social factors, it is said to foster strong sustainability. Significant environmental considerations, such as reducing global carbon emissions or achieving national benchmarks on biodiversity conservation, can limit the type of development that is approved. Alternatively, where science is ignored, or used to promote one criterion of SD, or overcome objections and obstacles to development, it fosters a weaker version of sustainability (Cashmore 2004, p. 417). In these cases, the outcome frequently equates with environmental degradation. Yet, this is not to say that SD is identical to environmental conservation, even though environmental protection is a relevant criterion (Kidd 2008, p. 102).

These observations lead to a conundrum posed by Kidd (2008). The socio-political contexts of decision-making, including the way regulators view criteria, can influence the weighting and balancing in ways that are not immediately apparent. Kidd (2008, pp. 91, 94), for example, questions whether decision-makers in South Africa examine socio-economic factors purely from a socio-economic perspective, or whether they consider these factors from an environmental or other perspective. If the evaluation is undertaken from an environmental perspective, then arguably the environment is given "double" weighting. At the same time, Bates (2013, para. 7.35) notes that policy and regimes in western neo-liberal societies already favour socio-economic outcomes. For this reason, decision-making in those jurisdictions often focusses on *how* development should proceed rather than *whether* it should proceed. The process is based on a tacit presupposition that in the absence of compelling reasons, development should ensue. This demonstrates a bias towards development where science is often regarded as a means of overcoming environmental objections, suggesting that environmental protection is not being optimised (Bates 2013, para. 7.35). In Australia, this tendency is exacerbated by the fact that references to SD are found in non-operative parts of legislation without direction on how to weigh and balance relevant criteria, (Stein 2000, pp. 3, 8) as well as the propensity of government to overrule and manipulate the science.

The problem is likely to be further complicated in New South Wales by moves to repeal the EPPA and replace it with the provisions of the Planning Bill (2013), which fosters a streamlined approach to planning and assessment. Part 4.18 of the Bill is equivalent to s 79C of the EPAA. The former sets out criteria for a consent authority to take into account when determining a development application and includes matters such as the content of planning instruments, environmental and socio-economic impacts it as well as public submissions and socio-economic impacts of regional or state significance. Significantly, the public interest criterion found in s 79C of the EPAA has been removed, although it still remains as an objective in part 1.2 of the Bill. As already discussed, the Land and Environment Court has determined that appraisal of the "public interest" is fully consistent with reviewing criteria that can help achieve SD (Taralga Landscape Guardians Inc v Minister for Planning and RES Southern Cross Pty Ltd, para. 3). The Environmental Defence Office (EDO) (2013, pp. 16, 57–58) of NSW is concerned that the potential impact of this change is to weaken the weighting given to principles of SD. In particular, by removing public interest from the list of relevant criteria it means that judicial review, which primarily focusses on the processes of the decision-making, is less able to consider the components of SD in the context of competing environmental and socio-economic claims. At the time of writing, the Bill had passed the lower house, but had stalled in the upper house and lapsed until the next sitting of parliament, later in 2015.

Decision-making in EIA already incorporates a strong political overlay as decision-makers try to balance competing factors. As a result, political influences often outweigh scientific considerations, leading Morrison-Saunders and Bailey

(2003, p. 685) to conclude that EIA is akin to a "political process involving trade-offs rather than a purely scientific undertaking". The authors do not suggest that EIA should be purely science-based, but they lament the prominence that politicisation enjoys in the process. Partly, for these types of reasons, the NSW EDO has recommended that decision-makers incorporate "objective decision making criteria (such as a 'maintain or improve environmental outcomes' test)" (EDO 2013, pp. 16, 57). This suggestion embodies an eco-centric approach, based on science, but does not necessarily equate with decision-making based on "applied" science that is the mainstay of data collection and risk identification. Nor does it mandate ignoring social and community concerns. Indeed, the consideration of social factors is strong enough to include community well-being and is one of the core objectives of Australia's *National Strategy for Ecologically Sustainable Development*.

EIA practitioners explain that civic science can improve stakeholder participation and also improve the "scientific integrity" of EIA (Morrison-Saunders and Bailey 2003, p. 693). In their view, an important preliminary step, which is part and parcel of civic science, is to obtain input from stakeholders in the planning stages of proposals to give stakeholders a voice (Morrison-Saunders and Bailey 2003, p. 693). On one reading of these viewpoints, the outcome of EIA procedures could therefore depend on the type of stakeholders that become involved, potentially skewing the EIA process in accordance with the group that is the most outspoken. Such an example has been discussed in the Bald Hills wind farm case, where environmental concerns were twisted for political motives. Yet, in other cases, environmental protection is regularly afforded a lower level of consideration than say, economic interests. Hence, where stakeholders highlight the detrimental environmental impacts of a project, this produces an adjustment, that prior to the exercise of discretion, allows environmental matters to be placed on at least an equivalent footing to other concerns.

Although, lack of appropriate consultation may in fact be a trade-off that developers tacitly adopt if they consider the EIA process is too slow, consultation in the early stages of proposals is vital, especially for those developments that generate complex, polycentric disputes. Polycentric disputes involve multiple criteria that are so interrelated it is neither possible to work through them consecutively, nor is it feasible to balance them objectively (Eisenberg 1978, p. 425). Preston *Chief Justice* of the NSW Land and Environment Court notes that in such cases, administrative decision-makers exercise a type of managerial role to determine subjectively the relevancy and weighting of matters (Warkworth's case, para. 35–36). Accordingly, the determination involves an "intuitive synthesis" (Warkworth's case para. 41; Norm Fletcher and Associates Pty Limited v Strathfield Municipal Council 2014, para. 67). Even where decision-makers deal with criteria sequentially, they can still consider interactions and dependencies, but the decision-maker is able to regard one or more issues as determinant of the matter, without intentionally synthesizing them (Barrak Corporation Pty Ltd v Parramatta City Council 2014, para. 94). Irrespective of which approach the decision-maker favours, it still involves the exercise of abundant and subjective discretion.

Accordingly, while polycentric and sequential approaches may be consistent with legislation that underpins the decision-making process, this is not necessarily decisive as to whether the approaches promote SD. In particular, Bond and Morrison-Saunders (2011, p. 2) maintain that if decision-makers assess a proposal piece by piece, this equates to a form of reductionism that does not capture "the relationship between important variables that contributing to effectiveness". The benefit of obtaining early input from stakeholders is that it allows the science to target those areas where it most needed. It can, for example, provide the decision-maker with alternatives and sufficient information to make a fairer, more equitable decision. It can also help to identify whether projects are in reality consistent with SD. Yet, at the end of the day the extent to which science and early stakeholder input are deemed critical to achieving SD depends on the political will of the government of the day, otherwise difficulties arise for the judiciary to ensure that environmental values receive equivalent consideration as socio-economic values.

6 Conclusion

It is a truism that to be effective, the concept of SD needs to be functionally operational. Although EIA provides such an opportunity, SD is frequently seen as an optional consequence of decision-making rather than as a mandated outcome. This is a deficiency of both process and substance that is aggravated by lack of effective guidelines on prioritizing competing claims. Indeed, in NSW, where legislative guidance is provided, it frequently prioritizes economic imperatives. In one sense, the vast range of criteria relevant to SD makes compromise inevitable. Yet, as decision-makers bargain values against each other, it may foster one imperative dominating, leading to compromise and trading-off that weaken SD. These deficiencies are exacerbated where decision-making is politicised.

Given the fact that in practice, the environment is often not afforded an equivalent regard to economic and social matters, it is important that decision-makers use science to pinpoint environmental problems and how they can be managed. Indeed, by tightening legislative guidance so that decision-makers apply a science-based approach, EIA can place environmental protection at the forefront of decisions. Significantly, the type of science proposed by this paper is based on civic science as identified by Cashmore. Consequently, while science plays a key role in identifying whether a proposed development is sustainable, social values and economic factors still have a role to play. The economic and political value of a development, however, would not be the overriding factor.

It is a matter of some concern that notwithstanding the shortcomings of the current system the NSW government is planning to introduce legislation that will facilitate poor regard to principles of SD, further weakening environmental protection. As Bates (2013, para. 7.33) points out, EIA currently weighs and balances criteria to determine to what extent SD should be achieved—and this is simply the wrong approach. Rather than trading-off SD against economic and social factors, the weighing and balancing should be done to achieve SD and legislative guidance in this regard is essential. To put it succinctly, the prioritisation of criteria should not be done to "assess the relative priority of SD to other factors. SD should stand alone as *the* object of the legislation".

References

Australian Bureau of Statistics (2013) Information paper: towards the Australian environmental-economic accounts, Chap. 3 Energy (27 Mar 2013), 4655.0.55.002. Retrieved 15 Mar 2015. http://www.abs.gov.au/ausstats/abs@.nsf/Products/4655.0.55.002~2013~Main +Features~Chapter+3+Energy?OpenDocument

Barrak Corporation Pty Ltd v Parramatta City Council (2014) New South Wales land and environment court 177

Bates G (2013) Environmental law in Australia, 8th edn. LexisNexis Butterworths, Australia, pp 1–964

Bond AJ, Morrison-Saunders A (2011) Re-evaluating sustainability assessment: aligning the vision and the practice. Environ Impact Assess Rev 31(1):1–7

Brownlie S, King N, Treweek J (2013) Biodiversity tradeoffs and offsets in impact assessment and decision making: can we stop the loss? Impact Assess Project Appraisal 31(1):24–33

Bulga Milbrodale Progress Association Inc v Minister for Planning and Infrastructure and Warkworth Mining Limited (2013) NSWLEC 48

Cashmore M (2004) The role of science in environmental impact assessment: process and procedure versus purpose in the development of theory. Environ Impact Assess Rev 24(4):403–426

Colyvan M, Justus J, Regan HM (2011) The conservation game. Biol Conserv 144(4):1246–1253

Demidova O, Cherp A (2005) Risk assessment for improved treatment of health considerations in EIA. Environ Impact Assess Rev 25(4):411–429

Ecosystem approach (22 June 2000) Adopted by the conference of the parties to the CBD in decision V/6 set out in report of the fifth meeting of the conference of the parties to the convention on biological diversity 103 UNEP/CBD/COP/5/23

EDO (June 2013) NSW white paper submission, 1-93. Retrieved 15 Mar 2015. http://d3n8a8pro7vhmx.cloudfront.net/edonsw/pages/180/attachments/original/1380534662/130628NSWPlanningWhitePaper_EDONSWsubmission.pdf?1380534662

Eisenberg MA (1978) Participation, responsiveness, and the consultative process: an essay for Lon Fuller. Harvard Law Rev 92:410–432

Environmental Planning and Assessment Act 1979 (NSW) Retrieved 20 Apr 2015. http://www5.austlii.edu.au/au/legis/nsw/consol_act/epaaa1979389/

Filming Approval Act 2004 (NSW) Retrieved 20 Apr 2015. http://www5.austlii.edu.au/au/legis/nsw/consol_act/faa2004177/

Fisher D (2013) Legal reasoning in environmental law: a study of structure, Form and Language. Edward Elgar, Cheltenham

Gibson RB (2013) Avoiding sustainability trade-offs in environmental assessment. Impact Assess Project Appraisal 31(1):2–12

Grumbine RE (1997) Reflections on what is ecosystem management? Conserv Biol 11:41–47

H & J Standen Pty Ltd v Minister for Planning and Infrastructure (2014) New South Wales land and environment court 113

IUCN (1980) World conservation strategy: living resource conservation for sustainable development, gland: IUCN, 1-77. Retrieved 15 Mar 2015. https://portals.iucn.org/library/efiles/documents/WCS-004.pdf

Kidd M (2008) Removing the green-tinted spectacles: the three pillars of sustainable development in South African environmental law. S Afr J Environ Law Policy 15:85–102
Local Government (Town and Country Planning) Amendment Act (1945) (NSW) Retrieved 20 Apr 2015. http://www5.austlii.edu.au/au/legis/nsw/num_act/lgacpaa1945n21561.pdf
MacDonald I (2004) Minister for primary industries second reading speech, filming approval bill 2004 (NSW), NSW hansard articles: LC: 02/06/2004: #42. Retrieved 15 Mar 2015. http://www.parliament.nsw.gov.au/prod/parlment/nswbills.nsf/0/87669a32fd2d64f5ca256e8b0010b09e/$FILE/C3804.pdf
Makmor M, Ismail Z (2014) A comparative study on EIA process in Malaysia, West Australia, New Zealand and Canada. Jurnal Teknologi 70(1):15–22
Martin P, Williams J (2014) Science, hubris and insufficient legal safeguards? Environ Plann and Law J 31:311–326
Mauerhofer V (2008) 3-D sustainability: an approach for priority setting in situation of conflicting interests towards a sustainable development. Ecol Econ 63:496–506
Mauerhofer V (2012) A 'legislation-check' based on '3-D Sustainability'—addressing global precautionary land governance. Land Use Policy 29:652–660
McShane TO, Hirsch PD, Trung TC, Songorwa AN, Kinzig A, Monteferri B, Mutekanga D, Thang HV, Dammert JL, Pulgar-Vidal M, Welch-Devine M, Brosius JP, Coppolillo P, O'Connor S (2011) Hard choices: making trade-offs between biodiversity conservation and human well-being. Biol Conserv 144(3):966–972
SEPP Mining (2013) State environmental planning policy (mining, petroleum production and extractive industries) amendment (resource significance). amended state environmental planning policy (mining, petroleum production and extractive industries [2007-65]. Retrieved 20 Apr 2015. http://www.austlii.edu.au/au/legis/nsw/num_epi/seppppaeia201320135811004.pdf
Morrison-Saunders A, Bailey J (2003) Practitioner perspectives on the role of science in environmental impact assessment. Environ Manage 31(6):683–695
Morrison-Saunders A, Retief F (2013) Editorial. Impact Assess Project Appraisal 31(1):1
National Parks and Wildlife Act (1974) (NSW). Retrieved 20 Apr 2015. http://www5.austlii.edu.au/au/legis/nsw/consol_act/npawa1974247/
National Strategy for Ecologically Sustainable Development (1992) Australian government, department of the environment. Retrieved 20 Apr 2015. http://www.environment.gov.au/about-us/esd/publications/national-esd-strategy
Norm Fletcher & Associates Pty Limited v Strathfield Municipal Council (2014) New South Wales land and environment court 157
Palmer C (1997) Environmental ethics. ABC-CLIO, California
Planning Bill (2013) (NSW) Retrieved 20 Apr 2015. http://www.parliament.nsw.gov.au/prod/parlment/nswbills.nsf/0/33E72AD6EA1238B5CA257C0C0014134D
Prest J (2007) The Bald Hills Wind Farm debacle. In: Bonhady T, Christoff P (eds) Climate law in Australia. Federation Press, Australia, pp 230–261
Retief R, Morrison-Saunders A, Geneletti D, Pope J (2013) Exploring the psychology of trade-off decision-making in environmental impact assessment. Impact Assess Project Appraisal 31(1):13
Rio Declaration on Environment and Development (1992) Report of the United Nations conference on environment and development, Rio de Janeiro, 3 June-14 June A/CONF.151/26 (vol. I). Retrieved 20 Apr 2015. http://www.un.org/documents/ga/conf151/aconf15126-1annex1.htm
Schaffer v Hawkesbury (1992) New South Wales land and environment court 77
SEPP 55 (1998) State environmental planning policy No 55—remediation of land (SEPP 55). Retrieved 20 Apr 2015. http://www.planning.nsw.gov.au/assessingdev/pdf/sepp55_remediatio.pdf
State of NSW and Office of Environment and Heritage (2014) NSW biodiversity offsets policy for major projects. Office of environment and heritage for the NSW Government (2014). Retrieved 15 Mar 2015. http://www.environment.nsw.gov.au/biodivoffsets/bioffsetspol.htm

Stein P (2000) Are decision-makers too cautious with the precautionary principle? Environ Plann Law J 17(1):3–23

Taralga Landscape Guardians Inc v Minister for Planning and RES Southern Cross Pty Ltd (2007) New South Wales land and environment court 59

Telstra Corporation Limited v Hornsby Shire Council (2006) New South Wales land and environment court 133, para. 123, Preston chief justice

The Law Society of New South Wales, Young Lawyers, Environment and Planning Law Committee (2013) Submission on the state environmental planning policy (mining, petroleum production and extractive industries) amendment (resource significance), NSW young lawyers. Retrieved 15 Mar 2015. http://www.lawsociety.com.au/cs/groups/public/documents/internetyounglawyers/763290.pdf

Threatened Species Conservation Act 1995 (NSW) Retrieved 20 Apr 2015. http://www5.austlii.edu.au/au/legis/nsw/consol_act/tsca1995323/

United Nations (2002). Report of the World summit on sustainable development, Johannesburg, South Africa, 26 August 4 September. A/CONF.199/20, 20. Retrieved 20 Apr 2015. http://www.un.org/jsummit/html/documents/summit_docs/131302_wssd_report_reissued.pdf

Walsh V (2014) The future of land and environment court oversight of major project offsets. Environ Plann Law J 31:397–412

Wathern P (1988) An introductory guide to EIA. In: Wathern P (ed) Environmental impact assessment theory and practice. Unwin Hyman, London, pp 3–30

WCED (1987) Report of the world commission on environment and development: our common future (short "Brundtland-Report"), transmitted to the general assembly as an annex to UN Doc A/42/427 Part 2.1

Wilderness Act (1987) (NSW). Retrieved 20 Apr 2015. http://www5.austlii.edu.au/au/legis/nsw/consol_act/wa1987139/

Reframing Sustainability in Taiwan: Legal Challenges and Opportunities

Yao-Ming Hsu

Abstract This chapter mainly describes Taiwan's policies and legal regimes for sustainability and its future prospective in international participation. First of all, along with the economic development, Taiwan gradually established specific authority for environmental governance, being responsible from prevention of pollution to precautionary measures for guaranteeing sustainable use of natural resources. In the approaching future, a new Ministry of Environment and Resources will be installed for a centralized authority and comprehensive governance. Besides, national programs and strategies announced by the Cabinet in Taiwan, accompanying with related laws and regulations, mostly passed by the legislation, played a significant role in implementations of environmental affairs. In a word, Taiwan has shown its capacity and ambition to cope with the need of sustainability in our age. However, because of its ambiguous status in international arena, Taiwan still lacks opportunities to participate in real, official international cooperation for environmental protection and sustainable development. Some *sui generis* participation is nevertheless envisaged in a long run.

Keywords Taiwan · Sustainable development · Environmental protection · Environmental impact assessment · Climate change

1 Introduction

Sustainable development has been for a long time recognized as a supreme value for national and international balancing between economic development and environmental protection, especially declared by the Rio Declaration (Ke 1994) and reconfirmed by the Rio+20 conference (Li 2015, p. 226). However, its concise meaning and actual implementations significantly differ, either in domestically

Y.-M. Hsu (✉)
College of Law, National Cheng-Chi University, Tapei, Taiwan
e-mail: ymhsu@nccu.edu.tw

V. Mauerhofer (ed.), *Legal Aspects of Sustainable Development*,
DOI 10.1007/978-3-319-26021-1_15

policies and legal framework (Li 2015, p. 229), or in international stage for drafting, contracting or even monitoring/implementing of international instruments.

For a special national scenario, Taiwanese sustainability encounters some specificity. First of all, the natural and geographical uniqueness push Taiwan to a vulnerable status under threats from typhoons and earthquakes. Possible influences form these catastrophic disasters and relevant reactions should be taken into account for the sustainable development in Taiwan. Besides, even though Taiwan was not able to join many international treaties or organizations because of its *sui generis* nature, Taiwan still bestows its possible contributions to international sustainability. For example, a "National Council of Sustainable Development"[1] was established under the Administrative Yuan (the Cabinet) in 1997; further, for national actions, the "National Policy Program for Sustainable Development"[2] was declared in 2009 and an "Action Plan for Sustainable Development"[3] was also created and reviewed every year, all by the National Council of Sustainable Development. Besides, since 2010 discussions about the reorganization of governmental structure has been launched and in the approaching future, all the environmental affairs and related issues will be incorporated into the new Ministry of Environment and Resources. However, policies and legal framework for sustainable development in Taiwan still lacks of *de jure* international monitoring. Sometimes, Taiwanese environmental policies and regulations would still give their way to needs of economic development. This weak sustainability triggers domestic and international concerns. So, new challenges emerge, and new opportunities come.

Subsequently, this chapter tries to use international benchmarking for checking real achievement of Taiwanese sustainability. The policies in Taiwanese programs and action plans, the existing legal frameworks, and the approaching establishment of Ministry of Environment and Resources, will all be discussed below.

2 Taiwanese Policies for Sustainability

The development of environmental governance in second half of the last century could be broadly divided into two stages (Tseng 2008, p. 129). From 1950 to 1979, Taiwanese environmental governance was solely controlled by powerful government, accompanying with natural resources conservationism. In this period, environmental sanitation, exploitation and use of resources, and public nuisance prevention constituted main issues for sustainability, even though at that time such an idea was still in development (Tseng 2000, pp. 52–72). From 1980 to 2000, a compounded pluralism emerged in environmental governance, in which

[1]Retrieved May 15, 2015, from http://nsdn.epa.gov.tw/en/index.htm.

[2]Retrieved May 15, 2015, from http://nsdn.epa.gov.tw/20100203.pdf.

[3]Retrieved May 15, 2015, from http://nsdn.epa.gov.tw/CH/DEVELOPMENT/DEVELOPMENT.HTM.

non-governmental organizations (NGO) play a significant role, searching for distributive justice in environment and development issues.

Organizationally, early in 1982, an Environmental Protection Bureau (EPB) was set under the Health Administration in the Cabinet. Along with more and more environmental disputes, a Trans-ministerial Working Group for Environmental Protection (TWGEP) was also established in 1986 and finally an Environmental Protection Administration (EPA) was installed in the Cabinet in 1986. Accompanying with these models of governance, and under the international influence urging for global environment protection, especially the 1992 Earth Summit in Rio, Taiwanese government gradually noticed the organizational importance to have a specialized organ to deal with sustainability issue. That's why a National Council of Sustainable Development was established under the Cabinet in 1997 and in 2012 a project of establishment of Ministry of Environment and Resources was enshrined in the revision of the Organization Act of the Cabinet. Therein, all the former regulatory powers for natural resources management in the Ministry of Internal Affairs and for energy development and management in the Ministry of Economics will be also incorporated in this new Ministry. Yet, the reorganizational project is not detailed because of relative revisions are still under discussions in the Cabinet.

For the policy coordination, the "Taiwan 21st Century Agenda"[4] was announced in 2004; the "National Policy Program for Sustainable Development" was declared in 2009 and an "Action Plan for Sustainable Development" was also created and reviewed every year. Even though these policy announcements do not have real legal binding force, they still show the decisive orientations for Taiwanese environmental governance.

In the "Taiwan 21st Century Agenda", three main targets are set: sustainable society, sustainable economy and sustainable environment (ECC 2004, p. 4). Because of special natural environment as a small island and largely export-oriented economy policy, Taiwan faces a dilemma between economic development and environmental protection. Thus, some basic principles are set in this agenda: considerations of environmental burdens, prevention, intergenerational justice, technological innovation and institutional reform, public participation and international participation (ECC 2004, pp. 8–9). Concretely, the indicators of sustainability should be established both on central and local government level; the mechanism of decision and capacity-building for implementation should be improved, and a strategic outline is also pictured.

In this "Strategic Outline for Sustainable Development"[5] (EEC 2004, p. 9), annexed to this Agenda, three major orientations are concerned: sustainable society, sustainable economy and sustainable environment. First of all, for guaranteeing a sustainable environment, three targets are aimed on nature conservation, prevention of public nuisances and environmental planning. Besides, for developing a

[4]Retrieved May 17, 2015, from http://www.ndc.gov.tw/m1.aspx?sNo=0000533#.VVhbUvmqpBc.

[5]Retrieved May 17, 2015, from http://www.ndc.gov.tw/m1.aspx?sNo=0000604#.VVhdLfmqpBc.

sustainable economy, green industries, clean production and green consumption are encouraged. Finally, for establishing a sustainable society, principle of justice as fairness should be taken into account at all levels of public participations for community development and public health decisions.

Subsequently, in the "National Policy Program for Sustainable Development", the National Council of Sustainable Development has detailed all policies about sustainable society, sustainable economy and sustainable environment. In addition, in this program a new dimension was added for monitoring and implementation evaluation. In the beginning, for a sustainable development principles of inter-generational justice, equitable deliberation, environmental capacity, prevention, social justice, health care, technical innovation, public participation, policy integration and international participation are declared (NCSD 2009, p. 9). In the policy settings, for a sustainable environment, all issues relating to air, water, land, ocean and biodiversity are addressed. Besides, for a sustainable society, population and health, environment for habitats, social welfare, cultural diversity and disaster prevention and rescue are also detailed. Furthermore, for a sustainable economy, all issues about economic, industrial and traffic developments, sustainable energy and resources recycling are well programmed. However, all the above mentioned principles and policies are similar to those in the "Strategic Outline for Sustainable Development" in 2004 and all we need should be real implementation. Thus, in this program another dimension of mechanisms for implementation were additionally established for achieving all the goals for sustainable development (NCSD 2009, pp. 96–117). By using the strategies in education, technological research and development, information society, public participation, reorganization in governmental structure and international participation, consciousness about the sustainable development would be addressed in education system and civil society, in all industrial development through effective public participation. Most important is that Taiwan would actively search for possibilities to cooperate with other countries or international organizations.

Nevertheless, without effective implementation and monitoring, all principles and programs will remains just as slogans. Every year from 2007 on, an "Action Plan for Sustainable Development" was announced and every season or every half a year, an evaluation report would be issued for verifying the actual detailed implementations. For example, in the recent 2014 report of 330 pages, nine working groups (WG) reports are adopted: WG Climate Change and Mitigation Measures, WG land use and resources, WG biodiversity, WG Energy and Production, WG Traffic and Living, WG Technology and Evaluation,, WG Urban and Countryside Development, WG Health and Welfare and finally WG Education and Guidance (NCSD 2014b). For example in WG Climate Change and Mitigation Measures, for the goals of reduction of amounts of greenhouse gas emissions to the level of 2005 in 2020 and to the level of 2000 in 2025, the National Development Committee has launched a National Strategy for Mitigation, in which the Energy Bureau of the Ministry of Economics has tested an offshore wind power plant project and a solar power plant project. Besides, the Industry Bureau of the Ministry of Economics has also declared revised industrial policies for steel and petrochemical production, and also promoted advanced developments of solar panel, LED and electronic vehicle industries. Furthermore, the Agriculture Committee announced some carbon-neutralization strategy, in which organic agriculture and new methods of animal husbandry are welcomed. Finally, for the voluntary international compliance, the Environmental Protection Administration declared plenty of international monitoring criteria and recorded lots of local data for environmental implementation (NCSD 2014b, pp. 2–30).

Additionally, since 2003, every year the National Council of Sustainable Development will publish a "Sustainable Development Indicators" report (NCSD 2014a), in which every indicator for evaluating levels of sustainability based on the indicators system proclaimed by the United Nations are reported in very details of 18 dimensions. For example, within the environmental dimensions, the indicators of PM 2.5 and pollutants in the air, drinking water quality, recycling rate, ecological budget… etc. are well announced.

In sum, we could see that Taiwanese government has launched lots of programs and evaluation mechanisms for guaranteeing sustainability in Taiwan. However, all these policies need further legally binding laws and administrative regulations for assuring their real application.

3 Taiwanese Legal Frameworks for Sustainability

Before 1980, most of Taiwanese environmental norms concentrate in the pollution-prevention dimensions and are mainly promoted solely by governmental commands. After 1980, along with the emerging needs for cultural diversity and ecological sustainability, massive environmental laws and regulations boosted (Liu 1997, p. 39; Tseng 2008, p. 139; Yeh 2010a, p. 75). Based on principles of polluter-pays, polluter-solves, environmental prevention, technology-forcing, multilevel collaboration, public participation and access to information (Yeh 2010a, pp. 91–98; Chen 2012, pp. 307–338), Taiwanese environmental laws and regulations stepped into a new era for sustainability. Besides, a systemized, hierarchical legal regime for sustainable development and environmental protection was also gradually well-established, being pushed by the demands of legislative integrity, the inspiration of foreign legal systems, international prescriptions and the petitions from the people, including certain supports from certain political figures (Yeh 2010a, pp. 103–107).

Theoretically speaking, for environment governance, there are four possible regulatory instruments: direct administration, command and control, economic incentives and moral persuasion (Yeh 2010a, pp. 116–118). A mixed pluralism exists in all kind of respective laws and regulations, except for using the method of moral persuasion. However, the command and control method remains the mainstream in most laws, in which an administrative responsibility and even a criminal penalty are prescribed for guaranteeing the effectiveness of governance.

In a word, we could sketch all the laws and regulations for sustainability and environment protection in a systematic order by its function (Yeh 2010a, p. 121). At the top of system, the Basic Environment Act[6] serves as the foundation of all environmental laws and regulations, even though it was promulgated in 2002, later than more than half of the existing environmental norms. In the categories, the concept of prevention, control, remedy and organization serves as criterion for four types of norms.

At first, the Basic Environment Act solemnly declares the sustainable development as a fundamental value in its Article 1 that "this Act is formulated to raise the quality of the environment, advance the health and well-being of citizens, preserve environmental resources, and pursue *sustainable development* by promoting environmental protection." Besides, it also clearly defines that "sustainable development means satisfying contemporary needs without sacrificing the ability of future generations to satisfy their needs" (Art. 2.2). For the subjects of realizing these ideas, "citizens, enterprises and government entities at all levels shall jointly share the duties and responsibilities of protecting the environment" (Art. 4.1). As we have seen in the above discussions about the programs and action plans, several principles are clearly set in this Act, including the principle of polluter-pays, green production and consumption, international cooperation, prevention, no nuclear energy… etc. Besides, it also urges that education for sustainable development should be promoted at all levels of administration, in which regular budgets, functions of expertise, training for staff and relative planning at all levels of government should be ascertained. Especially, in Article 7.1 it is clearly announced that "the central competent authority shall draft environmental protection laws and regulations, draw up national environmental protection plans, establish sustainable development indicators and promote and implement such laws and regulations, plans and indicators." For a better implementation, "government entities at all levels shall fully enforce environmental laws and regulations and interdict and punish violations thereof in accordance with the law." (Art. 39). Last but not least, for environmental disputes "the courts may establish a dedicated court or designate dedicated personnel for environmental protection disputes." (Art. 14); furthermore, "the central government shall establish a mechanism for resolving environmental disputes, strengthen approaches to assess reasons for dispute and burden of proof, education, training, research and development, and provide an appropriate dispute resolution mechanism." (Art. 33.1); even, "if a government entity at any level is negligent in enforcement, persons or public interest groups may, in accordance with laws and regulations, name said competent authority as a defendant and directly file a lawsuit with the Administrative Court." (Art. 34.1), wherein a new type of "public interests group" action for sustainability and environmental protection is created.

[6]In Taiwan, in most of the official English translation for the title of laws, which are passed by our Legislative Yuan (the Congress), the word "Act" is used.

Nevertheless, this Basic Environment Act sounds like general clauses by its nature and still needs more detailed, concrete environmental laws and regulations for actual realisation.

Second, for the prevention and for creating procedural rules, the Environmental Impact Assessment (EIA) Act was promulgated in 2003 for "preventing and mitigating the adverse impact of development activity on the environment in order to achieve the goal of environmental protection" (Art. 1). According to Article 4.2, "environmental impact assessment means an environmental management plan based on scientific, objective and comprehensive surveys, forecasting, analyses and evaluations conducted prior to project implementation in order to determine the degree and scope of the potential impact of development activity or government policy on the environment (including the living environment, natural environment and social environment), economy, culture and ecology, and the public explanation and review of such a plan." Procedural rules set details for phase I and sometimes phase II EIA (Arts. 6–19). For violation of these procedural exigencies, penalties and fines are obliged (Arts. 20–24). In principle, all the development activities of the establishment of a factory or the development of an industrial park, of a road, railway, mass rapid transit system, harbour or airport, of the extraction of soil and rock or the exploration and extraction of minerals, of water storage, water supply, flood control or drainage projects, of the use of land for agriculture, forestry, fisheries or livestock, of recreational areas, scenic areas, golf courses or sports fields, of cultural, educational or medical facilities, of new municipal districts, construction of tall buildings or renovation of old municipal districts, of environmental protection projects, and of nuclear energy or other energies or the construction of radioactive waste storage or treatment facilities, should be examined by phase I EIA; if necessary, due to concern of a significant impact on the environment, a phase II EIA should be conducted by the developer (Art. 5). These procedural requirements assure the sound sustainability for development. However, till now more than 90 % of development activities are just approved by simplified phase I EIA in Taiwan (Yeh 2010b), because in the Committee of EIA under the EPA, part-time commissioners could not have enough time and capacity to examine all the details, and also because under the substantial needs of economic development, commissioners are sometimes selected just by their pro-development attitude, not by their expertise. For preventing the rebound effect (Binswanger 2001, p. 119)[7] in actual EIA regime (Mauerhofer 2012, pp. 654, 658), in the future revision, an independent committee outside the EPA, in which commissioners are selected by their expertise and not influenced by governmental preferences, is preferable.

[7]"Efficiency improvements also affect the demand for resources and energy, and often an increase in efficiency by 1 % will cause a reduction in resource use that is far below 1 % or, sometimes, it can even cause an increase in resource use."

Thirdly, for the command and control, in Taiwan there are several laws[8] concerning the pollution prevention, for example the Air Pollution Control Act (1975, 2012 last revised), the Noise Control Act (1983, 2008 last revised), the Water Pollution Control Act (1974, 2015 last revised), the Marine Pollution Control Act (2000, 2014 last revised) and the Soil and Groundwater Pollution Remediation Act (2000, 2010 last revised). In addition, for sustainable use of resources, the Resource Recycling Act (2002, 2009 last revised) and the Waste Disposal Act (1975, 2013 last revised) are prescribed. Furthermore, for environmental safety, the Toxic Chemical Substances Control Act (1986, 2013 last revised), the Environmental Agents Control Act (1997, 2006 last revised), Agro-pesticides Management Act (1972, 2014 last revised) and Drinking Water Management Act (1972, 2006) are also set. For sustainable development of agriculture, fisheries and forestry, the Agricultural Development Act (1973, 2010 last revised), the Water Act (1942, 2014 last revised), the Fisheries Act (1929, 2015 last revised), the Forestry Act (1932, 2004 last revised), and the Utilization and Transfer of Reserved Mountainous Land Act (1976, 2006) are installed. For the value-added utilisation of natural landscapes, the National Park Act (1972, 2010 last revised) and the Act for the Development of Tourism (1969, 2015 last revised) are also promulgated. Finally, for protection of biodiversity, the Wildlife Conservation Act (1989, 2013 last revised) and the Administrative Rules for Natural Protection Area (2005) are well set. Nearly all the above mentioned acts are effectively implemented, although some violations of these acts occasionally still happen. Besides, we could notice that some of these command and control preventive norms were dated long enough, especially in the domain of pollution control. So, after 2000, there are more and more revisions of these acts for emphasising the conservation effects and the sustainable use of resources.

Fourthly, for environmental disputes, two ordinary resolution processes are provided: for environmental damages caused by one person to others, civil chambers of ordinary courts will deal with this person's civil liability, and sometimes even with the state compensation mechanism if it's generated by faults of civil servants; for actors being called for administrative responsibilities, it is also possible for them to appeal against the administrative act for punishment in administrative courts. In addition, some special law about the public nuisance is also announced. According to the Public Nuisance Dispute Mediation Act (1992, 2009), its Article 2.1 defines that "public nuisance referred to in this Act means human activities which destroy the living environment, and damage or endanger public health. Such activities include water pollution, air pollution, soil pollution, noise, vibration, noxious odors, waste, toxic substance pollution, land subsidence, radioactive pollution and other activities designated and officially announced as public nuisances by the central competent authority". This Act creates an alternative procedure for parties to a public nuisance dispute; in its Article 14.1, it's stated that

[8]For those who are interested in the texts of these laws, an official website of Taiwanese Department of Justice provides most of them in English version, retrieved July 28, 2015, from Law and Regulations Database of the Republic of China Web site: http://law.moj.gov.tw/Eng/.

"a party to a public nuisance dispute *may* apply for mediation by submitting an application form to the mediation committee in the special municipality, county or city where the public nuisance dispute or damage arose." Certainly, parties to a dispute could still choose to file a suit directly in a district court; here the mediation procedure is not compulsory nor a prerequisite for filing a suit. For the effects of the mediation, its Article 28 stipulates that "when a mediation is achieved, a mediation accord shall be prepared within 7 days of the achieved mediation and submitted for approval to a court holding jurisdiction; If the mediation accord does not contravene applicable laws and regulations, the court shall approve the accord at its earliest convenience and return the accord to the mediation committee for delivery to the parties." A approved mediation "shall have the same effect as a final civil decision by a court of law" (Art. 30.1).

At last, for organizational adjustments, the Taiwanese EPA was established in 1986 and will be promoted to a Ministry of Environment and Resources in a couple of years. For now, EPA is a second-tier agency under the Cabinet in Taiwan. The competences of the EPA includes: air pollution and noise, water, waste, toxic substances, environment monitoring and information, ruling for public nuisance disputes, reduction of greenhouse gas, and especially sustainable development. In the Office of Sustainable Development under the EPA, three divisions are set: the Division of Sustainable Development mainly cooperates with the National Council of the Sustainable Development, the Division of technological development cooperates with the Ministry of Technology, and finally the Division of international cooperation focuses on possible development of substantial relationship with other states, regional organizations and international organizations.

4 New Establishment of Ministry of Environment and Resources

Basically, the approaching Taiwanese governmental reorganization has just some minor influence for developing sustainability because it will just combine different authorities under different Ministries in a new Ministry. However, the promotion of Taiwanese EPA to Ministry of Environment and Resources in the future signifies that the expanded competence of environmental Ministry, especially for coping with the needs of sustainable development and for officially including the management of natural resources, which is now still in the competences of Ministry of Economics and Ministry of Internal Affairs.

According to the "Draft of Act of Ministry of Environment and Resources" in 2012, its Article 2 stipulates that its competences include: (1) Programming, monitoring and drafting of laws and regulations for environmental and natural resources policies, environmental impact assessment and environmental education; (2) Programming, monitoring and drafting of laws and regulations for climate change, international cooperation and sustainable development and relevant technologies (it replaces the status of National Council of Sustainable Development);

(3) Programming, monitoring and drafting of laws and regulations for air pollution, indoor air quality, noise and vibration control, ionizing radiation, water and ocean pollution, quality of drinking water, sewer, waste and recycling, soil and groundwater, toxic substances, environmental agents, public nuisance disputes, environmental sanitation, pollution sources, environmental testing; (4) Programming, monitoring and drafting of laws and regulations for nature conservation, wetland, natural landscape and biodiversity; (5) Programming, monitoring and drafting of laws and regulations for water resources, running water, hot spring, forest, wild fauna and flora, conservation of soil and water, geology, mineral resources and national parks; (6) Programming, monitoring and drafting of laws and regulations for certificates for environmental education, experiments and researches in nature conservation, forest and biodiversity; (7) any other relevant affairs in environment and conservation of natural resources.

For its future possible functioning, Taiwanese scholars expect some magnificent prospective (Deng et al. 2012, pp. 5–18). First of all, not only the integration of competences in one single Ministry is indispensable, but also the objectives of environmental protection and low carbon economy should be promoted for sustainability. Second, because policy and management of energy is still not incorporated in this new ministry, in the future the harmonization of energy policy and sustainable development should be prudently evaluated. Especially, when Taiwan has announced it "no nukes" policy (even though the deadline is not set yet) after the Japanese Fukushima accident in 2011, and the Project of Act of Abolition of Nuclear Power is officially under discussion since 2015, how to balance the need of electricity and sustainable development would be a crucial issue in next decades. At last, scholars also propose that the National Council of Sustainable Development should be untouched for steering framework programming, or at least it should be incorporated into this new ministry, but as a guiding agency for directing all other agencies (Deng et al. 2012, p. 5).

5 Implications for a Sustainable Development in Taiwan

From the paragraphs mentioned above, we could clearly note that Taiwan has basically achieved a comprehensive regime for sustainability. However, some further steps are still needed. Till now, most of the laws and regulations were set, based on the principle of prevention, but not on the idea of precaution. In fact, as it was promoted in European environmental governance since the seventies in the last century, and again promoted in Rio Declaration in the nineties, Taiwanese scholars noticed this idea a little bit later in the first decade of this century (Niu 2004; Yeh 2010a, p. 91; Chen 2012, p. 309). Even though the content and possible concrete implementation of precautionary measure are still disputed, Taiwan, under the condition of its geographical vulnerability, still seems to take some advanced considerations for avoiding catastrophic natural disasters and also for eliminating artificial environmental degradation.

Concretely, Niu (2004, p. 45) proposed that when facing scientific uncertainties, governments should take some precautionary measures for preventing irreversible damages. That implies that for assuring sustainable development, we could not just install some prohibitions for certain definite acts, which are scientifically proved to be harmful to environment, but also it's our obligations to step further for preventing any possible environmental disasters, even though they are scientifically uncertain. For example, in recent five years, possible influences from climate change to Taiwan, mitigation and adaptation both included, attract Taiwanese scholars' and officials' attention. It is promoted that this precautionary principle should serve as a fundamental principle for a better governance of climate change, in which global, large-scale, multilevel governance constitute its main figures, accompanying with scientific uncertainties and plurality (Yeh 2015, p. 435). Furthermore, from perspectives of state responsibility to its people, Li (2000) also proclaimed for assuring sustainable development, which is—according to this author—composed of principles of fairness, durability and commonality in Taiwan, that Taiwanese government should also take responsibilities to act and react, to provide information to people, to create a financial mechanism for supporting the functioning, and to adjust all interests among different sectors and even between generations. Li's proclamation could be also accessed here in the view of the precautionary principle, because harmonization for all, present and future, lacks of scientific certainty but still needs our efforts for guaranteeing its realization.

Especially, Taiwanese experience also provides us a unique example for a non-state political entity in the international participation for sustainability. Till now, Taiwan could not officially participate in multilateral environmental agreements, or in relevant international organizations for environmental protection, except as a separate custom territory in World Trade Organization. The real scenario here is merely voluntary compliance of Taiwanese government to international norms and standards for sustainability. However, some prospective to join international organizations as for example a status of observer will be of great help for Taiwan to achieve sustainability. Besides, the effectiveness for these voluntary compliances also needs verification by international monitoring.

For example, for combatting negative effects arising from climate change, Taiwanese EPA has established a website for propaganda to join the United Nations Framework Convention on Climate Change (UNFCCC). [9] On the websites, it is clearly shown the will of Taiwan to join the UNFCCC and the Intergovernmental Panel of on Climate Change(IPCC). In September 2014, Taiwan also declared a voluntary commitment to the UNFCCC that Taiwan would reduce its greenhouse gas (GHG) emissions by at least 30 % by 2020. Taiwan also actively promotes the Greenhouse Gas Reduction and Management Act, which sets the official new target of reduction of GHG to 50 % of the 2005 level by 2050 and is just promulgated in July 2015, in combination with the Energy Tax Bill, which is still being formulated, and the Energy Management Act and Renewable Energy Act, which have already

[9]Retrieved May 17, 2015, from http://unfccc.epa.gov.tw/unfccc/english/index.html.

been promulgated in 2009, will form the legal basis for the management of GHG reductions in Taiwan. Similarly, in 2012, Taiwan specified six GHGs as air pollutants—including carbon dioxide—in accordance with the Air Pollution Control Act. On December 25 of the same year, the Regulations of Mandatory GHG Reporting were announced, as well as Public and Private Premises under Regulations of Mandatory GHG Reporting. Besides, to enhance Taiwan's GHG management based on a system of Measurement, Reporting and Verification (MRV), and to monitor the main sources of GHG emissions in Taiwan, it is estimated that 265 public and private institutions will be included in the reporting mechanism, allowing the government to keep track of 90 % of GHG emissions. The above mentioned laws and regulations, mechanism and reporting, are domestically binding to all actors in Taiwan, but constitute some voluntary compliance to international community. In addition, some substantial inter-city international accords and cooperation, which appear to be of international soft law, for example the Green Cities Declaration of San Francisco in 2005 and the Urban Environmental Accords in 2011, is also proposed by Taiwanese scholars for evading the disputes of Taiwanese statehood (Hsu 2014, p. 270; Yeh 2015, p. 72). That would be a realistic solution for now for Taiwanese participation in the international arena.

6 Conclusions

In principle, Taiwan has established sound policies and legal norms for sustainable development. For traditional pollution control, we could observe that since the Seventies of the last century, Taiwan has achieved obvious effects in air, water and land use, even though sometimes occasional pollution still occurs, but this will be immediately sanctioned by the competent authority. For conservation of forest, wetland and biodiversity, Taiwan also set an effective mechanism in both local and central level, and a national park system was well established. For procedural requirements of new developments, the EIA provides an essential demand, even though till now it seems too easy to pass the phase I assessment. Besides, the approaching promotion of Taiwanese EPA to a new Ministry would to some extent reinforce relative competences for sustainability, being much more capable to systemize all relevant procedural and substantial norms for promising sustainability. In short, all the traditional prevention-oriented methods for sustainable development appear to be well set. But, a precautionary approach is still needed for better avoiding irreversible catastrophic disasters to environment.

In addition, all these efforts should be linked to the international level because we now live in a global village and international monitoring would be necessary for examining the exact effects of Taiwanese voluntary compliance to international norms. For example, for the proclaimed targets for reducing the Greenhouse gas emission, if Taiwan could not reach the reduction of 50 % by 2050, there's no direct sanction or pressure from international society. Even, every year a national report for sustainable development is issued, but no review mechanism through outsiders exists. At most, if Taiwanese government fails to achieve some expected target, there's just protest and opposition from Taiwanese civil society. Interlinkages between Taiwanese policies and legal regime for sustainable development and relevant international mechanisms and regimes will be of help to both.

References

Note: All the references cited in this chapter are in Traditional Chinese, apart the ones where it is explicitly mentioned "in English"

Binswanger M (2001) Technological progress and sustainable development: what about the rebound effect? Ecol Econ 36:119–132 (in English)
Chen T-Y (2012) General introduction of environmental law. Angle Publishing, Taipei
Deng C-J, Lo S-L, Chang C-Y, Yang C-Y, Chang T-E (2012) Expectations to ministry of environment and resources. CTCI Foundations, Taipei
Economic Construction Committee of the Cabinet Taiwan (ECC) (2004) Taiwan 21st century agenda. Administrative Yuan, Taipei
Hsu Y-M (2014) Transnational governance model for climate change. In: Yeh JR (ed) Institutional responses to climate change: decision, finance and Law. National Taiwan University Press, Taipei, pp 261–293
Ke T-D (1994) Sustainable development: analyses for the formation and responses of International environmental law. Nat Taiwan Univ Law Rev 23–2:33–56
Li C-L (2000) The state liability system for sustainable development. Nat Taipei Univ Law Rev 46:79–112
Li C-L (2015) Sustainable development and due process. Taiwan Law Rev 240:221–247
Liu T-T (1997) Legal system and methods for environmental protection in Republic of China. Nat Cheng-Chi Univ Law J 57:37–67
Mauerhofer V (2012) A legislation check based on '3-D sustainability'—addressing global precautionary land governance change. Land Use Policy 29:652–660 (in English)
NCSD (2009) National council of sustainable development: National policy program for sustainable development. Administrative Yuan, Taipei
NCSD (2014a) National council of sustainable development: 2013 evaluation report of sustainable development indicators. Administrative Yuan, Taipei
NCSD (2014b) National council of sustainable development: report of action plan for sustainable development: from January to June 2014. Administrative Yuan, Taipei
Niu H-J (2004) Studies of precautionary principle. Nat Taiwan Univ Law Rev 34(3):1–71
Tseng H-P (2000) Humankind and environment. Chen-Chung, Taipei
Tseng H-P (2008) Environmental governance in Taiwan (1950–2000): an analytical study in the light of ecological modernization and eco state theory. Taiwan Hist Res 15–4:121–148
Yeh J-R (2010a) Environmental policies and laws. Angle Publishing, Taipei
Yeh J-R (2010b) Report on environmental impact assessment. Committee on Researches of the Cabinet, Taipei
Yeh J-R (2015) Climate change: governance and law. National Taiwan University Press, Taipei

Part VI
Sectorial Policies: Forest and Water Management

A Legal Butterfly Effect: Unexpected Twists and Turns of the Law in Costa Rica's Payment for Ecosystem Services Program

Pablo G. Peña

Abstract Costa Rica's Payment for Ecosystem Services program (PES) is one of the most studied exercises of its kind but closer examination of the program's legal framework and governance is still lacking. The PES did not occur on a vacuum; laws and policies outside the boundaries of the PES' regulations shape the way it evolved and functions. The supervisory checks and balances of the forestry regency system, the public funds laws that reduced the program's flexibility, and the administrative simplification process across the Costa Rican government are all examples of policies outside the PES that strongly influence its functioning. Foreign policies also shaped the PES. For example, the World Bank-sponsored structural changes of the Costa Rican economy during the 1980's helped shift the rationale from forest subsidies to payments for ecosystem services. In addition, a closer look at the PES on the ground provides interesting opportunities to reflect on the effects of this legal framework. For example, the way violations to forest laws occur and are dealt with by judges and PES officials most likely had an effect on the Costa Rican forest cover, which is missed in studies focused on the additionality of the program. Ultimately, however, people implement the PES and this paper suggests an interesting dynamic between two types of bureaucrats at the program, the 'technicians' and the 'lawyers'. The 'lawyers' seem to have displaced the 'technicians' in a process of 'rendering legal' nature, which has conflicting implications for the PES effectiveness. All these dynamics may suggest a legal 'butterfly effect' that policy-makers ought to be aware of when designing and implementing environmental institutions and mechanisms.

Keywords Payment for ecosystem services · Forest governance · Environmental law · REDD+ · Costa Rica

P.G. Peña (✉)
Environmental Law and Policy Specialist, Peruvian Society for Environmental Law, Lima, Peru
e-mail: ppena@spda.org.pe

P.G. Peña
School of Forestry and Environmental Studies, Yale University, New Haven, USA

V. Mauerhofer (ed.), *Legal Aspects of Sustainable Development*,
DOI 10.1007/978-3-319-26021-1_16

1 Introduction

Costa Rica has a rich history of regulating and experimenting with different incentives for forestry activities and, more recently, for forest conservation through the well-known Payment for Ecosystem Services Program (PES). According to the Forestry Act of 1996, the Costa Rican PES pays landowners for a bundle of four ecosystem services provided by their forests: CO_2 fixation and sequestration, water conservation, biodiversity protection and scenic beauty. FONAFIFO (Fondo National De Financiamiento Forestal) is the institution entrusted with managing the PES. As of 2012, FONAFIFO had awarded 11,378 payment for ecosystem services contracts in different categories, more than 4500 of those are currently in force (FONAFIFO 2012a, b).

Despite the publicity, the PES' efficacy in restoring Costa Rica's forests remains a contested issue but there are interesting opportunities to improve our understanding of the program's governance. Costa Rica's PES is considered the first major payment for ecosystem services grand scheme exercise in the world and, as such, it is one of the most studied (Arraigada et al. 2012; Daniels et al. 2010; Fletcher and Breitling 2012). However, the bulk of previous research has focused mainly on the program's effectiveness to curb deforestation or help poor *finqueros* (farmers) (Fletcher and Breitling 2012). Thus, some scholars have been advocating for additional research that focuses on explaining the context in which the PES developed and is governed to better understand the program's current and potential impacts (Daniels et al. 2010; Fletcher and Breitling 2012). This paper aims to bridge that gap and to provide a better contextual understanding of how Costa Rica's PES operates within the specific set of rules and institutions that govern it, in particular from the perspective of those who administer it on a daily basis. In doing so, I offer some insights into the role of law in the PES and the way it functions on the ground.

2 Methods

I did field research in Costa Rica between May and July 2012. During this time, I conducted interviews with key informants, engaged in participant observation during field visits and conducted legal research. In total, I conducted 18 semi-structured interviews of 30 min to an hour. In addition, I held casual conversations on the field with many of them during longer periods. With the help of an informant in San José, head of an international NGO, I prepared a list of potential informants that had diverse experience with the PES and added others recommended by some interviewees. The 18 interviewees comprised mainly current and past government officials from the Ministry of the Environment (MINAET), FONAFIFO and the Sistema Nacional de Areas de Conservación (SINAC), as well as conservation NGO workers, representatives from the Colegio de Ingenieros

Agrónomos and some prospective and current beneficiaries. The semi-structured interviews focused on the informants' perception of the PES and FONAFIFO as an institution, and included questions on how they thought it could be improved or adjusted. In some cases, depending on the expertise of the interviewee, questions centred on legal matters of the PES in order to understand the legal framework that regulates it and the perceptions these different interviewees had on their efficacy at different stages. Oral information on the purpose of the interview and research were provided in accordance with human research protection standards.

I visited two implementation areas, Limón (E Costa Rica) and Sarapiquí (N Costa Rica), to do participant observation and conduct interviews. In Limón, I accompanied NGO workers in field visits to current and prospective PES beneficiaries and their *fincas* (farms). In Sarapiquí, I joined government officials in a supervision visit to a PES *finca*. The field visits and the time spent with these informants provided insights into the nature of the PES work on the ground, and its relationship to the law and policies that support it.

3 Literature Review

Costa Rica's PES is in part a product of a new paradigm in conservation where market-based tools are regarded as key to achieving efficient and sustainable forest protection. Throughout the history of the conservation movement, academics and practitioners have advocated and used tools based on shifting paradigms. The idea of 'fortress conservation' behind the creation of the Yellowstone National Park in the United States, for example, was in vogue during the beginning of the movement, followed by the integrated conservation and development projects (ICDPs), and soon after by a critique of those who acknowledged the existence of trade-offs, and the move towards using market incentives such as PES to drive conservation (Adams and Hutton 2007; Wells and Brandon 1992; Robinson and Redford 2004; Wells et al. 2004; Wunder 2005; McShane et al. 2011). Costa Rica's PESP is a prime example of a policy intervention that, although led by the State, aspires at being driven by ecosystem services markets.

The PES has been subject to abundant analysis, particularly on the question of its additionality. Studies on the efficacy of the PES have focused on understanding to what extent the programme has been responsible to curb deforestation in the country (Sánchez-Azofeifa et al. 2007; Pagiola 2008; Daniels et al. 2010; Arraigada et al. 2012; Plaff et al. 2008; Morse et al. 2009; Sierra and Russman 2006; Wunder 2007). Most of these studies rely on complex economic and statistical models and the majority of them have found that the PES has had no additional impact on slowing down the country's deforestation rate, although the subject remains highly contested (Daniels et al. 2010; Arraigada et al. 2012). For example, in a recent study Arraigada et al. found that, unlike prior studies, in *fincas* under the PES in the Sarapiquí region 'there was a net increase in total forest cover' compared to those not in the program, thus signalling a modest impact of the PES on the forest (2012, p. 393). However, the authors explain that it is difficult to determine whether this difference is a result of forest regrowth or avoided deforestation (Arraigada et al. 2012).

Another focus of attention has been the social impact of the PES. Many have pointed out that the legal requirement to have a land title is a major barrier to access to the benefits of the PES (Camacho et al. 2000; Pagiola 2002, 2008). This criterion seems to have resulted in more *fincas* under PES owned by better-off *finqueros* than by their poorest counterparts (Zbinden and Lee 2005; Pagiola 2008). Moreover, laws regulating the use of public funds barred the government from using those funds to pay landholders without clear titles, thus, restricting the access of the poorest *finqueros* to the benefits of the PES (Pagiola 2008). Congress later approved legislation clarifying this situation and allowing FONAFIFO to sign up some informal landholders into the PES. Also, the government put forth ingenious schemes in protected areas such as the Osa peninsula allowing landholders without titles to enter the program. FONAFIFO enlisted the Institución de Desarrollo Forestal (IDA) as these informal landholders' proxy, with whom FONAFIFO officially signed the PES contract (Castro Salazar and Peña Chacón 2011). Transaction costs involved in applying and maintaining *fincas* in the PES may prove an additional barrier to entry for the poor (Pagiola 2008). This was especially true in the early years of the PES; since then, FONAFIFO has worked on decentralizing its operations, making the program more widely known and making the procedures simpler and less time-consuming for applicants (FONAFIFO 2005; Pagiola 2008). Finally, another constraint for the poor in the original setup of the program was its lack of focus on deprived regions, or on activities regarded as more economically beneficial like agroforestry (Cole 2010; FONAFIFO 2005). FONAFIFO has changed its policy and currently allows *fincas* under agroforestry systems into the PES. FONAFIFO awarded the first agroforestry PES contracts in 2003, and the time of the research they amounted to more than 4000 ha (FONAFIFO 2012a). There has also been a change in regards to the geographical focus of the program. The PES Procedural Manual (Government of Costa Rica 2009) now states in article 2.2.7 that *fincas* located in districts with a social development index of less than 40 % are a priority.

Academics have not similarly focused on analysing the PES's governance but there have been some interesting studies that provide a first look at how the legal framework of the PES—the governance backbone—works and how it operates on the ground. Early work by Peña Chacón (2004, 2006) described the basic legal features supporting the PES. More recently, Takacs (2009) explained the legal basis for Costa Rica's allocation of carbon rights to *finqueros* (farmers), which used a rather simple rule that whoever owns the land owns the carbon rights and other rights related to the ecosystem services provided by that land. Castro Salazar and Peña Chacón (2011) review the overall legal framework structure of the PES, concluding that countries that wish to replicate Costa Rica's PES need a strong regulations and a specialized agency such as FONAFIFO. An interesting research by Le Coq et al. (2010) applies a policy process perspective to the PES and manages to provide an inner look at the politics and history behind its creation and development. Fletcher and Breitling (2012) make an effort to explain the PES governance concluding that despite its market-based intent the program remains a subsidy policy. Finally, Daniels et al. (2010) provide examples of how an improved and contextual understanding of the PES governance helps better frame its results. The present research builds on these past studies in order to move the conversation on the broader context of the PES forward, particularly from a legal and policy point of view.

4 Governance of the PES on the Ground

4.1 Applying to the PES: A Functional Sieve

As a state-led program, FONAFIFO relies on its bureaucracy and official procedures to run the PES. The Forestry Act of 1996, its regulations and its Procedural Manual provide the basic legal framework for this. The procedure starts with the application, which has come a long way since the program began in 1997. An informant from FONAFIFO's main office in San José, a lawyer, said that, before, the application process took 'a long time' but now it only takes around 1 and a half months and 'between 6 and 7 months, worst case scenario'. This, he explained, is a result of laws that mandated the implementation of 'administrative simplification' throughout all government agencies. For example, he explained that in the past FONAFIFO required applicants to submit official proofs of title from the public registry office; now, FONAFIFO eliminated this requirement and its officials review the legal status of the *finca* using the public registry online system, an almost automatic process. This is only possible because other agencies have modernized their database systems as well, and make the systematized data available online.

Currently, FONAFIFO relies heavily on an online system for the application process and its overall work. Digital forms have replaced most of the internal paperwork that bureaucrats needed to move from office to office physically, avoiding unnecessary delay in the application process. Moreover, good communications and widespread Internet usage in Costa Rica has provided ground for FONAFIFO to move the application online. An informant from FONAFIFO's San José office explained that, since 2011, they decided to require most of their applicants to submit their applications online. Some people on the ground, he said, met this with scepticism at the beginning but now the policy has proven successful: he states that around 90 % of the applicants now submit their applications online. An informant from an NGO in Limón explained, however, that this is mostly a pre-application form and prospective beneficiaries still need to submit additional documents such as a map and copies of their national identity cards. After FONAFIFO's lawyers in San José verify that the application complies with the legal requirements (for example, land ownership and that the applicant is up to date with social security payments), the next step is to require the technical study by the *regente*, an officially certified forester. FONAFIFO officials on the regional offices analyse this study and grade each application according to the priority scale approved in yearly decrees signed by the President of Costa Rica.

Despite the implementation of online systems, PES officials and some documents still have to move physically. FONAFIFO personnel work and communicate well through the online system but they still need to meet for planning work or go to field visits. One of the informants from FONAFIFO's headquarters in San José explained that Costa Rica's small size and the fact that it is relatively well connected made FONAFIFO's work effective. He exemplified the benefits of this by saying that if he needed to talk personally to one of his officers on the ground in the farthest office he could call him by telephone and he would be in San José '*by the afternoon*'. Furthermore, institutional cooperation among different agencies seems to be key. For example, during my visit to Sarapiquí I participated in a supervision visit to a *finca* under the PES with an official from SINAC and another from FONAFIFO. These informants explained that joint visits are common. In this specific case, it was also out of a necessity because the regional office of SINAC had run out of gasoline to fuel their vehicles so the SINAC official needed a ride from the better-equipped FONAFIFO office. As for transporting documents, FONAFIFO's San José office needs to send to the regional offices the contracts that the *finqueros* will sign. A former SINAC official and current NGO worker in Sarapiquí said that, in the past when SINAC officials on the ground had some PES responsibilities, they also prepared the contracts themselves. He explained that the government later concentrated this responsibility in FONAFIFO's headquarters in San Jose, and transferred it to 'the lawyers'. 'The ones who decide now are the lawyers', he concluded with contempt.

4.2 Trade-Offs in 'Rendering Legal' the PES

Some legal requirements are subject of controversy among PES actors. Uniformly, interviewees from NGOs on the ground and some government officials in regional offices mentioned their particular frustration with what they saw as excessive rigor by 'the lawyers in San José' responsible for evaluating the proofs of landownership. Their objections go beyond those related to the exclusion of landholders without title. A common criticism was that FONAFIFO expects a complete match between land title certificates from the public registry and the information in the cadastre, which does not occur in many instances. Informants from an NGO in Limón mentioned that fixing this meant a big investment in time and money by hiring technicians and paying government fees, perhaps more than what the beneficiary can expect to receive from the PES. Two of those same informants said that, in one case, FONAFIFO turned down the application of a poor *finquero* whose registered title deviated from the map in the cadastre by 1 hectare. Another informant from the same NGO told me about a *finca* whose title had a 10 % overlap with an indigenous territory and, thus, FONAFIFO also rejected it. Stories like these seem widespread and some border on the absurd. In Sarapiquí, an NGO worker told me about a *finca* that was rejected by FONAFIFO because the area in the title and the information in the cadastre had a difference of a few square decimetres—an area 'occupied by a cow's shit', in the informant's words.

Yet, this restrictive way of implementing the PES is not absurd in eyes of 'the lawyers'. Most of the same informants who narrated these stories also explained that FONAFIFO officials defer many decisions to 'the lawyers' and refrain from making calls themselves. A FONAFIFO official on the ground explained that he sends his report to the San José headquarters and 'the lawyers' tell him what to do. 'They are the ones who have power', this informant reasoned. One of the Sarapiquí informants, a NGO worker, explained that in recent years the press has been looking for irregularities in the PES, so now bureaucrats are very cautious and before even considering a special request they just say 'no'. In addition, due to their legal training, 'the lawyers' probably feel constrained by the letter of the law in the PES regulations and the mandates of other laws that tightly regulate the use of public funds (Pagiola 2008). Recent scholarship has framed this as a problem of a PES system with a positivistic approach that gives little value to interdisciplinarity (Barreiro 2012).

In reality, of course, cadastre and public registry do not go hand by hand because different agencies administered each of them in the past and they have been subjected to institutional, legal and technical changes throughout the years. However, the law and its operators within the PES are unable to capture this complexity. In this sense, the PES is an example of a scheme that deems necessary to simplify reality in order to make nature and people more attainable to a State that wants better control, as described in other places (Scott 1999).

As a result, the current 'rule of the lawyers' is highly functional to the PES. So far, due mainly to financial constraints, the PES has been incapable of satisfying the total demand of *finqueros* wishing to be included in the program. One informant from a NGO in Sarapiquí stated that in past years FONAFIFO ended up approving only 30 % of the applications and, thus, 'the lawyers cheered' to the prospect of further regulating the admissions criteria. A fellow NGO worker agreed to this idea saying that FONAFIFO, seeing so much demand, said 'Let's become *espesos*'.[1] Thus, a big demand may have driven FONAFIFO to come up with more barriers to access in the form of additional or more stringent legal requirements, which are framed as objective and technically sound. As seen in other places, FONAFIFO, as a governmental and development structure, places a high regard into simplifying and rationalizing nature and people's use of nature by 'rendering it technical' (Li 2007; Scott 1999). In the particular case of Costa Rica's PES, 'the lawyers' play

[1]Espesos is Spanish for thick or dense, meaning also picky or difficult.

a higher role in this simplification and rationalization process, through what could be called a process of 'rendering legal'.

4.3 Administering and Supervising the PES

Despite the bulk of demand, finding potential beneficiaries does not seem to be an easy task. As one experienced informant from an NGO in Sarapiquí put it: 'you don't see lines of people [waiting] outside'. Some informants, NGO workers, said that there has been disbelief among some *finqueros* on the program's intentions. This experienced informant explained that many in rural areas, especially the poor, doubt that someone will pay them for conserving their forests or 'for doing nothing'. In Limón, the daughter of a beneficiary described that his father was 'the most reluctant' of his neighbours to enter the program because he believed that this was a taking in disguise by the government. The NGO informants from Limón confirmed that this belief had been somewhat common among the rural people in the area. This may be originated in upsetting past experiences of expropriations of private land that overlapped protected areas under the Wildlife Conservation Act of 1992, as one of the informants suggested.

In general, the PES on the ground looks a lot like the instructions provided in the Procedural Manual but with more sweating. During my time in Limón, I accompanied two NGO workers to La Estrella valley to visit a *finca* from a local school that had just signed up to the program. According to plan, the group comprised by NGO and school workers walked the limits of the property to verify the condition of the forest, see if the borders had been properly cleared, and hang signs that read 'Private property under conservation: Payment for Ecosystem Services program…'. It was a 5-hour hike through a hilly terrain, without pathways at various times and under dense tropical foliage.[2] One of the NGO workers was a certified *regente* and as such, he was responsible for subscribing the PES documents of the *finca* and conduct yearly visits to make sure the landowners comply with the program. Visiting potential beneficiaries and traversing their *fincas* is part of the everyday job of NGO workers and *regentes* who promote and sign up *fincas* to the PES.

The *regencia* system, as it turns out, is one of the most interesting features allowing the program to function properly, a system that is not exclusive to the PES but part of the broader forestry governance of Costa Rica. Under this system, a certified *regente* must verify and approve most forestry activities in the country. This system includes checks and balances ensuring reliance in the *regencia* work and the information it provides. The Forestry Act of 1996 sanctioned the regencia

[2]To a city-dweller like me this felt like a very strenuous feat, although, the rest of the group seemed mostly adept at it.

system and assigned the task to supervise the work of *regentes* to the Colegio de Ingenieros Agrónomos, an officially regulated professional association. An informant from the Forestry Prosecutor's Office at the Colegio de Ingenieros Agrónomos (also a creation of the Forestry Act) explained that their work is to 'control the practice of the profession'. Six officers from this agency supervise all forestry *regentes* through administrative and on-site audits. Some of them go on one field tour per month to pre-selected sites, which may or may not include *fincas* under the PES. The database run by the Colegio de Ingenieros Agrónomos is also a product of the Forestry Act of 1996, which mandates that *regentes* send copies of all the documents they sign to this institution. This allows this agency to conduct administrative reviews, which means that each year they select approximately 10 % of the *regentes* and go over their documents to look for inconsistencies that look suspicious, such as too many *regencias* done by the same *regente* at the same time. Interestingly, the same informant explained that it is not usual for *regentes* to oversee many *fincas* under the PES at the same time, implying that the *regencia* work in these *fincas* is more demanding than the *regencia* work for timber management in other *fincas*.

These informants also explained that, unlike the other offices at the Colegio de Ingenieros Agrónomos, the Forest Prosecutor's office has reliable and usually sufficient funds to do their work. A provision in the Forestry Act of 1996 that earmarked a portion of the forestry tax to this agency is important to provide sustained funds. Also, the fact that this agency monopolizes the production and selling of official stationary that *regentes* must use in the documents they sign helps provide sustainable funds for their supervisory work. Financial stability strengthens this office's autonomy and allows them to work well, thus, providing confidence in the *regencia* system.

SINAC and FONAFIFO collaborate in this supervisory task of the *fincas* under the PES. FONAFIFO only acquired additional supervisory duties later in the life of the program as part of the decentralization and strengthening process that led to the creation of eight FONAFIFO offices throughout the country (FONAFIFO 2005). Thus, supervision of PES *fincas* is constant and the work of the *regentes* in the PES is under continuous examination by different agencies. One informant from a Sarapiquí NGO argued that, in their case, this makes the PES more burdensome than it needs to be. He mentioned that sometimes '[the auditing agencies] go too far', and preparing the documentation and being subject to the auditing process takes too much time.

4.4 *Managing Illegality*

Non-compliance within the PES, however, seems rare. Most informants explained that, in general, overt deforestation in Costa Rica is extremely uncommon and, instead, some *finqueros* engage in *socola*. These informants described the process of *socola* as being a concealed and slow degradation of the forest in the margins of farmland, which starts by clearing the understory and then converting the forest gradually into agricultural land. A FONAFIFO lawyer explained that they have detected some *socola* in *fincas* under the PES but it has been minimal. He also mentioned that in the PES they have found 'only a couple of cases [of land use change] but because it's a crime then [people don't do it]'. Informants described how banning land use change and other provisions restraining *finqueros* from cutting down trees in protection areas deter them from engaging in illegal acts. An experienced NGO worker from Sarapiquí explained that many rural people are afraid of getting caught in illegal activities because they have to go to court and use lawyers; that is to say, *finqueros* are afraid of getting entangled in a criminal justice system they are unfamiliar with. Building on this informant's suggestion it appears that deterrence in Costa Rica's forest governance stems from the very idea of being drawn into strange governmental offices and courtrooms and unknown procedures rather than by the prospect of punishment by itself (i.e. prison time).

Despite this deterrence effects, sometimes crimes occur and some informants from an NGO in Sarapiquí expressed their frustration with the criminal justice system in these cases. According to them, judges are too lenient with the 'poor' *finqueros*, and think as if 'cutting a little tree is not a big deal'. Furthermore, one of these informants explained that judges in rural Costa Rica try to avoid conviction by raising the standard of proof to unreasonable levels. 'Have you seen him with the match in his hand?' is the type of question a judge would ask to prosecutors or witnesses in a case involving the burning of a forest, according to this informant. An informant from the FONAFIFO headquarters, a lawyer, saw this from a different perspective. In this informant's view, the criminal justice system, compared to other countries, works well and is another reason why rural people are deterred from damaging the forest. Furthermore, this informant ties an explicit link between the deterrence factor of the ban on land use change and the incentives provided by the PES. To him, the PES is a way for *finqueros* to at least make some money off the *finca* and avoid getting into trouble.

Violations from *finqueros* under the PES are uncommon but they do occur, which triggers the involvement of government officials. A FONAFIFO official from Sarapiquí exemplified what to him constituted a major—although rare—violation. He showed me a picture on his computer of a *finca* forest that had been deforested by less than a hectare for agriculture; 'this is very atypical', he said. Before the supervision visit described earlier in Sarapiquí, the official from SINAC had received information of illegal logging inside the PES *finca*. The *finca* owners had in fact cut down trees in the area, as it was apparent from the wood laying on site. The FONAFIFO and SINAC officials had no problem entering the property and verifying the state of the forest. These informants explained that they have the authority to do so under the Forestry Act and that only in a rare occasion one of them needed to enforce his authority by bringing along the police. One of the informants from the Forestry Prosecutor's office of the Colegio de Ingenieros Agrónomos attested to the fact that officials working in the PES are highly respected on the ground. When recently appointed, fellow officials told this informant to 'be aware now that [he]'ll be the new son of a bitch!' but despite that frightening forecast he says he hasn't run into much trouble while on the job. One informant, a FONAFIFO lawyer, explained that most violations in the PES occur from causes other than damage to the forest, such as landownership conflicts or invasions from *precaristas*.[3] In these cases, he said, they would put the payment on hold or exclude the specific area in conflict. This informant explained that FONAFIFO is not interested in cancelling PES contracts and, thus, gives opportunity to beneficiaries who are not complying by giving them a warning.

Conservation NGOs play an active role in the PES (Locatelli et al. 2008). In Sarapiquí and Limón, the NGOs where the informants worked are intermediaries between FONAFIFO and the *finqueros* wishing to enter the PES or currently inside the program. One informant from Limón, a NGO worker, mentioned that many *finqueros* are grateful for this work. This informant perceived that the work they perform is very important; to him, if it were not for the NGO he works for there would be much less *fincas* under PES in his region. An NGO worker from Sarapiquí corroborated this idea by saying that *finqueros* 'need a lot of help' filling out the PES applications and thus they provide this support. He gave an example of this by saying that 'there are many people that don't even know what *notificación*[4] means' in the application form.

The role of NGOs as intermediaries of the PES is not a de facto job; the Procedural Manual states that NGOs wishing to collaborate with the PES can sign a formal agreement with FONAFIFO. This agreement allows NGOs to receive payments from FONAFIFO and channel them to the beneficiaries. This is an important part of their role as it provides means for beneficiaries to have access to their funds more easily by distributing the funds to where the *finqueros* are located. The way FONAFIFO distributed payments had been an issue of controversy in the past with the government being slow in disbursing the amounts owed, which led to disputes (Castro Salazar and Peña Chacón 2011). This situation has improved since then. Different informants explained that because applying for the PES requires specialized work from a *regente* it could be expensive. Thus, most *finqueros* and *regentes* make an arrangement by which they pay *regentes* only after the *finca* enters the program, using the funds from the PES payments. In the case of NGOs, informants explained that they too charge a fee they claim is lower than what an independent *regente* would charge. These arrangements are an important part of the PES mechanics as they incentivize *regentes* to look for more *fincas* to sign them up to the PES, and provide poor *finqueros* with a way of navigating the relatively high transaction costs of the program.

[3]From the Spanish word *precario*, which means instable; in reference to squatters.

[4]*Notificación* is Spanish for legal notice. In the application form there is a space asking applicants to write down their *dirección para notificación*, i.e. their official mailing address.

5 Flapping Butterflies: The Unexpected Influence of Past, Present and Foreign Laws and Policies in the PES

5.1 Past Flaps

Other laws and policies—past, external and foreign—have constantly shaped Costa Rica's PES. Costa Rica built the PES upon an extensive experience with past forest policies (Daniels et al. 2010; Pagiola 2007). By 1997, the country already had a strong and experienced forest bureaucracy as well as lessons learnt from experimenting with the initial forest subsidies mainly aimed at timber activities. In addition, in the 1980's and 1990's the government was subject to institutional changes that took out forests from the agricultural sector and put them inside the newly created MINAET. Despite this relocation, the forest service, the wildlife agency and the protected areas agency were still separate semiautonomous agencies within the same ministry. A former high official at MINAET involved in these institutional changes explained that the division of forest issues in 3 semiautonomous agencies was inefficient and led to lack of coordination on the ground. In 1994, the government merged these agencies, giving rise to SINAC, agency in charge of all forest-related issues in Costa Rica. In line with past scholarship, this informant considered that the past experimentation with forest incentives and the consolidation of agencies that allowed for a coherent public forest policy were 'enabling conditions' to design the PES that we know today (Pagiola 2008; Daniels et al. 2010).

5.2 The Flaps of the Swarm

At present, there are concurring legal provisions within the Forestry Act of 1996 that have been key in the development of the PES. The *regencia* system explained earlier is an interesting example, but there are others. The ban on land use change is particularly important to the extent that it effectively made the PES politically viable. The Forestry Act of 1996 that created the PES also forbade clear cutting or land use change in the entire country; the convergence of both policies in the same law was not a coincidence. A bill introduced in 1994 named 'Ley CULPA'[5] effectively sought to ban not only land use change but also all timber cutting from natural forests (Aguilar 1995). This bill eventually failed to pass Congress but became part of the political discussion on solutions to reduce deforestation in Costa Rica (Le Coq et al. 2010). Informants who were part of these political processes explained that the PES was in part a negotiated agreement to allow for a lighter 'Ley CULPA' (i.e. a ban on land use change but not on timber extraction from

[5]CULPA is the acronym for 'Cortar Únicamente lo Producido Ahora', which is Spanish for 'Cut only the [timber] currently planted'. *Culpa* also means 'guilt' in Spanish.

natural forests) to pass as a provision within the Forestry Act of 1996. Thus, the PES was effectively negotiated as a compensation to those affected by the prohibition of clear-cutting the forest, namely those who used the forest for agricultural and cattle ranching purposes. This ban on land use change, along with a tax on gasoline earmarked to fund the PES, are strongly tied together with the creation of FONAFIFO and can be regarded as part of the same PES system. Furthermore, these 3 policies have been coexisting since 1997. As a result, any analysis on the effectiveness of the PES would be incomplete without accounting for this symbiosis (Daniels et al. 2010; Pagiola 2008).

Behind the PES's rationale was the growing idea that the government and its laws could not just mandate an environmentalism of the rural people, they had to provide them with compensation. One informant, the leader of an NGO, explained that in the past people in cities believed that 'the forests belonged to all Costa Rican people' but now they are beginning to understand that those forests actually have owners that ought to be compensated. Several other informants reiterated this compensatory discourse. Interestingly, the PES political process and its subsequent development might have helped induce a far more interesting cultural change in which rural landowners and their land rights are better acknowledged and recognized by city dwellers than before.

Laws and policies outside the forestry sector also influence the way the program functions. One example is the Treasury Inspector's Office intervention on FONAFIFO's autonomy and use of funds. Originally, the government designed FONAFIFO to be a semi-public institution, a trust under private law. Most informants agreed that this feature made FONAFIFO an efficient institution; for example, it had very low administrative costs, below 7 %, according to one of them. This changed, as FONAFIFO became 'bureaucratized', in one informant's words. An informant from FONAFIFO's legal team explained that the Treasury Inspector's Office decided that FONAFIFO was an agency of the government in full and thus had to comply with the laws regarding public employees, doubling FONAFIFO's operating costs to 14 %.[6] As a result, through this interpretation of the public funds laws, many informants argue that the Treasury Inspector's Office changed the agile and flexible nature of FONAFIFO.

Furthermore, the Treasury Inspector's Office stated that the funds that FONAFIFO obtained from donations were also public funds under the law and thus more stringent legal conditions applied. For example, FONAFIFO could not use those funds to pay people illegally occupying public lands (Pagiola 2008). This later changed after Congress passed the law approving the second World Bank loan for the PES that included a provision allowing informal landowners to enter the PES, as mentioned by an informant who was a former high official at MINAET. Also, another informant, former head of the national parks service, explained that public funding laws such as the Law for the Financial Equilibrium of the Public Sector of 1984 and the Law for Containing Public Expenditure of 1985 were the

[6]Under the current law, FONAFIFO's operating costs theoretically reach to 23 %.

basis of these policies. These laws were originally put together to deal with the 1980's economic crises and by applying them to FONAFIFO they reduced its flexibility of expenditure (for example, they provided labour stability but also made more cumbersome hiring new personnel). An informant, the former head of an NGO, suggested that the reason for these institutional changes was the fact that FONAFIFO was 'the rich cousin of the poor Ministry' and thus, MINAET and the rest of the government wanted more control over FONAFIFO's valuable assets.

A well-known legal principle in the civil law tradition is the 'first in time, better in right'[7] principle, and FONAFIFO used it as the default filter to access the PES. FONAFIFO applied this principle as a 'first-come, first-served' application process in the early years of the program, leading to a disproportionate representation of large landowners in the PES (Castro Salazar and Peña Chacón 2011). An informant from an NGO in Sarapiquí explained that they complained to FONAFIFO about this policy because the process focused on the ability of prospective applicants to put together an application faster rather than the suitability of each *finca* and applicant. This principle was dropped by FONAFIFO soon after, according to this informant. He mentioned this example to describe that, in his view, FONAFIFO is an institution that is open to listening to other's opinions and learn from its mistakes. Furthermore, to him, the use and abandonment of this principle signalled a normal trend of regulating through trial and error in what he saw as a rather good institution. The opinion of FONAFIFO as a good institution was widespread across all interviewees. When asked about the possible institutional improvements to FONAFIFO, a former head of an NGO said that to him there is no obvious need for change; he even went as far as to say that FONAFIFO 'is perfect'.

5.3 Flapping from Afar

International laws and policies have also influenced Costa Rica's forest policy and the PES. The World Bank adjustment plans of the 1990's are examples of foreign policies that influenced the PES (Daniels et al. 2010). Following World Bank directives, Costa Rica had to liberalize its economy and discontinue many of its subsidies. Around that time, Congress was discussing the new Forestry Act of 1996 that included the PES, effectively a program to transfer public funds to private hands. In order for this to be amenable to the World Bank, supporters cleverly framed the PES as a program that does not provide subsidies but pays for actual services—a market mechanism (Le Coq et al. 2010). In this sense, the PES was a result of neoliberal international forces (Fletcher and Breitling 2012). These forces were in line with conservation policies coming from the Rio Conference of 1992,

[7]In Latin: 'prior in tempore, potior in iure', from Roman Law.

and the already ongoing PES projects executed by the not-for-profit Fundación para el Desarrollo de la Cordillera Volcánica Central—FUNDECOR with funds from the United States Agency for International Development—USAID. Framing the PES as a scheme to pay for services and not to provide subsides remains largely in the minds of most interviewees, as was constantly stressed during our conversations. However, despite the PES' aspirations at being the flagship of a market-based scheme for conservation, some scholars contend that it remains a subsidy program at its core (Fletcher and Breitling 2012).

The international context in favour of tools like PES also had an influence in the World Bank's approval of two consecutive loans to Costa Rica explicitly aimed at funding the PES. These loans have proven key for the relatively widespread implementation of the program around the country. Informants from FONAFIFO's office in San José said that the second loan still active at the time of this research accounted for around half of FONAFIFO's budget. In fact, the loans' impact goes beyond this major economic contribution. International contracts formalize these loans, which Congress then approves. This high legal layer surrounding the international loans shielded the PES from political instability and changes in fiscal priorities. This suggests that the importance of the loans is as much economic (i.e. necessary funding) as it is legal (i.e. abating the possibility that political instability could reduce funding for the PES). This does not mean that supporters' defense of the PES from political instability is effortless, despite most informants' comments that the PES enjoys a good amount of public support. An informant who was a former head of an NGO explained that some years ago there was a sense that the government was starting to question the PES, so the environmental movement put together an ample forum to defend it as a precautionary move. However, even when asked about external factors possibly affecting the largely favourable attitudes towards the PES (i.e. higher international prices of oil threatening the support of the gasoline tax that funds the PES) most informants agreed that the PES would stay put.

Despite public support and past financial stability, most informants said that the PES was soon to be in financial distress. An informant, the head of an NGO, explained that the second World Bank loan was set to expire very soon with no options for renewal. Two informants from FONAFIFO's headquarters confirmed that the expiration of the loan would have an impact, although they showed less concern than the non-governmental informants familiar with this did. In our conversations, the informants from FONAFIFO's main office focused more on explaining their efforts to diversify FONAFIFO's financial sources. For example, in 2005, a new law imposed a water tariff specifically aimed at financing PES farms of the watershed where the fee was collected (Pagiola 2008). In addition, FONAFIFO has been reaching out to hydroelectric operators and other big companies to set up voluntary financial agreements to finance the PES and receive ecosystem services certificates in return (Pagiola 2008; Russo and Candela 2006). An informant from

FONAFIFO's main office explained that, more recently, FONAFIFO made an agreement with the National Bank of Costa Rica to introduce a new 'green debit card' for which a percentage of the bank's commission goes to a Fund for the PES. Despite these efforts, there was certainly a sense of anxiety among most of the interviewees familiar with the PES' financial structure, despite official discourse. As one informant head of an NGO said, 'the thinking of those who run FONAFIFO has to change; they have yet to realize that the loan is ending'.

The PES' financial constraints have pushed for seeking alternative sources of funding in the Reduction of Emissions from Deforestation and Degradation (REDD +) mechanism. Costa Rica has been actively engaging in the REDD+ international conversation through the Readiness for REDD+ country process sponsored by the World Bank's Forest Carbon Partnership Facility (FCPF). FONAFIFO takes this effort seriously and it appointed a person within the agency to coordinate the readiness process for the entire country. Costa Rica's incorporation into the FCPF process assumes that it can strengthen current forest conservation policies or develop additional ones to the ones already existing like the PES, particularly in an academic and policy situation where the alleged causal relationship between the PES and the country's reduction of deforestation is a contested issue. When asked about the difference between the PES and REDD+, the person in charge of the readiness for REDD+ at FONAFIFO described REDD+ as being 'more comprehensive'. However, as our conversation progressed it became hard to distinguish this difference.

Conceptually, it seems difficult to differentiate the PES from REDD+ as both aim at reducing deforestation by compensating for the provision of ecosystem services, specifically carbon storage. This similarity may have actually played in favour of Costa Rica, which seems to have been successful in arguing the case that it should become a 'REDD+ country', in the FCPF jargon. Even though Costa Rica currently has net positive annual forest change rate (0.95 % in the period of 2005 to 2010)[8] and thus is not immediately appealing for REDD+ in its traditional sense, it compensates it with good and effective forest governance embodied in the PES, a characteristic that others have suggested may be even more relevant (Phelps et al. 2010). In the assessment of Costa Rica's Readiness Preparation Proposal for REDD + the FCPF (2012, p. 8) states: 'Costa Rica is now seeking a new funding mechanism that would reward the carbon services it provides to the world. It is envisaged that REDD would allow Costa Rica to receive sustainable financial transfers from the international community while consolidating, improving and scaling-up the PES program.' Framing Costa Rican forest conservation policies like the PES to fit current international trends in conservation, such as REDD+, is still an ongoing effort.

[8]see: http://foris.fao.org/static/data/fra2010/FRA2010GlobaltablesEnJune29.xls accessed on July 3, 2015.

6 The Butterfly Effect: Understanding the Policy and Legal Context of Costa Rica's Pes

This paper focuses on the legal and policy context of the PES, the backbone of the program's governance. It reviewed the perceptions of bureaucrats at governmental agencies and workers of non-governmental organizations in two implementation areas and the administrative centre in San José. This allowed for a comprehensive study that includes contrasting descriptions of the same phenomena. Through the analysis of the PES on the ground and the law that surrounds it, this paper sheds light into how the legal and policy context mattered for designing and implementing this conservation programme. However, this study has limitations bound by the short time spent in the area, the relatively small space covered, and the number as well as type of informants.

A look at on-the-ground implementation of the PES provides interesting opportunities to reflect on the effects of the legal framework. For example, the way violations to forest laws occur and are dealt with by judges and PES officials as well as the criminalization of land use change by the Forestry Act of 1996, most likely had an effect in the quality and quantity of the Costa Rican forest cover that is missed in additionality studies. Also, some *finqueros*' impression that the government was attempting concealed takings through the PES has historical and legal basis on the expropriation processes mandated by the protected areas legislation on private lands overlapping these areas. This could help understand what types of *fincas* were most likely to enter the program in the first years and why.

The PES governance has other interesting characteristics, as well. The issue of access, for example, has been subject to prior research focusing on how landholders without title have been mostly left out, or on the high transaction costs involved (Zbinden and Lee 2005; Pagiola 2008). However, also important for the question of access is understanding how the process leading to exclusion or inclusion—the PES procedure—works. This procedure has been changing and now includes modern technological solutions such as online application forms supported by a notable intranet system that benefits from other agencies' online databases. People and documents still have to move, however, and Costa Rica seems to have a comparative advantage in its size and accessibility.

Interestingly, the procedure itself is not currently a subject of much criticism by the informants as it is the rigor applied in the evaluation of new applications. This critique showcased an interesting bureaucratic divide within the PES, which does not run across traditional lines of scale (i.e. regional offices versus San José headquarters) so much as across professions—'the lawyers' versus 'the technicians'. Informants on the ground seem to believe there is a shift in power within FONAFIFO, from the technical to the legal. The will of rendering everything technical (Li 2007) was embedded in the PES as implied by informants on the ground who longed for a more technical and less legal past but now 'the lawyers' seem to have moved further into 'rendering legal' the program. This goes beyond the efforts to simplify people's attitudes and conducts towards nature to benefit a controlling State, which are a staple of the contemporary State (Scott 1999). The animosity of 'the technicians' towards 'the lawyers' highlights a deeper significance of the role of law in conservation.

For a State to be able to fully make nature attainable to its control it is in need of both, a process of 'rendering technical' that simplifies the environment and a process of 'rendering legal' to codify it, or in fact translate it into the governmental lingua franca. For better or for worse, the law ends up being the common language spoken by all forest bureaucrats working in the PES, where technicians may be capable but lawyers are native-speakers and the official translators. It is important to understand, however, that the current extent of the process of 'rendering legal' in the PES has not been always the same; it has changed and evolved. This asserted rule of 'the lawyers' through the process of 'rendering legal' plays a role that is functional to a program that has a three to one demand ratio, by defining the PES application requirements in a more restrictive way. In the process, some bizarre results occur, which are mainly due to an inherent difficulty of the law and its implementers to incorporate the complexities of reality, as explained by previous scholarship (see for example Scott 1999). We still need to comprehend fully whether or how exactly 'rendering legal' the PES or other environmental institutions and mechanisms affects their performance.

Past, present and international laws and policies outside the boundaries of the PES regulatory regime also shape the way the PES evolved and functions. The PES did not occur in a vacuum, as it is a product of years of previous policy incentives, which provided a solid basis for its development (Daniels et al. 2010; Le Coq et al. 2010; Pagiola 2008). However, concurring laws and policies also affect the PES. A first set of these influencing policies and laws are the ban of land use change and the gasoline tax, both within the Forestry Act of 1996. Without them, the PES would have had different results by lacking sustainable funding and rural support, or, more importantly, it is likely Congress would have not passed it in the first place. Thus, these are not just complementary legal provisions but key elements of the entire governance of the PES; they should be accounted for when analysing the PES' additionality. A way to do this would be to include forest crime data, effectively a proxy for land use change, in the models.

There are other examples of non-PES laws that influence its impact on the ground. The *regencia* system with its supervisory checks and balances, the application of public funds laws that reduced FONAFIFO's flexibility, as well as the administrative simplification process across the Costa Rican government are prime examples of policies outside the PES that strongly influence the way it is implemented. Finally, the default application of the 'first in time, first in right' civil law principle during the early years of the PES may have influenced who was able to access the program, likely benefiting even more those *finqueros* who had the means and knowledge to submit an application faster to the detriment of their poorer counterparts. It is also possible that by dropping this principle in the application process, 'the lawyers' became more prominent as FONAFIFO was in need of looking for newer or additional legal requirements different to this simple rule.

The legal framework surrounding the PES is certainly a product of many conditions of its time, including cultural ideas, but it is possible that a feedback loop allowed it to impact social constructs and cultural beliefs as well. Informants repeated the mantra within the PES system that *finqueros* are not paid subsidies but compensated for services; this may have far-reaching consequences. Interestingly, framing the PES as a market-based program that pays for something in return (i.e. not a subsidy) developed in part as a way around the imposed liberal policies of the World Bank of the 1990's but now it seems also embedded in the mindsets of many Costa Ricans. As one of the interviewees explained, apparently now people in cities understand that they owe something to the rural people for their conservation efforts, that is to say, that forests are not a given and they belong to somebody who needs to be compensated, usually poor *finqueros*. The importance of this cultural shift could be major on other social aspects beyond the scope of this research like economic inequality or democratic representation.

As FONAFIFO and Costa Rica looks into the future, uncertainty and a will to adapt mark the PES. The most pressing issue was how to ensure reliable sources of funding after the last World Bank loan expired. FONAFIFO has been trying to diversify its financial sources through partnerships with energy companies and the Bank of Costa Rica, as well as water tariffs. Nevertheless, the program is in need of more substantial and reliable funds and eyed REDD+ as the most likely option. In this process, they have engaged in a similar past experience of framing their efforts in the terms set by the international legal and policy agenda, which in the past was one that discouraged the use of subsidies and now is the REDD+ trend.

Finally, it is important to acknowledge that although the governance of the PES relies on a web of legal mandates and institutional arrangements, people on the ground ultimately execute the program. Implementing the PES is hard work. It requires walking across hilly forests, talking to many people, measuring, sweating, convincing, and walking further. In this sense, borrowing from Barnes' idea of what 'makes' water in Egypt (2014), the PES, and the ecosystems services provision it seeks to secure, is in fact constantly 'made' by the daily work and decisions of implementers and beneficiaries on the ground, from the *fincas* to the offices. After I stopped exhausted in the middle of a field visit, my NGO informant turned to me and sentenced: '"this" is PES'.

Acknowledgments I would like thank various people for their contribution to this chapter. First, Mr Carlos Manuel Rodríguez of Conservation International—Costa Rica, for his support and guidance for this research. Also, to the students and instructors of the 'Writing in the Social Sciences' workshop at Yale F&ES, for extensively reviewing the first drafts. Finally, to my informants on the ground from Limón and Sarapiquí, for generously sharing their time and patiently waiting for me on the field to catch my breath.

I would like to acknowledge the financial support at Yale University provided by the Tropical Resources Institute, the Program of Agrarian Studies, the Jubitz Family Endowment for Research Internships Fund, and the Carpenter Sperry Fund, which allowed me to conduct this research.

Finally, special thanks should be given to Dr Amity Doolittle at Yale University, for her advice and continuous support to this research project.

References

Adams WM, Hutton J (2007) People, parks and poverty: political ecology and biodiversity conservation. Conserv Soc 5(2):147–183

Aguilar X (1995) Veda forestal ¿una llamada de atención? Rev Forestal Centroamericana 4 (11):40–43

Arraigada RA, Ferraro PJ, Sills EO, Pattanayak SK, Cordero-Sancho S (2012) Do payments for environmental services affect forest cover? A farm-level evaluation from Costa Rica. Land Econ 88(2):382–399

Barnes J (2014) Cultivating the Nile: the everyday politics of water in Egypt. Duke University Press Books, Durham

Barreiro P (2012) Opportunity analysis of payment for ecosystem services: policy design and implementation for coffee agroforestry systems in Costa Rica. Trop Resour 31:56–66

Castro Salazar M, Peña Chacón M (2011) The case of Costa Rica. In: Greiber T, Schiele S (eds) Governance of ecosystem services: lessons from Cameroon, China, Costa Rica and Ecuador. IUCN, Gland, pp 73–91

Camacho MA, Segura O, Reyes V, Aguilar A (2000) Pago por servicios ambientales en Costa Rica. Prisma: San Salvador

Cole RJ (2010) Social and environmental impacts of payments for environmental services for agroforestry on small-scale farms in southern Costa Rica. Int J Sustain Dev World Ecol 17:208–216

Daniels A, Bagstad K, Esposito V, Moulaert A, Rodriguez CM (2010) Understanding the impacts of Costa Rica's PES: are we asking the right questions? Ecol Econ 69(11):2116–2126

FCPF (2012) Forest carbon partnership facility Costa Rica forest carbon partnership facility REDD readiness: readiness preparation proposal (R-PP) assessment note. Retrived April 26, 2013, from https://forestcarbonpartnership.org/sites/forestcarbonpartnership.org/files/Documents/Costa%20Rica%20RPP%20Assessment%20Note.pdf

Fletcher R, Breitling J (2012) Market mechanism or subsidy in disguise? Governing payment for environmental services in Costa Rica. Geoforum 43(3):402–411

FONAFIFO (2005) FONAFIFO: over a decade of action. FONAFIFO, San José

FONAFIFO (2012a) Distribución de hectáreas contratadas en pago de los servicios ambientales, por año y por modalidad. Retrieved December 1, 2012, from http://www.fonafifo.go.cr/text_files/servicios_ambientales/Distribucion%20de%20hectareas%20contratadas%20PSA%20por%20A%C3%B1o%20y%20Modalidad%201997-2011.pdf

FONAFIFO (2012b) Emission reduction program idea note: Costa Rica. Retrieved November 10, 2012, from http://www.forestcarbonpartnership.org/sites/forestcarbonpartnership.org/files/Documents/PDF/Oct2012/Costa%20Rica%20FCPF%20ER%20PIN%20v3%2016%20Sept%202012-ENGLISH%20rev.pdf

Government of Costa Rica (2009) Manual de Procedimientos para el Pago de Servicios Ambientales. Retrieved July 3, 2015 from http://www.fonafifo.go.cr/documentacion/biblioteca/decretos_manuales/manualpsa_2009.pdf

Le Coq JF, Froger G, Legrand T, Pesche D, Saenz-Segura F (2010) Payment for environmental services program in Costa Rica: a policy process analysis perspective. Retrieved September 9, 2012, from http://www.serena-anr.org/IMG/pdf/04-2010_PES_costa_rica_-_PES_policy_process_V03_final.pdf

Li T (2007) The will to improve: governmentality, development, and the practice of politics. Duke University Press, Durham

Locatelli B, Rojas V, Salinas Z (2008) Impacts of payments for environmental services on local development in northern Costa Rica: a fuzzy multi-criteria analysis. For Policy Econ 10 (5):275–285

McShane TO, Hirsch PD, Trung TC, Songorwa AN, Kinzig A, Monteferri B, Mutekanga D, Thang HV, Dammert JL, Pulgar-Vidal M, Welch-Devine M, Brosius JP, Coppolillo P, O'Connor S (2011) Hard choices: making trade-offs between biodiversity conservation and human well-being. Biol Conserv 144(3):966–972

Morse WC, Schedlbauer JL, Sesnie SE, Finegan B, Harvey CA, Hollenhorst SJ, Kavanagh KL, Stoian D, Wulfhorst JD (2009) Consequences of environmental service payments for forest retention and recruitment in a Costa Rican biological corridor. Ecol Soc 14(1):23

Pagiola S (2002) Paying for water services in Central America: Learning from Costa Rica. In: Pagiola S, Bishop J, Landell-Mills N (eds) Selling forest environmental services: market-based mechanisms for conservation. Earthscan Publications Ltd, London, pp 37–61

Pagiola S (2008) Payments for environmental services in Costa Rica. Ecol Econ 65(4):712–724

Peña Chacón M (2004) El régimen económico y jurídico de los servicios ambientales. Medio Ambiente & Derecho, 10

Peña Chacón M (2006, n.d.) The legal and economic regime of environmental services in Costa Rica. Retrieved December 1, 2012, from http://cmsdata.iucn.org/downloads/newsletter_2006_en.pdf

Phelps J, Guerrero MC, Dalabajan DA, Young B, Webb EL (2010) What makes a 'REDD' country? Glob Environ Change 20(2):322–332

Plaff A, Robalino JA, Sanchez-Asofeifa GA (2008) Payments for environmental services: empirical analysis for Costa Rica. Duke University, Durham

Robinson J, Redford K (2004) Jack of all trades, master of none: inherent contradictions among ICD approaches. In: McShane T, Wells M (eds) Getting biodiversity projects to work: towards more effective conservation and development. Columbia University Press, New York, pp 10–34

Russo R, Candela G (2006) Payment of environmental services in Costa Rica: evaluating impact and possibilities. Tierra Trop 2(1):1–13

Sánchez-Azofeifa A, Pfaff A, Robalino JA, Boomhower JP (2007) Costa Rica's payment for environmental services program: intention, implementation, and impact. Conserv Biol 21 (5):1165–1173

Scott J (1999) Seeing like a State: how certain schemes to improve the human condition have failed. Yale University Press, New Haven and London

Sierra R, Russman E (2006) On the efficiency of environmental service payments: a forest conservation assessment in the Osa Peninsula, Costa Rica. Ecol Econ 59(1):131–141

Takacs, D. (2009) Forest Carbon: Law and Property Rights. Conservation International, Arlington

Wells M, Brandon K (1992) People and parks: linking protected area management with local communities. The World Bank, Washington

Wells M, McShane T, Dublin H, O'Connor S, Redford K (2004) The future integrated conservation and development projects: building on what works. In: McShane T, Wells M (eds) Getting biodiversity projects to work: towards more effective conservation and development. Columbia University Press, New York, pp 397–419

Wunder S (2005) Payments for environmental services: some nuts and bolts. CIFOR Occasional Paper 42. CIFOR, Bogor

Wunder S (2007) The efficiency of payments for environmental services in tropical conservation. Conserv Biol 21(1):48–58

Zbinden S, Lee D (2005) Paying for environmental services: an analysis of participation in Costa Rica's PSA program. World Dev 33(2):255–272

From River Basins to Landscapes—Holistic Legal Constructs and Their Differentiation

Henrik Josefsson

Abstract This chapter explores themes related to the differentiation of spaces of regulation, with regard to the river basin/body of water constructs in the EU Water Framework Directive and the landscape construct in the European Landscape Convention. The two spaces of regulation are described and followed by an analysis informed by the sustainability perspective of E. Ostrom. Ostrom provided a scheme for differentiating spaces for sustainable management, 'action areas', and emphasized that the general community, biophysical materiality, and instructive rules are variables that need to be considered when differentiating sustainable management units. The analysis indicates that by combining the spaces of regulation in the Water Framework Directive and the Landscape Convention a more sustainable space of regulation can be established that both reflect Ostroms variables and the multiplicity of the social and ecological dimensions. Similar to the Landscape Convention, the general community, as the public concerned, should be eligible to participate in the differentiation of spaces of regulation. Biophysically a site-specific adaptation seems to better fit with both Ostroms reasoning and scientific literature on the subject. Instructive rules, as the focus or objective of the space, are suggested as population self-maintenance as an indication of ecological functionality.

Keywords EU water framework directive · European landscape convention · Elinor Ostrom · Spaces of regulation · Management units

1 Introduction

This chapter analyses the establishment of spaces of regulation (also see Josefsson 2014), with regard to the EU Water Framework Directive (henceforth, 'Directive') and the European Landscape Convention (henceforth, 'Convention'). The Directive

H. Josefsson (✉)
University of Uppsala, Uppsala, Sweden
e-mail: Henrik.josefsson@jur.uu.se

V. Mauerhofer (ed.), *Legal Aspects of Sustainable Development*,
DOI 10.1007/978-3-319-26021-1_17

and the Convention share important aspects of today's legal regulations: they are holistic, and focus on the entire landscape or river basin, and not on single species, protected areas, or single activities. The Directive and the Convention complement one another; the Convention is attuned to the aspirations of the public in its differentiation, and the Directive uses a biophysical river basin/body of water focus in its differentiation of spaces of regulation. The aim of this chapter is to explore the possibility to develop a space of regulation based on these complementarities informed by the sustainability perspective of E. Ostrom. Ostrom provided a scheme for differentiating spaces for sustainable management, named 'action areas', and emphasized that the general community, biophysical materiality, and instructive rules are variables that need to be considered when differentiating sustainable management units (Ostrom 2005, p. 15). The questions I attempt to explore are: 'How should an area in which an expanse of water is a dominant feature be legally differentiated, to establish spaces of regulation that allow for sustainable management of aquatic natural resources?

First, the Directive is described and discussed, followed by the Convention, the two legal acts are then analysed based on Ostroms emphasized variables: general community, biophysical materiality, and instructive rules. A discussion and a concluding section follow the analysis.

2 The EU Water Framework Directive

The Directive is notable for the complexity of its methodological and ecological assessment, for which it has been criticized (Moss 2008; Hering et al. 2010; Josefsson and Baaner 2011; Josefsson 2012, 2014). Still, many see the Directive as stepping stone towards sustainable management of EU river basins, and it is regarded as laying the foundation for a modern, holistic, and ambitious water policy for the European Union (European Commission 2009). The Directive establishes environmental objectives that are commonly recognized within the European Union. Here, the target of 'good ecological status' by 2015, as an expression of the quality of the structure and functioning of aquatic ecosystems, is a focal point for assessment and management by all Member States (Josefsson 2015a). The ecological aspects have been advanced since the enactment of the Directive in 2000, and the Directive specifies a legal requirement for Member States to construct biological monitoring systems that are comparable through an intercalibration exercise, to provide a harmonious definition of 'good ecological status'[1] (e.g. European Commission 2013; Josefsson 2015a). To achieve a harmonized view of 'good ecological status' the Directive establishes a typological system that differentiates the river basins (found primarily in Annex II). The typological

[1]See Art. 4, Annex V 1.4.1.

differentiation highlights how the river basin is to be understood legally and ecologically, as ‘good ecological status’ space of regulation.

2.1 *From River Basins to Bodies of Water*

The Directive frames and differentiates by the use of a number of defined geographical units, the largest being the river basin or sub-basin, which describes an area of land from which all run-off flows through a sequence of streams, rivers, and lakes, into a sea, river mouth, lake, or river confluence (Art. 2). These large geographical units are categorized and organized into smaller units, such as ‘artificial body of water’ (created), ‘heavily modified body of water’ (physically altered with substantial changes in character), or ‘body of surface water’ (a discrete and significant ‘natural’ element of surface water, e.g. a lake, stream, or river, or transitional or coastal waters; Art. 2).

The first step in the differentiation (system A) is to divide the features of a river basin into various categories, for example, a lake or river.[2] The second step is to differentiate based on type, for example, if a lake has a depth <3 m, 3–15 m and >15 m, each area is designated as one of three types, and each type will be compared to’ a type-specific ‘high ecological status’ lake reference point.[3] For heavily modified and artificial bodies of water, the line drawn between types of bodies of water is primarily based on the changes in hydromorphological characteristics resulting from physical alternations caused by human activity, preventing the attainment of ‘good ecological status’ characteristics[4] (also see Common Implementation Strategy No 2 2003). For example, a dam affects the hydromorphological characteristics of a body of water, owing to its physical alteration of the river basin. For rivers, as one example, the defining features of this differentiation include ecoregion; altitude, size, and geology.[5]

A differentiation based on kinds of pressure and impact may be used to distinguish among ‘natural’ bodies of water that are affected by human factors, but are not classified as heavily modified or artificial. A differentiation is complementing this based on use (which frequently overlaps impact), protected areas, or physical or chemical features that determine the characteristics of a body of water, and hence, the biological quality elements (Common Implementation Strategy No 2 2003; Common Implementation Strategy No 3 2003). These complementary differentiation methods are part of the alternative differentiation system (system B), but must achieve at least the same degree of differentiation as would be achieved using system A, that is, ensure that type-specific biological reference conditions may be

[2]Annex II 1.1 (i).

[3]Annex II 1.2.2.

[4]See Art. 4 (3) a, Annex II 1.1. (v) and Annex V 1.2.5.

[5]See Annex II 1.1. (iii) and 1.2.1.

reliably derived.[6] For rivers, as one example, the defining features for a system B differentiation include altitude, latitude, longitude, geology, and size. Several optional features are also defined, such as form and shape of the main riverbed, river discharge, valley shape, and transport of solids.[7] Member states have primarily used system B for the development of a typology (European Commission n.d., p. 62).

There also exists the possibility of reversing the differentiation process, and grouping bodies of water into a single type. This is possible only for small bodies of water that share their discrete and significant water element. Furthermore, bodies of water should be aggregated' only when category and type overlap, influenced by the same pressure category and type, and when they influence one another. However, bodies of water with high ecological status (i.e. reference points) should not be grouped together (Common Implementation Strategy No 2 2003). Thus, when contiguous elements of surface water of a type share the same status, combining them into a single body of water still provides an accurate description of the surface water status. This aggregation may also be conducted if the differentiation seems illogical to the public—that is, a section of a river or lake is differentiated to better fit the understanding of a lay observer—but only if this still follows the aim of the differentiation[8] (Common Implementation Strategy No 2 2003). The Directive allows for other types of differentiation systems, as long as they are firmly in agreement with the aim of differentiation, which is to provide a precise topographical map of a river basin, which secures the achievement of the Directive's purpose, objectives, and provisions.[9]

This differentiation is essential for several reasons, as it maps out the typological units for the reference system, classification, assessment, and management. The body of surface water is the space of regulation to which the environmental objectives, 'good ecological status' or 'good ecological potential',[10] applies.

2.2 *Public Participation*

The Directive was partly a response to the increasing criticism of the technocratic and top-down framework of EU water laws and policy (Kaika 2003). Still, another critical view relates to how the differential planning and most of the Directive are based solely on narrow ecological/scientific thinking. It has been emphasized that in order to achieve sustainable river basin management, knowledge should be gathered from all relevant disciplines, to be integrated into the planning processes, something

[6]See Annex II 1.1. (iv) and 1.2.

[7]Annex II 1.1. (iv) and 1.2.1.

[8]See Annex II 1.1.(iv).

[9]See Art. 4 (1)(i)(ii)(iii), Annex II 1.1. (iv), and Annex V 1.2.

[10]See Art. 4 (1)(a)(ii)(iii).

the Directive fails to stipulate (Ludwig et al. 2001; Hedelin 2008; Hedelin and Lindh 2008).

The Directive emphasizes public participation as a method for informing the public and hearing their thoughts about the established assessment, and management differentiation and planning (Steyaert and Ollivier 2007; Valinia et al. 2012). There is a requirement for information and consultation in the planning processes and the implementation process,[11] but not in the fundamental differentiation or the assessment. In other words, in accordance with the Directive, public knowledge is used after the critical assessment and planning phases (Common Implementation Strategy No 8 2003).

There is a guidance document on public participation that emphasizes that pressure and impact assessment, classification and objective-setting, and reference conditions should involve public participation, since it is argued its knowledge may be just as valuable as expert opinion (Common Implementation Strategy No 8 2003). However, the Directive specifies that active involvement should be encouraged in the processes or production, reviewing and updating of river basin management plans,[12] but fails to mention any other type of knowledge base than expert knowledge in the establishment of reference conditions, classification, monitoring, or differentiation.[13]

2.3 *The 'Body of Water' Construct from a Biomonitoring Perspective*

A type-specific, established system for biomonitoring means that you compare and assess sites, by assuming that one site has certain desired attributes/elements/conditions, which are then compared to those of another site with undesirable elements or conditions. This reasoning is used experimentally, when two similar sites exist, and one is experimented on, to understand and predict how ecosystems may respond to the induced stressors. In the science of ecological assessment, much depends on the possibility of creating expectations, which site-specific conditions must meet' (Stoddard et al. 2006). In management, the reasoning is practical: if human stressors impact a stretch of river, for example, you compare this stretch to a similar reference stretch with higher ecological status, and through management measures, try to alter the impacted stretch, to make it similar to the reference stretch (Friberg et al. 2011). This is of practical value, since a number of comparable bodies of water may be similarly assessed and managed (Verdonschot 2006).

[11]Art. 14(1) and 14(2).

[12]Art. 14(1).

[13]See Annex II 1.3. (iii) and Annex V 1.3.4.

The facilitation of the measure-compare-classify process is the driving force behind the differentiation of a river basin into discrete and significant elements of surface water, the typological bodies of water. Such an assessment and management map is based on the assumption that the biological communities in a type of body of water will deviate only slightly from the reference body of water and its biological communities, when it achieves 'good ecological status'[14] (Karr and Dudley 1981; Angermeier and Karr 1994, p. 693).

One troubling aspect of the foregoing thinking is that an a priori differentiation assumes that types of bodies of water provide the templates that determine ecological quality, and if chosen aspects of this quality are absent, the assumption is that the body of water is degraded (this is also amplified by the Directive's 'one out–all out' principle[15]). However, this is not definite the case (Friberg et al. 2011; Josefsson and Baaner 2011). Instead, the biotic interactions are key drivers, which is demonstrated when otherwise identical environments present very different sets of species and traits (Woodward 2009; Friberg et al. 2011).

With regard to the Directive, there have been scientific suggestions for how the differentiation could be developed, either towards simplicity, aiming for manageability, or more sophisticated systems that are better adapted to addressing aquatic ecosystems (Moss et al. 2003; Hering et al. 2010). A recurring, critical view is that a priori typological differentiation, with its broad categories of map-derived variables, fails to recognize the site-specific aspects of a river basin or body of water (Hering et al. 2010). Or, that typological river basin differentiation is a questionable method for describing how a river basin is affected or unaffected by establishing reference points in different ecological environments, for example (Whittier et al. 2007; Friberg et al. 2011). It is argued that instead of a priori judgements, more site-specific ecological aspects, such as the composition of river basin materials on which organisms are dependent, makes more sense, and should be used (Moss et al. 2003; Davy-Bowker et al. 2006; Verdonschot 2006; Johnson et al. 2007). Another criticism is that the differential process loses some of the benefits of being type-specific, since the Directive demands a high degree of differentiation, which has negative consequences for the applicability of the typology system (Verdonschot 2006). The body of water construct make sense, not in terms of its own attributes, but based on the structural factors that direct its differentiation. This kind of thinking has been called the 'episteme of similarity' (Foucault 1971, pp. 136–137, 144–145) and has been criticised in legal research literature (e.g. Graham 2010).

In general, it may be said that other typological factors are more reliable, simply because they apply a differentiation that is based on variables that are ecologically significant in a given region (site-specific), and comply more closely with the character of aquatic ecosystems (Davy-Bowker et al. 2006; Hering et al. 2010). As parameters relevant to typology are among the major sources of uncertainty in ecological assessment, this scientific discussion is not surprising (Hering et al. 2010).

---

[14]See Annex V 1.2.

[15]Compare to Annex V Sect. 1.4.2.(i).

3 The European Landscape Convention

The Convention is the world's first regional treaty specifically concerning landscapes (at a sub-regional level there is the Mediterranean Landscape Charter, which also influenced the Convention). It was opened for signature by the member states of the Council of Europe in 2000, and enacted in 2004. By May 2015 out of the EU Member States, all but two—Austria and Malta—have signed and ratified the Convention (Malta only signed it), the EU has not become Party to the Convention. The Convention was enacted in response to a growing concern across Europe regarding the nature and scale of landscape change, increasingly seen to result in a loss of distinctive local character. To address this growing concern, a differentiation was established, wherein the concerned general community perceptually frames the differentiation, in stark contrast to the Directive's biophysical differentiation.

In comparison to the Directive, the Convention is more general in form, focusing on fundamental, but still general, aspects that give direction regarding how a landscape, should be understood, differentiated, assessed, and managed. The Directive, on the other hand, imposes obligations on the Member States, with respect to individuals, with or without direct effect, which requires implementation of the necessary measures to achieve these objectives (see Art. 288 TFEU). This kind of rights for individuals is not possible to derive from the Convention itself; instead there the parties to Convention have full discretion to establish, or not, such rights.

4 The Differentiation of Landscapes

Ecologically, 'landscape' may be defined as a geographic area in which variables of interest are spatially heterogeneous, where the boundary is based on geographic, ecological, or administrative units (e.g. river basin, county, urban area) that are relevant to the research questions and objectives (Wu 2013). This definition agrees with the Convention's, in terms of the human-scale landscape, and that human perception often coincides with geographic units such as river basins or urban areas (Gobster et al. 2007). It is emphasized that the human landscape (or region) may be considered a basic spatial unit for studying and maintaining sustainability, because it represents the smallest scale above which nature-society interactions may be meaningfully addressed (Wu 2006). The primary objective of the assessment of such landscapes is the systematic and methodical determination of the capacity of landscapes to meet various social requirements (Bastian et al. 2006). The presence of a correlation between the public and broader geo- and biophysical spatial factors is emphasized as an important aspect. It is argued that by a correlation it is possible to connect assessment and management to public landscape perception, connecting humans to ecological phenomena, as the perceived spatial scales of a river basin, for example, often form the basis on which humans intentionally affect ecosystem and ecological processes (Gobster et al. 2007; Brace and Geoghegan 2011).

Corresponding to ecological landscape perspective the Convention defines landscape (includes land, inland water, and marine areas[16]):

> Landscape means an area, as perceived by people, whose character is the result of the action and interaction of natural and/or human factors.[17]

Here, a landscape is an evolving entity, resulting from the actions of natural and human forces, a whole whose components should be assessed and managed together, rather than separately (Council of Europe 2000). Assessment, management, and planning should comprise the whole landscape, and refrain from dividing it into a number of component elements. A landscape is not the sum of its constituent parts; instead, a landscape is a qualitative space, and not fully quantifiable (Council of Europe 2008).

Experts, in conjunction with local inhabitants, differentiate landscapes, from one another. Differentiating among landscapes means understanding that each landscape is a combination of constituent segments and structures, the whole of which may be fundamental to local identity, and intertwined with social perceptions, but also involves constantly changing/evolving natural, social, and economic processes (Council of Europe 2000). The differentiation is not intended to generate a map of landscapes based on a priori considerations; instead, the differentiation is open to what people perceive as landscape units, a differential process based on landscape democracy (Council of Europe 2000). However, if there are conflicting perceptions there are no rules for solving or prioritising one specific view over another. There are examples where municipal voting has solved conflicting perceptions of the use of a river as either exploited for hydropower or protected for fishing or other recreational activities, and the democratic process resulted in a unexploited river (Angelstam 2009). This correspond to other literature that emphasize that on a general level of perception there is a correlation between human and non-human interests (e.g. Gobster et al. 2007; Brace and Geoghegan 2011).

Thus, within the Convention the landscape unit is not an infinite one; it is a space that is recognized and identified by the public concerned. A landscape is not a natural entity that may be located according to a structural pattern, nor is it a specific assemblage of physical objects that may be objectively analysed by the natural or social sciences, but a shared cultural and natural heritage, a reflection of the diversity of cultural, ecological, and socio-economic surroundings[18] (Olwig 2007). Thus, the people whose daily practices and perceptions shape the social and physical landscape should establish the basis for landscape planning and development, by differentiating one landscape from another (Olwig 2007).

---

[16]Art. 2.

[17]Art. 1 (a).

[18]See Art. 5 (a).

Active public participation should occur at all stages, including differentiation.[19] Knowledge should be accessible to all, that is, be understandable to non-specialists (Council of Europe 2008). In such instances, participation is intended to be non-discriminatory, in accordance with the Aarhus Convention (henceforth, 'Aarhus'),[20] and includes administrators, professionals, and the public, both as visitors and residents, emphasizing that visitors' perceptions may be important in areas undergoing urbanization (Jones 2007; Lowenthal 2007). The openness to the broader and local communities with regard to policy, quality objectives, programmes, and plans, is restricted in the assessment and only includes the population concerned,[21] which, in Aarhus, is defined as the public affected or likely to be affected, or having an interest.[22] Here, Aarhus provides a complementary definition of 'the public' as one or more natural or legal persons and their associations, organization, or groups, emphasizing that the public is more than the citizens[23] (also see Prieur 2006). Differentiating landscapes based on a multi-layered knowledge base is intended to give management the opportunity to ensure social, economic, and environmental processes, and bring about changes that ensure the maintenance of the landscape,[24] in keeping with the aspirations of the public.

As described above, this differentiation is intended to allow for an analysis of morphological, archaeological, historical, cultural, and natural characteristics and their interrelation, as well as an analysis of change, reflecting the landscape's multiplicity of social and ecological functioning (Prieur 2006; Council of Europe 2008). The differentiation establishes a heterogeneous frame of reference for the Convention.

5 Analysing the Directive and the Convention

The following sections attempt to conjoin the Directive and the Convention, and attempts to answer the question presented in the introduction: How should an area in which an expanse of water is a dominant feature be legally differentiated, to establish management units that allow for sustainable management of aquatic natural resources?

The assumption is that a differentiation based on the complementariness of the Directive and the Convention increases the likelihood of managing the river basins and water landscapes towards an improved status more reliably than the Convention or the Directive alone. In conjoining the Directive and the Convention, two different

[19]Art. 5 (c).

[20]Aarhus Art. 3 (9).

[21]Art. 6 (C).

[22]Aarhus Art. 2 (5).

[23]Aarhus Art. 2 (4).

[24]Art. 1.

variable clusters for differentiation are highlighted: the general community (the Convention), and the geo- and biophysical (the Directive). It has been emphasized that management units, or 'action areas' as E. Ostrom described them, should be differentiated based on the general community, biophysical materiality, and instructive rules (Ostrom 2005, p. 15). These three variable clusters serve as the basis for the analysis.

5.1 The General Community

The general community is defined by boundary rules that define who are eligible to participate in the process of defining the aspirations of the general community, with regard to differentiation, assessment, and management (for boundary rules see Ostrom 2005, p. 194).

Even if the Directive fulfils the Aarhus requirements there are inconsistencies between the wide interpretations of the 'public' envisages in Aarhus and the differentiation and assessment is not open for public participation (Howarth 2009). For sustainable water management, it has been emphasized that it is crucial to involve the public in the initial phases, and if the public and local actors are asked to comment only subsequent to the fundamental planning, this is not achieved (Hedelin 2008; Wright and Fritsch 2011).

In comparison to the Directive, the Convention has a more open differential structure that is not determined a priori, but in conjunction with experts and local inhabitants, which demands extensive participation of the public, local, regional authorities, and experts, to acquire knowledge and an understanding of the aspirations of the public before competent authorities differentiate one landscape from another (Jones et al. 2007).

Thus, the Convention tries to be more attuned to the general community then the Directive. Based on the Convention in the differentiation of landscapes those who should be eligible to enter into the process are administrators, the public (visitors and residents), natural or legal persons and their associations, organizations, or groups should be allowed to be involved.[25] With regard to the differentiation and establishment of spaces of regulation, a narrower boundary rule are implemented, and only the public (visitors and residents), and natural or legal persons affected or likely to be affected are allowed to participate, the narrowed focus being intended to give more room for situated or place-based knowledge.[26] In comparison to the Directive[27] this is a more open and precise boundary rule, than 'all interested parties' with respect to river basin management plans and programme of measures, and establishes the possibility of lay knowledge being used in all phases of the

[25] Aarhus Art. 3 (9), 2 (4).

[26] See the Convention Art. 6 (C) and Aarhus Art. 2 (5).

[27] Art. 14(1).

assessment and management, including differentiation of spaces of regulation, whereas the Directive only implement participation after assessment and differentiation.

5.2 Biophysical Materiality

In the Directive, differentiation follows the general assumption that if assessment and management are applied according to river basin features, the institutional systems will perform better (Young 2002, p. 20). However, for a comparison of what 'good ecological status' means for each Member State, a predetermination of the differentiation is outlined in the Directive, which restricts a site-specific adaptation of the bodies of water and the institutional system, for example towards relevant river basin ecological processes. The river basin/body of water differentiation combines ecological thinking with legal prerequisites, such as a harmonized understanding of 'good ecological status', and should be understood as such.

In a landscape in which an expanse of water is a dominant feature, the most apparent and telling (ecological) characteristic is the flow of water (even if there is exceptions to this, such as lakes sourced solely by rainwater). Water, or the hydrological connectivity, mediates the transfer of matter, energy, and organisms within or among elements of the hydrological cycle (Pringle 2001). Various concepts have been used to describe, present, and differentiate ecological water landscapes, including 'riverine landscape' or 'riverscape', which focus on the macro-level of aquatic ecosystems. These representational concepts indicate a macro-level approach, focused on the patterns and processes associated with fluvial ecosystems, integrating ecological processes and spatial complexity (Ward 1998; Ward et al. 2002).

At the macro-scale of riverine landscapes, differentiation may apply broad, differential categories, such as 'erosive river', 'depositional floodplain river', their associated shallow lakes and wetlands, and deep lakes (Moss et al. 2009). Or, flow-regimes, as ecological processes, may be used as a basis for specifying stream and river types that exhibit similar ecological conditions, and may have similar responses to flow alternations (Poff et al. 2010). This is similar to the Directive's type-specific differentiation, and the criticism of type-specific reference systems voiced above may be applicable to this differentiation; also, social contextualization for resource users, and general administrative or legal arrangements are absent (Pahl-Wostl et al. 2013). In relation to the Convention a type-specific differentiation stands in stark contrast to how the Convention emphasizes that the landscape unit are a qualitative space, and not quantifiable. Ecological context, such as site-specific interactions between flow-patterns, geomorphology, and temperature heterogeneity, which vary markedly among differing geologic, climatic, and topographic settings, should frame the differentiation of water landscapes (Ward and Stanford 1982; Moss et al. 2009; Olden and Naiman 2010; Brierley et al. 2013). A site-specific, riverine, macro-scale differentiation should be based on the hydrological-ecological network of site-specific flow-patterns, sedimentation, nutrients, and organisms, elements that differentiate one unit from another (McCluney et al. 2014). A riverine differentiation would construct units that are large enough for an assessment of the variations that produce changes, both downstream and upstream from the immediate area of rehabilitation site, for example (McCluney et al. 2014). Avoiding type-specific models, we find ecological thinking that fits with the way in which the Convention emphasizes that all landscapes are potentially unique and interconnected. Both the Convention and this macro-level, ecological perspective focus on what is present and why, rather than what could be there, theoretically (Poole 2002; Pedroli et al. 2006; Brierley et al. 2013).

There is potential for a site-specific, macro-scale differentiation that acknowledges riverine processes, and emphasizes, as the Convention does, that river basins or landscapes cannot be managed by scaling up information (Council of Europe 2008). The ecological field of reference becomes one of site-specific landscape patterns and processes, rather than individual, type-specific sites, a 'least disturbed condition' frame of reference, similar to the Habitats Directive (Ward et al. 2002). Framing the Convention's differentiation in this manner, units could be aimed at managing measures that focus on what marks an area as special and distinctive. Heterogeneity of biophysical habitat conditions, intrinsic connectivity between water landscapes, and temporal fluctuations of variables, such as population abundance and nutrient maintenance, are potential ecological metrics for such macro-units (McCluney et al. 2014; Thorp 2014).

5.3 Instructive Rules

By emphasizing biophysical and general community elements, the Directive and the Convention operationalize certain elements of the legal/social aim of sustainability, which is part of the general legal fabric of today.[28] The general community represents an intragenerational, perceptual realm, and socio-cultural and economic dimensions, and the biophysical materiality provides an ecological dimension. However, sustainability also emphasizes intergenerationality, as an obligation to provide future generations with equal ecological resource options (in terms of renewables) in comparison to previous generations (compare to Brown-Weiss 1989, pp. 38–45; D'Amato et al. 1990). In (environmental) law, future generations are an intrinsic element, owing to the recognized aims of sustainability and intergenerational justice; it is an essential legal obligation of any legal space (Philippopoulos-Mihalopoulos 2009, p. 208). Therefore, it is necessary to consider how legal management units operationalize sustainability and address intergenerational management elements of a river basin.

[28] E.g. Art. 3 (3) of the Treaty on European Union and Art. 11 of the Treaty on the Functioning of the European Union.

The meaning of equal ecological resource options is a question of perspective; here, it is interpreted as an intergenerational obligation in the present, to maintain an abiotic and biotic template that enables organisms, populations, and communities to *self-maintain,*[29] measured by the ability to *self-generate*, and thereby provide for ecological functioning across generations (Heiden et al. 1985). As a self-maintaining system is based on the on-going generation and disappearance of constituent sub-processes (e.g. organisms and communities), performance and functionality may be understood as intergenerational ecological elements, and could provide a framework for intergenerational ecological obligations. That is, if organisms are able to maintain the essential conditions required for their own intergenerational existence, they also have a capacity to persist intergenerationally (see also Josefsson 2015b, pp. 100–102).

Measuring ecological functioning through self-maintenance restricts assessment and management to those elements, such as certain species or traits, that contribute to the maintenance of the ecological organization of the site-specific waterscape/river basin, while they are maintained by that same organizational structure, the very existence of each assessed and managed element depending on their being involved in the self-maintenance of the ecological organization (Mossio et al. 2013, Nunes-Neto et al. 2014). That is, we articulate self-maintenance as an individual and intragenerational property that contributes to the autonomous organization of individual organisms and populations, as well as intergenerationally contributes to the self-maintenance of the ecosystems (Elmqvist et al. 2003; Hughes et al. 2011; Saborido et al. 2011; Hughes et al. 2012).

This could produce units that could be focused towards addressing potential impairments to ecosystem functionality before they occur, while corresponding to intergenerational properties, such as resilience. In comparison to Brown-Weiss (1989, pp. 39–42), we do not aim to define units that conserve a certain status or options, but construct units intended to produce these options. A necessary condition is that the river basins be perceived as spatially connected; how we manage the river basin significantly affects its constituent parts, and vice versa.[30]

This is a system perspective, in which spaces of regulation are governed by both their respective intrinsic properties, and by their being embedded in the river basin. Within this frame of knowledge, the river basin is understood as a macro-legal-ecological construct, and the spaces of regulation are relational, regional (meso) scale legal-ecological constructs. Thus, even if each space of regulation is differentiated, it is still related to the whole river basin but not as a discrete, isolated element, as the body of water, but a constituent part of the whole river basin (Koestler 1989, pp. 341–348; Ostrom 2005, pp. 11–12; Jørgensen et al. 2007, pp. 246–250).

---

[29]See Habitat Directive Art. 1 (i).

[30]The WFD has certain transboundary elements incorporated into it, such as a set of specific monitoring requirements, but does not provide obligations that force Member States to inform each other when there is a violation of the objectives that will have transboundary effects (Keessen 2008).

6 Discussion

Any differential and simplified model of a river basin must function in relation to its intended outcome, and if the ecological status is to be improved, it is necessary for the differentiated units to contextualize themselves in relation to both the natural and social elements. It has been emphasized that even a biophysical unit is not something objective or universal, but a matter of normative settings, worldviews, and societal decisions (Jax 2010, p. 83). Therefore, the differential boundaries for a body of water or an ecosystem, for example, are not objective; the boundaries are chosen based on preferences or perspective (Lefebvre 1991, for example pp. 33, 48–49, 170; Jax 2006).

Ostrom defined the first variable for differentiation of 'action areas' as the general community, and legally those that are eligible to participate within EU and the wider Europe are often based on the Aarhus Conventions definitions. Both the Directive and the Convention are based on Aarhus but they differ in that the Convention emphasizes a wider conception of the 'public' then the Directive. Another difference is that the Convention allows for public participation in all stages of the assessment/management cycle, and allows for public participation in both the differentiation of spaces of regulation and assessment of these spaces, which the Directive does not. There is also scientific literature that emphasizes that full participation in all stages of both assessment and management is required for sustainability to be achieved (e.g. Ludwig et al. 2001; Hedelin 2008; Hedelin and Lindh 2008). Thus, it can be considered that the Convention have a better definition of its boundary rule for public participation then the Directive, even if also the Directive succeeds in being compatible with Aarhus.

How the biophysical materiality should influence the differentiation of spaces of regulation differs between the Convention and the Directive. At first glance it would seem as the Directive focus on the river basin, as the point of departure for differentiation of bodies of water, provide a suitable biophysical differentiation. The problematic aspects of the Directives differentiation are that it does not acknowledge the biophysical materiality, with regard to the ecosystems of each body of water, of each space of regulation, as the system of differentiation is type-specific. This means that there are fundamental flaws in how the body of water construct is adapted towards the area that it regulates. In Ostroms studies the biophysical materiality of the action areas under study is site-specific and not type-specific, a perspective that can find support in the science of biomonitoring. The biophysical focus on the river basin as the point of departure in differentiation of spaces of regulation seems correct, but the body of water construct should had been biophysically site-specific and not type-specific. As the differentiation under the Convention is without a detailed scheme each landscape could be biophysically site-specific, however this demands that this is the perception of the public affected and the governmental organizations organizing the process, but this cannot be guaranteed. Thus, both the Directive and the Convention fail to properly safeguard the site-specific biophysical materiality in the differentiation of spaces of regulation.

The last variable in Ostroms differentiation of 'action areas' is instructive rules, which aim to provide the necessary focus or aim that the 'action area' is created for. The body of water construct is the space for 'good ecological status', and the landscape construct is the space for what the public find are their aspirations. What is a proper legal focus for a space of regulation focused on sustainable management of aquatic natural resources? Above population self-maintenance, as a measure of ecological functioning of a space of regulation, was emphasized as such a focus (also see Josefsson and Baaner 2011). This focus is a development of the 'ecological status' construct, partial, emphasis on ecosystem functioning, and self-maintenance is intended to emphasize the site-specific abiotic and biotic needs for self-generation and not undisturbed conditions in the differentiation of spaces of regulation.

Conjoining and developing important elements of the spaces of regulation of the Directive and the Convention, a space of regulation focused on sustainable management of aquatic natural resources should emphasize the perceptual landscape perspective found in the Convention, while maintaining the Directive's focus on water as the biophysical frame of reference. A unit differentiated according to the discussion above would support assessment and management that focus on negotiation of how site-specific, ecological process-patterns relate to social meaning, its ecological and social interrelations, and leave room for its history, practices, and rules-in-use that produce and reproduce the particular waterscape (Lefebvre 1991, pp. 33, 89; Mitchell 2008). A procedural differentiation of a conjoined space of regulation satisfies requirements such as stability, legitimacy, public participation, and social and ecological, periodical evaluation, all of which are important elements of a legal-ecological system for adaptive sustainable management (see for example Ruhl 2012).

7 Conclusion

A unit to be assessed and managed must be differentiated from a larger geographical area. What may be known of a differentiated space is delimited by the location of its borders. The spaces of regulation analysed here, river basin-body of water and landscape both fail to properly describe a landscape in which an expanse of water is a dominant feature; instead, each is a differentiated actualization of chosen ideas and aspects.

The 'body of water' and the 'landscape' are social spaces with legal meanings, and it is crucial that the legal meanings they provide are significant for the legal management of water, and give legitimacy to it, if we are to improve river basin status. How we legally understand the differentiation of units is essential, particularly when they are legally materialized, the legal units of 'body of water' and 'landscape' are potential forces that will influence the kinds of landscapes or river basins we leave to future generations, as the systems of law and ecosystems intertwine, and give rise to emergent conditions and properties (Delaney 2003, 2011, p. 5; Ruhl 2007, 2012).

The differentiation of spaces of regulation should be ecologically comprehensive while providing the legal system with an appropriate focus. Neither the concept of landscape nor that of river basin/body of water provides such a management unit. Both fail to specify a differentiation that allows for the complexity of the place-based legal arrangements and ecosystem interrelatedness, for example.

Conjoining the Directive and the Conventions elements for differentiation and informing these elements with the three variable clusters of Ostrom, general community, biophysical materiality, and instructive rules, a more sustainable space of regulation would be established. Thus, the general community should be the public concerned, biophysically the space of regulation should be site-specific and have instructive rules that focus the differentiation towards providing spaces of regulation that can assess and manage population self-maintenance (as an indication of ecological functionality). Thus, boundaries should relate to the intergenerational functionality of ecosystems, and be used to attempt to manage the status at the macro-level of ecological processes, and provide units that could emphasize the social and economic values that exist outside the boundaries of traditional permit processes or property, in general (Hirokawa 2010; Josefsson and Baaner 2011). Practitioners should be aware of the limitation and flaws of both the 'landscape' and 'body of water' construct to better work towards a better spaces of regulation adapted to the areas it govern.

References

Angelstam P (2009) The challenge of governing and managing a riverine landscape—towards integrative knowledge production. In: Breuste J, Kozov M, Finka M (eds) European landscapes in transformation: challenges for landscape ecology and management, European IALE conference. University of Salzburg, Salzburg, pp 256–259

Angermeier PL, Karr JR (1994) Biological integrity versus biological diversity as policy directives. Bioscience 690–697

Bastian O, Krönert R, Lipský Z (2006) Landscape diagnosis on different space and time scales—a challenge for landscape planning. Landsc Ecol 21(3):359–374

Brace C, Geoghegan H (2011) Human geographies of climate change: landscape, temporality, and lay knowledges. Prog Hum Geogr 35(3):284–302

Brierley G, Fryirs K, Cullum C, Tadaki M, Huang HQ, Blue B (2013) Reading the landscape integrating the theory and practice of geomorphology to develop place-based understandings of river systems. Prog Phys Geogr 37(5):601–621

Brown-Weiss E (1989) In fairness to future generations: international law, common patrimony, and intergenerational equity. United Nations University, Tokyo

Common Implementation Strategy No 2 (2003) Common implementation strategy for the water framework directive (2000/60/EC) Guidance Document No 2 identification of water bodies

Common Implementation Strategy No 3 (2003) Common implementation strategy for the water framework directive (2000/60/EC) Guidance Document No 3 analysis of pressures and impacts

Common Implementation Strategy No 8 (2003) Common implementation strategy for the water framework directive (2000/60/EC) Guidance Document No 8 Public participation in relation to the water framework directive

Council of Europe (2000) European landscape convention, florence, explanatory report, CETS No. 176. Council of Europe, Strasbourg

Council of Europe (2008) Recommendation CM/Rec(2008)3 of the Committee of Ministers to member states on the guidelines for the implementation of the European Landscape Convention. Council of Europe, Strasbourg

D'Amato A, Brown-Weiss E, Gundling L (1990) Agora: what obligation does our generation owe to the next? An approach to global environmental responsibility. Am J Int Law 84:190–212

Davy-Bowker J, Clarke RT, Johnson RK, Kokes J, Murphy JF, Zahrádková S (2006) A comparison of the European water framework directive physical typology and RIVPACS-type models as alternative methods of establishing reference conditions for benthic macroinvertebrates. In: Furse MT, Hering D, Brabec K, Buffagni A, Sandin L, Verdonschot PFM (eds) The ecological status of European rivers: evaluation and intercalibration of assessment methods. Springer, Dordrecht, pp 91–105

Delaney D (2003) Beyond the word: law as a thing of this world. In: Holder J, Harrison C (eds) Law and geography. Oxford University Press, Oxford, pp 67–84

Delaney D (2011) The spatial, the legal and the pragmatics of world-making: nomospheric investigations (Reprint edition). Routledge-Cavendish, New York; Abingdon, Oxfordshire

Elmqvist T, Folke C, Nyström M, Peterson G, Bengtsson J, Walker B, Norberg J (2003) Response diversity, ecosystem change, and resilience. Front Ecol Environ 1(9):488–494

European Commission (2009) Report from the Commission to the European Parliament and the Council in accordance with article 18.3 of the Water Framework Directive 2000/60/EU on programmes for monitoring of water status (Communication) COM (09) 156 final

European Commission (2013) Commission Decision of 20 September 2013 establishing, pursuant to Directive 2000/60/EC of the European Parliament and of the Council, the values of the Member State monitoring system classifications as a result of the intercalibration exercise and repealing Decision 2008/915/EC

European Commission (n.d.) Commission Staff Working Document (European Overview (2/2) Accompanying the documents Report form the Commission to the European Parliament and the Council in the implementation of the Water Framework Directive (2000/60EC) River Basin Management Plans (Com(2012) 670 final))

Foucault M (1971) The order of things: an archaeology of the human sciences. Pantheon Books, New York

Friberg N, Bonada N, Bradley D, Dunbar M, Edwards F, Grey J, Hayes R, Hildrew A, Lamaroux N, Trimmer M, Woodward G (2011) Biomonitoring of human impacts in freshwater ecosystems: the good, the bad and the ugly. Adv Ecol Res 44:1–68

Gobster PH, Nassauer JI, Daniel TC, Fry G (2007) The shared landscape: what does aesthetics have to do with ecology? Landsc Ecol 22(7):959–972

Graham N (2010) Lawscape: property, environment, law. Routledge-Cavendish, Abingdon

Hedelin B (2008) Criteria for the assessment of processes for sustainable river basin management and their congruence with the EU Water framework directive. Eur Environ 18(4):228–242

Hedelin B, Lindh M (2008) Implementing the EU water framework directive—prospects for sustainable water planning in Sweden. Eur Environ 18(6):327–344

Heiden UAD, Roth G, Schwegler H (1985) Principles of self-generation and self-maintenance. Acta Biotheor 34(2–4):125–137

Hering D, Borja A, Carstensen J, Carvalho L, Elliott M, Feld CK, Heiskanen A-S, Johnson R, Moe J, Pont D, Solheim A, Van De Bund W (2010) The European water framework directive at the age of 10: a critical review of the achievements with recommendations for the future. Sci Total Environ 408(19):4007–4019

Hirokawa KH (2010) Three stories about nature: property, the environment, and ecosystem services. Mercer Law Rev 62:541

Howarth W (2009) Aspirations and realities under the water framework directive: proceduralisation, participation and practicalities. J Environ Law 21:391

Hughes FMR, Stroh PA, Adams WM, Kirby KJ, Mountford JO, Warrington S (2011) Monitoring and evaluating large-scale, 'open-ended' habitat creation projects: a journey rather than a destination. J Nat Conservation 19(4):245–253

Hughes FMR, Adams WM, Stroh PA (2012) When is open-endedness desirable in restoration projects? Restor Ecol 20(3):291–295
Jax K (2006) Ecological units: definitions and application. Q Rev Biol 81(3):237–258
Jax K (2010) Ecosystem functioning. Cambridge University Press, Cambridge
Johnson RK, Furse MT, Hering D, Sandin L (2007) Ecological relationships between stream communities and spatial scale: implications for designing catchment-level monitoring programmes. Freshw Biol 52(5):939–958
Jones M (2007) The European landscape convention and the question of public participation. Landsc Res 32(5):613–633
Jones M, Howard P, Olwig KR, Primdahl J, Sarlöv-Herlin I (2007) Multiple interfaces of the European landscape convention. Norsk Geografisk Tidsskrift–Norwegian J Geography 61 (4):207–216
Jørgensen SE, Fath B, Bastianoni S, Marques JC, Muller F, Nielsen SN, Patten B, Tiezzi E, Ulanowicz RE (2007) A new ecology: systems perspective (1a uppl). Elsevier, Amsterdam
Josefsson H (2012) Achieving ecological objectives. Laws 1(1):39–63
Josefsson H (2014) Assessing aquatic spaces of regulation: key issues and solutions. Nordic Environ Law J 3:23–44
Josefsson H (2015a) Ecological status as a legal construct—determining its legal and ecological meaning. J Environ Law (in press)
Josefsson H (2015b) Good ecological status, advancing the ecology of law. Uppsala University, Uppsala
Josefsson H, Baaner L (2011) The water framework directive—a directive for the twenty-first century? J Environ Law 23(3):463–486
Kaika M (2003) The water framework directive: a new directive for a changing social, political and economic European framework. Eur Plan Stud 11(3):299–316
Karr J, Dudley D (1981) Ecological perspective on water quality goals. Environ Manage 5(1):55–68
Keessen A, Van Kempen J, Van Rijswick H (2008) Transboundary river basin management in Europe: legal instruments to comply with European water management obligations in case of transboundary water pollution and floods. Utrecht Law Rev 4:35
Koestler A (1989) The ghost in the machine. Arkana, London
Lefebvre H (1991) The production of space. Basil Blackwell, Oxford
Lowenthal D (2007) Living with and looking at landscape. Landsc Res 32(5):635–656
Ludwig D, Mangel M, Haddad B (2001) Ecology, conservation, and public policy. Annu Rev Ecol Syst 32(1):481–517
McCluney KE, Poff NL, Palmer MA, Thorp JH, Poole GC, Williams BS, Williams M, Baron JS (2014) Riverine macrosystems ecology: sensitivity, resistance, and resilience of whole river basins with human alterations. Front Ecol Environ 12(1):48–58
Mitchell D (2008) New axioms for reading the landscape: paying attention to political economy and social justice. In: Jr JLW, Johnston DM (eds) Political economies of landscape change. Springer, Dordrecht, pp 29–50
Moss B (2008) The water framework directive: total environment or political compromise? Sci Total Environ 400(1–3):32–41
Moss B, Stephen D, Alvarez C, Becares E, Van Bund W, Collings S, Van Donk E, Eyto E, Feldmann T, Fernandez-Alaez C, Fernandez-Alaez M, Franken R, Garcia-Criado F, Gross E, Gyllstrom M, Hansson L-A, Irvine K, Jarvalt A, Jensen J-P, Jeppesen E, Kairesalo T, Kornijow R, Krause T, Kunnap H, Laas A, Lill E, Lorens B, Luup H, Miracle M, Noges P, Noges T, Nykanen M, Ott I, Peczula W, Peeters E, Phillips G, Romo S, Russell B, Salujoe J, Scheffer M, Sievertsen K, Smal H, Tesch C, Timm H, Tuvikene L, Tonno I, Virro T, Vicente E, Wilson D (2003) The determination of ecological status in shallow lakes—a tested system (ECOFRAME) for implementation of the European water framework directive. Aquatic Conservation: Marine Freshw Ecosyst 13(6):507–549
Moss B, Hering D, Green AJ, Aidoud A, Becares E, Beklioglu M, Weyhenmeyer G (2009) Climate change and the future of freshwater biodiversity in Europe: a primer for policy-makers. Freshw Rev 2(2):103–130

Mossio M, Bich L, Moreno A (2013) Emergence, closure and inter-level causation in biological systems. Erkenntnis 78(2):153–178

Nunes-Neto N, Moreno A, El-Hani CN (2014) Function in ecology: an organizational approach. Biol Philos 29(1):123–141

Olden JD, Naiman RJ (2010) Incorporating thermal regimes into environmental flows assessments: modifying dam operations to restore freshwater ecosystem integrity. Freshw Biol 55(1):86–107

Olwig KR (2007) The practice of landscape 'conventions' and the just landscape: the case of the European landscape convention. Landsc Res 32(5):579–594

Ostrom E (2005) Understanding institutional diversity. Princeton University Press, Princeton

Pahl-Wostl C, Arthington A, Bogardi J, Bunn SE, Hoff H, Lebel L, Nikitina E, Palmer M, Poff L, Richards K, Schluter M, Schulze R, St-Hilaire A, Tharme R, Tockner K, Tsegai D (2013) Environmental flows and water governance: managing sustainable water uses. Curr Opin Environ Sustainability 5(3–4):341–351

Pedroli B, Pinto-Correia T, Cornish P (2006) Landscape—What's in it? Trends in European landscape science and priority themes for concerted research. Landsc Ecol 21(3):421–430

Philippopoulos-Mihalopoulos A (2009) Niklas Luhmann: law, justice, society. Routledge, London

Poff NL, Richter BD, Arthington AH, Bunn SE, Naiman RJ, Kendy E, Acreman M, Apse C, Bledsoe B, Freeman M, Henriksen J, Jacobson R, Kennen J, Merritt D, O'keeffe J, Olden J, Rogers K, Warner A (2010) The ecological limits of hydrologic alteration (ELOHA): a new framework for developing regional environmental flow standards. Freshw Biol 55(1):147–170

Poole GC (2002) Fluvial landscape ecology: addressing uniqueness within the river discontinuum. Freshw Biol 47(4):641–660

Prieur M (2006) Landscape and social, economic, cultural and ecological approaches. I landscape and sustainable development : challenges of the European landscape convention. Council of Europe, Strasbourg

Pringle CM (2001) Hydrological connectivity and management of biological reserves: a global perspective. Ecol Appl 11(4):981–998

Ruhl J (2007) Law's complexity: a primer. Georgia State Univ Law Rev 24:885

Ruhl J (2012) Panarchy and the law. Ecol Soc 17(3)

Saborido C, Mossio M, Moreno A (2011) Biological organization and cross-generation functions. Br J Philos Sci 62(3):583–606

Steyaert P, Ollivier G (2007) The European water framework directive: how ecological assumptions frame technical and social change. Ecol Soc 12(1), Art. 25

Stoddard JL, Larsen DP, Hawkins CP, Johnson RK, Norris RH (2006) Setting expectations for the ecological conditions of streams: the concept of reference conditions. Ecol Appl 16(4):1267–1276

Thorp JH (2014) Metamorphosis in river ecology: from reaches to macrosystems. Freshw Biol 59 (1):200–210

Valinia S, Hansen H-P, Futter MN, Bishop K, Sriskandarajah N, Fölster J (2012) Problems with the reconciliation of good ecological status and public participation in the water framework directive. Sci Total Environ 433:482–490

Verdonschot PFM (2006) Evaluation of the use of water framework directive typology descriptors, reference sites and spatial scale in macroinvertebrate stream typology. In: Furse MT, Hering D, Brabec K, Buffagni A, Sandin L, Verdonschot PFM (eds) The ecological status of European rivers: evaluation and intercalibration of assessment methods. Springer, Dordrecht, pp 39–58

Ward JV (1998) Riverine landscapes: biodiversity patterns, disturbance regimes, and aquatic conservation. Biol Conserv 83(3):269–278

Ward JV, Stanford J (1982) Thermal responses in the evolutionary ecology of aquatic insects. Annu Rev Entomol 27(1):97–117

Ward JV, Malard F, Tockner K (2002) Landscape ecology: a framework for integrating pattern and process in river corridors. Landsc Ecol 17(1):35–45

Whittier TR, Stoddard JL, Larsen DP, Herlihy AT (2007) Selecting reference sites for stream biological assessments: best professional judgment or objective criteria. J N Am Benthological Soc 26(2):349–360

Woodward G (2009) Biodiversity, ecosystem functioning and food webs in fresh waters: assembling the jigsaw puzzle. Freshw Biol 54(10):2171–2187

Wright SAL, Fritsch O (2011) Operationalising active involvement in the EU water framework directive: why, when and how? Ecol Econ 70(12):2268–2274

Wu J (2006) Landscape ecology, cross-disciplinarity, and sustainability science. Landsc Ecol 21(1):1–4

Wu J (2013) Landscape ecology. In: Leemans R (ed) Ecological systems. Springer, New York, pp 179–200

Young OR (2002) The institutional dimensions of environmental change: fit, interplay, and scale. MIT Press, Cambridge

Coherence Issues Between Climate, Energy and Water in the Case of European Hydropower: Can We Have It All?

Jonida Abazaj

Abstract The questions of renewable and sustainable energy have attracted increasing attention in the last decades, becoming salient topics on the national and European political agendas. In parallel, hydropower has made an incredible return to the global agenda, after having been absent due to heavy criticism because of its social and environmental impacts. The common denominator for this surge is represented by the need to mitigate climate change and support low-carbon development paths. Drawing on secondary data analysis the chapter investigates the conflict occurring between climate change mitigation through hydropower and the protection of good water quality in the implementation of two EU directives: the Renewable Energy Directive (RES) and the Water Framework Directive (WFD). After a presentation of the evolving policy and regulatory context around hydropower, the paper explores the link between energy-climate-water-sustainability and addresses the trade-offs, synergies and opportunities for policy integration and coherence. The analysis reveals that while synergies between climate and energy are straightforward and clearly stated at the EU level, and that it is possible to track this co-ordination between water and biodiversity, challenges persist in relation to the link between the Climate and renewable energy package and the Water Framework Directive.

Keywords Sustainability · Policy coherence · Water framework directive · Renewable energy · Hydropower

J. Abazaj (✉)
Department of Industrial Economics and Technology Management,
Norwegian University of Science and Technology (NTNU),
Alfred Getz veg 3 Sentralbygg II, Trondheim 7491, Norway
e-mail: jonida.abazaj@iot.ntnu.no

V. Mauerhofer (ed.), *Legal Aspects of Sustainable Development*,
DOI 10.1007/978-3-319-26021-1_18

1 Introduction

The issues of renewable and sustainable energy have attracted increasing attention in the last decades, becoming salient topics on the national and international political agendas (Lenschow 2002a; Lafferty and Hovden 2003; Pittock 2011). Their origins lie in two distinct issues (Frey and Linke 2002). On the one hand, renewability can be seen as a response to concerns about the security of energy supply and the possible depletion of some primary energy sources such as fossil fuels. Because of the lower greenhouse gas (GHG) emissions the mix of renewable sources composed by hydropower, wind, solar, geothermal, biomass and other sources has become—despite the differences in GHG levels—one of the key instruments for mitigating climate change and a primary goal of the EU renewable energy policy (Bilgen et al. 2004; Moomaw et al. 2011).

On the other, sustainability is an answer to the environmental degradation of the planet and the apprehension of leaving a legacy of a reduced quality of life to future generations. Although the concept of living in harmony with nature is as old as mankind, the concept of "Sustainable Development" rose only in the 1960s becoming the key theme of the United Nations (UN) Conference on the Human Environment in Stockholm in 1972 and was consolidated in the 1987 UN report *Our Common Future* (WCED 1987). Since then, sustainability has become a key term in environmental policy and a major overall goal of national governments and the European policy agenda, especially in relation to alternative energy solutions and natural resource management (EU 2014; WWAP 2015).

During the last decade, hydropower has made an incredible return to the global agenda, after having been absent due to heavy criticism because of its social and environmental impacts (Truffer et al. 2001; Fletcher 2010; van Ginneken 2015). The proponents of hydropower development claim that hydropower, as a technically and economically mature renewable source of electricity can contribute and support low-carbon economic and social development paths (Truffer et al. 2001; IHA 2003; UNESCO 2003; IPCC 2014). However, recent studies highlight the issue of emissions from hydropower plants both in the tropics and in Europe, reopening thus the discussion on the definition of hydropower as a "green and clean energy" (Fearnside and Pueyo 2012; IPCC 2013; Maeck et al. 2013). In addition, hydropower represents an important supporting instrument for intermittent renewable sources by providing rapid response back-up power, storage capacity and stabilizing services for the power system, ensuring in this way security of electricity supply (Oud 2002; IRENA 2012; Locker 2004; Kaunda et al. 2012). It is thus the potential contribution in fighting climate change, energy security concerns and economic aspects that has brought hydropower back to the centre of the scene.

However, the social and environmental impacts that made hydropower so unpopular in the past have not disappeared, undermining the sustainability claim of the sector and creating trade-offs between goals such as: (a) satisfaction of electricity needs versus conservation of natural resources; (b) needs of current generation versus opportunities for next generations; (c) global climate change mitigation versus local environmental impacts; and increase of renewable technologies with less water intensive paths such as solar and wind versus those with higher water footprint such as hydropower and biomass (Mekonnen et al. 2015).

In fact, the environmental, social, and economic benefits of this renewable source come with a price as hydropower infrastructures pose an immediate risk to biodiversity and to water resources (Rodriguez et al. 2006; Bunn and Arthington 2002; Rosenberg et al. 1995; McCully 1996; Egré and Milewski 2002; Sternberg 2008). Trade-offs therefore manifest not only between energy and environmental goals, but also between climate goals (greenhouse gas reduction) and water and biodiversity protection. While discourses related to win-win sustainable solutions have become common in issues related to conservation and development outcomes in order to describe the simultaneous achievement of goals are to the satisfaction of all parties (Groom et al. 2010; Henriksen et al. 2011), several studies question the basic assumptions behind win-wins especially regarding the situations that involve competing rather than complementary social, economic and ecological goals (Rodriguez et al. 2006; Jackson 2011; McShane et al. 2011; Howe et al. 2014). In order to achieve win-win situations, conflicts and trade-offs need to be clearly spelled out and managed. In the event of hard cases it is crucial to understand how the reconciliation of goals is carried on and if by reconciliation is implied that goals are equally important or that there should be a prioritization among them. And in this last case, should the legal provisions aimed at water and biodiversity protection be seen as untouchable and sacred, even if their application challenges and hinders policies aimed at addressing climate change? Or should goals and legal provisions aimed at pursuing ecological protection of water bodies and biodiversity be set aside or "relaxed" in order to accommodate "overriding" climate change goals?

This paper addresses the above issues focusing on the European hydropower sector in the light of the EU Climate and energy strategy, as well as water and biodiversity policies. Driven by the three imperatives of security of supply, sustainability, and economic efficiency, the water and energy sectors have undergone rapid and complex reforms in the last decades substantiated by the elaboration of the EU Climate strategy, the Renewable Energy Directive (RES), and the EU Water Framework Directive (WFD). Because of its characteristics hydropower has one foot planted in the area of water policy and environmental protection, and the other in the area of energy policy. As such it is subject to policy, politics and regulations in both these fields.

Drawing on secondary data analysis (policy reports, European Commission documents, legal documents) the paper investigates the trade-offs and potential synergies between the goals of water, energy, biodiversity and climate policies. After a presentation of the analytical framework (Sect. 2), the paper first analyses the evolving policy and regulatory context around hydropower by looking at the above-mentioned legislations (Sect. 3). Then, based on concepts of policy integration and policy coherence, Sect. 4 highlights the key trade-offs materializing between the goals of the RES Directive and WFD and addresses the way sectoral interests are taken into consideration by the WFD. Finally Sect. 5 discusses the degree of policy coherence and opportunities for synergies.

2 Analytical Framework: Sustainable Development, Policy Integration and Policy Coherence

In the last decades, numerous scholars have emphasized the challenges of coordination, coherence and integration between different policy areas in order to overcome conflicts and redundancy between policy goals and programs, at every level of policy-making.

As already recognized by the Brundtland Report, the cross-cutting challenges of sustainable development are handled by institutions that tend to be independent, fragmented and working to relatively narrow mandates with closed decision-making processes (WCED 1987). Lack of communication and coordination between different departments or ministries has been identified as the cause of contradictions and/or redundancy in their programmes (Peters 1998), calling for governments to "[…] strengthen national institutional capability and capacity to integrate social, economic, developmental and environmental issues" (UNCED 1992 paragraph 8.12).

Since then, the objective of policy integration has been expressed in several contexts such as: "coherence" in the OECD context for Sustainable Development strategies (OECD 1995, p. 73; 1996) and in the European Commission's work on good governance (2001); horizontal (policy) integration in the governance discourse (OECD 2001; OECD 2002; EC 2004); and environmental policy integration (EPI) among environmental policy scholars (see e.g. Lenschow 2002a; Nilsson and Persson 2003; Lafferty 2004; Meijers and Stead 2004; Persson 2004; Jordan and Lenschow 2008; Knudsen 2009). Emerged and further developed as a concept because of the perceived failure of sectoral environmental policy to address the environmental pressures posed by the society, the issue of EPI in the EU has been addressed intensively, especially since the Single European Act (1987) introduced the objective of ensuring that "environmental protection requirements shall be a component of the Community's other policies" (Single European Act 1987, Article 130r.2). In addition, the Amsterdam Treaty (European Communities 1997) outlined sustainable development as an overall objective of the EU and has committed the EU by stating that "environmental protection requirements must be integrated into the definition and implementation of the Union policies and activities" (Article 6 of the Amsterdam Treaty, now Article 11 of the Treaty on the Functioning of the European Union, TFEU). EPI has been defined and framed differently in time by different authors.

Collier (1994) for example relates the motivation for EPI to the sustainable development agenda and defines it as "an approach which requires the inclusion of the environment amongst the set of values being considered" (p. 35) aiming at

- "achieving sustainable development and prevent environmental damage;
- removing contradictions between policies as well as within policies; and
- realising mutual benefits and the goal of making policies mutually supportive" (p. 36).

Liberatore (1997) contributes to the EPI debate not only by discussing the integration of sustainable development objectives in the EU policy—making but also by calling the attention on the weighting and "reciprocity" issue of the sector and environmental concerns (p. 119). In this case, the different concerns are required to have similar weight in order to avoid a policy that is diluted.

Other authors that have intensively contributed to the EPI are Lenschow and Lafferty. Lenschow (2002b, p. 6) conceives EPI as "a first-order operational principle to implement and institutionalise the idea of sustainable development". In addition Lenschow suggests that, as a principle, EPI should encompass both policy processes and outputs highlighting the relationship between integration process and substantive output: "The integration process currently faces the challenge of ensuring that substance follows from procedure" (2002b, p. 7).

For Lafferty (2003), EPI is not just about removing contradictions and realizing mutual benefits. He argues that "the whole point of EPI is, at the very least, to avoid situations where environmental objectives become subsidiary, and, in the broader purview of sustainable development, to ensure that they become principal or overarching societal objectives" (p. 13). Hence EPI implies:

- "the incorporation of environmental objectives into all stages of policy making in non-environmental policy sectors, with a specific recognition of this goal as a guiding principle for the planning and execution of policy;
- an attempt to aggregate presumed environmental consequences into an overall evaluation of policy, and a commitment to minimise contradictions between environmental and sectoral policies by giving priority to the former over the latter" (p. 13).

The integration imperative highlights the need to deal with trade-offs and conflicts that emerge between the different goals (Collier 1994; Biermann et al. 2009). These can be "external goal conflicts" in case when environmental and other sectorial goals are involved, as well as "internal goal conflicts" when different environmental goals cannot be reached simultaneously. As Biermann et al. (2009, p. 355) highlight "boundaries are blurred, and some policies may be seen as having environmental as well as non-environmental elements."

Together with EPI, policy coherence is another important objective in governance and policy making at the EU level, the need for which is acknowledged in an increasing number of official EU documents (e.g., CEC 2011a,b; EEA 2010, EU 2014). Although in the field of environment most of the work related to coherence has been addressed under the concept of EPI, there is confusion between the two concepts and a distinction can also be made between policy integration and policy coherence (see e.g. Kivimaa and Mickwitz 2009; Nilsson et al. 2012). Policy coherence provides criteria for evaluating policy outputs, rather than processes, and is used to describe a condition of synergy between different policy areas, where incentives and signals to target groups do not conflict (see e.g. Van Bommel and Kuindersma 2008, p. 15; Nilsson et al. 2012, p. 2).

The goal of policy coherence is hence to reduce conflicts and enhance synergies between and within different policy areas, enabling processes through which both aims can be redefined in order for win-win situations to occur (Winship 2006; den Hertog and Stross 2011; Nilsson et al. 2012). In fact it concerns how well different laws and policies work together, and where ideally, the objectives of different laws and policies complement each other. For this reason, coordination is one of the means to achieve policy coherence. However, it is important to keep in mind that greater consultation and coordination in itself, does not address the problem of deciding which policy objectives should be given priority when interests conflict (Ellison 2010; Pataki et al. 2011; more broadly, see Jordan and Schout 2006). Following up the problematique of coordination, Mauerhofer (2013, p. 329) underlines the importance of substance in coordination as opposed to a formal coordination between different ministries in order to be able to assess "the hierarchy and chosen priorities, within a sustainable development, in the relationship between environmental, social and economic sustainability".

3 The Evolving Policy and Regulatory Context Around the Hydropower Sector

3.1 Climate Action and the Renewable Energy Directive

Under the Climate Action and Renewable Energy (CARE) Package, EU leaders have committed to transform Europe into a highly energy-efficient and low carbon economy. The package represents a set of binding legislations which aims to ensure that the EU meets its ambitious climate target of 20 % reduction in GHG emissions below the 1990 levels, and a 20 % share of renewable energy in the EU energy consumption by 2020 (EC 2009a). In addition, driven by the three imperatives of security of supply, sustainability, and economic efficiency, EU leaders have committed to even higher targets to cut greenhouse gas emissions by at least 40 % by 2030 compared to 1990 levels, an EU-wide binding target for renewable energy of at least 27 % and an indicative energy efficiency target of at least 27 % (EC 2013a).

Despite the great ambitions for the decades to come, the trends presented in the last "State of the Environment Report" (EEA 2015) highlight not only that the EU has become since 2000 more resource efficient but also that it stays very resource intensive. This would seem to support the "Jevons Paradox" according to which economically justified energy-efficiency improvements increase rather than reduce energy consumption, having thus implications for energy and climate policy (Saunders 1992; Sorrel 2009).

As part of this package the EU RES Directive 2009/28/EC (EC 2009b) provides a framework for increasing the share of energy from renewable sources, diversifying and improving the security of the energy supply, stimulating the energy sector economically, as well as aiming at reducing greenhouse gas emissions in order to combat climate change. In the light of this directive, hydropower is defined (Art. 2[a]) as a renewable source utilizable for the achievement of renewable energy targets.

According to the data provided by Member States in the mandatory National Renewable Energy Action Plans (NREAPs) delivered in 2010, hydropower generation represents the largest share of renewable electricity and it is expected to continue playing an important role, being second only to wind in 2020. Based on these projections, hydro capacity is expected to be installed at around 1.5 GW per year between 2010 and 2015 and at around 1.9 GW per year between 2015 and 2020, leading to a total installed capacity of over 135 GW (Eurelectric 2011a). Although the importance of hydropower in supporting intermittent renewable sources and the potential still available in Europe, the Union of the Electricity Industry (Eurelectric) finds the level of ambition for hydropower in the NREAPs disappointing and attributes the constraint in growth to environmental restrictions and regulation, in particular the WFD (Eurelectric 2011b).

3.2 *The Water Framework Directive*

At the EU level, the Water Framework Directive (European Union 2000) represents the most important piece of water legislation. Adopted in the year 2000, this unique regulatory framework sets as objectives the general protection of the aquatic ecology, specific protection of unique and valuable habitats, protection of drinking water and bathing water.

A key element of the WFD is that water resource management is taking place at the river basin level and that all water bodies must reach “good ecological status” and “good chemical status” by 2015 (Art. 2.9; Art. 4.1). The WFD requires a River Basin Management Plan (RBMP) to be made for each river basin every six years as a tool for: (a) recording the current status of water bodies within the river basin district, (b) setting out the environmental objectives for the different water bodies, justification for setting lower objectives, and a summary of the programme of measures to reach the objectives of the WFD (Art. 11), and (c) act as the main reporting mechanism to the EU and the public. In this process, Member States have to “encourage the active involvement of all interested parties” (Art. 14).

In addition, the WFD calls explicitly for a continued dialogue and for the development of strategies towards a “(…) further integration of protection and sustainable management of water into other Community policy areas such as energy, transport, agriculture, fisheries, regional policy and tourism (...).” (Preamble 16). Therefore, it implicitly requires coordination and cooperation between Member States in the case of shared river basins, as well as among the different stakeholders, sectors and government levels of Member States in order to implement the measures and achieve the goals.

Although not directly regulated by the WFD, the hydropower sector is affected for the reason that water storage for power generation has been identified, at the EU level, as one of the main drivers to hydro-morphological pressures. At present 57 % of EU waters do not have good status and hydropower is identified as the third most common water use for designating “heavily modified water bodies” (EC 2012).

The WFD uses this category for water bodies that have been subject to physical alteration as well as substantial changes in character as a result of human activity that cannot be removed because of the high economic and social cost (Art. 2.9). For heavily modified water bodies derogations are made allowing a less stringent objective to be met—Good Ecological Potential—and offering the possibility to postpone the deadlines up to 2027 (Art. 2.16; Art. 4.3).

In addition, the WFD provides for exemptions to the principle of non-deterioration exceptionally allowing for new modifications to the physical characteristic of water bodies (new infrastructure projects, including hydropower) or failure to achieve good water status provided certain strict conditions are met (Art. 4.7).

3.3 Directives Relating to Biodiversity and Natura 2000 Network

In parallel with emissions, renewable energy and water quality targets, there are also biodiversity targets. Both at the global and EU level these targets have been ambitiously set (UNCBD 2002), missed and revised. The EU goal is currently "to halt the loss of biodiversity and the degradation of ecosystem services in the EU by 2020, restore them in so far as feasible, while stepping up the EU contribution to averting global biodiversity loss" (Environment Council of the EU 2010, p. 2; EC 2011a).

The Birds Directive (European Council 2009b) and the Habitats Directive (European Council 1992) are the cornerstones of the EU's biodiversity policy and nature conservation policy framework. The protection provided by the Directives is divided between species protection measures—EU's Member States shall protect certain species wherever they are found in the wild—and site protection measures that require Member States to identify, designate and protect conservation areas for certain habitat types and for the habitats of certain species. The network of nature protection areas selected under the Habitat Directive and the Bird Directive is known as Natura 2000, and represents the "centrepiece" of EU biodiversity policy (EC 2009c). Despite the fact that the network remains incomplete, it is defined as the largest network of protected areas in the world, containing currently nearly 26,000 protected sites, corresponding to about 20 % of the EU's territory (EC 2011b).

Article 6 of the Habitats' Directive plays a crucial role in the management of Natura 2000 sites making provision for the establishment of the necessary conservation and restoration measures of the habitats and species for which the site has been designated (Art. 6.1), as well as the avoidance of damaging activities that could significantly disturb species or deteriorate their habitats (Art. 6.2). In addition, similarly to the provisions of WFD Article 4(7), Articles 6(3) and 6(4) of the EU Habitat Directive depict the procedure to be followed for plans and projects that might affect a Natura 2000 site and the exceptions to the non-deterioration principle (Art. 6.3 and 6.4).

3.4 Directives Relating to Environmental Assessment

In order to ensure that environmental implications of decisions are taken into account before the decisions are made, an important instrument is offered through the procedure of the environmental assessment and the two EU directives on Environmental Assessment. The Strategic Environmental Assessment (SEA) Directive of the European Council (2001/42/EC) represents a tool to determine the positive or negative impact that a proposed plan or programme may have on the environment. On the other hand, the Environmental Impact Assessment (EIA) Directive (2011/92/EU) is an instrument used for improving the basis on which decisions are taken for single projects. It consists of gathering, analysing and presenting information on the likely environmental impacts of a project.

Both Directives have the common goal to provide a high level of protection of the environment by ensuring that plans and projects that are likely to have significant effects on the environment are made subject to an environmental assessment, prior to their approval or authorization. In this way they aim to contribute to the integration of environmental considerations into the preparation of programmes, plans and projects.

If on the one hand both SEA and EIA lay down procedural requirements and their results have to be taken into consideration, on the other the Habitats Directive and the Birds Directive establish more substantial provisions by introducing obligatory environmental standards that determine the provisions (EC 2014).

4 Trade-Offs, Synergies and Opportunities for Enhanced Coherence Among EU Policy Goals

In relation to hydropower, it is possible to identify several potential trade-offs and synergies among the WFD, the Climate Change Strategy and RES Directive, the Birds and Habitats directives, as well as the two Directives on Environmental Assessment including several options for better policy integration.

4.1 Trade-Offs Among Goals

Although through the wording and statements of the CARE and the WFD it is not possible to directly infer any conflict, trade-offs among their goals do materialize when the operation of existing hydropower plants or the establishment of new ones in order to meet the renewable energy goals and mitigate climate change, make it necessary to accept less stringent environmental objectives and might put water resources at risk (Opperman et al. 2011; Pittock et al. 2011). Hydropower was in fact identified in the first implementation report of the WFD as one of the main drivers of hydro-morphological alterations, loss of connectivity and significant adverse effects on the survival of fish populations (EC 2007, 2012).

On the other hand, the goals of the WFD, by putting a ban on deterioration or by requiring measures aimed at improving the ecological status of water bodies (where goals have not been achieved) might decrease or limit hydropower generation, impact negatively on the revenues of the operators, and limit further exploitation of the available potential (Ecologic 2009). Indeed, the EU environmental legislation represents according to Eurelectric one of the hindering factors to new hydropower projects or major costs in the operation of the existing ones (Eurelectric 2011b).

As we can see (Fig. 1), the conflict at stake does not only involve action in order to reconcile two sectoral policies (*energy generation vs. water and biodiversity protection*) but also different environmental concerns (*climate change mitigation vs. water and biodiversity protection*).

In addition, it is important to underline the temporal, geographical and stakeholder perspectives of the trade-offs (*long term vs. short term*; *global vs. local*). In fact, while the benefits of mitigating global climate change through hydropower generation are likely to be felt in a very long term—if at all in some regions—the environmental impacts on water quality and biodiversity are immediate and localized. On the other hand, the relevance of environmental benefits deriving by the EU water and biodiversity policies is extended to the society and might become evident sometimes only in the longer run, while the costs of the required ecological improvements are immediate and weigh on the hydropower operators (social vs. economic).

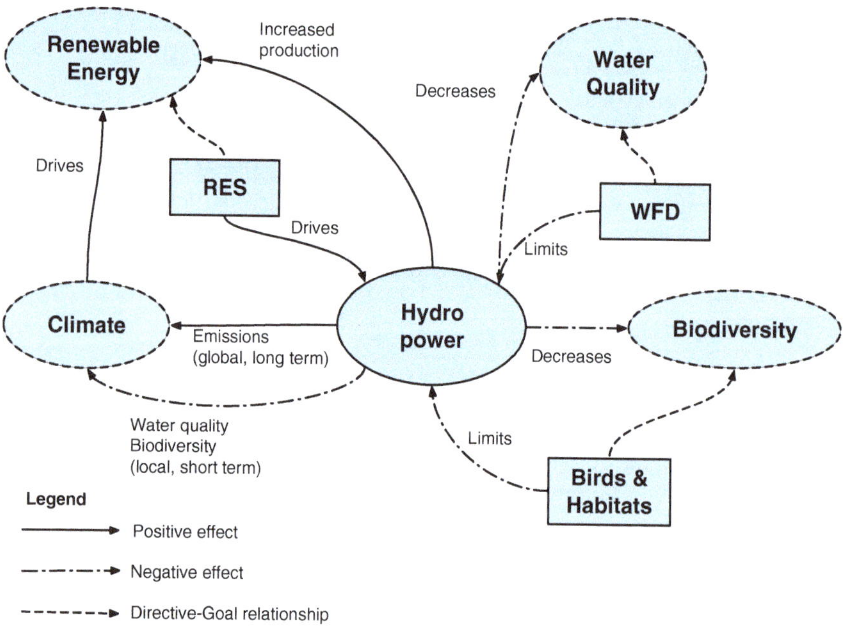

Fig. 1 Main relationships between the goals and directives related to hydropower

4.2 Consideration of Sectorial Interests in the Directives

The principle of sustainable development requires that resources should be managed in a holistic way, coordinating and integrating environmental, economic and social aspects. The WFD itself highlights the need for an integrated management and consequently provides through the use of exemptions a way to balance between water use and the protection of water bodies.

Thinking in terms of existing hydropower plants prior to the adoption of the WFD, the legislator has considered and accommodated the related social and economic interests by introducing the designation of Heavily Modified Water Bodies (Art. 4.3) and extension of deadlines up to 2027 (Art. 4.4). Member States may designate a body of surface water as heavily modified when:

(a) the changes to the hydromorphological characteristics which would be necessary for achieving good ecological status would have significant adverse effects on: (i) the wider environment; (ii) activities for the purposes of which water is stored, such as drinking-water supply, power generation or irrigation; (iii) water regulation, flood protection, land drainage, or (iv) other equally important sustainable human development activities;
(b) the beneficial objectives served by the modified characteristics of the water body cannot, for reasons of technical feasibility or disproportionate costs, reasonably be achieved by other means, which are a significantly better environmental option.

For this category of water bodies the directive allows to pursue the less stringent objective of Good Ecological Potential (Art. 4.5) when the Good Ecological Status is infeasible or disproportionately expensive, and when "The environmental and socioeconomic needs served by such human activity cannot be achieved by other means, which are a significantly better environmental option not entailing disproportionate costs" (Art 4.5[a]).

Thinking in terms of sustainable human development activities mentioned at the end of litera a above, the WFD does not give a definition of activities falling under sustainable development. As highlighted already in the CIS Guidance Document on Exemptions to the Environmental Objectives (EC 2009d) such activities require to be defined and framed in an iterative procedure in the course of the decision making process and ".... will thus depend on the time, scale, involved stakeholders and information available." (p. 24). Hence, although the WFD itself, the SEA and the EIA—guided by the polluter pays principle, the precautionary principle and preventive action and the principle of rectification of pollution at source—should provide the relevant process requirements in order to define if an activity falls under the sustainable development, the implementation of the directive depends on the interaction of the stakeholders.

The element of heavily modified water bodies serves therefore an economic function, highlighting the relevance of certain policy areas such as energy generation for member states, as well as the pressures exercised by the different stakeholders involved in the elaboration process of the WFD. It represents, in other words, the way sectorial interests and conflicts have been addressed in the new water policy (Acreman and Ferguson 2010).

On the other hand, when deterioration or failure of good water status is expected as a consequence of a planned hydropower project, Article 4.7 of the Directive enables the weighing process between the different interests (WD meeting, Segovia, 27–28 May 2010). Despite the obvious conflict with the non-deterioration principle, exemptions according to Article 4.7 WFD can occur when the benefits of the new infrastructure are of "overriding public interests", outweighing thus the benefits of achieving the WFD environmental objectives, and there are no significantly better environmental options which are technically and economically feasible. The concept of overriding public interest, also used in the Habitats Directive, has been related to situations where plans or projects prove to be indispensable for policies and actions aiming to protect fundamental value for citizen's lives such as health, safety and sustainable development. For the exemption to take place Article 4.7 requires also that the projects shall be reported in the River Basin Management Plans and that all practicable mitigation measures are taken to minimize negative effects on the aquatic ecology.

The Common Implementation Strategy (CIS) workshop "Water Management, Water Framework Directive & Hydropower" that was held in Brussels in 2011 found that a majority of Member States have yet to apply Article 4.7 for hydropower developments (Kampa et al. 2011). As confirmed in the European Commission Implementation Report of the WFD, out of the 116 plans assessed that refer to exemptions inter alia, only 12 plans refer to projects that fall under Article 4.7 (EC 2012, p. 13). Hence, although it is obvious that there are more planned and on-going projects that are likely to cause deterioration of status of water bodies, these are not mentioned in the RBMPs. This lack of references to new projects and programmes in most of the RBMPs indicates for the European Commission a "missed opportunity for the sustainable development of economic activities under a framework of real integrated water management." (EC 2012, p. 13).

Similarly to WFD Article 4.7, Article 6.3 and 6.4 of the Habitat Directive lay down the procedure to be followed when new developments like hydropower might affect a Natura 2000 site. Article 6.4 allows plans or projects that have been subject to a negative assessment to proceed only when: (1) an "absence of alternative solutions" is demonstrated; (2) "imperative reasons of overriding public interest" to proceed with the plan are demonstrated; (3) the Member State in question takes "all compensatory measures necessary to ensure that the overall coherence of Natura 2000 is protected".

5 Policy Coherence and Opportunities for Synergies

Although both the WFD and Birds and Habitats Directives allow for the use of exemptions in order to accommodate socio-economic needs and objectives, an effective way to reduce or eliminate conflicts would be to have coordination and integration of the different EU policies.

5.1 *Water and Biodiversity*

Looking at the WFD and Biodiversity strategy, their objectives not only are coherent, but they also complement each other towards the protection of natural resources, habitats and biodiversity. In fact, if plans and projects within the meaning or management measures of the Habitats Directive have an impact and are likely to cause a deterioration of the ecological status of water bodies, authorities also need to assess whether conditions of WFD Article 4.7 are fulfilled before they may authorize them. Equally, in case less stringent objectives are considered for *new* sustainable human development activity (e.g. a hydropower development) under the WFD, if the proposed development is likely to cause a failure to achieve the objectives for a Natura 2000 site, then the criteria for the use of exemptions of both the WFD and the Habitats Directive must be fulfilled.

In addition, Art. 4.9 WFD is clear in its obligation that when applying the exemptions of Art. 4, the same level of protection should be given as in existing Community legislation. This means that exemptions from the WFD environmental objectives cannot be used to deviate from objectives and obligations set by the Birds Directive and the Habitats Directive, and vice versa, otherwise infringements might occur. In such case it should be evaluated whether amendments can be made to the measure/project so that it satisfies the requirements of both directives.

5.2 *Biodiversity and Climate*

Biodiversity and climate change are certainly two areas of research that overlap. Large knowledge in climate science is based on studies of the changes and loss of biodiversity in various parts of the world, both in terms of species and habitats (Bertzky et al. 2010; Dupont 2010). In addition, scientific research documents that the direct impact of climate change on EU biodiversity exist and will become stronger in the future (Bertzky et al. 2010).

Given the links between climate change and biodiversity, discussions on the overlaps have come more and more to the fore in the policy documents, highlighting the importance of promoting synergies between the two policy areas. The statement in The Message from Athens (EC 2009e) makes it clear that "We cannot halt biodiversity loss without addressing climate change but it is equally impossible to tackle climate without addressing biodiversity". On the same line, a discussion paper prepared by the EU Ad Hoc Expert Working Group on Biodiversity and Climate Change (EU 2009) defines the conservation of biodiversity as a tool to both adapt to climate change, and mitigate the impacts of climate change as it protects the natural carbon capture and storage capacity of ecosystems.

Despite the need and call for major integration between climate and biodiversity policies goals the EU Ad Hoc Expert Working Group on Biodiversity and Climate Change—composed by Member State representatives, staff of the European Commission, scientists and civil society—stated that "true integration of climate and biodiversity policies remains the exception" (EU 2009, p. 5). In biodiversity action plans, the Working Group highlights that climate change aspects are often limited to adaptation measures for biodiversity without considering the role of biodiversity and ecosystem services in climate change mitigation and stabilization. Similarly, climate change programmes that include biodiversity sections are found to often tackle only climate change impacts on biodiversity, without considering the impact on biodiversity of other sectors such as energy in response to climate change, as well as the potential for other sectors to use ecosystem-based approaches to enhance the resilience of biodiversity (EU 2009).

In this matter, the European Commission Guidance on Integrating Climate Change and Biodiversity into Strategic Environmental Assessment sees the strategic environmental assessments carried out across the EU Member States as "an opportunity to systematically integrate climate change and biodiversity in a standardized approach into plans and programmes (PPs) across the EU" (EC 2013b, p. 12). In fact, by assessing the consistency and coherence between the proposed plans and programmes and the relevant policy objectives and targets for biodiversity protection and climate change, the SEA process should identify the policy objectives for biodiversity protection and climate change which may be relevant for the proposed PP and clearly describe whether it facilitates or contradicts their achievement.

Nevertheless, as the political attention to climate change and biodiversity goals integration represents a relatively recent phenomenon (Dypont 2010; Mant et al. 2014) time is needed in order to observe synergies between the two policies.

5.3 *Water, Energy and Climate*

Looking at the link between WFD and CARE, there would seem to be no immediate conflicts between them as they both aim at connected environmental objectives such as human environment, biodiversity conservation and greenhouse gas reduction. Nevertheless, as previously elaborated, the conflict materializes in the case of hydropower generation and plans for new hydropower development.

Given the importance of the WFD and the technical challenges that its implementation rises for the hydropower sector as well as other stakeholders, national governments and the Commission, the Common Implementation Strategy was proposed as a way to address the challenges in a co-operative and co-ordinated way. During the last 15 years many workshops and documents have been organized and elaborated in order to give guidance in the interpretation and implementation process of the WFD, as well as to manage the conflicts between the different stakeholders. Important reference points among others in this work are the CIS Guidance Documents N. 4 (Identification and Designation of Heavily Modified and Artificial Water Bodies) and N. 20 (Exemptions to the environmental objectives) and the two workshops on "WFD and Hydropower" where the need for policy coherence is continuously highlighted (Kampa et al. 2011). Although hydromorphological changes and hydropower have been the main themes of some of the activities under the CIS, more work remains to be done especially in relation to the integration of climate policy goals.

Climate change is indeed an additional pressure on many of the EU's already stressed water resources (EEA 2009; Wilby et al. 2006). The relationship between climate change and water policies, as well as the need for integration started coming to the attention of experts only in 2005 with the launch of the second European Climate Change Programme (Massey et al. 2010). The first effort to explore the relationship between the WFD and climate change adaptation came in 2007 with the constitution of a Strategic Steering Group (SSG) on Climate Change and Water under the umbrella of the Common Implementation Strategy (CIS) (EC 2008). The work of the SSG contributed to the elaboration of "Guidance document no. 24: "River basin management in a changing climate" (EC 2009f), where key guiding principles were provided on how to integrate climate change into the WFD planning and implementation processes. Climate change is not explicitly referred to in the text of the WFD and the European Commission sustains that the integration could occur through the same structures, objectives, timetables and river basin management cycles without revisions of the directive. This "defensive vision", as defined by Brower et al. (2013, p. 142), is not shared by some scholars that urge structural revisions of the WFD in order to consider climate change in the definition and evaluation of the status of water bodies (Wilby et al. 2006).

On the other hand, from the wording of the Guidance document no. 24 it is clear that the authors fear that climate considerations might be "used" to justify a relaxation of the WFD objectives (derogations under Art. 4): "There is a danger that anthropogenic climate change could be used as an excuse to set lower objectives for water bodies, even though formal attribution of a detected trend to anthropogenic climate change is unlikely at the scale of RBDs for several decades to come" (EC 2009f, p. 42). The frequency and way the derogations of Article 4 will be used has to be seen in the new RBMPs owed in 2015, depending as well on the views and reactions of the European Commission and European Court of Justice (Urwin and Jordan 2008; Brouwer et al. 2013). As today, the two policies continue to exist and operate in isolation (Henriksen et al. 2011; Brouwer et al. 2013).

There is however an important potential to increase the coherence between the CARE and the WFD. This involves combining the ecological restoration measures for improved status of the water (e.g. habitat measures, fish ways, increased minimum flow, discharge volumes in key periods) with extended refurbishments and modernization processes, which would allow for increased efficiency of the hydropower plants, thus higher production or limited losses of renewable electricity despite using a smaller share of water (ICPDR 2013; Forseth and Harby 2014).

High priority of the European Commission and Member States is to reduce the impact of existing hydropower plants on water bodies through modernization, refurbishment and ecological restoration measures, as well as to carefully evaluate the sustainability of new projects through the application of a strategic planning. As suggested by the report of the European Commission on the Implementation of the WFD "refurbishing and expanding existing installations should be given priority over new developments which should be underpinned by a strategic assessment at the river basin scale, selecting optimal locations in terms of energy production and lowest environmental impact." (EC 2012, p. 13).

In addition, opportunities for minimizing conflicts and integrating concerns are represented by: (a) major use of the guidance documents on hydropower developed under the WFD CIS; (b) sharing examples of good practices in hydropower planning and construction between member states; and (c) establishing platforms for sharing of insights and experiences from water basin management plans with public and private actors in the energy sector.

As the development of effective policies, processes and tools relies on understanding the connections, the trade-offs and their implications, it is critical for EU policy makers to understand the complex links between climate, energy, water and biodiversity sectors. Improved levels of information in all sectors accompanied by the discussion of interests and trade-offs in an open and transparent way, as already initiated inside the WFD Common Implementation Strategy workshops, can indeed represent important elements towards the achievement of major policy coherence (e.g. the elaboration of guidance documents, sharing of examples and good practices).

6 Conclusion

Strengthening interconnections between different economic, social and environmental policy areas, as well as improving the environmental sustainability of policies, is strictly related to the degree of coherence among policies. In this paper the link between energy-climate-water-sustainability has been explored, addressing the trade-offs, synergies and opportunities for policy integration and coherence.

Although the management of trade-offs requires an integrated response—hence major policy integration and coherence (Nilssen et al. 2012)—the existing policy frameworks that address climate and energy, water and biodiversity issues at the European level are largely independent of each other. In fact, despite the leading role of the EU in integrated policy development and the complex reforms of water and energy policies, a weak evidence of integration and coherence in water, energy and climate change policies has been pointed out on several occasions (Urwin and Jordan 2008; Opperman et al. 2011; Pittock 2011; Nilssen et al. 2012). An OECD report (2011) also underlines the lack of integration stating that overall "policies across the agriculture, water, energy and environment spheres are formulated without sufficient consideration of their interrelationship in any comprehensive manner, or their unintended consequences".

The search for co-ordination as a fundamental element for policy coherence is all good, but as highlighted by several scholars, in itself, leaves unsolved the problem of deciding which policy objectives should be given priority when interests conflict (Ellison 2010; Pataki et al. 2011; Jordan and Schout 2006). In the case of hydropower for instance, the importance of addressing climate objectives has complicated the trade-offs inherent in the EPI agenda, because it requires important water quality and biodiversity (environmental) policy targets to be reconsidered. As climate change is seen as an overriding environmental and political concern of this century there is a risk that the implementation and enforcement of existing water and biodiversity-friendly policies and measures may be reduced. As Pataki et al. (2011) have suggested, this situation has certainly not been improved by the creation of a new DG Climate Action to operate alongside DG Environment and their differences of emphases.

Thinking in terms of policy synergies, these are identified as the strategies that provide positive outcomes in terms of: reducing climate change, increasing or preserving levels of renewable energy, preserving/restoring good ecological status of water courses, as well as biodiversity. But can we really have it all? While synergies between climate and energy are straightforward and clearly stated at the EU level (an ambitious climate policy would contribute to the achievement of energy goals; an ambitious energy policy would contribute to the achievement of climate-policy goals), and it is possible to track this co-ordination between water and biodiversity, challenges persist in relation to the link between CARE and the WFD. As source of these challenges it is possible to trace the unintended consequences emerging from the increased use of renewables, especially with respect to their effects in the long run on other valuable natural resources. As new evidence is being brought on the different carbon, water and land footprint of the renewable energy sources (Hadian and Madani 2015) it is important to consider wisely what kind of energy mix should be pursued.

Addressing the impacts of hydropower opens up the debate on the sustainability of this technology and urges to address a fundamental question: should the concept of sustainability be seen as an absolute goal for a process, or should it be a more flexible concept where a process can have both positive and negative impacts, and what counts at the end is the balance between the two? Addressing this question corresponds to reflecting upon the need to make a choice between climate change and water protection, global needs versus local priorities, and long term versus short term/immediate impacts or benefits. As already highlighted by Peters coordination may inter alia fail "when policies with the same clients (including the entire society as the clients) have different goals and requirements (incoherence)." (1998, p. 303). Such incoherent goals and requirements seem to be archetypical to the political and administrative arenas reflecting the three different dimensions of sustainable development. The main question is therefore how to coordinate and integrate them through prioritisation in a more sustainable manner (Mauerhofer 2013, p. 334).

Defining, aiming, achieving and implementing sustainable hydropower has a considerable political relevance in many countries and decision-makers at the EU, national and sectorial level have the task to reconcile what Sternberg calls "the possible with the needed by means of the political–economic process that insures a viable environmental system" (Sternberg 2008, p. 1590). This implies for the decision-makers of our generation to decide whether the impacts created by hydropower facilities are a reasonable trade-off for the benefits generated according to the current value system and importance attached to both the positive and negative effects. Each future generation will have to make this judgment, according to the values it embraces.

7 Suggestion for Further Research

The implementation of the climate, energy, water and biodiversity policies is strictly related to synergies and trade-offs across climate mitigation and other policy objectives, which in turn depend on the national and local circumstances, as well as the national and international co-ordination and cooperation of the different stakeholders. There is a huge variation between countries in Europe as the share of hydropower varies from negligible quantities in Malta to very high values in Norway, Latvia, Austria and Sweden. It is in these peculiar contexts where policy integration and coherence should be observed and assessed in order to understand which are the processes and results of the balancing process, and what a win-win solution might signify.

References

Acreman MC, Ferguson AJD (2010) Environmental flows and the European water framework directive. Freshw Biol 55:32–48

Bertzky M, Dickson B, Galt R, Glen E, Harley M, Hodgson N, Keder G, Lysenko I, Pooley M., Ravilious C, Sajwaj T, Schiopu R, de Soye Y, Tucker G (2010) Impacts of climate change and selected renewable energy infrastructures on EU biodiversity and the Natura 2000 network: Summary Report. European Commission and International Union for Conservation of Nature, Brussels. Retrieved January 20, 2015, from http://ec.europa.eu/environment/nature/climatechange/pdf/study.pdf

Bilgen S, Kaygusuz K, Sari A (2004) Renewable energy for a clean and sustainable future. Energy Sour 26(12):1119–1129. doi:10.1080/00908310490441421

Biermann F, Davies O, van der Grijp N (2009) Environmental policy integration and the architecture of global environmental governance. Int Environ Agreem Polit Law Econom 9(4):351–369. doi:10.1007/s10784-009-9111-0

Brouwer S, Rayner T, Huitema D (2013) Mainstreaming climate policy: the case of climate adaptation and the implementation of EU water policy. Environ Plan C Gov Policy 31(1):134–153. doi:10.1068/c11134

Bunn SE, Arthington AH (2002) Basic principles and ecological consequences of altered flow regimes for aquatic biodiversity. Environ Manage 30(4):492–507. doi:10.1007/s00267-002-2737-0

CEC (2011a) A resource-efficient Europe—Flagship initiative under the Europe 2020 strategy, COM (2011) 571 final. European Commission, Brussels

CEC (2011b) SETIS Strategic Energy Technologies Information System/Hydropower: Technology Information Sheet. European Commission, Brussels

Collier U (1994) Energy and environment in the European Union. Avebury, Aldershot

Dupont C (2010) Political commitment to climate policy integration at EU Level: the case of biodiversity policy (October 13, 2010), Edinburgh Europa Paper Series 2010/05. Retrieved February 10, 2015, from http://ssrn.com/abstract=1691610

Egré D, Milewski JC (2002) The diversity of hydropower projects. Energy Policy 30 (14):1225–1230

den Hertog L, Stross S (2011, April) Policy coherence in the EU system: concepts and legal rooting of an ambiguous term. Paper presented at the conference "The EU as a Global Player", University Institute for European Studies, Madrid, Spain

EC (European Commission) (2001) European Governance, A White Paper. Retrieved January 24, 2015, from http://europa.eu.int/eur-lex/en/com/cne/2001/com2001_0428en01.pdf

EC (European Commission) (2004) National sustainable development strategies in the European Union: a first analysis by the European Commission. Commission Staff Working Document. Retrieved January 24, 2015, from http://europe.eu.int/comm/sustainable/docs/sustainable_development_strategies.pdf

EC (European Commission) (2007) Communication from the Commission to the European Parliament and the Council—Towards sustainable water management in the European Union —First stage in the implementation of the Water Framework Directive 2000/60/EC—[SEC (2007) 362] [SEC(2007) 363]

EC (European Commission) (2008) Climate change and water, CIS Policy Paper. Retrieved January 24, 2015, from https://circabc.europa.eu/sd/a/eada3719-bc27-4d4e-b3ef-12e21f473924/Final%20Policy%20Paper.pdf

EC (European Commission) (2009a) Communication from the Commission to the European Parliament, the Council, the European Economic and Social Committee and the Committee of the Region—20 20 by 2020 Europe's climate change opportunity. European Union, Brussels, Belgium. Retrieved January 24, 2015, from http://www.energy.eu/directives/com2008_0030en01.pdf

EC (European Commission) (2009b) Directive 2009/28/EC of the European Parliament and of the Council, of 23 April 2009, on the promotion of the use of energy from renewable sources and amending and subsequently repealing Directives 2001/77/EC and 2003/30/EC. European Union, Brussels (OJ L 140, 5.6.2009)

EC (European Commission) (2009c) Beyond 2010-options for EU biodiversity policy (MEMO/09/177, 23 April 2009). Retrieved January 24, 2015, from http://europa.eu/rapid/pressReleasesAction.do?reference=MEMO/09/177

EC (European Commission) (2009d) Common implementation strategy for the water framework directive. Guidance document no. 20. Guidance document on exemptions to the environmental objectives. Commission of the European Communities, Brussels. Retrieved February 22, 2015, from http://ec.europa.eu/environment/water/water-framework/objectives/pdf/Guidance_document_20.pdf

EC (European Commission) (2009e) The Message from Athens. 27–28 April 2009. Athens: European Commission Publications Office. Retrieved March 22, 2015, from http://ec.europa.eu/environment/archives/nature/biodiversity/conference/pdf/message_final.pdf

EC (European Commission) (2009f) Common implementation strategy for the water framework directive. Guidance document no. 24: river basin management in a changing climate, Commission of the European Communities, Brussels. Retrieved February 22, 2015, from https://circabc.europa.eu/sd/a/a88369ef-df4d-43b1-8c8c-306ac7c2d6e1/Guidance%20document%20n%2024%20-%20River%20Basin%20Management%20in%20a%20Changing%20Climate_FINAL.pdf

EC (European Commission) (2011a) Our life insurance, our natural capital: an EU biodiversity strategy to 2020 (3.05.2011). COM(2011) 244 final. Retrieved February 22, 2015, from http://ec.europa.eu/environment/nature/biodiversity/comm2006/pdf/2020/1_EN_ACT_part1_v7%5b1%5d.pdf

EC (European Commission) (2011b) Guidance document: Wind energy developments and Natura 2000. October 2010. Retrieved February 23, 2015, from http://ec.europa.eu/environment/nature/natura2000/management/docs/Wind_farms.pdf

EC (European Commission) (2012) Report from the commission to the European parliament and the council on the implementation of the water framework directive (2000/60/EC) River Basin Management Plans (Text with EEA relevance) Brussels, 14.11.2012 COM(2012) 670 final. Retrieved January 22, 2015, from http://eur-lex.europa.eu/legal-content/EN/TXT/PDF/?uri=CELEX:52012DC0670&from=EN

EC (European Commission) (2013a) Green paper A 2030 framework for climate and energy policies Brussels, 27.3.2013 (COM(2013) 169 final), Retrieved February 22, 2015, from http://eur-lex.europa.eu/legal-content/EN/TXT/PDF/?uri=CELEX:52013DC0169&from=EN

EC (European Commission) (2013b) Guidance on Integrating Climate Change and Biodiversity into Strategic Environmental Assessment, European Union. Retrieved February 22, 2015, from http://ec.europa.eu/environment/eia/pdf/SEA%20Guidance.pdf

EC (European Commission) (2014) Guidance Document "Streamlining environmental assessment procedures for energy infrastructure 'Projects of Common Interest' (PCIs)", European Commission. Retrieved February 22, 2015, from http://www.nepp.ie/wordpress/wp-content/uploads/20130919_pci-en-guidance.pdf

Ecologic (2009) Heavily Modified Water Bodies: 'Information Exchange on Designation, Assessment of Ecological Potential, Objective Setting and Measures'. Discussion paper prepared on behalf of the workshop "Common Implementation Strategy Workshop Brussels, 12–13 March 2009" co-organisers of Germany, UK and the EC. Ecologic: Berlin. Retrieved February 22, 2015, from http://www.ecologic-events.de/hmwb/documents/Discussion_Paper_Updated.pdf

Ellison D (2010) Addressing Adaptation in the EU Policy Framework. In: Keskitalo ECH (ed) Developing adaptation policy and practice in Europe: Multi-level governance of climate change. Springer, Netherlands, pp 39–96

Environment Council of the EU (2010) Environment Council conclusions of 15 March 2010. Retrieved February 26, 2015, from http://www.consilium.europa.eu/ueDocs/cms_Data/docs/pressData/en/envir/113373.pdf

EU (European Union) (2000) Directive 2000/60/EC of the European Parliament and of the Council of 23 October 2000 establishing a framework for Community action in the field of water policy. Official J Eur Communities L327: 1–73

EU (European Union) (2009) Towards a Strategy on Climate Change, Ecosystem Services and Biodiversity: A Discussion Paper Prepared by the EU Ad Hoc Expert Working Group on Biodiversity and Climate Change. Working with Nature. Retrieved February 28, 2015, from http://ec.europa.eu/environment/nature/pdf/discussion_paper_climate_change.pdf

EU (European Union) (2014) Decision No 1386/2013/EU of the European Parliament and of the Council of 20 November 2013 on a General Union Environment Action Programme to 2020 'Living well, within the limits of our planet', Text with EEA relevance. OJ L 354, 28.12.2013, pp. 171–200. Luxembourg: Publications Office of the European Union. doi:10.2779/66315

Eurelectric (2011a) National renewable energy action plans: an industry analysis. Retrieved January 13, 2015, from http://www.eurelectric.org/media/26710/resap_nreap_report_-_final_8-11-11-2011-135-0001-01-e.pdf
Eurelectric (2011b). Hydro in Europe: powering renewables, Eurelectric WG Hydro. Retrieved January 13, 2015, from http://www.eurelectric.org/media/26690/hydro_report_final-2011-160-0011-01-e.pdf
European Communities (1997) Treaty of Amsterdam amending the Treaty on European Union, the Treaties establishing the European Communities and certain related Acts. Luxembourg: Office for Official Publications of the European Communities. ISBN 92-828-1652-4
European Council (1992) Habitats Directive. Consolidated version of Council Directive 92/43/EEC of 21 May 1992 on the conservation of natural habitats and of wild fauna and flora (OJ L 206, 22.7.1992, p. 7)
European Council (2001) Directive 2001/42/EC of the European Parliament and of the Council of 27 June 2001 on the assessment of the effects of certain plans and programmes on the environment (OJ L 197, 21.7.2001)
European Council (2009b) Directive 2009/147/EC of the European Parliament and of the Council of 30 November 2009 on the conservation of wild birds (OJ L 20, 26.1.2010, p. 7), which is the codified version of the original Birds Directive, Council Directive 79/409/EEC of 2 April 1979 on the conservation of wild birds (OJ L 103, 25.4.1979)
EEA (European Environment Agency) (2009) Water resources across Europe—confronting water scarcity and drought. Copenhagen: EEA
EEA (European Environment Agency) (2010) The European Environment—State and Outlook 2010—Synthesis. EEA, Copenhagen
EEA (European Environment Agency) (2015) The European environment—State and Outlook 2015. EEA, Copenhagen
Fearnside PM, Pueyo S (2012) Underestimating greenhouse-gas emissions from tropical dams. Nat Clim Change 2:382–384. doi:10.1038/nclimate1540
Fletcher R (2010) When Environmental Issues Collide: Climate Change and the Shifting Political Ecology of Hydroelectric Power. Peace & Conflict Review, 5(1). Retrieved March 12, 2015, from http://www.review.upeace.org/pdf.cfm?articulo=106&ejemplar=20
Forseth T, Harby A (eds) 2014 Handbook for environmental design in regulated salmon rivers. NINA Special Report 53, 1–90
Frey GW, Linke DM (2002) Hydropower as a renewable and sustainable energy resource meeting global energy challenges in a reasonable way. Energy Policy 30:1261–1265. doi:10.1016/S0301-4215(02)00086-1
Groom B, Grosjean P, Kontoleon A, Swanson T, Zhang S (2010) Relaxing rural constraints: a "win-win" policy for poverty and environment in China? Oxford Econ Pap 62:132–156. doi:10.1093/oep/gpp021
Henriksen CB, Hussey K, Holm PE (2011) Exploiting soil-management strategies for climate mitigation in the European Union: maximizing "win–win" solutions across policy regimes. Ecol Soc 16(4):22. doi:10.5751/ES-04176-160422
Howe C, Suich H, Vira B, Mace GM (2014) Creating win-wins from trade-offs? Ecosystem services for human well-being: A meta-analysis of ecosystem service trade-offs and synergies in the real world. Glob Environ Change 28:263–275. doi:10.1016/j.gloenvcha.2014.07.005
ICPDR (International Commission for the Protection of the Danube River) (2013) Guiding Principles on Sustainable Hydropower Development in the Danube Basin, Vienna
IHA (International Hydropower Association) (2003) The role of hydropower in sustainable Development. IHA White Paper, February, 1–140
IRENA (International Renewable Energy Agency) (2012) Renewable energy technologies—cost analysis series, volume 1: power sector, IRENA Working Paper 3/5
IPCC (Intergovernmental Panel on Climate Change) (2013) Climate policymakers. Cambridge University Press, New York

IPCC (Intergovernmental Panel on Climate Change) (2014) Climate change 2014: mitigation of climate change. Contribution of Working Group III to the Fifth Assessment Report of the Intergovernmental Panel on Climate Change. Cambridge University Press, Cambridge/UK and New York/USA

Jackson A (2011) Renewable energy vs. biodiversity: policy conflicts and the future of nature conservation. Glob Environ Change 21(4):1195–1208

Jordan A, Schout A (2006) The coordination of the European Union. Oxford University Press, Oxford

Jordan A, Lenschow A (eds) (2008) Innovation in environmental policy? Integrating the Environment for Sustainability. Edward Elgar, Cheltenham UK

Kampa E, von der Weppen J, Dworak T (2011) Water management, Water Framework Directive & Hydropower, Common Implementation Strategy Workshop, Brussels, 13–14 September 2011. Issue Paper (final version). Retrieved January 15, 2015, from http://www.ecologic-events.eu/hydropower2/documents/IssuePaper_final.pdf

Kaunda CS, Kimambo CZ, Nielsen TK (2012) Hydropower in the context of sustainable energy supply: a review of technologies and challenges. International Scholarly Research Network ISRN Renewable Energy, Volume 2012, Article ID 730631

Kivimaa P, Mickwitz P (2009) Making the Climate Count: Climate Policy Integration and Coherence in Finland. Finnish Environment Institute, Helsinki

Knudsen JK (2009) Environmental policy integration and energy. conceptual clarification and comparative analysis of standards and mechanisms. Universiteit Twente/CSTM, Twente

Lafferty WM (ed) (2004) Governance for sustainable development. Edward Elgar, Cheltenham, UK

Lafferty WM, Hovden E (2003) Environmental policy integration: towards an analytical framework. Environ Polit 12(3):1–22

Lenschow A (ed) (2002a) Environmental policy integration: greening sectoral policies in Europe. Earthscan, London

Lenschow A (2002b) Greening the European Union: an introduction. In: Lenschow A (ed) Environmental policy integration: greening sectoral policies in Europe. Earthscan, London

Liberatore A (1997) The integration of sustainable development objectives into EU policy-making: Barriers and prospects. In: Baker S, Kousis M, Richardson D and Young S (eds) The politics of sustainable development: Theory, policy and practice within the European Union. Routledge, London

Locker H (2004) Environmental Issues and Management for Hydropower Peaking Operations. United Nations, Department of Economic and Social Affairs (UN-ESA), New York

Maeck A, DelSontro T, McGinnis DF, Fischer H, Flury S, Schmidt M, Fietzek P, Lorke A (2013) Sediment trapping by dams creates methane emission hot spots. Environ Sci Technol 47:8130–8137

Mant R, Perry E, Heath M, Munroe R, Väänänen E, Großheim C, Kümper-Schlake L (2014) Addressing climate change—Why biodiversity matters. UNEP-WCMC, Cambridge, UK. Retrieved March 25, 2015, from http://www.unep-wcmc.org/system/dataset_file_fields/files/000/000/221/original/IKI_report_2_accessible_version_20140530.pdf?1401884844

Massey E, Huitema D, Jordan A, Rayner T (2010) Adaptation in the water sector: will mainstreaming be sufficient? In: Jordan AJ, Huitema D, van Asselt H, Rayner T, Berkhout F (eds) Climate change policy in the European Union: confronting the dilemmas of mitigation and adaptation?. Cambridge University Press, Cambridge, pp 167–185

Mauerhofer V (2013) The 'Governance-Check': Assessing the sustainability of public spatial decision-making structures. Land Use Policy 30(1):328–336. doi:10.1016/j.landusepol.2012.03.016

McCully P (1996) Silenced rivers: the politics and the ecology of large dams. Zed Books, London

McShane TO, Hirsch PD, Trung TC, Songorwa AN, Kinzig A, Monteferri B, Mutekanga D, Hoang TV, Dammert JL, Pulgar-Vidal M, Welch-Devine M, Brosius JP, Coppolillo P, O'Connor S (2011) Hard choices: making trade-offs between biodiversity conservation and human well-being. Biol Conserv 144(3):966–972

Meijers E, Stead D (2004) Policy integration: what does it mean and how can it be achieved? A multi- disciplinary review. Paper presented at the Berlin Conference on the Human Dimensions of Global Environmental Change: Greening of Policies—Interlinkages and Policy Integration. Retrieved March 15, 2015, from http://www.fu-berlin.de/ffu/akumwelt/bc2004/download/meijers_stead_f.pdf

Mekonnen MM, Gerbens-Leenes PW, Hoekstra AY (2015) The consumptive water footprint of electricity and heat: a global assessment. Environ Sci Water Res Technol 1:285–297. doi:10.1039/c5ew00026b

Moomaw W, Yamba F, Kamimoto M, Maurice L, Nyboer J, Urama K, Weir T (2011) Introduction. In: Edenhofer O, Pichs-Madruga R, Sokona Y, Seyboth K, Matschoss P, Kadner S, Zwickel T, Eickemeier P, Hansen G, Schlömer S, von Stechow C (eds) IPCC special report on renewable energy sources and climate change mitigation. Cambridge University Press, Cambridge, United Kingdom and New York, NY, USA

Nilsson M, Persson A (2003) Framework for analysing environmental policy integration. J Environ Planning Policy Manage 5(4):333–359

Nilsson M, Zamparutti T, Petersen JE, Nykvist B, Rudberg P, McGuinn J (2012) Understanding policy coherence: analytical framework and examples of sector-environment policy interactions in the EU. Environ Policy Gov 22(6):395–423. doi:10.1002/eet.1589

OECD (1995) Planning for sustainable development: country experiences. OECD, Paris

OECD (1996) Building policy coherence: tools and tensions. Public Management Occasional Papers, No. 12. OECD, Paris

OECD (2001) Sustainable development: critical issues. OECD, Paris

OECD (2002) Improving policy coherence and integration for sustainable development: a checklist. OECD Policy brief October 2002. Retrieved February 12, 2015, from http://78.41.128.130/dataoecd/61/19/2763153.pdf

OECD (2011) Policy coherence between water, energy and agriculture. Working Party on Biodiversity, Water and Ecosystems, 10–11 March 2011

Opperman JJ, Royte J, Banks J, Day LR, Apse C (2011) The Penobscot River, Maine, USA: a basin-scale approach to balancing power generation and ecosystem restoration. Ecol Soc 16(3):7. URL: http://dx.doi.org/10.5751/ES-04117-160307

Oud E (2002) The evolving context for hydropower development. Energy Policy 30:1215–1223

Pataki G, High C, Nemes G (2011) Report on the policy and governance context for adaptation. CLIMSAVE deliverable. Retrieved February 12, 2015, from http://www.climsave.eu/climsave/outputs.html

Persson Å (2004) Environmental policy integration: an introduction. PINTS—Policy Integration for Sustainability. Background Paper. Stockholm Environment Institute—SEI, Stockholm (ISBN: 91 975237 1 2)

Peters G (1998) Managing horizontal government: the politics of co-ordination. Public Adm 76(2):295–311. doi:10.1111/1467-9299.00102

Pittock J (2011) National climate change policies and sustainable water management: conflicts and synergies. Ecol Soc 16(2):25. URL:http://www.ecologyandsociety.org/vol16/iss2/art25/

Rodriguez JP, Beard Jr. TD, Bennett EM, Cumming GS, Cork SJ, Agard J, Dobson AP, Peterson GD (2006) Trade-offs across space, time, and ecosystem services. Ecol Soc 11(1):28. URL:http://www.ecologyandsociety.org/vol11/iss1/art28/

Rosenberg DM, Bodaly RA, Usher PJ (1995) Environmental and social impacts of large scale hydroelectric development: who is listening? Glob Environ Change 5(2):127–148. doi:10.1016/0959-3780(95)00018-J

Saunders, Harry D (October 1992) The Khazzoom-Brookes postulate and neoclassical growth. Energy J 13(4):131–148. Retrieved May 17, 2015, from http://www.iaee.org/en/publications/ejarticle.aspx?id=1091

Sorrel S (2009) Jevons' Paradox revisited: the evidence for backfire from improved energy efficiency. Energy Policy 37:1456–1469

Sternberg R (2008) Hydropower: dimensions of social and environmental coexistence. Renew Sustain Energy Rev 12, 1588–1621. Retrieved March 28, 2015, from https://www.pik-potsdam.de/news/public-events/archiv/alter-net/former-ss/2009/working-groups/literature/w-sternberg.pdf

Truffer B, Markard J, Bratrich C, Wehrli B (2001) Green electricity from alpine hydropower plants. Mountain Res Dev 21(1):19–24, Retrieved March 22, 2015, from http://www.mtnforum.org/sites/default/files/publication/files/5954.pdf

UNCBD (2002) COP 6 Decision VI/26 (7–19 April 2002). Retrieved March 17, 2015, from http://www.cbd.int/doc/decisions/cop-06-dec-en.pdf

UNCED (1992) Agenda 21. United Nations, New York

UNESCO (2003) Water for people-water for life. UNESCO, Paris/New York

Urwin K, Jordan A (2008) Does public policy support or undermine climate change adaptation? Exploring policy interplay across different scales of governance. Glob Environ Change 18:180–191

Van Bommel S, Kuindersma W (2008) Policy integration, coherence and governance in Dutch climate policy: a multi-level analysis of mitigation and adaptation policy report 1799. Wageningen, Alterra

van Ginneken M (2015) A decade of sustainable hydropower development—What have we learned? HydroWorld.com article of 02/17/2015. Retrieved April 29, 2015, from http://www.hydroworld.com/articles/print/volume-23/issue-1/articles/a-decade-of-sustainable-hydropower-development-what-have-we-learned.html

Water Directors (2010) Final synthesis of informal meeting of water and marine directors of the European Union, Candidate and EFTA Countries, Segovia, 27–28 May 2010

WCED (World Commission on Environment and Development) (eds) (1987) Our common future. Oxford University Press, Oxford

Wilby RL, Orr HG, Hedger M, Forrow D, Blackmore M (2006) Risks posed by climate change to the delivery of Water Framework Directive objectives in the UK. Environ Int 32:1043–1055

Winship C (2006) Policy analysis as puzzle-solving. In: Moran M, Goodin RE, Rein M (eds) Oxford handbook of public policy (pp. 109–124). Oxford University Press, Oxford

WWAP (United Nations World Water Assessment Programme) (2015) The United Nations world water development report 2015: Water for a Sustainable World. UNESCO, Paris. Retrieved May 5, 2015, from http://unesdoc.unesco.org/images/0023/002318/231823E.pdf

Part VII
Sectorial Policies: Renewable Energy

Generating Renewable Energy for the Material Realization of Sustainable Development: What Do We Need from Multilateral Cooperation, the Climate Change and the International Trade Regimes?

Marco Citelli

Abstract This chapter investigates certain aspects of multilateral cooperation in the field of renewable energy as well as the role that renewable energy occupies within the climate change regime and the questions that both the generation of energy from renewable sources and the use of renewable energy related technologies are growingly posing to the agents of global trade. The premise is that production of energy from renewable sources is the key driver for the material realization of sustainable development. Inter State development cooperation can variously impact on the dynamics of the renewable energy sector. Different branches of public international law can also determine trends in this respect. This is why the case of the International Renewable Energy Agency (IRENA) will be analysed. International rules requiring States to foster the promotion of renewable energy within the climate change regime will then be addressed in light of some results of the Lima Summit (UNFCC-COP20). Finally, certain WTO disputes will be examined in order to identify the limits that international trade law imposes to States in relation to certain specific energy policy choices and to determine if, and to what extent, a space within the WTO system exists in order to accommodate WTO law inconsistent measures however meant to sustain the production of clean energy. The overall goal is to see what is required from international organizations, climate change and trade law to foster the dissemination of renewable energy thereby favouring the material realization of sustainable development.

Keywords International law · International organization · Renewable energy · International cooperation · Sustainable development

---

M. Citelli (✉)
Department of Legal Studies, "A. Staffa", Università Commerciale "L. Bocconi", Milan, Italy
e-mail: marco.citelli@phd.unibocconi.it; marco.citelli@gmail.com

V. Mauerhofer (ed.), *Legal Aspects of Sustainable Development*,
DOI 10.1007/978-3-319-26021-1_19

1 Introduction: Sustainable Development as a Principle of International Environmental Law: Why Renewable Energy Matters?

Renewable energy has been on the agenda of global environmental conferences in the last thirty years. However, the dissemination of its related technologies as well as the relationship between renewable energy and the principle of sustainable development were never fully addressed on those occasions. In fact, public international law has been lacking dedicated binding and non-binding rules on renewable energy due to the persistence of interests sustaining the exploitation of traditional energy sources as well as of market imperfections and technical constraints hampering a wider reliance on renewable energy (e.g. Redgwell 2008, pp. 85–108).

It has to be noted, though, that from the very first appearance of sustainable development as a principle of international environmental law, the linkage with renewable energy was acknowledged. The 1987 Brundtland Report of the World Commission on Sustainable Development, in fact, termed renewable energy as "the foundation of the global energy structure during the 21st century" (UN General Assembly 1987 Chap. 7, paragraph 88).

Despite this straightforward indication of the way to go, global environmental governance has just partially developed in this direction. Both the Principles agreed upon at 1992 Rio Declaration on Environment and Development and Agenda 21 (UNCED 1992) are of very limited relevance to renewable energy generation, signalling a comprehensive lack of State *consensus* which later crossed the decades. The 2002 Johannesburg Plan of Implementation (JPoI 2002), designed to guide States in further implementing the policy guidance provided by Agenda 21 had no clear focus on renewables and in 2007 the UN Commission on Sustainable Development (UNCSD 2007) witnessed a widespread resistance to the adoption of measurable targets for renewable energy.

The backdrop did not change in any meaningful fashion for the 2012 Rio + 20 UN Summit. The decisions taken there, in fact, do not entail any specific political commitment (let alone any legal development) on renewable energy, even if, it must be said, the link between access to energy and sustainable as well as human development was acknowledged and the goal of facilitating the access to energy services to the 1.4 billion people deprived thereof was set (UN General Assembly 2012, paragraph 125).

The limited support for renewable energy options emerging from a wide array of UN soft law instruments seems to neglect the theoretical premise of this analysis, according to which the use of renewable sources of energy can boost the material realization of sustainable development. However, it must be observed that the relationship between renewable energy and sustainable development has constantly been reaffirmed and that a full support from States has been prevented by a plurality of factors. From the international law standpoint, part of the problem is determined by the relativism of the sustainable development principle. The lack of a *consensus* among States on the widespread reliance on renewable energy itself mirrors the relativism around this concept. The nature of sustainable development is also scholarly debated. The complexity of defining this concept in legal terms has been evidenced by the doctrine (e.g. Pallemaerts 1996, pp. 630–634), which also highlighted the difficulties inherent to its implementation (Viñuales 2013, pp. 3–13). Global environmental law and governance, however, are currently devoid of alternative regulatory paradigms guiding States towards the adoption of clean energy choices. In an era where the human-environment interface becomes increasingly important due to pressing challenges, including energy security and the exhaustion of finite natural resources, that need to be addressed with renovated urgency (Kotzé 2014, pp. 121–156), sustainable development, growingly underpinned by the so-called ‘ecosystem approach’ shaping many international environmental law regimes, remains the predominant legal and policy “narrative” inspiring national and international action albeit competing with others, prominently eco-centric ones (De Lucia 2014, pp. 91–117).

2 Multilateral Cooperation in the Field of Renewables: The Rise of IRENA

Global governance on energy issues has traditionally been scattered and, to a certain extent, it continues to be so as many multilateral developmental agencies keep on playing a role conveying inter State cooperation. This is the case, for instance, of the World Bank, and of many specialized agencies and programmes under the aegis of the UN. A novelty, in this respect, is the creation of an International Renewable Energy Agency (IRENA) in 2009. The Conference leading to the adoption of its Statute was promoted by Germany and other European countries like Spain and Denmark, highly committed to favouring the dissemination of technologies for the exploitation of renewables. Furthermore, the initiative was backed up by NGOs and networks like REN21 and Eurosolar, working for fostering renewable energy cooperation also with the view to “counterbalancing” the prevalence of the International Atomic Energy Agency (IAEA) and the activities undertaken by the latter for the promotion of the civil uses of nuclear energy (Scheer 2007, pp. 306–310).

In order to assess the activities developed by IRENA in the first years after its appearance on the global governance scene, it is preliminarily necessary to refer to its Statute (IRENA 2009), one of the most rapidly ratified legal instrument in the recent history of major environmental agreements, with the view of understanding the mandate conferred to the Agency by its founding Members. The Preamble of the Statute explicitly acknowledges the positive role that renewable energy plays for the attainment of sustainable development. In the same vain, it recognizes the effects that the exploitation of renewable energy sources can have on the solution of geopolitical and economic issues, like energy security and the instability of energy prices on world markets. The significance of renewable energy for the transition to a low carbon economy is also underlined, as their contribution for reducing the concentration of GHGs in the atmosphere thereby stabilizing climate change. The Preamble further refers to the decentralized access to energy, the negative impact determined by the use of fossil fuels and by the inefficient utilization of biomass and the general need of increasing energy efficiency that, paired with an increased utilization of renewable energy sources, is identified as the key strategy to be adopted in order to satisfy an ever growing energy demand.

IRENA's main objective is to foster the use and the dissemination of renewable energy, taking into account, what can be referred to as national developmental priorities and positive environmental externalities deriving from the use of these energy sources (Article II). To this purpose, the Statute laid down a brief definition of renewable energy, which comprises «all forms of energy produced from renewable sources in a sustainable manner», including, among others geothermal energy, hydropower, ocean energy (i.e. tidal, wave, ocean thermal energy), solar energy and wind energy (Article III).

The activities of the Agency, which is intended to act as a centre of excellence in the field of renewables, are regulated by Article IV. According to this provision, the Agency shall primarily facilitate the circulation of information and knowledge relevant to the renewable energy sector by offering assistance to its Members in relation to domestic policies and their implementation. Furthermore, IRENA shall operate in order to assess, monitor and systematize the best practices, while interacting with other governmental organizations and NGOs, with the view of stimulating scientific and technical research on the socio-economic aspects connected to the utilization of renewable sources of energy. Besides this, the Statute requires the Agency to carry out assistance activities, upon request of its Members, aimed at supporting the development of local capabilities and interconnections. Assistance can involve the dissemination of information concerning national and international technical standards and can be aimed at increasing public awareness on the potential of renewables.

While displaying its activities pursuant to Article 4 of its Statutes, IRENA shall act in conformity to the objectives and principles of the United Nations Charter as well as to the policies adopted by the UN oriented to fostering sustainable development. Moreover, the activities of the Agency shall be aimed at addressing the needs of developing countries, least developed countries, small island states and remote regions, avoiding also unnecessary duplications with initiatives taken by other multilateral institutions, with which IRENA is required to cooperate.

In the first years after the entry into force of its Statute, IRENA has developed a practice in line with its objectives and has adopted a series of annual work programmes (e.g. IRENA 2012, 2013). Such instruments are proposed by the General Director of the Agency, are submitted to the IRENA's Council and are finally adopted by the Assembly, according to the procedures under Articles IX(g)2 and X (f)2 of the Statute. Each of these programmes has been based on three main sub-programmes concerning, respectively, the dissemination of information and knowledge, the dissemination of innovation and technology, capacity building and assistance. If in the first two years IRENA prominently focused on the structuring of its own staff and to the definition of internal administrative procedures, it subsequently started to develop and to implement its external initiatives in various sectors. The Agency's initiatives falling into the first sub-programme have primarily required the definition of a *Knowledge Management Strategy* aimed at guaranteeing access to all information and knowledge related to renewable energy to all IRENA's members and other public and private stakeholders alike. The Strategy has been then supported by the Global Atlas on solar and wind power, (IRENA 2015) two operative tools the elaboration of which has been coordinated by the Agency, which systematized already existing and new information provided by certain energy national authorities (e.g. the US Department of Energy) and by international organizations that entered a partnership with IRENA (e.g. the World Meteorological Organization). Another notable initiative underpinned by the Agency is the harmonization of statistical data on renewable energy concerning the countries that are not Members to the Organization for Economic Development in Europe (OECD), thus not covered by the information gathered and elaborated by the International Energy Agency (IEA), an intergovernmental agency established in the framework of the OECD, whose main goal is to facilitate the coordination of the energy policies pursued by its Members.

Innovation and technology, as another pillar of IRENA's action, saw the Agency engaged with a plurality of initiatives. First of all, the Agency began to elaborate criteria that could assist States in improving their domestic policies related to innovation. IRENA also started to work on the development of tools and methodologies for measuring the environmental impact of products and technologies connected to the generation of energy from renewable sources. The latter initiative was launched together with the United Nation Environmental Program (UNEP) and the Secretary of the 1979 Convention on migratory species (CMS). Furthermore, the cooperation between IRENA, the World Intellectual Property Organization (WIPO) and the European Patent Office (EPO) resulted in a joint project which aims at the creation of a database giving access to all information on patents in the field of renewables.

In the context of assistance and capacity building IRENA embarked on multiple activities. For instance, the Agency has started monitoring the best practices among the national policies of its Members in order to detect those that deliver the best results in terms of renewable energy dissemination. Moreover, at the international level, IRENA started putting forward its proposals to increase the weight of renewable energy in the framework of certain mechanisms provided for by the climate change (i.e. the Technology Mechanism and the Green Fund instituted by the 2009 Copenhagen Accord). In addition, IRENA contributed to a study of the International Labour Organization (ILO) on the employment opportunities in the renewable energy sector.

In the light of the activities briefly illustrated above, the Agency was described by scholars as an epistemic institution (Meyer 2013, pp. 15–44). As such, IRENA gathers and conveys policy-relevant scientific advice aimed at untapping the potential of renewables, given that the role played by governments and private economic operators alike in its sector has been rather limited so far. This function is certainly prominent for two reasons. Firstly, IRENA's Statute does not confer to the Agency any mandate for the establishment of standards and guidelines (soft law) either in the field of renewable energy or in relation to the definition of legally binding commitments in this sector (unlike, for instance, the case of the IAEA). Secondly, the assessment of the initial practice developed by the Agency reveals the peculiar connotation of its epistemic activities which are mostly oriented to assisting developing countries and, by virtue of the absence of legal and institutional connections with other processes underpinned in the framework of the UN and in other *fora*, most notably, the climate change regime (Uperlainen and Van de Graaf 2013, pp. 1–19) and the Global Environmental Facility, take place without the transfer of financial resources, on the basis of equality among States. In other words, these activities are conducted on the basis of a paradigm that, at least formally, seems to move away from the principle of common but differentiated responsibilities which permeates the global governance on environment and development from its inception in Principle 7 of the 1992 Rio Declaration. Many initiatives undertaken by IRENA, in fact, can be read in these terms. Certainly this is the case of the Renewable Readiness Assessments (RRAs) conducted by IRENA in various African, Middle Eastern, Latin American and Caribbean countries, in order to facilitate the dialogue between the subjects involved in the definition of national and sub-national strategies on renewable energy generation. Other partnerships and regional technical cooperation initiatives promoted by the Agency can be read in similar terms. In 2012, for instance, IRENA launched an initiative for the promotion of geothermal energy in the Andean region (with the support of two so-called "mentor" countries, Iceland and Mexico). In 2013, additionally, the Agency started cooperating with ASEAN with the view of studying the regional integration of energy markets in South East Asia and the expansion of the ASEAN Power Grid. What is becoming increasingly evident is the particular trajectory of the flow of information and knowledge shared by the Agency and of the assistance it conveys. In fact, most of these activities take place on a South-South axis, rather than between developed countries, on the one hand, and developing countries, on the other hand. This particular focus on developing countries' needs is also proved by the assistance provided by the Agency to improve the attractiveness of renewable energy-related projects both in the framework of supranational financial mechanisms like those administered by the World Bank (e.g. the Climate Technology Fund) and by the efforts that are being done for the elimination of technical, economic and financial barriers, that discourage the flow of foreign private investments from developed countries, and have historically favoured countries like China, India, Brazil and other emerging economic powers (e.g. Del Rio 2007, pp. 1361–1378).

3 The Post-2020 International Climate Change Regime: Which Place for Renewables?

Until 2020 the United Nations Convention on Climate Change (UNFCCC) and the Kyoto Protocol (KP) will be the main international legal instruments binding States to commit themselves in the fight against human-induced climate change. The UNFCCC is relevant to the renewable energy sector because the Contracting Parties to this Convention are required to control their sources of anthropogenic GHG emissions and to favour climate change mitigation by adopting programs to these ends while also streamlining climate change, *to the extent feasible*, in the preparation of their social, economic and environmental initiatives (Article 4, paragraph 1, (b) and (f)).

The KP, for its part, establishes that Annex I Parties shall 'implement and/or further elaborate policies and measures *in accordance with national circumstances*' on research and development of renewable energy technologies (Article 2, paragraph 1, (i) and (iv)). Furthermore, it must be noted that the KP provides its Contracting Parties with the faculty of undertaking additional efforts (supplementing domestic ones) directly involving the production of energy from renewable sources under some 'flexibility mechanisms', the functioning of which is regulated, in particular, by Article 6 ('Joint Implementation'), allowing an Annex I country (i.e. developed countries and economies in transition) to implement emission reduction projects in any other Annex I country and by Article 12 ('Clean Development Mechanism' or CDM), which permits Annex I countries to make investment aimed at achieving emissions reduction while fostering sustainable development in the territory of non-Annex I countries (i.e. developing countries). Although their operationalization is not obligatorily subjected to the use of renewable energy technologies, these instruments certainly hold the potential to sustain the increase of their share in the world energy supply.

Under any circumstances, however, neither the UNFCCC nor the KP do appear as sufficiently sustaining the dissemination, the adoption and the use of products and technologies for the generation of energy from renewable sources as the primary means to be used in order to curb GHG emissions. In the course of negotiations that followed the conclusion of the KP, various developing Contracting Parties expressed the view that renewable energies should have been specifically given priority, for instance, in relation to projects and activities under the CDM, but their proposals remained on paper. In any case, it must be admitted, the KP does not exclude investments in renewable energy either, but rather encourages them through its flexibility mechanisms designed to supplement the efforts undertaken by Annex I countries in achieving their national targets of emission reduction, particularly the CDM. In fact, 70 % of the total amount of the CDM projects from the start of the crediting period until the end of 2012 is related to renewable energies (Center on Energy, Climate Change and Sustainable Development 2015).

In spite of this, if the arguments on the scarce economic viability and technical feasibility of renewable energy options could hold and prevail at time of the inception of the KP and for several years after its adoption, today the cost efficiency of renewable energy products and technologies has greatly increased and giving more prominence to renewable energy utilization can no longer be viewed as another stumbling block in climate negotiations. Pursuant to the mandate established for the negotiating process launched in 2011 by COP17, known as the Durban Platform, by 2015 UNFCC Parties will have to consider for adoption either an international agreement, a protocol or another "agreed outcome with legal force". Such an arrangement will have to enter into force by the end of 2020, at end of the second commitment period of the KP.

At the current state of play it is not clear whether a new post-2020 arrangement will overcome what has been identified by scholars of various disciplines as one of the most evident failures of the present climate change regime, namely, its incapacity of shifting the world's energy base to renewable power instead of fossil fuels (e.g. Ferrey 2010, p. 68).

What is rather clear so far is that the principle of common but differentiated responsibilities will remain pivotal in any future arrangement on climate change, even thought it might be featured by more nuanced connotations (more complex and more differentiated), as progressively shown already by the most relevant decisions of the Conference of the Parties of the Convention (UNFCCC COP) in the last few years, from the adoption of the 2007 Bali Action Plan onwards (Brunnée and Streck 2013, pp. 590–607). In turn, this may theoretically encourage a process of *consensus* formation around obligations directly addressing the generation of energy from renewable resources, a choice that will ultimately depend, however, on the way in which States will perceive the seriousness of the challenges posed by climate change and natural resources management (De Cendra de Larragán 2012, pp. 6–27).

Certain aspects of the 2014 Lima Call for Climate Action (UNFCCC 2014), the core decision of COP 20 which marked the final stage of climate negotiations preceding the actual adoption of the text of an agreement in 2015, might be viewed as signs of an increasing *consensus* on renewable energy options. In fact, the so-called "Elements for a draft negotiating text", annexed to the decision at hand, refer widely to the dissemination and application of new and alternative technologies as the main way for achieving greater energy efficiency and combating the adverse effects of climate change.

Furthermore, amongst the proposals put forward in relation to the mobilization of financial resources, all premised by the recognition that leveraging private finance as a supplement of public funding should be the priority, the establishment of an international renewable energy and energy efficiency bond facility is envisaged, together with the phasing down of high-carbon investments and fossil fuel subsidies and the introduction of a tax on oil exports from developing to developed countries. It remains to be seen, however, the extent to which these proposals will be actually included in the future climate agreement and how they could actually contribute to realization of the social, economic and environmental aspects of sustainable development.

4 WTO Disputes on Renewable Energy: Signalling the Need for an International Trade Law Reform?

4.1 Measures for the Promotion of Renewable Energy Related Technologies

The Global trade in technologies which may allow a high level of environmental protection, hence the realization of sustainable development, has progressively increased, arousing interest around the consistency of energy policies with international trade law. Albeit the reference to the principle of sustainable development in the Preamble of the WTO Agreement, the WTO administered agreements do not lay down any rule concerning the so-called "energy-related products". Consequently, market access of such products is subjected to the non-discrimination principle permeating all WTO trade agreements and the general Lists of tariff concessions annexed to the Marrakesh Protocol. Energy products are not immune to unilateral and multilateral actions whenever they benefit from export subsidies, subsidies dependent on the use of national goods or contravening the anti-dumping discipline, as established, respectively, by the Agreement on Subsidies and Countervailing Measures (ASCM) and by the Agreement on the implementation of Article VI of the General Agreement on Tariffs and Trade 1994 (Anti-Dumping Agreement or ADA). However, besides the practices expressively prohibited by the WTO, uncertainty remains on the consistency of some either direct or indirect governmental interventions, which can be found as a steady trait of policies aimed at favouring the dissemination of technologies for the exploitation of renewable energy sources while counteracting the forces limiting the development of the renewable sector.

4.2 WTO Disputes the Promotion of Renewable Energy Related Technologies

The first Panel reports and Appellate Body (AB) decisions on the compatibility of renewable energy financial assistance schemes with WTO law were handed down in the cases *Canada—Renewable Energy* and *Canada—FIT Program* (WTO 2012b, 2013a). The challenged *Feed-In Tariff Program* (FIT Program) was based on a system of guaranteed prices aimed at increasing electricity generation from wind power and solar energy, with a two-fold objective: improving air quality while curbing fossil fuel dependence.

One requirement to satisfy for participating in the programme was the mandatory use of materials and technologies developed by enterprises operating in the Ontario Province ("local content requirement" or LCR). The claimants, Japan and the EU, pointed at the LCR inconsistency with the national treatment obligations *ex* Article III:4 of General Agreement on Tariffs and Trade 1994 (GATT 1994) and *ex* Article 2.1. of the Agreement on Trade Related Investment Measures (TRIMS Agreement).

On its part, Canada justified such measures invoking Article III:8 of GATT 1994 on governmental purchases. In their conclusions, however, both the Panel and the AB easily demonstrated the discriminatory character of the LCR and the inapplicability of the exception on governmental purchases due to the commercial character of the transactions occurring between the energy operators and the Ontario Province, ultimately aimed at phasing in electricity into the grid and at reselling it to final users. Furthermore, the AB excluded the application of Article III:8 of GATT 1994 as the product being purchased by the government (electricity) differed from the product being discriminated due to its origin (products for wind and solar power generation).

The claimants also claimed that the FIT Program was in breach of ASCM Articles 3.1(b) and 3(2) because, by imposing a LCR it subordinated the possibility of benefiting from a subsidy to the preferential use of national goods over imported ones. In the case at hand, although the Panel easily acknowledged the measure as constituting a financial contribution in the terms of ASCM Article 1.1(a)1, it did not manage to determine the conferral of a benefit as defined by ASCM Article 1.1(b). The Panel argued, in fact, that the difficulty encountered in making the determination, which ultimately prevented the identification of a benchmark to formulate economic and trade comparisons, were due to the non-perfectly competitive features of the electricity market, where public interventions are necessary to secure stable and long-term electricity supply. In this regard, the AB partially moved from the logic used by the Panel by rebutting the thesis according to which the entire electricity market was the reference to determine the benefit conferred by the subsidy. Taken as given the composition of the energy supply established by the Province of Ontario, the AB identified, as a specific market, the one of electricity generated by wind and solar power and claimed that a benchmark should have been found there. Notwithstanding this, neither the AB managed to detect an adequate benchmark and ultimately did not decide on the existence of measures fitting the definition of subsidy under ASCM Article 1.

4.3 WTO Disputes the Promotion of Biofuels

EU Directive 2009/28/EC concerning the promotion of the use of renewable energy sources, hereinafter RED (EU 2009) is a cornerstone of the so-called "EU Climate and Energy Package" (Kulovesi et al. 2009, pp. 829–891). The WTO law consistency of RED norms on biofuels and bioliquids is currently under the scrutiny of the WTO dispute settlement bodies. In particular, the alleged WTO inconsistency of the so-called *sustainability criteria* applicable to bioliquids and biofuels regardless from their country of origin, either within or outside the EU (RED Article 17), was targeted by Argentina. This country, which has developed one of the most flourishing biodiesel industries and has recently consolidated its position as the major world exporter of such product, lodged three different requests for consultations (WTO 2012a, 2013b, 2014) *ex* Article 4.4 of the Dispute Settlement Understanding (DSU).

RED Article 17 lays down a series of *sustainability criteria* (paragraphs 2–6). What matters under an international trade law perspective is the possibility that the commercial partners of the EU begin to challenge the consistency of the criteria with GATT 1994 and with the Agreement on Barriers to Trade (TBT).

The first complaint concerns the alleged breach of the national treatment provisions and of the discipline on internal regulations established by GATT 1994 Articles III.1, 4 and 5, Articles 2.1 and 2.2. of the TRIMS Agreement and by Article XVI:4 of the WTO Agreement. The issue revolves substantially on alleged exclusion of non-EU biodiesel from the calculation and the subdivision into various categories of produced and consumed biodiesel volumes for the purpose of achieving the "mandatory national objectives" introduced for each EU Member State by the Directive (Article 1). A possible discrimination seems at least to be indirect, in the sense that the sustainability criteria expressively exclude any differential treatment between domestic and foreign products. More precisely, RED Article 17 excludes all "non sustainable" biofuels, regardless of their country of origin, from the ones admissible for discounting the progress made to achieve national targets as well as from the range of products that can benefit from financial support. Therefore this non-tariff measure appears to be capable of altering the structure and the flows of trade in biodiesel without, however, breaching the GATT 1994.

The second complaint concerns certain measures on the importation and the commercialization of biodiesel and certain financial schemes supporting the production of biodiesel in the EU. Argentina, in particular, claims that the 35 % threshold of GHGs generated reduction to be satisfied by a product in order to be qualified as "sustainable" is ultimately arbitrary and unjustifiable. Moreover, the default value of 31 % assigned by the Directive to biodiesel generated from soy feedstock (the one mainly exported by Argentina) rules out this product from those eligible for achieving RED objectives. Argentina also brought claims under the TBT Agreement (Articles 2.1, 2.2, 5.1 and 5.2) pointing to certain administrative procedures prescribed by the Directive that would constitute an unfair burden for its domestic operators.

The present dispute, for which the Panel is expected to issue its final report by the end of 2015, was enriched with the filing of a third request for consultation by Argentina and Indonesia. The claim, this time, relates to the anti-dumping measures adopted by the Union against the biodiesel supposedly exported at a price inferior to its normal value. In particular, Argentina condemned the investigation methodology used by the European Commission following the complaint lodged in 2012 by the European Biodiesel Board concerning the determination of provisional and definitive anti-dumping measures imposed on imports of biodiesel by means of Regulation n. 490/2013 of the Commission and of Implementing Regulation n. 1194/2013 of the Council. The issue deals essentially with the inconsistency of such measures with Article VI, paragraphs 1 and 2 of GATT 1994, and with a plurality of ADA provisions, focusing on the calculation made to determine the minimal threshold of the tariff to be imposed in order to nullify the adverse effects caused by dumped imports. If, on the one hand, the EU and its biodiesel producers are worried by a threat of unfair competition and the measures adopted were justified on economic and industrial grounds, on the other hand, the reliance upon trade defence instruments is unlikely to appease with the tenets of the 2001 Doha Development Agenda and, above all, with the goal of liberalizing the market for the so-called "environmental goods and services" (EGSs). This could easily be one of the major challenges that the WTO judiciary would deal with in order to settle the dispute at issue.

4.4 Lessons from WTO Cases, Alternative Approaches and Prospects Ahead

The above cases show that the absence of methodological alternatives for individuating the market, in determining the existence of a subsidy runs the risk of limiting the scrutiny capacity of WTO dispute settlement bodies. Alternative comparisons could therefore be carried out, as suggested by the Panel on the basis of the remuneration of enterprises similar to the subsidized ones (i.e. also supplying goods and services to public authorities). By definition, this reference standard cannot be deduced by the market features of a certain country, recalled by ASCM Article 14.1 (d), but necessarily requires the adoption of a wider notion of market, enabling the search for comparable investment projects (Piérola 2013, pp. 293–296).

The discriminatory character of Canada's measures was not deemed justifiable on the basis of the public procurement exception under GATT 1994 Article III:8. In line of theory, however, the AB decisions could have been different. For instance, one could ask if and to what extent measures in breach of the ASCM discipline are justifiable under the general exceptions *ex* GATT 1994 by virtue of the positive environmental externalities stemming from the operation of the renewable energy sector, in the absence of specific rules on the trade of EGSs and given the lack of general exceptions to the ASCM Agreement. Hypothesizing the applicability of Article XX of the GATT 1994 to similar circumstances would necessarily entail the recognition of a hierarchy among WTO Agreements, with the GATT functioning as a *lex generalis* which, as such, would fill the void left by the lack of exceptions under the ASCM, in turn, describable as a *lex specialis* with a more limited and specialized scope of application (Farah and Cima 2013, pp. 707–740).

In the *EU – Biofuels* case the WTO dispute settlement body will analyse multiple international trade law issues and ultimately will prove how and to which extent the WTO system is committed to sustainable development (Lydgate 2012, pp. 157–180). Besides the issues raised by the extraterritorial application of unilateral domestic measures affecting international trade, it will also tackle issues relating to technical barriers and the determination of the "likeness" concept (between domestic and imported goods) under the Agreement on Technical Barriers to Trade (TBT Agreement). The space that within said determination can be reserved to measures affecting "process and production methods" (PPMs) unrelated to products might be addressed even though measures influencing PPMs not relating to the characteristics of a given product ("non-product related") were generally viewed as inconsistent with WTO core principles. However, the sustainability criteria might be portrayed as "climate policy-related PPMs" and as such their lawfulness might also be scrutinized in light of the general exceptions *ex* GATT Article XX (b) and (g), respectively regarding measures necessary to protect human, animal or plant life or health and those relating to the conservation of exhaustible natural resources.

Finally, it must be underlined, when challenging the WTO consistency of the measures enacted by some EU Member States to support European operators, Argentina pointed to the breach of ASCM obligations because Annex I to the Agreement on Agriculture (AoA), which defines its scope by relying on Chapters from 1 to 24 of the Harmonized System for the classification of products, adopted by the World Customs Organization (WCO) in 1988, does not encompass biodiesel. The case of biodiesel sheds light on the urgency of redefining, in the context of the wider debate on EGSs (Marceau 2010, pp. 83–93), the relationship between certain multilateral trade agreements, and the interaction between the trade and the climate change regimes (Low et al. 2012). Given that biodiesel is listed as an industrial product unlike ethanol, for instance, measures supporting its production are not admitted on the basis of an exception to the ASCM nor can fall amongst those exempted by reduction commitments ("green box") and allowed by the AoA pursuant to the requirements set under Annex II of said Agreement (Josling and Blandford 2009, pp. 530–568).

5 Conclusions

Spreading the use of products and technologies for the generation of energy from renewable sources is crucial to the materialization of sustainable development. Nevertheless, States have historically been slow in embracing *consensus* on renewable energy options as shown by many soft law developments in the framework of the UN.

Proofs of a reversal of this trend, apparently signalling the advent of a better future for renewable energy, are being provided by the epistemic activities undertaken by IRENA encouraging the dissemination of information and knowledge and, to a certain extent, by a few proposals on renewables put forward during the last round of climate negotiations.

However, it must be noted, without the inception of international legal instruments setting a specific discipline for renewable energy generation either within the climate change regime or in the context of the WTO, States will need to keep shaping their energy policies without departing from the binding rules of international trade law, as proven by recent and ongoing disputes. Any international legal development fostering renewable energy generation should integrate these concerns to the greatest extent possible.

References

Brunnée J, Streck C (2013) The UNFCCC as a negotiation forum: towards common but more differentiated responsibilities. Clim Policy 13:589–607
Center on Energy, Climate Change and Sustainable Development (2015) CDM projects by type, retrieved July 24 2015 from www.cdmpipeline.org)
De Cendra de Larragán J (2012) The future of international climate change law: a scenario-based perspective. Clim Policy 12:6–27
De Lucia V (2014) Competing narratives and complex genealogies: the ecosystem approach in international environmental law. J Environ Law 27(1):91–117
Del Rio P (2007) Encouraging the implementation of small renewable electricity CDM projects: an economic analysis of different options. Renew Sustainable Energy Rev 11:1361–1378
EU (2009) Directive 2009/28/EC of the European Parliament and the Council of 23 April 2009 on the promotion of the use of energy from renewable sources, OJ L 140, 5.6.2009, pp 16–62
Farah P, Cima E (2013) Energy Trade and the WTO: implications for renewable energy and the OPEC cartel. J Int Econ Law 707–740
Ferrey S (2010) The failure of the international global warming regulation to promote needed renewable energy. Environ Affairs 36:67–126
Josling T, Blandford D (2009) Biofuels subsidies and the green box. In: Meléndez-Ortiz R, Bellman C, Hepburn J (eds) Agricultural subsidies in the WTO green box. Cambridge University Press, Cambridge
JPoI (2002) Johannesburg plan of implementation, in UN report on the world summit on sustainable development, Johannesburg, South Africa, 26 Aug–4 Sept 2002, A/CONF/199/20, 6–72, Retrieved 24 July 2015, from www.iisd.ca/wssd/download%20files/wehab_energy.pdf
Kotzé LJ (2014) Rethinking global environmental law in the anthropocene. J Energy Nat Resour Law 32(2):121–156
Kulovesi K, Morgera E, Muñoz M (2009) Environmental integration and multifaceted international dimensions of EU Law: unpackaging the EU's 2009 Climate and Energy Package. Common Mark Law Rev 48:829–891
Low P, Marceau G, Reinaud J (2012) The interface between the trade and climate change regimes: scoping the issues. J World Trade 3:485–544
Lydgate E (2012) Biofuels, sustainability and trade-related regulatory chill. J Int Econ Law 157–180
Marceau G (2010) The WTO in the emerging energy governance debate. Glob Trade Customs J 83–93
Meyer T (2013) Epistemic institutions and epistemic cooperation in international environmental governance. Transnatl Environ Law 15–44
Pallemaerts M (1996) The future of environmental regulation: international environmental law in the age of sustainable development: a critical assessment of the UNCED process. J Law Commer 15:623–676
Piérola F (2013) The question of the benefit. Glob Trade Customs J 293–296
Redgwell C (2008) International legal responses to the challenges of a lower-carbon future: climate change, carbon capture and storage, and biofuels. In: Zillman DN, Redgwell C, Omorogbe Y, Barrera-Hernández LK (eds) Beyond the carbon economy: energy law in transition. Oxford University Press, Oxford, pp 85–108
IRENA (2009) Statute of the international renewable energy agency, Bonn (Germany), 26 Jan 2009, in force 8 July 2009. Retrieved 24 July 2015. www.irena.org
IRENA (2012) Decisions on the work programme and budget for 2012. Retrieved 24 July 2015, from www.irena.org
IRENA (2013) Decisions on the work programme and budget for 2013. Retrieved 24 July 2015, from www.irena.org
IRENA (2015) Global atlas on solar and wind power. Retrieved 24 July 2015, from www.globalatlas.irena.org

Scheer H (2007) Towards a solar proliferation treaty. Leaving the global atomic trap. In: Stockinger H, Van Dyke E (eds) Updating international nuclear law. Intersentia, Salzburg, pp 306–310

Viñuales JE (2013) The rise and fall of sustainable development. Rev Eur Commun Int Environ Law 22:3–13

UNFCCC (2014) Report of the conference of the parties on its twentieth session, held in Lima from 1 to 14 December 2014, "Lima Call For Climate Action", Decision 1/COP20, FCCC/CP/2014/10/Add.1 of 5 Feb 2015

UN General Assembly (1987) UN GA Resolution 427 (42), 4 Aug 1987, Report of the world commission on environment and development (Brundtland report), "Our Common Future"

UN General Assembly (2012) UN GA Resolution 288 (66), 11 Sept 2012, Report of the UN summit on sustainable development, "The Future We Want"

UNCED (1992) United nations conference on environment and development, the 1992 rio declaration on environment and development, UN Doc. A/Conf.151/5/Rev.1, of 14 June 1992

UNCSD (2007) United nations commission on sustainable development, E/CN.17/2007/15, Decision 9/1 on "Energy for sustainable development", para. 11, Retrieved 24 July 2015, un.org/ga/search/view_doc.asp?symbol=E/CN.17/2007/15(SUPP)&Lang=E

Uperlainen J, Van de Graaf T (2013) The international renewable energy agency, a success story in institutional innovation? Int Environ Agreem Politics Law Econ 1–19

WTO (2012a) WT/DS443/1, G/TRIMS/D/30, G/L/994, European Union and a member state—certain measures concerning the importation of biodiesels, request for consultation of 23 Aug 2012

WTO (2012b) WT/DS412/AB/R, WT/DS412/R, Canada—certain measures affecting the renewable energy generation sector, Canada—measures relating to the feed-in tariff program, reports of the panel of 19 Dec 2012

WTO (2013a) WT/DS426/AB/R, WT/DS426/R, Canada—certain measures affecting the renewable energy generation sector, Canada—Measures relating to the feed-in tariff program, reports of the appellate body of 6 May 2013

WTO (2013b) WT/DS459/1, G/L/1027, G/SCM/D97/1, G/TRIMS/D/36, G/TBT/D/44, European Union and Certain Member States—certain measures on the importation and marketing of biodiesel and measures supporting the biodiesel industry, request for consultation of 23 May 2013

WTO (2014) WT/DS473/1, G/L/1062, G/ADP/D101/1, European Union—anti-dumping measures on biodiesel from Argentina, request of consultation of 8 Jan 2014

The Fair and Equitable Treatment Standard and the Revocation of Feed in Tariffs—Foreign Renewable Energy Investments in Crisis-Struck Spain

Thomas Dromgool and Daniel Ybarra Enguix

Abstract This contribution explores a highly topical issue in international investment law—the protection of foreign investors' legitimate expectations through the Fair and Equitable Treatment standard (FET) in case of a repeal of renewable energy support schemes. On the grounds of the Spanish case of disruptive cuts, particularly regarding Feed in Tariff regulation supporting photovoltaic energy since 2008, a possible violation of legitimate expectations is being assessed. The investors may rely on the stability of the Spanish *régimen especial*, despite their own conduct and the State's right to regulate. Even in times of crisis the State may not justify changes in the regulation to the point of stripping away the very raison d'être of the initial investment. Such measures could constitute a breach of legitimate expectations. Regarding the current global trend of revoking renewable energy support schemes, the present case may initialize a cascade effect for a number of claims.

Keywords International investment law · Fair and equitable treatment · Legitimate expectations · Feed-in tariffs · Spanish FiT regime

1 Introduction

It is a seminal issue for the promotion of Sustainable Development—the search for convergence between investment law and Sustainable Development policies, or more specifically renewable energy policies. If renewable energy investment

T. Dromgool (✉)
Higher Regional Court of Berlin, Berlin, Germany
e-mail: thomas.dromgool@gmx.de

T. Dromgool
University of Freiburg, Freiburg im Breisgau, Germany

D.Y. Enguix
International Business Department, Spanish Engineering Company Specialised in Energy Efficiency and Renewable Energy, Alcoy, Spain
e-mail: dybarraenguix@gmail.com

V. Mauerhofer (ed.), *Legal Aspects of Sustainable Development*,
DOI 10.1007/978-3-319-26021-1_20

incentives are to be taken seriously, States have to assure a stable investment climate, as well as they have to consider the public interest. Renewable energy has become the buzzword behind the global climate change initiative (e.g. United Nations 1997, Art. 2[1]a] [iv]) and new national energy, industrial and environmental policies (e.g. Spanish national plan for the promotion of renewable energy 2000–2010 (Spain 1999), Spanish national renewable energy plan 2005–2010 (Spain 2005) and German Renewable Energy Act 2014). Furthermore, and as confirmed by the energy and climate package (hereinafter 2020 package) of the European Commission (2009),[1] renewable energy promotion is at the very heart of the European Union's environmental policy. Notwithstanding this global trend, during the current global financial crisis many Governments have revoked their commitments to renewable energy support and have undertaken major changes in their regulation. This has triggered a whole wave of claims in investor-state dispute settlement in a number of countries, particularly with international renewable energy investors in Spain, the Czech Republic and Italy having already made a start in filing for arbitration (Jha 2012; Peterson 2013). The largest claims are the ones in the Spanish case, which are probably the most drastic proceedings, especially as far as photovoltaic (hereinafter PV) energy is concerned. Since the Spanish Government started reducing incentives for renewable energy and as of July 26, 2015, 20 arbitration claims have been brought against Spain under the Energy Charter Treaty (hereinafter ECT), all of which involve solar energy (Energy Charter Secretariat 2015).[2] PV investors are likely to rely on the Fair and Equitable Treatment Standard (hereinafter FET) contained in Article 10 of the ECT (Energy Charter Secretariat 1984). The main question will be whether PV investors can successfully claim a breach of their legitimate expectations, due to cuts, including retroactive changes, in incentive programmes for PV plants the Spanish regulator has introduced since 2008. Particularly important for assessing a potential breach of

[1]Enacted in 2009, the 2020 package consists of a set of binding legislation establishing three objectives to be reached by all Member States by 2020: to reduce greenhouse gas emissions by 20 %, to raise the energy consumption produced from renewable resources to 20 % and to improve the energy efficiency by 20 %.

[2]These claims are (in chronological order): *The PV Investors v Spain, Charanne (the Netherlands) and Construction Investments (Luxembourg) v Spain, Isolux Infrastructure Netherlands B.V. v Spain, CSP Equity.*

Investment S.à.r.l. v Spain, RREEF Infrastructure (G.P.) Limited and RREEF Pan-European Infrastructure Two Lux S.à.r.l. v Spain, Antin Infrastructure Services Luxembourg S.à.r.l. and Antin Energia Termosolar B.V. v Spain, Eiser Infrastructure Limited and Energia Solar Luxembourg S.à.r.l. v Spain, Masdar Solar & Wind Cooperatief UA v Spain, NextEra Energy Global Holdings B.V. and NextEra Energy Spain Holdings B.V. v Spain, InfraRed Environmental Infrastructure GP ltd. et al. *v Spain, RENERGY S.à.r.l. v Spain, RWE Innogy GmbH and RWE Innogy Aersa S.A.U. v Spain, Stadtwerke München GmbH, RWE Innogy GmbH* et al. *v Spain, STEAG GmbH v Spain, 9REN Holding S.a.r.l v. Spain, BayWa r.e. Renewable Energy GmbH and BayWa r.e. Asset Holding GmbH v. Spain, Cube Infrastructure Fund SICAV and others v. Spain, Matthias Kruck and others v. Spain, KS Invest GmbH and TLS Invest GmbH v. Spain, JGC Corporation. v. Spain.* Cases No. 31, 34, 38, 39, 40, 41, 43, 50, 52, 53, 58, 61, 62, 63, 67, 68, 69, 70, 71 and 72, respectively.

legitimate expectations of PV investors in this case is the legal nature of the normative framework which the Spanish legislator revoked, being feed-in tariffs (hereinafter FiTs), which are incentives guaranteeing the payment of a fixed amount of money per unit of electricity supplied to the grid for a certain period of time (United Nations Environment Programme 2011, p. 226, European Commission 2005, points 4–5). Furthermore, the PV investors' own conduct involving the duty to reasonably assess the investment risks of the host country, as well as the Host State's right to regulate are limitations to the legitimate expectations of high relevance for a possible breach in the Spanish case. Interestingly, the central argument of the Spanish Government to justify the cuts has not so much been the financial crisis that the country faces today or the austerity measures imposed by Brussels, but the necessity to correct the country's energy tariff deficit (*déficit tarifario*) in order to ensure the financial stability of the Spanish energy system (Soria 2014).

The claims brought against Spain will serve as case study, by which this article will explore an increasingly important but until now uncommon issue of international investment law—the protection of foreign investors' legitimate expectations in a case of FiT revocation. To this end, the article will be after this introduction (Part 1) organized as follows. Part 2 will consider the background in which the Spanish FiT regime was set up and examine the regulations that have led to the rise and fall of support schemes for PV energy in the country. Part 3 will briefly introduce the two most important standards of investment protection for cases of renewable energy incentive revocation and will specifically assess the content and scope of FET, with special regard to the legitimate expectations of PV investors in Spain. In Part 4 the findings determining the chances of a successful claim based on a violation of FET in the Spanish case will be weighed and summed up.

2 Context and Evolution of the Photovoltaic Energy's Economic Regime in Spain

2.1 Introduction to the Spanish Energy Sector—Putting All Its Eggs in the Renewable Energy Basket

During the past decades, the Spanish energy sector has revealed three main and repeated features: a high level of dependency on energy imports, a heavy reliance on fossil fuels and a poor record on energy efficiency (National renewable energy plan 2011–2020 (Spain 2011, p. 10). Different than in other EU Member States, the energy consumption structure in Spain is dominated by imported oil products. This, added to the meagre contribution of indigenous resources to the national energy mix, is the main reason for the country's high rate of energy import dependence (close to 80 %) when compared to the average rate in the EU (54 %) (Eurostat 2013). Moreover, as in many other EU economies, the economic growth experienced by Spain before 2008 led to an increase in the country's energy consumption.

These features and incidents have made the Spanish energy regulator pursue three prime goals during the past twenty years, namely, (1) to secure energy supply (2) in a way that is affordable and (3) respectful of the environment (Spain 1997). The EU regulation has also played a major role in reshaping the energy sector in Spain, especially in competition and environmental issues. With regard to the latter, the 2020 package's objectives of reducing greenhouse gas emissions, raising the energy consumption produced from renewables and improving the energy efficiency (European Commission 2009) have been high on the agenda of the Spanish regulator.

In order to pursue these national and European objectives, renewable energy had to be taken forward. In fact, the progress of green energy sources in Spain in the last twenty years has been remarkable. While in the eighties energy production in Spain was mainly based on coal, in the mid-nineties nuclear energy became the most important domestic source and today renewable energies represent the prime energetic asset of the country (Sevilla et al. 2013, p. 37; Spanish Power Transmission Company 2014, pp. 5 and 10). In particular, two sources advanced considerably over the past decades in Spain in terms of installed capacity, namely wind and solar energy (Agosti and Padilla 2010, p. 518). Wind power started expanding first, however, since 2005 PV energy has been the fastest-growing energy source in the country, mainly because of the high level of public support for installations using this technology (Agosti and Padilla 2010, p. 521). Along with the national and European goals explained above, such incentives were motivated by the aim of the Spanish Administration to position Spain among the world leaders in PV energy, both in installed capacity and technology production [National renewable energy plan 2005–2010 (Spain 2005, p. 157)].

Notwithstanding the efforts put in promoting a more sustainable energy system, it is most likely that Spain will not meet its 2020 climate and energy targets in terms of greenhouse gas emissions reduction and increase of renewables' shares in total energy consumption (European Environment Agency 2014, p. 10). Furthermore renewable energies have had a significant impact on the sharp rise of energy prices in Spain in the last years (Sallé 2012, pp. 104–105). This is due to the large number of energy sources benefiting from subsidies and the above average support granted to renewable energies when compared to other EU countries (Sallé 2012, p. 105). What is more, the Spanish Government identifies the support for renewables as a main cause of the national energy tariff deficit (*déficit tarifario*) [e.g. Press conference following the Council of Ministers held on 27 January 2012 (Spain 2012a), first intervention of Minister of Industry Soria; EFE Economía 2013], that is the difference between the sum owed by the Administration to electrical companies and the amount the Administration receives from consumers.

These developments, which coincided in time with a severe financial crisis in the country and strict austerity measures imposed by the EU, prompted the Spanish Government to introduce major changes to the renewable energy regulation. These changes have ultimately led to serious damages for low-carbon investors. In order to provide enough detail to draw an informed opinion on whether Spain should be held accountable for the losses suffered by foreign PV energy investors, the focus of the next section is on the evolution of the PV energy's economic regime in Spain. However, before reaching any conclusion, political and economic issues surrounding the decisions to cut incentive programmes also need to be taken into consideration. While the section on the Host State's right to regulate will address the questions of the crisis and the austerity measures imposed by Brussels as possible arguments to limit Spain's responsibilities, it is important here to provide further detail on the tariff deficit.

Initially, the tariff deficit was designed to address temporary and small imbalances between revenues and costs in the Spanish electricity system. These occurred because electricity prices in Spain are revised only once a year according to cost predictions (at the end of each year for the next year) and the latter are most often inaccurate. In order to solve this problem, and motivated by the desire to maintain low energy prices for consumers, the Spanish Administration introduced a system for financing the gap through the contribution of five energy companies (Hidrocantábrico, Endesa, EON, Iberdrola and Gas Natural Fenosa) (Sallé 2012, p. 107). The tariff deficit first arose in the year 2000, reached the sum of € 5.6 billion in 2012 (Spanish Energy Commission 2013, p. 3) and today seems to be under control, given that it was reduced to € 2.9 billion in October 2014 (Spanish National Stock Market Commission 2014, p. 3).

2.2 *The Rise and Fall of the Support Regime for Photovoltaic Energy in Spain*

2.2.1 The Rise: The Development of Support Schemes for Photovoltaic Energy in Spain

In Spain, the generation of electricity from renewable sources has been regulated through statutes, such as laws (*leyes*) and royal decree laws (*reales decretos ley*), as well as through regulations, for instance national plans (*planes nacionales*), royal decrees (*reales decretos*) and ministry orders (*órdenes ministeriales*).[3] The 1997 Electricity Sector Law, which provided the basis for the liberalization of the electricity market in the country, initiated the process of regulating the special regime (*régimen especial*) (Spain 1997, Title IV, Chap. II), i.e. the set of rules that apply exclusively to renewable energies and that define their economic regime. During the period 1998–2007 two National Renewable Energy Plans established the goals to be reached for each renewable energy, and several decrees developed the regime applicable for each technology. Royal Decrees (hereinafter R.D.)

[3]In Spain, laws and royal decree laws have the same position in the hierarchy of legal sources (statues) but differ in procedural terms. Laws are initiated and approved by the Congress. Royal decree laws are issued by the Government for extraordinary and urgent reasons, but have to be approved by the Congress. Regulations rank below statues, serve to complete, specify and implement statues and the Administration controls their whole development process.

Table 1 Evolution of tariffs and premiums in the special regime (adapted from Agosti and Padilla 2010 p. 526)

Technology	R.D. 2818/1998	R.D. 436/2004		R.D. 661/2007	
	1998	2004		2007	
	Premium (€/MWh)	Tariff (€/MWh)	Premium (€/MWh)	Tariff (€/MWh)	Premium (€/MWh)
PV < 100 kWp	60.00	414.4	n/a	440.0	n/a
PV > 100 kWp	30.00	216.2	187.4	229-417	n/a
Solar thermal	20.00	216.2	187.4	269.0	254.0
Wind energy	31.6	64.9	36.0	73.0	29.0
Hydro power < 10 MW	32.8	64.9	36.0	78.0	25.0
Hydro power > 10 MW	0–35.8	57.7–64.9	28.8–36.0	66–78	13.0
Geothermal	32.8	64.9	36.0	69.0	38.0
Biomass	28.0	64.9	36.0	107–158	61–115

For the 2007 regime only average prices given

2818/1998, 436/2004 and 661/2007 were the key royal decrees for the sharp rise in PV energy subsidies in that period.

In line with the 1997 Electricity Sector Law, R.D. 2818/1998 classified renewable energy plants according to the technology used and the installed capacity [Electricity production through renewable sources, cogeneration and waste (Spain 1998), art. 2]. Article 26 provided that the remuneration mechanism for plants will consist of a fixed premium being placed on the electricity market price dependent upon the type of the plant, and articles 27–31 specified the premium applicable to each type of plant (see details in Table 1). According to the decree, the regulator had to reassess and adjust the premiums every four years (Electricity production through renewable sources, cogeneration and waste 1998 art. 32).

One year later, the Spanish Government approved the National Plan for the Promotion of Renewable Energy 2000–2010. As part of the strategy for attaining the three main national energy objectives (energy import independence, affordable prices and an environment-friendly system),[4] the Plan set investment objectives for increasing the share of renewable energy in primary energy consumption to 12 % by 2010 (Spain 1999, p. 1). Furthermore, the Plan established objectives of installed capacity for each renewable energy source and included information on the technological and environmental aspects, the investment and construction costs, the barriers and possible support measures associated with each source (Spain 1999, Chap. IV). When tackling the question of incentive measures for PV energy, the Plan stated:

[4]See Part 2.1.

> On the premises of the high solar radiation in Spain, the favourable environmental effects of PV energy and the particular characteristics of this source, it is necessary to undertake measures and incentives to increase the presence of this energy in the territory of the State. The measures proposed are designed to […] conduct an integrated plan that strengthens the Spanish photovoltaic sector […] (Spain 1999, p. 116).[5]

R.D. 436/2004 included important modifications into the economic regime of renewables with a view to facilitating the achievement of the 2010 renewable energy objectives [Methodology for updating and structuring the special regime (Spain 2004)]. It allowed the majority of renewable energy producers (depending on the technology) to choose between two options for selling their energy: (1) to sell the energy directly on the national market, on the futures market, or through a bilateral contract, each time at the market price plus a premium; (2) to sell the energy to distribution companies at a fixed tariff (Spain 2004, art. 22). As far as the PV energy is concerned, this decree set a considerably higher remuneration for small plants, but only allowed their owners to sell the energy through FiTs (Spain 2004, art. 33) (see Table 1).

Having realized that the target set for 2010 of raising the share of renewables in primary energy consumption to 12 % was not going to be achieved, in August 2005 the Spanish Government adopted the National Renewable Energy Plan 2005–2010. This new text did not only reaffirm the 2010 target, but it also recommended increasing renewable incentives and fixed one new objective: by 2010 electricity production from renewable sources in the country had to increase to 29.4 % (Spain 2005, pp. 7 and 9). In what constituted an even more exhaustive report than the 2000–2010 National Plan, this Plan presented for each type of renewable energy (1) a comparative analysis of installed capacity across EU countries, (2) a summary on its evolution and the state of the technology in Spain, (3) a new target in terms of installed capacity, (4) the measures needed to meet this target, and (5) possible lines for technological innovation (Spain 2005, Chap. III). Furthermore, an entire chapter was dedicated to the funding of the Plan, which included different technical and financial hypotheses based on the specific features of each renewable source, a detailed evaluation of the investment envisaged, the nature of this investment and the public aid needed to meet the targets (Spain 2005, Chap. IV). With reference to PV energy, while the Plan recognized the development of this energy source in the country, it nevertheless acknowledged that its progress was insufficient (Spain 2005, p. 157). In this context, the Plan proposed a series of measures with a view to raising the share of PV energy in the national energy mix and to furthering the national PV energy industry.

Following the recommendations of the 2005–2010 Plan, R.D. 661/2007 introduced two main changes to the special regime. First, it divided the technologies into new categories [Law regulating the activity of electricity production under the special regime (Spain 2007) art. 2] and set the premiums and/or tariffs according to each category, the installed capacity and the age of the installation (Spain 2007,

[5]Translation by the authors.

Tables 1, 2 and 3). Second, it provided higher incentives for certain technologies, particularly for PV energy (Spain 2007, cf. Tables 1, 2 and 3).[6] Despite the fact that the regulator fixed higher tariffs and premiums for the first 15–25 years of each installation (depending on the technology), no lifetime caps on production were imposed (Spain 2007, Tables 1, 2 and 3). On the other hand, this R.D. stipulated that once the 85 % of the objectives set for a technology in the National Renewable Energy Plan 2005–2010 were reached, the regulator would fix a date for changing the tariffs and/or premiums applicable to that technology (Spain 2007, art. 22).

Due to the incentives introduced by R.D. 661/2007 for small PV installations (€ 440.0/MWh, i.e. 567 % above the reference average tariff for that year), in May 2008 the total installed capacity of PV energy in Spain reached 1000 MW and in October of the same year exceeded 2200 MW (Sevilla et al. 2013, p. 44). The 2005–2010 Plan had fixed the target of 400 MW for the year 2010. This boom of PV installations, seen as financial investment products by national and international investors, led to the adoption of R.D. 1578/2008.

2.2.2 The Fall: Cuts and Retroactive Changes in the Spanish Support Schemes for Photovoltaic Energy

R.D. 1578/2008 can be seen as the turning point in Spain's approach to renewable energy support schemes. For the first time since 1997, Spain reduced incentives for renewable energy sources. Having surpassed the target set in the 2005–2010 National Plan for installed capacity of PV energy, the regulator decided to lower PV energy tariffs with a view to not discouraging technological innovation [Remuneration for photovoltaic installations created after the deadline set in the R. D. 661/2007 (Spain 2008)]. This cut only affected PV energy plants installed after 29 September 2008, applying the R.D. 661/2007 regime to all plants built before that date (Spain 2008, art. 2).

Nevertheless, since the approval of R.D. 1578/2008, the regulator has adopted countless regulations that introduced drastic changes to the economic regime of renewable energy sources, also for plants that were operating before 29 September 2008. The Preambles of those regulations (e.g. Preambles of R.D.L. 14/2010, R.D. L. 2/2013, R.D.L. 9/2013 and R.D. 413/2014 in this section) and the statements made by the Government (e.g. Soria 2014) refer to two interrelated reasons for such changes: the necessity to reduce the tariff deficit and to guarantee the financial stability of the electricity system. Regulators must be in the position to control the effects and costs of their national support schemes (European Parliament and Council 2009, 25th recital in the Preamble); the pertinent question here is, how far regulatory changes with retroactive effects can go. While Part 3 of this contribution will examine in detail that question through the lense of international investment law, the focus of this section is on the retroactive changes in PV energy incentive

[6]PV energy tariffs defined in Table 3, under energy type b.1.

programmes, which are central for the claims of international PV investors against Spain. It is noteworthy saying that retroactivity refers in this section to rules applying in the future to regimes introduced in the past.[7]

Three years after the adoption of R.D. 661/2007, R.D. 1565/2010 aimed at adjusting and correcting the special regime [Law regulating and modifying certain aspects relating to the production of energy production under the special regime (Spain 2010a)]. Among other changes, the decree removed administrative barriers for new installations and added clarifications on existing regulation (Spain 2010a, p. 97429). However, the most polemical aspect of this decree was the elimination of all the benefits for PV projects operating under the R.D. 661/2007 regime after the twenty-fifth year (Spain 2010a, art. 1.10). As explained in the previous section, R. D. 661/2007 did not establish lifetime caps for PV energy installations.

Despite the fact that the Government had approved in 2009 a new regulation for tackling the tariff deficit [Law approving extraordinary measures in the energy sector (Spain 2009)], in 2010 the regulator passed new urgent measures in order to correct it. The reason for this was that the forecasts which accompanied the 2009 regulation did not materialize and the tariff deficit continued rising. Royal Decree Law (hereinafter R.D.L.) 14/2010 included new measures so that all players in the energy industry contributed with an additional and shared effort to the deficit reduction [Law establishing urgent measures to correct the tariff deficit (Spain 2010b)]. Hence, the R.D.L. introduced a new toll for accessing the transport and distribution networks for all electricity generation companies, obliged the companies operating under the ordinary regime to finance the Energy Savings and Efficiency Plans 2004–2012 and set two limitations on the operating hours of PV installations operating under the R. D. 661/2007 regime (Spain 2010b, pp. 106387–106388). The First Additional Disposition fixed production-hour caps based on five climatic solar zones. Additionally, the Second Transitory Disposition introduced further hour production restrictions for the period 27 December 2010–31 December 2013.

In 2012 the Spanish Parliament approved Law 15/2012, which sought internalizing the environmental costs linked to energy production to energy producers [Fiscal reform in order to ensure the sustainability of the energy system (Spain 2012c)]. This bill imposed seven new taxes on the energy sector (Spain 2012c, p. 88081), including a tax on electricity production with a single tax rate of 7 % for all power plants (Spain 2012c, cf. art. 1, art. 6.1 and art. 8). It is striking that, even if this bill aimed at promoting Sustainable Development, the same tax was imposed on fossil-fuel and renewable energy plants.

Notwithstanding the efforts of the Government, the tariff deficit continued growing and reached the sum of € 5.6 billion at the end of 2012 (Spanish Energy Commission 2013, p. 3). In an attempt to alleviate the ever-increasing sum owed by the Administration to electric companies and to avoid raising energy prices, R.D.L.

[7]The Spanish Supreme Court refers to this type of retroactivity as "improper retroactivity" and considers it permissible because it only affects expectations, e.g. Spanish Supreme Court (2012) fifth legal basis.

2/2013 introduced a new system for updating the prices of energy products [Urgent measures in the electricity system and in the financial industry (Spain 2013a)]. Since its entry into force, energy prices adjusted annually for inflation are not based on the Consumer Price Index, but use instead a special index that does not include non-elaborated food products and energy products (Spain 2013a, art. 1). This measure further harmed (renewable) energy producers, because these two types of products usually present the biggest changes occurring in the price level of consumer goods and services throughout a year.

The period July 2013–June 2014 might have witnessed the definitive step from the Spanish Government towards eliminating the tariff deficit and adjusting the remuneration for renewable energy sources. This last stage was initiated on 12 July 2013, with the approval by the Government of R.D.L. 9/2013, and ended on 16 June 2014, once the Government passed the new remuneration mechanism for renewables in R.D. 413/2014 and Ministerial Order IET/1045/2014.

R.D.L. 9/2013 had three major aims: (1) to reduce extra costs arising from the tariff deficit accumulated during the first semester of 2013; (2) to set the basis of a new regulatory framework that would ensure the financial stability of the electricity system; (3) to guarantee energy supply at the lowest cost possible [Urgent measures to guarantee the financial stability of the electricity system (Spain 2013b)]. For tackling the tariff deficit's extra costs of 2013, the regulator sought to implement "balanced, proportionate and wide-raging measures" across the different stakeholders in the energy sector (Spain 2013b, p. 52110). In the end, energy producers operating under the special regime and energy distribution companies assumed the bulk of those costs (Noceda 2013). Drastic modifications to the economic regime of renewable energy production included: First, R.D. 661/2007 and R.D. 1578/2008 were repealed (Spain 2013b, sole Repeal Provision), what amounted to a de facto revocation of the special regime. Second, even if the specific remuneration for each technology was to be set in subsequent regulations, this R.D.L. revealed that instead of tariffs, the new remuneration of renewable energy plants will consist in the income from the sale of the energy produced paid at market price plus, if applicable, a complementary retribution aimed at ensuring a reasonable profitability (Spain 2013b, art. 1.2). This complementary retribution could include the remuneration for investment and operating costs which are not covered by the market price (Spain 2013b, art. 1.2). In connection with the remuneration for installations operating between 14 July 2013 (date of entry into effect of the decree) and the date of the approval of the definite regime, the reformed 661/2007 regime continued applying in an interim basis, but the sum received for the energy produced between those dates had to be credited against the remuneration to be received under the new regime (Spain 2013b, Third transitory disposition).

The new regime for renewable energy plants was finally unveiled in June 2014. R.D. 413/2014 set the methodology to be used for calculating the remuneration of plants [Law regulating the generation of electricity using renewables, cogeneration, and waste (Spain 2014a)]. According to article 13 of the decree:

1. A Ministerial Order will establish a classification of installation types according to the kind of technology, the installed capacity, the age, the electrical system, and any other criteria deemed necessary for implementing the remuneration system (…).
2. The remuneration of each installation type will be calculated taking into account a fixed set of criteria, (…). The most relevant criteria include:

   a) Remuneration on the investment
   b) Remuneration on the operation
   c) Regulatory lifetime
   d) The number of minimum and maximum operating hours
   e) Operating threshold
   f) The average market price (…).

Ten days after the approval of R.D. 413/2014, the Government passed the Ministerial Order IET/1045/2014, which detailed in more than 1700 pages the specific remuneration parameters applicable to standard renewable facilities [Ministerial Order validating the remunerative parameters for installations using renewables, cogeneration, and waste (Spain 2014b)]. Annex I listed the installation types existing under the new regime, showed how they corresponded to the categories in the 661/2007 regime and established a code for each installation type. While the 661/2007 regime classified PV installations in six categories, the new regulation includes 91 categories (cf. art. 2 and Table 3 in R.D. 661/207 with Annex I in Ministerial Order IET/1045/2014). Annex II established the concrete valuation of the remuneration parameters for installation types with a right to a feed-in tariff prior to 14 July 2013, differentiating between the parameters applicable in 2013 from those applicable in 2014-2016. Remuneration parameters will be revised every three years (Spain 2014b, p. 46431).

The radical overhaul of the energy system has certainly contributed towards the reduction of the tariff deficit and the financial stability of the system, as evidenced by the fact that the tariff deficit was reduced to € 2.9 billion in October 2014 (Spanish National Stock Market Commission 2014, p. 3). Moreover, this has been achieved amidst the most severe financial crisis experienced by Spain in the last century. However, such changes have had devastating effects on the country's renewable energy industry,[8] have put at stake the Spain's 2020 climate and energy targets (European Environment Agency 2014, p. 10) and have led to massive losses in revenue for PV system owners. Due to the numerous remuneration parameters introduced in the Ministerial Order IET/1045/2014, conclusions cannot be generalised to all renewable energy sources. But if we take the small PV energy installations built before September 2008 as reference, their remuneration has been cut in more than half in a period of only seven years (cf. art. 2 and Table 3 in R.D. 661/207 with Annexes I and II in Ministerial Order IET/1045/2014). Previously, in the time frame 2010–2013, PV investors had already suffered important shortfalls in

[8]Since 2013, major Spanish PV panels producers such as Isofotón, T-Solar, Siliken and 3S Soluciones have entered into state of bankruptcy, liquidation and cessation or suspension of activities.

Table 2 Retroactive measures affecting PV installations introduced in the period 2010–2014 (author's elaboration)

Regulation	Objective	Retroactive measure(s)
R.D. 1565/2010	Introduction of new technical requirements and specifications for renewable energy installations	– Set up of lifetime caps (25 years) for PV installations
R.D.L. 14/2010	Reducing the tariff deficit	– Introduction of production-hour caps for PV installations based on five climatic solar zones throughout the country – Establishment of limitations on production hours for PV plants for the period 27 December 2010–31 December 2013
Law 15/2012	To ensure the stability of the electricity system	– Introduction of a 7 % tax on the value of the electricity produced by all plants, including renewable energy plants
R.D.L. 2/2013	Reducing the tariff deficit	– Introduction of a new actualisation method for energy prices which is not linked to the Consumer Price Index, but instead to an index that does not include unprocessed food and energy products
R.D.L. 9/2013	To ensure the stability of the electricity system	– Revocation of R.D. 661/2007 and R.D. 1578/2008 (de facto revocation of the special regime) – Despite the fact that the new remuneration mechanism for renewable energy plants had to be set in subsequent regulations, this regime had to apply from 14 July 2013 (date of entry into effect of R.D.L. 9/2013)
R.D. 413/2014 and IET/1045/2014	To ensure the stability of the electricity system	– Introduction of the new remuneration mechanism for renewable energy plants applying since 14 July 2013

the returns of their investments as a result of the retroactive norms passed by the Government (see Table 2). Major changes on the revenue of PV plants have forced many investors to renegotiate the loans for their installation (Cerrillo 2014). They accumulate now higher costs due to loan repayments and installations' maintenance costs than gains. Even worse, some investors have entered bankruptcy and their plants now belong to banks. What level of protection does international investment law offer in such cases? The next part analyses this question in detail.

3 Fair and Equitable Treatment as Protective Standard and the Spanish Revocation of Renewable Energy Incentives

The FET standard is the most frequently invoked protective standard in international investment law (Dolzer and Schreuer 2012, p. 130). FET is part and parcel of nearly every bilateral investment treaty (hereinafter BIT), but it is also included in other international investment agreements (hereinafter IIAs), such as the North American Free Trade Agreement (hereinafter NAFTA) and the ECT. In this line, in the proceedings of the PV investors against Spain and in other cases of revocation of renewable energy investments, Article 10(1) ECT, containing the Host State's obligation to ensure a stable investment environment and including the protective standard of "fair and equitable treatment", is likely to be in the claimants' focus (Alfonso 2011; Freshfields Bruckhaus Deringer 2013[9]).

3.1 The Fair and Equitable Treatment Standard—Attractive and Promising

The main reasons for the FET's attractiveness among investors lie in its flexibility [*Waste Management, Inc. v United Mexican States* (2004)][10] and absoluteness (Dralle 2011, p. 6).[11] This makes FET the most successfully invoked but also the most heavily criticized standard in investment arbitration. For greater legal certainty, it is at first necessary to identify the situations in which the FET standard may operate (*Total S.A. v The Argentine Republic*, Decision on Liability (2010) para. 107),[12] which in line with the interpretation by Dolzer and Schreuer, would be: the stability and the protection of the investor's legitimate expectations; compliance with contractual obligations; procedural propriety and due process; good faith; freedom from coercion and harassment (Dolzer and Schreuer 2012, p. 145). These categories also apply to FET as it is found in Article 10 ECT, which by itself does not provide any explicit specifications regarding the content of FET.[13] In the

---

[9]For the Romanian Government's Emergency Ordinances of 2013.

[10]Para. 99: "[…] the standard is to some extent a flexible one which must to some extent be adapted to the circumstances of each case".

[11]As opposed to the relative standards of "national treatment" and "most favored nation treatment".

[12]Para. 107: "[…] there cannot be a single definition of FET, but that although its exact content is not predefined, except in cases where a treaty provides additional specifications".

[13]Energy Charter Treaty, Article 10(1) on the Promotion, Protection and Treatment of Investments:

> Each Contracting Party shall, in accordance with the provisions of this Treaty, encourage and create stable, equitable, favourable and transparent conditions for Investors of other Contracting Parties to make Investments in its Area. *Such conditions shall include a*

following it is the element of the investor's legitimate expectations that will be shed light upon and that will prove FET to be the most promising standard of protection against the revocation of green energy incentives, respectively FiTs (Boute 2012, p. 613; Boute 2009, p. 333; Kasolowsky 2011). In line with some views, invoking the non-expropriation standard, contained in Article 13 ECT, also seems possible (Cf. *Nykomb Synergetics Tech. Holding AB v Latvia*, Arbitration (2003); Wälde and Hobér 2004, p. 15. For an overview of the views on expropriation, such as police powers doctrine, sole effects and economic effects test see Dolzer and Schreuer 2012, p. 112). Yet, absent of a deprivation of a foreign investor's acquired rights and the transfer of ownership rights to the state or a third person (Reinisch 2008, p. 408) through the revocation of FiTs, a direct expropriation or nationalization can be ruled out here. An indirect expropriation, which leaves the investor's title untouched and which is by far the more common form of an expropriation in international investment law (Dolzer and Schreuer 2012, p. 101) would only be thinkable in case of a "partial expropriation" (Kriebaum 2007, pp. 69, 83; opposing this view is Boute 2012, p. 635). In the end, according to none of the dominant views, the revocation of FiTs will amount to an indirect expropriation, since the investors usually still retain control of their power plants and will receive the profits of the electricity output (Boute 2009, pp. 333, 363).

3.2 Legitimate Expectations of Renewable Energy Investors

Acknowledging the dependence of the renewable energy sector on private foreign investments and the subsequent support of this branch of industry through specialized economic incentives,[14] it is not surprising that an investor expects and relies upon the predictability and stability of these mechanisms. However, the individual factual background has to be assessed when judging the legitimacy of the investor's expectations. Could, for example, a reliance on a tariff regime granting more than 500 % above the reference average tariff for that year[15] still be assumed to be a reasonable rate of return and thus constitute a legitimate expectation?

(Footnote 13 continued)

commitment to accord at all times to Investments of Investors of other Contracting Parties fair and equitable treatment. Such Investments shall also enjoy the most constant protection and security and no Contracting Party shall in any way impair by unreasonable or discriminatory measures their management, maintenance, use, enjoyment or disposal. In no case shall such Investments be accorded treatment less favourable than that required by international law, including treaty obligations. [...] Each Contracting Party shall observe any obligations it has entered into with an Investor or an Investment of an Investor of any other Contracting Party [...].

[14]For an explanation of Feed-in Tariffs (FiTs) see Part 1—Introduction.

[15]Such was the remuneration for small photovoltaic plants regulated through R.D. 661/2007.

3.2.1 Functional Importance of Legitimate Expectations to the FET Standard

Since the tribunal in *Tecmed v. Mexico* was the first to explicitly consider investment protection through the "basic expectations" of the investor, today, no real dispute remains as to the existence of this sub-element [*Técnicas Medioambientales Tecmed S.A. v United Mexican States* (2003, para. 154)]. The tribunal in *Saluka v Czech Republic* even considered the legitimate expectations to be "the dominant element" of FET (*Saluka Investments BV (The Netherlands) v Czech Republic* (2006) para. 302). Recently, the tribunal in *Electrabel S.A. v Hungary* stated that it was "widely accepted" that the legitimate expectations were the "most important function" of the FET standard (*Electrabel S.A. v Hungary* 2012, para. 7.75). It is also named "one of the major components" of FET (*Ulysseas Inc. v Ecuador* (2012) paras. 248–249). Such a qualification within FET is convincing, because the closely related principle of good faith even qualifies as a "general principle of law" in the sense of Article 38(1)(c) of the Statute of the International Court of Justice (*Total* para. 111, in reference to Forsyth 1988 p. 242).[16]

3.2.2 Definition and Scope of Legitimate Expectations in Cases of Renewable Energy Investments in Spain

Concerning the general definition of legitimate expectations the decision in *Tecmed v. Mexico* is predominantly considered as being a landmark award. It has served as a sample for almost identical FET provisions in most IIAs:

> The foreign investor expects the host State to act in a consistent manner, free from ambiguity and totally transparently in its relations with the foreign investor, so that it may know beforehand any and all rules and regulations that will govern its investments, as well as the goals of the relevant policies and administrative practices or directives, to be able to plan its investment and comply with such regulations […] The investor also expects the State to use the legal instruments that govern the actions of the investor or the investment in conformity with the function usually assigned to such instruments. (*Tecmed* para. 154).

However, the *Tecmed* approach has been criticized as "not being a standard at all [but] rather a description of perfect public regulation in a perfect world, to which all states should aspire, but few [if any] will ever attain" (Douglas 2006, pp. 27–28). Opposing the broad and subjective definition by the *Tecmed* tribunal, yet, not shaping the term of legitimate expectations equally precise, was the tribunal in *Saluka* which found that "[…] in order for [investor expectations]to be protected, [they] must rise to the level of legitimacy and reasonableness in light of the circumstances." (*Saluka* para. 304). In the view of some authors, therefore, FET has already been devaluated to be merely subject to the distinctive views of single arbitral tribunals (Hobér 2010, pp. 153, 158).

---

[16]Considering the concept of legitimate expectations to have emanated from German law where it is extensively applied in the function of "Vertrauensschutz".

The most precise definition of legitimate expectations has, so far, been provided in the case of *Thunderbird Gaming v. Mexico*: "[…] The concept of 'legitimate expectations' relates […] to a situation where a Contracting Party's *conduct creates reasonable and justifiable expectations on the part of an investor* (or investment) to act in reliance on said conduct, such that a failure by the [Host State] Party to honour those expectations could cause the investor (or investment) to suffer damages." (*International Thunderbird Gaming Corporation v The United Mexican States* (2006) para. 147 *emphasis added*).

With regard to the Spanish case, the ECT has been assumed to include the concept of legitimate expectations (Hobér 2010, p. 158). Yet, although explicitly embodying FET in its Article 10, the ECT is lacking a reference to legitimate expectations. Such lacuna can be found in basically every existing IIA.[17] This makes the exact content of legitimate expectations difficult to assess through the ordinary meaning of FET in accordance with Article 31(1) of the Vienna Convention on the Law of Treaties. However, the terms in Article 10(1) ECT to "encourage and create stable, equitable, favourable and transparent conditions for investors", requiring a positive commitment of the contracting parties, suggest a stronger obligation than in other IIAs (See e.g. Article 2(2) of the German Model BIT (2008): "Each Contracting State shall in its territory in every case accord investments by investors of the other Contracting State fair and equitable treatment […]"; Wälde 2006, paras. 31, 113). The imperative language in sentence 2 of Article 10(1) ECT, such as "shall", "commitment", "at all times", supports this finding (Wälde 2006, para. 114). The tribunal in *Electrabel v. Hungary* also made a reference to a particular feature of Article 10 ECT, adding the obligation of the Host State to establish "favourable and transparent conditions" to the general understanding of FET. The tribunal deducted from this the obligation to be "forthcoming with information about intended [investment-relevant] changes in policy and regulations" and, thus, provide the possibility for the investor to "engage the host state in dialogue about protecting its legitimate expectations" (*Electrabel* para. 7.79).

After all, the main purpose of the ECT is to promote conditions for profitable investments in energy projects and to ensure a high level of legal security and to provide a stable and transparent framework.[18] This purpose already comprises the main aim of the general concept of legitimate expectations, "to enable the foreign

---

[17]Few IIAs, and mainly concerning the question of indirect expropriation, merely include the term of 'investment-backed expectations' such as the US Model BIT (2012), Annex B(4)(a)(ii). But also see the consolidated text of CETA (2014), stating in Article X.9, para. 4: "When applying the above fair and equitable treatment obligation, a tribunal may take into account whether a Party made a specific representation to an investor to induce a covered investment, that created a legitimate expectation, and upon which the investor relied in deciding to make or maintain the covered investment, but that the Party subsequently frustrated [...]'.

[18]Furthermore, within a contextual interpretation, the ECT preamble's wording, with its aim to "liberalize investment" and by its explicit labelling the of the ECT's commitments as legally binding, suggests that the State is under an obligation to provide a high level of investment protection.

investor to make rational business decisions relying on the representations made by the host State" (Schreuer and Kriebaum 2009, p. 265).

The ECT, thus, offers a higher level of protection of the investor's legitimate expectations than other IIAs would. In the following, these findings, relevant for renewable energy investors in Spain, will be assessed against the background of the general understanding of legitimate expectations by investment law tribunals outside the ECT context.

3.2.3 Legitimate Expectations Based on Contractual Arrangements, Specific Representations or Even Less?

The main critique regarding the legitimate expectations concept is that it is of such breadth that it may cover an infinite number of situations (Sornarajah 2010, p. 355).

Hence, many tribunals have begun narrowing down its scope by introducing new criteria. In this way, the situations in which legitimate expectations may arise can be systematized into three groups (see Schill 2006, p. 16; Hirsch 2011, p. 8):

(1) Contractual arrangements (*Parkerings Compagniet AS v The Republic of Lithuania* (2007) para. 334; *Total* para. 117; *CME Czech Republic B.V. v Czech Republic* (2001) para. 611);
(2) Representations, specific commitments (*Ulysseas Inc. v The Republic of Ecuador* (2012) para. 249, *CMS Gas Transmission Company v The Republic of Argentina* (2005) para. 277) or assurances which were reasonably relied upon by the investor (*Waste Management* para. 98; *CME* para. 611);
(3) Legitimate expectations deriving from the general regulatory framework the Host State has put in place, as long as the confidence that the framework generates is sufficiently specific (*Total* para. 122, describing this as being "the most difficult case"; *CMS* paras. 266-284; *Electrabel* para. 7.78; *Occidental v Ecuador* (2004) para. 196; *Suez and Vivendi Universal S.A. v Argentine Republic* (2010)[19]). Yet, in the last group the Host State must have acted in order to induce the investment (*Glamis Gold v United States* (2009) para. 766; *Total* para. 121; *Suez* para. 208[20]).

Of further importance in each group is the interplay of the specificity of the representations and the legitimacy of expectations in the stability of the legal

[19]Para. 226: "In examining the various cases that have justifiably considered the legitimate expectations of investors and the extent to which the host government has frustrated them, this Tribunal finds that an important element of such cases has not been sufficiently emphasized: that investors, deriving their expectations from the laws and regulations adopted by the host country, acted in reliance upon those laws and regulations and changed their economic position as a result".

[20]Para. 208: "Argentina through its laws, the treaties it signed, its government statements, and especially the elaborate legal framework which it designed and enacted, deliberately and actively sought to create those expectations in the Claimants and other potential investors in order to obtain the capital and technology that it needed to revitalize and expand the Buenos Aires water and sewage system.".

framework. The more formal and specific and clear a representation is, the more legitimate the investor's expectation will be and the less regulatory space will remain for the Government. Or as Wälde (2006, para. 31) has put it: "[...] [T]he less formal "personal communications", the less likely is the emergence of a legitimate expectation; this means that the greater the formality of an assurance, the greater its ability to trigger a legitimate expectation." (Cf. *Total* para. 121).

Addressing the group of "contractual arrangements", as well as the one of "specific representations" is the tribunal in *Parkerings*. Here the ICSID tribunal found that an expectation can only be legitimate if there is as a basis a "received [...] explicit promise or guaranty from the host state" and that in absence of an agreement "in the form of a stabilisation clause or otherwise [...]" the amendment of the original regulatory framework was lawful (*Parkerings* paras. 331, 332; Cf. *CME* para. 611; Tudor 2008, p. 165).

More recently, and relating to Article 10 and 13 ECT, the award of *AES v. Hungary* demanded a specific stabilization agreement or stabilization clause as a condition for the investor's legitimate expectations (*AES Summit Generation Limited and AES—Tisza Erőmú Kft. v Hungary* (2010) para. 9.3.18, where the tribunal ultimately found no breach of FET had occurred).

However, the majority of awards did not require specific contractual stabilization clauses,[21] yet, they are unclear and inconsistent in their terminology of what should form the basis of legitimate expectations. Some tribunals named "specific representations" (*Ulysseas* para. 249; *Glamis Gold* para. 627) others required "commitments" as a basis for a claim (*Duke Energy Electroquil Partners and Electroquil SA v Ecuador* (2008) para. 340; *Continental Casualty v The Argentine Republic* (2008) para. 252).

Few tribunals even regarded a modification of the general regulatory framework at the time the investment was made, not specifically addressed to the investor, as sufficient for a breach of legitimate expectations (e.g. *Occidental* para. 196; *Suez* para. 226). In this way the tribunal in *Electrabel v. Hungary* concluded that specific assurances are "not always indispensable", hence, not always absolutely necessary for the investor's legitimate expectations and continued "specific assurances will simply make a difference in the assessment of the investor's knowledge and of the reasonability and legitimacy of its expectations [...]" (*Electrabel* para. 7.78).

Specifically relating to an economic crisis and absent a contract or specific representation, the tribunal in the case of *Total* qualified a specific stabilization clause to be "undoubtedly" sufficient. Interestingly, the same tribunal identified that, as a rule, the general regulatory framework would not suffice for legitimate expectations, however, as an exception, the change of regulation of "inherently prospective nature [which is] aimed at providing a defined framework for future operations [...]" may constitute a breach of FET (*Total* paras. 122, 129). To

[21]The high standards for legitimate expectations set in the awards in line with *Parkerings* will have to be regarded as exceptions, especially because there was each time a contract between the investor and the Host State.

determine the threshold for a breach of such prospective regulation, the tribunal distinguished between a general promise of legislative stability (*Total* paras. 297, 308–314) and, absent such a promise, a commitment to provide for an "economic equilibrium" based on the principle of "regulatory fairness" (*Total* paras. 309, 122). The economic equilibrium had been violated by Argentina's energy price regulation, which thus constituted a breach of FET (*Total* paras. 327, 330, 333). Importantly, the case indicates that the claimant can only reasonably rely on a diminished protection, when the regulatory framework was of unilateral and general character and not specifically addressed to him (*Total* paras. 119–124).

One can conclude that, in absence of an investor-state contract, but in presence of specific representations and assurances, there is a firm basis for the protection of the investors' legitimate expectations. Yet, many investment law tribunals and scholars assume the above-mentioned unspecified "general regulatory framework" as being too low a threshold for a breach of legitimate expectations. Wälde and Kolo (2001, pp. 824–825) clarify that: "One cannot postulate that the environmental regime should be absolutely frozen […]. The question is rather to identify the threshold of an unexpected regulatory change and its impact on the investor's legitimate expectation […]".

However, as the case of *Total* and other awards have shown, the option to rely on the stability of the general regulatory framework is not completely barred, but diminished, in cases of regulation of a "prospective nature", as long as the investor could "reasonably" rely on it and the government acted "to induce" the investment.

For the cases of foreign PV investors in Spain no investor state contracts have been publicized containing a stabilization clause in the sense of the first group. Concerning the second group, of specific representations reasonably relied on by the investor, at first sight, the special regime,[22] containing foremost general laws, would be lacking a specific addressee. As the tribunal in *Total* pointed out: "Representations made by the host State are enforceable and justify the investor's reliance only when they are specifically addressed to a particular investor." [*Total* para. 119; Cf. *Thunderbird Gaming* para. 147; *Merrill & Ring Forestry L.P. v Canada* (2010) para. 242; *El Paso Energy International Company v Argentine* Republic (2011) para. 375; Cf. on the necessarily individualized character of representations in order to form a basis for legitimate expectations, which is not present in abstract administrative decisions such as decrees: Diehl (2012, pp. 398–402)].

While the Electricity Sector Law was aiming at renewable energy investments, noticeably, it was neither only addressed to foreign investors nor was it restricted to PV energy production only (Electricity Sector Law 1997, Title IV, Chap. II). The same holds true for the National Renewable Energy Plan of 1999 which aimed at promoting renewable energies, among others the PV sector, yet, remained very general in its statements (National Plan for the Promotion of Renewable Energy 2000–2010 (1999) p. 116). Hence, the special regime by itself is not yet a specific representation in the above stated terms. However FiTs, as those affecting the PV

[22]See above Part 2.2.1.

investors in Spain, set the exact amount of support for a specific group of investors with the particular aim to incentivize investments in an innovative industry with an environmental purpose for a minimum timeframe. The decree R.D. 2818/1998 was not yet directly favouring solar energy producers as a group among all renewable energy producers and only offered premiums as remuneration mechanism. However, R.D. 436/2004 divided the technologies into categories and now allowed energy producers to choose between premiums and/or tariffs, while for small PV plants only FiTs were foreseen (R.D. 436/2004, Table 3). Then R.D. 661/2007 stands apart as it further specified and privileged renewable energy as opposed to conventional energy suppliers, and, within this group of FiT recipients, strongly promoted solar installations. Table 3 of this decree contained specific FiT provisions not expressly naming one addressee, yet, evidently offering far higher sums to PV and solar thermal suppliers as opposed to any other renewable energy form. Thus, de facto, the addressees of the significantly elevated tariff were only solar investors. This may not make R.D. 436/2004 and R.D. 661/2007, as most FiT regimes, eligible to be "specific representations". However, as seen, this regulatory framework was precise in its conditions, geared at a specific group, being domestic and foreign solar investors, contained an inducement to invest and, thus, formed the basis of the investors' legitimate expectations. These R.D.s were, moreover, embedded into the broader regulatory framework of National Energy Plans, which through their continuous renewal and evolution produced an increasingly beneficial and specific framework for solar energy promotion. The two R.D.s taken together with the 1999 and 2004 National Plans constitute a firm basis for legitimate expectations to emerge on the side of any PV investor having invested within the timeframe between the enactment of R.D. 436/2004 and R.D. 661/2007 and the retroactive changes initiated by R.D. 1565/2010 (Real Decreto 1565/2010 Art. 1). Lastly, this also fulfils the prerequisite set up by the *Total* decision, of regulation of forward-looking, hence, prospective, nature.[23] Yet, as a caveat, one may add that due to the low threshold of the general regulatory framework, being neither a formal representation nor a contractual arrangement, the test for the reasonableness of the legitimate expectations will be a stricter one and may lead to a diminished protection. Whether the Spanish FiTs in fact render the expectations legitimate depends on other factors such as the investor's own conduct or possible defences on the Government's side. These will be examined next.

3.2.4 Legitimate Expectations Limited by the Renewable Energy Investors' Own Conduct

There is a tendency among tribunals that the business risk inherent in an investment is to be borne by the investor, as shown by *Consortium RFCC v. Morocco*: "C'est le lieu de rappeler qu'un Traité de protection des investissements ne peut servir à

[23]Until R.D. 1565/2010, the regulatory framework included no lifetime caps, see above.

compenser un investisseur déçu du résultat financier de l'opération réalisée,[…]." (*Consortium RFCC v Royaume du Maroc* (2002, para. 108); See also *Parkerings* para. 333, where the tribunal required the investor to "anticipate that the circumstances could change, and thus structure its investment in order to adapt it to the potential changes of legal environment"; Yannaca-Small 2008, p. 127). This will be true, in particular, in case of the investor's own "bad business judgments" (*Maffezini v Spain* [2000] para. 64). Thus, for determining the scope of legitimate expectations, the investor's *own conduct* has to be taken into account in order to conclude what is "fair and equitable" (Alvarez 2011, pp. 185, 383, 385; Potestà 2013, pp. 88, 119).

Muchlinski categorizes that the investor has the *duty* (1) to avoid unconscionable conduct, (2) to assess the investment risks of the host country reasonably and (3) to operate the investment reasonably (Muchlinski 2006, p. 527). Particularly relevant for the present case seems to be the second duty of due diligence to reasonably assess the risk, meaning "*all* circumstances, including not only the facts surrounding the investment, but also the political, socioeconomic, cultural and historical conditions prevailing in the host State." (*Duke Energy* para. 340; *Bayindir Insaat Turizm Ticaret Ve Sanayi AS v Pakistan* (2009) para. 195; *Maffezini* paras. 64–71; *Methanex Corporation v USA* (2005) Part IV, Chap. D, para. 10). This seems necessary before deciding to invest in the first place (Tudor 2008, p. 217; *Tecmed* para. 154), especially when committing oneself to a long-term investment covering years of energy supply. Thus, the investor's own conduct constitutes a general limitation to his legitimate expectations.

Regarding PV energy investors in Spain, it is thus a critical question as to how intensely they had to assess the socio-economic and regulatory circumstances in the pre-investment phase in order to maintain the legitimacy of their expectations in a stable and predictable business environment. Concerning the ranking of different risk factors, to the detriment of the solar investors, *Boute* has fittingly remarked that public support schemes are, almost always, a "conditio sine qua non" for the initial decision to invest in the renewable energy sector (Boute 2009, p. 637). The UNCTAD (2010, p. 30) states similarly express: "Foreign investment into new low-carbon industries may not be competitive in the start-up phase and may therefore need government support, such as feed-in tariffs for renewable energy or public procurement". The Spanish regime reveals the common characteristic of such supportive legislation to be already in place before the investment, because it should incentivize the renewable energy investments. As in any renewable energy investor's pre-assessment of the investment climate, the financial stability of the Spanish support system, thus, had to be considered as the most important risk factor (Cf. European Commission 2005, pp. 16–17).

A further argument for a stronger weight of the investors' own conduct in the Spanish case may be the unreasonableness of an expectation in case of tariffs being offered at a rate of more than 500 % above the reference average tariff. However, specific rates cannot be expected from an investment, which has been undertaken in awareness of the common business risk of loss. A point in favour of photovoltaic energy investors would, again, be the fact that the Spanish Government applied detailed market research and empirical analysis of respective renewable energy costs to establish FiT payment levels in the first place. This was supposed to ensure that the FiTs would allow competently operated projects to be profitable (Couture et al. 2010, pp. 7–8). Still, considering each investor's professional "businessman" background (*Parkerings* para. 332), it seems reasonable to apply a stricter test to the investor's own conduct. The large sum of investments totalling € 2–4 billion (Hepburn 2011), on the basis of a very generous FiT scheme, in the particular claims of PV Investors in Spain could, after all, indicate some naïveté. A possible point on the side of the solar investors would exist, if, at the time the investment was made, the tariff deficit was not foreseeable to serve as the Government's main argument for cutting the support for renewables. Notably, the expectations must be assessed at the time of making the investment (*Duke Energy* para. 340; *LG&E v Argentine Republic* (2006) para. 130: "[expectations] are based on the conditions offered by the host State at the time of the investment."). However, the investors had access to the tariff deficit and this very deficit was already eight years old when the Spanish Government started cutting support for renewables (Royal Decree 1578/2008). But until April 2009, the date when the first regulation adopting measures for tackling the tariff deficit was approved (Royal Decree Law 6/2009), the regulator did not seem concerned about this problem. Quite the opposite, in 2007, the regulator introduced higher incentives for plants operating under the special regime. Thus, the tariff deficit has not been a foreseeable argument for the cuts at the time the investments were made.

To sum up, the impact of the photovoltaic energy investors' own conduct in Spain on their legitimate expectations is subject to a stricter test as they would have to assess whether there is at all a supportive regime in place that, as a "conditio sine qua non", is eligible to let them, as renewable energy investors, compete with conventional energy producers. Yet, the scope of the investor obligation to due diligence does not cover miscalculations, hence, the question in what way the supportive regime may properly operate. This risk will remain within the sphere of the legislator setting up the regime, as in the Spanish case. Ultimately, as will be unfolded, the investors' conduct by itself will not be the decisive factor for the tribunal's decision, on whether there was a breach of legitimate expectations.

3.2.5 Legitimate Expectations Limited by the Right to Regulate of the Spanish State in Crisis

Notably, the legitimate expectations and the implied requirement of stability of the legal framework do not per se affect the Host State's right to exercise its sovereign regulatory powers—its right to regulate (Dolzer and Schreuer 2012, pp. 148–149; More recently the European Parliament (2011) called on the Commission to include in all future EU investment agreements specific areas for a right to regulate. In the ECT Article 18 also takes into consideration a right to regulate, specifically addressing energy resources).

The tribunal in the *Saluka* case made it clear that the investor cannot reasonably expect that "the circumstances at the time the investment is made *remain totally unchanged*" and that the Host State's right "to regulate domestic matters in the public interest" must not be neglected (*Saluka* para. 305; Tudor 2008, p. 167). A regulatory measure would not have to be compensated, if States adopted laws within "the normal exercise of regulatory powers, […] in a non-discriminatory manner […] bona fide, that are aimed at the general welfare […]" (*Saluka* para. 255; Cf. *Feldman v Mexico* [2002] para. 112). Similar formulations with respect to the indirect expropriation of investments are to be found in later IIAs and in Article 13 ECT. Moreover, the right to regulate requires a "weighing" of the investor's legitimate expectations on the one hand and the host State's legitimate regulatory interest on the other hand (*Saluka* para. 306). The tribunal in *EDF v Romania* established that in the case of the revocation of business licenses the aim to fight corruption within the Romanian state was found to be legitimate, as long as there was a "reasonable relationship of proportionality between the means employed and the aim sought to be realized" (*EDF v Romania* (2009) para. 293; Similarly *LG&E* para. 189). While tribunals have adopted the systematization that public policy objectives need to be "bona fide", "non-discriminatory" and in the "public interest" in reference to indirect expropriation (*Methanex* Part IV, para. 7) they have not as clearly applied such categories to cases in which the right to regulate conflicts with FET and legitimate expectations (The tribunal in *El Paso* recently applied the *Saluka* test for the non-discriminatory, bona fide regulation of the Host State in the public interest to the FET claim of the investor and in this way pointed out that "legitimate expectations necessarily vary with the surrounding circumstances […]" *El Paso Energy v Argentine Republic* (2011) paras. 358–359). If the State's measure should not fulfil these conditions, rather than as a clear prerequisite for illegality in case of an expropriation, for the case of FET, this would serve as strong indicator for a violation. Except for a case of obvious discrimination or documented "mala fides" of the Host State the absence of a public purpose would not automatically entail a breach of FET, as the weight of the conflicting public and investors' interests depends strongly on the surrounding circumstances. (Yet, for the criterion of "bona fide" *Schreuer* contends that: "[…] it may be regarded as established that action against the investor that is demonstrably in bad faith would be a violation of the fair and equitable treatment standard." In his view, FET may also be violated in absence of "mala fides" on the side of the host State and the burden of proof in so far does not lie with the claimant, Schreuer (2005, pp. 383–385); Cf. *Glamis Gold* para. 22).

The first question to ask is whether Spain with its revoking and retroactive measures, in particular visible in the regulation following R.D. 1578/2008, has pursued a legitimate public interest aim. The public interest receives an increased importance in cases in which an abnormal socio-economic situation is present. Thus, the financial crisis of Spain since 2008 will be probably brought to the fore as an argument by the Spanish State and may give stronger weight to the right to regulate in the public interest to reduce public debt or to avoid the tariff deficit.[24] Argentina faced similar problems resulting from its severe economic crisis from 1999 to 2002. Comparably, for example in the case of *CMS*, the Argentine Republic invoked the "right to regulate" as a defence against an alleged breach of the energy investors' legitimate expectations (*CMS* para. 93; *CMS v Argentina*, Argentina's Annulment Reply 2007, para. 43; Alvarez 2011, p. 247). Notably, in the award of *National Grid v. Argentina* FET was acknowledged to be defined differently "in case of an economic and social crisis" (*National Grid v Argentina* 2008, para. 180; Alvarez 2011, pp. 264–265). However, for the case of Spain, the weight of the crisis should not be overestimated because the economic conditions are far less severe than they were in Argentina,[25] and already the clear wording of most energy norms passed in Spain since 2008 such as R.D.L. 14/2010, R.D.L. 2/2013, R.D.L. 9/2013 and R.D. 413/2014 reveals that they pursue one major goal: the elimination of the tariff deficit.[26] However, the aim of reducing appears legitimate as Spain, being an EU Member State, is bound by the aims set forth in the EU Stability and Growth Pact which, inter alia, has fixed annual objectives per country in terms of deficit reduction since 1999 (European Commission 1999). Spain has not met these aims since the emergence of the crisis and would thus face sanctions if it not drastically changed its spending policies, which can be pursued, among other ways, by reducing subsidies in the electricity sector. This, taken together with the fact that the deficit is also a burden for the public, makes its reduction a legitimate aim, still, with a different weight than a severe socio-economic crisis such as the one experienced in Argentina.

The next question is whether the measures have been applied in a proportionate manner. The measures visible in the regulation following R.D. 1578/2008 were appropriate to reach the aim, as a large amount of public spending on PV energies would have been saved and the deficit presumably reduced. Yet, the measure was only necessary if there had not been any equally efficient means at hand for the Spanish State less violating for FET, to reduce the deficit. While higher taxes, in particular for other conventional energy sources, may have been thinkable, their outcome is just as uncertain. Moreover, it must not be overseen that, by 2012, Spain has implemented the most severe economic policies ever since the democracy has

---

[24]As a legal basis for such an argument the exceptions of Article 24(3)(c) ECT may be invoked, as it is not explicitly inapplicable to Article 10 ECT. However, the Spanish measures cannot be construed as "for the maintenance of public order", for in this case the latter did not seem at stake as a consequence of the economic crisis.

[25]At least three elements support this claim. First, while the Argentine Government passed the *Corralito* in 2001 for "fencing in" withdrawals and prohibiting international transfers, no limits on bank withdrawals have been imposed in Spain. See Argentine Decree No. 1570/01 of 1 December 2001 (Argentina 2001). Second, contrary to Argentina, Spain has not defaulted on its debts. Third, unlike what happened in Argentina, no international institution such as the International Monetary Fund has intervened in Spain. Furthermore, not even in the Argentine Gas Cases was financial crisis alone a sufficient argument for a permitted breach of international standards of investment protection with *National Grid* being the only exception.

[26]See Part 2.2.2.

been introduced to the country and that these now affect basically every sector, not only PV energy (see Spain 2012a, b). Thus, the measure was necessary.

Most importantly, the balancing of interests must have been proportionate. As shown, the own conduct of the renewable energy investors and their obligation to diligently pre-assess the investment climate appear stricter as the financial stability of the support system represents “the most important risk factor.” (European Commission 2005, pp. 16–17). Yet, the Spanish Government failed to act against the tariff deficit for nine years. It also cannot be expected of the investor to react to the revocation, becoming gradually visible, by autonomously removing the investment himself: an investment in renewable energies is not just a simple financial product which could be easily distributed or removed again but involves a complex and long-term construction, i.e. setting up a PV plant. Moreover, the bona fides of the Government remains doubtful because of three reasons. First, despite the difficulty to understand the precise cause of the tariff deficit, all the Parliament Members of the *Partido Popular* (party heading the current Government) rejected a parliamentary motion presented by a left-wing coalition in June 2013 for conducting an audit to reflect the true costs of energy production and distribution in Spain (Economía 2013). Second, the tariff deficit was initiated in the year 2000 and escalated to the amount of € 4 billion in the year 2005, yet tackled by Spain for the first time as late as the year 2009. Third, Spain was urged by the European Commission to proceed in its promotion of renewable energies and even surpassed the intermediate goal set out in the directive 2009/28/EC of raising the share of renewable energy in the final energy consumption to 10.9 % by 2.9 % (in the year 2010 this share amounted to 13.8 %) (European Commission 2013, p. 175, Annex 1). Thus, the legislator should have been more cautious in the implementation of support schemes for renewables from the beginning in order to avoid miscalculations.

The Spanish Supreme Court’s has found that the reforms in the special regime qualify as an “improper” legal retroactivity[27] and this indicated investors in the Spanish case could not expect the regulatory framework to remain unchanged, (an analysis of the question of relevance of national jurisprudence for international investment arbitration would surpass the scope of this article. Extensive research on such interplay of national and international legal orders has already been conducted by Kjos (2013). While it is clear that the standard of legitimate expectations is not identical with existing standards of legal certainty as laid down in many Constitutions[28] and that the principle of legal certainty cannot be understood as an “immutable right”, however, the principle of legitimate expectations must not be deprived of its main function: to provide protection by “enabl[ing] the foreign investor to make rational business decisions relying on the representations made by the host State” (see above Part 3; Schreuer and Kriebaum 2009, p. 265). In this way the “economic equilibrium” of the investment may have been distorted. According

[27]See above Part 2.2.2.

[28]E.g. Article 9.3 of the Spanish Constitution; Article 20(3) of the German Constitution.

to the tribunal *Total* a breach of the "economic equilibrium", hence of legitimate expectations, is reached when the investors are no longer able to cover their costs and make a reasonable return on their investment (*Total* paras. 313, 327.) The tribunal in *Total* already found that this could already be the case when not renegotiating the electricity tariff regime representing "regulation of general nature" (*Total* paras. 312, 327). In the Spanish case, the Government guaranteed high profit margins through R.D. 661/2007, as shown above. However, the changes introduced through R.D. 1565/2010, R.D.L. 14/2010, Law 15/2012, R.D.L. 2/2013, R.D.L. 9/2013, R.D. 413/2014 and Ministerial Order IET/1045/2014 have resulted in greater costs than revenues for many PV investors.[29] In this context, the very *raison d'être* of the initial PV investment becomes doubtable. With a special regard to the fact that the FiTs in place qualify as inducements to investment[30] their revocation can be seen as a heavy distortion of the economic equilibrium.

It can be concluded that the right to regulate may establish a limitation to the protective scope of FET. In relation to renewable energy investments, States have the right to adapt their support regimes in order to avoid overcompensation (Boute 2012, p. 648). Yet, the State cannot justify the withdrawal of a support scheme as regulating in the public interest when, in fact, it is primarily aiming to reduce an energy tariff deficit which is due, at least in part, to the Government's mishandling.

In this way, it has to be said that the interest of the PV investors outweighs the regulatory interest of the Spanish State and the revocation measures constitute a breach of legitimate expectations and, hence, a violation of FET.

4 Conclusions for Legitimate Expectations in Photovoltaic Energy Investments in Spain

Renewable energy investors in Spain, in particular PV investors, will be able to successfully claim a breach of FET due to the far-reaching revocation measures[31] put in place from 2008-2014. Regarding the scope of legitimate expectations in the case of Spain, as shown, PV investors can reasonably rely on the general regulatory framework the Host State has put in place. Particularly R.D. 661/2007,[32] can be qualified as being part of the general regulatory framework formed by the special regime, as part of which the FiTs have generated a confidence that was sufficiently specific (cf. *CMS*) and have served as an inducement to invest (cf. *Total* para. 110; *Glamis Gold* para. 766) in the Spanish solar industry.[33] This underscores the State's

[29] See above Part 2.2.2.

[30] See above Part 3.2.3.

[31] See Part 2.2.2.

[32] R.D. 661/2007.

[33] First, Article 2 in conjunction with Table 3 of R.D. 661/2007 promised a certain sum of payment for an unlimited period of time, offering significantly higher sums to PV energy suppliers, hence,

inducement ambitions once more (see e.g. National renewable energy plan 2005–2010, pp. 183–184). What is more, the protective scope of the legitimate expectations under the applicable ECT reaches further than it does under most IIAs.[34] Yet, in no regime, the FET protection is capable of reaching as far as a stabilization clause would (Cf. Hirsch 2011, p. 24). While international investment law is not supposed to force countries to keep in place subsidy programs that are inefficient and unintended in their consequences, renewable energy investors may legitimately expect the maintenance of an "economic equilibrium", at least in terms of the viability of their business, as shown in *Total* at para. 313: "The respect for economic equilibrium principle entails that, in normal situations and from a long term perspective, the private generators are able to cover their costs and make a return on their investment, while providing their services to the market and consumers as required under the Electricity Law", (also see Crockett 2012, pp. 516, 523). In the end, the interest of the PV investors outweighs the regulatory interest of the Spanish State and the revocation measures constitute a breach of legitimate expectations and, hence, a violation of FET. The provided evidence suggests that the conflict between the Spanish State and foreign investors could be best resolved through a compensation for the investors that takes into account the losses of the investors, the inappropriate regulation of the State and the financial limitations of the country due to the crisis.

In order to fulfil Sustainable Development goals States will have to promote renewable energies. Today there is a broad consensus that not only economic development, but all "three pillars", the economic, the ecological and the social aspects of each regulatory or private measure have to be seen as one, in order to guarantee the sustainability of the regulatory measure (Principles 5, 7, 8 and 10 of the Rio Declaration [United Nations Conference on Environment and Development 1992]). However, as the concept of Sustainable Development cannot, yet, be considered as a binding norm in international law, also in investment law it, nonetheless, serves as an aid to interpretation. In the present disputes the claimants cannot and do not rely on Sustainable Development in a general manner. Yet, as Sustainable Development divides itself into several subprinciples an invocation of the latter, may be helpful. In this way, the Principle of Sustainable Use of Natural Resources demands of the actors to use all of their natural resources sustainably and efficiently, set out for a long-term maintenance for a general utility for society [Principle 1 of the New Delhi Declaration (International Law Association 2002); Principle 8 of the Rio Declaration (United Nations Conference on Environment and Development 1992)]. Clean energy investments, such as solar energy investments,

---

(Footnote 33 continued)

de facto specifying the addressees of the tariff. Second, the progression of implementation over a long time generated a more reliable, more credible regime, of a prospective nature. Third, the purpose of the FiT regime in Spain comprised environmental goals, as well as the industrial policy objectives of fostering the development of domestic innovative technology. This underscores the State's inducement ambitions once more.

[34]See Part 3.2.2.

fulfil the aims of this subprinciple as they constitute a surrogate for conventional energy sources, such as coal or gas, which will deplete over time and do far greater harm to the environment. However, only in sectors for which a specific international regime exists, the Principle of Sustainable Use has acquired some normative content (Birnie et al. 2009, p. 200). For international investment law, Sustainable Development has, so far, not attained a normative form, yet, its relevance is increasing rapidly, as can be seen e.g. in the broad-based Investment Policy Framework for Sustainable Development (UNCTAD 2012). Finally, it becomes visible in the proceedings of the international PV investors in Spain: For a credible pursuit of Sustainable Development goals States may be legally held to maintain a stable investment climate for the international renewable energy projects they have lured into the country.

References

Agosti L, Padilla J (2010) Promoción de las energías renovables: la experiencia de España. In: Moselle B, Padilla J, Schmalensee R (eds) Electricidad verde: energías renovables y sistema eléctrico. Marcial Pons, Madrid, pp 518–541

Alfonso J (2011) El gobierno guarda silencio ante la reclamación de inversores extranjeros por el cambio legislativo en la FV. Interview with Jeffrey Sullivan, counsel from Allen & Overy. Retrieved 14 Apr 2015. http://www.energias-renovables.com/articulo/ldquo-el-gobierno-guarda-silencio-ante

Alvarez JE (2011) The public international law regime governing international investment. Hague Academy of International Law, The Hague

Argentina (2001) Decree No. 1570/01 of 1 December 2001 (Corralito). Retrieved 15 Apr 2015. http://infoleg.mecon.gov.ar/infolegInternet/anexos/70000-74999/70355/norma.htm

Birnie PW, Boyle AE, Redgwell C (eds) (2009) International law and the environment. Oxford University Press, New York

Boute A (2009) The potential contribution of international investment protection law to combat climate change. J Energy Nat Resour Law 27:333–376

Boute A (2012) Combating climate change through investment arbitration. Fordham Int Law J 35:613–664

Cerrillo A (2014) El sueño solar se apaga. Retrieved 14 Apr 2015. http://www.lavanguardia.com/natural/20141101/54418479670/sueno-solar-apaga.html

Couture T, Cory K, Kreycik C, Williams E (2010) A policymaker's guide to feed in tariff policy design. National Renewable Energy Laboratory—NREL. Retrieved 14 Apr 2015. http://www.nrel.gov/docs/fy10osti/44849.pdf

Crockett A (2012) Stabilisation clauses and sustainable development: drafting for the future. In: Miles K, Brown C (eds) Evolution in investment treaty law and arbitration. Cambridge University Press, Cambridge, pp 516–539

Diehl A (2012) The core standard of international investment protection: fair and equitable treatment. Wolters Kluwer Law & Business, Alphenaan den Rijn

Dralle, T. M. (2011) Der fair and equitable treatment-standard im Investitionsschutzrecht am Beispiel des Schiedsspruchs Glamis Gold v. United States. Beiträge zum Transnationalen Wirtschaftsrecht, vol 115. Retrieved 14 Apr 2015. http://telc.jura.uni-halle.de/sites/default/files/BeitraegeTWR/Heft115.pdf

Dolzer R, Schreuer C (2012) Principles of international investment law. Oxford University Press, Oxford

Douglas Z (2006) Nothing if not critical for investment treaty arbitration: occidental, Eureko and Methanex. Arbitr Int 22:27–51

Economía (2013) El Congreso rechaza la petición de IU-ICV-CHA de auditar el déficit de tarifa previo a la reforma energética. Retrieved 14 Apr 2015. http://www.europapress.es/economia/energia-00341/noticia-economia-congreso-rechaza-peticion-iu-icv-cha-auditar-deficit-tarifa-previo-reforma-energetica-20130627200420.html

EFE Economía (2013) Soria cree que el problema del déficit está muy encauzado pero no solucionado. Retrieved 14 Apr 2015. http://economia.elpais.com/economia/2013/02/05/agencias/1360061444_785068.html

Energy Charter Secretariat (1984) Energy charter treaty and related documents. Retrieved 14 Apr 2015. http://www.encharter.org/fileadmin/user_upload/document/EN.pdf

Energy Charter Secretariat (2015) Investor-state dispute settlement cases. Retrieved 14 Apr 2015. http://www.encharter.org/index.php?id=213

European Commission (1999) Stability and growth pact. Retrieved 15 Apr 2015. http://ec.europa.eu/economy_finance/economic_governance/sgp/index_en.htm

European Commission (2005) Communication on the support of electricity from renewable energy sources COM (2005) 627 final. Retrieved 15 Apr 2015. http://eur-lex.europa.eu/legal-content/EN/TXT/PDF/?uri=CELEX:52005DC0627&from=EN

European Commission (2009) The climate and energy package (The 2020 package). Retrieved 14 Apr 2015. http://ec.europa.eu/clima/policies/package/index_en.htm

European Commission (2013). Renewable Energy Progress Report COM (2013) 175. Retrieved 15 Apr 2015. http://eur-lex.europa.eu/LexUriServ/LexUriServ.do?uri=COM:2013:0175:FIN:EN:PDF

European Environment Agency (2014) Trends and projections in Europe 2014: tracking progress towards Europe's climate and energy targets for 2020. Retrieved 14 Apr 2015. http://www.eea.europa.eu/publications/trends-and-projections-in-europe-2014

European Parliament (2011) Resolution of 6 April 2011 on the future European international investment policy 2010/2203(INI). Retrieved 15 Apr 2015. http://www.europarl.europa.eu/sides/getDoc.do?type=TA&reference=P7-TA-2011-0141&language=EN

European Parliament and Council (2009) Directive 2009/28/EC of 23 April 2009 on the promotion of the use of energy from renewable sources and amending and subsequently repealing Directives 2001/77/EC and 2003/30/EC. Retrieved 14 Apr 2015. www.eur-lex.europa.eu/legal-content/EN/TXT/PDF/?uri=CELEX:32009L0028&qid=1423661368386&from=EN

Eurostat (2013) EU27 energy dependence rate at 54 % in 2011. News release 23/2013. Retrieved 14 Apr 2015. http://ec.europa.eu/eurostat/documents/2995521/5162466/8-13022013-BP-EN.PDF/67cc4ff6-81bb-4387-ac05-16862e822593?version=1.0

Freshfields Bruckhaus Deringer (2013) Romania cuts renewable energy subsidies: remedies under the Energy Charter Treaty. Retrieved 14 Apr 2015. http://www.freshfields.com/uploadedFiles/SiteWide/Knowledge/36315.pdf

Forsyth CF (1988) The provenance and protection of legitimate expectations. Camb Law J 47:238–260

German Model BIT (2008) German model bilateral investment treaty from 2008. Retrieved 27 January 2015. http://www.italaw.com/sites/default/files/archive/ita1025.pdf

Germany (1949) Grundgesetz für die Bundesrepublik Deutschland (Constitution for the Federal Republic of Germany)

Germany (2014) Gesetz zur grundlegenden Reform des Erneuerbaren-Energien-Gesetzes und zur Änderung weiterer Bestimmungen des Energiewirtschaftsrechts (German Renewable Energy Act)

Hepburn J (2011) Renewable energy arbitration claims on horizon, but states take differing approaches to public disclosure. International Arbitration Reporter, vol 4

Hirsch M (2011) Between fair and equitable treatment and stabilization clause: stable legal environment and regulatory change in international investment law. International Law Forum vol 8, Hebrew University of Jerusalem Research Paper No 07-13. Retrieved 27 January 2015. http://papers.ssrn.com/sol3/papers.cfm?abstract_id=2272952

Hobér K (2010) Investment arbitration and the energy charter treaty. J Int Dispute Settl 1:153–190
International Law Association (2002) The 70th conference of the international law association, held in New Delhi, India, 2–6 Apr 2002 (The New Delhi Declaration). Retrieved 15 Apr 2015. http://www.ila-hq.org
Jha V (2012) Trends in investor claims over feed-in tariffs for renewable energy. Investment treaty news quarterly (International Institute for Sustainable Development), vol 2(4), pp 14–5. Retrieved 14 Apr 2015. http://www.iisd.org/pdf/2012/iisd_itn_july_2012_en.pdf
Kasolowsky B (2011) Tariff changes for photovoltaic solar power in Spain. Mondaq. Retrieved 14 Apr 2015. http://www.mondaq.com/x/126100/Renewables/Tariff%2BChanges%2BFor%2BPhotovoltaic%2BSolar%2BPower%2BIn%2BSpain
Kjos HE (2013) Applicable law in investor-state arbitration: the interplay between national and international law. Oxford University Press, Oxford
Kriebaum U (2007) Partial expropriation. J World Investment Trade 8:69–84
Muchlinski P (2006) Caveat investor? The relevance of the conduct of the investor under the fair and equitable treatment standard. Int Comp Law Q 55:527–558
Noceda MA (2013) La reforma eléctrica supondrá una subida de la luz del 3,2 % de inmediato. Retrieved 14 Apr 2015. http://economia.elpais.com/economia/2013/07/12/actualidad/1373628661_266334.html
Peterson LE (2013) Solar investors file arbitration against Czech Republic; intra-EU BITs and energy charter treaty at the center of dispute. International Arbitration Reporter, 6(10)
Potestà M (2013) Legitimate expectations in investment treaty law: understanding the roots and the limits of a controversial concept. ICSID Rev 28:88–122
Reinisch A (2008) Expropriation. In: Muchlinski P, Ortino F, Schreuer C (eds) The Oxford handbook of international investment law. Oxford University Press, Oxford, pp 407–459
Sallé C (2012) El déficit de tarifa y la importancia de la ortodoxia en la regulación del sector eléctrico. Papeles de Economía Española 134:104–116
Schill SW (2006) Fair and equitable treatment under investment treaties as an embodiment of the rule of law. International law and justice working papers, New York University School of Law, vol 6, pp 1–37. Retrieved 15 Apr 2015. http://iilj.org/publications/documents/2006-6-GAL-Schill-web.pdf
Schreuer C (2005) Fair and equitable treatment in arbitral practice. J World Investment Trade 6:357–386
Schreuer C, Kriebaum U (2009) At what time must legitimate expectations exist? In: Werner J, Ali AH (eds) A liber amicorum: Thomas Wälde—law beyond conventional thought. Cameron May Publishing, London, pp 265–277
Sevilla M, Golf E, Driha O (2013) Las energías renovables en España. Estudios de Economía Aplicada 31(1):35–58
Soria JM (2014) Sostenibilidad sostenible. Opinion letter from the current Spanish Minister of Industry, Tourism and Energy to El País. Retrieved 14 Apr 2015. http://www.economia.elpais.com/economia/2014/06/23/actualidad/1403551157_315628.html
Sornarajah M (2010) International law on foreign investment. Cambridge University Press, Cambridge
Spain (1997) Ley 54/1997, de 27 noviembre, del sector eléctrico (Electricity sector law). Official State Gazette n. 285, 28 November 1997, p. 35097. Retrieved 14 Apr 2015. http://www.boe.es/boe/dias/1997/11/28/pdfs/A35097-35126.pdf
Spain (1998) Real Decreto 2818/1998, de 23 de diciembre, sobre producción de energía eléctrica por instalaciones abastecidas por recursos o fuentes de energía renovables, residuos y cogeneración (Electricity production through renewable sources, cogeneration and waste). Official State Gazette n. 312, 30 December 1998, p. 44077. Retrieved 14 Apr 2015. http://www.boe.es/boe/dias/1998/12/30/pdfs/A44077-44089.pdf
Spain (1999) Plan de fomento de las energías renovables 2000–2010 (National plan for the promotion of renewable energy 2000-2010). Retrieved 14 Apr 2015. http://idae.es/index.php/mod.documentos/mem.descarga?file=/documentos_4044_PFER2000-10_1999_1cd4b316.pdf

Spain (2004) Real Decreto 436/2004, de 12 de marzo, por el que se establece la metodología para la actualización y sistematización del régimen jurídico y económico de la actividad de producción de energía eléctrica en régimen especial (Methodology for updating and structuring the special regime). Official State Gazette n. 75, 27 March 2004, p. 13217 . Retrieved 14 Apr 2015. http://www.boe.es/boe/dias/2004/03/27/pdfs/A13217-13238.pdf
Spain (2005) Plan de energías renovables 2005–2010 (National renewable energy plan 2005–2010). Retrieved 14 Apr 2015. http://idae.es/uploads/documentos/documentos_PER_2005-2010_8_de_gosto-2005_Completo.(modificacionpag_63)_Copia_2_301254a0.pdf
Spain (2007) Real Decreto 661/2007, de 25 de mayo, por el que se regula la actividad de producción de energía eléctrica en régimen especial (Law regulating the activity of electricity production under the special regime). Official State Gazette n. 126, 26 May 2007, p. 22846. Retrieved 14 Apr 2015. http://www.boe.es/boe/dias/2007/05/26/pdfs/A22846-22886.pdf
Spain (2008) Real Decreto 1578/2008, de 26 de septiembre, de retribución de la actividad de producción de energía eléctrica mediante tecnología solar fotovoltaica para instalaciones posteriores a la fecha límite de mantenimiento de la retribución del Real Decreto 661/2007, de 25 de mayo, para dicha tecnología (Remuneration for photovoltaic installations created after the deadline set in the R.D. 661/2007). Official State Gazette n. 234, 27 September 2008, p. 39117. Retrieved 14 Apr 2015. http://www.boe.es/boe/dias/2008/09/27/pdfs/A39117-39125.pdf
Spain (2009) Real Decreto-ley 6/2009, de 30 de abril, por el que se adoptan determinadas medidas en el sector energético y se aprueba el bono social (Law approving extraordinary measures in the energy sector). Official State Gazette n. 111, 7 May 2009, p. 39404. Retrieved 14 Apr 2015. http://www.boe.es/boe/dias/2009/05/07/pdfs/BOE-A-2009-7581.pdf
Spain (2010a) Real Decreto 1565/2010, de 19 de noviembre, por el que se regulan y modifican aspectos relativos a la actividad de producción de energía eléctrica en régimen especial (Law regulating and modifying certain aspects relating to the production of energy production under the special regime). Official State Gazette n. 283, 23 November 2010, p. 97428. Retrieved 14 Apr 2015. www.boe.es/boe/dias/2010/11/23/pdfs/BOE-A-2010-17976.pdf
Spain (2010b) Real Decreto Ley 14/2010, de 23 de diciembre, por el que se establecen medidas urgentes para la corrección del déficit tarifario del sector eléctrico (Law establishing urgent measures to correct the tariff deficit). Official State Gazette n. 312, 24 December 2010, p. 106387. Retrieved 14 Apr 2015. http://www.boe.es/boe/dias/2010/12/24/pdfs/BOE-A-2010-19757.pdf
Spain (2011) Plan de energías renovables 2011–2020 (National renewable energy plan 2011–2020). Retrieved 14 Apr 2015. http://www.idae.es/uploads/documentos/documentos_11227_PER_2011-2020_def_93c624ab.pdf
Spain (2012a) Press conference following the Council of Ministers held on 27 January 2012. Retrieved 14 Apr 2015. http://www.lamoncloa.gob.es/consejodeministros/ruedas/Paginas/2012/cmrp20120127.aspx
Spain (2012b) Las Reformas del Gobierno de España—Determinación frente a la crisis. (Reforms by the Government of Spain—combating the crisis with determination). Retrieved 15 Apr 2015. http://www.lamoncloa.gob.es/Documents/lasreformasdelgobiernodeespa%C3%B1aespa%C3%B1olcongr%C3%A1ficos.pdf
Spain (2012c) Ley 15/2012, de 27 de diciembre, de medidas fiscales para la sostenibilidad energética (Fiscal reform in order to ensure the sustainability of the energy system). Official State Gazette n. 312, 28 Dec 2012, p. 88081. Retrieved 14 Apr 2015. http://www.boe.es/boe/dias/2012/12/28/pdfs/BOE-A-2012-15649.pdf
Spain (2013a) Real Decreto Ley 2/2013, de 1 de febrero, de medidas urgentes en el sistema eléctrico y en el sector financiero (Urgent measures in the electricity system and in the financial industry). Official State Gazette n. 29, 2 Feb 2013, p. 9072. Retrieved 14 Apr 2015. http://www.boe.es/boe/dias/2013/02/02/pdfs/BOE-A-2013-1117.pdf
Spain (2013b) Real Decreto Ley 9/2013, de 12 de julio, por el que se adoptan medidas urgentes para garantizar la estabilidad financiera del sistema eléctrico (Urgent measures to guarantee the financial stability of the electricity system). Official State Gazette n. 167, 13 July 2013, p. 52106. Retrieved 14 Apr 2015. http://www.boe.es/boe/dias/2013/07/13/pdfs/BOE-A-2013-7705.pdf

Spain (2014a) Real Decreto 413/2014, de 6 de junio, por el que se regula la actividad de producción de energía eléctrica a partir de fuentes de energía renovables, cogeneración y residuos (Law regulating the generation of electricity using renewables, cogeneration, and waste). Official State Gazette n. 140, 10 June 2014, p. 43879. Retrieved 14 Apr 2015. http://www.boe.es/boe/dias/2014/06/10/pdfs/BOE-A-2014-6123.pdf

Spain (2014b) Orden IET/1045/2014, de 16 de junio, por la que se aprueban los parámetros retributivos de las instalaciones tipo aplicables a determinadas instalaciones de producción de energía eléctrica a partir de fuentes de energía renovables, cogeneración y residuos (Ministerial order validating the remunerative parameters for installations using renewables, cogeneration, and waste). Official State Gazette n. 150, 20 June 2014, p. 46431. Retrieved 14 Apr 2015. http://www.boe.es/boe/dias/2014/06/20/pdfs/BOE-A-2014-6495.pdf

Spanish Constitution (Constitución Española) (1978)

Spanish Energy Commission (2013) Nota resumen del saldo de la deuda del sistema eléctrico a 10/05/2013 (Debt balance sheet of the electricity sector 10/05/2013). Retrieved 14 Apr 2015. http://www.cne.es/cne/doc/publicaciones/cne64_13.pdf

Spanish National Stock Market Commission (2014) Informe sobre la liquidación provisional 10/14 del sector eléctrico: análisis de resultados y seguimiento mensual de la proyección anual de los ingresos y costes del sistema eléctrico (Interim balance sheet 10/2014 for the electricity sector). Retrieved 14 Apr 2015. http://www.cnmc.es/Portals/0/Ficheros/Energia/Informes/Liquidaciones_Electricidad/20141218_Informe_Liquidaci%C3%B3n_10_2014_energia_electrica.pdf

Spanish Power Transmission Company (2014) Avance del informe del sistema eléctrico español 2014 (2014 advanced report on the Spanish electricity system). Retrieved 14 Apr 2015. http://www.ree.es/sites/default/files/downloadable/avance_informe_sistema_electrico_2014b.pdf

Spanish Supreme Court (2012) Ruling STS 2586/2012 on appeal for judicial review 35/2011. Fuente Álamo Fotoparque 1 et al. Retrieved 14 Apr 2015. http://www.poderjudicial.es/search/doAction?action=contentpdf&databasematch=TS&reference=6352769&links=%2235/2011%22&optimize=20120504&publicinterface=true

Tudor I (2008) The fair and equitable treatment standard in the international law of foreign investment. Oxford University Press, New York

United Nations (1997) Kyoto protocol to the United Nations framework convention on climate change. Kyoto. Retrieved 14 Apr 2015. http://unfccc.int/resource/docs/convkp/kpeng.pdf

United Nations Conference on Environment and Development (1992) Rio declaration on environment and development. Retrieved 15 Apr 2015. http://www.unep.org/Documents.Multilingual/Default.asp?documentid=78&articleid=1163

United Nations Conference on Trade and Development (2010) UNCTAD world investment report 2010: investing in a low-carbon economy. Retrieved 15 Apr 2015. http://unctad.org/en/Docs/wir2010_en.pdf

United Nations Conference on Trade and Development (2012) UNCTAD investment policy framework for sustainable development 2012. Retrieved 15 Apr 2015. http://unctad.org/en/PublicationsLibrary/diaepcb2012d5_en.pdf

United Nations Environment Programme (2011) Renewable energy: investing in energy and resource efficiency. Retrieved 14 Apr 2015. http://www.unep.org/greeneconomy/Portals/88/documents/ger/GER_6_RenewableEnergy.pdf

US Model BIT (2012) U.S. model bilateral investment treaty from 2012. Retrieved 15 Apr 2015. http://www.state.gov/e/eb/ifd/bit/

Wälde TW (2006) International Thunderbird Gaming Corporation v United Mexican States. NAFTA/UNCITRAL arbitration, separate opinion of Thomas Wälde of 26 January 2006. Retrieved 15 Apr 2015. http://italaw.com/sites/default/files/case-documents/ita0432.pdf

Wälde TW, Hobér K (2004) The first energy charter treaty arbitral award. Trans Dispute Manage 4:1–15. Retrieved 15 Apr 2015. http://www.transnational-dispute-management.com/article.asp?key=285

Wälde TW, Kolo A (2001) Environmental regulation, investment protection and 'regulatory taking' in international law. Int Comp Law Q 50:811–848

Yannaca-Small K (2008) Fair and equitable treatment standard: recent developments. In: Reinisch A (ed) Standards of investment protection. Oxford University Press, Oxford, pp 111–131

Arbitral Awards

AES Summit Generation Limited and AES—Tisza Erőmú Kft. v Hungary, ICSID Case No ARB/07/22, Award of 23 September 2010

Bayindir Insaat Turizm Ticaret Ve Sanayi AS v Islamic Republic of Pakistan, ICSID Case No ARB/03/29, Award of 27 August 2009

CME Czech Republic B.V. v Czech Republic, UNCITRAL, Partial Award of 13 September 2001

CMS Gas Transmission Company v The Argentine Republic, ICSID Case No ARB/01/8, Award of 12 May 2005

CMS Gas Transmission Company v The Argentine Republic, Argentina's Annulment Reply of 22 January 2007

Consortium RFCC v Royaume du Maroc, ICSID Case No ARB/00/6, Award of 22 December 2002

Continental Casualty v The Argentine Republic, ICSID Case No ARB/03/9, Award of 5 September 2008

Duke Energy Electroquil Partners and Electroquil SA v Ecuador, ICSID Case No ARB/04/19, Award of 18 August 2008

EDF (Services) Limited v Romania, ICSID Case No ARB/05/13, Award of 8 October 2009

Electrabel S.A. v Republic of Hungary, ICSID Case No ARB/07/19, Decision on Jurisdiction Applicable Law and Liability of 30 November 2012

El Paso Energy International Company v The Argentine Republic, ICSID Case No ARB/03/15, Award of 31 October 2011

Emilio Agustín Maffezini v The Kingdom of Spain, ICSID Case No ARB/97/7, Award of 13 November 2000

Glamis Gold Ltd. v United States of America, NAFTA/UNCITRAL, Award of 8 June 2009

International Thunderbird Gaming Corporation v The United Mexican States, NAFTA/UNCITRAL Award of 26 January 2006

LG&E Energy Corp., LG&E Capital Corp., and LG&E International, Inc. v Argentine Republic, ICSID Case No ARB/02/1, Decision on Liability of 3 October 2006

Marvin Feldman v The United Mexican States, ICSID Case No ARB(AF)/99/1, Award on Merits of 16 December 2002

Merrill & Ring Forestry L.P. v The Government of Canada, UNCITRAL/ICSID Award of 21 February 2010

Methanex Corporation v United States of America, NAFTA/UNCITRAL Award of 3 August 2005

National Grid v The Argentine Republic, UNCITRAL Award of 3 November 2008

Nykomb Synergetics Tech. Holding AB v Republic of Latvia, Arbitration Institute of the Stockholm Chamber of Commerce Case No 118/2001, Award of 16 December 2003

Occidental v Ecuador, LCIA Case No UN3467, Award of 1 July 2004

Parkerings Compagniet AS v The Republic of Lithuania, ICSID Case No ARB/05/8, Award of 11 September 2007

Saluka Investments B.V. (The Netherlands) v The Czech Republic, UNCITRAL Partial Award of 17 March 2006

Suez, Sociedad General de Aguas de Barcelona, S.A. and Vivendi Universal, S.A. v The Argentine Republic, ICSID Case No ARB/03/19, Decision on Liability of 30 July 2010

Técnicas Medioambientales Tecmed S.A. v The United Mexican States, ICSID Case No ARB(AF)/00/02, Award of 29 May 2003

Total S.A. v The Argentine Republic, ICSID Case No ARB/04/01, Decision on Liability of 27 December 2010

Ulysseas Inc. v The Republic of Ecuador, UNCITRAL Award of 12 June 2012

Waste Management, Inc. v The United Mexican States ("Number 2"), ICSID Case No ARB(AF)/00/3, Award of 30 April 2004

Horizontal and Vertical Integration of Sustainability into Policymaking, Planning and Implementation of Renewable Energy Projects—The New Zealand Model

David Grinlinton

Abstract While sustainable development is now widely accepted as an international normative principle guiding human interaction with the natural environment, it can be little more than an aspirational platitude unless incorporated in a practical and enforceable way in domestic regulation and processes. This paper addresses the vertical and horizontal integration of sustainability into policy-making, planning and decision-making with a particular focus on renewable energy developments. New Zealand is unique in the way it has incorporated the principle of sustainability as an enforceable concept in domestic legislation. The approach incorporates a hierarchical model with an environmental sustainability objective at the apex. This influences policy-making and "macro-planning" at the national and regional levels, which in turn influences lower level planning and operational decision-making. The system is integrated both vertically between different levels of government (central, regional and municipal), and horizontally between central and local government and resource management agencies, corporations, public interest groups and individuals. The system is a result of an exhaustive administrative and law reform process in the late 1980s-early 1990s that restructured central and local government agencies, and implemented major legislative reform. The underlying conceptual model may be described as "integrated environmental management" (IEM). The system has now been in place for 25 years, and has been continually amended and refined in that time. It provides a useful model of a considered and coherent approach that facilitates sustainable management of the environment and natural resource development, including encouraging greater uptake of renewable energy.

D. Grinlinton (✉)
Faculty of Law, The University of Auckland, Auckland, New Zealand
e-mail: d.grinlinton@auckland.ac.nz

D. Grinlinton
Faculty of Law, University of Western Ontario, London, ON, Canada

V. Mauerhofer (ed.), *Legal Aspects of Sustainable Development*,
DOI 10.1007/978-3-319-26021-1_21

Keywords Renewable energy • Sustainable development • Sustainable management of natural and physical resources • Integrated environmental management

New Zealand Legal Terms and Abbreviations Used in this Part

EC or EnvC or Env Ct	Environment Court
DC	District Court
HC	High Court
CA	Court of Appeal
SC	Supreme Court
NZSC	New Zealand Supreme Court
NZLR	New Zealand Law Reports
NZRMA	New Zealand Resource Management Appeals
Judge	Judge of the Environment Court or District Court
Justice	Judge of the High Court, Court of Appeal or Supreme Court
J	Abbreviation for Judge of the High Court, Court of Appeal or Supreme Court
JJ	Abbreviation for two or more Judges of the High Court, Court of Appeal or Supreme Court
CJ	Chief Justice

1 Introduction

This paper addresses the integration of sustainability into policy-making, planning and decision-making in regard to the management and use of natural resources with a particular focus on renewable energy projects.

Sustainable development is now widely accepted as an international normative principle providing governance principles for human interaction with the natural environment (Bosselmann 2008). However, the principle of sustainable development is little more than an aspirational platitude unless incorporated in a practical and enforceable way by states in domestic environmental and natural resource development regulation, and in relevant administrative and legal processes.

At the national level, governments develop policy, enact legislation and manage activities impacting upon the environment through ministries and agencies, and through control of activities on government owned land. Local government agencies and municipalities usually exert the most immediate environmental and natural resource management, often through town and country planning rules and the management of water and atmospheric emissions.

While internationally agreed principles such as "sustainable development" (Brundtland Report 1987, at 27)[1] and "the precautionary principle" (Rio Declaration 1992, Cameron 2006)[2] find expression in many international instruments, it is far more difficult to incorporate them into domestic regulation in a meaningful way. New Zealand is unique in the way it has incorporated the principle of "sustainable management of natural and physical resources" as an enforceable concept in domestic legislation (Resource Management Act 1991, Sect. 5). The approach incorporates a hierarchical model placing sustainability at the apex, which influences policy-making and "macro-planning"[3] at the national and regional levels, which in turn influences lower level planning and operational decision-making. The system is integrated both vertically between different levels of government (Central, Regional and Municipal), and horizontally between local government and other resource management agencies. The system requires integration of administrative bodies and government agencies, along with complementary regulatory reform, and enforcement agencies such as the "environment court". The underlying conceptual model may be described as "integrated environmental management" (IEM) (for development of the concept, see Mitchell 1986, pp. 13–26; Rabe 1986; Bartlett 1990, pp. 235–254; Grinlinton 1992).

The New Zealand system has been in place since 1991, and has been continually developed, refined and interpreted by the courts in specific environmental and resource development cases. A recent decision of the Supreme Court of New Zealand [*Environmental Defence Society Inc v The New Zealand King Salmon Co Ltd* (2014) NZSC 38] has clarified the way in which the sustainability principle is to be implemented through policy and planning instruments in specific cases. This paper will review that decision, and other decisions of the Courts, and apply them to renewable energy developments that are ongoing. In New Zealand, currently around 37 % of total primary energy supply is made up of renewable energy sources (primarily hydro and geothermal), and 75 % of electricity is generated from renewables (NZ Govt. 2014, pp. 3, 55). The government has an objective to achieve 90 % renewable electricity production by 2025 (NZ Govt. 2011, pp. 6, 9). The system therefore provides a useful model for other jurisdictions that hope to increase the uptake of renewable energy.

---

[1]The widely accepted definition of "sustainable development" is development that "meet[s] the needs of the present without compromising the ability of future generations to meet their own needs" (Brundtland Report 1987, 27).

[2]The "precautionary approach" was defined in the Rio Declaration (1992): "where there are threats of serious or irreversible damage, lack of full scientific evidence shall not be used as reason for postponing cost-effective measures to prevent environmental degradation". The application of the principle in the context of environmental risk management in New Zealand is discussed in Cameron (2006).

[3]In this context "macro-planning" includes national and sub-national policy statements and regulatory instruments dealing with higher-level environmental and resource management aims and objectives, as opposed to conventional town planning ordinances and rules.

2 The IEM "Model"—An Overview

The interrelationships between the various elements of the biosphere are complex and environmental and natural resource management should reflect this interdependence. An IEM approach to natural resource and energy developments should provide for policy-making, regulation and decision-making that accommodates related issues, flow-on consequences, and the cumulative effects of actions and activities. It must be applied not just through isolated reactive statutory measures, but across the full spectrum of administration, regulation and implementation, including (Grinlinton 2013, pp. 26–32):

- administrative structures,
- policy-making and planning,
- legislation and regulation,
- processes of participation and decision-making, and
- operational implementation including environmental monitoring, impact assessment and enforcement of actions and responsibilities.

Such a structure requires a sound philosophical foundation upon which the policy and regulatory system can be developed in an integrated and coherent way. As already mentioned, the international consensus on principles such as sustainability and the precautionary approach in environmental protection and energy development provide such a foundation. If these normative principles are incorporated into domestic policy and government action at the national level, they provide strategic guidance for specific environmental legislation and sub-national policy and regulation of environmental protection and resource management. Under this structure, local government normally has the primary role in the implementation of higher-level policy and regulation through specific planning instruments containing rules and procedures for resource use, and for enforcement of the system. The model is broadly illustrated in Fig. 1 (Grinlinton 2013, p. 33):

As with any "model" the divisions are not mutually exclusive, and there is some overlap in the purpose and content of policy and regulation at the normative, strategic and operational levels.

At the *normative level* the structure integrates the normative principle of sustainability through recognition of value premises and priorities that underlie environmental management systems, including policy-making, planning and decision-making processes (Grinlinton 1992, pp. 5–7). The progressive development of international instruments such as the *Stockholm Declaration* (1972), the *World Conservation Strategy* in 1980, the *Brundtland Report* (1987) and the *Rio Declaration* (1992) are illustrations of developing normative global principles of environmental management. The system is dynamic as it accommodates reconsideration of higher-level objectives and re-definition of desired goals and values on an ongoing basis to accommodate developments in science and technology, and increasing understanding of natural processes and ecosystem dynamics. The successes and failures of policy, regulation and other management tools are also

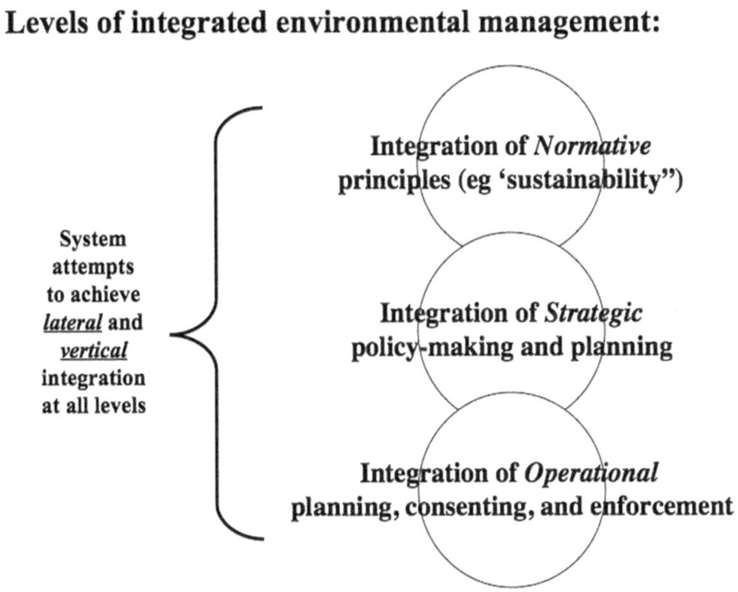

Fig. 1 Levels of Integrated Environmental Management (adapted from Grinlinton 2013, p. 33)

relevant in this context. As mentioned, the various levels of IEM overlap with the setting of objectives and policies at the national level for environmental management and natural resource use falling at the transition between normative and strategic levels of governance.

At the *strategic level* alternative goals and objectives, and the selection of the policy and regulatory means to achieve these, are addressed (Grinlinton 1992, pp. 8–11). This process may include administrative reform to provide for an integrated administrative and management framework to drive law reform and provide a management structure capable of implementing these elements. Legislation, policies and plans may be progressively introduced to provide the framework for implementation of the management structure. Macro-planning in the form of national policy statements and environmental standards, and sub-national policy instruments are appropriate, and fall at the transition between strategic governance and operational management.

At the *operational level* there is the allocation of specific responsibility for the various elements of resource management, usually to agencies of local government including regional and municipal levels (Grinlinton 1992, pp. 11–15). Micro-planning includes the preparation of regional and municipal planning instruments containing rules governing the use of natural resources such as water, land, and atmosphere, and providing procedures for obtaining consents to use those resources.

Such a model has been implemented in New Zealand over the last 25 years, and provides something of a case study for integrating sustainability into environmental governance, and specifically in relation to renewable energy development, at both institutional and regulatory levels.

3 Integrating Sustainability into Environmental Management and Natural Resource Development in New Zealand

New Zealand embarked on an ambitious process of environmental reform in the mid-1980s (Grinlinton 1995, pp. 14–23). The main features of this reform were:

- Institutional restructuring and rationalization of administrative governance through central and local government reform;
- Legislative reform providing for the management of land air and water through the new governance structure under the central guiding principle of "sustainable management of natural and physical resources" (RMA, Sect. 5).

A. **Administrative governance reforms 1986–90**

The following administrative reforms took place in New Zealand in the late 1980s to early 1990s (Palmer 2012, pp. 773–781):

- Establishment of a "Ministry for the Environment" under the Environment Act 1986;
- Creation of the Office of the Parliamentary Commissioner for the Environment (or "Environmental Ombudsman") under the same Act;
- Establishment of a Department of Conservation under the Conservation Act 1987 to manage New Zealand's 'conservation estate'; and
- Local government reform with the rationalization and restructuring of regional government and municipal authorities.

(i) *Central government restructuring*

The Environment Act 1986 established the Ministry for the Environment, and the separate Parliamentary Commissioner for the Environment (PCE), sometimes referred to as the "Environmental Ombudsman" (Environment Act 1986, Sect. 4).[4] The term "environment" was given an expansive meaning, encompassing ecosystems and their constituent parts, including all natural and physical resources, and the physical, social, economic, cultural and aesthetic aspects of an area

[4]The PCE was intended to be an independent "system guardian" for the environment responsible to Parliament rather than the Executive, and not subject to direction by a particular minister.

(Environment Act 1986, Sect. 2).[5] Further, the Act recognized in the management of natural and physical resources that a full and balanced account should be taken of the intrinsic values of ecosystems, all values placed by people on the quality of the environment, the rights of the Maori (the indigenous people of New Zealand), the sustainability of natural and physical resources, and the needs of future generations (Environment Act 1986, Preamble).

The Conservation Act 1987 established a new Department of Conservation, to have responsibility for administration of national parks and public (Crown) conservation lands. The Department has particular functions in advocating conservation and sustainable management of approximately 30 % of New Zealand's land area (Conservation Act 1987).[6]

(ii) *Local government restructuring*

Between 1988 and 1989, the Local Government Commission reviewed all existing local authorities, resulting in a substantial reduction of the number of public bodies covering 12 regions, and 74 districts. The geographic boundaries of the regions followed catchment areas, with the intent that comprehensive integrated management of water and soil conservation would be achieved. This biogeographical "catchment" approach reflects similar approaches in the EU under the Water Framework Directive (Directive 2000/60/EC), although as New Zealand is an island nation trans-border complications inherent in the EU measure do not arise.

Regional Councils were given responsibility for regional water planning, and to provide broad policy directions for land use planning, which would guide planning at the district level (Palmer 2012, pp. 36–47, 776–779).

B. **Law and policy reforms 1988–91**

In the late 1980s the newly created Ministry for the Environment developed and implemented a range of new policies and legislation. Underlying these environmental reforms was the desire to incorporate the normative principle of sustainability under a single integrated system of resource management (Palmer 2013 pp. 14–20; Grinlinton 2013 pp 26–39). The concept was consistent with the Brundtland Report of 1987 which gave general recognition to the objective of sustainable development recognizing intra-generational equity by redistribution of wealth, and

[5]The extent to which this "expansive" definition of environment influences the outcome when weighing sustainability concerns against other, more anthropocentric, interests is discussed below in 4. D.

[6]For further detail, see the Department of Conservation website: www.doc.govt.nz. Accessed 17 March 2015.

inter-generational equity through maintaining the viability of the ecosystem for the benefit of future generations. The primary mechanism[7] was the Resource Management Act 1991 ("RMA"), which came into force on 1 October 1991.

C. The Resource Management Act 1991

The RMA attempted to integrate into one statute the law relating to the management of land, air and water and replaced over 50 other Acts dealing with these matters. The overriding thrust of the legislation is to provide for integrated environmental and natural resource management (*Environmental Defence Society Inc v The New Zealand King Salmon Co Ltd* [2014] NZSC 38, [2014] 1 NZLR 593 (SC) at [9–11], [24] and [64] per Elias CJ, McGrath, Glazebrook and Arnold JJ).[8] The Act requires a holistic approach to planning and administration. It recognizes the balance required between environmental objectives, social and cultural objectives, and economic objectives.

(i) *The purpose and principles of the RMA*

The RMA has as its central purpose "…the sustainable management of natural and physical resources" (Sect. 5[1]). "Sustainable management" as defined in Sect. 5(2) contemplates enabling communities to provide for their social, economic and cultural wellbeing, while protecting the life-supporting capacity of the biosphere, and sustaining resources for the foreseeable needs of future generations:

(2) In this Act, sustainable management means managing the use, development, and protection of natural and physical resources in a way, or at a rate, which enables people and communities to provide for their social, economic, and cultural well-being and for their health and safety while—

- (a) sustaining the potential of natural and physical resources (excluding minerals) to meet the reasonably foreseeable needs of future generations; and
- (b) safeguarding the life-supporting capacity of air, water, soil, and ecosystems; and
- (c) avoiding, remedying, or mitigating any adverse effects of activities on the environment.

(Emphasis added).

[7]Other enactments have been passed which also incorporate the sustainability principle, including: The Forests (Amendment) Act 1993 (incorporating sustainable management of indigenous forest on private land); the Fisheries Act 1996 (incorporating a "sustainable utilization" principle in management of commercial fisheries); and the Energy Efficiency and Conservation Act 2000 (sustainable use of energy).

[8]See also *Falkner v Gisborne District Council* [1995] 3 NZLR 622 at 632 (integrated holistic system); *Auckland Regional Council v North Shore City Council* [1995] 3 NZLR 18 (CA) at 22–23 (District Plans must not be inconsistent with regional policies); *Canterbury Regional Council v Banks Peninsula District Council* [1995] 3 NZLR 189 (CA) (integration of instruments).

All functions and decision-making carried out under the Act are guided by this purpose, and must actively promote it. In this sense the Act itself provides a powerful statement of government policy.

The "sustainable management" purpose is possibly unique in domestic legislation. However, the definition has given rise to some difficulties in interpretation. The balance between the "management purpose" of providing for the wellbeing of communities appears to be qualified by so-called ecological "bottom lines" in Sect. 5(2)(a)–(c). However, the courts have taken the view that the words should be given a wide meaning of purpose and principles, rather than strictly subjugating the management purpose to the ecological bottom lines.[9] The prevailing view has been stated by the Environment Court in *North Shore City Council v Auckland Regional Council* [1997] NZRMA 59 at 94 as follows:[10]

> The method of applying Sect. 5 then involves an *overall broad judgment* of whether a proposal would promote the sustainable management of natural and physical resources. That recognizes the Act has a single purpose .... Such a judgment allows for comparison of conflicting considerations at the scale or degree of them, and their relative significance or proportion in the final outcome. (italics added by author).

In that case the Environment Court approved restricting the metropolitan urban limit around Auckland City in a location to protect an estuary from pollution.

This pragmatic view of the statutory purpose of sustainable management sees the concept, purpose, or ethic of sustainable management as establishing the *prime objective* of the measure, rather than requiring a narrow legalistic approach to the particular words.

Section 5 of the RMA was recently examined in some detail by the Supreme Court in *Environmental Defence Society Inc v The New Zealand King Salmon Co Ltd* [2014] NZSC 38, [2014] 1 NZLR 593 (*King Salmon*). The case concerned the effect of provisions in the *New Zealand Coastal Policy Statement 2010* ("NZCPS") in relation to aquaculture development in an area designated as being of outstanding natural character and having an outstanding natural landscape. The NZCPS is a statutory instrument (subordinate legislation) promulgated under the RMA. The appellants had applied for a change to the relevant planning instrument to reclassify salmon farming from a "prohibited activity" to a "discretionary activity" under the

---

[9]For example, in *New Zealand Rail Ltd v Marlborough District Council* [1994] NZRMA 70 at 86, Grieg J in the High Court upheld a consent to construct an export wharf in a natural part of a coastal area, as more important than conservation of the coastline.

[10]See also *New Zealand Rail Ltd v Marlborough District Council* [1994] NZRMA 70 (HC) at 86; *Royal Forest and Bird Protection Society of New Zealand Inc v Manuwatu-Wanganui Regional Council* [1996] NZRMA 241 (PT) at 269; *North Shore City Council v Auckland Regional Council* [1997] NZRMA 59 (EnvC) at 93–94 (the Environment Court uses the words "overall broad judgment" at 194); *Man O'War Station Ltd v Auckland Council* [2013] NZEnvC 233 at 35–47. See also: *Mangakahia Maori Komiti v Northland Region* [1996] NZRMA 193 (PT) at 215; *Genesis Power Ltd v Franklin District Council* [2005] NZRMA 541 (EnvC) at 228; and *Coromandel Watchdog of Hauraki Inc v Chief Executive of the Ministry of Economic Development* [2008] NZRMA 77 (CA) at 50.

RMA that may be given consent by the local authority if certain criteria are met. The Minister of Conservation had directed that a Board of Inquiry determine the application in the first instance.[11] The Board found that the proposed salmon farm would have significant adverse effects on the natural character and landscape of that area and that, as a consequence, policies 13(1)(a) and 15(a) of the NZCPS (which were concerned with preserving and protecting the coastal environment)[12] would not be complied with if the plan change were granted. Nevertheless, the Board considered that those policies, while carrying considerable weight, were not determinative. It decided that it was required to give effect to the NZCPS "as a whole", and applied the "overall broad judgment" approach relating to pt 2 of the RMA to grant the plan change (*King Salmon*, para [5]).

On appeal, the Supreme Court undertook a detailed analysis of the RMA, in particular pt 2 (purpose and principles) (*King Salmon*, paras [8]-[30]). It confirmed that the correct interpretation of "while" in Sect. 5(2) of the RMA, is that it means "at the same time as" (para [24(c)]). The Court further stated that the various elements of the Sect. 5(2) definition of sustainable management should be read as an "integrated whole", and that the elements in Sect. 5(2) (a)–(c) do not constitute a strict "environmental bottom line" in themselves (para [24(c)]). However, in considering the overall judgment approach, the Court noted that the approach could not be used to undermine or veto clear directive requirements of policies, plans and rules that have been prepared in accordance with the RMA (paras [106]-[149] for full discussion). The Court held that policies 13(1)(a) and 15(1)(a) and (b) of the NZCPS "provide something in the nature of a bottom line", and this was consistent with the definition of sustainable management in Sect. 5(2) of the RMA (para [132]). The Court disallowed the plan change as it did not give effect to policies 13 and 15 in the NZCPS.

The decision in *King Salmon* has much wider application than simply interpreting the NZCPS. The judgment provides useful clarification of the meaning of "sustainable management" as contained in Sect. 5 of the RMA, and its implementation through the hierarchy of policies and plans put in place by central and local government. It also confirms that such instruments may indeed contain limitations in the nature of "environmental bottom lines" that take effect in the planning and consenting process, and in the decisions of the courts on appeal.

---

[11]Acting on the advice of the Environmental Protection Agency pursuant to Sect. 147(1) of the RMA.

[12]Department of Conservation, *New Zealand Coastal Policy Statement 2010* (November 2010), policy 13 (preservation of natural character) is expressed as: "To preserve the natural character of the coastal environment and to protect it from inappropriate subdivision, use and development: … (a) avoid adverse effects of activities on natural character in areas of the coastal environment with outstanding natural character." Policy 15 (natural features and natural landscapes) is expressed as: "To protect the natural features and natural landscapes (including seascapes) of the coastal environment from inappropriate subdivision, use, and development: … (a) avoid adverse effects of activities on outstanding natural features and outstanding natural landscapes in the coastal environment …".

The RMA also includes a number of other "supplementary" purposes to guide policy and decision-makers. For example, matters in Sect. 6, RMA, include preservation of the coastal environment, wetlands lakes and rivers, and their margins, outstanding landscapes, and indigenous flora and fauna. Matters in Sect. 7, RMA, include the efficient use and development of natural and physical resources, the intrinsic values of ecosystems, the efficiency of the end use of energy, the effects of climate change and benefits of renewable energy, and maintenance and enhancement of environmental quality.

(ii) *The policy and planning structure under the RMA*

The RMA creates a vertically and laterally integrated structure for environmental management. It provides for central government policies, regional government policies and planning instruments, and territorial (city/municipal) level planning instruments. Each level of government has differing, but sometimes overlapping resource management responsibilities.

Vertical integration is achieved by the requirement that lower level plans and policies must "give effect to" higher level policies and plans (RMA, Sects. 67[2], [3], 75[3]).[13] Lateral integration is achieved by the requirement to consult with neighbouring councils, central government agencies, some NGOs and other interest groups regarding the effects on them of proposed policies and plans when preparing such instruments (RMA, Schedule 1, clause 3 [consultation]). Such consultation is guided by the purpose and principles in ss 5–8, RMA, and on the respective functions and responsibilities of those government agencies as provided for in Part 4 of the RMA.

Central government may promulgate "National Policy Statements" (NPSs) and "National Environmental Standards" (NESs) pertaining to various aspects of environmental protection and natural resource management.

Strategic planning and operational management of land air and water resources is largely devolved to regional councils and "territorial" authorities (Local Government Act 2002, Sect. 5[1] [meaning of "territorial authority"]; Palmer 2012 para [1.2]), with regional councils primarily responsible for managing water use and discharges into water, and district and city councils primarily responsible for land use.

(iii) *The "resource consent" permitting system*

People wishing to undertake activities with environmental effects are required to apply for "resource consents" (planning permissions).[14] Often a number of different

[13]For judicial discussion of the integration of policy and planning instruments see: *Application by the Canterbury Regional Council* [1995] NZRMA 110 at 111, and *North Shore City Council Application* [1995] NZRMA 74 at 86, where the planning Tribunal held that "regional council function must be able to impose some measure of restraint on management decisions made in exercise of territorial authority function".

[14]"Resource consents" include land use consents, subdivision consents, water permits, coastal permits and discharge permits under the RMA: Sects. 2, 87.

resource consents may be required for a particular activity. For example, renewable energy developments such as hydro, tidal and geothermal may require a range of planning permissions, including: land use permits for structures, transmission lines and access roads; and water use permits and discharge permits during the construction phase where there is water diversion, concentration, sedimentation or contamination, and for ongoing operation. Wind farms and solar developments would certainly require land use permits to operate, and possibly land use, water use and discharge permits during the construction phase. Department of Conservation approval may be required if the activity is on conservation land, or may have a significant effect on vegetation, wildlife and natural habitat.[15] Coastal permits may also be required from the Department of Conservation or Regional Council if the activity is in the coastal marine area (RMA, Sect. 12, 28, 56–58A, 89, 117, 119A). Marine consents may also be required for marine energy activities in the Exclusive Economic Zone or the extended continental shelf of New Zealand (Exclusive Economic Zone and Continental Shelf (Environment Effects) Act 2012, Sects. 3, 13).[16] The RMA provides detailed time schedules in Part 6 for processing resource consent applications.

The horizontally integrated nature of the system is illustrated by the resource consent application procedure. Applications for resource consents may be made on a publicly notified or non-notified basis in accordance with statutory notification criteria. The Council hears the application and, where notified, *any person* may make submissions. The decision must be made in accordance with the statutory purpose of "promoting sustainable management" and in accordance with the objectives and criteria in the Plan. The Plan, in turn, must not be inconsistent with any higher level regional or Government Policy statements, and is also subject to the sustainable management purpose.

Further horizontal integration of decision-making is providing for by "joint hearing committees" made up of representatives of the various consent authorities, and which can conduct hearings and grant all resource consents required in one hearing and decision-making process (see *AFFCO New Zealand Ltd v Far North District Council (2)* [1994] NZRMA 224, 233–234). Only when publicly notified are hearings for applications open to objections and submissions by any person without the need to have *locus standi* ('standing').[17]

---

[15]Such activities and effects are governed under both the Conservation Act 1987 and the RMA: Conservation Act 1987, Part 3B, and esp. Sect. 17P.

[16]The purpose of this Act is "to promote the sustainable management of the natural resources of the exclusive economic zone and the continental shelf": Sect. 10(1).

[17]Traditionally under the common law, "standing" requires the litigant to have a property interest or some special interest greater than the general community. While "any person" can theoretically make submissions and objections to proposed plans, and also to resource consent applications that are publicly notified, in reality less than 5 % of resource consent applications are notified, so "open participation" is very limited. Criteria for notification/non-notification are contained in RMA: Sects. 95A-95G.

In considering an application, the consent authority must have regard to the purposes and objectives of the RMA under Part II of the Act as outlined earlier and any relevant NPSs or NESs, regional policies.

Decisions at the council level can be appealed to the specialist Environment Court by both the applicant and any objectors.[18] This appeal can be on both law and merits issues. The Environment Court is also bound by the sustainable management purpose of the Act. Further appeals to the High Court. Court of Appeal and, in certain cases to the Supreme Court, can only be on matters of law such as procedural compliance, jurisdiction, and correct application of legal rules and principles, as opposed to merits issues (RMA, Sects. 299, 308; Supreme Court Act 2003). Judicial review is another avenue for redress in the higher courts.

In summary, the planning and resource consent system works as follows:

- Regional Councils make ‘Regional Plans’ dealing with issues such as water use, coastal area, and major land use management;
- District/City Councils make ‘District Plans’ dealing with subdivision, developments and land uses;
- Regional and District Plans contain ‘Rules’ for use of land, air and water;
- People must normally apply for ‘Resource Consents’ for activities involving the use of, or impact of activities upon, land, water and air;
- A hearing committee representing the Council(s) who must grant consent considers the application(s) and makes a decision;
- If the application will have a significant environmental and/or social effect, there may be a public hearing and objections;
- The principle of “sustainable management” guides decision-making;
- The resource consent is then granted (or declined);
- Parties can appeal to the Environment Court de novo (from the beginning; afresh);
- Appeal can be made to the High Court (and higher courts) on legal issues only, or via an application for judicial review.

(iv) *Enforcement under the RMA regime*

Failure to comply with the RMA, plans and rules made under it, or the conditions of resource consents, may constitute offences under the Act. Penalties include the possibility of heavy fines of up to $NZ 300,000 for individuals or $600,000 for corporations. Imprisonment for up to 2 years is also an option for individuals and corporate officers. Liability for the most serious offences is strict, and the Act provides for vicarious corporate liability (RMA, Sects. 338–341). Most of the prosecutions under the Act relate to pollution of waterways and groundwater, or removal of protected indigenous vegetation. Although rare, some prosecutions have resulted in

---

[18]The Environment Court (previously the Planning Tribunal) is a specialist judicial body set up to arbitrate and adjudicate on environmental disputes (Part 11, RMA).

sentences of imprisonment (Grinlinton 2009).[19] The Act also provides for pre-hearing conferences (RMA Sects. 99, 267), caucusing of experts (Bollard 2007), other forms of alternative dispute resolution (ADR) (RMA, Sects. 268, 356; Clapshaw 2009), and the use of imaginative sentencing options such as community service.[20]

4 Applying IEM to Renewable Energy Development in New Zealand

A. **Background**

Renewable energy uptake has experienced a dramatic increase in the last decade. The *Renewables 2014 Global Status Report* (Ren21 2014) indicates that by the end of 2013 renewables comprised more than 26 % of global electric generating capacity, and supplied around 22 % of electricity generated (Ren21 2014, p. 25). While the majority source was hydropower, modern renewables such as wind and solar (both photovoltaic [PV] and thermal) have shown strong growth. The US, China, Germany, Spain, Italy, Turkey, Brazil, and India accounted for the majority of generation growth in 2013, and New Zealand led the growth in geothermal power generation (Ren21 2014, p. 16).

Renewable resources accounted for 38.2 % of New Zealand's primary energy supply, and over 75 % of electricity production in 2013 (NZ Govt. 2014, p. 50). This is the fourth highest proportion of renewable sources to non-renewable sources for electricity production in the OECD. Hydro power accounted for 54.5 %, geothermal 14.5 %, wind 4.8 % and bioenergy (and other sources) 1.5 % of New Zealand's electricity generation in 2013. Gas contributed 19.4 % and coal 5.3 % (NZ Govt. 2014, pp. 56). While much of the hydro generation is from dams that were built in the mid to latter part of the twentieth century and therefore were consented under less rigorous planning and environmental protection regimes,[21] recent activity has centred on geothermal and wind energy developments. Many such developments have had to obtain consent under the RMA and therefore the

---

[19]Examples include: *Franklin District Council v McCollum* Unreported, District Court, CRN 3057005960, 14 February 1994 (pig farmer sentenced to 6 months imprisonment for polluting a waterway); *R v Conway* [2005] NZRMA (sentence of 3 months for pollution of waterways with oil and fuel upheld).

[20]For example, in *Smith v Auckland City Council* [1996] NZRMA 274 a sentence of 6 months community service/periodic detention was imposed for mortally damaging a landmark tree as a political protest.

[21]The transitional provisions of the RMA provide, in most cases, "deemed" consent for existing water uses for hydro dams, but these expired after 10 years, after which time hydro operators had to obtain water permits under the RMA, and consent authorities were able to impose more stringent conditions of consent at that time if appropriate: RMA, Sect. 386 (transitional provisions for water permits, including for hydro dams). See also: Sects. 123 (duration of resource consents), 128–132 (revision of consents).

planning and consenting process has been subject to the sustainable management purpose of the Act.

B. **Policy measures relevant to renewable energy development**

New Zealand is an active participant in many international instruments relevant to environmental protection and climate change (NZ Govt. 2015).[22] It has also incorporated the international principle of sustainable development in a number of statutes as a statutory objective or purpose. Such measures are themselves a statement of policy reflecting normative principles such as sustainability at a national level. For example, Sect. 5 of the RMA, can be viewed as both a statement of government policy enshrined in legislation, as well as an operative provision with binding effect in the national and sub-national policy-making, planning and decision-making framework contained in that Act.

The RMA also provides for National Policy Statements and National Environmental Standards to be promulgated on various matters. These NPSs and NESs have the effect of subordinate legislation, and are therefore binding on central government, local government and on the courts. Between themselves, NPSs and NESs are at the same level of subordinate legislation although are used for different purposes. NPSs are more general guidance policy documents containing matters such as objectives and higher-level policies that are relevant to achieving the purpose of the Act.[23] NESs contain more quantified technical performance standards for matters such as contaminants, water and air quality, soil quality, and noise standards.[24]

In 2011 the *National Policy Statement on Renewable Electricity Generation* (NPSREG) came into force. This measure recognizes as "matters of national significance" (NPSREG, p. 4):

(a) the need to develop, operate, maintain and upgrade renewable electricity generation activities throughout New Zealand; and
(b) the benefits of renewable electricity generation.

The Policy requires decision-makers to recognize the benefits of renewable electricity generation activities and to facilitate their implementation where possible (NPSREG, pp. 5–6). It also specifically requires local authorities to incorporate—within 24 months of the measure—objectives, policies and methods in regional policies and plans, and in district plans, to increase the uptake of renewable energy generation from sources such as solar, hydro, wind, tidal, geothermal and biomass (NPSREG, Policies [E1]–[E4]). Provision for small and community-scale generation is also provided for (NPSREG, Policy [F]). In *Meridian Energy Ltd v Canterbury*

[22]For example, New Zealand was one of 185 developed and developing countries that adopted the Framework Convention on Climate Change at the Rio Earth Summit in 1992. New Zealand also ratified the Kyoto Protocol to the UNFCCC in December 2002. While New Zealand has withdrawn from the Kyoto commitment period from 2013–2020, it has agreed to remain bound to take its next emissions reduction commitment directly under the UNFCCC.

[23]RMA: Sect. 45(1).

[24]RMA: Sect. 43.

Regional Council [2013] NZEnvC 70, the Environment Court examined the effect of the NPSREG on the decision-making process for granting resource consents for activities that may impact upon renewable energy. In that case Meridian Energy—a major hydro generator—had appealed against the grant of a water permit by the Canterbury Regional Council to a vineyard owner for water off-take for irrigation purposes from Lake Aviemore which Meridian relied upon for its hydro generation. The hearing commissioners who granted the consent had not considered the NPSREG in their deliberations, and the Court gave the following clear guidance of the importance of the measure in the decision-making process (at paras [9–10]):

> [9] The NPSREG was gazette on 14 April 2011. The Commissioners' Decision is dated five months later, but does not refer to it. That is of concern because the NPSREG contains policies which should have been had regard to. [Section 104(1)(b)(iii), RMA]
>
> … The NPSREG makes the continued availability of the upper Waitaki water a matter to which particular regard should be had– first by the Hearing Commissioners, and now on appeal, by this court.
>
> [10] It is curious that the Hearing Commissioners did not refer to the NSPREG at all. As a matter of law it should have been had regard to. …[25]

Clearly the NPSREG contains policies and considerations that must be given effect to in policies and planning instruments of local authorities, by decision-makers determining resource consents, and by the courts in determining appeals.

C. **Statutory measures relevant to renewable energy development**

The RMA contains a number of sections that address renewable energy. Section 5 (2) refers to promoting resource efficiency, sustaining resources for future generations, and "safeguarding the life-supporting capacity of air, water, soil, and ecosystems". These elements generally favour renewable energy proposals even where there may be some negative effect on other environmental or amenity values. Section 6 sets out a number of "matters of national importance" that must be recognized and provided for in achieving the purpose of the Act. They include matters such as the protection of the natural character of the coastal environment, and the protection of outstanding natural features and landscapes from "inappropriate" development. In recent decisions the courts have stated that more localized landscape and amenity effects of renewable energy developments will often be subordinated to national and global considerations favouring renewable energy, which are implicit in the sustainable management purpose in Sect. 5.[26] This is not

---

[25]See also NPSREG, paras [11] and [18–20].

[26]See, for example, *Genesis Power Ltd v Franklin District Council* [2005] NZRMA 541 at [228], *Unison Networks Ltd v Hastings District Council* [2007] NZRMA 340, *Meridian Energy Ltd v Wellington City Council*, Environment Court W031/2007, 14 May 2007 (Project West Wind). *Meridian Energy Ltd v Central Otago District Council* [2011] 1 NZLR 482, *Mainpower New Zealand Ltd v Hurunui District Council* [2011] NZEnvC 384, [2012] NZEnvC 21. *Final Report and Decision of the Board of Inquiry into the Hauauru ma Raki Wind Farm and Infrastructure Connection to Grid* (13 May 2011) (application by Contact Energy Ltd). *Final Report and Decision of the Board of Inquiry into the Turitea Wind Farm Proposal* (6 September 2011) (application by Mighty River Power Ltd).

always the case with the Court having some discretion (albeit guided by the s 5 "sustainable management" purpose), which has resulted, in some cases, in local property rights or indigenous cultural sensitivities being prioritized over the broader national and global benefits of renewable energy.[27] In Sect. 7 of the RMA decision-makers are required inter alia to *have particular regard to*:

(b) the efficient use of natural and physical resources; and
(ba) the efficiency of the end use of energy; and
(i) the effects of climate change; and
(j) the benefits to be derived from the use and development of renewable energy.

These sections have also been applied to favour renewable energy developments over other land uses in a number of cases over the last decade.[28]

D. **The wind energy cases**

A precursor to the line of wind energy cases discussed below was the decision in *Environmental Defence Society Inc v Auckland Regional Council* [2002] NZRMA 492. The Court had to consider whether greenhouse gas emissions and New Zealand's obligations under the UNFCCC and Kyoto Protocol were relevant considerations when considering an application for an air discharge permit under the RMA for a gas-fired power station. Although not part of domestic law, the Court determined that such international obligations could be taken into account under Sect. 104 of the RMA (at para [28]):

> The weight we give to the instruments is dependent on the nature of New Zealand's obligations under them, and the extent to which New Zealand government policy has crystallized, so as to give an indication of how New Zealand's obligation under the instruments will be given effect in New Zealand law.

In 2004 the RMA was amended to limit the extent to which such international instruments may be taken into account by decision-makers when considering applications for activities resulting in discharges of GHGs.[29] Nevertheless a line of decisions, starting in 2005, have weighed such international obligations against the effects on property owners and communities of wind energy developments. In some case the international obligations under such instruments as the UNFCCC have influenced the outcome in favour of the development; while in others private property and cultural interests have prevailed (Grinlinton 2007).

---

[27]See, for example, *Outstanding Landscape Protection Society v Hastings District Council*, unreported, Environment Court W24/07, 13 April 2007, where the Environment Court favoured the protection of Māori cultural values related to a ridgeline over the desirability of an extension to a windfarm.

[28]*Genesis Power Ltd v Franklin District Council* [2005] NZRMA 541 at 220–224, *Unison Networks Ltd v Hastings District Council* [2007] NZRMA 340 at 74, and *Meridian Energy Ltd v Wellington City Council*, Environment Court W031/2007, 14 May 2007 at 582.

[29]See discussion in *Greenpeace New Zealand v Northland Regional Council* [2007] NZRMA 87 at 39–57 per Williams J.

Genesis Power Ltd v Franklin District Council [2005] NZRMA 541 was an appeal against the refusal of the council to grant land use consent to construct and operate up to 18 wind turbines at a coastal site on the West Coast of the North Island of New Zealand. After a full examination of the evidence, and local objections from property owners and the proprietors of a horse-breeding operation, the Court upheld the appeal. In its decision it accepted that the proposal would have significant adverse effects on the natural character and landscape of the coastal environment contrary to Sect. 6 of the RMA, and those effects could not be mitigated. However, it noted that Sect. 6 was subordinate to the primary Sect. 5 purpose of sustainable management (para [55]), and this purpose would be best served by granting consent than refusing it. This was also supported by the various specific references to encouraging efficiency in the use of energy, addressing the effects of climate change, and the benefits of renewable energy in Sect. 7 of the RMA (para [228]):

> We find that the benefits of the proposal, when seen in the national context, outweigh the site-specific effects, and the effects on the local surrounding area. To grant consent would reflect the purpose of the Act as set out in Sect. 5.

Ultimately the development did not go ahead for economic reasons. The *Genesis* case was followed by *Unison Networks Ltd v Hastings District Council* [2007] NZRMA 340 which concerned appeals and cross-appeals by developers of two wind farm projects in a rural area on the East coast of the North Island of New Zealand. Approving modified consents for the developments, the court elevated the global and national objectives to reduce GHGs over the objections of property owners and local Māori—the latter objecting on cultural and spiritual grounds. The Court did, however, make the point that renewable energy generation would not always be favoured, stating (at para [82]):

> We make this decision on a site-specific basis. It may well be that other sites, perhaps for example more iconic in character, or closer to houses or clusters of population, will call for a different result. [Affirmed in *Unison Networks Ltd v Hastings District Council* [2011] NZRMA 394 (HC)]

This comment foreshadowed the decision in *Outstanding Landscape Protection Society Inc v Hastings District Council* (unreported, Environment Court W24/07, 13 April 2007) where objections by local Māori against granting consent to an extension of a windfarm along a ridge-line that had spiritual significance, were upheld. The court stated (at para [116]):

> Important as the issue of climate change and the use of renewable sources of energy unquestionably are, they cannot dominate all other values. The adverse effects of the proposal on what is undoubtedly an outstanding landscape, and its adverse effects on the relationship of Māori with this land and the values it has for them, clearly brings us to the conclusion that the tipping point in favour of other values has been reached.

Again, this case illustrates the discretion that the Courts have exercised to apply the overall judgment approach in balancing the many social, economic, cultural, ecological and intergenerational factors—many of which are often in tension—that are included in the s 5, RMA, purpose of sustainable management.

The decision in *Meridian Energy Ltd v Wellington City Council* (unreported, Environment Court, Wellington W 31-07, 14 May 2007) followed a month later with the Court approving a 70 turbine wind farm development on the West Coast of the North Island near Wellington. This case had raised a storm of protest from local landowners based on many matters including noise effects, effects on vegetation and habitat, amenity and visual effects, recreation and public access, and heritage effects. In granting consent, albeit modified in some cases as to location of turbines, the Court stated (para [459]):

> [W]e accept that there are concerns about a good number of turbines on various adverse effect grounds. In our judgement however the benefits to be gained from the project in terms of the promotion of sustainable management of natural and physical resources in terms of Sect. 5 [RMA], as informed by Sect. 6 and Sect. 7 factors we have reviewed, outweigh those concerns in respect of the great majority of turbines.

In 2009 a major wind farm development (Project Hayes), comprising some 176 turbines to be erected along the iconic and largely unspoiled Lammermoor Range in Central Otago in the South Island of New Zealand, was declined by the Environment Court: *Maniototo Environmental Society Inc v Central Otago District Council* (unreported, Environment Court, Christchurch, C103/2009, 6 November 2009). The Court considered the development would have a major adverse visual impact, and held that the applicant had failed to conduct a thorough analysis to show alternative sites were not available (at para [757], per Judge Jackson summarising the Court's reasoning). Meridian Energy—a Crown-owned energy generation company—successfully appealed the decision to the High Court: *Meridian Energy Ltd v Central Otago District Council* [2011] 1 NZLR 482. While acknowledging that the availability of alternative sites could be considered by the decision-maker when assessing the application, the Court held that the applicant should not be required to undertake a full economic cost-benefit analysis of alternative sites (para [123]). Despite its success on appeal, Meridian Energy ultimately did not proceed with the development.

A number of other wind farm applications have gained consent, or modified consent, sometimes following appeals, or consideration by a Board of Inquiry.[30] In 2011 another major development of up to 168 turbines—the Hauauru Wind Farm Project—received approval despite objections from local landowners regarding visual amenity and detraction from landscape views, and objections based on the ecological effects of the development: see *Final Report and Decision of the Board of Inquiry into the Hauauru ma Raki Wind Farm and Infrastructure Connection to Grid* (13 May 2011). In this case local Māori groups had entered into environmental compensation side agreements with the developer to address their concerns. Local landowners did not receive financial compensation for detraction from their views.

[30]This is an alternative procedure for dealing with proposals of "national significance": see RMA, Part 6AA, and esp. Sect. 142(2)(a) and (3).

Many conditions were imposed in the consent, including measures to control sediment, dust, glare and noise, relocation of a bat colony, measures to address the impact on birds and terrestrial invertebrates, road and air safety, radio and TV interference avoidance, and cultural effects (see Volume 2, Conditions and Schedules).

Also in 2011 an application by Mainpower New Zealand Ltd to construct 67 turbines at Mt Cass in the central South Island high country was also successful despite objections on landscape and ecological grounds: *Mainpower New Zealand Ltd v Hurunui District Council* [2011] NZEnvC 384, and see [2012] NZEnvC 21. The Environment Court imposed a number of conditions to address those issues, including undertaking pest and weed control, and assisting regeneration of indigenous species ([2011] NZEnvC 384, paras [485–488] and attached Conditions).

In *Contact Energy Ltd v Manawatu-Wanganui Regional Council* [2011] NZRMA 155 the Environment Court approved a 52–58 turbine development on a farm in a relatively unique karst landscape. Objectors were concerned with adverse effects on water quality and supply due to construction effects, visual impact and the effects on a nearby horse stud. Again the positive benefits of renewable energy overrode the negative effects on property rights and amenity.

Also in 2011, consent was granted for 60 turbines in the Turitea Wind Farm development. Although the applicant had sought approval for 122 turbines, the Board of Inquiry reduced this by 50 % to reduce the adverse effects on an outstanding natural landscape, and to avoid excessive clearance of indigenous vegetation of high ecological value. The applicant and local Māori had also entered into a side agreement which provided various financial benefits: *Final Report and Decision of the Board of Inquiry into the Turitea Wind Farm Proposal* (6 September 2011).

Further developments have been approved near Wellington in the North Island (*Meridian Energy Ltd v Wellington City Council* [2011] NZEnvC 406), and Mt Cass in the central South Island (*Re Meridian Energy* [2013] NZEnvC 59).

One interesting paradox is that, although the RMA, and subordinate legislative instruments such as the NPSREG, require positive consideration in the planning and decision-making process for renewable energy developments, the converse does not apply. In *Genesis Power Ltd v Greenpeace New Zealand Inc* [2008] 1 NZLR 803 the Court of Appeal held that climate change issues may not be considered as a negative factor militating *against* applications for fossil fuel driven power stations (see paras [39–44] per William Young P (President of the Court)).

E. **Other renewable energy cases**

Several recent decisions for hydro, tidal and geothermal generation further highlight the integrated environmental management approach to renewable energy in New Zealand.

Although hydro generation has provided the backbone of New Zealand's electricity generation for many decades, and still provides the majority of generation, it is unlikely that any large-scale hydro development will take place in the foreseeable future. There are a number of reasons for this including: the withdrawal of government from building and operating electricity generation schemes; the stricter limitations on environmental impact that flow from the RMA regime; the many demands placed on rivers and lakes by recreational and agricultural users; the establishment of a number of *Water Conservation Orders* under the RMA restricting damming or diversion of major rivers (RMA, Sects. 199–217; Nolan 2015 paras [8.56]–[8.80]); and Māori claims to water bodies (Palmer and Grinlinton 2014, pp. 257–258).

The one major development to be proposed since the RMA was enacted was "Project Aqua" involving several hydro dams on a major South Island river system which would take 73 % of the water flow. This met with substantial opposition from farmers who took water for irrigation, recreational users and local Māori. The Government, under the Resource Management (Waitaki Catchment) Amendment Act 2004, set up a Board of Inquiry which recommended a scheme of allocation between competing users, leaving significantly less for hydro generation. Although the hydro scheme did not proceed it did lead to a more integrated approach to fresh water management and the promulgation under the RMA of a National Policy Statement on Freshwater in 2011.[31] It is likely that in the future only smaller hydro developments—of which there is considerable potential—will be successful in the resource consent approval process.

More recently a proposal was floated to harness the energy from the tidal flow of water in and out of the Kaipara Harbour on the West Coast of the North Island north of Auckland. The applicants Crest Energy Ltd wished to place 200 × 1 MW turbines on the seabed in the main channel at the mouth of the harbour. Objectors included fishermen concerned with effects on spawning grounds and net obstruction, environmentalists concerned about the effects on rare species of dolphin and other marine life, and local Māori concerned with both ecological and cultural effects. An interim decision was issued by the Environment Court in 2011 (*Crest Energy Kaipara Ltd v Northland Regional Council* (unreported), EnvC Auckland A 132/2009, 22 December 2009; Wright and Leary 2011). A final decision granting a full 35 year consent term was given in 2011 (*Crest Energy Kaipara Ltd v Northland Regional Council* [2011] NZRMA 420), but the development remains on indefinite hold mainly due to technical and economic factors. However, the application and appeal process further illustrates the integrated system of environmental management incorporating consideration by central and local government, and accommodating objections and submissions from interested parties, all under the guiding principle of sustainable management.

Geothermal energy generation has experienced very significant growth in recent years, with New Zealand in 2013 having the highest growth globally in the utilization of this resource (NZ Govt. 2014, p 50). In 2013 almost 15 % of New Zealand's electricity was provided from geothermal generation, and this figure has

[31]See: www.mfe.govt.nz. Accessed 17 March 2015. The NPS focuses on best use of fresh water and prevention of overallocation.

been growing by around 2–4 % per annum in recent years (NZ Govt. 2014, pp 55–56). The *Te Mihi* geothermal proposal in 2008 provides a good example of the application of the IEM approach for this type of renewable energy. The application was dealt with by a Board of Inquiry due to its national significance. A number of matters were considered, including land stability, subsidence, discharge of contaminants into the air, and reinjection of surplus water and steam. The project was approved in 2008 on a similar basis to the wind energy cases already discussed; i.e. that the project would contribute to more sustainable and efficient use of energy and address climate change issues in line with New Zealand's commitments under the UNFCCC and the Kyoto Protocol: see *Final Report and Decision of the Board of Inquiry Te Mihi Geothermal Power Station Proposal* (3 September 2008).[32] The 250 MW *Tauhara 2* geothermal development project being developed by Contact Energy and the Tauhara Moana Trust in the central North Island was also approved in December 2010: *Final Report and Decision of the Board of Inquiry into the Tauhara II Geothermal Development Project* (10 December 2010).[33]

Geothermal energy development has also involved joint development between generating companies and local Māori. The *Ohaaki* station was built on land leased from local Māori, and also utilizes surplus energy for heating for an adjacent glasshouse horticultural operation. Although the station's electricity production has not met original expectations, it does provide an example of cooperative developments integrating local communities and indigenous groups in the business side of the development (Waikato Regional Council 2015). The *Tauhara 2* development is also a cooperative venture with a local Māori landowning trust.

5 Conclusions and Implications for Sustainable Development

A. Conclusions

Effective environmental governance requires an integrated system of environmental and natural resource management. Integration must occur at a number of levels. First, and probably foremost, the system must have strong normative guiding principles. The sustainability principle provides this. Secondly, these principles must be fully integrated into every level of administration, policy-making, regulation and implementation of the system. This may require significant reform of governance and administration organs. Thirdly, the system itself must be part of an

[32]The station is sited 5 km west of the 1958 Wairakei station which will in time be phased out of production (except for an existing binary station commissioned in 2005): http://www.mfe.govt.nz/rma/call-in-temihi/. Accessed 17 March 2015.

[33]See: http://www.epa.govt.nz/Publications/thii-boi-report-vol-01.pdf (vol 1), and http://www.epa.govt.nz/Publications/thii-boi-report-vol-02.pdf (vol 2). Accessed 17 March 2015.

integrated environmental regulatory and management structure reflecting the interrelatedness of all ecological elements within the biosphere.

New Zealand has implemented such a system. Integration of administrative structures has occurred with restructuring of environmental, conservation and natural resource management functions between central and local government. Integration of environmental and resource legislation has also occurred with environmental resource management policy-making, planning and decision-making now largely under a unified legislative regime. The RMA, while not without some flaws, provides a useful example of integrated environmental and natural resource management based on, and guided by, its sustainable management purpose.

Renewable energy developments are implemented through rules in planning instruments, and the grant of resource consents (planning permissions) by local authorities allocating rights in the necessary natural resources (e.g., land, water, geothermal steam). Vertical integration is provided by the overarching legislative purpose in the RMA of sustainable management of natural and physical resources, which permeates the entire policy, planning and consenting process at central and local government levels, and through the appeals process in the Environment Court. Policies promulgated under the RMA (such as the NPSREG) also guide the contents of plans and the decision-making process, and regional and district plans contain the criteria and rules that govern the grant of resource consents. Horizontal integration is provided by requirements to consult with neighbouring authorities and government agencies when preparing policies and plans, and with the joint hearing approach that allows combined decision-making on related consent applications when consents are required from different agencies. Broad rights of public participation in policy and plan preparation, and in consent applications, allow further integration of the interests of other stakeholders and interested parties.

There are of course some failings in the system, and these include (Grinlinton 2013, pp 39–46):

- A lack of coordinated planning and policy-making at local government level;
- Lack of detailed requirements for environmental impact assessment and independent auditing;
- Limitations on full public participation in resource consent hearings;
- Variable and light-handed enforcement by local authorities for breaches of the RMA or conditions in resource consents. Private citizens and environmental organizations are discouraged from bringing enforcement actions because of the risk of large costs and damages awards if unsuccessful.

The broad discretion allowed decision-makers and courts—albeit within the boundaries of the sustainable management purpose and principles in ss 5–8, RMA, and polices and rules in plans—may be a failing if inappropriately exercised. On the other hand the flexibility provided by such discretion is a strength if appropriately used to advance sustainable management of natural and physical resources. There are of course examples both ways, but on balance it is this writer's view that the advantages outweigh the disadvantages especially if the focus of the decision-makers is on "effects" of activities rather than the rigid "zoning" approach of traditional town planning.

Another criticism may be the lack of specific protections for species and other elements of the biosphere at the national level in the Act itself. The RMA was not intended to provide absolute prescriptive critical limits for species and natural resources. Rather it provides a framework for establishing national policies and standards, and regional and local limits and rules that may deal with those matters in more detail. Other legislation also provides specific protection for species and elements of the biosphere (e.g., Conservation Act 1987, Biosecurity Act 1993, Trade in Endangered Species Act 1989, Wildlife Act 1953, Fisheries Act 1996).

B. Implications for Sustainable Development

This chapter argues that an effective and workable theoretical IEM model incorporating sustainable management of natural and physical resources can be implemented domestically by states through structural governance change, and law reform. Such a system can effectively promote and facilitate greater renewable energy development, although the trans-boundary implications must be considered for countries that share land borders, or have close maritime borders, with other countries.

Despite some shortcomings noted above, the New Zealand environmental management regime provides a useful IEM model, not only to encourage greater uptake of renewable energy, but as a system to facilitate the sustainable management of natural and physical resources generally. In many ways New Zealand is a fortunate country. It has a secure and effective democratic structure, a capable and impartial judiciary, a reasonably affluent and well-educated society, a low level of corruption, and a sparsely populated and unspoilt natural environment. As these characteristics are often lacking in other jurisdictions the extent to which the IEM structure may be applied elsewhere will depend upon many factors, including:

- The political and governmental structure—both central and local government levels;
- The ability to implement coherent, comprehensive and integrated legislation and regulation;
- Clear allocation of responsibilities and the vesting of appropriate powers of regulation and enforcement in central and local government environmental and resource management agencies;
- Strong leadership and the impartial and effective exercise of those functions and powers;
- The expertise and independence of the legal profession and the judiciary;
- The level of consultation and participation allowed to public and private agencies, private individuals, landowners and developers;
- Availability of full information on the state of the environment when making policies, plans or decisions of specific developments; and
- The ability to consider cumulative and "cocktail" effects of activities in any decision-making processes.

References

Bartlett R (1990) Comprehensive environmental decision-making: can it work? In: Vig NJ, Kraft ME (eds) Environmental policy in the 1990s: toward a New Agenda. CQ Press, Washington, DC, pp 235–254

Brundtland Report (1987) World commission on environment and development our common future. Oxford, New York

Bollard J (2007) Caucusing of experts. Butterworths Resour Manag Bull 7:141

Bosselmann K (2008) The principle of sustainability. Ashgate, Aldershot

Cameron L (2006) Environmental risk management in New Zealand—Is there scope to apply a more generic framework? New Zealand treasury policy perspectives paper 06/06. New Zealand Government, Wellington

Clapshaw D (2009) Restorative justice in resource management prosecutions—A facilitator's perspective. Butterworths Resour Manag Bull 8:53–55

Grinlinton D (1992) Integrated resource management: a model for the future. Environ Plan Law J 9:4–19

Grinlinton D (1995) Natural resources law reform in New Zealand—Integrating law, policy and sustainability. Aust J Nat Resour Law Policy 2:1–37

Grinlinton D (2007) Property Rights, the "public interest" and global considerations: The case of wind energy development. Butterworths Resour Manag Bull 7:62–66

Grinlinton D (2009) Sentencing under the RMA. Butterworths Resour Manag Bull 8:33–38

Grinlinton D (2013) Integrating sustainability into environmental law and policy in New Zealand. In: Bosselmann K, Grinlinton D, Taylor P (eds) Environmental law for a sustainable society, 2nd edn. NZCEL, Auckland, pp 21–47

International Union for Conservation of Nature and Natural Resources, United Nations Environment Programme, World Wildlife Fund, Food and Agriculture Organization of the United Nations, and UNESCO (1980) World conservation strategy: living resource conservation for sustainable development. Gland, Switzerland: IUCN (World Conservation Strategy, 1980)

Mitchell B (1986) The evolution of integrated resource management. In: Lang R (ed) Integrated approaches to resource planning and management. University of Calgary Press, Calgary, pp 13–26

NZ Govt. (2014) New Zealand government, ministry of business, innovation and employment. Energy in New Zealand 2014. MBIE, Wellington. Retrieved March 17, 2015, from http://www.med.govt.nz/sectors-industries/energy/energy-modelling/publications/energy-in-new-zealand

NZ Govt. (2011) New Zealand government, ministry of economic development. New Zealand energy strategy 2011–2021. Issued under the Energy Efficiency and Conservation Act 2000. Wellington, MED. Retrieved March 17, 2015, from http://www.med.govt.nz/sectors-industries/energy/pdf-docs-library/energy-strategies/nz-energy-strategy-lr.pdf

NZ Govt. (2015) New Zealand government, ministry for the environment. United Nations framework convention on climate change. Wellington, MfE. Retrieved March 17, 2015, from http://www.mfe.govt.nz/climate-change/international-forums-and-agreements/united-nations-framework-convention-climate

Nolan D (ed) (2015) Environmental and resource management law. LexisNexis, Wellington

Palmer KA (2012) Local authorities law in New Zealand. Brookers, Wellington

Palmer KA (2013) Origins and guiding ideas of environmental law. In: Bosselmann K, Grinlinton D, Taylor P (eds) Environmental law for a sustainable society, 2nd edn. NZCEL, Auckland, pp 3–20

Palmer KA, Grinlinton D (2014) Developments in renewable energy law and policy in New Zealand. J Energy Nat Resour Law 32(3):245–272

Rabe BG (1986) Fragmentation and integration in state environmental management. The Conservation Foundation, Washington, DC

Ren21 (2014) Renewables 2014 global status report. Paris: REN21 Secretariat. Retrieved March 17, 2015, from http://www.ren21.net/ren21activities/globalstatusreport.aspx

Rio Declaration (1992) Rio declaration on environment and development. UN Doc. A/CONF.151/26 (vol. I); 31 ILM 874 (1992)

Stockholm Declaration (1972) Declaration of the United Nations conference on the human environment. U.N. Doc. A/Conf.48/14/Rev. 1(1973); 11 ILM 1416 (1972)

Waikato Regional Council (2015) Waikato Regional Council, Ohaaki. Hamilton, NZ. Retrieved March 17, 2015, from http://www.waikatoregion.govt.nz/Environment/Natural-resources/Geothermal-resources/Geothermal-systems-map/Ohaaki/

Wright G, Leary D (2011) Marine energy. N Z Law J 227

Tackling Climate Change Through the Elimination of Trade Barriers for Low-Carbon Goods: Multilateral, Plurilateral and Regional Approaches

Christopher Frey

Abstract While the trade and environment debate has long focused on the adverse impact of environmental measures on free trade and vice versa and on securing policy space to regulate in the environmental interest, the focus has only recently started to shift to the potentially reinforcing relationship between trade and sustainable development. The present contribution explores this notion by discussing the liberalization of trade in low-carbon goods as a contribution to the fight against climate change. It provides an overview over the existing trade barriers and the efforts to eliminate these barriers for renewable energy goods and energy efficient technologies. The contribution furthermore identifies the liberalization potentials and main challenges in the multilateral, plurilateral and regional contexts. Perhaps surprisingly, mega-regional trade agreements are identified as the most promising venue to pursue ambitious liberalization efforts for low-carbon goods, if trade negotiators were willing to make this a central part of the agreements.

Keywords Environmental goods • Renewable Energy • Non-tariff barriers • Trade facilitation • Climate change

1 Introduction

The relationship between sustainable development and international trade can be characterized as ambiguous. On the one hand, most of the contemporary environmental problems are associated with economic growth and increased industrial output. Growing volumes of global trade contribute significantly to rising greenhouse gas emissions (Peters and Hertwich 2008). On the other hand, the increas-

C. Frey (✉)
Dresden University of Technology, Dresden, Germany
e-mail: christopher.frey@tu-dresden.de

C. Frey
University of Canterbury School of Law, Christchurch, New Zealand

V. Mauerhofer (ed.), *Legal Aspects of Sustainable Development*,
DOI 10.1007/978-3-319-26021-1_22

ingly global nature of environmental threats requires a range of responses including the exchange of mitigation technology and associated services. As the outcome document of the 2012 United Nations Conference on Sustainable Development "The Future We Want", recalls, international trade can assist countries in their advance towards sustainable development, among others through trade in environmental goods and services (UN 2012, para. 281). In a similar vein, the Johannesburg Plan of Implementation of the 2002 World Summit on Sustainable Development called for establishing and strengthening trade agreements with a view to achieving sustainable development and for supporting initiatives for the creation and expansion of domestic and international markets for environmentally friendly goods and services (UN 2002, para. 99).

Climate Change is a prime example of this ambiguous but ultimately reinforcing relationship between trade and sustainable development. While it is true that the transportation of goods around the world is responsible for a large part of global CO_2 emissions, trade in goods and services associated with the abatement of climate change helps countries to replace fossil fuels and energy-intensive technologies and consequently reduce their carbon emissions.[1] This potential contribution of the international trade regime to the fight against climate change has gained increasing attention in different fora in recent years.[2] At the outset of the "Doha Development Round" of multilateral trade negotiations in 2001, the Members of the World Trade Organization (WTO) agreed on a mandate for the reduction or elimination of barriers to trade in environmental goods and services (EGS). In line with the general pace of the Doha Development Agenda, progress on environmental goods trade liberalization has been slow. In July 2014, a group of 14 WTO Members including the U.S., the EU and China formally launched negotiations on a plurilateral "Environmental Goods Agreement" (EGA) which aims at the elimination of tariffs on a range of goods deemed beneficial for the environment (European Commission 2014a). Similar negotiations in the framework of the Asia-Pacific Economic Cooperation (APEC) have progressed even further as APEC Members have agreed on reducing applied tariffs on 54 categories of environmental goods to five percent or less until the end of 2015 (APEC 2011).

At present, world trade relations are being fundamentally altered through the rise of trans-regional or "mega-regional" trade agreements of which the Transatlantic Trade and Investment Partnership (TTIP) and the Trans-Pacific Partnership (TPP) are two prominent examples. This recent trend prompts the question what potential the self-proclaimed "21st century trade agreements" (USTR 2015) offer for committing to ambitious liberalization of trade in environmental goods.[3]

This contribution seeks to give an overview over the current state of affairs of climate-friendly goods trade liberalization as well as to outline the potentials and stumbling blocks for reaching agreement in the different fora. The definitional

---

[1]See below, part 2.

[2]See below, part 4.

[3]See below, part 4.5.

question of which types of goods should be labelled as "environmental" for the purpose of the elimination of trade barriers has been one of the most contentious points during negotiations. While such questions of coverage are ultimately for the negotiating parties to be decided, it is submitted here that low-carbon goods would be a good starting point for liberalization because of the common (if differentiated) responsibility of the community of states to fight climate change.[4] Hence, this contribution focuses on two subsets of environmental goods—those related to the generation of energy from renewable sources and energy-efficient goods. As environmental goods and associated services are often provided on an integrated basis (e.g. installation and maintenance of wind turbines), there is a strong case to be made for combining efforts to facilitate trade in low-carbon goods and services. Services liberalization will however not explicitly be addressed here as this would go beyond the scope of this publication (on environmental services liberalization see Monkelbaan 2013).

2 The Case for Liberalizing Trade in Low-Carbon Goods

The commitment to cut carbon emissions, stabilize the concentration of greenhouses gases in the atmosphere and reduce the effects of global warming is a responsibility which all but two of the currently 161 WTO Members share. With the exceptions of Hong Kong and Macao, to which the United Nations Framework Convention on Climate Change (UNFCCC) provisionally not applies, all WTO Members have acceded to the UNFCCC. A slightly lesser number of WTO Members has committed to more concrete emission reductions through ratification of the Kyoto Protocol (KP). During the 21st Conference of Parties to be held in December 2015 in Paris, the UNFCCC Members will likely agree on a text to succeed the Kyoto Protocol.

The UNFCCC and the KP both call on their members to design their policies in accordance with the obligations they have entered into in the context of international trade. Article 3 of the UNFCCC lays down the commitment to an "open international economic system" as one of the fundamental principles. The mechanisms foreseen in the climate treaties themselves are all market-based. The approach of the international climate regime is not to restrict international trade because of its negative environmental externalities. Rather, it is to encourage using international trade as an enabling environment for reaching climate targets. The most obvious contribution would be an agreement to facilitate trade in those goods and services that are directly associated with the abatement of climate change. Renewable energy technologies are a good case in point. The economic opportunities of a shift towards a low-carbon economy have been recognized by many countries and policies have been implemented at the national level as well as in some regional settings to initiate this transition. (European Commission 2011).

---

[4]See below, part 2.

While the international community has been struggling to agree on a meaningful and urgently needed Post-Kyoto policy framework, the global economy is witnessing a radical shift towards cleaner and more climate-friendly technologies. Global markets for renewable energies are growing at a rapid pace. Reflecting the increasing economic significance of low-carbon technologies, renewable energies have recently taken center-stage at the WTO, as a number of disputes involving solar and wind power have reached the dispute settlement mechanism.[5] The WTO Dispute Settlement Body (DSB) has been engaged by Members to deal with trade remedies for solar power and alleged violations of the WTO Agreements brought about by support measures for renewable energy. While the DSB has always been a prominent forum for addressing environmental questions with trade implications (Carlarne 2006, p. 53), it is increasingly confronted with issues of green energy and climate change. The recent cases parallel a development in international investment law, where a number of cases concerning support measures for renewable energy are currently pending before investment tribunals (Dromgool and Ybarra 2015). The rising trade tensions surrounding solar and wind energy demonstrate that there is a need to address specific non-tariff barriers which impede free trade in these sectors. Agreement on the reduction or elimination of such trade barriers would send a strong signal that trading partners are willing to give up their protectionist stance and allow goods associated with the shift to a low-carbon economy to circulate freely. Importantly, a glance at the trade flows of clean energy goods reveals that there is strong theoretical potential for climate abatement through targeted trade liberalization because the main greenhouse gas emitting countries are also those which have a strong manufacturing base in renewable energy generation equipment (Sugathan 2013, p. 4). For example, China is not only by far the principal emitter of CO_2, but also the top exporter of solar panels and continuously gaining market share in wind energy (REN21 2014). The economic and climate benefits of opening markets and phasing out barriers to trade in environmental goods have been highlighted in several studies and reports (UNEP 2013; OECD 2009).

3 Barriers to Trade in Low-Carbon Goods

As the row of recent trade disputes over solar panels has illustrated, a free global market in solar energy has not yet emerged (for a recent overview of these trade disputes see Kasteng 2013) and trade barriers are often prohibiting innovative producers to sell their products in important emerging markets. For wind energy,

---

[5]These are Canada—Renewable Energy, WT/DS412; China—Measures Concerning Wind Power Equipment, WT/DS419; Canada—Feed-in Tariff Program, WT/DS426; US—Countervailing Measures (China), WT/DS437; US—Countervailing and Anti-Dumping Measures (China), WT/DS449, European Union and Certain Member States—Certain Measures Affecting the Renewable Energy Generation Sector, WT/DS452; India—Certain Measures Relating to Solar Cells and Solar Modules, WT/DS456.

the situation is only marginally better (Jha 2013). Despite the growing significance of climate-friendly technologies for the low-carbon economy, a variety of important barriers to trade in climate-related environmental goods remain in place throughout the globe (Balineau and De Melo 2013). These barriers create unnecessary additional costs which impede on the worldwide dissemination of technology. According to a study convened for the OECD, the main problems encountered by companies trading in environmental goods relate to product testing, certification requirements and customs procedures (Fliess and Kim 2008, p. 535). Exporters furthermore experience problems in relation to intellectual property protection, government procurement procedures and technical regulations and standards (Fliess and Kim 2008, p.535).

3.1 Tariffs

Tariff levels for renewable energy goods appear to follow the general pattern of tariffs in other sectors—they are low on average, but prohibitively high in exceptional circumstances (Hufbauer and Kim 2012). Most developed countries including the EU and the U.S. apply tariffs of five percent or less to most climate-friendly goods (Sugathan 2013, p. 34). In important emerging markets like China and India, tariffs for most of these goods range between 5 and 20 % (Sugathan 2013, p. 34). Analysis of specific tariff data for renewable energy goods does not deviate from this general picture. Taking wind turbines as an example, most of the applied tariffs range between 0 and 10 %, with only a few exceptions. Brazil, which is one of the most important markets for wind energy, applies an average tariff of 14 % for wind-powered generating sets[6] (WTO Tariff Database 2015). While tariffs overall do not seem to be major trade barriers, they are considered to be "low-hanging fruit"—a good starting point for the negotiation of concessions with less potential for disagreement than other trade barriers. Nevertheless, as important as the elimination of remaining tariffs on clean energy goods may be, a variety of non-tariff barriers (NTBs) exist for low-carbon goods which should be tackled in parallel with tariffs.

3.2 Domestic Regulation with Respect to Standards and Procedures

Technical standards and procedural requirements for product testing rank among the foremost non-tariff barriers to trade in climate-friendly goods. In the RE sector, standards usually relate to safety, performance or installation (Rai and Payosova

[6]Under tariff line 850231 of the Harmonized System of the World Customs Organization (WCO).

2015). They are also of special significance for energy-efficient goods, as most of these are distinguished solely by way of labelling and are otherwise physically indistinguishable (Sugathan 2015, p. 28). Worldwide standards applying to all the main markets of renewable energy and energy-efficient goods have not yet been agreed upon. As a result, exporters of clean energy goods are confronted with a number of differing standards and technical regulations. Complying with all of these at the same time induces considerable administrative costs for producers. A greater degree of harmonization or even mutual recognition of standards and procedures would facilitate innovation, allow for economies of scale and ultimately lead to the optimization of costs. In the solar PV industry, several initiatives by international bodies are underway to harmonize standards (Rai and Payosova 2015).

In international trade law, a distinction is usually made between voluntary and mandatory standards. WTO law treats standards which are mandatory throughout the Agreements as "technical regulations". The Agreement on Technical Barriers to Trade (TBT) requires that technical regulations "are not prepared, adopted or applied with a view to, or with the effect of, creating unnecessary obstacles to international trade."[7] It furthermore requires WTO Members to resort to international standards when adopting national regulations that include standards.[8] The TBT Committee, which is tasked with reviewing the operation and implementation of the Agreement, has been examining the TBT-compatibility of Korean standards for thin-film solar panels, the use of which is subject to certification by a Korean authority (Hufbauer and Kim 2012, p. 10). At the time of writing (April 2015), the issue is still pending in the Committee. Some WTO Members have begun to tackle the issue of differing product standards and safety and testing requirements through regulatory cooperation. An institutionalized system of regulatory cooperation will be one of the cornerstones of the TTIP agreement.[9] "Expanding regulatory cooperation and advancing regulatory coherence" is also one of the objectives of the APEC's Committee on Trade and Investment (APEC 2015).

3.3 Protection of Intellectual Property

The inadequacy of intellectual property (IP) protection in many countries importing environmental goods is conceived by producers in exporting countries as a major burden (Maskus and Okediji 2010). The most important categories of intellectual property rights (IPRs) are patents, copyrights and trade secrets. For companies in the renewable energy sector, patents are especially important because technological innovation is a major part of success. Statistics provided by the World Intellectual Property Organization (WIPO) show that in the five years from 2006–2011 the

[7]Art. 2.2 TBT Agreement.

[8]Art. 2.4 TBT Agreement.

[9]See below, part 4.5.

increase in the number of patent filings in the field of renewable energy was four times greater than general patenting activity (WIPO 2014). Manufacturers of renewable energy equipment have complained about treatment of their intellectual property in important emerging markets. As the experience of the leading German manufacturer of wind turbines ENERCON suggests, patent disputes may prompt companies to completely abandon a market they have previously invested in. ENERCON has been fighting before Indian courts over some of its patents related to wind energy generation allegedly worth up to 1$ billion, without much success (Basheer 2014; Lath 2014).

While exporters from many developed countries want to ensure that they are rewarded for their investments in innovation through protection of their patents and copyrights, importing developing countries without a significant domestic production sector take the view that a high level of IP protection constitutes a barrier to effective technology diffusion. Whether a high degree of protection of IPRs in fact drives or hinders technology diffusion overall is subject to an intense empirical debate (Littleton 2008, p. 4). Therefore, it is far from clear whether a stronger protection of IP will increase overall exports of low-carbon goods, especially to developing countries (cf. Glachant et al. 2013). On the other hand, the extent to which developing countries will benefit from trade liberalization depends largely on the degree of IP protection that developed countries will retain for themselves in the negotiations. Discussions in the framework of the UNFCCC have therefore focused, among other things, on mechanisms to facilitate technology transfer in climate-friendly technologies to developing countries. The UNFCCC Members have agreed to promote, facilitate and finance access to and transfer of climate-friendly technology (Littleton 2008, p. 4). The Clean Development Mechanism (CDM) which was established by the KP can also help to enhance the transfer of know-how and technology relating to energy-efficient and renewable energy goods to developing countries when CDM projects involve such arrangements. The 16th session of the Conference of Parties to the UNFCCC in December 2010 in Cancun saw the establishment of a "Technology Mechanism" which aims to "facilitate the implementation of enhanced action on technology development and transfer in order to support action on mitigation and adaptation to climate change" (UNFCCC COP 16 2010, para. 117).

The protection of IPRs has found multilateral recognition in international Conventions since the late nineteenth century and more recently through the TRIPS Agreement which has been negotiated in the framework of the WTO. The Agreement sets global minimum standards for the protection of IPRs and subjects violations of the provisions to the binding system of WTO dispute settlement. In its Article 7, the TRIPS agreement, without specifically mentioning developing countries, identifies "the promotion of technological innovation" and "the transfer and dissemination of technology" as objectives of IPR protection and enforcement. One of the most important means to achieve this objective is the mechanism of "compulsory licensing".[10] Another option is the so-called "patent pooling", by which patent holders agree to make patents available amongst each other or to third parties. Whereas compulsory licensing is specifically addressed under the TRIPS (and allowed as long as a number of conditions are fulfilled),[11] there is no corresponding provision for patent pooling.

3.4 Government Procurement

The procurement of "green" goods by government institutions plays an increasingly important role in driving demand for these goods as governments are looking to replace existing energy-intensive or inefficient technologies in public infrastructure with state-of-the-art technologies. In this respect, there is considerable potential for climate mitigation in government procurement (van Asselt et al. 2006). Governments are often inclined to design their policies in a way that promotes domestic industry development and creates employment (Monkelbaan 2014a, p. 117). When government procurement policies are not designed in a transparent and non-discriminatory manner, foreign producers find it hard to successfully compete for public tenders and obtain contracts.

In the framework of the WTO, a number of Members have agreed on common rules governing public procurement which have been inscribed in the Government Procurement Agreement (GPA). The GPA provides for transparent and non-discriminatory procurement practices and condemns procedures which create unnecessary obstacles to international trade. Technical specifications prescribed for in procurement tenders must be based on performance and not on design or descriptive characteristics.[12] Rules on Government Procurement are increasingly finding their way into bilateral and regional free trade agreements (FTAs). A number of FTAs include provisions which go further than what has been agreed on in the WTO framework ("WTO-plus" commitments) (on EU practice on government procurement see the Commission's DG Trade 2014).

3.5 Discriminatory Design of Support Measures for Renewable Energy

While most of the trade barriers discussed here also exist in relation to other specialized high-technology goods, some trade barriers are specific to the RE

[10] Compulsory licensing denotes a mechanism under which the use of a patent can be authorised without the explicit consent of the patent holder when overriding public interests are at stake.

[11] Compare Article 31 TRIPS.

[12] Art. X GPA.

industry and relate to the competitive opportunities of climate-friendly technology. Generally, to be competitive, the RE industry relies on a policy framework which supports their position in the marketplace and helps pioneering companies become established players in the market. This usually means that laws have to be put in place which provide for the financial support of renewable energy generators and ultimately allow for grid parity of electricity generated from renewable sources.[13]

The most widely used mechanism is a Feed-In Tariff (FIT) (REN 21, p. 77). While FITs have greatly contributed to increasing the share of renewable energy in the electricity consumption of many countries, their policy design often leads to distortions in the market and in many cases disadvantages foreign over domestic producers (Kuntze and Moerenhout 2013; Dromgool and Ybarra 2015). The reason is that governments aim to combine the—often costly—support of renewable energy electricity producers with positive effects on the domestic economy and job creation. (Kuntze and Moerenhout 2013). Local content requirements (LCRs) are among the most widely used means to achieve the aforementioned goals (UNCTAD 2014). LCRs in the renewable energy sector usually make the benefits of the FIT contingent on the domestic or local sourcing of equipment used for the generation of electricity. These considerations were allegedly also underlying the FIT scheme of the Canadian Province of Ontario, which has successfully been challenged by the EU and Japan before the Dispute Settlement Body of the WTO. The Appellate Body in its decision upheld the findings of a Panel that the Ontario FIT, because of its LCR, is in violation of Article III GATT. A violation of the Agreement on Subsidies and Countervailing Measures, however, was not established because the FIT did not amount to a subsidy (for a critical account of this judgment see Rubini 2014).

4 Approaches to Liberalizing Trade in Environmental Goods

Negotiations have been initiated in different fora with an aim to reduce or eliminate barriers to trade in certain defined categories of environmental goods. These initiatives have taken place at the multilateral and plurilateral levels as well as in preferential trade agreements between two or more countries. The approaches differ considerably with respect to their constituency, coverage of goods and the envisaged scope of trade facilitation. The present part provides an overview over these initiatives and their current state of affairs, with a special emphasis on low-carbon goods.

[13]Grid parity is reached when the costs of production for renewable electricity equal the cost of conventional electricity fed into the grid.

4.1 Doha Round Negotiations

The Members of the WTO agreed during the Fourth Ministerial Conference (which marked the beginning of the Doha Round) in November 2001 that negotiations should be commenced on the reduction or elimination of tariff and non-tariff barriers to environmental goods and services (WTO 2001, p. 31). This specific mandate of the Doha Round is part of a wider approach to "enhance the mutual supportiveness" of trade and environment which has since been discussed in the Committee on Trade and Environment in Special Session (CTE-SS) (on the CTE see Tarasofsky 1999). Aside from the liberalization of EGS trade, the negotiations focus on the relationship between multilateral environmental agreements and the WTO. Article 31 of the Ministerial Declaration does not give any conclusive guidance as to what the outcome of the negotiations should be, especially whether this will result in a self-standing sectorial agreement (contra: Vikhlyaev 2004, p. 93). In line with the general spirit of the Doha Round and the "single undertaking" approach, the sought after outcome should be a multilateral solution.

Discussions in the CTE-SS on EGS trade liberalization have mainly focused on four issues: The preambular language of an agreement or the envisaged scope of liberalization, the exact coverage of goods and services, modalities of treatment and development elements (WTO CTE-SS 2011, para 11). WTO Members expect a "triple win" outcome with benefits for trade, the environment and development (WTO CTE-SS 2011, p. 13.). The facilitation of trade is expected to lead to increased dissemination of products which have a positive impact on the environment while at the same time increasing welfare effects from increased trade (Balineau and De Melo 2013, p. 2). Importantly, in line with the broader goal of the DDA, the liberalization of trade in environmental goods and services is intended to be beneficial for developing countries by increasing the availability of environmental technologies and reducing their costs. However, whether this "third win" scenario is going to materialize remains doubtful. Considering the expectedly unequal distribution of economic gains it is questionable whether the "development" part of the Doha EGS liberalization agenda will convince developing countries in the WTO (cf. Wu 2014).

With respect to coverage, different perceptions among WTO Members have existed throughout the negotiations and positions are converging slowly. The Doha mandate goes further than most subsequent initiatives by including services and non-tariff barriers. It does not, however, provide any further classification or specification of what environmental goods and services entail. Discussions in the CTE-SS have therefore focused on definitional aspects, on which consensus has proved hard to reach. Members were invited to submit to the CTE-SS proposals for environmental goods, taking as a reference HS-6 lines of the Harmonized System.[14]

[14]The Harmonized System, as explained by the WTO, is an international nomenclature developed by the World Customs Organization, which is arranged in six-digit codes allowing all participating countries to classify traded goods on a common basis. Beyond the six-digit level, countries are free

The outcome of this process is a compiled list of goods covering 408 tariff lines (Hufbauer and Kim 2012). The goods which have been proposed were categorized broadly into air pollution control, renewable energy, waste management and water treatment, environmental technologies, carbon capture and storage and others.[15] However, the submissions show a wide divergence of positions about what an "environmental" good is. In fact, as Balineau and De Melo have shown, there is not a single good which has been proposed by all Members and two thirds of the goods which have been proposed appear only on one list of submissions (Balineau and De Melo 2013, p. 6). Hence, the challenge now lies in structuring the discussions and finding common ground. A group of five WTO Members has proposed a "core" list of 26 goods in 2011, which were taken as a starting point for future discussions (Balineau and De Melo 2013, p. 7). As regards treatment of environmental goods, the focus is on tariffs but Members have also briefly touched upon reducing and eliminating NTBs. While some NTBs which are specific to the environmental goods and services industry have been identified by the Members, the discussions have generally been rather limited on the NTB side (CTE-SS 2011, p. 17).

The environmental goods and services discussions of the Doha Round must be appreciated in the broader context of the Doha mandate for negotiations on non-agricultural market access (NAMA). As environmental goods in principle fall under the category of non-agricultural goods, the relationship between the general and the more specific mandates are not entirely clear and require further examination, especially if the members in the CTE adopt a broad approach to the classification of environmental goods (see Cottier et al. 2009, p. 396; Vikhlyaev 2004, p. 105).

4.2 Environmental Goods Agreement (EGA)

The World Economic Forum in Davos in January 2014 provided the ground for the birth of an agreement among 14 WTO Members aiming at targeted trade liberalization for environmental goods (Cosbey 2014).[16] The negotiations, which are taking place outside the Doha Mandate of the WTO, were formally launched in July 2014. The material scope of the envisaged trade liberalization is only a fraction of what the Doha Ministerial Declaration had foreseen as the negotiating parties only focus on goods and leave out services entirely. While some members have indicated their willingness to consider discussing NTBs and the agreement has been

---

(Footnote 14 continued)

to introduce national distinctions for tariffs and other purposes. Whereas a customs code with low number of digits stands for a broad product category, 6-digit lines are the most detailed standard product definitions.

[15]Compare Annex III of the March 2010 Report to the Trade Negotiations Committee.

[16]The Joint Statement Regarding Trade in Environmental Goods of 24 January 2014 is available online at the Commission's DG Trade Website.

designated as a "living agreement" possibly containing a revision clause on the inclusion of services and a future work program (European Commission DG Trade 2015), the initial focus is on tariffs only. The original participants of the plurilateral initiative include the European Union, the United States, China, Japan and Australia, Canada, Costa Rica, Hong Kong, Korea, New Zealand, Norway, Singapore, Switzerland and Chinese Taipei. Israel, Turkey and Iceland have joined the negotiations in the meantime, bringing the total number to 17. Whether the EGA will be embraced by the COP21 in Paris and possibly made part of a package of measures for post-2015 climate change cannot be said with certainty at the time of writing.

The negotiations are progressing on the basis of a list comprising 54 environmental goods which is modelled on a list drawn up by the members of the APEC. The climate benefits of a comprehensive and ambitious agreement could potentially be significant, as the negotiating parties comprise not only 86 % of world trade (ICTSD BIORES 2014), but also some of the largest emitters of CO_2. While the first steps are promising, the tariffs-only approach has been criticized as lacking ambition (Cosbey 2014). It is expected that the EGA will follow the Information Technology Agreement and operate on the basis of the most favoured nation (MFN) principle Sugathan 2014). This would mean that advantages and concessions granted to other EGA Members would in principle have to be granted in relation to all other WTO Members as well. The Agreement will only enter into force once a "critical mass" of countries has acceded to it. While the exact threshold is not known, the only time the concept of "critical mass" was previously applied by WTO Members was in the context of the Information Technology Agreement. The Agreement entered into force in 1997 once the international trade volume in IT products represented by the signatories reached 90 % (WTO 2012). In the long run, the decisive question for the EGA will be whether the Agreement will have the dynamism to evolve into a multilateral agreement involving most or all of the WTO Members or whether it will stay on the sidelines of the WTO like the Government Procurement Agreement. From a climate-protection point of view, the agreement will only be effective if the negotiations ultimately move beyond goods tariffs to services liberalization and NTBs.

4.3 Bilateral and Regional Free Trade Agreements

Many FTAs include chapters on environmental protection or sustainable development and stress the importance of mutual supportiveness of trade and environment (Gehring et al. 2013). Whereas FTAs for a long time focused on including general provisions on environmental protection to secure policy space, some more contemporary FTAs contain positive commitments on the liberalization of trade in EGS. The EU has also included more specific provisions on green technology and renewable energy in some of its recent FTAs. EU practice has traditionally been to include a chapter on trade and sustainable development in its FTAs, while U.S. and

Canadian FTAs usually contain environmental side agreements. The NAFTA members negotiated the North American Agreement on Environmental Cooperation as a "framework (…) to facilitate effective cooperation on the conservation, protection and enhancement of the environment." (NAFTA 1993). To coordinate environmental activities between NAFTA members and to mediate disputes in the environmental field, the North American Commission for Environmental Cooperation (CEC) was set up. Explicit positive obligations on the liberalization of environmental goods trade are absent from the NAFTA itself as well as from the NAAEC. Some of the NAFTA members, notably Canada and the US, have started including provisions on EGS in subsequent FTAs. The environmental side agreement of the Canada-Peru FTA (the Canada-Peru Environment Agreement), which entered into force in August 2009, provides in its Article 2.6 that "the parties shall encourage the promotion of trade and investment in sustainable goods and services."[17] Furthermore, Annex I of the agreement lists the "promotion of production and trade of environmentally-friendly goods and services" as one of the priority areas for cooperation between the parties.

In some of its more recent FTAs, the European Union has similarly included provisions encouraging trade in environmental goods and services. In the EU-Colombia-Peru FTA, for example, "the parties (…) stress the need to enhance the mutual supportiveness between trade and environment." More specifically, "the parties shall strive to facilitate and promote trade and foreign direct investment in environmental goods and services".[18] The agreement also includes specific provisions on climate change. Taking the commitments under the UNFCCC as a starting point, the parties to the Agreement agree to

> promote trade and investment measures that promote and facilitate access, dissemination and use of best available technologies for clean energy production and use, and for mitigation of and adaptation to climate change.[19]

Furthermore, according to Article 275 (5) the parties agree to consider actions to contribute to achieving climate change mitigation and adaptation objectives through their trade and investment policies, inter alia, by:

> (a) facilitating the removal of trade and investment barriers to access to, innovation, development, and deployment of goods, services and technologies that can contribute to mitigation or adaptation, taking into account the circumstances of developing countries; and
> (b) promoting measures for energy efficiency and renewable energy that respond to environmental and economic needs and minimize technical obstacles to trade.[20]

[17] Text of the Agreement retrieved 15 May, 2015, from https://www.ec.gc.ca/caraib-carib.

[18] EU-Colombia-Peru FTA, Art. 271 (2), retrieved at 15 May, 2015, from http://trade.ec.europa.eu/doclib/docs/2011/march/tradoc_147704.pdf.

[19] EU-Colombia-Peru FTA, Art. 275 (4).

[20] EU-Colombia-Peru FTA, Art. 275 (5).

The EU-Korea FTA has been described as "a new approach on trade and sustainable development" (European Commission DG Trade 2011, p. 3). The agreement, which ranks among the most comprehensive free trade agreements ever negotiated by the EU, places a special emphasis on green technology, renewable energies and energy efficiency. Parties agree to facilitate trade in these goods sectors, including through addressing related non-tariff barriers.

The EU-Singapore FTA, in addition to containing a chapter on trade and sustainable development, also includes a sector-specific chapter which deals with non-tariff barriers to trade and investment in renewable energy generation. The Parties agree to refrain from adopting trade distorting measures like local content requirements and mandatory joint ventures. Furthermore, they require authorization and licensing procedures to be objective, transparent and non-arbitrary (European Commission 2011). This quite progressive approach represents the negotiating partner's ambition to draft a "comprehensive 21st century trade agreement" (European Commission DG Trade 2013, p. 15) which would provide a reference point for future FTAs of the two countries.

4.4 Mega-Regionals

While the approaches described above have the potential to further cut or even completely eliminate tariffs on low-carbon goods, important non-tariff barriers will remain in place. As some of the major trading powers in the world are in the process of negotiating a new kind of preferential trade agreements covering unprecedented trade flows, these "mega-regional" trade agreements could possibly be harnessed for further facilitation of trade in low-carbon goods. Currently under negotiation are, among others, the Transatlantic Trade and Investment Partnership (TTIP) between the EU and the U.S., the Trans-Pacific Partnership (TPP) between 12 Pacific countries and the Comprehensive Economic and Trade Agreement (CETA) between the EU and Canada. Furthermore, China is negotiating a Regional Comprehensive Economic Partnership (RCEP) with the Member States of ASEAN and with India. The most notable elements of the envisaged agreements include "deep integration" through the elimination of a range of non-tariff barriers and greater regulatory coherence. Furthermore, most of the draft texts of the "mega-regional" trade agreements include substantive provisions on investment protection and a mechanism for investor-state dispute settlement (ISDS) (see European Commission 2014b; Wikileaks 2015). Rising public opposition against mega-regionals, e.g. in Europe and New Zealand, has been directed against the anticipated negative effects on environmental and consumer protection and a perceived constraint on regulatory space. A commonly voiced fear is that regulatory leeway on the side of democratically elected governments might be undermined through the agreements, especially the mechanisms of regulatory cooperation and ISDS (see Euractiv.com 2015). While the public debate on the environmental side has so far focused almost entirely on negative effects and the need for regulatory space, the potential positive contribution of these agreements to certain areas of environmental protection has not been addressed sufficiently. It is submitted here that the agreements offer interesting opportunities from the point of view of environmental goods trade liberalization.

On a general note, the agreements aim by their very purpose at across-the-board trade facilitation for goods and services between the Parties. This means an elimination of tariffs for virtually all goods on which tariffs are still in place. More importantly, the agreements provide for market opening through the facilitation of customs procedures and the mutual recognition of standards. As only a limited number of countries with similar interests and objectives are involved in the negotiation process, the likelihood of achieving agreement is much higher than in the multilateral environment of the WTO. Especially as regards coverage, which is one of the most contentious items on the negotiating table in Geneva, one can expect more common ground between regional trading partners. By forging a strong "regional consensus" first, WTO members could then openly bring their results to the multilateral negotiating table and create positive spillovers. Another reason for pursuing the EGS liberalization agenda through mega-regionals is that the countries currently negotiating such agreements are among the main emitters of CO_2 (Leal-Arcas 2013).

One of the preconditions for an ambitious reduction of barriers to trade in climate-friendly goods would be a common understanding of and commitment to the low-carbon economy. Considering the efforts that the EU has so far undertaken in the domestic sphere and the high level of environmental protection it has pledged to adhere to in its Treaties, this should be feasible at least in the case of TTIP and CETA. As regards the TPP and especially the RCEP, Members could be more reluctant to make special commitments in the low-carbon sector.

While the final texts of the different agreements have not yet been carved into stone, the negotiating mandates and the draft texts which have leaked since the start of the negotiations reveal the position of the negotiators towards issues such as environmental protection, sustainable development and the green economy. In line with recent trade policy, and the environmental integration principle of Article 11 TFEU, the Agreements to which the EU is a party will include a chapter on sustainable development. Notably, the CETA draft contains both a chapter on sustainable development and a separate chapter on the environment. However, so far, specific commitments to advance a low-carbon economy through sectorial liberalization are missing from the texts of the agreements (Frey 2015). CETA and TTIP do not follow the approach taken in the EU-Singapore FTA, which contains a chapter on renewable energy. This is regrettable, as including far-reaching sectorial provisions (e.g. front-loading the elimination of tariffs for defined climate-friendly goods and committing to eliminate local content requirements for the RE sector) would be an important contribution to reaching climate change mitigation targets (Frey 2015; De Ville and Gheyle 2014). A common commitment to a low-carbon economy would also have the potential to enhance public legitimacy of the agreements and set a counter-mark to the manifold apprehensions of environmental NGOs.

5 Conclusions

Recent years have witnessed a rising awareness of the various interlinkages between the international climate and trade regimes. A post-2015 climate regime needs to recognize the important role that world trade can play in flanking other policy areas. Binding targets to be agreed upon in the framework of the UNFCCC should therefore be combined with efforts within the trade regime. The Work Program of the Doha Development Agenda, by providing a mandate for discussions on the liberalization of trade in environmental goods and services, is a step in the right direction. However, considering the current stalemate of the multilateral Doha talks, the WTO seems to be the most unlikely avenue for timely action. This is certainly regrettable, as the WTO is the only multilateral forum for trade negotiations which brings together the UNFCCC parties, includes developing countries and provides for a binding system of dispute settlement. Moreover, the Doha Mandate sets out the most ambitious roadmap by including environmental services and non-tariff barriers. Common rules in the framework of the WTO would therefore have by far the biggest impact on emission reductions. A serious attempt to move the agenda forward has been initiated by a 17-country bloc which is currently in the midst of negotiations on a plurilateral Environmental Goods Agreement. Negotiations have already progressed further than on the multilateral level and final agreement seems within reach. Importantly, most of the main emitters of CO_2 are on board of the EGA talks. While the initiative certainly provides fresh impetus to the EGS liberalization agenda, the ambition stays behind the Doha Mandate as the negotiations currently focus on tariffs only.

While tariff reduction or elimination for EGS seems feasible through multilateral and plurilateral efforts, other barriers will remain. In this respect, FTAs and especially mega-regionals could take the agenda further by specifically tackling the trade barriers that exist with respect to renewable energy and energy-efficient goods. Contemporary EU FTAs include specific sectorial provisions on the elimination of trade barriers and market access for environmental goods and services. To date, such provisions have not been agreed on with major greenhouse gas emitting countries. The currently ongoing mega-regional negotiations will show whether such an approach is feasible. Mega-Regionals will by their very purpose and design tackle many of the barriers which have been identified for free trade in RE Goods. Furthermore, they are set to include common rules on regulatory issues and in their most far-reaching form a mechanism for regulatory cooperation.

An analysis of the published negotiating mandates, draft agreement texts and the results of the negotiating rounds which have already taken place reveals that the mega-regionals do not built forth on the latest trends in drafting FTA's, i.e. including sectoral provisions on tariffs and NTBs for low-carbon goods. This would be a missed opportunity, especially since the COP 21 conference in December 2015 provides important political momentum for fighting climate change. While the climate cannot wait on the Doha negotiations to be wrapped up, all of the non-multilateral efforts to facilitate EGS trade could ultimately spill-over to the multilateral level and thus have a positive impact. Lessons learned from the plurilateral and FTA experience can facilitate the finding of common ground during multilateral negotiations.

References

APEC (2011) Leaders' declaration—toward a seamless regional economy. Retrieved 15 May 2015, from http://www.apec.org/Meeting-Papers/Leaders-Declarations/2011/2011_aelm.aspx

APEC (2015) Committee on trade and investment. Retrieved 25 July 2015, from http://www.apec.org/Groups/Committee-on-Trade-and-Investment

Balineau G, De Melo J (2013) Removing barriers to trade on environmental goods: an appraisal. World Trade Rev 12(4):693–718

Basheer S (2014) Enercon and positive patent perception. SpicyIP Blog. Retrieved 25 July 2015, from http://spicyip.com/2014/06/enercon-and-positive-patent-perception.html

Carlarne C (2006) The Kyoto protocol & the WTO: reconciling tensions between free trade & environmental objectives. Colorado J Int Environ Law Policy 17(1):45–48

Commission for Environmental Cooperation (1993) North American agreement on environmental cooperation (NAAEC), 14 Sept 1993, 32 ILM 1480, 1482. Retrieved 25 July 2015, from http://www.cec.org/Page.asp?PageID=1226&SiteNodeID=567

Cosbey A (2014) The green goods agreement: neither green nor good?. International Institute for Sustainable Development, IISD Commentary, Geneva

Cottier T, Nartova O, Bigdeli SZ (2009) International trade regulation and the mitigation of climate change. Cambridge University Press, Cambridge

De Ville F, Gheyle N (2014) The transatlantic trade agreement and the green economy: which way will the balance (T) tip? Retrieved 25 July 2015, from http://www.eutradepolitics.ugent.be/?q=content/transatlantic-trade-agreement-and-green-economy-which-way-will-balance-ttip

Dromgool T, Ybarra D (2015) The fair and equitable treatment standard and the revocation of feed in tariffs—Foreign renewable energy investments in Crisis-struck Spain. In: Mauerhofer V (ed) Legal aspects of sustainable development: horizontal and sectorial policy issues, (pp. -). Springer, Dordrecht

Euractiv.com (2015) Brussels, Washington seizing new regulatory powers through TTIP. Retrieved 15 May 2015, from http://www.euractiv.com/sections/trade-society/brussels-washington-boost-regulatory-powers-through-ttip-312005

European Commission (2011a) Free trade agreement between the European Union and Singapore (not yet in force). Initialled text retrieved 15 May 2015, from http://trade.ec.europa.eu/doclib/press/index.cfm?id=961

European Commission (2011b) Communication a roadmap for moving to a competitive Low-Carbon Economy in 2050. Retrieved 15 May 2015, from http://eur-lex.europa.eu/legal-content/EN/TXT/?uri=CELEX:52011DC0112

European Commission (2014a) Joint statement regarding trade in environmental goods of 24 Jan 2014 at Davos. Retrieved 15 May 2015, from http://trade.ec.europa.eu/doclib/docs/2014/January/tradoc_152095.pdf

European Commission (2014b) Investment provisions in the EU-Canada free trade agreement (CETA). Retrieved 15 May 2015, from http://trade.ec.europa.eu/doclib/docs/2013/November/tradoc_151918.pdf.European

European Commission Directorate-General Trade (2011) The EU-Korea free trade agreement in practice. Retrieved 15 May 2015, from http://trade.ec.europa.eu/doclib/docs/2011/October/tradoc_148303.pdf

European Commission Directorate-General Trade (2013) The Economic Impact of the EU-Singapore free trade agreement. Retrieved 15 May 2015, from http://trade.ec.europa.eu/doclib/docs/2013/September/tradoc_151724.pdf

European Commission Directorate-General Trade (2014) Public procurement in a nutshell. Retrieved 25 July 2015, from http://ec.europa.eu/trade/policy/accessing-markets/public-procurement/

European Commission Directorate-General Trade (2015) Minutes of the civil society dialogue's meeting on the trade and sustainability impact assessment of the environmental goods agreement. Retrieved 25 July 2015, from http://trade.ec.europa.eu/doclib/docs/2015/may/tradoc_153411.pdf

Fliess B, Kim JA (2008) Non-tariff barriers facing trade in selected environmental goods and associated services. J World Trade 42(3):535–562

Frey C (2015) Mega-regional trade agreements and post-2015 climate protection. JEEPL 12 (3–4) (forthcoming)

Gehring MW, Cordonier Segger MC, de Andrade Correa F, Reynaud P, Harrington A, Mella R (2013) Climate change and sustainable energy measures in regional trade agreements (RTAs): an overview. ICTSD, Geneva

Glachant M, Dussaux D, Ménière Y (2013) Promoting the international transfer of low-carbon technologies: evidence and policy challenges, report for the commissariat général à la stratégie et à la prospective. Retrieved 13 Oct 2001, from http://personal.lse.ac.uk/dechezle/Promoting_the_international_transfer_of_low_carbon_techs.pdf

Hufbauer G, Kim JA (2012) Issues and considerations for negotiating a sustainable energy trade agreement, ICTSD global platform on climate change, trade and sustainable energy. ICTSD, Geneva

ICTSD (2014) The road ahead for the environmental goods agreement talks. BIORES 8(7):14–17. Retrieved 25 July 2015, from http://www.ictsd.org/sites/default/files/review/biores8-7_0.pdf

IPCC (2000) Methodological and technological issues in technology transfer: summary for policymakers. Retrieved 15 May 2015, from http://www.ipcc.ch/ipccreports/sres/tectran/index.php?idp=0

Jha V (2013) Removing trade barriers on selected renewable energy products in the context of energy sector reforms: modelling environmental and economic impacts in a general equilibrium framework. ICTSD, Geneva

Kasteng J (2013) Trade remedies on clean energy: a new trend in need of multilateral initiatives. Geneva: E 15 Initiative

Kuntze JC, Moerenhout T (2013) Local content requirements and the renewable energy industry—a good match?. ICTSD, Geneva

Lath A (2014) Breaking news: SC streamlines patent revocation procedures in Enercon Litigation. SpicyIP Blog. Retrieved 25 July 2015 from http://spicyip.com/2014/06/breaking-news-sc-streamlines-patent-revocation-proceedings-in-enercon-litigation.html

Leal-Arcas R (2013) Climate change and international trade. Edward Elgar, Cheltenham

Littleton M (2008) The TRIPS agreement and transfer of climate-change-related technologies to developing countries. UN DESA Working Paper No. 71. Retrieved 15 May 2015, from http://www.un.org/esa/desa/papers/2008/wp71_2008.pdf

Maskus KE, Okediji RL (2010) Intellectual property rights and international technology transfer to address climate change: risks, opportunities and policy options, ICTSD's programme on iprs and sustainable development, Issue Paper No. 32. ICTSD, Geneva

Monkelbaan J (2014a) The benefits of a sustainable energy trade agreement (SETA). In: Cahoy DR, Colburn JE (eds) Law and the transition to business sustainability. Springer, Dordrecht, pp 103–123

Monkelbaan J (2014b) Trade preferences for environmental goods and services, ICTSD global platform on climate change working paper. ICTSD, Geneva

North American Agreement on Environmental Cooperation (1993) (signed 8 September 1993, entered into force 1 January 1994) (1992) 32 ILM 1482

OECD (2009) The Economics of Climate Change Mitigation: Policies and Options for Global Action Beyond 2012. OECD, Paris
Peters GP, Hertwich EG (2008) CO2 embodied in international trade with implications for global climate policy. Environ Sci Technol 42(5):1401–1407
Rai S, Payosova T (2015) Selling the sun safely and effectively: Solar photovoltaic (PV) standards, certification, testing and implications for trade policy. Geneva: ICTSD (forthcoming)
REN21 (2014) Renewable energy policy network for the 21st century (2014) Renewables 2014 global status report. Retrieved 15 May 2015, from http://www.ren21.net
Rubini L (2014) What does the recent WTO litigation on renewable energy subsidies tell us about methodology in legal analysis? the good, the bad, and the ugly. EUI Robert Schuman Centre for Advanced Studies Research Paper No. 2014/05
Sugathan M (2013) Winds of change and rays of hope: how can the multilateral trading system facilitate trade in clean energy technologies and services? Geneva: the E15 initiative. Retrieved 25 July 2015, from http://www.e15initiative.org/publications
Sugathan (2014) The road ahead for the environmental goods agreement talks, Biores Vol. 9 No. 8, ICTSD, Geneva
Sugathan M (2015) Addressing energy efficiency products in the environmental goods agreement, Issue Paper No 20. ICTSD, Geneva
Tarasofsky RG (1999) The WTO committee on trade and environment: is it making a difference? Max Planck Yearb United Nations Law 3:471–488
UNCTAD (2014) Local content requirements and the green economy. Retrieved 25 July 2015, from http://unctad.org/en/PublicationsLibrary/ditcted2013d7_en.pdf
UNEP (2013) Green economy and trade—trends, challenges and opportunities. Retrieved 25 July 2015, from http://www.unep.org/greeneconomy
UNFCCC (2010) Report of the conference of the parties on its sixteenth session: decisions adopted. Retrieved 25 July 2015, from http://unfccc.int/resource/docs/2010/cop16/eng/07a01.pdf#page=18
United Nations (2002) Plan of implementation of the world summit on sustainable development ("Johannesburg Plan of Implementation"). Retrieved 25 July 2015, from http://www.un.org/esa/sustdev/documents/WSSD_POI_PD/English/WSSD_PlanImpl.pdf
United Nations (2012) The future we want, outcome document of the united nations conference on sustainable development (Rio + 20). Retrieved 25 July 2015, from https://rio20.un.org/sites/rio20.un.org/files/a-conf.216l-1_english.pdf.pdf
USTR (2015) Trans-Pacific partnership (TPP): unlocking opportunity for Americans through trade with the Asia Pacific. Retrieved 25 July 2015 from https://ustr.gov/tpp
van Asselt H, van der Grijp N, Oosterhuis F (2006) Greener public purchasing: opportunities for climate-friendly government procurement under WTO and EU rules. Clim Policy 6 (2):217–229
Vikhlyaev A (2004) Environmental goods and services: defining negotiations or negotiating definitions? J World Trade 38(1):93–122
Wikileaks (2015) Leaked draft of the TPP Investment chapter on Wikileaks. Retrieved 15 May 2015, from https://wikileaks.org/tpp-investment/WikiLeaks-TPP-Investment-Chapter/page-1.html
WIPO (2014) Global challenges report, renewable energy technology: evolution and policy implications—evidence from patent literature. Retrieved 25 July 2015, from http://www.wipo.int/export/sites/www/policy/en/climate_change/pdf/ccmt_report.pdf
WTO (1994) Agreement on Government Procurement, (signed 15 April 1994, entered into force 1 January 1996) 1915 UNTS 103
WTO (2001) Doha Ministerial Declaration, adopted on 14 Nov 2001. Retrieved 25 July 2015 from https://www.wto.org/english/thewto_e/minist_e/min01_e/mindecl_e.htm
WTO (2011) Committee on trade and environment in special session report by the chairman to the trade negotiations committee of 21 April 2011, WTO Document TN/TE/20

WTO (2012) 15 Years of the information technology agreement: trade, innovation and global production networks. Retrieved 25 July 2015, from https://www.wto.org/english/res_e/publications_e/ita15years_2012_e.htm
WTO (2015) Tariff database. Retrieved 25 July 2015, from http://tariffdata.wto.org/ReportersAndProducts.aspx
Wu M (2014) Why developing countries won't negotiate? the case of the WTO environmental goods agreement. Trade Law Dev 6:93–176

Part VIII
Sectorial Policies: Cities, Waste and Material Management

Environmental Integration in China's Eco-city Development—From an Institutional Perspective

Ying Yin

Abstract In the last decade, Eco-City development has been emerging as test beds in China, not only for environmental technologies but also for a new integrated approach of urban planning to achieve better environmental performance in the urban areas development. The study aims to build knowledge, from legal aspects, on how environmental integration in sustainable urban planning is achieved or hindered in China's eco-city practice. The lessons are as well supposed to give some implications for sustainable urban development in future. The study is based on an overview of China's national institutional condition for sustainable urban development and a case study of two examples of Tangshan Bay Eco-city and the Sino-Swedish Low-Carbon Eco-City in China. Methods of literature review, document studies, study visits and interviews were applied for gathering relevant knowledge and collecting empirical data on the cases. To conclude, in spite of some progress, the absence of a holistic perspective and effective guidance and constraints in China's environmental and planning formal rules at different levels appeared as an institutional weakness, hindering environmental integration in urban planning process. Whether or not environmental requirements and their status was clearly stated in the formal rules, such as regulatory detailed plans and land use agreements, also considerably affected the enforcement of environmental integration in urban planning process. In addition, the results indicate an important role of powerful actors, such as local leaders and governments, in issuing an improved legitimacy of high environmental requirements in sustainable urban development through changing formal institutional conditions.

Keywords Environmental integration · Sustainable urban planning · Eco-city development · Legal aspects · China

---

Y. Yin (✉)
Royal Institute of Technology, Drottning Kristinas Väg 30, Plan 1,
10044 Stockholm, Sweden
e-mail: ying.yin@abe.kth.se

V. Mauerhofer (ed.), *Legal Aspects of Sustainable Development*,
DOI 10.1007/978-3-319-26021-1_23

1 Introduction

In most developed countries, the built environment accounts for an increasing share of environmental problems, such as pollution (Rogers 1998) and energy consumption (Birkeland 2002). In recent years, due to a fast urbanization, environmental degradation in China's cities has become a considerable challenge to the built environment in China. Eco-Cities have been described as test beds in China, not only for environmental technologies but also for a new integrated approach of urban planning to achieve better environmental performance in the urban areas development. However, to implement this holistic approach to integrate environmental requirements in the eco-city development often faces a number of barriers in urban planning process, since the holistic approach requires changes of the current formal institutional conditions, such as planning regulations and policies.

In this study, the overall aim is to build knowledge, from legal aspects, on how environmental integration in sustainable urban planning is achieved or hindered in China's eco-city practice. The lessons are supposed to give some implications as well for sustainable urban development in future. To investigate how legal aspects affect environmental integration in eco-city development could fill the gap between high environmental standards or technologies and weak local environmental performance in urban areas. It may help to find out more resilient ways to support applying environmental technologies in sustainable urban development. The study has evolved from two research questions:

- What kinds of legal aspects, such as planning regulations, policies and instruments at both national and local levels, could have crucial impacts on environmental integration in urban planning process?
- How do those formal institutional conditions promote or hinder environmental integration in sustainable urban planning? Why?

The investigation of environmental integration in sustainable urban planning is based on an overview of China's national institutional condition for sustainable urban development and a case study of two examples of Tangshan Bay Eco-city and the Sino-Swedish Low-Carbon Eco-City in China. Four primary methods were selected for gathering relevant knowledge and collecting empirical data on the cases. These include literature review, document studies, study visits and interviews. This study is qualitative; i.e., the majority of the data collected and presented is qualitative rather than quantitative.

2 Theoretical Basis

Literature review on institutional theories and Environmental Policy Integration (Nilsson and Persson 2003; Persson 2007; Jordan and Lenschow 2010) provide some insights to exam and analyse the ways in which formal institutional conditions, namely formal rules, hinder or promote environmental integration in the selected Eco-Cities. Here the concept of institution follows the definition that North (1990) and Olsson (2008) refer to—rules or norms that shape the interaction among actors. The construction of stable institutions is intended to reduce uncertainty about such things as actors' behaviour, practices and their outcomes, and to provide constraints regarding what actors are and are not permitted to perform (North 1990; Scott 2001; Ostrom 2005). Formal rulemaking (e.g. political systems, administrative structure, policy, regulation, mandates and contracts) is one of the well-recognized categories of institutions (North 1990; Ostrom 1990; Williamson 1998). Formal rulemaking is often used as written and legal sense for framing individuals' behaviour and political and legal monitoring, and enforcement may shape actors' cognition and habits in the long-term (Scott 2001), such as administrative commands, sanctions and penalties.

2.1 Formal Rulemaking for Environmental Integration - Regulations, Policies, Plans and Agreements

Environmental integration in policy-making and planning practice, as a trend without a long history or formal institutionalization, demands a reformulation of e.g. rationale rules in policy-making, overarching regulations, objectives and planning systems in the longer term (Persson 2007). As discussed above, the vague definition of the environmental priority and trade-offs leads to unclear directives, guidance and enforcement of environmental integration in policy-making and practice (Nilsson and Persson 2003; Nilsson 2005; Persson 2007; Storbjörk et al. 2009; Runhaar et al. 2009; Jordan and Lenschow 2010). The priorities of sector objectives are expected to be addressed in the political process during regulation- and policy-making in order to make civil servants at the administrative level obliged to deal with conflicts between environmental objectives and their own sectors' policy objectives (OECD 2002; Wandén 2003; Flipse 2007). This might be the most effective approach—'*that the objectives should be weighed up against one another* '*at the sharp end*'' (Wandén 2003, p. 39). To overcome the challenge, as they also argue, strong legislative and judicial systems regulating adherence to prescribed standards are necessary, as well as strict monitoring and enforcement of sanctions and penalties.

Maintaining consistency in formal rules at different levels, such as national and local environmental policies, local regulations, city master plans, detailed plans and development agreements is crucial to ensure the achievement of high-level environmental goals in planning practices (Jordan 2002; Jordan and Lenschow 2010; Wheeler 2004). Offering incentives for improving coherence and integration of sector policies across different levels demands longer-term financial and fiscal mechanisms, i.e. budgeting instruments (OECD 2002). In addition, early incorporation of environmental concerns, such as energy efficiency targets, into policy-making, regulations and planning (Owens 1990; Runhaar et al. 2009) is considered important for reducing controversies in the operational stages of development.

Support from local politicians or city commissions for the approval of plans or passing of local regulations is required in order to ensure consistency at different scales of plans and the early involvement of environmental concerns (Wheeler 2004). A low degree of political will can hinder the adoption of policies and regulations that enforce environmental integration (Raydan and Steemers 2006).

2.2 *Institutional Maintenance or Change*

Wheeler (2004) stresses the role of institutions and the need for institutional changes to help shape urban settings to support sustainability goals. What, then, are the conditions that affect the maintenance or transformation of current institutions? North (1990) argues that the stability of institutions results from the existence of a large number of specific constraints, and that these affect a particular choice, including both legal constraints and norms of behaviour. The transformation of institutions requires changes to a set of these formal and informal constraints (e.g. norms, culture, values and traditions). According to his idea, whether or not institutions could be incrementally altered or maintained also depends on the 'pay-offs' associated with conforming to the current and alternative institutions, i.e. new ideas/logics can show/approve its efficiency. However, the institutional constraints may not always or necessarily be ideal or efficient for all the sets of choices by individuals, but rather that they could be designed by governments to guard common interests. In this sense, major changes within the formal institutional framework also demand considerable interference from *'those with sufficient bargaining strength of the individuals and organizations'* (North 1990: 68), such as authorities in power. This means that either acknowledgement of efficiency or the involvement of key stakeholders with strong power is one precondition for institutional change. Powerful actors such as professionals and leaders may play important roles in recreating or transforming institutional forms by arguing and articulating their legitimacy (DiMaggio and Powell 1991).

To summarize from above, to realize environmental integration in practice, regulations, policies and plans need to be formulated or reformulated to provide normative and legal legitimacy to the public, in order to control and guide performance in the right direction. A legal enforcement to ensure early consideration of environmental issues and consistency in regulations and plans at different levels could contribute to the formal institutionalization of environmental integration in other sectors.

3 China's Institutional Conditions for Sustainable Urban Planning at the National Level

3.1 Urban Planning Policies, Regulations and Instruments

The regulatory framework for urban planning in China contains three parts: national acts, administrative regulations and local regulations (Fig. 1).

The Urban and Rural Planning Law of the P.R.C. (2008) is the central national law; it states the main principles, actors and tasks for the adoption, implementation and management of plans. Administrative regulations such as Measures for Formulating Urban Planning and Detailed Rules for Formulating Urban Planning clarify planning procedures and content and prescribe concrete rules and indexes. These provide an important basis for the formulation, assessment and approval of plans, as well as monitoring and evaluation of implementation. There are also a number of regulations for urban residential buildings, transportation and urban green space. For example, the Code of Urban Residential Areas Planning and Design defines the requirements for residential planning and design. Local regulations, such as the Technical Regulation of Urban Planning and Management of Beijing, are developed by local governments based on specific local conditions, but must be in line with national law.

For the urban planning instruments in general, a comprehensive plan of national land use made by the Ministry of National Land Resource defines and allocates the basic national interests on land resources. Besides this, regulatory plans defined in the Urban and Rural Planning Law of the P.R.C. (Table 1) encompass (1) urban

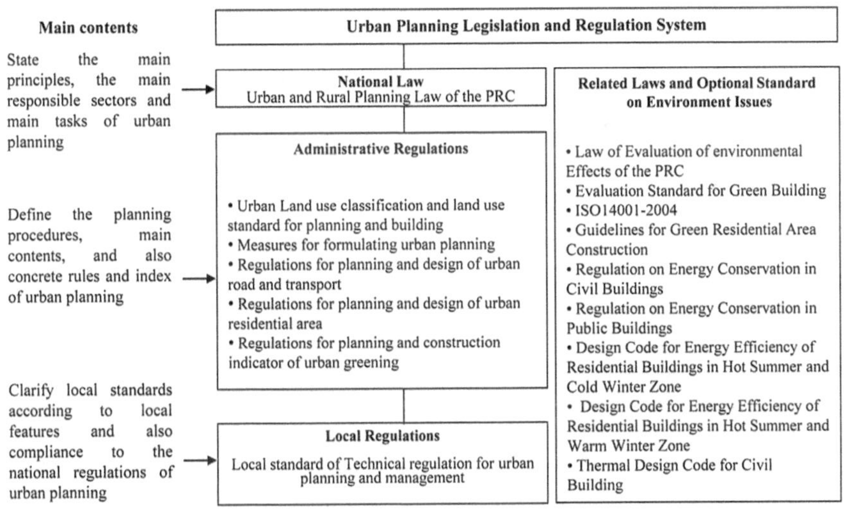

Fig. 1 The structure of the urban planning law and regulation system in China (Note: only a number of examples of regulations are listed in this figure)

Table 1 Regulatory plans defined in the Urban and rural planning law of the P.R.C

Various levels of urban plans	Names of regulatory urban plans
Urban system plans	National urban system plan Provincial urban system plan Municipal and County urban system plan
Local comprehensive plans	Municipal comprehensive plan District plan
Local detailed development plans	Regulatory detailed plan Construction detailed plan

system plans at national and provincial levels, (2) local comprehensive plans and (3) local detailed development plans (The NPC of P.R.C. 2007).

For example, Municipal Comprehensive Plans, legally binding, include a series of sector plans to provide specific arrangements from individual sectors' areas of concern, such as ecological conservation, land use, transport systems, tourism or water conservation.

Municipal Comprehensive Plans require approval by the upper level of the Provincial People's Congress and usually have a term of one to two decades after ratification. Planning instruments, regulating and guiding concrete issues, such as Regulatory Detailed Plan and Construction Detailed Plan, are organized by corresponding City Planning Departments and approved by their local governments.

3.2 Environmental Objectives/Concerns in National Plans, Policies and Programs

The Five-Year Plan is one of the most important governmental frameworks for economic and social development. The environmental programs in the 11th and 12th Five-Year Plans contribute to environmental progress by stating environmental objectives, environmental plans and implementation instruments, including for example financial distribution (OECD 2007). A number of policies and programs have been adopted and issued by several national authorities and ministries to improve environmental performance in sustainable urban planning and encourage developments of eco-provinces/-cities/-districts or low-carbon cities (Table 2).

The Environmental Protection Agency of P.R.C. initiated and launched, in 1995, the first eco-demonstration program—the Demonstration Projects of Ecological Provinces, Cities and Counties. To promote its implementation, the MEP of P.R.C. (2003) adopted The Construction Indicators of Ecological Counties, Cities and Provinces, and then in 2007 revised it to guide the local actions and monitoring.

In 2010, the National Development and Reform Commission (NDRC) of P.R.C. initiated the development of the Pilot Projects of Low-Carbon Ecological Provinces

Table 2 National pilot projects on improving environmental performance in urban development

Year	Names of the pilot projects	Authorities issuing the projects
1995	The Demonstration Projects of Ecological Provinces, Cities and Counties	The Environmental Protection Agency of P.R.C.
2010	The Pilot Projects of Low-Carbon Ecological Provinces and Cities	National Development and Reform Commission of P.R.C.
2011	The Pilot Projects of Low-Carbon Eco-City (Town)	Ministry of Housing and Urban-Rural Development of P.R.C.
2012	The second term of Pilot Projects of Low-Carbon Ecological Provinces and Cities	National Development and Reform Commission of P.R.C.
2013	Developing Green Houses and Eco-city Districts during the period of the 12th five-year plan	Ministry of Housing and Urban-Rural Development of P.R.C.
2013	The Pilot demonstration area for National Ecological Civilization Construction	Ministry of Environmental Protection of P.R.C.

and Cities by signing agreements with five provinces and eight cities (The NDRC of P.R.C. 2010). In 2012, the Second Term of Pilot Projects of Low-Carbon Ecological Provinces and Cities was launched, this time including one province and twenty-eight cities (The NDRC of P.R.C. 2012). Besides the pilot projects initiated by NDRC, there is also a parallel project called the Pilot Projects of Low-Carbon Eco-City (Town), developed by the Ministry of Housing and Urban-Rural Development of P.R.C. (The MHURD of P.R.C.) in 2011.

In 2013, the MHURD of P.R.C. also issued the 12th Five-Year Plan for Development of Green Houses and Eco-City Districts to complement the implementation of energy conservation and emission-reduction goals from the 12th Five-Year Plan-period (The MHURD of P.R.C. 2013a). The MHURD listed several focus tasks in the 12th Five-Year Development Plan of Green Buildings and Green Eco-Cities (2013). Three of the tasks are institutionally relevant and need to be mentioned here: firstly, the establishment of regulations and incentive policies; secondly, the formulation of an indicator system for green eco-cities (e.g. ratios of space usage, green coverage, renewable energy usage, green transport, material and waste recycling, usage of non-traditional water resources), which provides the basis for making Regulatory Detailed Plans for eco-city areas. Thirdly, to incorporate green requirements as pre-conditions for issuing initiatives, plans and land use agreements for development projects in eco-city areas. Moreover, the plan states tasks of evaluation and monitoring of eco-city development, as well as announces the intention to plan and develop 100 eco-city pilot projects in the 12th Five-Year Period. In 2012, the Sino-Swedish Low-Carbon Eco-City and Tangshan Bay Eco-City were included as two of the first eight Green Eco-Cities by MHURD (The MHURD of P.R.C. 2013b).

In 2013, the MEP of P.R.C. also issued a new document, Indicators of The Pilot Demonstration Area for National Ecological Civilization Construction[1] (trial), wherein a new concept of Ecological Civilization Construction, emerging from the new central politician agenda was introduced. The document adds new and crucial indicators to guide, monitor and evaluate the pilot demonstration areas; these include carbon emission intensity, efficiency of resource output, renewable resource recycling and public transport. In addition, besides technique-relevant indicators, the document also lists a number of quantitative criteria from aspects of public participation, governmental subsidy, the evaluation of officials' performance, sharing of information and environmental education, which already were included in the previous document of indicators.

Inspired by a series of demonstration projects, the development of eco-cities or low-carbon cities has become increasingly popular all over China. More and more cities are using eco-city models or low carbon-initiatives as development strategies to brand their cities, both out of consideration for an attractive built environment and to increase politicians' and officials' credibility; the environmental concerns are supposed to be gradually integrated into the performance evaluation system for promotion and considered crucial issues by the public (The MEP of P.R.C. 2009, 2013). According to a study by the Chinese Society for Urban Studies, by 2011 there were 259 cities with eco- or low-carbon city construction development objectives, accounting for almost 90.2 % of China's municipalities (Liu 2011).

3.3 Environmental Requirements in Planning Regulations

Energy matters, identified as a crucial aspect of sustainable development, have been raised and expanded to include not only industry, but also areas such as urban planning, housing and the construction sectors. For example, the EPL of P.R.C. requires local governments to submit energy consumption plans for buildings and public transportation to the state. Furthermore, the Regulation on Energy Conservation in Civil Buildings (2008) states the standards and requirements for energy efficiency in existing and new civil buildings (Seligsohn et al. 2009). However, environmental concerns have not been comprehensively integrated into planning regulations.

To further explore the general situation of the integration of environmental requirements/indicators in current Chinese urban planning regulation systems. A number of legislations and regulations (see Fig. 1) are looked at. Seven environmental perspectives are selected for review; i.e. to determine if they exist in the current system. These include ecological environment, effective and compact urban

[1]The concept of Ecological Civilization Construction was proposed as an important national development strategy to face the current problems of environmental pollution, ecosystem degradation and resources shortage in the 18th National Congress of Chinese Communist Party in 2012 (People's Daily 2012).

function, green transportation, green buildings, energy production and conservation, water utilization, waste treatment and reuse. These seven perspectives were generated based on Environmental Indicators in Low-Carbon Eco-City Development developed by the Chinese Society for Urban Studies (2009) as well as the indicator systems planed in Tangshan Bay *Eco-City (Tangshan), Sino-Swedish Low-carbon Eco-City (Wuxi) and Sino-Singapore Eco-City (Tianjin).*

In the regulatory framework, the rules relevant to **ecological environment** mostly focus on the planning of green landscape structure and green land use layout, which concern morphology rather than ecological/environmental aspects. The indicators frequently listed in the regulations are for example average public green area per person, green land rate and green coverage of the completed area; thus, the indicators focus on quantitative requirements, but not qualitative improvements. Thus, other aspects of indicators, such as native vegetation, species diversities and microclimate created need to be incorporated in the planning regulatory framework, and all requirements should be in quantitative as well as qualitative forms. The principle of **Effective and compact urban function** is stipulated in national regulations. The sub-categories of compact layout and effective public facilities are also included as general principles for urban planning, such as the Floor Area Ratio (FAR) of construction land and the coverage of municipal pipeline networks, but the regulations do not offer guidelines as to how these should be approached. For instance, mixed function of land is considered to contribute to efficient use of land resource, less vehicle travel and more active neighbourhoods, yet it is not mentioned in the regulations. Instead, the principle of clear function zoning is still predominant. Principles of *Green transportation*, such as improving public transport facilities and making the transition to a green travel model are covered in Code of Transport Planning on Urban Roads, with the exception of the requirement regarding renewable energy use. Although green modes of travel such as walking, bicycling and using public transport are encouraged in some items, they sometimes conflict with other standards. For instance, in the Code of Urban Residential Areas Planning and Design, there is an item stating that 'crossing vehicles should be avoided in residential neighborhoods' to ensure a safe residential environment. However, many large-scale neighborhoods are planned without bus accessibility, potentially increasing the demand and need for private vehicles in the area (Hai 2009). In addition, items regarding modes of travel and commuting times exist in the regulations, but concrete guidelines for implementation are absent. Meanwhile, provisions in regulations, for instance regarding parking space, seems to be focused on facilitating the use of private vehicles rather than constraining them. For example, in the Code of Urban Residential Areas Planning and Design, the parking space ratio for private cars in residential areas may be no less than 0.1/householder, but no maximum limitation is provided. **Green building design and green material use** are seldom mentioned in current planning regulations, and the Code of Urban Residential Areas Planning and Design only mentions daylight utilization and harvesting natural wind power to improve microclimate in a small number of standards. Although several standards refer to green buildings established in China, the combination with the planning sector needs to be clarified in the regulations, as they are closely connected during sustainable urban area development. The **energy production and conservation** target is clearly stated in the regulations, but there are no concrete suggestions available. **Water utilization**, water conservation, sewage treatment and water quality and health are mentioned, but not presented in concrete terms in the current regulations, while the new subject of utilization of non-traditional water resources, has not been integrated. **Waste management**, which is a new area considered important for the building of an environmentally friendly society, is only mentioned in terms of waste collection, without recommendation or mention of recycling.

Many of the principles, standards and indicators regarding environmental urban development are absent. Mentions of planning principles such as green space coverage, landscape design, green transportation, waste treatment, effective and compact land use and energy conservation and utilization are fragmented and ambiguous in the current regulations. In addition, some standards and indicators in the regulations are inconsistent with sustainable urban planning principles, such as the contradiction between clear function zoning and mixed land use, increasing parking ratios and reducing private vehicle use.

4 Summary and Discussion of China's General Institutional Conditions

Environmental requirements have not been systematically integrated in the urban planning regulations, and many important aspects concerning environmental sustainability are absent. The integration of environmental requirements into formal regulations in the sector of urban planning, housing and construction exists, but predominantly with regard to energy use in civil buildings (Cai et al. 2013). Policy-making and capacity building in other scopes (e.g. new material, reuse and recycling of resources, indoor environments) thus need to be fostered to promote environmental concerns in sustainable urban development. Conflicts also exist between some propositions in the planning regulations and general sustainable development principles, such as the contradiction between zoning and mixed land use, increasing parking ratios and reducing the use of private vehicles.

Several pilot projects initiated by different national authorities and ministries have been adopted during the last two decades, and thereby accelerated the incentives and actions of eco-city or low-carbon city development at local level. However, each of them uses different key concepts. For instance, the concepts of eco-city, low-carbon eco-city and ecological civilization construction appear in different pilot projects. A consistent standard- and indicator system for rewarding and monitoring qualified projects seems to be lacking. Relevant government authorities have not identified differences or correlations between the pilot projects, at least not in their official websites. Fragmentation, overlaps and inconsistency within the pilot projects bring about confusion and difficulties for implementation in the long run. Understanding, communication and coordination for promoting environmental integration in sustainable urban development among authorities and ministries appear to be weak in China. Generally, China is in urgent need of holistic and operative guidelines and standards for eco-city development.

The adoption of indicator systems for pilot projects is in progress, though certain inadequacies remain, especially for implementation. For the latest pilot project, the Pilot Demonstration Area for National Ecological Civilization Construction (trial), the indicator system has a much wider view on environmental matters. For example, rather than focusing solely on pollution prevention and treatment in the industry sector, the system integrates aspects of resource efficiency and recycling, renewable energy and green transport as well some social aspects, such as public participation. In spite of this, there are weaknesses such as a lack of instructions on how to successfully achieve the indicators, as well as the clarification of authorities' and agencies' responsibilities to fulfil the indicators (Lin and Tian 2011).

The description and discussion in this section of the general institutional conditions in China outlines the basic context in which the following two eco-city cases are embedded. This context helps to improve the understanding, at higher levels, of various legal conditions that might affect the integration of environmental issues in the local sustainable urban development practice, which is presented in the next sections.

5 Case Study—Tangshan Bay Eco-city, Tangshan, China

The Caofeidian New Area, Tangshan Bay Eco-city (previously called Caofeidian New City; called 'Tangshan Bay Eco-City' thereafter) is located in the Coastal New City, which is in the southern part 80 km from the inner city of Tangshan.[2] Tangshan Bay Eco-City is a cooperative project for sustainable urban development between the City of Tangshan, the Swedish consultant company Sweco and the Centre for Environment Technology (CENTEC) within the Swedish Embassy in Beijing. Due to its strategic position as an important port in the Bohai region, located 80 km from Tangshan city and 220 km from Beijing (Fig. 2), Caofeidian is planned as an area with heavy industry development. In its entirety, Tangshan Bay Eco-City has a total area of 150 km^2, planned to provide residential support and services for a new international deep-water harbour and a large industrial area to be developed in the Coastal New City (Ranhagen 2009; Lin and Tian 2011; Tian 2012). It is estimated for completion in 2020, and its 15.8 km^2 residential area will accommodate 800,000 inhabitants. Besides residential land use, Tangshan Bay Eco-City also incorporates urban functions such as an administrative area, a commercial area, culture and recreational area, sports facilities, an educational and research area as well as medical service facilities. The initiative and planning began

[2]Tangshan is a city situated in the Urban Agglomerations of Tianjin, Beijing and Hebei in China.

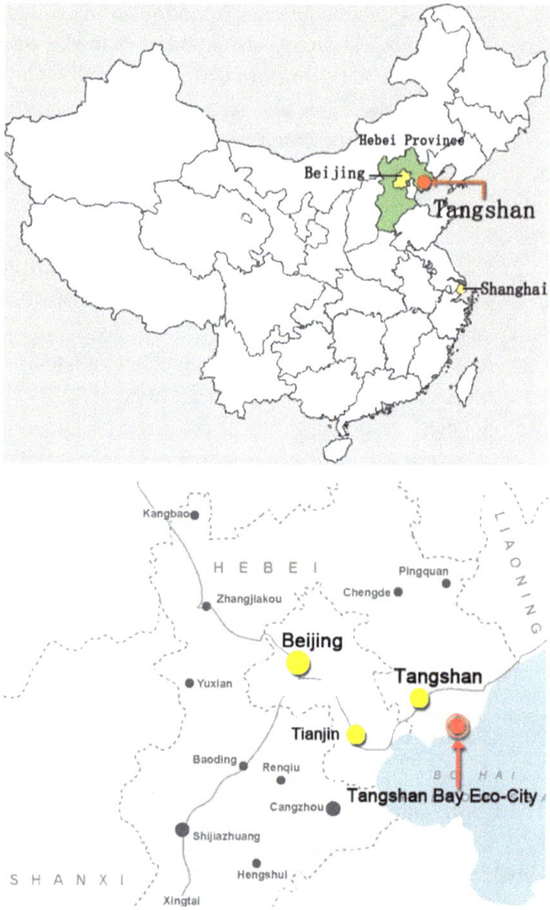

Fig. 2 a, **b** The location of the Tangshan Bay Eco-city

in 2008, and construction began in 2009 (Lin and Tian 2011). The first tenants (employees of the on-site Administration Committee) moved into the staff dormitories during the first phase in 2011. Figure 3 of on-site pictures of Tangshan Bay Eco-City shows its development status until 2013.

5.1 An Indicator System for Guiding Sustainable Urban Planning

In 2008, in one of its planning tasks, Sweco proposed a planning indicator system, sustainability principles, integrated spatial planning as well as an eco-cycle model. Ultimately, Sweco adopted the indicator system for the eco-city that covered 52

Fig. 3 On-site pictures of Tangshan Bay Eco-city in 2013 (Photos taken by Tian, Xinxin)

items, and which was later adopted by the Administration Committee[3] as the planning indicator framework for the project. However, the indicators needed occasional adjustments to be suitable to the local conditions of Tangshan Bay Eco-City and the national standard[4] (Lin and Tian 2011). Thus, based on the Swedish indicator system, the Planning Group in the Administration Committee formulated an overall indicator system for implementation in the eco-city (Sweco 2008). The final overall indicator system covers 7 sub-systems, namely eco-industry, green building, renewable energy, green transport, water management, waste management, and urban landscape. They are described by 141 indicators, including urban function, architecture, transportation, energy, waste, water and urban landscape and space. To clarify and ensure the implementation of the indicators, a corresponding handbook for the overall indicator system was made from four perspectives: content, controlling types, timescale and responsible actors (Lin and Tian 2011) (Fig. 4). These indicators provided a basis for formulating planning indicators in the Regulatory Detailed Plan in the next step. This indicator system provided not only specific objectives but also instruments to support implementation (Lin and Tian 2011).

[3]Tangshan Bay Eco-City Administration Committee was established by Tangshan City as a temporary but formal administrative organization, to be located on-site at Tangshan Bay Eco-City. The Administration Committee is an administrative model to closely combine all relevant sectors both geographically and administratively.

[4]The national standard referred to is The Construction Indicators of Ecological Counties, Municipalities and Provinces (pilot), developed by Minister of Environmental Protection of P.R.C in 2003 and revised in 2007. It is mainly used as a standard for the government's general evaluation of 'ecological counties, municipalities and provinces'.

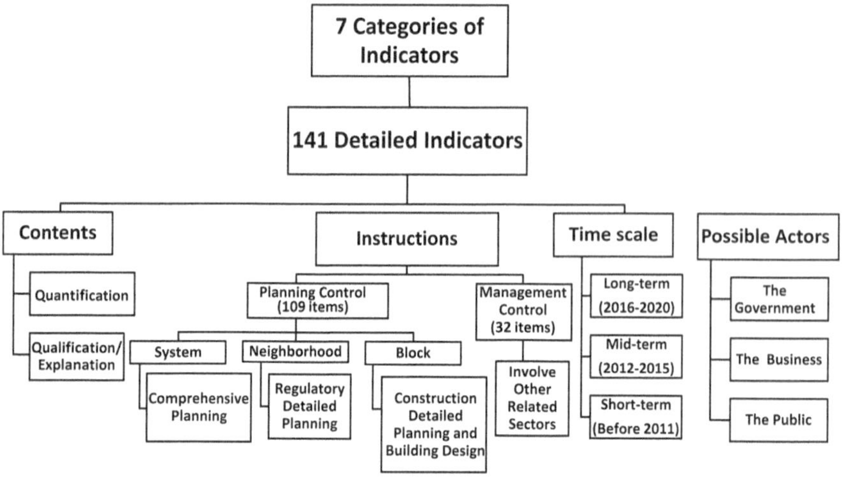

Fig. 4 The structure of adjustment and classification of the eco-indicator system (*Source* based on Lin and Tian 2011)

5.2 *Urban Planning Instruments and Land Use Agreement*

In a typical urban development project, it usually requires adoptions of several plans, including Conceptual Detailed Plans, Regulatory Detailed Plans and Construction Detailed Plans (Fig. 5). In this eco-city project, to integrate environmental concerns, two more types of plans were added: the indicator system (see the previous section) and a series of eco-subject plans (Fig. 5).

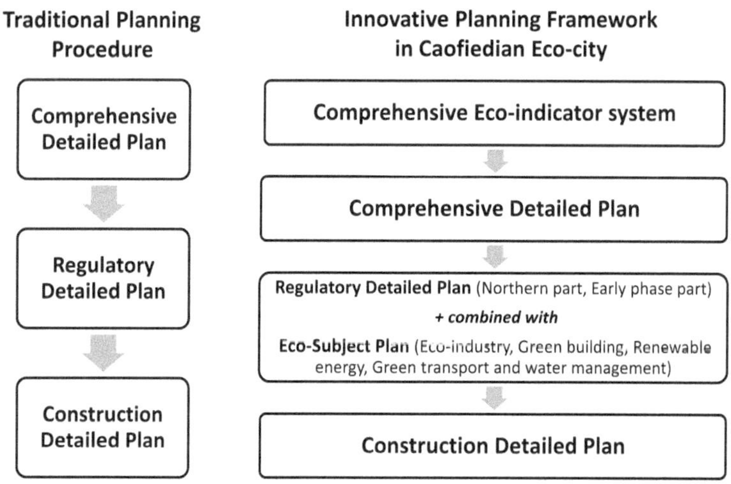

Fig. 5 The comparison of traditional and adjusted planning frameworks in Tangshan Bay Eco-City (*Source* based on Tian 2012)

Eco-subject Plans were intended to cover the seven sub-systems in the overall indicator system. Consultants from each sub-field developed eco-subject plans to adjust and specify indicators, and based on the indicators they proposed design and operative guidelines for the eco-city implementation. For example, regarding the sub-system of green building, several eco-subject plans have been developed, e.g. the Green Building Implementation Handbook and Green Design Guidelines for Public Buildings. The planning and environmental requirements in eco-subject plans also helped the formulation of the handbook of the overall indicator system as well as the Regulatory Detailed Plans (Tian 2012). The combination of environmental indicators in the Regulatory Detailed Plan is crucial for environmental integration (Lin and Tian 2011) (Box 1), because it provides legally-binding references for formulating Land Use Agreements in the area.

BOX 1: Regulatory Detailed Plan and Land Use Agreement: *In accordance with the typical Chinese planning procedure, the Regulatory Detailed Plan is an important regulatory planning instrument, which mainly specifies aspects offloor area ratio, building density, building height, ratio of green space and other elements concerning urban spatial form. This plan then provides the baseline for the creation of the Land Use Agreement, which is another legally binding instrument of importance for ensuring the quality of urban area planning and development. The Land Use Agreement, which is issued by the Land Resource Bureau, is one of the most important legally- binding instruments, specifying planning and construction requirements before developers purchase land. The Land Use Agreement has well-established routines and a standardized template that all local land resource bureaus in China must follow strictly, but the standardized template does not accommodate binding environmental requirements. It is therefore difficult to bind the environmental standards in the same way as other regular planning requirements.*

Box 1: Regulatory Detailed Plan and Land Use Agreement in China's urban planning system (Source: Yin 2014) Despite this innovative planning framework applied, problems showed up with the—also compulsory—land use agreement (Box 1). Planners from the Administration Committee indicated that although the Land Use Agreement was considered an important legal instrument for ensuring implementation that followed planning requirements, it was difficult to formally integrate the environmental requirements into the Land Use Agreement in this case. The explanation given by the planners was that the contents and format of the Land Use Agreement should strictly follow the national model, which is inflexible and hardly concerns environmental requirements in sustainable urban development.

5.3 Summary and Discussion of the First Case Study

The relative absence of national and local rules, regulations, policies and an urban planning system for guiding and enforcing environmental integration in sustainable urban development were a weak precondition. Members of the Administration Committee made efforts to reformulate or re-organize the Regulatory Detailed Plans and Land Use Agreements to include binding environmental indicators. However, the hierarchical structure and the well-established routines of the Land Use Administration increased the difficulties of introducing legally-binding environmental concerns.

There seems to be a gap between the imported sustainable planning ideals and regular formal rules in the Chinese urban planning system with regard to the integration of environmental requirements in sustainable urban development. For example, legally- binding environmental requirements were absent from the Regulatory Detailed Plan and the Land Use Agreement, which limited the regulatory legitimacy and enforcement of environmental integration. The Planning Group in the Administration Committee made progress in adding environmental requirements to Regulatory Detailed Plans. To strengthen enforcement, the Regulatory Detailed Plan and Land Use Agreement were reformulated and re-organized to include the environmental indicators in legally-binding planning instruments. However, according to the interview with the planning manager involved in the project, the barrier for environmental integration in the Land Use Agreements emerged: the strict national format of Land Use Agreements limited possibilities to add environmental indicators as standards for enforcement in the agreement. This barrier is also related to the hierarchical structure within China's Land Use Administrations, where local land use departments must strictly adhere to sector routines from the national level. Therefore, even though the environmental requirements were incorporated into Land Use Agreements, they were supplied as recommendations in attachments and provided limited enforcement of environmental implementation in the area development. The following sections prevents and analyses another case of Sino-Swedish Low-Carbon Eco-City, Wuxi, which shares some similarities but also difference with Tangshan Bay Eco-city, with a regard of how legal aspects providing impacts on environmental integration in the practices.

6 Case Study—Sino-Swedish Low-Carbon Eco-city, Wuxi

In 2010, the Wuxi local government and the Swedish Ministry of Environment signed the document Sino-Swedish Low-Carbon Eco-City Cooperation Memorandum, which designated the Sino-Swedish Low-Carbon Eco-City (2.4 km^2) as a cooperative, demonstrative project between Sweden and China. The Sino-Swedish Low-Carbon Eco-City is located in the south central area of

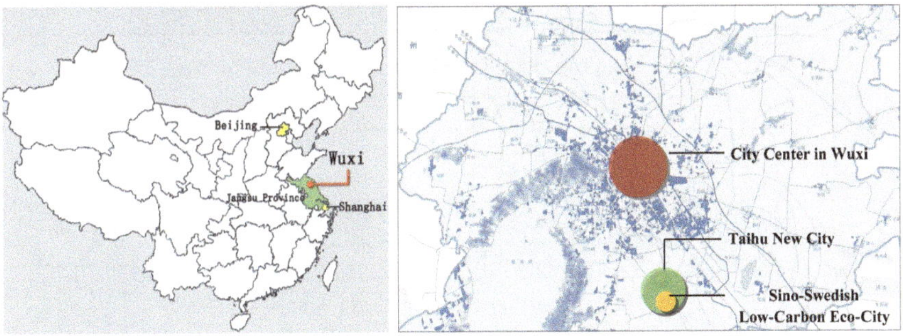

Fig. 6 a, b The location of Sino-Swedish Low-Carbon Eco-City in Wuxi, formulated plans in Taihu New City and Sino-Swedish Low-Carbon Eco-City

Taihu New City's total 150 km^2, approximately 6 km from the old city center of Wuxi city[5] (Fig. 6). According to the local government's ambition, the Sino-Swedish Low-Carbon Eco-City is the pioneer and representative area for Taihu New City's eco-city construction practice, as well as an eco-city model for the Wuxi city. Planned with the urban functions of a residential area with municipal and community public facilities, it will accommodate about 20,000 inhabitants. The public facilities areas will cover 1.04 km^2, including commercial area (0.10 km^2), a hotel industry area (0.12 km^2), a cultural and recreational area (0.30 km^2), and sports facilities (0.52 km^2). Planning of the area started in 2009, and construction began in 2013 and is due to be completed in 2018 (Gao 2013).

After the decision to launch the Sino-Swedish Low-Carbon Eco-City pilot project, Wuxi engaged the Swedish architecture firm Tengbom in late 2009 to help develop the Comprehensive Detailed Plan for the area. As the Sino-Swedish Low-Carbon Eco-City is located inside the Taihu New City, the plan would have to adhere to the requirements in the Comprehensive Plan for New Taihu City (Fig. 7). The Comprehensive Detailed Plan aimed to provide general guidelines for the area from aspects of land use, urban function, environmental requirements, road systems and public facilities. Besides the physical design, the environmental indicators were also incorporated in the Comprehensive Detailed Plan, and provided insights and framework for planning at other levels in the area, as well as in Taihu New City (Fig. 7). Some environment-technical solutions from Sweden were also introduced in the plan. For example, decisions were made to implement a vacuum waste management system in the Sino-Swedish Low-Carbon Eco-City.

Inspired by the Swedish contributions of an integrated planning approach and environmental indicators, the indicator systems and implementation guidelines for both Taihu New City and Sino-Swedish Low-Carbon Eco-City were developed in 2010 by the Chinese Academy of Building Sciences and Wuxi Institution of Urban

[5]Wuxi city, situated in the southern part of Jiangsu Province, is included in the metropolitan region of Yangtze River Delta, China.

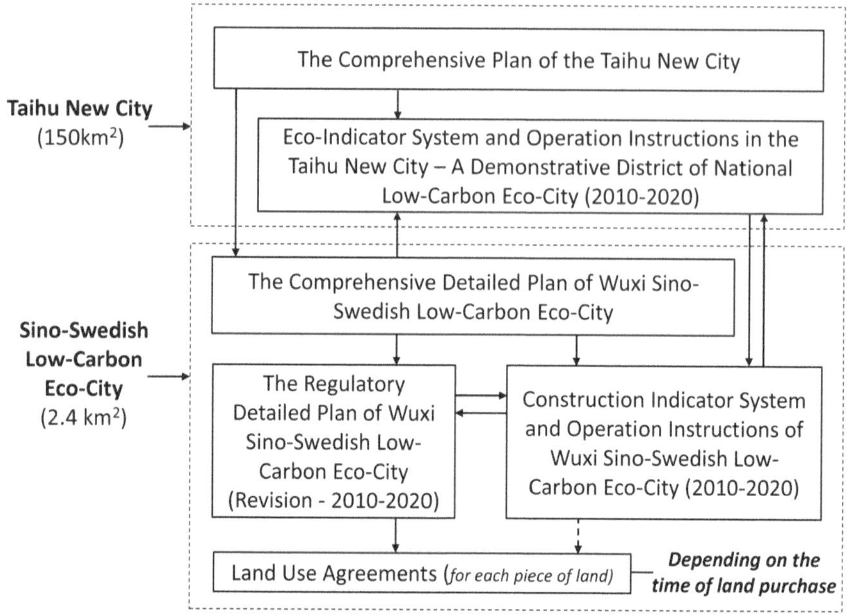

Fig. 7 Formulated plans in Taihu New City and Sino-Swedish Low-Carbon Eco-City (2009–ongoing)

Planning and Design (Fig. 7). Since the Sino-Swedish Low-Carbon Eco-City is located inside the Taihu New City, their planning and development are closely interrelated. The Planning Indicator System and Implementation Guideline of Taihu New City—The National Low-Carbon Eco-Demonstration Area (2010–2020) includes six dimensions: urban function, green transportation, energy and resources, ecological environment, green building and social harmony. It provides comprehensive guidelines and requirements for improving environmental performance in Taihu New City. Meanwhile, it also provides the baseline for the Construction Indicator System and Implementation Guideline of Sino-Swedish Low-Carbon Eco-City, Wuxi (2010–2020), the environmental profile of which was supposed to be stricter than in Taihu New City. The environmental indicator system in the Sino-Swedish Low-Carbon Eco-City contains 7 categories (47 indicators), which are urban functions and management, green transportation, energy use, water use, solid waste management, landscape and green buildings. In the planning for Taihu New City, the larger scale urban functions, green transport systems, ecological infrastructures (e.g. energy supply, water supply) were taken into consideration from a holistic point of view. The indicator system for the Sino-Swedish Low-Carbon Eco-City, tended to reinforce operative guidance, for example by proposing higher standards of water reuse, waste collection infrastructure and detailed operative measures.

To enforce environmental integration, the environmental requirements needed to be specified in the Regulatory Detailed Plan for Sino-Swedish Low-Carbon Eco-City, Wuxi (Box 1). After introducing the environmental profile for the area, the Regulatory Detailed Plan of Sino-Swedish Low-Carbon Eco-City was revised in 2010 based on Swedish planning experts' comments. The Construction Indicator System and Operation Instructions of Sino-Swedish Low-carbon Eco-City, Wuxi, in particular was subject to revision (Fig. 7). The indicator system included mandatory indexes and regulations aimed at promoting green transport, improving efficient utilization of land resources, protecting ecological environment and promoting resource and energy use as regulatory planning requirements to be considered when adopting the Land Use Agreements in the area *(Box 1)*.

6.1 Driving Forces—Local Policies and Regulations

According to the respondents from the City Planning Bureau and the Construction Headquarters, the issuing of local policies and regulations played crucial roles in overcoming obstacles and promoting the project's progress and development. And it was the local leaders of the Party Secretary and the Mayor who reacted quickly and positively to upcoming barriers and conflicts during the process, and thereby considerably facilitated the formulation of these local formal rules.

According to project leaders and planners from the Construction Headquarters, resistance from the relevant administrative departments to undertake the tasks necessary for fulfilling the environmental requirements became a significant obstacle for the implementation of plans with high environmental requirements. The City Planning Bureau coordinated the project in its early stages; i.e. before construction began in 2013, but the responsibilities for relevant issues such as water management were not under its jurisdiction. For example, when the City Planning Bureau communicated with the Water Supply and Treatment Bureau about the new objective of water standard in the area, the water department was reluctant to apply the higher standard for drinking water. The main reason, noted by a planner from the Construction Headquarters, was that the existing national and local regulations regarding environment requirements have lower standards than in the Sino-Swedish Low-Carbon Eco-City. Officials in the Water Supply and Treatment Bureau were concerned that applying these higher standards would entail extra work during the management process.

In addition, the high environmental profile required that an increased number of sector authorities were involved in the project, which in turn led to more disagreements and conflicts among sectors than in regular urban development projects. For example, a planner from the Construction Headquarters noted that the City Planning Bureau and the Water Supply and Treatment Bureau had different objectives regarding the form of riverbanks: the City Planning Bureau's wishes were to increase public access to nature, such as waterfronts, through the landscape design, while the Water Supply and Treatment Bureau aims were to build higher concrete barriers to protect citizens from the rivers and flooding and ensure their security.

Reacting to these obstacles, in 2010, the official document named "Three-Year Action Plan for Accelerating the Implementation of the Sino-Swedish Low-Carbon Eco-City in Taihu New City" was issued by the Wuxi City Government Office (2010) as an administrative command. The action plan was based on the contents in a number of plans, such as the Conceptual Detailed Plan, the indicator systems, and the Regulatory Detailed Plan. It designated the tasks of planning, construction and management to corresponding public administrations through both political and administrative enforcement. Then in 2012, the Eco-City Regulation in Taihu New City, Wuxi was issued by Wuxi City People's Congress[6] (2012), as the first local regulation on eco-city development in China to address legal constraints for the eco-city implementation. It included the contents of general principles, eco-city planning, eco-city construction, eco-city management, responsibilities and supplementary provisions. The purpose of the regulation was to provide regulatory enforcement for the integration of environmental requirements throughout the entire development process. For example, the regulation appointed specific authorities or administrative departments as responsible for corresponding tasks. The adoption of the Three-Year Action Plan and the Eco-City Regulation in Taihu New City provided formal enforcements to ensure the implementation of environmental standards, especially when the formal basis, such as planning regulations and procedures, for sustainable urban development are lacking at the national level.

6.2 Summary and Discussion of the Second Case Study

Formal rules in sector regulations sometimes contradicted other sectors' preferences for reaching environmental profile goals in the project, e.g. the Planning Bureau and the Water Supply and Treatment Bureau's different proposals for the area's riverbanks. The conflicts between different sectors' objectives that emerged in the project were difficult to resolve during the implementation, as a clear priority of different objectives had not been stated. In addition, neither national regulations nor the planning system provided guidance and constraints for environmental integration in urban planning practice. In a typical Chinese urban development project, environmental integration between legally binding plans such as Regulatory Detailed Plans and Land Use Agreement is absent, limiting the regulatory capacity of enforcing environmental implementation in eco-city development. In this case, the formats of the Regulatory Detailed Plans and Land Use Agreements were rearranged to include the environmental requirements in order to strengthen the

[6]Wuxi City People's Congress is the legislative body, authorized by the state government office, in Wuxi city. It can issue local regulations that can only apply to specific local areas within the city's administrative boundary rather than in the national dimension.

legally-binding enforcement in planning and construction. The rearrangement of the Regulatory Detailed Plans and Land Use Agreements also provided the platforms for generating inter-organizational coordination to ensure environmental integration in the eco-city development.

7 Final Discussion

7.1 A Holistic Perspective of Policies, Regulations and Plans

The evidence provided in this study indicates that an overarching structure of formal rules needs to be in place to guide and enforce environmental integration in sustainable urban planning. In recent years, the number of national programs and plans for guiding and encouraging sustainable urban development has increased, but their effectiveness seems uncertain, since they are fragmentarily distributed in various regulations and policies, and the relationships between environmental issues and urban planning are vague. The concepts used in the various pilot programs/plans—such as eco-city, low-carbon eco-city and ecological civilization construction– are only vaguely defined. In addition, almost each pilot program/plan was led by different ministries and authorities (Table 2), whose responsibilities to environmental integration in urban planning are overlapping and indistinct. This fact led to failures in the guidance and control of local activities regarding sustainable urban planning, where the indicators in those programs were not used as guidelines during the planning process. In addition, the national planning legislations and regulations appear to be weak in combining environmental concerns within them, and some of the requirements are evenly contradicted with sustainable goals. Thus, to summarize, the national policies' vague definition and contents, together with the lagging-behind regulations, created confusion and failed to provide effective guidance for environmental integration in urban planning at local levels.

At the local level, generally, the local governments appear to be more interested in environmental indicator systems rather than in overall environmental programs. Although the formulation of concrete indicators would provide practical guidance and constraints for environmental performance in the two eco-city cases presented, the lack of a holistic view in overall policy- and regulation-making might not offer an appropriate formal institutional environment for ensuring the implementation of environmental objectives.

Besides the weakness, there exists progress as well at the local level. In the Sino-Swedish Low-Carbon Eco-City, the City of Wuxi adopted the action plan for tasks clarification and responsibility distribution among relevant local authorities and agencies and Wuxi City People's Congress issued the local eco-city regulation in order to fill the gap of normative and regulatory legitimacy for the local eco-city development. Comparatively, in Tangshan Bay Eco-City, local regulations or other mandated documents specifically for the promotion of environmental integration in the project have not been developed. This finding indicates that although the lack of policies and regulation at the national level increased difficulties for environmental integration in planning practice, it is not necessarily a hindering institutional condition per se.

7.2 Regulations, Policies, Plans and Agreements

Besides basic guidance and constraints at national and local levels, the findings in the study indicate that concrete standards and instructions for environmental requirements also play an important role in the projects. In both case studies presented, efforts have been made to combine environmental requirements in detailed plans and development agreements in order to legally enforce them in the implementation. However, according to the planners in Tangshan Bay Eco-City, it was difficult to use Land Use Agreements to enforce environmental requirements due to the well-established and strict format of the current administration in the land resource sector. In other words, in Tangshan Bay Eco-City, the Administrative Committee succeeded in realizing environmental integration in regulatory detail plans, but not in land use agreements. This problem appears to have been less pronounced in the Sino-Swedish Low-Carbon Eco-City than in Tangshan Bay Eco-City. The environmental requirements could be listed as an appendix of environmental indicators to the Land Use Agreements in the project, and *'the appendix has the same legality as the regular provisions in the Land Use Agreement'*, according to a planner in the Sino-Swedish Low-Carbon Eco-City. The strong political support, such as the issuing of the local action plan and the local eco-city regulation, seemed to have given the necessary political and regulatory legitimacy for processing binding environmental requirements in the Land Use Agreements. Thus, strong political support is indicated here to be crucial for driving the achievement of environmental integration in eco-city development, referring to what Wheeler (2004) and Raydan and Steemers (2006) argue.

7.3 Concluding Reflections

Generally, more attention is being paid to environmental matters in China's national policy-making. However, the formal rules in current urban planning system have yet to provide legitimacy for implementation of the integrative planning approach for sustainable urban development.

The evidence in this study indicates that the absence of the overarching perspective and effective guidance and constraints in China's environmental and planning formal rules at different levels appeared as an institutional weakness, hindering environmental integration in urban planning process. In a resource book compiled by Dalal-Clayton and Bass (2002), they point out that strong legislative systems adhering to a holistic perspective are necessary for introducing experimental approaches or changes in order to implement an overarching sustainable policy framework in practice. Further, the lack of regulatory guidance and constraints at the national level actually could be complemented by adopting local policies and regulations, if local politicians have strong intentions to fulfil a commitment of sustainable urban development, for example in the case of Sino-Swedish Low-Carbon Eco-City. This evidence also proves what North (1990) and DiMaggio and Powell (1991) argue that active involvement of and considerable interference from powerful actors play a key role in making institutional changes to legitimate supported actions.

Whether or not environmental requirements and their status were clearly stated in the formal rules also played a crucial role in ensuring the local regulatory legitimacy and enforcement of environmental integration in the both cases. In other words, regulations and policies for enforcing the integration of environmental requirements in regulatory detailed plans and land use agreements is crucial for providing legitimacy for ensuring sustainable urban development. There was a rearrangement of legally-binding planning tools such as Regulatory Detailed Plans and Land Use Agreements to fill the regulatory gaps. Generally, the common approach employed in the cases is to incorporate environmental requirements in programs, regulatory plans and agreements in order to ensure their implementation. The foundation of regulatory or normative rules in Tangshan Bay Eco-City appears weaker than the Sino-Swedish Low-Carbon Eco-City. To some extent, the new local eco-city regulation in Wuxi overcame the barrier of weak regulatory legitimacy for environmental integration in the Sino-Swedish Low-Carbon Eco-City, despite China's lacking national guidance and control. The different attitudes and actions to use Land Use Agreements between the two case studies indicate that the Wuxi city government provided more effective support than the Tangshan government by establishing other formal institutional conditions to offer norms and regulatory enforcement. This indicates again the considerable influence of local leaders and governments' intention on driving an achievement in sustainable development.

References

Birkeland J (2002) Design for sustainability: a sourcebook of integrated ecological solutions. Earthscan, London

Cai ZC, Yin Y, Wennerstern R (2013) From energy efficiency to integrated sustainability in housing development in China: a case study in a hot-summer/cold-winter zone in China. J Hous Built Environ 28(2):329–344

Dalal-Clayton B, Bass S (2002) Sustainable development strategies: a resource book. Organisation for economic co-operation and development, Paris and United Nations Development Programme, New York

DiMaggio JD, Powell WW (1991) Introduction. In: Powell WW, DiMaggio JD (eds) The new institutionalism in organizational analysis. University of Chicago Press, Chicago
Flipse L (2007) Environmental policy integration and spatial planning: an evaluation of the MILO and LOGO spatial planning. Master thesis, Utrecht University
Gao SM (2013) The construction of the Sino-Swedish Low-carbon Eco-city, Wuxi, starts off, and will be accomplished within five years. Wuxi Daily. Retrieved 20 May 2015, from http://www.wxrb.com/node/kandian/2013-4-9/F25H44ECG429542.html
Hai XP (2009) Report III: Strategy research for sustainable urban planning. In: China's research council of urban science, China's low carbon eco-city development strategy. China City Press, Beijing
Jordan A, Lenschow A (2010) Environmental policy integration: a state of the art review. Environ Policy Gov 20(3):147–158
Jordan A (2002) Efficient hardware and light green software: environmental policy integration in the UK. In: Lenschow A (ed) Environmental policy integration: greening sectoral policies in Europe. Earthscan, London, pp 35–56
Lin P, Tian XX (2011) The implementation and specification of the indicator system in Cao Feidian eco-city planning. Constr Sci Technol 13:40–44
Liu Y (2011) The overall development and typical cases of China's eco-cities. Constr Sci Technol 13:35–39
National Development and Reform Commission of P.R.C. (NDRC of P.R.C.) (2010) The pilot projects of low-carbon ecological Provinces and Cities. Retrieved 1 Dec 2012, from http://www.ndrc.gov.cn/
National Development and Reform Commission of P.R.C. (NDRC of P.R.C.) (2012) The second term of pilot projects of low-carbon ecological Provinces and Cities. Retrieved 1 Dec 2012, from http://www.ndrc.gov.cn/
Nilsson M, Persson A (2003) Framework for analyzing environmental policy integration. J Environ Plann Policy Manage 5(4):333–359
Nilsson M (2005) Learning, frames, and environmental policy integration: the case of Swedish energy policy. Environ Plann C Gov Policy 23(2):207–226
North DC (1990) Institutions, institutional change and economic performance (Political economy of institutions and decisions). Cambridge University Press, Cambridge
Olsson RA (2008) Planning metropolitan regions—Institutional perspectives and the case of spare. Doctoral Thesis, Royal Institute of Technology, Stockholm
OECD (Organization for Economic Co-Operation and Development) (2002) Improving policy coherence and integration for sustainable development—A checklist
OECD (Organization for Economic Co-Operation and Development) (2007) Environmental performance reviews, China.
Ostrom E (1990) Governing the commons: the evolution of institutions for collective action. Cambridge University Press, Cambridge
Ostrom E (2005) Understanding institutional diversity. Princeton University Press, Princeton
Owens SE (1990) Land-use planning for energy efficiency. In: Cullingworth JB (ed) Energy, land, and public policy. Energy policy studies, 5. Transaction Publishers, Piscataway
Persson Å (2007) Different perspectives on EPI. In: Nilsson M, Eckerberg K (eds) Environmental policy integration in practice: shaping institutions for learning. Earthscan, London
Ranhagen U (2009) Introduction of planning work for Caofeidian International Eco-City. World Archit 06:17–27
Raydan D, Steemers K (2006) Environmental urban design. In: Santamouris M (ed) Environmental design of urban buildings: an integrated approach. Earthscan, London
Rogers R (1998) Cities for a small planet. Basic Books, New York
Runhaar H, Driessen PPJ, Soer L (2009) Sustainable urban development and the challenge of policy integration: an assessment of planning tools for integrating spatial and environmental planning in the Netherlands. Environ Plan 36(3):417–431
Scott WR (2001) Institutions and organizations. Sage Publications, Beverly Hills

Seligsohn D, Heilmayr R, Tan XM, Weischer L (2009) China, the United States, and the climate change challenge. Wri policy brief. Retrieved 1 Dec 2011, from http://www.ccchina.gov.cn/en/column.asp?ColumnId=25

Storbjörk S, Lähteenmäki-Smith K, Hilding-Rydevik T (2009) Conflict or consensus: the challenge of integrating environmental sustainability into regional development programming. Eur J Spat Dev 34:1–22

Sweco (2008) Caofeidian international eco-city, Tangshan—Sustainability Guidelines

The MEP of the P.R.C. (Ministry of Environmental Protection of the P.R.C.) (2003) The construction indicators of ecological country, municipality and province (The pilot)

The MEP of the P.R.C. (Ministry of Environmental Protection of the P.R.C.) (2009). Capacity building of national environment supervision: Mid-Term evaluation of the 11th Five-Year Plan

The MEP of the P.R.C. (Ministry of Environmental Protection of the P.R.C.) (2013) Indicators of the pilot demonstration area for national ecological civilization construction (Trial)

The MHURD of P.R.C. (Ministry of Housing and Urban-Rural Development of P.R.C.) (2013a) Developing green houses and eco-city districts during the period of the 12th five-year plan

The MHURD of P.R.C. (Ministry of Housing and Urban-Rural Development of P.R.C.) (2013b) Evaluation report of energy saving of buildings consumption in 2012

The National People's Congress of P.R.C (The NPC of P.R.C) (2008) Urban and Rural Planning Law of the P.R.C. Retrieved 12 July 2012, from http://www.gov.cn/flfg/2007-10/28/content_788494.html

Tian XX (2012) Case study of eco-city urban planning—the case of Tangshan Bay Eco-city. Conference paper in Annual national planning conference

Wandén S (2003) Society, systems and environmental objectives. Swedish Environmental Protection Agency, Sweden

Wheeler SM (2004) Planning for sustainability—creating livable, equitable, and ecological communities. Taylor and Francis, Abingdon, UK

Williamson OE (1998) Transaction cost economics: how it works, where it is headed. De Economist 146(1):23–58

Wuxi City Government Office (2010) Three-year action plan for accelerating the implementation of the Sino-Swedish low-carbon eco-city in Taihu New City

Wuxi City People's Congress (2012) Eco-city regulation in Taihu New City, Wuxi

Yin Y (2014) Environmental integration in sustainable urban planning from an institutional perspective. Doctor thesis. Royal Institute of Technology, Stockholm, Sweden

Identifying the Interaction Between Landfill Taxes and NIMBY. A Simulation for Flanders (Belgium) Using a Dynamic Optimization Model

Rob Hoogmartens, Maarten Dubois and Steven Van Passel

Abstract In the past, legally backed landfills were emerging at an increasing pace in order to deal with growing waste generation. The negative externalities that are caused by these landfills however, together with the emergence of what is nowadays called the NIMBY (Not In My Back Yard) syndrome, led to the awareness that volumes of landfilled waste had to decrease. As a result, restrictions on remaining landfill capacities emerged which causes remaining capacity to be regarded as a non-renewable, scarce resource. In this paper, a dynamic optimization model is constructed to assess the evolution of landfill volumes and landfill prices in time. Carrying out a simulation for Flanders (Belgium), landfill paths and price paths were constructed for two different scenarios. In the first scenario, landfill taxes are taken up in the model, whereas these taxes were omitted from the model in scenario two. As the results show, when landfill taxes are legally levied, it takes 42 years for landfill exhaustion to occur. When no landfill taxes are being used, this period would be shortened to only 20 years. Therefore, it is clear that a legally introduced landfill tax has the effect that yearly landfilled volumes decrease considerably, managing the remaining landfill capacity in a more sustainable way. In addition, when landfill taxes are used, discounted total welfare increases significantly. So we can conclude that, from a broad societal perspective, the added value of a legally introduced landfill tax is considerable in terms of welfare gains.

Keywords Exhaustible resources · Landfilling · Landfill tax · NIMBY · Scarcity

R. Hoogmartens (✉) · S. Van Passel
Research Group of Environmental Economics, Centre for Environmental Sciences, Hasselt University, Agoralaan—Building D, 3590 Diepenbeek, Belgium
e-mail: rob.hoogmartens@uhasselt.be

S. Van Passel
e-mail: steven.vanpassel@uhasselt.be

M. Dubois
Policy Research Centre for Sustainable Materials, KU Leuven, Kasteelpark Arenberg 44, 3001 Louvain, Belgium
e-mail: maarten.dubois@kuleuven.be

V. Mauerhofer (ed.), *Legal Aspects of Sustainable Development*,
DOI 10.1007/978-3-319-26021-1_24

1 Introduction

In the 1960s, as a result of mass production and growing consumption that led to a steep incline in waste generation, landfills were popping up everywhere. Starting in the 1970s, public attitude towards waste started to change as people realized the negative environmental externalities caused by landfilling and the valuable space it occupies (Strasser 1999; Walsh 2002, pp. 4936–4942; Van Passel et al. 2013, pp. 92–102). A well-known and often used instrument to internalize external effects such as noise, odor, groundwater pollution and air emissions, is a legally introduced landfill tax.

In Europe, as it is moving towards an open market for waste management, calls are made to harmonize waste policies and laws across borders. Considering that several front runners in waste management have high landfill taxes, the arguments for high and harmonized landfill taxes in Europe seem strong. In addition, landfill taxes directly target the lowest level of the Waste Hierarchy, inducing more sustainable waste treatment practices. Indeed, the landfill tax increases the cost of landfilling such that other waste treatment methods become more attractive (Bio 2012; Dinan 1993, pp. 242–256; IVM 2005). Although there is some evidence on the effectiveness of a landfill tax to reduce landfilling (Bio 2011; OECD 2012; Oosterhuis et al. 2009), not all economic scholars are convinced that high landfill taxes are justified (Dijkgraaf and Vollebergh 2004, pp. 233–247; Dubois 2014; Eshet and Shechter 2005, pp. 487–504; Kinnaman 2006, pp. 219–232). These scholars argue that external costs of modern sanitary landfills with methane extraction are rather low (5–30 euro). A Pigovian[1] landfill tax would therefore be positive, but typically lower than landfill taxes such as in Belgium or in the UK (50–90 euro) (CEWEP 2014). The question that then arises, is how to reconcile the economic perspective with policy discourses.

A historic element that has determined waste policies is the 'Not In My Back Yard' syndrome (NIMBY), which is mainly caused by spatial scarcity, environment related and health impact concerns (Levinson 1999, pp. 31–51). Although this syndrome has not been a conscious policy decision, its occurrence has actually avoided new permits for landfills. Considering that NIMBY has had the same effect as a moratorium on new landfill zones, landfill volumes have become a scarce good in densely populated regions in Western Europe. Economic theory predicts that scarce non-renewable goods will increase in price such that consumption will be restrained. Remarkably, this effect is also the aim of a legally introduced landfill tax, so attention has to be paid to the interaction between both instruments.

The purpose of the paper is to model optimal landfill and price paths (to be included into legal frameworks), and to investigate when landfill capacity will be exhausted. In view of the above, this papers handles following research questions:

[1]A Pigovian tax is applied to a market activity that is generating negative externalities. By setting the tax equal to the social cost of these negative externalities, social costs are covered by the private costs of an activity and inefficient market outcomes are corrected for.

- What are optimal landfill and price paths, and when will landfill capacity be exhausted?
- Knowing that NIMBY has made new permits for landfills impossible, what has been the added value of the landfill tax in terms of social welfare gains?
- Can a landfill tax above the landfill externalities be justified?

To answer these research questions, a dynamic economic model was developed to identify the relationships between all relevant variables such as landfill volumes, landfill prices and landfill taxes. As the model should identify the best allocation of landfill volume over time, the model was set up as a dynamic optimization problem. Focusing on the simulation of economic models, the paper does not address non-economic factors other than the potential impacts of applying landfill taxes. When supporting decision-making in practice, also non-economic and societal factors should normally be included, such as social structures, employment and technological developments. As a case study, the Flemish region of Belgium was chosen to illustrate the theoretical approach.

2 Theoretical Framework

In this part of the paper, all elements of the dynamic optimization model are discussed and a theoretical model is built up. Numerical optimization problems are known to serve at least two functions. First, they make theory and methods less abstract and more meaningful and secondly, they can serve as a bridge from theory and general models to actual analyses of real-world allocation problems (Conrad 1999). Furthermore, optimization problems are capable of providing exploratory numerical insights that might be included into legal frameworks and in this way support legal decision making.

2.1 Landfill Demand

Figuring out optimal landfill and price paths, and effects of influencing variables, requires a dynamic optimization model to be elaborated. A first element of this model is the inverse linear demand function.

To map the aggregate landfill volume demanded to market price, an inverse demand curve is used. In general, we write $p_t = D(S_t)$, where p_t is the price per ton landfilled in period t given that an aggregate landfill volume of S_t is supplied to the market. We will assume that price decreases with increases in S_t (so $D'(S_t) < 0$). In our model, a linear inverse demand function is used, which is given by:

$$p_t = D(S_t) = A - bS_t \tag{1}$$

with:

p_t = price in year t (euro/ton)
S_t = volume landfilled in year t (million ton)
A = choke-off price, intercept on price axis (euro/ton)
b = slope of the inverse demand function

An important characteristic of the linear demand curve is the implied maximum choke-off price A. When this choke-off price is reached, the equilibrium quantity on the landfill market falls to zero. Such an upper bound may result from the existence of a substitute, available at constant marginal cost MC = A. In scheduling landfill volumes, each competitive firm is assumed to know about this backstop substitute and to know what price will reach the intercept when the full remaining landfill volume has been exhausted. As substitutes for Flemish landfilling, one can think of an increase in waste export or an increase in the waste recycling rate. In the course of time, as the landfill tax gradually reduces the volumes that are landfilled, these substitutes are becoming more attractive. In the model, we assume that exhaustion occurs in t = T and that by then, remaining landfill volume falls to zero ($S_T = 0$) so that only substitutes can be used. The date of exhaustion, T, is unknown and must be determined along with the competitive landfilling and price paths.

2.2 Competitive Landfill Companies

In the model, it is assumed that there exists a competitive landfill industry facing a linear inverse demand curve for aggregate landfill volume, S_t. The landfill companies are maximizing their profits, so they will try to offer landfill volume so as to:

$$\text{Maximize}_{S_t}\ \pi = \sum_{t=0}^{T} \beta^t * \pi_t = \sum_{t=0}^{T} \beta^t * [p_t - c - l]S_t \tag{2}$$

$$\text{s.t.} \quad \sum_{t=0}^{T} S_t = \overline{S}_0$$

with:

π_t = profit in year t (euro)
c = landfill cost (euro/ton)
l = landfill tax (euro/ton)
$\overline{S}_0$ = Remaining landfill capacity in year 0 (million ton)
$\beta = 1/(1 + \delta)$ and δ is the discount rate

In mathematical optimization problems, usually the method of Lagrange is used as a strategy for finding optimal solutions of an objective function subjected to equality constraints. In our case, the equality constraint is formed by the limited remaining landfill capacity, while the objective for landfill operators is to maximize their profits. The Lagrangian for our problem may be written as:

$$\text{Maximize}_{S_t} L = \sum_{t=0}^{T} \beta^t * [p_t - c - l]S_t - \lambda \left[\sum_{t=0}^{T} S_t = \overline{S}_0 \right] \tag{3}$$

The first-order-conditions require:

$$\begin{aligned} \frac{\partial L}{\partial S_t} &= \beta^t * [p_t - c - l] - \lambda = 0 \text{ or:} \\ \beta^t * [p_t - c - l] &= \lambda \end{aligned} \tag{4}$$

Considering that we work with an inverse linear demand, it is straightforward to determine the choke-off price A. The price in period T (the last period of landfilling) will be equal to this choke-off price. So:

$$p_T = A \tag{5}$$

Equality (5) can then be inserted into (4) to determine λ, the shadow price of volume restriction (NIMBY). λ can be regarded as an economic measure of resource scarcity which is different from standard measures based on physical abundance. From an economic point of view, scarcity should reflect marginal value net of the marginal costs associated with landfilling. Filling in (5) into (4) gives us:

$$\beta^T * [A - c - l] = \lambda \tag{6}$$

In order to assess the yearly volumes landfilled, we insert (1) into (4) and we get:

$$\beta^t * [A - bS_t - c - l] = \lambda. \tag{7}$$

If we rewrite (7), we get:

$$S_t = \frac{A - c - l - \lambda\beta^{-t}}{b} \tag{8}$$

Again, rewriting (8) results in:

$$S_t = \frac{\left[1 - \beta^{T-t}\right] * [A - c - l]}{b} \quad \text{if} \quad \lambda > 0, \text{ i.e. NIMBY} \tag{9}$$

With:

$$\overline{S}_0 = \sum_{t=0}^{T-1} \frac{\left[1 - \beta^{T-t}\right] * [A - c - l]}{b} \tag{10}$$

Making use of Eq. (10), we can calculate a value for T, so we can estimate how long it takes before all remaining landfill volume will be exhausted.

2.3 Societal Point of View

From a societal point of view, there are two major problems related to landfilling. First of all, landfilling has some negative externalities, like for example noise, odor, groundwater pollution and air emissions. As these externalities have an impact on society, they carry a cost with them, which is called an externality cost. In the model, the unit externality of landfilling is presented by parameter e, whose value is strictly larger than zero. The second problem relates to the Marginal Cost of Public Funds (MCPF). Typically, taxes are a destructive nuisance that can penalize workers and distort economic decision-making. Moreover, governmental revenues from such taxes are typically expensive for society because of tax dodging and administration (Barrios et al. 2013; Glomm et al. 2008, pp. 19–32; Schob 1997, pp. 167–176). Using common sense, many nations are starting to improve their tax structures by reducing distorting taxes on for example labour and drawing more revenue from environment related activities. Following this line of reasoning, landfill taxes are way easier to monitor, especially in developed regions such as Flanders. Indeed, there are only a few landfill sites and landfill monitoring would happen regardless of the fact whether taxes apply or not. The Marginal Cost of Landfill Taxes (MCLT) is therefore lower than the marginal cost of more conventional taxes (MCPF), indicating the comparative advantage of using landfill taxes. This gives:

$$\chi = MCPF - MCLT > 0 \tag{11}$$

With χ representing the relative benefit of using landfill taxes instead of other, more expensive taxes like for example a labour tax.

Based on all preceding equations, total welfare can be calculated with B = 0 as the residual benefit of landfill volume after year T:

$$W = \sum_{t=0}^{T} \beta^t * \left[\int D(S_t) - c - e + \chi l \right] S_t + B \tag{12}$$

In the next section, all of the foregoing formulas will be used in an illustrative simulation. With these formula, values are defined for T, pt, St, πt and W. All other variables are defined exogenously.

3 Case Study: Landfilling in Flanders

Based on the given theoretical underpinnings, simulations are made using data from the Flemish part of Belgium. Making use of Eq. (10) and using exogenously determined values for parameters A, c, l, b, e, δ, χ and $\overline{S}_0$, a value can endogenously be determined for parameter T. The exogenous values that were used, are presented in Table 1. These figures were chosen in such a manner that they reflect reality as closely as possible.

Using the choke-off price and taking into account landfill volumes and prices from previous years (Briffaerts et al. 2011), the linear inverse demand function was estimated to be:

$$p_t = 100 - 50\ S_t$$

Based on a study carried out by OVAM (2013), the remaining landfill volume in period 0 was taken to be 12 million tons, so the value of parameter $\overline{S}_0$ was set to 12. Below, the simulation results are given for two scenarios, one with and one without using landfill taxes.

3.1 *Scenario 1: Simulation with Landfill Taxes*

By assigning parameter l a positive value, a landfill tax is directly taken up in the simulation exercise. By solving the dynamic maximization problem and using Eq. (10) to define a value for parameter T, results like presented in Table 2 can be obtained.

As can be deduced from Table 2, the value of T satisfying Eq. (10) is T = 41.33. In a discrete-time problem such as this, where T must be an integer, we round T up to 42. This means that it takes 42 years for exhaustion to occur. The bottom row in Table 2 shows the total landfilled volume, the discounted total profit of the landfill companies and the discounted total welfare. Remember that λ can be regarded as an economic measure of resource scarcity. In the above model, λ is the value of marginally loosening the constraint, that is, increasing the landfill capacity. It can be

Table 1 Exogenous parameters

Parameter	Value	Parameter	Value
A	100	e	50
c	15	δ	0.05
l	60	χ	0.2
b	50	$\overline{S}_0$	12

A = choke-off price; c = landfill cost; l = landfill tax; b = slope inverse demand function; e = externality cost; δ = discount rate; χ = relative advantage of using landfill tax; $\overline{S}_0$ = remaining landfill capacity

Table 2 Simulation with landfill taxes

t	S_t	p_t	π_t (discounted)	W_t (discounted)	λ
0	0.43	78.33	1,442,349.28	15,675,116.13	3.33
1	0.43	78.49	1,431,276.22	14,848,151.02	3.49
2	0.43	78.67	1,419,649.51	14,060,024.74	3.67
3	0.42	78.85	1,407,441.46	13,308,860.79	3.85
4	0.42	79.04	1,394,623.00	12,592,870.68	4.04
5	0.42	79.25	1,381,163.63	11,910,349.67	4.25
…					
38	0.08	96.25	249,674.83	530,221.44	21.25
39	0.05	97.31	178,968.04	366,227.93	22.31
40	0.03	98.43	104,725.92	206,592.46	23.43
41	0.01	99.60	26,771.69	50,934.97	24.60
42	0	100	0	0	25
$\sum_{t=0}^{T}$	12		39,931,451.13	239,109,115.1	

seen that λ rises at the rate of interest, reflecting the increasing opportunity cost as the remaining landfill capacity diminishes. This phenomenon is also known as the Hotelling Rule (Hotelling 1931, pp. 137–175; Perloff 2011). Given the results in Table 2, one could numerically plot the time paths for landfill volume and landfill price. These paths are respectively shown in Figs. 1 and 2 and show how landfill volumes and landfill prices change in time.

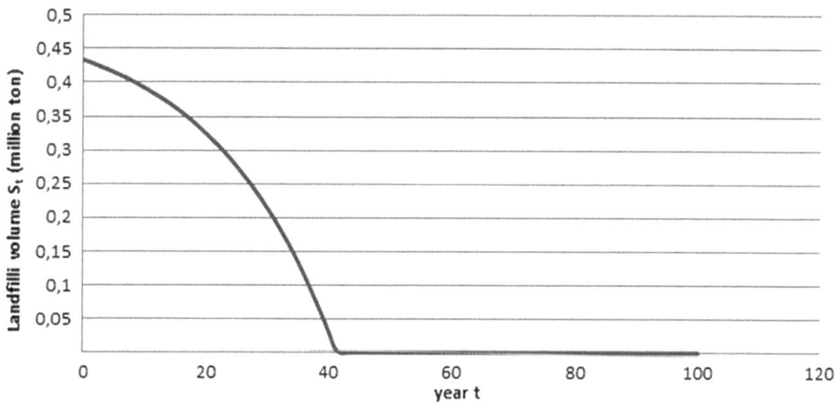

Fig. 1 Landfill path when using landfill taxes

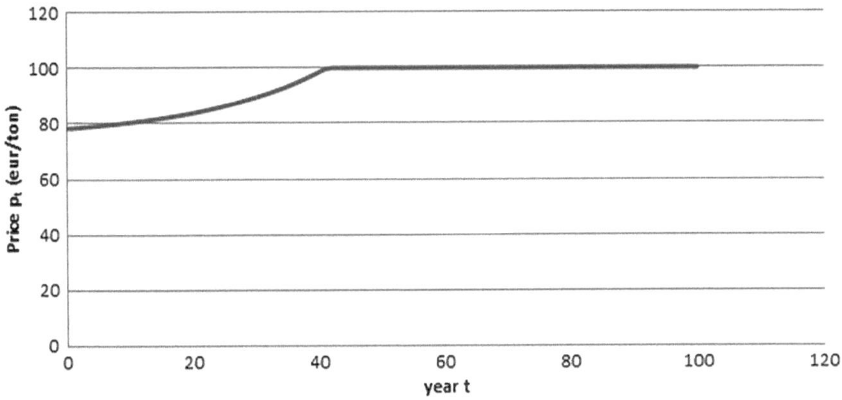

Fig. 2 Price path when using landfill taxes

3.2 *Scenario 2: Simulation Without Landfill Taxes*

By setting the value of parameter l equal to zero, landfill taxes are left out of the model. When we then solve the dynamic optimization problem and use Eq. (10) to define a value for parameter T, results like presented in Table 3 are obtained.

Looking at Table 3, we see that the value of T satisfying Eq. (10) has decreased from T = 41.33 in the case of using landfill taxes, to T = 19.23. As before, we round T up to 20. This means that it takes 20 years for exhaustion to occur. When we compare Tables 2 with 3, we see that a landfill tax substantially reduces the volumes that are landfilled each year. For example, when a landfill tax would be applied, the total volume landfilled in the first six years would add up to only 2,547,316 tons, whereas this total volume landfilled would add up to 5,674,436 tons when no landfill tax is levied. By reducing these volumes, the remaining landfill capacity is managed in a more sustainable way. When landfill taxes are used, prices start at a much higher level than when no landfill taxes would be applied, but they move less rapidly to the choke-off price level. As expected, the landfill companies' discounted total profits are much higher when no landfill taxes have to be paid. However, when we look at discounted total welfare, we see that from a societal point of view, the use of landfill taxes is preferable. Discounted total welfare is much higher in the scenario where landfill taxes are applied. This difference is mainly related to the parameter χ, which takes into account that the MCPF is larger than the MCLT. The parameter λ can still be regarded as the value of marginally increasing the available landfill capacity. In this scenario where no landfill taxes are applied, the value of this scarcity indicator is higher than in the case where landfill taxes were used. This is quite logical, taking into account that the remaining landfill capacity is depleted at a higher rate when no landfill taxes are being used, making the remaining stock more scarce and valuable.

Table 3 Simulation without landfill taxes

t	S_t	p_t	π_t (discounted)	W_t (discounted)	λ
0	1.03	48.27	34,419,977.41	9,450,015.63	33.27
1	1.00	49.93	33,313,294.35	9,503,816.96	34.93
2	0.97	51.68	32,151,277.15	9,501,039.65	36.68
3	0.93	53.51	30,931,159.08	9,441,677.08	38.51
4	0.89	55.44	29,650,035.11	9,325,587.93	40.44
5	0.85	57.46	28,304,854.94	9,152,495.77	42.46
…					
16	0.25	87.62	8,238,632.16	3,268,421.10	72.62
17	0.18	91.25	5,822,881.85	2,338,685.08	76.25
18	0.10	95.06	3,286,344.01	1,335,310.58	80.06
19	0.02	99.06	622,979.29	255,908.62	84.06
20	0	100	0	0	85
$\sum_{t=0}^{T}$	12		399,202,148.3	132,563,771.7	

Given the results in Table 3, the time paths for landfill volume and landfill price can numerically be plotted. These paths are respectively shown in Figs. 3 and 4 and show how landfill volumes and landfill prices change in time, when no landfill taxes are applied. Based on these paths, and by comparing them to Figs. 1 and 2, the same conclusions can be drawn as were discussed in the description of Table 3.

4 Discussion and Conclusions

As remaining landfill capacity is scarce, care has to be taken to deplete the remaining capacity in the most sustainable way. Responsible for putting a sustainable way of doing into practice, are both law and policy makers, as well as

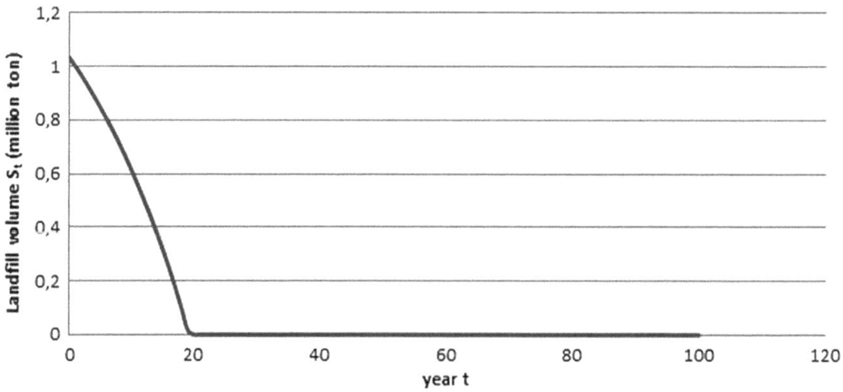

Fig. 3 Landfill path without using landfill taxes

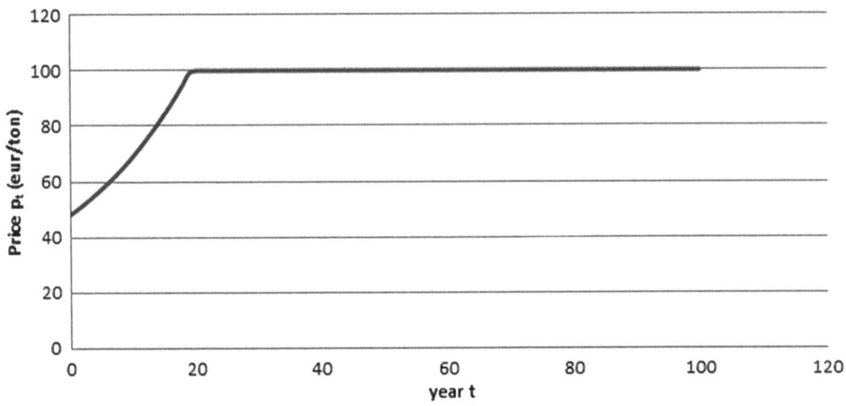

Fig. 4 Price path without using landfill taxes

private companies. As an instrument, landfill taxes can be used to increase the cost of landfilling such that other waste treatment methods become more attractive. This way, these taxes directly target the lowest level of the Waste Hierarchy, inducing more sustainable waste treatment practices. Next to landfill taxes, there is another, more indirect instrument that can cause an increase in the price of landfilling. This is the so called NIMBY syndrome, whose existence avoids new authoritative permits for landfills to be created. Given this, landfill prices rise through the fact that landfill volume can be seen as a scarce, non-renewable good. A phenomenon that also emerges when using landfill taxes.

Optimization models are capable of providing exploratory numerical insights that might be included into legal frameworks and in this way support legal decision making. Also, such models make theory and methods more meaningful and they serve as a bridge from theory and general models to actual analyses of real-world allocation problems. In this paper, by elaborating a dynamic optimization model and using data of Flanders, landfill paths and price paths were defined for both a scenario where landfill taxes are being used and a scenario without landfill taxes being applied. Starting with a remaining landfill capacity of 12 million tons in year zero, the scarcity of landfill capacity, which can partly be explained by the NIMBY syndrome, was taken into account. As could be seen, when landfill taxes are used, the volume that is landfilled each year will drastically decrease. When landfill taxes are used, it takes 42 years for full exhaustion to occur, whereas this period would be shortened to 20 years would no landfill taxes be levied. Besides a sustainable management of remaining landfill capacity, although discounted total profit falls when landfill taxes are used, discounted total welfare increases considerably (from 132,563,772 to 239,109,115 euro). This difference is mainly achieved by the comparative cost advantage of using landfill taxes instead of other, more expensive taxes. To conclude we can say that, from a broad societal point of view and knowing that NIMBY has made new permits for landfills impossible, the added value of a landfill tax is considerable in terms of welfare gain.

Another research question that could be answered using the dynamic optimization model, is whether a landfill tax above the landfill externalities would be justified. One could expect, looking at the MCPF, that a landfill tax above the landfill externalities can be justified. As the simulation model showed, this effect is not as large as expected. Only when minor tax increases are implemented, discounted total welfare increases. However, a turning point will be achieved from where discounted total welfare starts to decrease. Eventually, discounted total welfare will be lower than the value initially started from. It should be obvious that an increase in the parameter χ (MCPF-MCLT) increases discounted total welfare and prolongs the period until the turning point is reached. When improving tax structures and refining legislation, nations should be aware of these factors as they can have a large impact on the effectiveness of the imposed tax structures. Although the optimization model shows that the added value of a landfill tax in terms of social welfare is considerable, there are some aspects that could deserve more in-depth analysis in future research. Firstly, it could be examined whether other methods that prevent the creation of waste also become attractive with regard to favouring waste treatment methods other than landfilling. Extended producer responsibility systems complemented by additional product taxes on non-collected waste fractions for example can have the same effect. As a consequence, it could be interesting to investigate the efficiency of different instruments and to analyze the impacts this could have on the used assumptions and model. Moreover, it would be interesting as well to investigate whether all these instruments lead to incentives for waste prevention and green product design incentives. A second aspect that deserves more attention is the inclusion in the model of non-economic and societal factors like technological developments and employment that might influence the results of the analysis. Especially in the case that the results would be used in supporting decision-making in practice, these non-economic factors should be added. A third aspect is related to the substitutes for Flemish landfilling, like an increase in waste export or an increase in the waste recycling rate. In this paper, it is not explicitly stated to which substitute people switch when landfilled volumes are phasing out. In future research however, it would be very interesting to be able to predict how the demand for each substitute changes when landfill volumes are phasing out and how these substitutes interrelate. The fourth and last aspect is formed by the fact that politicians can be under pressure to create extra landfilling capacity. Especially at the end, when the initial landfill capacity has practically fully been used. This way, it is possible that eventually, a release of additional permits is enforced. Thereby, the NIMBY effect can be postponed or attenuated, which has an effect on the functioning of the tax and on the evolution of the yearly landfill volumes that are gradually being used.

Landfill taxes target the lowest level of the Waste Hierarchy by increasing the cost of landfilling such that other waste treatment methods become more attractive. Looking at the added value of the tax, policy makers could handle remaining landfill capacity and waste more sustainably and at the same time make society better off.

References

Barrios S, Pycroft J, Saveyn B (2013) The marginal cost of public funds in the EU: the case of labour versus green taxes. Retrieved 30 Jan 2013. http://ec.europa.eu/taxation_customs/resources/documents/taxation/gen_info/economic_analysis/tax_papers/taxation_paper_35_en.pdf

Bio IS (2011) Implementing EU waste legislation for green growth, Report prepared for European Commission DG ENV. Bio Intelligence Service, Paris

Bio IS (2012) Use of economic instruments and waste management performances, Report prepared for European Commission DG ENV. Bio Intelligence Service, Paris

Briffaerts K, Claes K, D'Haese A, Dubois M, De Groof M, Putseys L, Umans L, Van Acker K, Vandeputte A, Van der Linden A, Vander Putten E, Wille D (2011) Environmental report for flanders, managing waste materials. (Translated from: Milieurapport Vlaanderen, thema beheer afvalstoffen.). Retrieved 28 Jan 2013. http://www.milieurapport.be/Upload/main/0_achtergronddocumenten/2011/AG2011_Afval_TW.pdf

CEWEP (2014) Landfill taxes and bans. Retrieved on 31 Feb 2014. http://www.cewep.eu/media/www.cewep.eu/org/med_557/955_2012-04-27_cewep_-_landfill_taxes__bans_website.pdf

Conrad JM (1999) Resource economics. Cambridge University Press, United States of America

Dijkgraaf E, Vollebergh H (2004) Burn or bury? A social cost comparison of final waste disposal methods. Resour Energy Econ 50:233–247

Dinan TM (1993) Economic efficiency effects of alternative policies for reducing waste disposal. J Environ Econ Manage 25:242–256

Dubois M (2014) Environmental taxes for efficient waste and materials management. Chapter in book: Taxes and the economy: government policies, macro-economic factors and impacts on consumption and the environment. NOVA Science Publishers, New York

Eshet T, Shechter M (2005) A critical review of economic valuation studies of externalities from incineration and landfilling. Waste Manage Res 23:487–504

Glomm G, Kawaguchi D, Sepulveda F (2008) Green taxes and double dividends in a dynamic economy. J Policy Model 30:19–32

Hotelling H (1931) The economics of exhaustible resources. J Polit Econ 39:137–175

IVM (2005) Effectiveness of landfill taxation. Institute for Environmental Economics, Amsterdam

Kinnaman TC (2006) Policy watch: examining the justification for residential recycling. J Econ Perspect 20:219–232

Levinson A (1999) NIMBY taxes matter: the case of state hazardous waste disposal taxes. J Pub Econ 74:31–51

OECD (2012) Sustainable materials management: making better use of resources. OECD Publishing, Paris

Oosterhuis FH, Bartelings H, Linderhof V, van Beukering P (2009) Economic instruments and waste policies in the Netherlands: inventory and options for extended use. Institute for Environmental Studies, Amsterdam

OVAM (2013) Tariffs and capacities for landfilling and incineration, actualization till 2012. (Translated from: Tarieven en capaciteiten voor storten en verbranden, actualisatie tot 2012.). Retrieved 28 Jan 2013. http://www.ovam.be/jahia/Jahia/cache/offonce/pid/176?actionReq=actionPubDetail&fileItem=2952

Perloff JM (2011) Microeconomics with calcalus, 2nd edn. Pearson Education Limited, England

Schob R (1997) Environmental taxes and pre-existing distortions: the normalization trap. Int Tax Public Finance 4:167–176

Strasser S (1999) Waste and want. German Historical Institute, Washington DC

Walsh DC (2002) Urban residential refuse composition and generation rates for the 20th century. Environ Sci Technol 36:4936–4942

Van Passel S, Dubois M, Eyckmans J, De Gheldere S, Ang F, Jones PT, Van Acker K (2013) The economics of enhanced landfill mining: private and societal performance drivers. J Clean Prod 55:92–102

Pressing Forward—Developments in the Transition Towards Sustainable Materials Management in EU Environmental Law

Thomas J. de Römph

Abstract The European Union is experiencing a transition towards the Sustainable Material Management. Ultimately, radical changes are necessary to break down existing strong path-dependencies and lock-ins. Policy and legislation play important roles in this process, as they can either stimulate or obstruct such developments. A variety of constructive policies has been launched in the past fifteen years that form the basis for legislative action to stimulate the sustainable use of materials. Despite numerous fruitful attempts to adjust particular laws, the legal framework as a whole is still lagging behind policy. Nevertheless, several developments in law-making and law can be identified that enhance the legal transition, i.e.: a strategic approach is being adopted, the Better Regulation Package is intended to create coherence and simplicity, and the entire life-cycle of a material is increasingly emphasized in legislation. Especially the latter development, however, must be further deepened and better applied to press forward. In this regard, opportunities lie in the alignment of legislation, the integration of policies, the broadening of the scope of laws and the introduction of novel concepts into the legal framework. All things considered, the European Union has just started its legal transition. The revised Circular Economy Package, including its legislative proposal, might accelerate the transition.

Keywords EU environmental law · Sustainable materials management · Circular economy package · Life-cycle perspective · Better regulation

T.J. de Römph (✉)
Department of European and International Law, KU Leuven, Tiensestraat 41, 3000 Leuven, Belgium
e-mail: thomas.deromph@law.kuleuven.be

T.J. de Römph
Centre for Government and Law, Hasselt University, Martelarenlaan 42, 3500 Hasselt, Belgium

V. Mauerhofer (ed.), *Legal Aspects of Sustainable Development*,
DOI 10.1007/978-3-319-26021-1_25

1 Introduction

Some believe we live in the so-called Anthropocene (Rockström et al. 2009, p. 2). Human-caused pressures on the Earth's system are widespread. In this new geological epoch (i.e. from the Holocene to the Anthropocene), materials underpin to a great extent the quality of human life and are driving forces of the economies. Many people around the world use smart phones, drive cars, want to wear the latest fashion clothing and trade money for numerous other products on a daily basis. The other side of the coin is that pressures on natural resources, the environment and human health have increased severely by the use of materials over the last decades and will continue to rise if the existing routines endure. On top of that, due to the desire in developing countries to attain the same welfare as in developed countries and because of the expected dramatic growth of the world population in the coming years [from 7 billion now to over 9 billion people in 2050 (OECD 2012a, pp. 46, 50)], we put more and more pressure on the environment and human health. Hence, a change in many of our current manufacturing, production, consumption, usage, waste treatment and trade patterns is required. Despite the awareness of the need to radically change these processes and behaviours when drafting, amending and implementing European Union (EU) policy and legislation, many shortcomings remain in place. In fact, the EU is facing a *legal* transition towards the sustainable management of materials.

2 Objective and Scope

Although there are various approaches on how to characterize a transition, all of which highlight different dynamics, it can generally be assumed that 'sustainability transitions' take place in interconnected and interdependent socio-technical systems, such as in the fields of energy, water, food and materials. Fundamental and structural changes are necessary to transform these socio-technical systems. Generally, transitions need incremental steps to come to such radical changes (Loorbach and Rotmans 2010, pp. 144–147, with emphasis on p. 147). The challenges identified are commonly linked with and stimulated by the strong path-dependencies and lock-ins in established sectors and regimes that show unsustainable symptoms. (Markard et al. 2012, pp. 955–956; Paredis 2013, pp. 2, 10–11) In the context of materials these are—amongst many others—concerns about natural resource scarcity, the short life of products, materialisation and chemical contamination. Existing policy, legislation and institutional structures are just three possible causes of these challenges.

Even though law could also play an important role in guiding the direction of change for the better (Loorbach 2007, p. 125; Paredis 2013, p. 7), it is generally only briefly touched upon in literature, if mentioned at all. A difficulty would probably be that no sustainability approach comes with a blueprint. For example, the legal potentials for Sustainable Materials Management (SMM) are endless and the final result is far from clear. After all, sustainability comes in all shapes and sizes. Moreover, legislation as such is a continuous process of determining norms that are deemed right in a certain period of time and in a certain society. Therefore, legislation is just a tool and is always 'under construction'. On the other hand, one of the characteristics of law is that it ought to provide stability and fixity, which seems to be contrasting with a transitional approach.

Since it is impossible to pinpoint the exact stage in the transition towards SMM, the aim of this chapter is to get an impression of where the EU stands at the moment and of where it is heading to. This may at least indicate where the Union could improve its practice and legislation applicable to materials. This chapter therefore briefly explains the legal framework of Sustainable Materials Management in the EU. Subsequently, it sheds light on some important developments in EU law-making and law in the SMM transition, whereupon related opportunities are put forward, illustrated by examples. Evidently, each particular law and life-cycle stage has their own SMM challenges. To discuss these challenges one by one and in more detail, however, is not the aim of this study; this has already been done elsewhere (e.g. for waste-specific issues see Van Calster 2014). The examples are merely put forward to show the opportunities from a wider perspective.

3 Sustainable Materials Management and EU Law

In 2010, the Environment Council defined Sustainable Materials Management as '*an integrated approach that seeks to reduce environmental impacts of materials use and conserve natural resources throughout the life-cycle, a prerequisite for a more resource-efficient Europe.*' (EU Environment Council 2010, p. 1). SMM has been developed in the first place by the OECD (since 2004) and has been fleshed out by two major reports in 2005 and 2012. The Environment Council definition has thus largely been inspired by the OECD's working definition, according to which SMM is '*an* [policy] *approach to promote sustainable materials use, integrating actions targeted at reducing negative environmental impacts and preserving natural capital throughout the life-cycle of materials, taking into account economic efficiency and social equity.*' (OECD 2012b, p. 15). The 2012 report further clarifies that 'materials' include all the extracted or derived from natural resources at all points throughout their life-cycles, and that 'life-cycle of materials' includes all the activities related to materials (e.g. extraction, transportation, production, consumption and waste treatment). In addition, it highlights the multitude of actors and [governmental] levels involved (OECD 2012b, pp. 15–16). On the whole, Sustainable Materials Management addresses the transition towards the sustainable use of materials, as the approach not only tries to preserves resources, it also intends to reduce waste and to minimize the environmental impacts of the materials we use. Moreover, it is based on the idea to not merely shift the burdens to another stage of the material's life-cycle or to another generation, or across geographical and political borders.

The explanation eminently reflects the broad framework in which legislators need to operate when dealing with materials. Achieving SMM is further challenged by the strong connection with other socio-technical systems, most of which also undergo sustainability transitions, e.g. the energy system. A particular change in one system may cause great effects in other systems.

Despite the OECD's work on practical guidance for policy makers to put in place SMM policies, it has not been able to fully address the legislative challenges attached to it. It only briefly touched upon this issue by formulating a guiding principle that encourages policy makers to apply the full diversity of policy instruments (including regulation) to stimulate and reinforce sustainable environmental, economic and social outcomes (Principle 3) (OECD 2012b, pp. 61–62), which indeed relates to the well-known tripartite nature of Sustainable Development (SD). According to the OECD, combining and integrating these instruments must be stimulated. In the light of increasing stakeholder engagement and sharing of responsibility, governments can also consider shifting from command-and-control models of regulation toward more policy governance by setting long-term goals (OECD 2012b, p. 66). In addition, the 2012 report specified additional principles and subprinciples that can be carried out *through* law, such as economic incentives, increase of reuse and recycling, detoxification, eco-innovation, trade stimulation, the gathering and sharing of information, and many more. Overall, the OECD emphasizes that an effective SMM framework must embrace a life-cycle perspective and a high level of co-ordination between economic actors and different policy areas and ministries (OECD 2012b, pp. 22, 24, 28, 51–81, see specifically pp. 61–62).

It is not the purpose of this chapter to elaborate upon all the legal measures applicable to materials separately, but it may be useful to explain the relationship between SMM and EU primary law.[1] After all, it is best when the legal transition towards Sustainable Materials Management establishes a legal basis in the broader legal framework, which is not so much subject to external pressures, such as political instabilities or financial uncertainties.

As from the Treaty of Lisbon (European Union 2007), which amends the two Treaties and entered into force in December 2009, Article 3(3) TEU (European Union 2012a) has been one of the primary provisions when it comes to Sustainable Materials Management, as it aims for Sustainable Development.[2] Building on the 2005 Draft Declaration on Guiding Principles for Sustainable Development (European Commission 2005a) and the comprehensive 2006 Sustainable

[1]In brief, EU primary law consists of the founding Treaties which form the constitutional basis of the EU (e.g. Treaty on European Union [TEU] and Treaty on the Function of the European Union [TFEU]). EU secondary law consists of legislative acts (e.g. regulations, directives & decisions) and non-legislative acts (e.g. communications and recommendations), which are based on primary law.

[2]See also Recital 9 and Article 21(2)(d)(f) TEU, and Article 11 TFEU.

Development Strategy (SDS) (European Council 2006), the Commission adopted the Review of EU SDS in 2009 (European Commission 2009), which underlines the link between SD and SMM. It sums up policy achievements as well as major challenges for the EU and follow-up action. It clearly states that unsustainable trends persist in many areas and the efforts to address these trends should for that reason to be scaled up. Amongst other things, the Strategy identifies trends such as ongoing unsustainable consumption and production patterns, more production of toxic chemicals and the fast growing global demand for natural resources. As regards the latter, it therefore stresses the need for more and better recycling in the European Union.

Article 3(3) TEU further aims at achieving a high level of protection and the improvement of the quality of the environment, and highlights the need for stimulating scientific and technological progress and the establishment of an internal market in the Union. As a general rule, measures concerning the internal market are based on Article 114 TFEU (European Union 2012b) and measures reflecting the environmental objectives in Article 191 on Article 192 TFEU. Hence, most environmental measures are based on Article 192, for example when it concerns waste. Article 191(1) TFEU prescribes that EU environmental policy shall contribute to pursuit of the following objectives: *preserving, protecting and improving the quality of the environment; protecting human health; prudent and rational utilisation of natural resources;* and *promoting measures at international level to deal with regional or worldwide* environmental problems, *and in particular combating climate change*. These objectives unmistakably underline the most important aspects of Sustainable Materials Management.

Because Articles 191 and 192 TFEU do not contain any precise prohibitions or obligations, nor do they lead to other concrete requirements for legislative action, the objectives need to be operationalized in EU secondary law. In the same way, Article 191(2) TFEU contains several environmental legal principles (i.e. the precautionary and preventive principles, the source principle and the polluter pays principle) that merely function as guidance for further policy and legislation (Krämer 2012, p. 14), despite the fact that the 7th Environment Action Programme (EAP) adheres to the principles by enumerating them once more. This might be important, because EAPs are currently adopted under Article 192(3) TFEU (as a decision) and are thus to a certain extent legally binding and official sources of EU law. This may have given greater legal weight to the principles.

On the whole, the rather flexible and open nature of EU primary law makes it theoretically easy to embed Sustainable Materials Management into the environmental legal framework. By the same token, it does not give much practical guidance either. This chapter therefore continues by looking at several developments in law-making and secondary laws.

4 Developments and Opportunities

To avoid repetition of things which have already been extensively explained in textbooks on EU environmental law, this chapter refrains from merely enumerating all the landmark dates for Sustainable Materials Management in environmental policy and law. Instead, it sets out three developments in Sect. 4.1 that potentially facilitate the transition towards SMM: (a) a more strategic approach; (b) the Better Regulation Package; and (c) a life-cycle perspective. Meanwhile, discussing these developments also shows the relatively short period in which SMM has emerged and the contextual setting of the Circular Economy Package, which in itself is a fine example of the current state-of-the-art and will for that reason be explained in Sect. 4.2. Since the transition towards SMM is currently taking place and is far from finished, several opportunities arise that mainly relate to the continuing development of the life-cycle perspective; these will be put forward in Sect. 4.3.

4.1 Developments in Law and Law-Making

Since the 70s, environmental concerns have been gaining ground in policy and legislation, both on State and EU level. The first legal measures were a reaction to large-scale accidents with great impacts on the environment. Over the last forty years, environmental policy-makers and law-makers have changed from taking an ad hoc approach to a more strategic approach. This evolution is very much reflected in the communications of the Commission, particularly the ones drafted in the past fifteen to ten years. The Commission—the engine of SMM influenced EU policy and law—launches green papers, action plans, strategies and general communications (legally speaking they are all communications) and roadmaps on a regular basis, which increasingly include 'horizon targets'. These horizons have moreover been shifting from no date at all (e.g. European Commission 2001a), to ten years (e.g. European Commission 2010) and even to thirty-five years (e.g. European Commission 2011). However, there are some exceptions to this observation: e.g. the Strategy for Sustainable Development targeted 2020 in 2001 (European Commission 2001b). Consistent with the horizon of 35 years, the EU institutions as a whole engaged itself to a so-called circular economy '*where nothing is wasted and where natural resources are managed sustainably*' in 2050 in the 7th EAP, which was adopted by the appropriate name of 'Living well, within the limits of our planet' in 2013 (European Union 2013a, Annex I). Although legislators have also increased the use of targets over time, they have not been following the same pattern as regards the expanding periods. Arguably, that would have been counterproductive: while communications may be changed rather easily, often by the Commission alone, it is more difficult to adapt regulations and directives owing to the legislative procedure(s). This is why these legal acts should not include horizons in a faraway future. In order to balance both extremes (too short and too long periods), several waste laws fixed incremental deadlines up to fifteen years: for example up to 2020 in the 2008 Waste Framework Directive (WFD) (European Union 2008) and up to 2015 in the 2000 End-of life Vehicles Directive (European Union 2000) (see e.g. Article 11(2) WFD and Article 7(1) End-of life Vehicles Directive). All things considered, the increasing use of horizons in policy and law perfectly fits into the notion of a sustainability transition.

The second development in the interest of Sustainable Materials Management is the growing formalised engagement of non-State and non-EU actors in EU law-making and the assessment of legislation at every stage (i.e. from planning to implementation and revision). Under the guise of 'better regulation', the Commission—assigned by the current and the previous two regimes (Juncker Commission, Barroso Commission and Prodi Commission, respectively)—has introduced various overlapping and supplementing horizontal initiatives over the past fifteen years: e.g. Better Regulation, Smart Regulation and REFIT (Regulatory Fitness and Performance programme) (see for instance the Commission communications and brochures: European Commission 2001c, 2002, 2014a, b, 2005b). Together, these initiatives—from now on referred to as the Better Regulation Package, in compliance with the 2015 communication on better regulation and the EU website created by the Juncker Commission to promote the Better Regulation Package (European Commission 2015a)—promote the use of roadmaps that describe the planned follow-ups and the examination of potential economic, social and environmental consequences before the Union takes any action. The first element ties in with the observation on the increasing use of horizons, whereas the latter element puts emphasis on Sustainable Development and the Integration Principle (see below). Furthermore, once the EU has adopted secondary legislation, the Commission evaluates the legislation's implementation and executes a so-called 'fitness-check' to identify opportunities for simplification and regulatory burdens reduction. An initial but significant step would be to improve the overall coherence of the legal framework by aligning definitions, deleting overlaps and filling gaps. Also, the Commission asks industries, citizens, administrations, research bodies and organisations at various stages of this process to express their views and to lower costs of information collection, not least to gain public support for the particular measure. In general, it seems that this development also corresponds to the transitional approach. It is as if the inherently inflexibility of law is slightly decreased by the fitness-checks and public consultation, because the Commission is able to spot the potential flaws of a legal measure easier and right at the beginning, in the course of its drafting or after its implementation. When there is greater clarity and support as to what changes are required, the Commission could start a legislative procedure to adapt the draft. In addition, according to the Commission, better regulation also entails the greater use of different (legal) policy tools; this has also been advocated by the OECD in the context of SMM.

A third development concerns the growing use of a life-cycle approach, even though the term has not been well-defined in EU policy and law. The life-cycle logic can be interpreted very broadly: one has to take the environmental concerns into account during the entire life-cycle of a material, because decisions made in one phase of a life-cycle may influence other life-cycle(s) phases. In accordance with the work of the OECD, the EU laid strong emphasis on *waste* treatments in the past. In recent years, however, there have been calls not only to focus on the waste stage, but to broaden the scope and change the direction of European waste legislation more towards the management of *materials* in general, regardless of the material's status (e.g. waste, product or resource). In this regard, the 2008 Waste Framework Directive contains an expanded five-stage waste hierarchy (Article 4(1) WFD). As a general rule for all waste streams, the prevention of waste is the first to tick off on the priority list and takes place entirely *outside* the scope of the waste stage. In addition, EU waste policy and legislation stresses the importance of re-use and recycling. These two waste treatments follow the prevention of waste in the hierarchy and place former waste back into the user stage (preparation for reuse) or add an entire new life-cycle to the previous one (recycling). Recycling is really at the heart of current EU waste policy; binding engagements can be found in Recitals (28), (29) and (41), and Article 11(2) WFD. In order to deliver the best overall environmental outcome, Member States may however customize the hierarchy where this is justified by 'life-cycle thinking' on the overall impacts of the generation and management of such waste (Article 4(2) WFD). Another indication of a life-cycle perspective is the introduction of the Extended Producer Responsibility concept and the Polluter Pays Principle in the framework directive in 2008, as they further highlight the significance of the design, production and user stages of a material's life-cycle. The life-cycle perspective has also been incorporated in other waste legislation targeting specific waste streams. This is for example given in the WEEE Directive (European Union 2012c), as it contains a provision on the design and production of electrical and electronic equipment (Article 4 WEEE). Additionally, the objective of the End-of life Vehicles Directive is *inter alia* to improve the environmental performances of all the economic operators involved in the life cycle of vehicles and especially the operators directly involved in the waste treatment of end-of life vehicles (Article 1). All in all, it seems that EU waste legislators have already undertaken a number of endeavours when it comes to the actual application of the life-cycle approach in law.

4.2 The Circular Economy Package

As all Commission administrations tried to improve EU waste legislation in accordance with the views accepted and the technologies available at the time of application, the previous regime (Barroso 2004–2014) dedicated much effort to the overall coherence of EU waste legislation. An example is the Waste Framework Directive of 2008: the directive annexed the Hazardous Waste Directive (European Union 1991) and the Waste Oils Directive (European Union 1975), and it was 'officially' called a *framework* directive, whereas its predecessors were not (legally speaking, however, there is no difference between directives and framework directives). Despite this development, the legal waste framework was to a large extent still fragmented. Indeed, this can be explained by the historic ad hoc approach for environmental issues. Under the second presidency of Barroso (2009–2014), the Commission launched a new and comprehensive policy package mid-2014 in order to install a '*common and coherent EU framework for promoting the circular economy*', also known as the Circular Economy Package (European Commission 2014c). The programme was really meant to be the apotheosis of a whole range of policy initiatives (some of which are touched upon above) and fleshes out the notion of 'circular economy', which had initially been introduced in EU policy and legislation by the 7th EAP and can be understood as the European interpretation of SMM. Several approaches have been developed internationally that support SMM. The policy differences are not straightforward, however; they can be explained in various political and historical ways [for a general overview of the approaches, including their similarities see Happaerts (2014, pp. 1–49)]. By the time this study was published, 'circular economy'—let alone the Circular Economy Package—was not yet a commonly used term. Hence, it does not include any reference to the circular economy whatsoever.

The Circular Economy Package consisted of a communication and a legislative proposal (European Commission 2014d), which amends six waste directives, including the Waste Framework Directive and the Landfill Directive (LFD) (European Union 1999). Importantly, it was the result of a long period of drafting roadmaps, performing assessments, consulting (e.g. the Targets Review Project), carrying out public consultations... *et cetera* so as to enhance simplicity, clarity and predictability, and thereby to reduce costs and efforts. Indeed, the Circular Economy Package is a prime example of the Better Regulation programme. One of the outcomes of those procedures was the need to align definitions in EU waste law. For instance, the definitions of recycling used in waste laws regulating specific waste streams are slightly different than the WFD recycling definition, which causes confusion and conflicting calculation methods for the recycling targets throughout the EU. As regards targets, the legislative proposal also gradually increased the majority of the reuse and recycling targets; generally, they were initially aimed at 2030. In addition, the Circular Economy Package aimed for the phasing-out of landfilling by 2025 of recyclable wastes in non-hazardous landfills. Overall, it clearly endorsed a transitional approach by sticking to a long-term vision, which has already been adopted by the 2008 WFD.

At first glance, it looked as though the Circular Economy Package was to boost the transition towards SMM. Not before long, however, there were clouds on the horizon. The newly appointed European Commission officially killed the Package in December 2014 after just having been in office for less than two months. Its withdrawal shows the political vulnerability of the legal transition towards Sustainable Materials Management in the EU. While the Package had been proposed in times of economic crisis, which could had restrained the proposal's drafting in the first place, it was eventually withdrawn due to a wind of change that blew through EU politics. Of course, this state of affairs is not a new phenomenon: a newly elected President of the European Commission usually announces withdrawals of legislative proposals in his Work Programme at the start of his mandate [the Juncker Work Programme 2015 is officially called 'A New Start' (European Commission 2014f)]. Yet, this time there was a broad-based engagement for the Circular Economy Package by the Member States, the European Parliament (EP) and industries alike, for they already invested quite some time, effort and money in 'greening' their policies and activities over the past few years. Moreover, a recent decision of the Grand Chamber of the Court of Justice of the EU (Court of Justice of the European Union 2015) considerably questions the legitimacy of the withdrawal. In a nutshell, the Court ruled that the power of withdrawal cannot '*confer upon a right of veto* [for the Commission] *in the conduct of the legislative process, a right which would be contrary to the principles of conferral of powers and institutional balance*' (para 75). It further declared that the Commission '*must state to the Parliament and the Council the grounds for the withdrawal, which, in the event of challenge, have to be supported by cogent evidence or arguments*' (para 76). These findings are important in the context of the Circular Economy Package, since the Commission merely stated that the legislative proposal for a circular economy was under scrutiny to cut red tape, without giving much detail on the reasons for withdrawal (European Commission 2014f, pp. 5–6). Ironically, the original proposal had been the *result* of the Better Regulation Package (e.g. see European Commission 2014 g). Hence, the Commission's own assessments showed the opportunities for business and job growth, as well as better environmental and human health protection. On top of that, the preparation and announcement of the withdrawal was done by Timmermans, who currently serves as the First Vice-president of the Commission. Part of his responsibility is the Better Regulation programme.

The Commission assured to come up with a new and 'more ambitious' proposal by the end of 2015 (European Commission 2014f, Annex II, p. 8, European Commission 2015b) and even initiated a new public consultation round and a 'Circular Economy Conference' mid-2015 (European Commission 2015d). It should nevertheless be stressed that such a promise is by no means legally binding in a strict sense, leaving a lot of wiggle room for the Commission in terms of time, content and type of legal measure. It might well be that the reanimated proposal will be more economic-driven or environment-driven, or both. Additionally, it could also be more comprehensive in the sense that it covers more than just incremental adjustments to 'only' six waste laws. All in all, the sustainable use of materials is a much debated and important issue in the Union's quest to enhance its ecological resilience and to transform into a sustainable society.

4.3 Opportunities in the Application of the Life-Cycle perspective

The Circular Economy Package neatly illustrates the difference between policy and law on EU level. Efforts have been made to consolidate policies into one coherent policy on SMM. The policy-makers took into account a long-term vision and the public's opinion, and the communication really tried to promote a *circular* line of reasoning instead of a linear. Whereas waste had been the centre of gravity for EU policy for many years, the importance of design and production processes, products and services were also addressed by the Package.

Although the legislative proposal had been the result of the Better Regulation Package (which can be understood as a challenge just as well, as discussed below) and contained several upgraded and new targets, it did not broaden the scope of the legislation addressed, nor did it incorporate legislation regulating other life-cycle stages. It was only intended to amend waste legislation by means of familiar and overall effective legal tools. As a consequence, if the proposal had been adopted, there would still be a patchwork of legislation in place regulating the use of materials. The Package's title is therefore somewhat misleading.

Below, four interconnected flaws are put forward that should be addressed in the future (in the new Circular Economy Package). They all relate to the crucial role life-cycle perspective plays in the legal transition towards Sustainable Materials Management.

4.3.1 Integrate Policies—The Integration Principle

As stated above, the OECD sees the coordination and integration of policies and institutional structures as something different than the life-cycle approach. In my view, however, the integration of policies is a fundamental aspect of the life-cycle perspective, because the sustainable use of material is a highly complex and comprehensive issue, like the environment as a whole, involving many actors and policies throughout the life-cycle. It requires an integrated approach and a shared responsibility to address the entire material life-cycle—or at least as much as possible. Basically, this boils down to yet another touchstone for SMM in EU primary law: the Integration Principle. According to Article 11 TFEU environmental protection requirements are to be integrated into other policy fields, in particular in view of promoting Sustainable Development (although the practical legal link between these concepts is still quite unclear, because it did not have any notable legal effects in terms of judicial reasoning or legal action brought to confirm or interpret the Article's meaning [Fisher et al. 2013, p. 430]). The provision recognizes that the environment is always affected by other policies, the obvious examples being policies addressing socio-technical systems other than the material system, such as transportation, trade, energy and industry. Indeed, the Integration Principle captures very well the 'sustainability transition' approach and the OECD's support for policy cohesion. Because EU policies are in principle equally important and the Union is obliged to try to achieve them all (Article 7 TFEU), the Union is not mandatory to give priority to environmental protection requirements. Instead, the EU institutions have broad political discretion in putting the environmental principle into practice (Krämer 2012, p. 418). In addition, after the Treaty of Lisbon, not only Articles 7 and 11 TFEU refer to a form of integration: there are several other provisions referring to different policy integration (or better: 'policy coherence'). Like Article 11 TFEU, Articles 8–10 and 12–13 TFEU apply on particular topics, e.g. consumer protection and the promotion of a high level of employment. Notably, it seems that the Integration Principle has forfeited at least some of its visibility and status due to the proliferation of integration provisions (Lee 2014, p. 69, Jans 2011, pp. 1543–1547). At the same time, this change in significance can be regarded as being more in line with the broader scope of SD (Lee 2014, p. 69, European Commission 2005a, p. 5).

In general, within the Commission the integration of policies, as expressed in Article 11 TFEU, cannot be considered successful, because the principle is still poorly understood and implemented (Lee 2014, pp. 67–69). For example, administrations typically work vertically, without many institutional or regulatory interdepartmental meetings (Krämer 2012, pp. 368–369). Moreover, political interest, including a 'frame of reference', is lacking or unstable (Jordan and Lenschow 2010, pp. 151–152). The SMM transition will therefore most likely require stronger commitments, new partnerships and communication channels between the departments. In fact, this is precisely the reason why the Directorate-General (DG) for the Environment (DG ENV) allied with different departments for the launch of the Circular Economy Package, e.g. DG for Employment, Social Affairs and Inclusion (which also presented a new initiative in mid-2014, called the Green Employment Initiative) and the DG for the Internal Market, Industry, Entrepreneurship and SMEs (small and medium sized enterprises) also called DG GROWTH. In addition to the aim of policy integration, working together with other DGs was politically speaking also the best option, because DG ENV alone would probably not had been able to get enough support for the package. For this reason, this strategy is frequently used.

Another issue relating to the integration of policies concerns the assessments made in view of the Better Regulation Package. As it happens, these assessments are generally made in economic terms, while environmental impacts can often not be expressed in terms of money. For this reason, arguably more weight will be given to trade and industrial considerations instead of environmental considerations, although not necessarily intended. The assessments can thus lead to less environmental protection caused by the watering down of proposals (Krämer 2012, p. 433). Notably, the activity of other EU institutions is to a large extent dependent on those proposals. The institutions—the Commission in particular—must therefore ensure that the Better Regulation Package is not misused in a way to disconnect policies. Moreover, better regulation and deregulation should not be mixed up, especially in cases when it concerns environmental protection.

4.3.2 Align Legislation—the Waste Framework Directive and REACH

The measures in place do not necessarily correspond or connect with one another, which could cause miscommunications, legal uncertainty and other obstacles for innovation. Aligning secondary legislation would thus be a great opportunity for the transition towards SMM in the EU. Let us take the Waste Framework Directive (DG ENV) and the Regulation on Registration, Evaluation, Authorisation and Restriction of Chemicals (REACH; European Union 2006; administered by DG GROWTH) as an example, since both measures regulate the recycling process of wastes containing chemical substances and thus play significant roles in the SMM transition. After all, chemical recycling, for example in the case of plastics, generally offers lower environmental impacts than other waste management options provided by the waste hierarchy (e.g. Bio Intelligence Service 2013a, pp. 33–34; European Commission 2014e, pp 138–140; Huysman et al. 2015).

It is worth repeating that EU waste policy and legislation generally stresses the importance of recycling. REACH, on the other hand, stimulates above all a clean and safe material cycle,[3] meaning that as little as possible impurities and hazardous substances are contained in substances and products ('articles' in REACH terminology) containing chemicals. Both objectives can be defended in the context of SMM. However, current practices show that chemical recycling is arguably excessively hindered by some obligations under REACH.

As for the purity of the chemical cycle and the safety for humans and the environment, REACH contains several mechanisms relating to the registration, authorisation and restriction of substances, which is generally in the form of information assessment and supply. In this context, it could be observed that there are some exceptions to the registration obligation, one of which is most important for recyclers and is expressed in Article 2(7)(d) REACH.[4] It is however extremely difficult for recyclers to fulfil the requirements when they receive post-consumer waste (e.g. household waste), because it is generally delivered without any information on the composition of the substances. Consequently, recyclers have to do their own (laboratory) analysis of the constituents in order to proof the 'sameness' of the already registered substances, requiring an in-depth research that is time-consuming and costly. In addition, the registration fees are quite high. In fact, the same difficulties exist with regard to the determination of the hazard profile of the recycled substances and products (i.e. for the restriction and authorisation requirements). Admittedly, regulatory compliance rarely is free of challenges and of the need to devote financial resources to it. On the other hand, it should not stop the recycling industry from pursuing their activities. The question that arises is why taking the innovation path, which enquires so many efforts, if there is an easy way that equally takes you to success? In other words, why recycle when the use of new

[3]Article 1(1) REACH further expresses the Regulation's purpose as to ensure the free circulation of substances on the internal market while enhancing competitiveness and innovation.

[4]Article 2(7)(d) REACH: '*(d) substances, on their own, in mixtures or in articles, which have been registered in accordance with Title II and which are recovered in the Community if: (i) the substance that results from the recovery process is* ***the same as the substance*** *that* ***has been registered*** *in accordance with Title II; and (ii)* ***the information required*** *by Articles 31 or 32 relating to the substance that has been registered in accordance with Title II* ***is available*** *to the establishment undertaking the recovery.*' Emphasis added.

chemicals may lead to a similar or financially even better reward (Van Calster and de Römph 2015, pp. 12–18)? It looks as if REACH is not well equipped to stimulate recycling.

The Commission already responded to this flaw through the Better Regulation Package: it reduced the registration and authorisation fees for SMEs (although registration fees for larger companies increased) (European Union 2013b). In addition, the Commission is working on Union-wide End-of-Waste (EoW) criteria for plastics, which might improve the overall coherence between REACH and the Waste Framework Directive. The incorporation of existing definitions of hazard profiles and lists of hazardous substances, such as the ones listed and referred to in REACH, in the criteria could be a binding factor between waste legislation and product legislation (European Commission 2014e). The question is whether these changes do the trick in this specific case.

4.3.3 Broaden the Scope—the Ecodesign Directive

Another difference between waste legislation and product legislation is the transitional stage in which they seem to be. Although the implementation of the life-cycle approach in waste legislation is far from perfect (e.g. Lazarevic et al. 2012) and despite the fact that significant steps have already been made in REACH (there are several obligations in place to pass on information through the chemical cycle, which ideally also includes the waste stage), not all laws are that progressive to date. Generally speaking, there is a lack of environmentally orientated product legislation, despite the good intentions of the 2008 Action Plan on Sustainable Consumption and Production and the 2001 Green paper on the Integrated Product Policy, which are both administered by DG ENV (European Commission 2008, 2001a, respectively).[5]

The Ecodesign Directive (European Union 2009) administered by DG GROWTH is primarily aimed at the energy efficiency of energy-related products, such as kitchen equipment, air conditioning products and washing machines, as well as windows and insulation material. However, since the importance of resource use increases in the Union, the Directive's focus on energy efficiency standards has often been explained, criticized and counter criticized (e.g. Dalhammar 2015; Ecofys 2014, pp. 29–33; Arditi and Toulouse 2012). The Commission contracted a study on material-efficiency for the 'Ecodesign Methodology' (called: MEErP) to address this issue, which was finalised in December 2013 (Bio Intelligence Service 2013b). The report suggested amongst other things that the following parameters could at least be added to the Ecodesign Directive: recyclability benefit rates and recycled content. These conclusions show the willingness of the Commission to expand the scope of the Directive, e.g. by

[5] In fact, the first and biggest leap forward for the life cycle perspective in EU law had been the adoption of the Integrated Product Policy.

considering including more material-related considerations to the valuation. As a matter of fact, the Ecodesign Directive already refers to the life-cycle perspective in two Recitals (7 and 13) and Annexes I and II specify several parameters that can be used in this respect. These yardsticks, however, have been largely neglected so far. Hence, another possibility would be to expand the Directive's scope in the context of products, meaning that the eco-design requirements would not 'only' apply to all energy-related products but also to non-energy related products. This would probably cause a tremendous transformational change, especially because the Ecodesign Directive is a horizontally based directive and can thus be a perfect platform to encourage the use of the life-cycle approach (comparable to the Waste Framework Directive). Incentives for eco-design in law—for any product whatsoever—could be seen as a key function for the application of the life-cycle approach (Dalhammar 2015).

These potentially fundamental changes should not only be very carefully examined, taking the critical comments into account, it should also be nuanced and viewed in the light of the integration of policies. In point of fact, the Directive is—like most product-related laws—adopted on the basis of Article 114 TFEU. The Article on the internal market is most relevant in the case of product-related (environmental) measures, because products usually require uniform rules. For that reason, these measures are generally not in the Environment Commissioner's portfolio. Instead, DG GROWTH pursues this matter. While DG ENV puts more and more emphasis on the life-cycle stages apart from the waste stage in waste legislation, DG GROWTH is somewhat slower in experimenting in cross-border policies in product legislation. In addition, the main responsibility for the Ecodesign Directive is shared with the DG for Energy; it is no surprise that both DGs tend to focus more on energy issues than on additional environmental concerns (Bundgaard et al. 2015, pp. 45–46; Dalhammar 2015).

4.3.4 Introduce New Concepts—Enhanced Landfill Mining

The Circular Economy Package continued to cling to well-known and well-working legal tools, such as setting new targets. Disappointingly, it did not introduce any innovative ideas to improve the life-cycle perspective in legislation, for example with regard to landfilling. Landfills have long been a final 'solution' for waste; due to a whole host of environmental and social problems, the EU now carries out a policy that severely restrains this type of disposal. Nonetheless, there are still numerous closed and/or operational landfills in place on the Union's Territory and many problems continue to exist. The idea of mining landfills has been developed to address these issues and can be defined as '*a process for extracting minerals or other solid natural resources from waste materials that previously have been disposed of by burying them in the ground*' (Krook et al. 2012, p. 513). The Flemish Enhanced Landfill Mining (ELFM) Research Consortium—which is now endorsed by a European consortium—developed a comprehensive concept that builds upon this idea by integrating in an integrated and systematic resource recovery practice.

ELFM can be understood as the safe conditioning, excavation and integrated valorisation of landfilled waste streams as materials and/or energy, using innovative technologies and respecting the most stringent social and ecological criteria (Jones et al. 2013, p. 48). Phrased differently, the collected and pretreated non-recyclable waste is temporary stored, during which gas is being valorised. When it is technologically and economically viable to recover the waste, the materials are excavated, dried, washed, further processed… *et cetera*, whereupon the actual recycling or energy recovery takes place. The overall aim is to maximize the valorisation of the waste in all possible ways.

It appears, however, that the concept of ELFM—or landfill mining in general—does not sit well with the current Waste Framework Directive and Landfill Directive, mainly due to certain terminology constrains (de Römph 2014). The biggest obstacle is the absence of a 'temporary storage' status for these activities in waste law: after more than three years storage pending recovery, the storage changes its status from a normal waste treatment (Article 3(14) WFD) to the particular treatment of disposal (i.e. landfilling, Article 2(g) LFD) (Wante 2010, p. 7). This change results in the application of the Landfill Directive, which is without doubt not the aim and the activity of ELFM: that is recovery. For this reason, legislators may want to consider abolishing the three-year limit that triggers the application of the Landfill Directive. What is most important is that if this is actually implemented, ELFM could cause a real revolution in current waste management practices, because landfilling would not be a 'final solution'. In fact, landfilling as such will not even exist anymore. After all, ELFM adds a new life-cycle to almost all waste materials or valorises them as energy. No precious materials will get lost in the treatment processes, which makes ELFM an excellent example of the application of the life-cycle approach. It seems that the incorporation of transformational initiatives like ELFM would not only pave the way for SMM in EU legislation, it may also contribute to the integration of policies and to other sustainability transitions, because ELFM also addresses energy considerations.

5 Concluding Remarks

Whether the EU is on the right track is hard to tell: in any case, the EU is experiencing a transition towards the sustainable use of materials. Fundamental changes are however still necessary to break down strong path-dependencies and lock-ins in established sectors and regimes. Secondary environmental laws play important roles in this: they could either keep unsustainable practices and structures in place or they could promote sustainable alternatives instead. A variety of constructive policies has been installed in the past fifteen years that form together with several provisions in the primary legislation the basis for legislative action in the field of Sustainable Materials Management. Despite the numerous regularly fruitful attempts to adjust particular regulations, directives and decisions, legislation is generally still lagging behind policy. This of course, characterizes law in general, since policy generally serves as the kick off for legislative proposals.

Several developments in secondary law and law-making can however be identified that (potentially) enhance the transition to SMM. First, following a similar development in policy, waste legislation too includes an increasing amount of targets and other less binding long-term goals, which stimulate Member States to continuing their practices. Moreover, the period of many of these targets has been extended in the past fifteen years, however not to the same extent as in policy. The latest landmark development in the transition was the launch and withdrawal of the Circular Economy Package. The legislative proposal contained a number of target upgrades and targets for new waste streams. The use of long-term goals is seen as a characteristic of sustainability transitions. Second, the Better Regulation Package is becoming more important in EU law-making than before. The Commission provides full assessments before and after legislation becomes into force. Public consultation and so-called fitness-checks are part of the Package and the use of them indicates that legislators are aware of the necessity to respond faster and more accurate to revealed flaws in certain legal measures. The programme's downside is that it could also be used to merely deregulate, which may turn out to be all wrong in case of environmental legislation. In addition, the assessments are generally made in economic terms, which may exclude environmental consideration to be taken into account. Moreover, the withdrawal of the Circular Economy Package illustrates that the use of the Better Regulation Package is rather arbitrary. Last, the EU institutions have been taken account of the life-cycle perspective for some time now. Ambitious words were particularly uttered in policy, some of which have already been turned into legislative deeds: the 2008 Waste Framework Directive, for example, includes an enhanced waste hierarchy and the directive also aims at life-cycle stages other than the waste stage. Another example is the REACH Regulation, as it contains various obligations to pass on information throughout the chemical chain, including the waste stage.

Nevertheless, still lots of opportunities regarding the life cycle approach exist that were not seized by the previous Commission in the Circular Economy proposal. Key is that SMM requires greater integration of policies. At EU level, achieving this requires cooperation across different Commission departments. The appreciation and implementation of the Integration Principle should thus be deepened and strengthened. Furthermore, it is important to align the different stages of a material life-cycle in legislation. Challenges lie for instance in the consistency between REACH and the Waste Framework Directive. Additionally, the scope of the Ecodesign Directive should be broadened in the sense that it should be applicable to all products and that it includes more material-related parameters, covering all life-cycle stages. Another reflection is that the Commission should not back away from the introduction of new concepts. Quite the contrary: fostering them and finding a legal basis stimulates the necessary innovations. For example, Enhanced Landfill Mining maximizes the valorisation of waste in materials and energy, but there seems to be no legal basis for it. Significant changes to current practices and law are required. In the meantime, we should not see SMM as a stand-alone transition. The introduction of new concepts may be used to build bridges between the materials system and other socio-technical systems. As these opportunities for an improved application of the life-cycle approach in legislation show, they are interrelated and overlap with one another. Furthermore, they make crucial links to other policies and systems in transition.

All things considered, the question of whether the EU is on the right track in respect of the legal transition towards Sustainable Materials Management remains unanswered to date, because the future of the Circular Economy Package, which is a clear benchmark in the transition, is currently uncertain. On the on hand, the withdrawal of the Package could be seen as an impediment, because the legislative proposal contained solutions to several problems identified, e.g. by the Better Regulation programme. On the other hand, the proposal had only been loosely inspired by a life-cycle approach. Although there is a possibility that the Commission will continue along those same lines, it may be equally quite the reverse. As a matter of fact, there are positive signals from the Commission showing that the incorporation of the life-cycle approach has been placed high on the political agenda, linking production and consumption to the waste stage (European Commission 2015c, d). How well the life-cycle perspective will be incorporated remains to be seen.

Acknowledgements This chapter is based on a paper that is included in the conference proceedings of the 21st Annual Conference of the International Sustainable Development Research Society (ISDRS), held on 10–12 July 2015 in Geelong (Australia). In addition, it must be seen in the context of the Policy Research Centre on Sustainable Materials Management, which is a multi-disciplinary consortium coordinated by the Flemish Public Waste Agency (OVAM).

References

Arditi S, Toulouse E (2012) Priorities for the improvement of the EU Ecodesign & Energy Labelling policies. Recommendations from environmental NGOs. Retrieved 10 June 2015, from http://www.eceee.org/ecodesign/Horizontal-matters/ENGO%20Ecodesign%20improvement%2005-03-12.pdf

Bio Intelligence Service (2013a) Study on an increased mechanical recycling target for plastics. Final report prepared for Plastics Recyclers Europe. Retrieved 17 June 2015, from http://www.plasticsrecyclers.eu/sites/default/files/Study%20on%20an%20increased%20mechanical%20recycling%20target%20for%20plastics_BIOIS.pdf

Bio Intelligence Service (2013b) Material-efficiency Ecodesign Report and Module to the Methodology for the Ecodesign of Energy-related Products (MEErP). Final report to the European Commission—DG Enterprise and Industry. Retrieved 10 June 2015, from http://ec.europa.eu/enterprise/policies/sustainable-business/ecodesign/methodology/index_en.htm

Bundgaard AM, Remmen A, Zacho KO (eds) (2015) Ecodesign directive version 2.0: from energy efficiency to resource efficiency. København K Miljøstyrelsen. Retrieved 10 June 2015, from http://vbn.aau.dk/files/208981632/eco_design_2.0.pdf

Court of Justice of the European Union (2015) Council of the European Union vs. European Commission, Case of 14 April 2015, C-409/13. Retrieved 16 June 2015, from http://curia.europa.eu/juris/document/document.jsf;jsessionid=9ea7d2dc30dd6828654bcea741e2bc008c865d7ba5c9.e34KaxiLc3qMb40Rch0SaxuQaxf0?text=&docid=163659&pageIndex=0&doclang=EN&mode=lst&dir=&occ=first&part=1&cid=360922

Dalhammar C (2015) The application of 'life cycle thinking' in European environmental law: theory and practice. J Eur Environ Plann Law 12(2):97–127

de Römph T (2014) Enhanced landfill mining: legal barriers and opportunities for sustainable materials management in the EU. Presentation at the 2nd European environmental law forum conference, Brussels

Ecofys (2014) Evaluation of the energy labelling directive and specific aspects of the ecodesign directive. Final technical report to the European Commission. Retrieved 10 June 2015, from http://www.ecofys.com/files/files/final_technical_report-evaluation_eld_ed_june_2014.pdf

EU Environment Council (2010) Council conclusions on sustainable materials management and sustainable production and consumption: key contribution to a resource-efficient Europe. 3061st ENVIRONMENT Council meeting, Doc. 17495/10. Retrieved 17 June 2015, from http://register.consilium.europa.eu/doc/srv?l=EN&f=ST%2017495%202010%20INIT

European Commission (2001) Green paper on Integrated Product Policy. COM(2001) 68 final. Retrieved 17 June 2015, from http://eur-lex.europa.eu/legal-content/EN/TXT/PDF/?uri=CELEX:52001DC0068&from=EN

European Commission (2001b) A sustainable Europe for a better world: a European union strategy for sustainable development. COM(2001)264 final. Retrieved 17 June 2015, from http://eur-lex.europa.eu/LexUriServ/LexUriServ.do?uri=COM:2001:0264:FIN:EN:PDF

European Commission (2001c) European Governance—a white paper. COM(2001) 428 final. Retrieved 17 June 2015, from http://eur-lex.europa.eu/legal-content/EN/TXT/PDF/?uri=CELEX:52001DC0428&rid=2

European Commission (2002) Action plan "Simplifying and improving the regulatory environment". COM(2002) 278 final. Retrieved 17 June 2015, from http://eur-lex.europa.eu/LexUriServ/LexUriServ.do?uri=COM:2002:0278:FIN:en:PDF

European Commission (2003) Communication on integrated product policy building on environmental life-cycle thinking. COM 2003(302) final. Retrieved 17 June 2015, from http://eur-lex.europa.eu/legal-content/EN/TXT/PDF/?uri=CELEX:52003DC0302&from=EN

European Commission (2005a) Draft declaration on guiding principles for sustainable development. COM(2005) 218 final. Retrieved 17 June 2015, from http://eur-lex.europa.eu/legal-content/EN/TXT/PDF/?uri=CELEX:52005DC0218&from=EN

European Commission (2005b) Communication: better regulation for growth and jobs. COM (2005) 97 final. Retrieved 17 June 2015, from http://eur-lex.europa.eu/legal-content/EN/TXT/PDF/?uri=CELEX:52005DC0097&from=EN

European Commission (2008) Communication on the sustainable consumption and production and sustainable industrial policy action plan. COM(2008) 397 final. Retrieved 17 June 2015, from http://eur-lex.europa.eu/legal-content/EN/TXT/PDF/?uri=CELEX:52008DC0397&from=EN

European Commission (2009) Mainstreaming sustainable development into EU policies: 2009 review of the european union strategy for sustainable development. COM(2009) 400 final. Retrieved 17 June 2015, from http://eur-lex.europa.eu/LexUriServ/LexUriServ.do?uri=COM:2009:0400:FIN:en:PDF

European Commission (2010) A European strategy for smart, sustainable and inclusive growth. COM(2010) 2020 final. Retrieved 17 June 2015, from http://eur-lex.europa.eu/LexUriServ/LexUriServ.do?uri=COM:2010:2020:FIN:EN:PDF

European Commission (2011) Roadmap to a resource efficient Europe. COM(2011) 571 final. Retrieved 17 June 2015, from http://ec.europa.eu/food/safety/food_waste/library/docs/com2011_571_en.pdf

European Commission (2014a) brochure REFIT—making EU law lighter, simpler and less costly. Retrieved 10 June 2015, from http://ec.europa.eu/smart-regulation/docs/refit_brochure_en.pdf

European Commission (2014b) Regulatory fitness and performance programme (REFIT): state of play and outlook. COM(2014) 368 final. Retrieved 17 June 2015, from http://ec.europa.eu/smart-regulation/docs/com2014_368_en.pdf

European Commission (2014c) Towards a circular economy: a zero waste programme for Europe. COM(2014) 398 final/2. Retrieved 17 June 2015, from http://eur-lex.europa.eu/resource.html?uri=cellar:aa88c66d-4553-11e4-a0cb-01aa75ed71a1.0022.03/DOC_1&format=PDF

European Commission (2014d) Proposal for a Directive amending directives 2008/98/EC on waste, 94/62/EC on packaging and packaging waste, 1999/31/EC on the landfill of waste, 2000/53/EC on end-of-life vehicles, 2006/66/EC on batteries and accumulators and waste batteries and accumulators, and 2012/19/EU on waste electrical and electronic equipment. COM(2014) 397 final. Retrieved 17 June 2015, from http://ec.europa.eu/environment/waste/pdf/Legal%20proposal%20review%20targets.pdf

European Commission: Villanueva A, Eder P (2014e) End-of-waste criteria for waste plastic for conversion, Technical proposal, JRC Technical Reports. Retrieved 17 June 2015, from https://ec.europa.eu/jrc/sites/default/files/2014-jrc91637_.pdf

European Commission (2014f) Commission Work Programme 2015: a new start. COM(2014) 910 final. Retrieved 17 June 2015, from http://ec.europa.eu/atwork/pdf/cwp_2015_en.pdf

European Commission (2014g) Commission staff working document: impact assessment. Accompanying the document proposal for a directive of the European Parliament and of the Council amending Directives 2008/98/EC on waste, 94/62/EC on packaging and packaging waste, 1999/31/EC on the landfill of waste, 2000/53/EC on end-of-life vehicles, 2006/66/EC on batteries and accumulators and waste batteries and accumulators, and 2012/19/EU on waste electrical and electronic equipment. SWD(2014) 207 final. Retrieved 17 June 2015, from http://eur-lex.europa.eu/legal-content/EN/TXT/PDF/?uri=CELEX:52014SC0207&from=EN

European Commission (2015a) Better regulation for better results—An EU agenda. COM(2015) 215 final. Retrieved 10 June 2015, from http://ec.europa.eu/smart-regulation/better_regulation/key_docs_en.htm

European Commission (2015b) Roadmap: circular economy strategy. Retrieved 17 June 2015, from http://ec.europa.eu/smart-regulation/impact/planned_ia/docs/2015_env_065_env+_032_circular_economy_en.pdf

European Commission (2015c) Exchange of views with committee on the environment, public health and food safety (european parliament) 24 Feb 2015. Speech Commissioner Vella. Retrieved 17 June 2015, from https://ec.europa.eu/commission/content/exchange-views-committee-environment-public-health-and-food-safety-european-parliament-24_en

European Commission (2015d) Website European Commission. Retrieved 17 June 2015, from http://ec.europa.eu/environment/circular-economy/index_en.htm

European Council (2006) Review of the EU sustainable development strategy (EU SDS)—renewed strategy. 10917/06. Retrieved 17 June 2015, from http://register.consilium.europa.eu/doc/srv?l=EN&f=ST%2010917%202006%20INIT

European Union (1975) Directive 75/439 of 16 June 1975 on the disposal of waste oils. OJ L 194/23. Retrieved 17 June 2015, from http://eur-lex.europa.eu/LexUriServ/LexUriServ.do?uri=CELEX:31975L0439:EN:HTML

European Union (1991) Directive 91/689 of 12 December 1991 on hazardous waste. OJ L 377/20. Retrieved 17 June 2015, from http://eur-lex.europa.eu/legal-content/EN/TXT/PDF/?uri=CELEX:31991L0689&from=EN

European Union (1999). Directive 1999/31 of 26 April 1999 on the landfill of waste. OJ L 182/1. Retrieved 17 June 2015, from http://eur-lex.europa.eu/legal-content/EN/TXT/PDF/?uri=CELEX:31999L0031&from=EN

European Union (2000) Directive 2000/53 of 18 September 2000 on end-of life vehicles. OJ L 269/34. Retrieved 17 June 2015, from http://eur-lex.europa.eu/LexUriServ/LexUriServ.do?uri=CONSLEG:2000L0053:20050701:EN:PDF

European Union (2006) Regulation No 1907/2006 of 18 December 2006 concerning the Registration, Evaluation, Authorisation and Restriction of Chemicals (REACH), establishing a European Chemicals Agency, amending Directive 1999/45/EC and repealing Council Regulation (EEC) No 793/93 and Commission Regulation (EC) No 1488/94 as well as Council Directive 76/769/EEC and Commission Directives 91/155/EEC, 93/67/EEC, 93/105/EC and 2000/21/EC. OJ L 396/1. Retrieved 17 June 2015, from http://eur-lex.europa.eu/legal-content/EN/TXT/PDF/?uri=CELEX:02006R1907-20140410&from=EN

European Union (2007) Treaty of Lisbon 2007/C. Amending the treaty on european union and the treaty establishing the european community. OJ C 306/01. Retrieved 16 June 2015, from http://eur-lex.europa.eu/legal-content/EN/TXT/PDF/?uri=CELEX:12007L/TXT&from=EN
European Union (2008) Directive 2008/98 of 19 November 2008 on waste and repealing certain Directives. OJ L 312/3. Retrieved 17 June 2015, from http://eur-lex.europa.eu/legal-content/EN/TXT/PDF/?uri=CELEX:32008L0098&from=EN
European Union (2009) Directive 2009/125 of 21 October 2009 establishing a framework for the setting of ecodesign requirements for energy-related products. OJ L 285/10. Retrieved 17 June 2015, from http://eur-lex.europa.eu/legal-content/EN/TXT/PDF/?uri=CELEX:32009L0125&from=EN
European Union (2012a) Consolidated version of the treaty on European Union. OJ C 326/13. Retrieved 16 June 2015, from http://eur-lex.europa.eu/legal-content/en/TXT/PDF/?uri=CELEX:12012M/TXT
European Union (2012b) Consolidated version of the treaty on the functioning of the European Union. OJ C 326/47. Retrieved 16 June 2015, from http://eur-lex.europa.eu/legal-content/EN/TXT/PDF/?uri=CELEX:12012E/TXT&from=EN
European Union (2012c) Directive 2012/19 on waste electrical and electronic equipment (WEEE). OJ L 197/38. Retrieved 17 June 2015, from http://eur-lex.europa.eu/legal-content/EN/TXT/PDF/?uri=CELEX:32012L0019&from=EN
European Union (2013a) Decision 1386/2013 on a general union environment action programme to 2020 'living well, within the limits of our planet'. OJ L 354/171. Retrieved 17 June 2015, from http://eur-lex.europa.eu/legal-content/EN/TXT/PDF/?uri=CELEX:32013D1386&from=EN
European Union (2013b) Regulation No 254/2013 of 20 March 2013 amending Regulation (EC) No 340/2008 on the fees and charges payable to the European Chemicals Agency pursuant to Regulation (EC) No 1907/2006 of the European Parliament and of the council on the registration, evaluation, authorisation and restriction of chemicals (REACH). OJ L 79/7. Retrieved 17 June 2015, from http://eur-lex.europa.eu/LexUriServ/LexUriServ.do?uri=OJ:L:2013:079:0007:0018:en:PDF
Fisher E, Lange B, Scotford E (2013) Environmental law: text, cases and materials. Oxford University Press, Oxford
Happaerts S (2014) International discourses and practices of sustainable materials management. Policy research centre for sustainable materials management, pp 1–55. Retrieved 10 June 2015, from http://steunpuntsumma.be/docs/research%20paper%205.pdf
Huysman S, Debaveye S, Schaubroeck T De, Meester S, Ardente F, Mathieux F, Dewulf J (2015) The recyclability benefit rate of closed-loop and open-loop systems: A case study on plastic recycling in Flanders. Resour Conserv Recycl 101:53–60
Jans JH (2011) Stop the integration principle? Fordham Int Law J 33(5):1533–1547
Jones PT, Geysen D, Tielemans Y Van, Passel S, Pontikes Y, Blanpain B, Quaghebeur M, Hoekstra N (2013) Enhanced landfill mining in view of multiple resource recovery: a critical view. J Clean Prod 55:45–55
Jordan A, Lenschow A (2010) Environmental Policy Integration: a State of the Art Review. Environmental Policy and Governance 20:147–158
Krämer L (2012) EU environmental law. Sweet & Maxwell, London
Krook J, Svensson N, Eklund M (2012) Landfill mining: a critical review of two decades of research. Waste Manag 32:513–520
Lazarevic D, Buclet N, Brandt N (2012) The application of life cycle thinking in the context of European waste policy. J Clean Prod 29–30:199–207
Lee M (2014) EU environmental law, governance and decision-making. Hart Publishing, Oxford
Loorbach D (2007) Transition management. New mode of governance for sustainable development. Dissertation, Erasmus University Rotterdam. Retrieved 17 June 2015, from http://repub.eur.nl/pub/10200/

Loorbach D, Rotmans J (2010) Towards a better understanding of transitions and their governance: a systemic and reflexive approach. In: Grin J, Rotmans J, Schot J (eds) In collaboration with Geels F and Loorbach D, Transition to sustainable development. New directions in the study of long term transformative change. Routledge, New York/Abington, pp 105–223

Markard J, Raven R, Truffer B (2012) Sustainability transitions: an emerging field of research and its prospects. Res Policy 41:955–967

OECD (2012a) OECD environmental outlook to 2050: the consequences of inaction. OECD Publishing, Paris. Retrieved 17 June 2015, from http://www.oecd-ilibrary.org/docserver/download/9712011e.pdf?expires=1434563587&id=id&accname=ocid49025850&checksum=703368D7936418F6448BC9C5F9D6056F

OECD (2012b) Sustainable materials management: making better use of resources. OECD Publishing, Paris. Retrieved 17 June 2015, from http://www.oecd-ilibrary.org/docserver/download/9712061e.pdf?expires=1434563625&id=id&accname=ocid49025850&checksum=AD136E2A5DE1E1EEB3CC976013DF3F18

Paredis E (2013) A winding road. Transition management, policy change and the search for sustainable development. Dissertation, Ghent University. Retrieved 17 June 2015, from http://www.cdo.ugent.be/drupal-7.15/sites/default/files/publicatie_pdf/A%20Winding%20Road%20PAREDIS%20online.pdf

Rockström JS, Noone W, Persson K, Chapin Å, III Lambin FS, Lenton E, Scheffer TM, Folke M, Hans Schellnhuber C, Nykvist J, de Wit B, Hughes CA, van der Leeuw T, Rodhe S, Sörlin H, Snyder S, Costanza PK, Svedin R, Falkenmark U, Karlberg M, Corell L, Fabry RW, Hansen VJ, Walker J, Liverman B, Richardson D, Crutzen K, Foley J (2009) Planetary boundaries: exploring the safe operating space for humanity. Ecol Soc 14

Van Calster G (2014) Opportunities and pitfalls for sustainable materials management in EU waste law. In: Panoussis I, Post H (eds) Waste management in European law. Eleven International Publishing, The Hague, pp 97–105

Van Calster G, de Römph T (2015) Regulating opportunity and innovation in the EU. The case of sustainable materials (plastics) management. Social science research network. Retrieved 10 June 2015, from http://papers.ssrn.com/sol3/papers.cfm?abstract_id=2588562

Wante J (2010) A European legal framework for enhanced waste management. Paper presented at the 1st international symposium on enhanced landfill mining, Houthalen-Helchteren. Retrieved 17 June 2015, from http://www.elfm.eu/Uploads/ELFM/FILE_21227270-9625-404b-8b5d-f9cd744b40b8.pdf

Part IX
Sectorial Policies: Biodiversity, Nature Conservation, Oceans and Spatial Planning

Sustainable Development: New Thoughts, New Policy, New Law?

Frederik Hendrik Kistenkas

Abstract New thoughts and new policy on sustainable development have been brought forward and widely discussed and accepted, but law is still lagging behind. This paper aims to fill up that gap and tries to put some new light on how legislation and jurisprudence could meet up with modern sustainability insights. Much nature and planning legislation predates our common understandings of sustainability and might be able to obstruct sustainable development. However, sustainable growth, usually seen in terms of development for which assets and impacts for ecology, economy and society are brought in balance, should be facilitated by future law rather than being obstructed. The paper will focus on European and domestic nature and planning law and their shortcomings and possibilities in terms of sustainable development. The concept of ecosystem services will show the likely obstructions of current law and will help to alter these law provisions as some new improvement directions will be presented.

Keywords Ecosystem services · EU law · European nature conservation law · Policy · Spatial planning law · Sustainable development

1 Introduction

Sustainable development is mostly understood as a development for which assets and impacts for ecology, economy and society are in balance. Probably the best and less disputed definition of sustainable development is given by the 2005 United Nations resolution A/Res/59/227 (2005): 'the continuing need to ensure a balance between economic development, social development and environmental protection as interdependent and mutually reinforcing pillars of sustainable development'. Again and above all, this definition indicates a balance. This balancing approach

F.H. Kistenkas (✉)
Wageningen University and Alterra Wageningen UR, Wageningen, The Netherlands
e-mail: fred.kistenkas@wur.nl

V. Mauerhofer (ed.), *Legal Aspects of Sustainable Development*,
DOI 10.1007/978-3-319-26021-1_26

might also be called the *triple P* approach (people, planet, profit), or in short: 3P balancing. Such a 3P balancing might of course be based on the well-known 1987 Brundtland-report (Brundtland 1987).

In the meantime also the International Court of Justice (ICJ) in *Gabcikovo-Nagymaros* case (Hungary v. Slovakia) clearly accepts 'the need to reconcile economic development with protection of the environment' as this 'is aptly expressed in the concept of sustainable development' (ICJ 1997, para. 197). Once perhaps started as a political slogan or some kind of soft law in *status nascendi* the principle of sustainable development seems indeed rapidly hardening and moving towards a normative principle (De Sadeleer 2002; Nollkaemper 2011, p. 381) as the International Court of Justice still recalls this principle in more recent case law like the 2010 case of *Pulp Mills on the river Uruguay* (ICJ 2010, para. 177).

In terms of sustainable development and triple P-demands, regulating and supporting ecosystem services (ES) might be regarded as typical planet-services whereas provisioning services are mainly profit-linked. Cultural services obviously are largely people-related ES. Thus, sustainability could well be linked to the concept of ES (Termorshuizen and Opdam 2009; Kistenkas 2014). Whereas ES could become a valuable tool to reach sustainable development through balancing, up till now ES are not *verbatim* mentioned in most law instruments, especially not in European nature conservation and planning law.

Sustainable development then, is in fact a three scale balance: a triangle basically holds up by environmental (natural) capacity, social capacity and economic capacity. It is all about balancing the conflicts of interest between social (people), environmental (planet) and economic (profit) sustainability. These three dimensions of sustainability are improvingly and quite rightly depicted in what I would call the *cone of Mauerhofer* (Mauerhofer 2008 and 2013). His concept of 3-D Sustainability can be represented by a cone with on the base concentric rings of economic capital embedded in social capital, both embedded in environmental capital as the outside ring. Based on the outside ring is the column of environmental capacity and based on the ring of social capital is the social capacity column. In the middle the economic capacity column arises from the inner ring of economic capital. The three columns are bottom–up fed by their capital and top–down influencing and balancing the 3-D Sustainability triangle spread among and hold up by environmental (natural) capacity, social capacity and economic capacity. The diagonal cone side represent the limits of the environmental system and the column based on the outer ring will of course reach the (physical) limits of this environmental carrying capacity first.

Each of the three capacities is influenced by six criteria: eco-efficiency, eco-effectiveness, socio-efficiency, socio-effectiveness, sufficiency and ecological equity. Sustainable development measures related to sufficiency and eco-effectiveness should be addressed first, as they have a direct influence on the environmental capacity, which is as said above the first column inside the cone to reach the environmental carrying capacity limit (Mauerhofer 2008).

Such a 3-D sustainability triangle does indeed reflect and address decision making that produces an ecologically sustainable scale, a socially fair and just distribution and a market-based allocation mechanism (Costanza et al. 1997; Mauerhofer 2008).

It can also be used as a critique to traditional nature conservation and spatial planning legislation. Domestic forestry legislation for instance tends to give priority to economy or use impact and profit considerations only (e.g. *Nutzwirkung* in Austrian forest law and *bos-economie* or forest economy in Dutch forest law) blocking away the other columns and denying the concentric circles (Mauerhofer 2008; Kistenkas 2012). Also European nature conservation law a priori blocks away columns and entire clusters of ES, as the Natura 2000 appropriate assessment of Article 6 EU Habitats Directive only gives de iure weight to planet desiderata, thus ruling out provisional services and cultural services as well as most regulating and supporting ES. Also jurisprudence, i.e. court law of the European Court of Justice (ECJ), recently upholds this, confirming limited testing on only a few planet desiderata (one and a half hectare of limestone pavement within a subarea of 85 ha) in obstructing the Galway bypass road scheme around the busy city centre (ECJ 2013 C-258/11). Here, as to law, only limestone (being a limited planet desideratum) counts, not traffic safety, logistics or other dimensions (Kistenkas 2014). Reaching a sustainable development three scale balance might be far away then.

Likewise in domestic spatial planning law entire ES clusters may be ruled out when it comes to zoning in local and regional spatial plans with its traditional mono-functionality. A good example is the Dutch Ecological Main Structure (*Ecologische Hoofdstructuur*) throughout the land, being a by subordinated legislation (a Royal Decree called *Besluit algemene regels ruimtelijke ordening*) protected network of non-multifunctional countryside zoning again a priori denying the social and economic capacity. So, new thoughts on sustainability easily lead to an unmasking of traditional nature, forest and spatial planning legislation, on a supranational level as well as domestically. It is remarkably ominous the concept of ES still up till now is not explicitly mentioned in European nature conservation directives.

2 Ecosystem Services in Law

ES definitions do range from conditions and processes through which natural ecosystems, and the species that make them up, sustain and fulfil human life to the definition of *The Economics of Ecosystems and Biodiversity* (TEEB): ES are the direct and indirect contributions of ecosystems to human well-being (TEEB 2010). A widely accepted and quite workable definition comes from the famous *Millennium Ecosystem Assessment* (MEA) report, which defines ES as benefits people obtain from ecosystems (MEA 2005), distinguishing four categories of benefits to be obtained from ecosystems. Though there may be various alternatives (Fisher et al. 2009), the original version of the following categories tends to be a classic and widely accepted classification:

1. provisioning services (food, crops, water, minerals, industrial products, energy etc.),
2. supporting or habitat services (nutrient cycling, seed dispersal etc.),

3. regulating services (carbon sequestration, crop pollination, pest control etc.),
4. cultural services (leisure, spiritual inspiration etc.).

As early as in 2007 Ruhl and Salzman (2007) concluded that the status of ES in law is that they simply have none. Which is slightly regrettable because regulation, even the conventional command-and-control variety of prescriptive regulation, can indeed provide effective and efficient solutions for management of public goods. Legislation already could have introduced, prescribed and mainstreamed the ES approach, but remarkably both international as national and perhaps even subnational legislators seem to have turned a blind eye to the concept of ES. Basically there has not been changed much ever since. Notably European environmental legislation as well as European court law neglected this concept up till now or even have lost some good momentum to engage with this concept (Kistenkas 2014).

However, in law literature a very small amount of authors has occasionally and slightly mentioned the ES concept. Thus, also law doctrine never elaborated the introduction of ES into law profoundly. Reid (2011) reminds us that conservation today is increasingly adopting an ecosystems approach rather than focusing narrowly on specific species or habitats and therefore concludes that alternative approaches may have a lot to offer, notably 'an ambitious attempt to charge for ecosystem services' but he does not make or even start up this attempt. Also Monteduro (2013), seeking to overcome the distinction and dualism between agricultural law and environmental law as well as de iure facilitating the transition from mono-functional to multifunctional land use, ends with a call for a radical shift claiming the whole environmental law field should be re-thought and reconstructed as a law of ecosystems, ecosystemic law or just 'law of ecosystem services', concluding this could become a future challenge to legal doctrine. According to this author law scholars should inaugurate a new line of research.

Hauck et al. (2013) see an adjacent and complementary role of the ES concept for planning and decision making, as this concept could, in their view, only complement existing instruments that focus solely on a specific task or sectoral interest. In contrast, Mertens et al. (2012) and Kistenkas (2014) advocate for explicitly building in the ES concept into legislation replacing the existing instruments like for instance the infamous habitats assessment in EU nature conservation law which only addresses a few limited conservation objectives and thus ruling out entire clusters of provisional, cultural, regulating and supporting services. Mertens et al. (2012) propose to adopt a so called 'ecosystem services check' into EU legislation. They too, conclude legal research has been lagging behind and their final remark that ES protection by means of legislation is still at its infancy might well be seen as an understatement.

Whereas in other scientific disciplines the ES concept is quite widely adopted as a useful tool, the most salient missing formulation in the ongoing policy assessment for ES still is the legal and regulatory component (Ruhl and Salzman 2007).

3 Sustainable Developments: New Policy

Whilst legislation and its jurisprudence may not be fully compatible with modern academic sustainable development thoughts, novel policy initiatives may. While the ES concept is key in recent EU policy documents like Green Infrastructure (EC 2013) and EU Biodiversity Strategy to 2020 (EC 2011), European legislation still seems to ignore this concept. Ecosystem services are scarcely mentioned in traditional EU environmental legislation and therefore de iure still rather unknown or even seen as slightly irrelevant where they might de facto be relevant or urgent for sustainability reasons. Here, interestingly, law may lag way behind policy and governance or perhaps even collide with each other like a *contrapunctus*. The EU Green Infrastructure (GI) (EC 2013) as well as the EU Biodiversity Strategy (EC 2011) for instance both aim at serving interests of people, profit and planet (triple P) by the delivery and balancing of ecosystem services or, in other words, an infrastructure generating value for society and the diversity of outputs (private, public and social goods). One of the key attractions of GI is its multi-functionality, which allows unlocking several benefits on the same spatial area. In short, it is about triple P, whereas the very essence of traditional European nature conservation law is only about a few planet desiderata: it basically is about a one criterion test.

However, GI should be built up by identifying multi-functional zones where compatible land uses that support healthy ecosystems are favoured over other more destructive single-focus developments (so called grey infrastructure). In contrast with Natura 2000 legislation, GI does emphasize on holistic solutions: not only on (some) environmental benefits, but also on socio-economic aspects and advantages. It should likewise be included into spatial planning. Clearly and explicitly the European Commission (EC) in its GI links the ES concept to sustainability and triple P balancing (EC 2013, p. 1). Here policy is apparently far ahead of legislation and jurisprudence. Nevertheless, the EC claims that the GI strategy can be implemented within the context of existing legislation (EC 2013 at p. 10). I think this could well become quite difficult when it comes to nature conservation legislation (Borgström and Kistenkas 2014). Especially the one criterion law finding method of the *habitats assessment* appears not to be fully designed for new GI purposes. It may perhaps be useful to reconsider this legal instrument as part of the recently announced EU Regulatory Fitness (REFIT) program (COM(2012) 746 final and SWD(2013) 401 final).

Also domestic policy initiatives may mention the ES concept and link it with sustainable development. For instance the recent Dutch Nature Policy Vision Document (*Natuurvisie*) does (Natuurvisie 2014), whilst domestic and European nature conservation and spatial planning legislation still do not. In fact here a very interesting *contrapunctus* could indeed be noticed: European and national policy documents as well as law literature increasingly opt for triple P balancing (or, in short, 3P-weighing) and the acknowledgement of ES, whereas European and domestic nature and spatial planning legislation and jurisprudence still focus on 1P-testing. Designing new environmental policy we should avoid that this colliding *contrapunctus* eventually turns into a counterproductive governance dissonant. In a successful policy mix of sticks and carrots also classic coercive command-and-control regulatory instruments need to be in line with other governance instruments and policy documents (Gunningham et al. 1998).

4 Drafting New Nature Conservation Legislation

Whilst the GI initiative is about the delivery of ecosystem services and balancing of social, economic and environmental benefits, the habitats assessment is not designed for a triple P balancing but it is merely an one criterion test able to ignore entire clusters of ecosystem services like provisioning and cultural services. In contrast, recently drafted new EU nature legislation, such as the EU Timber Regulation (EUTR), which came into force in 2013 (Regulation 995/2010/EU2010) already shows new legal law finding methods and instruments and seems to move away from an one criterion test towards a balancing approach, as it adopts a so called Due Diligence System (DDS) instead of a rigid and dogmatic assessment. EUTR aims at sustainable forest management by combatting illegal logging basically through risk mitigation efforts of DDS (Article 6 EUTR). This Article 6 is in fact slightly different than Article 6 of the Habitats Directive: it appears to introduce a mere balancing approach of socio-economic and ecological aspects instead of an ecological test only (Kistenkas 2013b). It might also give way to introduce the ecosystem services concept, whereas the *habitats assessment* up till now has not given any legal relevance to this concept.

When it comes to accommodating diverging interests and flexibility the habitats assessment might very well be an obstruction, as it serves some limited planet aspects (the conservation objectives of the site) (McGillivray 2011). The concept ecosystem services, being key to GI, still is not explicitly referred to or made de iure relevant in the wording of the Habitats Directive. This surely might become a problem when GI must be integrated into EU nature conservation or landscape law and perhaps we should therefore reconsider the wording of (at least Article 6 of) the Habitats Directive. Court law is not likely able or willing to solve this problem as the recent 2013 Galway-case shows us.

In its judgement on the Galway bypass the ECJ seems to confirm the rigid dogmatic approach, strongly linking the criterion of ‘integrity of the site’ to the site’s conservation objectives. In this judgement the court considers that the constitutive characteristics of the site are connected to the presence of the natural habitats whose conservation was the objective justifying the designation of that site (ECJ, April 11, 2013, C-258/11 at paragraph 48). Then, despite the different language versions the natural characteristics with its limited conservation objectives do prevail. This also gives an a priori preference to some narrow planet-considerations above people-, profit- and other remaining planet-considerations. Sustainable land use and development, as said above, merely demand a 3P-weighing and an ES balance. It is quite questionable whether this can be reached by the legal method of assessing limited ecological criteria only. Some pre-set conservation objectives do a priori have a preferred position, as only they are within the criteria to be assessed. Not only during the first step of the habitats assessment (significancy test) but also during the second step of appropriate assessment of Article 6(3) the rigid conservation objectives seem to be the only criteria to be tested. Such an assessment might be argued then as an instrument not really compatible to sustainable land use and an ES-balance. Nevertheless, the ECJ confirmed this dogmatic approach again in the more recent A2 Motorway case (ECJ, May 15, 2014, C-521/12 at paragraph 28): the integrity of the site is only about the conservation objectives of the particular site. So Article 6(3) keeps a narrow focus on a few planet desiderata only, ruling out other ecosystem services. Again logistics, traffic safety and other interests at stake could not play a proper role and again a road scheme had to be halted.

It may be true socio-economic aspects might play a role in the exemption regime of Article 6(4) further on. Basically Article 6(4) is an exceptive clause: "If, in spite of a negative assessment of the implications for the site and in the absence of alternative solutions, a plan or project must nevertheless be carried out for imperative reasons of overriding public interest, *including those of a social or economic nature*, the Member State shall take all compensatory measures necessary to ensure that the overall coherence of Natura 2000 is protected" (emphasis added). At that second and exceptive stage the habitats assessment is not entirely blind for socio-economic aspects, but then they must qualify as imperative reasons of overriding public interest. It is not clear whether all relevant ES could be given adequate attention then. Is, for instance, a leisure cycling path or an outdoor walkway an imperative reason of overriding public interest? Probably not. It is however a people-related ecosystem service we are talking about. Nevertheless, it cannot play a proper role: not in the appropriate assessment of Article 6(3), nor in the exemption regime of Article 6(4).

The appropriate assessment could then be a good opportunity to determine all relevant ES at an early stage already. Why not broaden up the 'integrity of the site' criterion? This 'integrity of the site' could well be a lot more than only a few conservation objectives, as the two ECJ cases demonstrate. Therefore we might consider the help of the concept of ES already in the stage of the appropriate assessment. The traditional conservation objectives then could be related to other objectives and perhaps be balanced in a way it adds simultaneously value to both socio-economic and ecological ES;. Then, ES might turn out to be very helpful. Nature conservation law could benefit from an analysis and weighing of different values and their trade-offs based on a ES concept and the three main clusters of ES it distinguishes: (a) provisioning, (b) cultural and (c) ecological services being the two planet-related ES of regulating and supporting services.

Reconsidering Article 6 of the Habitats Directive, the habitats assessment could be brought back to a significancy test on the site's conservation objectives followed by an appropriate assessment in which the integrity of the site is not only a matter of, again, the conservation objectives but also of other ecosystem services. Already here, compensation and alternatives might play a role to reach a sustainable development. Relevant (other) ecosystem services do no longer have to qualify as imperative reasons of overriding public interest. Thus, an exemption regime of Article 6(4) will no longer be necessary, as the significancy test and the integrity of the site test of Article 6(3) may already provide sustainable development then. This calls for new legislation, perhaps within REFIT, but also for a new balancing jurisprudence, as the ECJ adopted a strong rigid dogmatic approach as we have seen above in the Galway and A2 Motorway cases of 2013 and 2014.

I think, the ECJ lost momentum by not adopting a broadened view of the integrity of the site, as it easily could have adjusted this criterion and thus giving more weight to sustainable development demands. Now, the legislator, probably through REFIT, is able to alter the habitats assessment towards an instrument facilitating sustainable development.

5 Drafting New Spatial Planning Legislation

Spatial planning has become increasingly part of European Union governance. The days of only some soft law documents like the 1999 European Spatial Development Perspective (ESDP) seem to be over (Tiefenthaler 2011). Lots of sectoral policies do or will indeed have spatial planning effects on a domestic member-state level. The Habitats Directive for instance is not solely nature conservation but spatial planning as well as it is focused on network zoning of Natura 2000 sites. Transnational protected areas like Natura 2000 are a shared competence between the European Union and its member-states (Article 4(2)(e) TFEU).

A new remarkable step was made in 2013, when a proposal for a new EU directive establishing a framework for maritime spatial planning and integrated coastal management (COM(2013) 133 final) directly and *sans scrupules* appears to pave the way for a genuine European spatial planning law, though be it, for the time being, only spatial planning at sea.

In this proposal interestingly Article 5 reads as follows:

> Maritime spatial plans and integrated coastal management strategies shall apply an ecosystem-based approach to facilitate the co-existence and prevent conflicts between competing sector activities in marine waters and coastal zones, and shall aim to contribute to:
>
> (a) securing the energy supply of the Union by promoting the development of marine energy sources, the development of new and renewable forms of energy, the interconnection of energy networks, and energy efficiency;
> (b) promoting the development of maritime transport and providing efficient and costeffective shipping routes across Europe, including port accessibility and transport safety;
> (c) fostering the sustainable development and growth of the fisheries and aquaculture sector, including employment in fisheries and connected sectors;
> (d) ensuring the preservation, protection and improvement of the environment as well as the prudent and rational use of natural resources, notably in order to achieve good environmental status, halt the loss of biodiversity and degradation of ecosystem services and reduce marine pollution risks;
> (e) ensuring climate resilient coastal and marine areas.

So according to Article 5 of this draft directive spatial plans should apply an ecosystem-based approach to facilitate the co-existence of and prevent the conflicts between competing sector activities. Co-existence indicates a multi-functionality at sea: energy supply, transport, fisheries, aquaculture but also preservation, protection and improvement of the environment. Here, even the concept of ES is mentioned explicitly in this article. Basically, this is one of the first times in European environmental legislation (Borgström and Kistenkas 2014). At last, ES are becoming part of EU legislation too and its legal framework can no longer deny the ES-concept.

The explanatory memorandum underlines that this novel directive is about sustainable growth: here spatial planning needs to achieve not only environmental but also social and economic objectives and to ensure this sustainability spatial planning will have to employ an ecosystem-based approach protecting natural resources that provide the basis for carrying out the various activities. For the time being the proposal is only on maritime spatial planning and thus it does not interfere with member state's prerogatives for town and country planning (terrestrial planning) yet, but the novel views of Brussels on future spatial planning might be made clear by this draft. Spatial planning is seen as a practical tool to comply with sustainable growth and 3P-balancing. What strikes the most is that this ES-concept is explicitly brought forward in a coercive legislative text, both Article 5 and in the Fourteenth Recital in the Preamble to this draft directive; now, also legally, due regard should be given to ecosystem services in spatial planning matters.

Although the final text of Article 5 of this novel directive (finally called the framework directive for maritime spatial planning; Directive 2014/89/EU of 23 July 2014) does no longer seem to mention ecosystem services such explicitly, it even stronger underlines an ecosystem-based approach, 3P weighing and 3D sustainable co-existence: 'Member States shall consider economic, social and environmental aspects to support sustainable development and growth in the maritime sector, applying an ecosystem-based approach, and to promote the coexistence of relevant activities and uses'. Besides, the Thirteenth Recital in the Preamble to this directive still brings forward the concept of ecosystem services explicitly. So, in my view, this new piece of legislation, albeit only on a subsectoral matter like maritime spatial planning, might well grow into a breakthrough directive.

Currently one dimension testing (like the habitats assessment or, for instance, the Dutch Ecological Main Structure regime) eventually leads to one dimension zoning and is likely to obstruct three dimensional sustainability. Earlier case studies showed that it inevitably implied rigid spatial segregation of functions as a kind of spatial *apartheid* denying other relevant ecosystem services and not complying with sustainable co-existence (Kistenkas 2014). However, this newly proposed ES-based approach might prelude a breakthrough in European spatial planning law. At last traditional spatial planning approaches like rigid monoculture zoning seem to be questioned as being too static (Ebbesson 2010) and potentially obstructing a weighing of socio-economic and ecological interests or three scale balancing and, hence, sustainable growth.

In sum, ecosystem services are about to be mentioned in a legislative text and thus they are made de iure relevant for spatial planning, be it only at sea initially, facilitating co-existence and spatial multi-functionality. In near future, it might perhaps encourage altering traditional domestic spatial planning legislation with its often rigid zoning of monocultures towards a modern co-existence of activities.

6 Discussion

New thoughts on sustainable development have led to new policy, but strange to say as yet not to new law. We, lawyers, seem to lag behind. Conventional and coercive command-and-control public law always needs to be part of smart governance (Gunningham et al. 1998), also of sustainability governance arrangements. Basically, the concepts of ES, 3-D Sustainability and 3P-balancing are well echoed in novel policy documents like Green Infrastructure, EU Biodiversity Strategy 2020 and No Net Loss. However, besides perhaps the maritime spatial planning directive as the only specific exception, up till now they are given no binding force by environmental law legislation. Integration into general environmental law is needed and more novel legislation should follow and enforce these policy initiatives.

Even a contradiction could now be noticed: European and national policy documents as well as law literature increasingly opt for triple P balancing (or 3P-weighing) and the acknowledgement of ES and 3-D Sustainability, whereas European and domestic nature and spatial planning legislation and jurisprudence still focus on 1P-testing (only planet or only profit). Such a new diagnosis may be the beginning of curation.

As to spatial planning we could perhaps follow the ideas on novel spatial planning as published in the draft and final version of the directive on Maritime Spatial Planning: the ES concept as a tool towards multi-functionality instead of rigid monoculture zoning. Then, instead of focusing solely on a specific task or sectoral interest, spatial planning truly helps us towards sustainable use of ecosystems and their services (Hauck et al. 2013). Here, through the ES-concept, also Ebbesson (2010) is put in the right as traditional zoning and designation of nature reserves might be less adequate means to promote sustainable management.

As to European nature conservation REFIT might perhaps questioning the traditional habitats assessment as it is currently rigidly being applied by the ECJ. The ‘integrity of the site’ criterion could perhaps be broadened up during the appropriate assessment. Not only the site’s preset and limited conservation objectives but also other ecosystem services could then be taken into account properly. Relevant (other) ecosystem services do no longer have to outstand or (over)qualify as ‘imperative reasons of overriding public interest’. Article 6(4) appears no longer to be necessary then, as the significancy test and the integrity of the site test of Article 6(3) may already provide sustainable development.

Domestic forestry legislation tends to give priority to economy or use impact and profit considerations (forest economy) blocking away entire ES clusters and denying the concentric circles in the 3-D *Cone of Mauerhofer*. Modern forest law must pay attention to both natural and socio-economic capital. Nowadays, on a EU level the new EU Timber Regulation with its Due Diligence Systems (DDS) already may be aware of this need for a 3P-balancing aiming at sustainable forest management by promoting environmentally appropriate (planet), socially beneficial (people) and economically viable (profit) forestry management (Kistenkas 2013b), but there still will be several domestic legislative provisions in member states which certainly may not (Mauerhofer 2008).

Here again, the ES-concept, three-dimensional sustainability and triple P balancing might serve as a critique to traditional nature conservation and spatial planning legislation, both European and domestic. Critical change could also be backed and inspired by growing recognition by environmental lawyers of the principle of sustainable development. As already explained above this principle is moving from a merely political principle to a normative principle of law (in *status nascendi*). Recognizing and applying this principle as an environmental law principle surely might support redirecting and re-interpreting traditional 1P-regulations towards more 3P-balancing. Environmental law principles make it possible to resolve hard cases and bear heavily on the weighing of interests (De Sadeleer 2002). Such law principles may indeed play a helpful role in balancing interests and ecosystem services (De Sadeleer 2002; Van Rijswick and Salet 2012; Kistenkas 2013a) or to put in the words of Ebbesson (2010): the development of general legal principles may be much more adequate in terms of promoting sustainable development than traditional one dimension zoning and static nature reserve designation. Also Van Rijswick and Salet (2012) advocate a more, what they call *principal* use of legal norms, as environmental law principles, notably the international principle of sustainable development, can have a guiding nature and simultaneously give a novel normative sense of (re)direction. They argue such a principal approach may enable a recombination of policy-making and legislation.

At least the principle could trigger a re-interpretation of traditional norms. Already courts could start with this re-interpretation; not only the International Court of Justice—who, as described above, already does so—but also the European Court of Justice. The latter already may use the written version of the principle of sustainable development as anchored in Article 3 of the Treaty on the European Union (TEU) and Article 37 of the Charter of Fundamental Rights of the European Union. Both articles of the EU *acquis communautaire* basically aim at a 3P-balancing (Kistenkas 2013b). Here, especially EU legislation and jurisprudence could easily meet up with (post-)modern sustainability insights and take the lead to give way to unhindered sustainable growth and alter traditional 1P nature conservation and spatial planning law.

7 Conclusions

Investigating new thoughts on sustainable development the ES-concept may de iure grow into a helpful tool in reaching the three scale balance as described above. Law finding techniques should therefore be a weighing of economic (profit), social (people) and environmental (planet) capital rather than a 1P-assessment as in traditional European nature conservation legislation and (domestic) spatial planning zones.

However, as to law, the ES-concept still has no true status in general environmental legislation and is barely mentioned in European and national statute and court law. This increasingly contrasts with novel EU policy documents such as Green Infrastructure, EU Biodiversity Strategy and No Net Loss, which are all basically about triple P and sustainable multi-functionality. Here, a sharp *contrapunctus* could be noticed: EU policy documents and doctrine increasingly opt for 3P-balancing acknowledging the ES-concept, whereas European and domestic nature, forest and spatial planning law, i.e. both statute and court law still focus on a one dimension test.

This will become problematic when policy initiatives like Green Infrastructure must be integrated into EU nature conservation law and perhaps we should therefore reconsider the wording of the Habitats Directive, as court law is not likely able or willing to solve this problem (the EU court is strongly linking the criterion of 'integrity of the site' to the site's conservation objectives). Now, the EU legislator, probably through the REFIT-program, could be the one to alter the habitats assessment towards an instrument facilitating sustainable development.

As to spatial planning law, however, a newly made maritime EU-directive recently gives a future role to ecosystem services and sustainable multifunctionality, being the first time ever in EU environmental legislation and as yet the only tiny and subsectoral example. So new thoughts may have led to new policy initiatives, but as yet scarcely to new legislation. Nevertheless, also on terrestrial spatial planning we could perhaps follow the ideas on novel spatial planning at sea as published in the draft and final version of the directive on Maritime Spatial Planning and on European nature conservation REFIT might alter the traditional habitats assessment into a more 3P weighing. Likewise on traditional one dimension domestic forestry legislation the new EU Timber Regulation with its Due Diligence Systems (DDS) might give rise to a new awareness in (post-)modern forest law of this need for a 3P-balancing aiming at sustainable management by promoting environmentally appropriate (planet), socially beneficial (people) and economically viable (profit) forestry management. But above all change could finally be backed and triggered now by de iure recognition of the principle of sustainable development as an environmental law principle, for recognizing and applying this principle in European and domestic law in my opinion might support redirecting and re-interpreting traditional one dimension regulations towards more three dimensional balancing.

Acknowledgements This paper was funded by the Dutch Statutory Research Tasks Unit for Nature and the Environment (*WOt Natuur & Milieu*), a division of Wageningen UR, and the EU 7th framework project OpenNESS. The author wishes to thank the contact persons of these institutions for their input.

References

Borgström S, Kistenkas FH (2014) The compatibility of the habitats directive with the novel EU green infrastructure policy. Eur Energy Environ Law Rev 23:36–44

Brundtland GH (1987) Our common future. Oxford University Press, Oxford

Costanza R, d'Arge R, De Groot R, Farberk S, Grasso M, Hannon B, Limburg K, Naeem S, O'Neill RV, Paruelo J, Raskin RG, Suttonkk P, Van den Belt M (1997) The value of the world's ecosystem services and natural capital. Nature 387:253–260

de Sadeleer N (2002) Environmental principles. From political slogans to legal rules. Oxford University Press, Oxford

Ebbesson J (2010) The rule of law in governance of complex socio-ecological changes. Glob Environ Change 20:414–422

EC (2011) EU biodiversity strategy to 2020. COM (2011) 244

EC (2013) Communication on green infrastructure (GI). Enhancing Europe's natural capital. COM (2013)249 final

Fisher B, Turner RK, Morling P (2009) Defining and classifying ecosystem services for decision making. Ecol Econ 68:643–653

Gunningham N, Grabowsky P, Sinclair D (1998) Smart regulation. Designing environmental policy. Clarendon Press, Oxford

Hauck J, Schweppe-Kraft B, Albert C, Görg C, Jax K, Jensen R, Fürst C, Maes J, Ring I, Hönigova I, Burkhard V, Mehring M, Tiefenbach M, Grunewald K, Schwarzer M, Meurer J, Sommerhäuser M, Priess J, Schmidt J, Grêt-Regamey A (2013) The promise of the ecosystem services concept for planning and decision making. GAIA 22:232–236

ICJ (1997) Gabcikovo-Nagymaros (Hungary v. Slovakia). ICJ Rep 1997

ICJ (2010) Pulp mills on the river Uruguay (Argentina v. Uruguay). ICJ Rep 2010

Kistenkas FH (2012) Recht voor de groenblauwe ruimte. Wageningen Academic Publishers, Wageningen

Kistenkas FH (2013a) Rethinking European nature conservation legislation: towards sustainable development. J Eur Environ Plan Law 10(1):72–84

Kistenkas FH (2013b) Concurring regulation in European forest law. Forest certification and the new EU timber regulation. GAiA 22(3):166–168

Kistenkas FH (2014) Innovating European nature conservation law by introducing ecosystem services. GAiA 23(2):88–92

Mauerhofer V (2008) 3-D sustainability: an approach for priority setting in situation of conflicting interests towards a sustainable development. Ecol Econ 66:496–506

Mauerhofer V (2013) The 'Governance-Check': assessing the sustainability of public spatial decision-making structures. Land Use Policy 30:328–336

McGillivray D (2011) Mitigation, compensation and conservation: screening for appropriate assessment under the EU habitats directive. J Eur Environ Plan Law 8(4):329–352

Mertens K, Cliquet A, Vanheusden B (2012) Ecosystem services. What's in it for a lawyer?. Eur Energy Environ Law Rev 21:31–40

Millennium Ecosystem Assessment (MEA) (2005) Ecosystems and human well-being: synthesis. Island Press, Washington/DC

Monteduro M (2013) Environmental law and agroecology. Transdisciplinary approach to public ecosystem services as a new challenge for environmental legal doctrine. Eur Energy Environ Law Rev 1:2–11

Natuurvisie (2014) Rijksnatuurvisie 2014 Natuurlijk verder. Sdu, The Hague
Nollkaemper A (2011) Kern van het internationaal publiekrecht. Boom Juridische Uitgevers, The Hague
Reid CT (2011) The privatisation of biodiversity? Possible new approaches to nature conservation law in the UK. J Environ Law 23(2):203–231
Ruhl JB, Salzman J (2007) The law and policy beginnings of ecosystem services. J Land Use Environ Law 22(2):157–172
TEEB (2010) The economics of ecosystems and biodiversity (TEEB): ecological and economic foundations. Earthscan, London
Termorshuizen J, Opdam P (2009) Landscape services as a bridge between landscape ecology and sustainable development. Landscape Ecol 24:1037–1052
Tiefenthaler V (2011) Spatial planning in Europe. The impact of European Union law on national planning systems and territorial transnational cooperation. J Eur Environ Plan Law 8:115–140
van Rijswick M, Salet W (2012) Enabling the contextualization of legal rules in responsive strategies to climate change. Ecol Soc 17(2):18

Sustainable Development of the Oceans: Closing the Gaps in the International Legal Framework

Glen Wright, Julien Rochette and Thomas Greiber

Abstract The world's oceans are critical providers of ecosystem services and they are under increasing pressure from expanding and intensifying human activities. A range of international instruments and institutions aim to regulate maritime activities, though some legal gaps in the international framework remain. In particular, areas beyond national jurisdiction (ABNJ) lack an overarching regulatory framework, with no provisions for marine protected areas, environmental impact assessment, or access and benefit sharing in relation to marine genetic resources. There are also gaps and weakness in the international framework for the exploitation of offshore oil and gas resources. In this chapter, we highlight these gaps, outline relevant ongoing processes to fill them, and propose ways forward.

Keywords Areas beyond national jurisdiction · Marine protected areas · Environmental impact assessment · Access and benefit sharing · Marine genetic resources · Offshore oil and gas

1 Introduction

A growing population and appetite for resources, coupled with innovation and rapid technological advancement, is driving unprecedented exploitation of the marine environment. This is driving a new 'industrial revolution' of the oceans (Charter 2007; Salcido 2008), encouraged by a growing policy focus on 'Blue Growth', i.e.

G. Wright (✉)
International Marine Policy, Institute for Sustainable Development and International Relations (IDDRI), Paris, France
e-mail: glen.wright@iddri.org

J. Rochette
Institute for Sustainable Development and International Relations (IDDRI), Paris, France

T. Greiber
Institute for Advanced Sustainability Studies (IASS), Potsdam, Germany

V. Mauerhofer (ed.), *Legal Aspects of Sustainable Development*,
DOI 10.1007/978-3-319-26021-1_27

development of policy aimed at leveraging the natural resources of the marine environment to achieve economic growth (European Commission 2012).

Traditional maritime activities such as fishing, navigation, or offshore oil and gas drilling are intensifying: fishing vessels target deeper as less fish remains in shallower waters; ship traffic is increasing, fuelled by booming world trade; and oil and gas drilling is moving ever deeper and further from shore. In addition, new activities are in varying stages of implementation: aquaculture is developing rapidly to ensure food security in the context of declining fish stocks (FAO Fisheries and Aquaculture Department 2014); marine genetic resources (MGRs) are attracting the interest of researchers and developers due to their unique biochemical properties (Leary 2011; Broggiato 2013); various options for offshore renewable energy (REN21 2013; Wright 2015) and geoengineering (Rayfuse et al. 2008; Boyd 2013) are being developed in pursuit of carbon reduction goals; and seabed mining is on the brink of becoming a commercial reality after a long gestation period.[1] Climate change, ocean acidification and various forms of pollution also threaten the oceans.

The more that we learn about marine ecosystems, the more apparent their importance becomes. The world's oceans are critical providers of the ecosystem services on which humanity depends. Given this importance, there exist a great number of international instruments and institutions aimed at regulating many maritime activities, though some legal gaps in the international framework still remain.

2 Sustainable Development, Implementation Gaps and Legal Gaps

The outcome document of the United Nations Conference on Sustainable Development 2012 ('Rio + 20'), *The Future We Want* (UN 2012), recognised that "oceans, seas and coastal areas form an integrated and essential component of the Earth's ecosystem and are critical to sustaining it, and that international law, as reflected in the United Nations Convention on the Law of the Sea, provides the legal framework for the conservation and sustainable use of the oceans and their resources".

Indeed, the United Nations Convention on the Law of the Sea (UNCLOS 1982) is generally considered to be the 'Constitution of the Sea' setting out "the legal framework within which all activities in the oceans and seas must be carried out" (UNGA 2010). As such, it does institute some specific obligations and responsibilities in relation to the marine environment. PART XII contains a general obligation to "protect and preserve the marine environment",[2] an obligation to take

[1]See The International Seabed Authority (ISA) has entered into 26 exploration contracts in the Atlantic, Indian and Pacific Oceans and is currently in the process of developing regulations regarding the exploitation of deepsea mineral resources.

[2]Article 192.

measures to prevent, reduce and control pollution,[3] and an obligation to assess the potential effects of activities that may "cause substantial pollution of or significant and harmful changes to the marine environment".[4]

In addition to the UNCLOS (1982) environmental provisions, a range of international instruments and institutions has been developed, including: the Convention on Biological Diversity (CBD 1992), which provides a framework for the conservation and sustainable use of biodiversity in both terrestrial and marine environments; the UN Fish Stocks Agreement (UN 2001), which regulates migratory fish stocks; and the International Convention for the Prevention of Pollution from Ships (MARPOL 1973) and London Convention (1972), which regulate pollution from ships and dumping respectively.

States have also recently agreed to develop a set of Sustainable Development Goals (SDGs) (UNGA 2014) building upon the previously agreed Millennium Development Goals (MDGs). The SDGs will incorporate the MDGs' primary aim of alleviating poverty, but will also focus more concretely on environmental concerns. The proposed goal 14 on oceans is to: "Conserve and sustainably use the oceans, seas and marine resources for sustainable development" (UNGA 2014).

A number of discrete targets are proposed, to be met by 2020, including to: "sustainably manage and protect marine and coastal ecosystems to avoid significant adverse impacts, including by strengthening their resilience, and take action for their restoration, to achieve healthy and productive oceans"; and "conserve at least 10 per cent of coastal and marine areas, consistent with national and international law and based on best available scientific information" (UNGA 2014).

The proposed oceans goal also calls on states to "ensure the full implementation of international law, as reflected in UNCLOS (1982) for States parties thereto, including, where applicable, existing regional and international regimes for the conservation and sustainable use of oceans and their resources by their parties" (UNGA 2014). This implies an increased focus on, and role for, existing regional arrangements, as well as better implementation of existing international and sectoral measures.

The plethora of existing international legal instruments suggests that, in general, efforts to ensure sustainable development of the oceans are hampered by weaknesses in implementation, rather than a lack of legal coverage. The international community has expended a huge amount of time and effort on establishing policy commitments that aim to protect the oceans, but there is still a considerable difference between the commitments formally expressed by States in these policy documents and their subsequent willingness or capacity to fully implement them.

To take only one example, the 2002 World Summit on Sustainable Development committed States to establish representative networks of marine protected areas (MPAs) by 2012. At the 2012 Rio + 20 conference, it was clear that little progress had been made towards meeting this target, which had subsequently been recast

[3]Article 194.

[4]Article 206.

under the auspices of the CBD as one of the 'Aichi Targets'. Aichi Target 11 now foresees that "at least (…) 10 % of coastal and marine areas (…) are conserved through effectively and equitably managed, ecologically representative and well connected systems of protected areas and other effective area-based conservation measures" by 2020. As of today, less than 3 % of the ocean has been designated as MPAs, and only around 0.5 % of Areas Beyond National Jurisdiction (ABNJ) are protected (IUCN and UNEP-WCMC 2013), despite these commitments. Extrapolations made in the Global Biodiversity Outlook 4 therefore suggest that the marine target is still not on course to be met (CBD Secretariat 2014, p. 83).

While implementation of existing agreements and commitments must be accorded a high priority, it should not be forgotten that there remain some significant gaps in the international legal framework for sustainable development in the oceans. Firstly, in relation to ABNJ, there is no coherent and comprehensive framework for the implementation of important conservation tools or for the regulation of research and development on MGRs. Secondly, there is currently no legally binding agreement regulating liability and compensation for pollution damage resulting from offshore oil and gas activities.

3 Conservation and Sustainable Use of Marine Biodiversity in ABNJ

Marine areas beyond national jurisdiction represent approximately half of the planet's surface. Consisting of both the 'high seas',[5] and the 'Area' (UNCLOS 1982[6]), ABNJ host a significant proportion of the earth's biodiversity (Census of Marine Life 2011, p. 16). As with the oceans at large, pressure on ecosystems in ABNJ has been mounting, and further pressure is on the horizon as scientific discoveries and technological developments now make it possible to exploit new resources, particularly marine generic resources and deep-sea minerals.

3.1 A New International Legally Binding Instrument on Marine Biodiversity in ABNJ

Current discussions on marine biodiversity in ABNJ have their origins in the Ad Hoc *Open-ended Informal Working Group to study issues relating to the conservation and sustainable use of marine biological diversity beyond areas of national*

---

[5]The High Seas cover the water column beyond the Economic Exclusive Zone of Coastal States.

[6]Defined as 'the seabed and ocean floor and subsoil thereof, beyond the limits of national jurisdiction'. Art. 1(1).

jurisdiction ("BBNJ Working Group") established by the UNGA in 2004 (see Druel et al. 2013).

Since 2011, States have worked on a 'Package Deal' of issues, specifically: (i) marine genetic resources (MGRs), including questions on the sharing of benefits; (ii) area-based management tools (ABMTs), including marine protected areas (MPAs); (iii) environmental impact assessments (EIA); and (iv) capacity-building and the transfer of marine technology.

At the Rio + 20 conference, States committed to urgently address the issue of the conservation and sustainable use of marine biological diversity in ABNJ, including by taking a decision on the development of an international instrument under UNCLOS (UN 2012, para. 162). Between 2014 and 2015, three meetings of the BBNJ Working Group were held in order to discuss whether or not negotiations should commence. This process culminated in January 2015 with the historic step of States agreeing to open negotiations for a new international legally binding instrument under UNCLOS (Rochette et al. 2015). This will presumably take the form of an implementing agreement (IA) to UNCLOS (1982) on the conservation and sustainable use of marine biodiversity in ABNJ.

We briefly discuss below the key challenges and options for the development of such an IA in relation to three of the four Package Deal areas: MPAs, MGRs and EIA.[7]

3.2 *Elements of a New UNCLOS IA*

3.2.1 Marine Protected Areas

There is no universally agreed definition of 'marine protected area', though various organisations and institutions have developed a number of definitions (Druel 2011). The basic idea shared between all definitions is that MPAs will have "a special status in comparison with the surrounding area due to their more stringent regulation of one or more human activities […] by one or more measures […] for one or more purposes" (Molenaar and Elferink 2009, p. 6).

In MPAs, the level of protection may vary depending on the pressures on the area to be protected and on the needs for conservation. Some MPAs may be entirely or partly marine reserves, including so-called 'no-take' zones, while in others only

[7]Technology transfer and capacity building are undoubtedly a crucial part of the overall sustainable development framework, though the legal gaps are arguably less pronounced than in the other areas. Article 144 of UNCLOS (1982) outlines the principles of technology transfer in relation to the Area—these could be extended to all of ABNJ—while the Intergovernmental Oceanographic Commission (IOC) has developed non-legally binding Criteria and Guidelines on Transfer of Marine Technology (2003). UNCLOS (1982) also promotes the strengthening and establishment of national and regional centres for marine science and technology (Articles 275 and 276).

certain activities such as fishing or tourism will be regulated and not necessarily prohibited.

There is currently strong interest in the establishment of multi-purpose MPAs in ABNJ, i.e. MPAs which aim to regulate a large variety of human activities with the ultimate objective to address different, cumulative pressures on marine biodiversity. However, there is no global mechanism for the establishment of such multi-purpose or multi-sectoral MPAs. Instead, the prevailing approach to conservation and sustainable use at the international level is sectoral.

Several international and regional organisations are already able to establish what can be called ‘sectoral MPAs’ or ‘area-based management tools’ in ABNJ, e.g.:

- The International Maritime Organization (IMO) can designate Particularly Sensitive Sea Areas (PSSA) to protect areas that, for recognised ecological, socio-economic or scientific reasons, may be vulnerable to damage by international shipping activities (IMO 2005). No PSSAs have been designated in ABNJ to date.
- The International Seabed Authority (ISA) can designate Areas of Particular Environmental Interest (APEI) and preservation reference zones to address impacts from deep seabed mining activities (ISA 2013). The ISA has declared 9 APEIs in the Clarion-Clipperton Zone (North Central Pacific) (ISA 2012).
- Regional Fisheries Management Organisations (RFMOs) can designate closures of certain fisheries to protect or restore the stocks they manage, or to protect the vulnerable marine ecosystems (VMEs) located on the seabed (pursuant to relevant UNGA resolutions, in particular UN 2006). Approximately 30 such closures have been made in the North-East Atlantic, North-West Atlantic, and South-East Atlantic (Wright et al. 2014a, b).

In parallel, a scientific process has been ongoing under the auspices of the CBD to identify ecologically or biologically significant marine areas (EBSAs). However identification of an EBSA does not have any immediate legal effect, and the management of these marine areas remains in the hands of the competent authorities.

While a number of international agreements and institutions thus have mandates to establish area-based management measures in ABNJ, there is currently no institution with an explicit mandate to establish cross-sectoral MPAs, nor a global procedure to bring together the various organisations and coordinate their activities in order to achieve this. The need to establish an international framework for the creation and management of multi-sectoral, well connected and internationally-recognised MPAs in ABNJ has been considered in detail within the BBNJ Working Group (Druel et al. 2013).

In order to consider what the substantive content of a new UNCLOS IA may be in relation to MPAs, it is helpful to think of the establishment of an MPA as a process requiring a number of steps to be taken. These include: (i) the description of a suitable area according to determined scientific criteria; (ii) the proposal of an MPA; (iii) official designation by a competent authority; and (iv) the adoption of a management plan and management measures aimed at meeting the objectives of the MPA.

A number of sets of scientific criteria for identifying MPAs, or similar areas, have already been developed, such as EBSAs, VMEs, and PSSAs, mentioned above. A new IA could utilise any one of these approaches, establish a new set of criteria inspired by them, or both. There is even the possibility that the criteria could go beyond merely scientific factors so as to include areas of socio-economic, cultural, and educational importance.

This is already the case for PSSAs. The IMO criteria for identification of PSSAs lists 'social, cultural, and economic criteria' and 'scientific and educational criteria' as two of the three categories for designation of a PSSA. An area can be designated as a PSSA on the basis of one criterion alone (IMO 2005 Sect. 44).

As to the manner in which an MPA might be proposed, options include proposal by one or a number of States, by a specific body convened under the auspices of the IA, or by NGOs or organisations with State support. Provision may be needed to ensure that a dedicated scientific body considers proposals and that they are officially endorsed by a Conference of the Parties (COP) or relevant organisational meeting.

There are also many potential structures that could be implemented for the adoption of management plans and management measures for meeting the objectives of an MPA. Indeed, adoption of a management plan may not even be necessary; the focus being placed instead on the adoption of specific management measures. Alternatively, a proponent may be required to submit a management plan when proposing an MPA, or one could be subsequently developed and adopted by an organ of the IA, or a regional organization.

In any event, management measures will be an essential part of ensuring the effectiveness of the MPA and mechanisms for their adoption will be needed. Such mechanisms could include proposal along with the MPA, or development by States cooperating directly and through competent international, regional, and sectoral organisations. To this end, regional working groups or advisory bodies could be established to bring together States, competent organisations, scientists, and other stakeholders in order to consider the management of MPAs in a given region.

3.2.2 Marine Genetic Resources

The use of MGRs found in ABNJ is currently not specifically regulated by an international legal instrument. The international regime for access to genetic resources and the sharing of benefits from their utilization (ABS) established by the CBD and its Nagoya Protocol is limited to genetic resources over which States have sovereign rights, thus not applicable to MGRs from ABNJ.[8] MGRs are also not

[8]Article 15.1 of the CBD and Article 3 of the Nagoya Protocol on Access to Genetic Resources and the Fair and Equitable Sharing of Benefits Arising from their Utilization (Nagoya Protocol).

mentioned in UNCLOS (1982), and the 'resources' regulated by Part XI are explicitly defined as mineral (non-living) resources, thus excluding MGRs.[9]

This leaves a situation where either full implementation of UNGA Resolution 2749 is still pending, according to which all resources from the Area including also MGRs should be considered as Common Heritage of Mankind (CHM);[10] or where the freedom of the high seas principle applies to MGRs.[11] Regardless of the approach taken, and the possible monetary benefit-sharing obligations this may or may not entail, use of MGRs remains subject to provisions on Marine Scientific Research (MSR). While these MSR provisions already include non-monetary benefit-sharing obligations,[12] no mechanism currently exists to ensure their complete and coherent implementation.

Again there are several potential structures that could be implemented for the adoption of an ABS regime for ABNJ. For example, the existing multilateral ABS approach under the FAO International Treaty on Plant Genetic Resources for Food and Agriculture (ITPGRFA) provides some ideas that could be adjusted to ABNJ. The ITPGRFA establishes a common pool of resources designed to facilitate access to genetic resources (and thereby further research), and to ensure the benefits derived from their use are fairly and equitably shared based on previously agreed terms.

A common pool could also be created comprising MGRs from ABNJ, drawing on established elements of the ITPGRFA's multilateral ABS process; in particular the development of standard material transfer agreements, differentiated and flexible access rights and benefit-sharing obligations, and the regulation of intellectual property rights. Regardless of the particular structure, approach and ideas followed, an ABS regime will have to be based on three main pillars: access to the resources, fair and equitable sharing of benefits, and compliance.

In terms of regulating access to MGRs, a distinction is generally made between in situ, ex situ, and in silico access. In situ refers to samples of MGRs collected in their natural setting, while ex situ *refers* to samples previously collected in ABNJ and subsequently stored in 'biorepositories'. In silico refers to access to any knowledge associated with the MGRs, such as observational or experimental data and other findings.

The regulation of in situ access raises questions of geographic scope, as well as sustainability: sampling takes place in both the 'Area' but also in the water column. Some of the resources are even 'transboundary', i.e. existing in and migrating between both maritime areas. To avoid loopholes MGRs from both spaces should be covered by an ABS system. Furthermore, the question of sustainability of access

---

[9]Article 133(a) UNCLOS.

[10]Resolution 2749 (XXV): Declaration of Principles Governing the Sea-Bed and the Ocean Floor, and the Subsoil Thereof, beyond the Limits of National Jurisdiction, 12 December 1970, A/RES/25/2749. In contrast to the Part XI regime of UNCLOS, Resolution 2749 does not define the term 'resources of the Area' and thus does not limit it only to minerals/non-living resources.

[11]Article 87–88 UNCLOS.

[12]Articles 242, 244 and Articles 143.3, 144.2 UNCLOS.

should not be neglected. While the environmental impacts of collecting MGRs are most of the time minimal or non-existing, ultimately this is highly site-specific, depending on the fragility of the area, the amount of resources taken and the frequency of taking, as well as the sampling techniques used and the standards applied by the scientists.

In contrast, the regulation of ex situ and in silico access will become a question to be addressed in the context of benefit-sharing, as facilitating such access will provide a clear benefit for the international scientific community, while also promoting further scientific research. Two broad categories of benefits are usually distinguished: monetary and non-monetary.[13] The high cost of obtaining MGRs in ABNJ and the long route to developing a commercial product (Broggiato et al. 2014, p. 177) mean that the most secure and direct benefits that can be shared are non-monetary. As mentioned before, the UNCLOS (1982) provisions related to MSR already envisage international cooperation in MSR,[14] publication and dissemination of results,[15] and promoting data flow and knowledge transfer.[16] These basic provisions under UNCLOS (1982) could provide the starting point for the further development of non-monetary benefit-sharing obligations.

Though less common, monetary benefits may also flow from the development of MGRs into commercially viable products, and in particular developing states will be keen to ensure that these benefits are distributed appropriately. Key questions here concern the trigger for monetary benefit sharing, and the blurred distinction between commercial and non-commercial research and development. An upfront payment for access could be appropriate where there is a clear commercial intent, but it would further increase the costs of non-commercial research and thus create a financial problem for such activities. In practice, sampling cruises in ABNJ tend to be non-commercial, or at least their objectives are not solely or primarily commercial. This makes them difficult to distinguish and therefore difficult to ensure that the appropriate remunerations are sought at the point of access.

As an alternative to upfront payments, a multilateral ABS system under a new UNCLOS IA could provide for payments at various stages along the chain of MGR research and development. Payment could become due upon reaching certain milestones (e.g. an exclusivity fee when an intellectual property right is granted), or when a commercial product is created and sold. At the same time, fees could be charged to acquire MGR samples from ex situ collections, or for access to in silico knowledge.

Some form of trust fund for ABNJ could be established to administer the monetary benefits on behalf of the international community. These resources could be used to support further non-monetary benefit-sharing (e.g. capacity-building and

[13]See also the Annex to the Nagoya Protocol and its indicative lists of monetary and non-monetary benefits.

[14]Articles 242 and 143.3(a).

[15]Articles 244.1 and 143.3(c).

[16]Articles 244.2 and 144.2.

technology transfer)). They could also be used to support activities related to conservation and sustainable use of marine biodiversity in ABNJ, thereby linking a benefit-sharing regime with the other elements of the Package Deal.

3.2.3 Environmental Impact Assessment

EIA is a key tool of environmental law globally, and its application to activities in the marine environment has been endorsed by many international legal instruments and policy documents (Morgan 2012). Although an obligation to conduct environmental assessment is well established in both customary and conventional international law, including the obligation in Article 206 of UNCLOS (1982), implementation in ABNJ is fragmented across different sectoral and regional bodies. There is no overarching international process or agreement for the implementation of EIA in ABNJ: a few Regional Seas programs have specific environmental protection responsibilities for ABNJ,[17] while the ISA has comprehensive environmental protection powers for seabed mining activities affecting the Area, but not in relation to other activities or parts of the marine environment. There are also no mandatory EIA regulations at the international level to govern new or emerging activities such as geo-engineering and sampling of MGRs.

This lack of an integrated system of environmental governance for ABNJ presents a challenge for effectively implementing EIA in these vast expanses of ABNJ. The predominant form of jurisdiction in ABNJ is flag State jurisdiction; for shipping and fishing vessels in ABNJ, responsibility for enforcement falls largely to flag States, rather than any global body. This results in varying levels of compliance with environmental standards, and a lack of auditing and sanctions for those falling short. Many stages in an EIA process require a strong lead agency to play a coordinating role, but this is lacking in the fragmentary system of governance applicable to most activities in ABNJ. These stages include the initial screening process (selecting which activities are subject to EIA), the scoping process (deciding the terms of reference for an EIA), the public notification and consultation process, and ongoing monitoring of impacts.

Including EIA within an UNCLOS IA can provide best practice standards for EIA in ABNJ, setting out an EIA process that is biodiversity inclusive, transparent and subject to international scrutiny, with associated powers to impose conditions on any activities that may negatively impact marine ecosystems in ABNJ.

[17]The scope of application of the 1986 Convention for the Protection of the Natural Resources and Environment of the South Pacific Region (Noumea Convention), the 1992 Convention for the Protection of the Marine Environment of the North-east Atlantic (OSPAR Convention) and the 1995 Convention for the Protection of the Marine Environment and the Coastal Region of the Mediterranean (Barcelona Convention) extend to ABNJ.

4 Next Steps

Based on the recommendations made by the BBNJ Working Group in January 2015 (UNGA 2015) the process for the development of a new UNCLOS IA will take a two-step approach:

- A preparatory commission ("PrepCom") will first be established and work in 2016 and 2017 in order to make substantive recommendations on the elements of a draft text. The PrepCom will report to the UNGA by the end of 2017;
- Before the end of its 72nd session (i.e. September 2018), the UNGA will decide on the convening and on the starting date of an intergovernmental conference to consider the recommendations of the PrepCom and elaborate the new instrument.

Importantly, the PrepCom will be open not only to Member States of the United Nations, but also to "members of specialised agencies (…) and others invited as observers" in accordance with past practice (UNGA 2015, §5a). The process will therefore be transparent and authorise the participation of the civil society.

Even though consensus was reached on the opening of negotiations, a few States, primarily the US, Canada, Japan and Russia, remain "unconvinced" on the absolute need to elaborate a new instrument (IISD 2015). They therefore may continue to express their doubts and concerns during the PrepCom meetings, slowing down the process by returning to the perennial debate on whether or not there are gaps in the current legal framework.

Moreover, the explicit reference to the elements of the Package Deal does not mean that delegations share a common vision on the content of the future instrument. Negotiations on MGRs, particularly on an ABS mechanism, are likely to be complex. Likewise, though States agreed not to "undermine existing relevant legal instruments and frameworks and relevant global, regional and sectoral bodies" (UNGA 2015, §7), the practical ways of integrating biodiversity conservation and enhancing coordination through the future instrument with the existing legal and institutional framework will be one of the key challenges.

5 Liability and Compensation for Pollution Damage Resulting from Offshore Oil and Gas Activities

Another gap in the international legal framework relates to offshore oil and gas activities, which have developed considerably in recent decades. Due to increasing energy demand and technological advancements, drilling has moved further into deep and ultra-deep areas. Today, almost a third of the oil and a quarter of the natural gas consumed in the world come from offshore sources, and forecasts predict continued growth for the foreseeable future (Pike 2013). Intensification of offshore oil and gas exploitation means increasing threats to marine ecosystems, as well as potential consequences for the human activities dependant on those ecosystems.

Recent disasters have demonstrated that the environmental risks of offshore drilling are widespread, affecting all regions and all types of company. As such disasters have had transboundary impacts, discussions have recently resurfaced on the suitability of the current international framework for the regulation of offshore oil and gas activities (Rochette et al. 2014a, b).

In this regard, it is clear that there is a regulatory gap, in that there are currently no global rules regulating liability and compensation for pollution damage resulting from offshore drilling activities (Rares 2011). Regional agreements have not addressed this issue and are very limited (in the Mediterranean and the European Union, for instance), while the offshore pollution liability agreement (OPOL), a private regime, is limited in its geographical coverage and capped at a rather low level with regard to the compensation of damages. As a result, "it is left to national laws to deal with this matter. Such laws vary enormously both in the way that the law itself deals with it and with the way contractual indemnities are interpreted and enforced, or not as the case may be" (Cameron 2012, p. 211).

An Indonesian proposal to elaborate a specific international convention was considered by the IMO Legal Committee from 2010 onwards, however there is no evidence of the necessary political will amongst other States. Reluctant States reject the idea of a global regulation and argue in favour of regional regulation.

Several risks can be highlighted if the legal *status quo* remains: (i) a risk of legal uncertainty and therefore a risk of political dispute between States; (ii) a risk of partial or nonpayment of damages because of the absence of clear rules; (iii) a risk of insolvency: indeed, "the international oil industry is now populated with a combination of big oil companies such as BP and ExxonMobil, medium to large oil companies such as Anadarko and many National Oil Companies, and numerous "new entrant" companies, including service companies, which certainly do not have the access to capital to pay the kind of large claims which BP faced following the Macondo oil spill" (Cameron 2012, p. 213).

It is therefore urgent to fill the regulatory gaps. In this regard, a potential global agreement on liability and compensation would implement the polluter-pays principle. It could be based on the following elements:

- **Type of damage covered**. The loss and damage covered by the regime should be as broad as possible and include, beyond economic losses, the ecological damage. "Without defining pollution damage to include these non-economic, abstract claims, [a treaty] regulating oil pollution from fixed platforms will still fail to leave victims fully recovered after massive oil disasters" (Smith 2011).
- **Strict liability**.[18] A future regime should be based on the strict liability of operators, for three main reasons. First, it is the most pragmatic regime: "strict liability would avoid argument about whether some other criterion of

[18]Strict liability holds a company liable for all damages, regardless of whether or not they arise from activities carried out within the permit provided.

responsibility, such as negligence or other fault, has occurred before someone is required to pay compensation. Strict liability offers certainty both in fixing immediate responsibility on an identified person to pay compensation as soon as a casualty occurs and, generally, in identifying what is payable" (Rares 2011). Second, "the various sources of customary international law reflect the emergence of the doctrine of strict liability and support its application in transnational offshore oil accidents" (Cates 1984). Third, it is the regime adopted in many national legal systems for abnormally dangerous activity.

- **Joint liability**. Liability should be shared between all license holders and their subcontractors. Indeed, "there will always be a risk that insurance, bank guarantees, or protection and indemnity arrangements may fail to respond, due to the insolvency of the person with the obligation to indemnify the controller. Thus, a wider range of persons involved in the ownership operation or control of an off-shore rig should be made responsible" (Rares 2011).
- **Financial capacity of operators**. States should ensure that operators have adequate financial capacity to pay possible compensation. To that purpose, compulsory third party insurance should become a necessary requirement for all companies.
- **Liability cap**. A cap may be set on the level of compensation, in order to balance the strict liability regime. However, such cap must be set at a level that can ensure the recovery of costs associated with environmental remediation and compensation and losses born by public and private entities. It also needs to take into account lessons learned from the level of costs incurred by recent accidents as specified by activity and/or technology and the sensitivity of the environment and ecosystem services it provides. Furthermore, a compensation fund, either publicly or privately funded, or both, could be set up to address major disasters that are likely to exceed the liability cap.
- **Judicial settlement**. First, any regime should allow the widest range of persons and States affected by pollution damage to make claims for compensation. Second, as with the Bunker Oil Convention and International Convention on Civil Liability for Oil Pollution Damage (UN 1969), jurisdiction should be given to the courts of any State Party in which the damage occurred and judgments given by that courts should be recognised by the courts of other States Parties.

6 Conclusion

The discussion above highlights two of the most important remaining legal gaps in the international framework for sustainable development in the marine environment: conservation and sustainable use of marine biodiversity in ABNJ and liability and compensation for oil pollution damage resulting from offshore oil and gas activities.

The consensus reached within the last BBNJ Working Group meeting is undoubtedly a historical one, paving the way for a global legally-binding instrument specifically dedicated to filling the governance and regulatory gaps in ABNJ. Nonetheless, pitfalls and challenges remain in the road ahead, and these will need to be carefully navigated. The opening of the negotiations for an international instrument does not diminish the need to advance sectoral and regional initiatives to conserve and sustainably use marine biodiversity in ABNJ. These must be strengthened hand in hand with the development of the new agreement as any new IA will depend on strong and well-coordinated action at multiple levels (Rochette et al. 2014a, b).

As for offshore oil and gas, this sector remains, for the moment, one of the least regulated maritime industries. Given the current growth of offshore activities and the recent accidents that highlighted considerable risks to the environment, it is time to reconsider international regulations in this area.

References

Boyd P (2013) Ocean fertilization for sequestration of carbon dioxide from the atmosphere. In: Lenton T, Vaughan N (eds) Geoengineering responses to climate change SE, vol 5. Springer, New York, pp 53–72. Retrieved 23 July 2015, from http://dx.doi.org/10.1007/978-1-4614-5770-1_5

Broggiato A (2013) Exploration and exploitation of marine genetic resources in areas beyond national jurisdiction and environmental impact assessment 429(178):237–241

Broggiato A, Arnaud-Haond S, Chiarolla C, Greiber T (2014) Fair and equitable sharing of benefits from the utilization of marine genetic resources in areas beyond national jurisdiction: bridging the gaps between science and policy. Mar Policy 49:176–185. Retrieved 23 July 2015, from http://doi.org/10.1016/j.marpol.2014.02.012

Cameron P (2012) Liability for catastrophic risk in the oil and gas industry. Int Energy Law Rev 6:207–219

Cates M (1984) Offshore oil platforms which pollute the marine environment: a proposal for an international treaty imposing strict liability. San Diego Law Rev 21:691–708

CBD (1992) Convention on Biological Diversity of 5th June 1992, 1760 UNTS 79. Retrieved 23 July 2015, from https://www.cbd.int/doc/legal/cbd-en.pdf

CBD Secretariat (2014) Global Biodiversity Outlook 4. Montréal

Census of Marine Life (2011) Scientific results to support the sustainable use and conservation of marine life—a summary of the census of marine life for decision makers

Charter R (2007) Life on the edge: the industrialization of our oceans. In: Proceedings of coastal zone 07, Portland, Oregon, 22–26 July 2007

Druel E (2011) Marine protected areas in areas beyond national jurisdiction: the state of play IDDRI working paper 7/2011. Retrieved 23 July 2015, from http://www.iddri.org/Publications/Marine-protected-areas-in-areas-beyond-national-jurisdiction-The-state-of-play

Druel E, Rochette J, Billé R, Chiarolla C (2013) A long and winding road: international discussions on the governance of marine biodiversity in areas beyond national jurisdiction, IDDRI study 07/13, IDDRI, Paris. Retrieved 23 July 2015, from http://www.iddri.org/Publications/A-long-and-winding-road

European Commission (2012) Blue growth: opportunities for marine and maritime sustainable growth, European Commission, Luxemburg. Retrieved 23 July 2015, from http://ec.europa.eu/maritimeaffairs/policy/blue_growth/documents/com_2012_494_en.pdf

FAO Fisheries and Aquaculture Department (2014) The state of world fisheries and aquaculture 2014. FAO, Rome
IISD (2015) Summary of the 9th meeting of the working group on marine biodiversity beyond areas of national jurisdiction, 20–23 January 2015. Retrieved 23 July 2015, from http://www.iisd.ca/vol25/enb2594e.html
IMO (2005) Revised guidelines for the identification and designation of Particularly Sensitive Sea Areas A.982(24)
IOC (2003) Advisory body of experts on the law of the sea, IOC criteria and guidelines on transfer of marine technology. Retrieved 12 October 2015, from http://unesdoc.unesco.org/images/0013/001391/139193m.pdf
ISA (2012) Decision of the Council relating to an environmental management plan for the Clarion-Clipperton Zone, ISBA/18C/22. Retrieved 23 July 2015, from http://www.isa.org.jm/files/documents/ EN/18Sess/Council/ISBA-18C-22.pdf
ISA (2013) Decision of the Council of the International Seabed Authority relating to amendments to the regulations on prospecting and exploration for polymetallic nodules in the area and related matters, ISBA/19/C/17
IUCN and UNEP-WCMC (2013) The World Database on Protected Areas (WDPA). Retrieved 23 July 2015, from http://www.protectplanetocean.org/official_mpa_map
Leary D (2011) Marine genetic resources: the patentability of living organisms and biodiversity conservation. In: Jacquet LTP, Pachauri RK (eds) Oceans: the new frontier—a planet for life 2011, TERI Press, pp 183–193
London Convention (1972) Convention on the prevention of marine pollution by dumping of wastes and other matter 1972, 1046 UNTS 120. Retrieved 23 July 2015, from http://www.austlii.edu.au/au/other/dfat/treaties/1985/16.html
MARPOL (1973) International Convention for the Prevention of Pollution from Ships 1973, MARPOL 73/78, 1340 UNTS 184. Retrieved 23 July 2015, from http://www.jus.uio.no/english/services/library/treaties/06/6-05/ships-pollution.xml
Molenaar EJ, Elferink AGO (2009) Marine protected areas in areas beyond national jurisdiction: the pioneering efforts under the OSPAR convention. Utrecht Law Rev 5(1):5–20
Morgan RK (2012) Environmental impact assessment: the state of the art. Impact Assess Proj Appraisal 30(1):5–14. Retrieved 23 July 2015, from http://www.tandfonline.com/doi/abs/10.1080/14615517.2012.661557. Accessed 3 Apr 2014
Pike WJ (2013) High crude oil prices sustain stable production, spur new discoveries. World Oil Mag 234:1. Retrieved 23 July 2015, from http://www.worldoil.com/magazine/2013/january-2013/features/deepwater-high-crude-oil-prices-sustain-stable-production-spur-new-discoveries
Rares S (2011) The need for an international convention to deal with offshore hydrocarbon leaks. Paper presented at the 2011 Biennial mini conference of the maritime law association of Australia and New Zealand (NSW branch), Lilianfels, Katoomba on 11 Mar 2011
Rayfuse R, Lawrence MG, Gjerde KM (2008) Ocean fertilisation and climate change: the need to regulate emerging high seas uses. Int J Marine Coast Law 23(2):297–326. Retrieved 23 July 2015, from http://openurl.ingenta.com/content/xref?genre=article&issn=0927-3522&volume=23&issue=2&spage=297. Accessed 1 Mar 2012
REN21 (2013) Renewables 2013 global status report, REN21, Paris
Rochette J, Wemaëre M, Chabason L, Callet S (2014a) Seeing beyond the horizon for deepwater oil and gas: strengthening the international regulation of offshore exploration and exploitation, IDDRI Study 01/14, IDDRI, Paris. Retrieved 23 July 2015, from http://www.iddri.org/Publications/En-finir-avec-le-bleu-petrole-Pour-une-meilleure-regulation-des-activites-petrolieres-et-gazieres-offshore
Rochette J, Unger S, Herr D, Johnson D, Nakamura T, Packeiser T, Proelss A, Visbeck M, Wright A, Cebrian D (2014b) The regional approach to the conservation and sustainable use of marine biodiversity in areas beyond national jurisdiction. Marine Policy Spec Segment 49:109–117

Rochette J, Wright G, Gjerde K, Greiber T, Unger S, Spadone A (2015) A new chapter for the high seas? Historic decision to negotiate an international legally binding instrument on the conservation and sustainable use of marine biodiversity in areas beyond national jurisdiction, IDDRI issue brief 02/15, IDDRI, Paris. Retrieved 23 July 2015, from http://www.iddri.org/Publications/A-new-chapter-for-the-high-seas

Salcido RE (2008) Offshore federalism and ocean industrialization. Tullane Law Rev 82:1355–1445

Smith M (2011) The Deepwater horizon disaster: an examination if the spill's impact on the gap in international regulation of oil pollution from fixes platforms. Emory Int Law Rev 25 (3):477–1516

UN (1969) International Convention on Civil Liability for oil pollution damage (CLC) 1969. http://www.imo.org/en/About/Conventions/ListOfConventions/Pages/International-Convention-on-Civil-Liability-for-Oil-Pollution-Damage-(CLC).aspx

UN (2001) Agreement for the implementation of the provisions of the convention on the law of the sea of December 1982 relating to the conservation and management of straddling fish stocks and highly migratory fish stocks, entered into force on 11th December 2001, 2167 UNTS 88. Retrieved 23 July 2015, from http://daccess-dds-ny.un.org/doc/UNDOC/GEN/N95/274/67/PDF/N9527467.pdf?OpenElement

UN (2006) Resolution 61/105: sustainable fisheries, including through the 1995 agreement for the implementation of the provisions of the United Nations convention on the law of the sea of 10 December 1982 relating to the conservation and management of straddling fish stocks and highly migratory fish stocks, and related instruments, A/RES/61/105

UN (2012) The future we want, A/CONF.216/L.1, Rio de Janeiro, Brazil, 19 June 2012

UNCLOS (1982) United Nations Convention on the Law of the Sea (LOSC). 10 Dec 1982, entered into force 16 Nov 1994, 1833 UNTS 3

UNGA (2010) UNGA resolution 65/37. Oceans and the law of the sea, UN doc, A/RES/65/37, of 7 December 2010 [Preambular Paragraph 4]. Retrieved 23 July 2015, from http://daccess-dds-ny.un.org/doc/UNDOC/GEN/N10/514/76/PDF/N1051476.pdf?OpenElement

UNGA (2014) See United Nations 2014 open working group proposal for sustainable development goals. Retrieved 23 July 2015, from http://sustainabledevelopment.un.org/?menu=1300

UNGA (2015) Recommendations of the ad hoc open-ended informal working group to study issues relating to the conservation and sustainable use of marine biological diversity beyond areas of national jurisdiction to the 69th session of the general assembly, A/69/780, 23 Jan 2015

Wright G (2015) Marine governance in an industrialised ocean: a case study of the emerging marine renewable energy industry. Mar Policy 52:77–84. Retrieved 23 July 2015, from http://linkinghub.elsevier.com/retrieve/pii/S0308597X14002838

Wright G, Ardron J, Gjerde K, Rochette J (2014a). Advancing marine biodiversity protection through regional fisheries management: a review of high seas bottom fisheries closures, IDDRI working paper 14/14, IDDRI, Paris. Retrieved 23 July 2015, from http://www.iddri.org/Publications/Advancing-marine-biodiversity-protection-through-regional-fisheries-management-a-review-of-high-seas-bottom-fisheries-closures

Wright G, Rochette J, Unger S, Gjerde K, Ardron J (2014b) The scores at half time: an update on the international discussions on the governance of marine biodiversity in areas beyond national jurisdiction, IDDRI issue brief 02/14, IDDRI, Paris. Retrieved 23 July 2015, from http://www.iddri.org/Publications/The-Scores-at-Half-Time-An-update-on-the-international discussions-on-the-governance-of-marine-biodiversity-in-areas-beyond-nati

Knowledge in Sustainable Resource Management in Australia

Judith Preston

Abstract Sustainable development (SD) balances conservation and exploitation of natural resources. The concept mandates that biodiversity conservation and ecological integrity are fundamental considerations in environmental decision-making. Loss of biodiversity is a serious global issue and is one of Australia's most pressing environmental problems. Increasing recognition of the role of indigenous knowledge (IK) is part of effective solutions to biodiversity loss and ways to achieve benefits for industry and agriculture. Research on IK provides information for understanding the role and importance of customary livelihoods within SD. The importance of combining indigenous and non-indigenous knowledge for conservation and resource management is reflected in the adoption of international principles and instruments which link SD, biodiversity conservation and poverty alleviation. There are a number of innovative tools being developed to embed IK principles into the legal framework for natural and cultural resource management. In Australia and globally, the focus remains on legal protection for IK through intellectual property, contract and trade practices law. There are important policies, programs and projects that value and use IK in Australia but they are piecemeal and generally unsupported by law. More comprehensive constitutionally-based and specific purpose-built legislative models are being developed in Europe, Asia and South America to incorporate IK into environmental decision-making which could be used as a blueprint for Australia. These legislative models should include provisions for the custodians of that IK, to have efficient and inexpensive access to courts and tribunals to enforce duties and protect rights in relation to sustainable resource management.

Keywords Indigenous knowledge · Sustainable development · Traditional knowledge · Indigenous protected areas · Environmental decision-making

J. Preston (✉)
The University of Western Sydney (UWS), Sydney, Australia
e-mail: judith.preston@mq.edu.au

V. Mauerhofer (ed.), *Legal Aspects of Sustainable Development*,
DOI 10.1007/978-3-319-26021-1_28

1 Introduction

The concept of sustainable development is one that seeks to balance socio-economic development with protecting Earth's ecosystems and its services for present and future needs. International assessments of the health of the world's biodiversity and ecosystems such as by the International Union for Conservation of Nature (IUCN) and the United Nations Educational, Scientific and Cultural Organisation (UNESCO) reveal that the application of sustainability has not achieved an effective balance between development and environmental protection. This leads to a state beyond Earth's tipping point where consumption of natural resources exceeds the sustainable limit of the natural ecosystem services. Lester Brown (2006) argues that the first signs of economic collapse appear in the environment due to factors such as over-consumption of natural capital, pollution, use of non-renewable fuels and over-population. Maffi and Woodley (2010, p. 193) have recognised that culture, cultural diversity and its concomitant knowledge is important for sustainability paving the way for humanity to maintain and extend the quality of life for humans and other species.

This chapter intends to demonstrate how indigenous knowledge ('IK') can be more effectively incorporated into legal and policy frameworks in Australia for sustainable development. Indigenous Protected Areas ('IPAs') will be used as an example of how IK is practically and effectively applying sustainable development objectives and can be part of a solution to rebalancing the tipping point. The chapter will firstly overview the concept of sustainable development and briefly explain how it has been incorporated into Australian legal and policy frameworks. Secondly, selected Indigenous Protected Areas will be used as case studies to demonstrate how this conservation model can effectively use IK for sustainable development objectives.

2 The Concept of Sustainable Development

The genesis of the concept of sustainable development is found in the first major international environmental conference convened by the United Nations General Assembly in December 1968, in Stockholm, Sweden ('The Stockholm Conference'). An important outcome of the Stockholm Conference was the connection between socio-economic development and environmental protection.

> Since that time, the inseparability of environment and development has become an increasingly important theme of international law … (Godden and Peel 2010, p. 42).

In 1987, a report entitled "Our Common Future" ('the Report') produced by the World Commission on Environment and Development ('the Commission') assisted in efforts of the international community to link issues of environment and development (World Commission on Environment and Development 1987, p. 43).[1] This report identified three main factors in sustainability: environment, society and economy.

Sustainable development was defined in the Report as "development that meets the needs of the present without compromising the ability of future generations to meet their own needs". Maffi notes that the sustainable development concept inspired a wide push for "a three-pillar" form of balance between economy, society and environment (Maffi and Woodley 2010, p. 191). She argues that there is sufficient evidence to include culture as another basic consideration of sustainability. Moreover, the concept itself has become obscure or skewed and not as useful as it was originally intended (Maffi and Woodley 2010, p. 192). Maffi and Woodley consider that sustainable development requires a biocultural dimension, "sustaining life", therefore comes to mean "sustaining life in nature and culture—the biocultural diversity of life" (Maffi and Woodley 2010, p. 193). Consequently "to achieve sustainability we need to explicitly incorporate an expanded understanding of 'diversity of life' in this definition [of sustainability]" (Maffi and Woodley 2010, p. 193). This has been addressed to an extent in the outcomes of the 2012 RIO+20 conference (United Nations 2012, p. 11, para 58[j]).

This involves public and private partnerships to achieve genuine, rights-based, equal and equitable collaborations between conservationists and indigenous and local communities—"[to] fully recognise the interdependence of biological and cultural diversity, and deploy all means necessary to support both" (Maffi and Woodley 2010, p. 195). These partnerships would necessarily require the involvement of development agencies, corporations and individuals to achieve a balanced approach to sustainable development. Maffi and Woodley (2010, p. 193) observe that the concept of resilience is also important in thinking about sustainability:

> … when we refer to resilience we should be thinking not only of the ecological resilience of the biosphere … but also of the cultural resilience of the 'ethnosphere' and the 'logosphere'—the planetary webs of people and language…and the reference of the very interconnectedness among all three…This realisation must now become central to the agenda of sustainability.

Losses in any part of the integrated biocultural ecosystem have effects in other parts of that system—"a loss of resilience anywhere in an integrated biocultural system is likely to contribute to a loss of resilience elsewhere" (Maffi and Woodley 2010, p. 193).

Recognition of the connections within the ecosystem and the need to support the resilience in each part is acknowledged at the international level. A report by the International Fund for Agricultural Development ('IFAD'), entitled "Indigenous People and Sustainable Development" (IFAD 2003, p. 8), notes:

[1]See World Commission on the Environment and Development (1987, p. 41) for a definition of "sustainable development".

> [c]ulturally sensitive approaches to addressing the specific needs of indigenous peoples have shown that strengthening cultural identity and promoting sustainable socio-economic development are mutually reinforcing, rather than mutually exclusive objectives.

The World Bank has been funding projects to integrate IK for development goals and helping to achieve Millennium Development Goals which include poverty alleviation and environmental sustainability. One example is rural communities in Mozambique that have addressed coastal forest exploitation using "myths and traditional rites maintaining these resources for the next generations" (The World Bank 2004, p. 51).

The next section considers the development of the concept of sustainable development in Australia.

3 Sustainable Development in Australia

The tension between socio-economic development and environmental protection has played a pivotal role in international environmental law since the 1970s. This is due in part to concerns of developing countries that environmental protection should not place unreasonable impediments on their right to develop. International policy documents such as the World Conservation Strategy promoting a balance of development with environmental protection inspired Australian public policy. The Federal Government worked proactively "to influence arguments about the environment and the economy" (Papadakis 1993, p. 132 as cited in Godden and Peel 2010, p. 135). Working groups were established to develop the concept of sustainable development which was originally known in Australia as "ecologically sustainable development" (ESD).

The concept was considered in detail and later adopted in the National Strategy for Ecologically Sustainable Development—NSESD (Department of Environment 1992). Godden and Peel (2010, p. 135) observe that ESD was readily accepted in Australia as it was consistent with "dominant understandings of the environment drawn from philosophical, scientific and economic traditions". Further, ESD had the potential to incorporate responsibilities such as to future generations, and integrative environmental concepts such as those reflected in IK (Godden and Peel 2010, p. 136).

The five working groups produced a number of recommendations which culminated in several final reports including the National Strategy for the Conservation of Australia's Biological Diversity (ANZECC 1996). The recommendations of the working groups formed the basis for the NSESD which was accepted by the Council of Australia Governments in December 1992. By its acceptance by the Council of Australian government from the federal to the local level, NSESD became the central policy document concerning ESD in Australia (Godden and Peel 2010, p. 136). Other reports were Australia's Ocean Policy (Commonwealth of

Australia 1998a) and the National Greenhouse Strategy (Commonwealth of Australia 1998b).

The concept of ESD is that which improves the quality of life in a way that maintains ecological processes. Godden and Peel (2010, p. 136) explain that the function of ESD is to provide for specific environmental rules and a framework for decision-making that balances environmental and development considerations. A number of principles reflecting those found in the United Nations Convention on Biological Diversity underpin the goal of ESD. These principles include balanced and integrative decision-making, the precautionary principle, intergenerational equity, biodiversity conservation, improved valuation and pricing including the polluter pays principle, and the promotion of public participation in governance. Decision-making conducted on the basis of ESD is a holistic consideration of short and long term impacts of proposed developments, takes into account scientific uncertainty, and encourages public participation (Godden and Peel 2010, p. 137).

4 Regulatory and Legal Implementation of ESD

The 1960s and 1970s for Australia, and the United States of America particularly, were watershed periods for environmental law and policy "in response to an upsize in public interest and concern about environmental issues" (Macintosh 2015, p. 3). Macintosh (2015, p. 6) comments that "[Australia's] adoption of ESD led to a rebranding of pre-existing institutions and provided a rhetorical tool to attract support for their reform". Public concern about the effects of modern life on the environment together with the outcome of the Stockholm Conference stimulated the Australian governments to make changes to environmental laws and policies including more rigorous environmental impact assessment regimes for proposed developments. Natural resource management governance was strengthened and native plant and animal protection laws reformed. This included designing processes to protect vulnerable species such as listing of threatened species of animals and plants and recovery plans together with environment assessment processes such as bilateral assessment (Macintosh 2015, p. 5). In the present legal and political system, Australia would require substantive regulatory and institutional change to fully incorporate ESD into public decision-making and implementation (Dovers and Connor 2006, pp. 21, 38 as cited in Godden and Peel 2010, p. 137).

In Australia, regulatory action is usually reactive to environmental problems as they arise. Australian environmental legislation distinguishes legal responsibility for different environmental issues to separate government departments. ESD contemplates designing policies and programmes that are integrated to deal with natural resource management with its environmental challenges. Within environmental legislation, ESD principles are found in objects clauses which relate to a duty for decision-makers to adopt the interpretation that best advances ESD in the decision-making process. Examples include section 5 of the *Environmental Planning and Assessment Act* 1979—NSW) and section 5 of the *Waste Management and Pollution Control Act* 2014 NT). Statutes may also require decision-makers to take ESD into account or apply ESD or its principles. Examples include section 136 of the *Environment Protection and Biodiversity Conservation Act* (1999) (Cth) and section 2A(2) of the *National Parks and Wildlife Act* (1974) (NSW) (Godden and Peel 2010, p. 139) note that "…the tendency has been for ESD to be given effect, not as a goal or outcome but as a consideration which goes into the mix of factors considered in the decision-making process".

The understanding and application of the concept of ESD has been enlivened by the judiciary. Cases like *Walker v Minister for Planning* [2007] NSWLEC 741; (2007) 157 LGERA 124 (Walker), which went on appeal to the NSW Court of Appeal (*Minister for Planning v Walker* [2008] NSWCA 224; (2008) 161 LGERA 423), demonstrate how ESD can become a mandatory consideration for decision-making as it is integral to the public interest. In the primary proceedings in the Land and Environment Court of NSW, Justice Biscoe held that the procedural role of ESD is an integral part of the decision-making process. The legislation encourages decision-makers "to take [ESD] … along with other considerations so as to ensure an environmentally informed decision-making process takes place" (at para. 74).

Preston also notes that "the principles of ESD are relevant matters to be considered in determining applications for approvals to carry out development under [the *Environmental Planning Assessment Act* 1979 (NSW)] … other statutes and statutory provisions" (Preston 2012, p. 7).

Although interpretation and application of ESD through policy, governance and case law are not always clear, Macintosh (2015, p. 11) considers that "while ESD may have been a useful rhetorical tool, its impact on Australia's institutions has been limited". Godden and Peel (2010, p. 143) argue that "ESD is the path on which environmental law and policy in Australia seem committed to travel …".

The next section considers how IK can be useful in achieving sustainable development outcomes for environmental protection.

5 Indigenous Knowledge and Sustainable Development

There are many definitions of IK which have been applied in many settings from anthropology and ecology to legal frameworks. IK includes knowledge, beliefs, traditions, practices, institutions and world views developed and continued by indigenous, and local communities as a result of their interaction with the natural environment. Mead (Tobin 2014, p. 157) explains that "[t]raditional knowledge is the knowledge that we're born with, that we've inherited, that we contribute to in our lifetime and pass onto future generations. Its whole function is survival and the development of a culture of a people".

IK is held and developed by indigenous people who are defined as groups and communities who have occupied land and water from times existing from pre-invasion/pre-colonial societies. These indigenous communities often consider themselves distinct and form non-dominant sectors of society, intending to pass IK onto present and future generations (Craig and Davis 2005). J. Martinez Cobo, former UN special Rapporteur defines 'Traditional Knowledge' as "Indigenous communities, peoples and nations are those which, having a historical continuity with pre-invasion and pre-colonial societies that developed on their territories, consider themselves distinct from other sectors of the societies now prevailing on those territories, or parts of them. They form at present non-dominant sectors of society and are determined to preserve, develop and transmit to future generations their ancestral territories, and ethnic identity, as the basis of their continued existence as peoples, in accordance with their own cultural patterns, social institutions and legal system." (Martinez Cobo 1986).

This knowledge is often referred to as 'traditional' and 'local' or 'community' knowledge which has been interpreted as outdated, inflexible and narrow-scaled. This is not consistent with the research:

> … the perception of [Traditional Ecological Knowledge] … is shifting from one as existing in a rather essentialised and static form … to [one] seen as having a hybrid and dynamic nature, more capable of adapting to new ecological and socio-economic conditions than previously assumed (Gomez-Baggethun et al. 2013).

Research has also indicated that despite factors such as globalisation, land appropriation and unsustainable developments that have eroded IK, it has persisted in what has been described as 'biocultural refugia' which can be used to build resilience in the face of adverse effects of modernisation to evolve and adapt to change (Barthel et al. 2013 as cited in Gomez-Baggethun et al. 2013). Additionally IK systems are being "increasingly acknowledged for their contributions to sustaining biodiversity and ecosystem service, and to building resilience in the face of global change" (Gomez-Baggethun et al. 2013, p.76).

One of the ways IK contributes to biodiversity protection and building resilience is by preserving and promoting biocultural diversity. IK is based on long term observation of the dynamism of local ecosystems and learning through trial and error. When IK is lost there is a corresponding loss of knowledge and experience to respond effectively to natural disturbance, conflict and global change (Gomez-Baggethun et al. 2013).

6 IK Protection and the Legal Framework

The role and value of IK has been acknowledged at the international level in international environmental and human rights reports and instruments.

The report entitled "Our Common Future" (World Commission on Environment and Development 1987, pp. 114–115) declared that "[indigenous communities] are the repositories of vast accumulations of traditional knowledge and experience that links humanity with its origins [and] larger society … could learn a great deal from their traditional skills in sustainably managing very complex ecological systems" (see also Richardson 2008, p. 1).

Examples of such instruments include the Convention for Biological Diversity (1992) ('CBD') and the United Nations Declaration on the Rights of Indigenous Peoples (2007) ('UNDRIP'). The CBD recognises the important role of IK in a number of articles, including Articles 8(j), 10(c), 17(2) and 18(4). For example, Article 8j of the CBD encourages ratifying States to respect, preserve and maintain IK relevant for the sustainable use and conservation of biodiversity. Article 8j recommends that this should be done with the free prior and informed consent of the owners and custodians of the IK with an equitable sharing of benefits flowing from the IK.

Protocols like the Nagoya Protocol, CBD targets such as the Aichi Targets and guidelines such as the Akwé: Kon Guidelines (Secretariat of the Convention on Biological Diversity 2004) have "enhanced the participation of indigenous people in biodiversity protection and strengthened the legal and practical protections afforded to indigenous people" (Godden and Peel 2010, pp. 37–38).

UNDRIP is an internationally recognised standard acknowledging the need for States to take action about the disadvantage and mistreatment of indigenous people. It also supports the need to realise the cultural, socio-economic and political aspirations of indigenous people. Article 31 provides that indigenous people have a right to maintain, control and protect their IK which includes cultural expressions, heritage, and language. It mandates States to take effective measures to recognise and protect the exercise of these rights (University of Technology Sydney and North West Local Land Services 2014, p. 12).

7 Australia's Legal Protection of IK

Godden and Peel (2010, p. 47) observe that Australian laws regulating the environment have begun to reflect changes in understanding of the complexity and interconnectedness of the environment. Further that:

> [O]ne of the most important changes in this regard has been the recognition of environmental interconnectedness flowing from ideas in ecological science and their interaction with particular strands of environmental philosophy. (Godden and Peel 2010, p. 47).

The legal system has reconstructed the self-contained sectors of the environment within legislative definitions, administrative regulatory practices and judicial interpretations of the law. Current legal notions of the environment expand the scope of environmental law and play a key part in regulating society's response to environmental problems (Godden and Peel 2010, p. 47). A part of a more sophisticated understanding of the complexity of the environment has been "towards more culturally inclusive notions … [leading] to greater involvement of Indigenous peoples in many aspects of environmental management and the promotion of sustainability" (Godden and Peel 2010, p. 36).

Despite commitments to more involvement of indigenous peoples in natural resource management in the National Strategy for the Conservation of Australia's Biological Diversity to implement the provisions of the CBD, there is a weak and disjointed legal framework underpinning that objective (Davis 1999 as cited in Godden and Peel 2010, p. 38). In Australia's latest national report to the CBD (Australian Government 2014), reference is made to inclusion of IK in biodiversity conservation and natural resource management. This inclusion is achieved through policy and programmes rather than within the substantive legal framework (Australian Government 2014, p. 2, sub-priority 1.2). Moreover, Aboriginal involvement in joint/co-managed protected areas can generally be described as passive.

The *Environment Protection and Biodiversity Conservation Act* (1999) (Cth) (EPBC Act) is the central piece of the Commonwealth's environmental legislation. Protection of IK is not achieved directly under the EPBC Act but incidentally considers indigenous interests, especially tangible heritage items, in the environmental impact assessment process. Other protection is achieved under a number of other laws such as the *Native Title Act 1993* (Cth) by considering IK as evidence to prove native title claims and the protection of IK in cultural heritage areas and objects under the *Aboriginal and Torres Strait Islander Heritage Protection Act 1984* (Cth).

8 Indigenous Protected Areas (IPAs)

From 1996, the Indigenous Protected Areas ('IPA') program was initiated to include large tracts of the Indigenous Estate (defined as rights and interests flowing from diverse indigenous systems of land ownership (Altman et al. 2007, p. 5)) with high environmental and cultural heritage values into the National Reserve System. Australia's National Reserve System includes more than 10,000 protected areas of almost 18 % of the continent—over 137 million hectares (Australian Government 2013). These areas must be compliant with criteria established by the IUCN (Altman and Kerins 2012, p. 13).

IPAs are areas of indigenous owned land or sea which are the subject of the agreement between the traditional owners and the Australian Federal Government. The agreement is to ensure that these areas are protected for biodiversity and cultural values. "IPAs make a significant contribution to Australian biodiversity conservation" (Australian Government 2013).

The first IPA, which was declared in South Australia in 1996, was Nantawarrina. The IPA system has grown to 51 IPAs covering over 36 million hectares of land and represents around one third of Australia's National Reserve System. IPAs are found in most Australian states and territories (Hill et al. 2011).

IPAs have been described as a ground breaking initiative which combines environmental management and indigenous cultural survival principles. The developments of IPAs have paralleled similar global initiatives such as the Indigenous and Community Conserved Areas (Hill et al. 2011, p. 2).

The goals of the IPA system are consistent with the aims of sustainable development. Ngurrara Traditional Owners note that "[b]eing sustainable on country means making decisions locally that do not harm country, now or in the future. It means providing future generations with the rights and the abilities to enjoy their responsibilities to country." (Australian Government 2012).

IPAs are declared in response to a process initiated by the indigenous owners and/or custodians as part of a consultative and participatory process using community controls and customary decision-making (Hill et al. 2011, p. 4). For the IPA to be recognised by the Federal Government, it must be based on a management plan consistent with an IUCN Protected Area Management Category, describe measurable objectives and have management approach objectives for conservation, cultural maintenance and other community benefits.

> Many IPAs describe how their Management Plans aim to be 'two way' bringing together management based on Indigenous ecological knowledge, practices, beliefs and tradition, with that based on western scientific knowledge and practices. (Gilligan 2006 as cited in Hill et al. 2011, p. 5)

Examples of combining the two styles of management are demonstrated in the content, description of the management aims and objectives and presentation of the material in the plans.

IPAs can provide many social, economic and other benefits such as improved health outcomes for local communities. In 2006, an evaluation found that 95 % of communities within IPAs reported improved economic participation and development benefits for communities (Hill et al. 2011, p. 37). Other benefits included better school engagement, reduction in substance abuse and improved family life. IPAs contribute to the Federal Government's 'Closing the Gap Policy' which aims to reduce differences in health, education and employment opportunities between indigenous and non-indigenous Australians (Hill et al. 2011, p. 37).

The Angas Downs IPA (Angas Downs IPA 2009 as cited in Hill et al. 2011, p. 6) describes both non-indigenous and indigenous objectives in English and indigenous language. The content of the plans clearly set out objectives that are important for indigenous and scientific approaches such as: "First way is Indigenous law, customs, culture and Indigenous ecological knowledge—Second way is non-Indigenous conservation and western science" (Hill et al. 2011, p. 6). The format is presented in various ways to demonstrate the aims of the IPA through diagrams, symbols, paintings and poster style. Statements of intent vary from simple to complex such as "Keep the land alive. Keep the story alive too" (Hill et al. 2011, p. 6) which is an example from the Paruku IPA (Paraku IPA 2004).

A governance objective for IPAs is to integrate sustainable development principles. Principles of IPA management include connection to place, stewardship, community participation, effective and transparent decision-making, IK and its role, the precautionary and intergenerational equity principles (Hill et al. 2011, p. 27). IPAs recognise the value of IK particularly Indigenous Ecological Knowledge (IEK). Many IPA managers support the recording of IEK "through indigenous use and occupancy mapping, ethno-ecology, digital databases, I-tracker and other projects" (Hill et al. 2011, p. 21). These processes give effect to the obligations arising out of recording, protecting and use of IK in UNDRIP (Hill et al. 2011, p. 21).

IPAs may provide business and commercial opportunities which are consistent with ESD goals including tourism, sustainable harvesting, payment for ecosystem services such as improved fire control, invasive species management and through enhanced carbon capture and storage process. There may be opportunities for joint ventures and commercial partnerships for major capital initiatives including road-house management, mining and other developments provided these developments are consistent with conservation and looking after country objectives (for example Jawoyn Association Aboriginal Corporation n.d.).

The next section provides more detailed examples of IPAs in one territory—Northern Territory, and two states—South Australia and Victoria. These three case studies have evolved differently but demonstrate various successes and challenges for ecologically sustainable developments which integrate IK effectively and achieve good social and economic outcomes.

8.1 Case Study 1—Jawoyn IPA (Northern Territory)

The Jawoyn Indigenous people (the Jawoyn people) have proposed a three-stage plan over a number of years to develop their land (of 1,830,000 ha) in the Nitmiluk National Park areas. The Jawoyn Association Aboriginal Corporation ('the Association') was established in 1985 as the representative body of the Jawoyn people. It is the business structure to allow the Jawoyn people to "[work] towards achieving sustainable improvements in economic social cultural health, education and other indicator of overall quality of life." (Jawoyn Association Aboriginal Corporation n.d.).

The Association has established a range of businesses in the tourist, real estate and aviation sectors. Nitmiluk Tours runs all the tourist operations in Nitmiluk National Park and is fully owned by the Jawoyn people. The Association operates real estate acquisition and management enterprises as well. An aviation company has been established to service its land management and tourism operations (Jawoyn Association Aboriginal Corporation n.d.). Unfortunately some aspects of the Association's business enterprises have not been successful and have led to a substantial loss of profits together with an investigation into its corporate governance activities. In 2013, that investigation was conducted by the Registrar of Indigenous Corporations, due to complaints about the behaviour of a number of officers and employees of the Association. It was concluded that there was insufficient evidence to prove breaches of the legislation administered by the Registrar were committed by the Association employees and no further action was taken (Jawoyn Association Aboriginal Corporation 2013; Office of the Registrar of Indigenous Corporations 2013). However, financial and management difficulties which face the Association do not undermine the positive outcomes of achieving sustainable development projects based on a combination of IK, western environmental management and customary law.

Indigenous Business Australia (IBA) has been established for over 25 years to invest in Aboriginal and Torres Strait Islander business ventures. This investment is only undertaken after a rigorous assessment process. IBA has financially backed a tourist accommodation project called Cicada Lodge in the Nitmiluk National Park (Indigenous Business Australia n.d.). Cicada Lodge is a luxury accommodation lodge on the edge of the Katherine River and is achieving success. It has been positively reviewed by tourists who have stayed there.

Nitmiluk Tours is owned and operated by the Jawoyn Association for tours and cruises to Nitmiluk Gorge as well as promotion of its restaurant facilities. Profits from these ventures support social welfare outcomes and cultural and spiritual renewal for Jawoyn people. Sharing cultural experiences may also lead to an improved understanding between indigenous and non-indigenous people.

8.2 Case Study 2—Paruku Wetlands IPA (South Australia)

For a number of years, the World Wildlife Foundation ('WWF') has been working with the Walmajarri people and the communities of Mulan, Billiluna and Balgo to help manage the Paruku wetland and establish the Paruku IPA. The Paruku wetlands are located in the Tanami Desert, 300 km south of Halls Creek in the Northern Territory and are of great cultural significance to the Walmajarri Traditional owners. "The IPA combines important traditional knowledge with western scientific techniques to train local rangers and better plan and manage country" (WWF Australia n.d.).

Through the Paruku Wetlands IPA, there are indigenous ranger training programs. These rangers assist with IPA management including biodiversity surveys to build knowledge about local species' conditions and scientific techniques. This IPA work is a successful partnership between Traditional Owners Indigenous organisations, the WWF and the Kimberley Land Council. The WWF intends to develop and extend this program to other rich biodiversity sites (WWF Australia n.d.).

8.3 Case Study 3—Budj Bim Within the Lake Condah IPA (Victoria)

This IPA covers the area contained in the Budj Bim National Heritage landscape in Western Victoria which has been managed by the Gunditj Mara people (the Gunditj Mara) for over 30,000 years. The eruption of Mt Eccles changed the drainage pattern creating large wetlands. The wetlands were used and modified by the Gunditj Mara to raise and harvest eels. This sustainable aquaculture provided the economic basis for a densely populated and settled community. A development by Alcoa Pty Ltd (Alcoa) for an aluminium smelter at Portland was proposed to the Victorian government in 1984. In 1980, two women who were traditional owners of the proposed smelter site started a legal action in the Victorian Supreme Court to prevent Alcoa from potentially damaging their cultural heritage-in particular sacred sites (Onus and Frankland v Alcoa (1981) 149 CLR 27). In 1984, the then Victorian Premier, John Cain proposed a settlement to end the litigation against Alcoa. This was that in exchange for indigenous title over the land in the Lake Condah mission area that the Gunditj Mara community should not object to the smelter proposal. The Victorian government was unable to pass appropriate legislation to give effect to that proposal. However, the Federal government using relevant constitutional powers, passed the *Aboriginal Land (Lake Condah and Framlingham Forest) Act 1987* (Cth) to return the mission land to the Aboriginal people (Australian Heritage Places Inventory n.d.) The Alcoa smelter development proceeded despite the strong objections of the local indigenous community by way of a joint venture agreement between the Victorian government and Alcoa to produce alumina at Portland which is still operational.

The Budj Bim National Heritage Landscape overlaps part of the area contained in the Lake Condah IPA and the outstanding values of this unique area are included in the Management Plan. The short-finned eels in Lake Condah migrate between Australia and the South Pacific and highlight the inter-linkages between local IPAs, Australian and international environmental management (Framlingham Aboriginal Trust and Winda Mara Aboriginal Corporation 2004).

9 International Models for IK Protection

There are a number of regional initiatives that are developing in external jurisdictions that may be instructive for Australian legal and regulatory reform to enable the inclusion of IK in environmental decision-making and natural resource management for achieving beneficial outcomes such as improved biodiversity protection and sustainable development. The Andean Community, through a series of Community Decisions have created a legal framework where indigenous people may promote the application of customary law as a precondition for access to and use of their IK (Andean Community of Nations 1996, 2000, art. 27(i) as cited in Tobin 2014, pp. 170–171).

In the South Pacific, the Pacific Islands Forum Secretariat published detailed guidelines for developing model legislation for the protection of IK based on the *Traditional Biological Knowledge Innovations and Practices Act*. This model law requires a licence to be "obtained for use of IK prior to use conforming with customary law principles" (Tobin 2014, p. 172). Another model for protecting rights of indigenous people over the use and access of their IK is the Peruvian Law 27811 (Peru Law No. 27811 of July 2002) for protecting the collective rights over traditional knowledge relating to biodiversity (Tobin 2014, p. 172).

Tobin (2014, p. 172) notes that "[t]he law recognizes traditional knowledge to be the cultural patrimony of Indigenous people, thereby recognizing both intergenerational rights and responsibilities relating to such knowledge". That author also argues that there is a general need for a more effective sui generis (of its own kind) approach for development frameworks and mechanisms for protecting IK for, inter alia, meeting needs for changing conditions of the environment. This is necessary in Australia and this position has been confirmed by the opinion of a prominent member of the Australian indigenous community, Special Rapporteur Mick Dodson, "that a completely new and customized approach is needed" (Dodson 2007 as cited in Tobin 2014, p. 157).

There have been proposals in Australia for incorporating the principles of biocultural protocols into the legal and regulatory framework but there are few incentives or pressure for governments to adopt measures to ensure that access and use of IK is obtained with free informed and prior consent of indigenous people (Tobin 2014, p. 161).

10 Conclusion

The concept of sustainable development is increasingly embedded within the Australian legal and regulatory framework. Whilst not sufficiently acknowledged and protected in Australian legislation, IK is proving to be beneficial in the long term for achieving positive environmental and cultural outcomes. Ideally, Australia could develop a sui generis legal framework to acknowledge, respect and give effect to use of IK for environmental objectives and the thriving of indigenous cultural heritage. Within the IPA system, indigenous people are demonstrating their active commitment to sustainable development which is proving to be a positive force in support of self-determination goals as well.

References

Altman J, Kerins S (2012) People on country, vital landscapes, indigenous futures. The Federation Press, Sydney

Altman J, Buchanan G, Larsen L (2007) The environmental significance of the indigenous estate: natural resource management as economic development in remote Australia. Centre for Aboriginal Economic Policy Research Paper No 286/2007, Canberra, Australian National University, Australian Capital Territory

Andean Community of Nations (1996) Decision no. 391 establishing the common regime on access to genetic resources. Retrieved 29 Mar 2015. http://www.wipo.int/wipolex/en/details.jsp?id=9446

Andean Community of Nations (2000) Decision no. 486 establishing the common intellectual property regime. Retrieved 29 Mar 2015. http://www.wipo.int/wipolex/en/details.jsp?id=9451

Angas Downs IPA (2009) Progress with the management plan, conservation of biodiversity, sustainable production and training needs. Presentation to IPA managers workshop 2009. Northern Territory, Angas Downs IPA, Australia

ANZECC Australian and New Zealand Environment and Conservation Council (1996) National strategy for the conservation of Australia's Biological diversity. Retrieved 29 Mar 2015. http://www.environment.gov.au/archive/biodiversity/publications/strategy/pubs/national-strategy-96.pdf

Australian Government Department of the Environment (2012) About indigenous protected areas. Retrieved 29 Mar 2015. http://www.environment.gov.au/indigenous/ipa/background.html

Australian Government Department of the Environment (2013) Indigenous protected areas. Retrieved 29 Mar 2015. http://www.environment.gov.au/indigenous/ipa/index.html

Australian Government Department of the Environment (2014, May) Australia's fifth national report to the convention on biological diversity. Retrieved 29 Mar 2015. http://www.cbd.int/doc/world/au/au-nr-05-en.pdf

Australian Heritage Places Inventory (n.d.) Budj Bim national heritage landscape—Mt Eccles lake Condah Area. Retrieved 4 June 2015. http://www.heritage.gov.au/cgi-bin/ahpi/record.pl?NHL105673

Barthel S, Crumley CL, Svedin U (2013) Biocultural refugia: combating the erosion of diversity in landscapes of food production. Ecol Soc 18(4):71

Brown LR (2006) Plan B 2.0: rescuing a planet under stress and a civilization in trouble. Earth Policy Institute Bookstore, Washington DC

Commonwealth of Australia (1998a) Australia's oceans policy vol 1: caring, understanding using wisely. Retrieved 29 Mar 2015. http://155.187.2.69/archive/coasts/oceans-policy/publications/pubs/policyv1.pdf

Commonwealth of Australia (1998b) National greenhouse strategy. Australian Government Publishing Service, Canberra

Craig D, Davis M (2005) Ethical relationships for biodiversity research and benefit-sharing with indigenous peoples. Macquarie J Int Comp Environ Law 2:31–74

Davis M (1999) Indigenous rights in traditional knowledge and biological diversity: approaches to protection. Aust Indigenous Law Rep 4(4):1–32

Department of Environment (1992, December) National strategy for ecologically sustainable development. Australian Government, Canberra. Retrieved 29 Mar 2015. http://www.environment.gov.au/about-us/esd/publications/national-esd-strategy

Dodson M (2007) Report of the secretariat on indigenous traditional knowledge. E/C.19/2007/10. United Nations Permanent Forum on Indigenous Issues, Geneva Switzerland

Dovers S, Connor R (2006) Institutions and policy change for sustainability. In: Richardson B, Wood S (eds) Environmental law for sustainability, 1st edn. Hart Publishing, Oxford, pp 21–60

Environment Protection and Biodiversity Conservation Act 1999 (Cth) Retrieved 29 Mar 2015. http://www.comlaw.gov.au/Details/C2015C00265

Environmental Planning and Assessment Act 1979 (NSW) Retrieved 29 Mar 2015. http://www.legislation.nsw.gov.au/maintop/view/inforce/act+203+1979+cd+0+N

Framlingham Aboriginal Trust & Winda Mara Aboriginal Corporation (2004) Kooyang sea country plan. Framlingham Aboriginal Trust and the Winda Mara Aboriginal Corporation, Victoria. Retrieved 29 Mar 2015. http://www.environment.gov.au/system/files/resources/4fc3aad8-47f4-4d90-b58e-bbadbd9e369d/files/kooyang-plan05.pdf

Gilligan B (2006) The indigenous protected areas programme: 2006 evaluation. Australian Government Department of the Environment and Heritage, Canberra

Godden L, Peel J (2010) Environmental law: scientific, policy and regulatory dimensions. Oxford University Press, South Melbourne

Gomez-Baggethun E, Corbera E, Reyes-Garcia V (2013) Traditional ecological knowledge and global environmental change: research findings and policy implications. Ecol Soc 18(4):72

Hill R, Walsh F, Davies J, Sandford M (2011) Our country our way: guidelines for Australian indigenous protected area management plans. CSIRO Ecosystem Sciences & Australian Government Department of Sustainability, Water, Environment, Population and Communities, Cairns

IFAD (2003, February) Indigenous peoples and sustainable development. Roundtable discussion paper for the twenty-fifth anniversary session of IFAD's governing council. Retrieved 29 Mar 2015. http://www.ifad.org/gbdocs/gc/26/e/ip.pdf

Indigenous Business Australia (n.d.) Our investment philosophy. Retrieved 29 Mar 2015. http://www.iba.gov.au/investments/our-investment-philosophy/

Jawoyn Association Aboriginal Corporation (2013, July 29) Jawoyn Association welcomes ORIC findings [press release]. Retrieved 29 Mar 2015. http://www.jawoyn.org/uploads/media-releases/MEDIA%20RELEASE%20issued%2029%20July%202013-1.pdf

Jawoyn Association Aboriginal Corporation (n.d.) Retrieved 29 Mar 2015. http://jawoyn.org/

Macintosh A (2015) The impact of ESD on Australia's Environmental Institutions. Aust J Environ Manage 22:33–45

Maffi L, Woodley E (2010) Biocultural diversity conservation: a global sourcebook. London and Washington DC, Earthscan

Martinez Cobo JR (1986) Study of the problem of discrimination against indigenous populations. United Nations Economic and Social Council Commission on Human Rights, New York. Retrieved 29 Mar 2015. http://undesadspd.org/IndigenousPeoples/LibraryDocuments/Mart%C3%ADnezCoboStudy.aspx

National Parks and Wildlife Act (1974) (NSW). http://www.legislation.nsw.gov.au/maintop/view/inforce/act+80+1974+cd+0+N

Office of the Registrar of Indigenous Corporations (2013, July 29) Registrar completes investigation into Jawoyn Association Aboriginal Corporation. Retrieved 29 Mar 2015. http://www.oric.gov.au/publications/media-release/registrar-completes-investigation-jawoyn-association-aboriginal, (press release)

Papadakis E (1993) Politics and the environment: the Australian experience. Allen & Unwin, St Leonards

Paruku IPA (2004) Paruku IPA management plan for community. Paruku IPA, Western Australia

Preston B (2012) Jurisprudence on ecologically sustainable development: Paul Stein's contribution. Environ Plann Law J 29:3–15

Richardson B (2008) The ties that bind: indigenous people and environmental governance. Comparative research in law & political economy. Research paper no. 26/2008. Retrieved 29 Mar 2015. http://digitalcommons.osgoode.yorku.ca/clpe/197

Secretariat of the Convention on Biological Diversity (2004) Akwé: Kon: voluntary guidelines for the conduct of cultural, environmental and social impact assessment regarding developments proposed to take place on, or which are likely to impact on, sacred sites and lands and waters traditionally occupied or used by indigenous and local communities. Montreal. Retrieved 31 Mar 2015. https://www.cbd.int/doc/publications/akwe-brochure-en.pdf

The World Bank, Knowledge and Learning Group (2004) Indigenous knowledge: local pathways to global development, Washington DC. Retrieved 29 Mar 2015. http://www.worldbank.org/afr/ik/ikcomplete.pdf

Tobin B (2014) Indigenous peoples, customary law and human rights—why living law matters. Routledge, London and New York

United Nations (2012) Report of the United Nations conference on sustainable development. A/CONF.216/16. Rio de Janeiro, Brazil. Retrieved 29 Mar 2015. http://www.uncsd2012.org/content/documents/814UNCSD%20REPORT%20final%20revs.pdf

United Nations Convention on Biological Diversity (1992) 1760 UNTS 79; 31 ILM 818. Retrieved 29 Mar 2015. http://www.cbd.int/default.shtml

United Nations Convention on Biological Diversity (2010) Aichi biodiversity targets. Retrieved 29 Mar 2015. http://www.cbd.int/sp/targets/

United Nations Declaration on the Rights of Indigenous Peoples (2007). A/RES/61/295; 46 ILM 818. Retrieved 29 Mar 2015. http://www.un.org/en/ga/search/view_doc.asp?symbol=A/RES/61/295

University of Technology Sydney & North West Local Land Services (2014) Recognising and protecting Aboriginal knowledge associated with natural resource management. White Paper for the Office of Environment and Heritage, NSW. Retrieved 29 Mar 2015. http://www.indigenousknowledgeforum.org/components/com_content/models/forms/white_paper.pdf

Waste Management and Pollution Control Act 2014 (NT) Retrieved 29 Mar 2015. http://notes.nt.gov.au/dcm/legislat/legislat.nsf/d989974724db65b1482561cf0017cbd2/17a7c0c09216ff1c69257d9300113d11?OpenDocument

World Commission on Environment and Development (1987, March) Our common future. Oxford University Press, Oxford and New York. Retrieved 29 Mar 2015. http://www.un-documents.net/our-common-future.pdf

WWF Australia (n.d.) Paruku wetlands—indigenous protected area (IPA). Retrieved 9 Apr 2015. http://www.wwf.org.au/our_work/partnership_with_indigenous_communities/working_with_indigenous_communities_in_the_kimberley/paruku_wetlands/

Part X
Sectorial Policies: Agriculture and Rural Policy

The Sustainable Use of Biodiversity and Its Implications in Agriculture: The Agroforestry Case in the Brazilian Legal Framework

Marcia Fajardo Cavalcanti de Albuquerque

Abstract Brazil is one of the biggest exporters of farm products and one of the most biologically diverse countries in the world. Nevertheless, agricultural expansion and the constant search for higher productivity are main factors leading to biodiversity loss. Thus, the promotion of rural sustainable development is considered a major challenge for Brazilian agricultural policy. In this context, agroforestry system of production might represent a viable solution, since it promotes biodiversity conservation without jeopardizing farm output. However, the complexity of the system may represent an obstacle to its legal implementation. Hence, this paper aims to analyse whether the Brazilian legal framework is well enough structured to implement this system in all its complexity. More widely, this research is seeking solutions to the conflict between traditional farming methods and the biodiversity conservation imperative in Brazil. In order to assess these issues question applied research was implemented and supported by an in-depth literature and legislation review. Amongst the results, this paper shows that according to the green economy's theory of strong sustainability, the agroforestry way of production represent a sustainable agricultural system because the quality and the quantity of each component of the natural stock is maintained in its original state. Nevertheless, this paper concludes that, regarding the Brazilian context, the potential of the agroforestry systems is constraint by the lack of supportive regulatory framework.

Keywords Agroforestry system · Sustainable agriculture · Brazilian legislation

M.F.C. de Albuquerque (✉)
University Paris 1-Sorbonne and Mackenzie, Paris, France
e-mail: Marcia.Fajardo-Cavalcanti-Albuquerque@malix.univ-paris1.fr

M.F.C. de Albuquerque
University of São Paulo, São Paulo, Brazil

V. Mauerhofer (ed.), *Legal Aspects of Sustainable Development*,
DOI 10.1007/978-3-319-26021-1_29

1 Introduction

Currently humanity faces a major issue: combining economic, environmental and social goals. That conciliation covers the rational management of natural resources and human knowledge as well as the search of economic development which is respectful of the environment, meeting the essential needs of present populations without compromising the needs of future generations. Sustainable development is difficult to achieve, especially for developing countries[1] like Brazil, where public policies are oriented almost exclusively to economic growth, frequently neglecting the need to preserve nature.

The Brundtland Report (WCED 1997) named "Our Common Future" defines sustainable development as the "development that meets the needs of the present without compromising the ability of future generations to meet their own needs" (pp. 16). According to Voigt (2013), sustainable development can be considered at the same time as an objective to be achieved and as a principle that should guide citizens' action and political, economic and judicial decisions. Thus, this paper will analyse it as goal.

Sustainable development is pursued differently, depending on the social, economic, and environmental context of each country and depending on the level of sustainability that one wants to reach. According to economic science, there are two paradigms of sustainable development: weak sustainability or "substitutability paradigm" (Neumayer 2003), based on "the assumption that natural capital is either abundant or substitutable both as an input into the production of consumption goods and as provider of direct utility" (Neumayer 2003, p. 22), so the natural capital can be safely explored "as long as enough man-made capital is built up in exchange" (Neumayer 2003, p. 22); and strong sustainability or "non-substitutability paradigm" (Neumayer 2003), based on the theory of ecological economics, which holds that there are environmental limits restricting the expansion of economic activity and natural resources are not considered interchangeable (Taýra 2006). According to Neumayer (2003), supporters of the strong sustainability paradigm are not against the achievement of weak sustainability, but they consider this goal as "an important first, but insufficient, step into the right direction" (Neumayer 2003, p. 24). Based on the fact that "natural capital loss is often irreversible" (Neumayer 2003, p. 26), this paper will be built on the strong sustainability paradigm. Though, it doesn't mean that it is necessary to "keep nature as it is" (Neumayer 2003, p. 25), but rather that the use of natural capital cannot exceed its capacity to regenerate, "so that their environmental function remains intact" (Neumayer 2003, p. 25).

Concerning more specifically the Brazilian context, one can consider integrating agricultural activities and forest protection as one of the elements for achieving sustainable development. This statement is based on the country's paradoxical position in the world. While Brazil is considered the "farm of the planet", being the

[1]These countries are defined according to their Gross National Income (GNI) per capita per year.

fifth largest agricultural producer (IPEA 2011), it is also seen as the "lungs of the planet", having the largest extension of tropical forest in the globe (FAO 2010).

According to Leff (2001) the achievement of sustainable development requires three types of rational behaviour. The social rationality, grounded in principles of equity, diversity and democracy, once community's participation increases the reflection on local problems and strengthens local governance; the productive rationality which consists of sustainable natural resources' management practices better adapted to the ecological conditions of each region and to the cultural values of each community; and the environmental rationality, which represents "the expression of the conflict between the use of the law (market's) by a class, the search for the common good with State's intervention and the participation of civil society in a process of nature re-appropriation, guiding its values and potential for a sustainable and democratic development" (Leff 2001, p. 143).[2]

Raman (2006) asserts that the concept of sustainable development represents an appropriate grounding for the definition of future development goals for agriculture. In this sense, the Brundtland Commission Report encouraged the implementation of agricultural systems "that focus as much attention on people as they do on technology, as much on resources as on production, as much on the long term as on the short term" (Raman 2006, p. 31). Raman also affirms that "this was the harbinger of a new paradigm of sustainable agriculture" (Raman 2006, p. 31). The same author defines the basic principles of sustainable agriculture as: (1) "plant and animal productivity adequate to meet the needs of a growing population"; (2) "ecological security"; (3) "economic viability" and (4) "social responsibility and acceptability" (Raman 2006, p. 32). Regarding the last element, Raman explains that sustainable agriculture should promote social justice and equity, taking into account the welfare of farming communities and he also asserts that the farming method applied should be socially accepted (Raman 2006). About the same subject, Gitau et al. (2008) affirm that sustainable agriculture is aimed at satisfying the changing human needs with little or no secondary effects on the ecosystem. The author stresses that in terms of resilience, the sustainability of an agricultural system could be measured by its ability to maintain its productivity when it is subject to perturbation. They also claim that sustainability is enhanced by natural diversity and by an efficient use and recycle of nutrients and energy (Gitau et al. 2008).

The concept of rural sustainable development is addressed in a slightly different way. Dutu (2013) stresses that its ecological element is composed by the rational, efficient and sustainable use of resources, whilst not damaging the environment; its economic element consists of the promotion of economic instruments that don't allow the natural capital do degrade; and finally, its social component comprehends the establishment of a more equitable system for production and distribution of agricultural products between North and South, the conservation of family farms and the strengthening of local communities' land use. According to Moreira and Carmo (2004) sustainable rural development reflects the need to strengthen the

[2]Unofficial translation by the author of this article.

collective social action of resistance to large-scale agriculture. The same authors attest that sustainable rural development strategy should be established in a participatory manner, according to the socio-cultural specificities of each region, providing for a transformation from inside out.

In sum, rural sustainable development should be built on the participation, democracy and equity principles, promoting the rational and sustainable use of natural resources, by the adoption of ecological practices which are adapted to each region and culture. In this context, the agroforestry system could represent a major ally to the promotion of rural sustainable development. By taking into account the local particularities and by engaging the local stakeholders, it promotes the rational use of biodiversity, one of the objectives announced by the Convention on Biological Diversity[3] (CBD). In this direction, Kings and Chandler (1978, p. 2) define agroforestry as "a sustainable land management system which increases the overall yield of the land, combines the production of crops (including tree crops) and forest plants and/or animals simultaneously or sequentially, on the same unit of land, and applies management practices that are compatible with the cultural practices of the local population."

Hence, this paper seeks to study the importance of agroforestry for achieving sustainable rural development, analysing the practice according to the concept of strong sustainability. Furthermore it aims to examine the way agroforestry is interpreted by the Brazilian legal framework and if the Brazilian legislation is well suited to implement it. To do so, in the first part, the paper will analyse the complexity of Brazilian agricultural systems that foster both the exploitation and the conservation of agrobiodiversity (I). The first part will be divided into two sections: section A, which is aimed at the study of some historical aspects of Brazilian agriculture and agricultural policy and section B, where conceptual aspects regarding agroforestry systems will be clarified. In the second part, the author will investigate how agroforestry systems can be implemented in the country (II) by examining the way agroforestry is interpreted by Brazilian law (A) and finally by pointing out the obstacles and suggestions for its legal implementation (B).

I. The complexity of agricultural systems in brazil: between exploitation and conservation of agrobiodiversity

Brazilian agriculture is well diversified and it represents one of the most important income sources of the country, figuring among the major producers and exporters of commodities in the world (FAOSTAT 2012). To better understand Brazil's farming systems it is necessary to analyse the history of Brazilian rural development (A). Later, this paper will examine the agroforestry system as an effective method of biodiversity conservation (B).

[3]The Convention on Biological Diversity (CBD) entered into force on 29 December 1993, signed by 150 government leaders at the 1992 Rio Earth Summit.

A. The Brazilian rural development: a reflection of the country's colonization process

The existence of previous crops upon the arrival of the Portuguese is undeniable. Before the 1500s, Native Indians practiced subsistence agriculture, mainly cultivating cassava, peanuts, tobacco, sweet potatoes and corn and also practicing foraging. However, the development of agriculture in Brazil is deeply rooted in the country's colonization process. As the purpose of the Portuguese Kingdom was to occupy the territory and exploit it economically to the most, the Portuguese managed to introduce in the country a large-scale agriculture production mainly addressed to export (Reifschneider et al. 2010).

The Portuguese colonization led to the formation of rural properties as the Brazilian territory was divided into large tracts of land, called hereditary captaincies. These were distributed to the people who had conditions to explore them, especially members of the Portuguese nobility, as a way to ensure an accelerated territorial occupation (Lima 2008).

The Portuguese agricultural model was based on export-oriented monoculture of exotic species, large estates and slave labour. As such, even though the Brazilian soil was rich enough, there weren't incentives to diversify the production. Wood products or sugar were intended for export and only the essential food for the reproduction of labour-work was grown. At that time, agriculture was characterized by the near absence of domestic demand for agricultural products on the one hand, and the strong European demand for certain exotic agricultural products on the other (Espirito Santo 2001).

The basic features of the primarily export-based economic model remained until the early twentieth century, when the industrialization process of the Brazilian economy began. To achieve it, a large proportion of state incentives were given to the industrial sector. Nevertheless, this fact led to the increased demand for food, accelerating the agricultural modernization process. The modernization of the Brazilian agricultural production happened parallel to the global trend of production mechanization and through heavy use of fertilizers known as the "Green Revolution". Aiming at increasing agricultural output, production practices were mechanized and monoculture fields were established, generating a devastation of the country's biodiversity (Teixeira 2005).

The modernization of the agricultural sector was essentially based on applied research development, the extension of fertile land and the granting of public credit. However, the provision of subsidized credit mainly favoured large producers and certain products, such as wheat and soybeans, marginalizing small farmers (Dal Soglio et al. 2009).

The consequences of agriculture industrialization were numerous. Among the environmental impacts are the loss of fertility, contamination of soil and water from excessive use of pesticides, deforestation and biodiversity loss (Balsan 2006). The socio-economic impacts include the increase of social inequality, the difficulty to access land-related rights, the decline in income, and unemployment (Balsan 2006). According to Balsan (2006, p. 135) "the survival capacity of small producers was determined by inter-capitalist competition in product and input markets, in which many were forced to abandon the race, confirming the exclusive nature of capitalist modernization in the field".[4]

Despite the devastation caused by agribusiness in Brazil, we can still identify the existence of sustainable agricultural practices, mostly used by small-scale farmers, also called family farmers. Article 3, I of Act 11.428/2006 defines smallholder farmer or family farmer as the farmer who lives in the countryside, holds the possession of a rural farmland not bigger than 50 acres, exploring it by its personal work, with the help of his family or eventually with the help of third parties. 80 % of the farmer's gross income must be derived from agricultural, livestock, forestry or extraction activities. If the individual fraction of the land in collective land holdings does not exceed 50 acres, it could also be considered smallholder farming.

According to the Agricultural Census conducted in 2006 (MPA 2010), 84.4 % of Brazilian farms are within the family farm profile, occupying 24.3 % of the territory. Agribusiness represents 15.6 % of the rural establishments and occupies 75.7 % of the territory. Surprisingly, research shows that family farming produces 70 % of the food consumed in Brazil. Smallholder farmers are the key actors for ensuring food security in the country.

Therefore, small-scale farming not only ensures food security, but could represent an opportunity for the development of alternative farming practices in Brazil. Family farming is highly versatile, it has a diversified production structure and it is concerned with the durability of natural resources, and with the community's welfare. The relationship between the family farmer and the land is not only one of productivity ratio, but one of identification with the working and living place (Finatto and Salamoni 2008).

In this sense, as a means of resistance to the capitalist mode of production, the first principles of agroecology were introduced in Brazil in the 1970s, during the explosion of agribusiness (Dal Soglio et al. 2009). According to Altieri (2005), the agroecological guidelines consist of optimizing nutrient cycles, ensuring a good condition of the soil, minimizing energy losses in the system, increasing biodiversity and encouraging beneficial biological interactions.

It is noteworthy that the Federal Government hasn't been completely unconcerned about this matter. To encourage agroecological practices, the National Program for Strengthening Family Agriculture—PRONAF (Programa Nacional de Fortalecimento da Agricultura Familiar), was established by the Decree 1.946 on 28, June, 1996, aimed at promoting sustainable development of family farming by providing technical and financial support. The PRONAF also has a Technical Assistance and Rural Extension sub-program that aims to support farmers in adopting sustainable agricultural practices.

In 2012, the Federal Government issued the Decree 7.794 on 20, August, establishing the National Policy for Agroecology and Organic Production—PNAPO (Política Nacional de Agroecologia e de Produção Orgânica), which aims to "integrate, coordinate and adapt policies, programs and actions to induce a

---

[4]Unofficial translation by the author of this article.

transition to agroecological, organic and agroecological production contributing to sustainable development and people's quality of life through sustainable use of natural resources and the supply and consumption of healthy foods"[5] (Article 1). The main instrument to carry out the policy is the National Plan for Agroecology and Organic Production—PLANAPO (Plano Nacional de Agroecologia e Produção Orgânica). It seeks to implement agroecological transition actions and programs and it contributes to sustainable rural development. Therefore, the plan provides for rural credit, technical assistance, incentives for research and agroecological education, support during the certification process and during the environmental regularization of the property. Decree 7.794/2012 also establishes two agencies responsible for PNAPO's management. Those are the National Agroecology and Organic Production Committee (Comissão Nacional de Agroecologia e Produção Orgânica), a body composed of government members and civil society, responsible for promoting social control through participation; and the Interministerial Chamber of Agroecology and Organic Production (Câmara Interministerial de Agroecologia e Produção Orgânica), responsible for establishing and implementing the PLANAPO. The first PLANAPO was launched in October 2013, remaining in force until December 2015. Furthermore, Act 12.854, established on 26, August, 2013, provides that the Federal Government should encourage, through public policies, the implementation of agroforestry systems in expropriated or degraded rural areas.

In sum, although the Brazilian agricultural model is historically built on the monoculture system, one can observe the development of sustainable ways of production, essentially associated with family farming. In Brazil's context, alternative farming practices are very important to solve the dilemma created between the agricultural model based on monoculture and environmental concerns of biodiversity protection.

It is important to highlight the fact that the exploration method to be adopted by family farmers should be the one that ensures greater environmental sustainability over the years, because generally, any farmer wants to obtain the maximum economic income per cultivated area. Thus, agroforestry can be a good option for small farmers (Abdo et al. 2008). In this sense, Bolfe and Bergamasco (2010) point out that, in Brazil, the agroforestry systems have been disseminated mainly among family farms. Thus, next the paper will analyse the agroforestry systems as an agroecological practice that promotes biodiversity conservation (B).

B. The agroforestry system as a means to conserve biodiversity

The term agriculture is derived from latin *agri cultura*, which means the culture of the land or the soil (Flobert 2000, p. 93). Agriculture is inherent to human history. It is not only a way of living, but also a means of social affirmation. Agriculture is the result of an integration and exchange relationship between man and the environment.

[5]Unofficial translation by the author of this article.

Agricultural activity is closely linked to natural biological processes, being part of the ecosystem. At the same time, agriculture can compete with the natural environment, as it artificializes and explores ecosystems to produce agricultural commodities through different practices that reorder natural biochemical cycles (OECD 2001). Agriculture has several functions and the market alone is not able to manage them all (Caron and Cairol 2008). It's not surprising that France has officially recognized the multifunctionality of agriculture and its promotion has become one of the objectives of the National Agricultural Policy (Agricultural Orientation Act 1999—La Loi d'Orientation Agricole de 1999).

In 1998, CBD's Parties had already emphasized the importance of agricultural biodiversity in ensuring a good quality of life (COP4 Report 1998). On the occasion of the 4th Conference of the Parties (COP), held in Bratislava, the parties stressed the importance of sustainable agriculture in order to promote the harmonization of production goals and conservation. It also suggested the promotion and encouragement of sustainable agricultural practices aimed at increasing the ecological services provided by agricultural biodiversity (COP4 Report 1998).

Within this logic, the agroforestry system is a practice that enhances different aspects of agriculture. It reconciles the livestock, the agricultural cultivation, and forest cultivation; representing an alternative to the problems of environmental degradation and loss of soil productivity. The agroforestry system is an agroecosystem that has characteristics similar to those of a natural forest. It is a high-productivity system created through the appreciation of nature (Ewert 2014). Santos (2007) demonstrates that some families are capable of producing over 10.000 Kilos of food per hectare per year.

Lundgren (apud Clarke and Thaman 1993, p. 9) affirms that the World Agroforestry Center defines agroforestry as "a collective name for all land-use systems and practices in which woody perennials are deliberately grown at the same land management unit as crops and/or animals. This can be either in some form of spatial arrangement or in a time sequence. To qualify as agroforestry, a given land-use system or practice must permit significant economic and ecological interactions between the woody and non-woody components". The definition of the Brazilian National Environmental Council—CONAMA (Conselho Nacional do Meio Ambiente) is not that different. According to Article 2, IV of the Resolution of the Brazilian National Environmental Council—CONAMA 429/2011, implemented on February 28, 2011, agroforestry systems are "use and land use systems in which woody perennial plants are managed in association with herbs, shrubs, trees, crops, and forage in the same management unit, according to spatial and temporal arrangements, with diversity of native species and interactions between these components."[6]

Clarke and Thaman (1993) explain that there are two different approaches to agroforestry: the modern or institutional which relies on modern agronomic science and field experimentation; and the traditional or indigenous which emerged from

[6]Unofficial translation by the author of this article.

"cultural geography and ecological anthropology" (Clarke and Thaman 1993, p. 2). In Brazil, both approaches can be identified, as there is a growing dissemination of agroforestry practices through different agronomy research centres[7] and, at the same time, indigenous peoples often practice agroforestry. One can quote as an example the "Agroforestry Indigenous Actors", belonging to different indigenous ethnicity from the Acre State in the north of Brazil, who are chosen from each village to help implement agroforestry practices based on their own traditional practices (Bianchini 2006).

As follows, agroforestry refers to a relationship of continuity, integration and support between humans and nature. Rather than forcing the dynamic nature of technology to specialized production and the maximum output, as seen in modern agriculture, man seeks here to fit into and accelerate the dynamics of nature, in order to optimize production to meet a wide range of economic and social needs (Ewert 2014). This makes sustainability an intrinsic characteristic of the agroforestry system. The alternation of production during the year ensures greater profits per acreage unit and greater economic stability since the earnings of certain seasonal products are balanced by others, reducing market risks for the farmer (Müller 2006). Moreover, agroforestry has the social role of maintaining the population in rural areas by increasing the labour offer throughout the year. It also holds an important ecological role, since agroforestry systems provide several environmental services such as erosion control; retention of organic matter and improvement of the physical and chemical structure of the soil; an increase of nitrogen fixation and the promotion of efficient nutrient cycling; maintenance of biodiversity at levels similar to natural ecosystems; increase of agrobiodiversity; and the reduction of greenhouse gases. It also has the capacity to recover and rehabilitate degraded land (Müller 2006).

According to Schroth et al. (2004), agroforestry practices can enhance both planned and unplanned diversity. Usually, agroforestry systems contain more planned diversity such as the planted and selected species (trees and crops), but it also increases the unplanned diversity, such as plants and animals that colonize and use the structure formed by the plants species, providing more niches for native flora and fauna. The soil rich in humus could also provide a good habitat for a diverse soil fauna and microflora (Schroth et al. 2004, p. 8).

Since one of the most important benefits from agroforestry practices is the enhancement of biodiversity, it is important to elucidate its direct and indirect use value. According to Pascual and Perrings (2009), "managed on-farm biodiversity can be represented as a stock or economic asset. The asset represents the mix of species and communities that supply a flow of ecological services on-farm that can directly benefit farmers by maintaining and enhancing agricultural productivity" (Pascual and Perrings 2009, p. 158). The direct use value of agrobiodiversity could be expressed by the ability of some species to "provide pollination and biological

[7]The courses about agroforestry systems offered by the Brazilian Enterprise for Agricultural Research (EMBRAPA) are an example (G1 Journal 2014).

control services against pests and invasive species" (Pascual and Perrings 2009, p. 159) and its indirect value could be represented by its ability to help regulating ecosystem services as Pascual and Perrings stress that "an increase in on-farm species richness and the diversity of overlapping functional groups of species enhance the level of functional diversity, which, in turn, increases ecological stability and resilience" (Pascual and Perrings 2009, p. 159).

In order to analyse the sustainability of agroforestry systems in the Pacific Islands, Clarke and Thaman (1993) conducted a study acknowledging that the agroforestry systems implemented by traditional Pacific Island societies had a high degree of stability, fitting into the concept of "sustained-yield". According to the authors, the systems (1) "did not depend on external energy subsidies or extra-system nutrient sources"; (2) did not receive applications of poisonous agricultural chemicals" or other substances that polluted the environment; (3) "had strongly positive net energy yields"; (4) "used only renewable sources as inputs"; (5) "were structured so that the resources supporting agriculture were equitably spread throughout the community rather than being concentrated in the hands of a few or in urban areas"; (6) "contained resources that were looked upon as productive capital to be preserved"; and (7) "were based on polyculture and a diversity of tree and non-tree crops, wild plants, and animals rather than on monoculture or on specialized animal production" (Clarke and Thaman 1993, pp. 17 and 18).

Through another study aiming to measure the sustainability of agroforestry systems implemented in the south region of Brazil, Lopes and Almeida (2002) established a few indicators related to the production system management; to the productivity of the land and of the labour; the economic resilience; the relation with the markets; the net income; the quality of the soil; the impacts caused in other systems; the decision-making process; and the institutional and community participation. They compared the sustainability of agroforestry systems composed by different species arrangements and institutional arrangements such as the individual arrangement, where the producers work alone or with their family, without any bond to enterprises, associations or cooperatives; the integrated arrangement, where the agribusiness provides inputs and technical assistance, getting the agroforestry production in return; and the associative arrangement, where the producers are gathered in associations or cooperatives, aiming at improving commercialization, exchanging technical assistance, increasing the acquisition of credit and social assistance (Lopes and Almeida 2002). The authors concluded that the systems reaching a higher level of sustainability were those represented by the associative arrangement, because the exchange of knowledge and the producers' participation in decision-making processes strengthened the practice methods. The authors also concluded that the systems whose indicators were in balance showed a greater degree of sustainability; nevertheless, they stressed the fact that local factors can also have influence on the sustainability level (Lopes and Almeida 2002).

It is interesting to highlight that each agroforest is unique because its management depends on many local, social and environmental factors. Agroforestry can answer both small farmers' and large companies' needs, adapting to the size of the property and to the amount of economic investment (Vivan 2001). Hence, some Brazilian companies, such as the cosmetics company Natura, are already practicing agroforestry. In 2012, the global initiative TEEB—The Economics of Ecosystems and Biodiversity (Bishop 2012), launched a study project on business initiatives that promote biodiversity conservation in Brazil, the "TEEB for business Brazil project". The study compared the environmental value associated to the palm oil harvest from a monoculture system and another harvest from an agroforestry system. The environmental value obtained from agroforestry palm oil was three times higher than the monoculture one. The agroforestry system also contributed to the increase of the farmer's income and reduced the risk of pests and diseases in crops (Bishop 2012).

Moreover, in order to verify if agroforestry systems could fit into the strong sustainability concept, Seoane et al. (2014) analysed its potential to generate forest restoration. They concluded that the agroforestry systems could indeed fit into the strong sustainability paradigm, because of the environmental services provided, being a good alternative for restoration of Brazilian forests. However, they stressed the fact that the implementation of agroforestry systems to forest restoration can face some legal and conceptual obstacles, once the agroforestry system is a productive system, not a restoration methodology (Sloane et al. 2014).

From all the arguments above, it can be concluded that the agroforestry system represents a successful model for fostering rural sustainable development as it relies only on renewable resources, not depending on artificial inputs and it maintains or even increases the quality and quantity of natural capital over time, once the practice can be used for recovering damage areas. Thus, it clearly fits into the concept of strong sustainability. However, we need to analyse if the Brazilian legal framework is capable of taking into account all the agroforestry features (II).

II The arduous task of implementing agroforestry in Brazil

In this section, we will analyse agroforestry systems under Brazil's legislation perspective (A). Then, we will identify the obstacles to its implementation, in order to propose solutions (B).

A. The agroforestry system within Brazilian Legislation

First, it is important to clarify a key aspect of Brazilian legislation. In environmental topics, sometimes in order to enforce the law, it may be necessary to detail some administrative or practical procedures referring to the legal instrument's content. The regulatory agencies[8] are responsible for the detailing through the elaboration of normative instructions and ordinances.

A normative instruction is an administrative act issued by the "head of the service" to his subordinates establishing disciplinary rules that should be adopted for the functioning of public service (Acquaviva 1999). Ordinances are acts by which the competent authorities determine administrative character arrangements,

[8]Such as the Brazilian Institute of Environment and Renewable Natural Resources, the Chico Mendes Institute for Biodiversity Conservation and the state and municipality's environmental bodies.

give instructions on execution of laws and services, define functional situations and apply disciplinary measures (Kaspary 2004).

Environmental boards (federal, state or municipal) are responsible for drafting resolutions to regulate issues not directly treated by the law (Ewert 2014). No legal instrument should be in conflict with the Brazilian Constitution.

Below we will examine the agroforestry systems according to the Brazilian Constitution (1) and the Brazilian Forest Code (2).

1. The agroforestry system within the Brazilian Constitution

According to Article 5°, XXIII, all property must fulfil its social function, since property is not only an individual right but a social right, subject to a purpose. The Brazilian Civil Code defines this right and duty of the owner in Article 1228, §1, which provides that "the right of ownership must be exercised in accordance with its economic and social purposes and in a way to preserve the flora, fauna, natural beauty, ecological balance and the historical and artistic heritage, and avoid air and water pollution."[9] Therefore, all property must also fulfill an environmental function, ensuring the protection of the environment and the rational use of natural resources. It could not be any different since the guarantee to a healthy environment is a right of citizens and a duty to be protected by the government and throughout society. Thus, the environmental function authorizes the imposition of positive obligations to the owner and not just a duty to refrain.

As for the rural property, the Constitution established some guidelines for satisfying the social function. The farmer must make use of the rural property in a rational and appropriate way, properly using the natural resources available, ensuring the preservation of the environment, observing the regulations of labour relations, and assuring a form of usage that favours the well-being of the owners and workers (Article 186, I, II, III and IV). In this sense, the rural property meets its social function when it is productive without damaging the environment.

Non-compliance with the regulations concerning the socio-environmental function may provoke the property's expropriation. The property will then be destined to land reform purposes, in order to reduce social inequalities (Article 184). The severity of the penalty demonstrates the importance given by the law to compliance with the socio-environmental function.

That said, agroforestry systems are in complete harmony with the constitutional impositions above. As seen previously, agroforestry systems have a threefold vocation: economic, social, and ecological. It is a sustainable system of land-use that aims to optimize production through the appreciation of nature.

In addition to Articles 5 and 186, the Federal Constitution disposes about the environment in different titles and chapters. Chapter VI however, is devoted specifically to it. Within this chapter, Article 225 provides that "everyone has the right to an ecologically balanced environment, good for common usage and essential to a healthy quality of life, imposing both the Government and the

[9]Unofficial translation by the author of this article.

community the duty to defend it and preserve it for present and future generations".[10] The Government is responsible for ensuring the effectiveness of such a right and it has the duty to preserve and restore the essential ecological processes and promote the ecological management of species and ecosystems (Article 225, §1, I).

The International Union for Conservation of Nature (IUCN 2012) defines ecological restoration as the care process for the recovery of degraded, damaged or destroyed ecosystems. It is a deliberate intervention to accelerate recovery of the ecosystem with respect to its structure and its functional properties, taking into account the different exchanges with the environment. According to this concept, the agroforestry system is a viable alternative for the recovery of degraded areas for it seeks a greater functional similarity with the composition and vegetation structure of a native forest. Therefore, the species to be introduced in the system are chosen according to the ecophysiological[11] characteristics of those belonging to the local vegetation. Hence, the regeneration of environmental functions is made possible through the observation and attempt to reproduce the natural ecosystem dynamics, that allows the increase of biodiversity and makes the environment more complex (Silva 2008). In this sense, as seen previously, Sloane et al. (2014) concluded that agroforestry systems could represent a viable way to restore degraded areas.

Regarding the concept of ecological management, the dictionary of environmental law defines it as "the use of natural resources by man, based on principles and methods that preserve the integrity of ecosystems, reducing human interference in the self-regulatory mechanisms of living organisms and the physical environment"[12] (Krieger et al. 2008). Within this logic, agroforestry systems favour the interactions between the different natural components in order to reduce the use of external inputs and to reduce the environmental impacts caused by agricultural activities.

According to Paulo Affonso Machado (2012), what brings efficiency to Article 225 is the right to propose an environmental lawsuit. There are two types of lawsuits aimed at guarantying the protection of the environment: the popular lawsuit that can be proposed in justice by any citizen in order to cancel a harmful administrative act against the environment or to repair an environmental damage (Article 5°, LXXIII of the Federal Constitution) and the civil public lawsuit proposed by the federal prosecutors in order to protect the environment and to repair an environmental damage. (Article 129, III of the Federal Constitution). Both lawsuits are based on the collective interest in protecting nature and any person or institution responsible for an environmental damage, private or public, can figurate as a defendant.

[10]Unofficial translation by the author of this article.

[11]Physiological ecology is the study of how organisms function and respond to changes in their natural environments.

[12]Unofficial translation by the author of this article.

Therefore, according to the Brazilian Constitution, the implementation of agroforestry systems is not only possible but it is legitimate, fulfilling the socio-environmental function of the land. Nevertheless, we will analyse further on if the overall Brazilian legislation is capable of prescribing solutions to the existing conflicts regarding its implementation.

2. The agroforestry systems within the Brazilian Forest Code (Act 12.651/2012)

In 2012 a modification of the Forest Code gave rise to many debates. On the one hand, large farmers sought the reduction of their obligations, claiming excessive severity of the 1965 Forest Code, prevailing at that time. On the other hand, scientists warned for the damages to the environment that could occur if the text was adopted. Despite a heated debate, the act was approved in May 25, 2012.

The new text may have been considered a major setback by environmentalists, because among its changes, the text determined the decrease of the delimitation of certain protected areas such as the areas of permanent preservation (APP)[13] and the legal reserve areas (LR).[14] It was however a great breakthrough for the agroforestry cultivators, once the new code favoured the implementation of agroforestry systems for forest restoration in the same quoted protected areas above. From a general perspective, it is believed that the legislative change will encourage the environmental regularization of consolidated areas.[15] The environmental regularization should follow the guidelines put forth by the Environmental Adjustment Program (EAP), established by the Federal Government or the States. As an incentive to the regularization, the plan exempts landowners of infractions committed before July 22, 2008 from sanctions, if they are registered in the Rural Environmental Registry (Article 59).

Regarding agroforestry systems specifically, the Code considers sustainable agroforestry as a casual activity or of low environmental impact (Article. 3°, X), and classifies it as an activity of social interest (Article 3°, IX). As a result, the removal of vegetation in APP or LR for agroforestry activities on small farms is

[13]According to Article 3, II of the Forest Code, a permanent preservation area is a "protected area, covered or not by native vegetation, with the environmental function of preserving water resources, landscape, geological stability and biodiversity, facilitating gene flow of fauna and flora, ensuring soil protection and the well-being of human populations." Unofficial translation by the author of this article.

[14]According to Article 3, III of the Forest Code, a legal reserve is an "area located within a property or rural possession, defined under Art. 12, with the task of ensuring the sustainable economic use of the property's natural resources assisting the conservation and rehabilitation of ecological processes and promoting biodiversity conservation." Unofficial translation by the author of this article. Article 12 prescribes the demarcation parameters of the legal reserve in accordance with the type of native vegetation in the region.

[15]The Forest Code defines the consolidated areas as those where human occupation was established until July 22, 2008, with buildings, improvements or agrosilvopastoral activity. The law allows the continuation of agroforestry activities, ecotourism and rural tourism if they comply with certain terms of commitment. The law also provides for the obligation of recovery of vegetation removed without authorization.

possible, but depends on certain conditions. The implementation is permitted as long as it does not harm the ecological function of the area and the cutting of native vegetation[16] is prohibited. In addition, the activity depends on the development of a sustainable forest management plan approved by the competent environmental authority and on a simple statement issued by the same authority once the property is registered in the Rural Environmental Registry (Article 31).

The Rural Environmental Registry is a public record with effects for third parties. The registry has the objective of defining the borders of the preserved area and composing a database for the control; monitoring; environmental and economic planning and fighting against deforestation. Regarding the registration of a small farm, registering is free and the owner shall present his identification, a property or possession proof and the LR's field of identification data (Article 8° Decree 7.830/2012).

The Forest Code also encourages the implementation of agroforestry through the Support and Encouragement of the Preservation and Restoration of the Environment Program (Article 41). The program provides for the possibility of payment or incentives for environmental services (Article 41, I) and will be primarily focused on small farmers (Article 41, § 7°). It will help them obtain compensation for nature conservation measures through tax reliefs, concessions of agricultural credit (Article 41, II), or through the allocation of resources for scientific research (Article 41, §1, I). The program is rooted on the social function of property, to ensure its appropriate use and the farmer's welfare (Machado 2012).

The Forest Code highlights the importance of the different functions of agroforestry as it encourages its implementation for the purpose of forest restoration, environmental conservation, sustainable management and provision of environmental services. However, there are some obstacles to the implementation of agroforestry systems, as we shall see below (B).

B. Obstacles and suggestions for the implementation of agroforestry in Brazil

The Brazilian environmental legislation analysed so far encourages the implementation of agroforestry systems, but reality is somewhat different regarding the subordinated norms, and obstacles abound. First, there are certain legal instruments that penalize practices often necessary for the implementation of agroforestry systems. Act 11.428/2006, which regulates the use and protection of the native vegetation of the Atlantic Forest biome, determines the types of management allowed in the different stages of regeneration of vegetation and prohibits the cutting of vegetation in an advanced stage of regeneration. However, for the proper functioning of the agroforestry system, pruning or cutting of vegetation for the renewal of the system is necessary. The farmer that does so can then be framed in the Environmental Crimes Act (Act 9605/1998) for practice of deforestation

[16]Native vegetation is the natural vegetation from a particular region, without human interference.

(Article 38). Moreover, Decree 660/2008 that regulates such an act forbids the removal of endangered species.[17] The cut of endangered vegetal species is prohibited, even when planting was done by the farmer himself and in situations that could enhance its occurrence, as in agroforestry systems (Ewert 2014).

Furthermore, according to Ewert (2014) legislation is demanding regarding some bureaucratic procedures and technical regulations are often difficult to interpret, a fact that discourages compliance with environmental standards by small producers. For instance, the Forest Code authorizes the implementation of agroforestry systems in APP's and LR's if the property is registered in the Rural Environmental Registry. However, several small farmers cannot present the required documents for registration. Many of them don't hold a title deed, sometimes not even an identity card. Hence, it is necessary to establish a legal framework better adapted to the specificities of the agroforestry system and to the Brazilian rural reality.

The lack of information and technical assistance is also a great burden. There is no point in creating legal instruments that promote the implementation of agroforestry systems without proper information and capacity building of farmers, because agroforestry systems may not adequately perform their duties and the sustainability exalted by Brazilian law will not materialize (Martins and Ranieri 2014).

Another major obstacle is Brazil's current political situation, since the main goal of the Brazilian agricultural policy is the expansion of agribusiness, focused primarily on the export of commodities. Because of that, there are few credit programs focused on the development of agroforestry systems, or even small rural properties. The rural sector has undeniable influence on public policy making. The Agriculture Parliamentary Front[18] has the support of 191 deputies and 11 senators, the equivalent to half of the National Congress (Camara dos Deputados 2015). Given that this interest group does not represent small farmers, it is clear that the country's present political situation is not conducive to the development of agroecological practices. The guarantee of more incentives and credit could bring economic viability to the first stage of implementation of the agroforestry system, which usually has low profitability.

An additional factor burdening the establishment of agroforestry systems is the difficult access to markets. To minimize this problem, the Education Ministry, Deliberative Council of the National Education Development Fund Resolution No. 38/2009 established the National School Meal Program providing that all public schools must buy at least 30 % of the food used in school meals directly from farmers, primarily opting for organic and agroecological products (Act 11.947/2009). The program ensures a guaranteed source of income for small farmers and

[17]According to Article 3°, I of the Normative Instruction n° 6/2008 from the Ministry of the Environment, endangered species are "those at high risk of extinction in a near future, as recognized by the Ministry of the Environment, based on scientific documentation available". Unofficial translation by the author of this article.

[18]A Parliamentary Front is composed by members of the legislative power following a common interest.

promotes the development of agroforestry and agroecological practices (Rorro and Miccolis 2011). The Food Acquisition Program, established on 2, July, 2003, by Act 10.696, also encourages the purchase of food coming from the family farm, representing an opportunity for agroforestry cultivators.

Difficulties in certifying agroforestry products may also be seen as a barrier to marketing. However, Brazilian law provides for the possibility of certification of agricultural products through participatory systems, which can be more accessible to small farmers. According to Act 10.831 established on 23, December 2003 (regulated by Decree 6.323/2007), environmental certification can be done in two different ways: certification by audit and certification by Participatory Systems of Organic Quality Assurance. In both cases, the certification organizations should be previously accredited by the Ministry of Agriculture, but the certification by audit is made by an independent certifying agency and the certification by Participatory Systems of Organic Quality Assurance is done through a mutual control of farmers in compliance with agroecological production standards. Furthermore, certification is optional in case of direct commercialization from farmers to consumers, once the farmers are previously registered under the surveillance authority and attached to an association. The participatory system could represent a great opportunity for the opening of new markets, as it is a certification mechanism based on farmers' credibility and relies on the participation of all stakeholders willing to ensure the quality of the final product.

Moreover, an additional way for increasing the farmer's income could be the adoption of the Payment for Environmental Services mechanism.

According to UNEP (2008), there are four types of ecosystem services: the environmental goods, the regulating services, the supporting services and the cultural services. The key characteristics of payment for ecosystem services is that "the focus is on maintaining a flow of a specified ecosystem "service" (UNEP 2008, p. 3) and that "the payment causes the benefit to occur where it would not have otherwise" (UNEP 2008, p. 3).

As seen previously, the Brazilian Forest Code establishes the Support and Incentives for Conservation and Environment Recovery Program, which introduces the mechanism into Brazilian legislation. In this context, Bill 792/2007 and Bill 5487/2009 are currently ready to be voted by the National Congress. Both aim to implement the National Policy on Payment for Environmental Services. The country does not yet have a federal regulatory instrument to regulate the PES. Hence, many states and municipalities issued laws establishing their own PES.

There are already some successful PES schemes in Brazil. In 2012, the State of Espírito Santo created the Reforestation Program (Act 9.864/2012 regulated by Decree 3.182-R/2012). Its goal is to compensate certain land use methods (which include agroforestry systems) that generate environmental services through a contract between the landowner and the Department of the Environment and Water Resources (Article 3, d of Decree 3.182-R/2012). The economic rewards for the maintenance/generation of environmental services can be granted by providing financial support for the development of technical projects or through financial compensation made directly to the owner (Tejeiro and Stanton 2014).

Albeit the several remaining obstacles to agroforestry's implementation, Brazilian legislation is starting to take into account alternative agriculture practices that could foster a more sustainable development of the rural sector.

2 Conclusion

Brazil's colonization process deeply marked the development of agricultural activity in the country. Several features of the current rural system were inherited from the time of the Portuguese Kingdom. The dominance of agribusiness for commodities export, the supremacy of the monoculture system and the monopoly of big landowners are strong features of the Brazilian reality. Yet, we observe the dissemination of agroecological practices, mainly linked to family farming.

Hence, family farming has the potential to provoke important changes in the Brazilian way of agricultural production, which is basically composed by export-oriented monoculture fields. Family farming is generally represented by an intensive practice, once the farmer seeks to obtain the maximum profit per acre cultivated. However, the sustainability of the system is equally important to the farmer, since agriculture is often his only income source. Thus, the family farmer should diversify the production in order to ensure the quality of the soil and the adequate replacement of nutrients. In this context, agroforestry is a practice that enhances different aspects of small-scale agriculture, since it promotes biodiversity conservation without jeopardizing farm output. This is one of the reasons why, in Brazil, agroforestry is mostly disseminated amongst family farmers (Bolfe and Bergamasco 2010).

This paper also demonstrated that agroforestry practices can perfectly fit into the concept of strong sustainability, once it is a completely renewable system, without the use of artificial inputs and it maintains and even increases the biodiversity level, as it can also be implemented to recover damage areas. Such practices may contribute to the promotion of the three pillars of sustainable development, as they are environmentally friendly, socially fair and economically viable, representing the conciliation of the development of agricultural production and environmental protection.

Nevertheless, this potential is constrained by the lack of supportive regulatory framework and the lack of a public policy specially aimed at developing the agroforestry systems. Although Brazil already has several instruments that legitimize and encourage the practice, such instruments need to be harmonized, since the legislation is often contradictory, encouraging the implementation of agroforestry systems while criminalizing certain practices necessary for its management.

In conclusion, we observe that it is necessary to draw up a legal instrument better adapted to the complexity of the agroforestry system, capable of presenting solutions to the present and future conflicts that may arise.

References

Abdo MTVN, Valeri SV, Martins ALM (2008) Sistemas Agroflorestais e Agricultura Familiar: Uma Parceria Interessante. Revista Tecnologia e Inovação Agropecuária. Retrieved 5 April 2015. http://www.dge.apta.sp.gov.br/Publicacoes/T&IA2/T&IAv1n2/Artigo_Agroflorestais_5.pdf

Acquaviva MC (1999) Dicionário Acadêmico de Direito. Editora Jurídica Brasileira, São Paulo

Agricultura, do subsidio à política agrícola (2011) Year 8, 68 edn. Retrieved 12 Dec 2015. http://www.ipea.gov.br/desafios/index.phpoption=com_content&view=article&id=2599:catid=28&Itemid=23

Altieri MA (2005) Agroecology: principles and strategies for designing sustainable farming systems. Retrieved 5 Jan 2014. http://www.biosafety-info.net/article.php?aid=291

Balsan R (2006) Impactos decorrentes da modernização da agricultura brasileira. CAMPO-TERRITÓRIO: Revista de Geografia Agrária, vol 1(2), pp 123–151

Bianchini PC (2006) Agroflorestas e Agentes Agroflorestais no Acre. Dissertation. Federal University of Santa Catarina

Bishop J (ed) (2012) TEEP—the economics of ecosystems and biodiversity in business and enterprise. Earthscan, London

Bolfe APF, Bergamasco SPP (June) Desvendando a Relação: Agricultores Familiares e Sistemas Agroflorestais. Paper presented at the 48° Congress Brazilian Society of Economy, Administration and Rural Sociology, Campo Grande

Brazil. Federal Constitution of 1988. October, 5th, 1988. Retrieved10 Feb 2015. http://www.planalto.gov.br/ccivil_03/constituicao/constituicaocompilado.htm

Brazil. Presidency of the Republic. Decree 1.946/1996 that establishes the National Program for Strengthening Family Agriculture. 28 June 1996. Retrieved 10 Feb 2015. http://www.planalto.gov.br/ccivil_03/decreto/D1946.htm

Brazil. Federal Act 9605/1998. Environmental Crimes Act. 12 Feb 1998. Retrieved 10 Feb 2015. http://www.planalto.gov.br/ccivil_03/leis/l9605.htm

Brazil. Civil Code. Act 10.406/2002. 10 Jan 2002. Retrieved 10 Feb 2015. http://www.planalto.gov.br/ccivil_03/leis/2002/l10406.htm

Brazil. Federal Act 10.696/2003 creates the food acquisition program. 3 July 2003. Retrieved 10 Feb 2015. http://www.planalto.gov.br/ccivil_03/leis/2003/l10.696.htm

Brazil. Federal Act 10.831/2003 provides guidelines for the organic agriculture. 23 Dec 2003. Retrieved 10 Feb 2015. http://www.planalto.gov.br/ccivil_03/leis/2003/l10.831.htm

Brazil. Federal Act 11.428/2006 regulates the use and protection of native vegetation of the Atlantic Forest biome. 22 Dec 2006. Retrieved 10 Feb 2015. http://www.planalto.gov.br/ccivil_03/_ato2004-2006/2006/lei/l11428.htm

Brazil. Bill 792/2007 defines the mechanism of payment for ecosystem services. 19 Apr 2007. Retrieved 10 Feb 2015. http://www.camara.gov.br/proposicoesWeb/fichadetramitacao?idProposicao=348783

Brazil. Presidency of the Republic. Decree 6.323/2007 regulates Act 10.831/2003. 27 Dec 2007. Retrieved 10 Feb 2015. http://www.planalto.gov.br/ccivil_03/_ato2007-2010/2007/Decreto/D6323.htm

Brazil. Ministry of the Environment. Normative Instruction no. 06 regulates about endangered species. September 2008. Retrieved 10 Feb 2015. http://www.mma.gov.br/estruturas/179/_arquivos/179_05122008033615.pdf

Brazil. Presidency of the Republic. Decree 660/2008 regulates Act 9605/1998. 24 Nov 2008. Retrieved 10 Feb 2015. http://www.planalto.gov.br/ccivil_03/_ato2007-2010/2008/decreto/d6660.htm

Brazil. Federal Act 11.947/2009 sets out guidelines on the establishment of the National School Meal Program. 16 June 2009. Retrieved 10 Feb 2015. http://www.planalto.gov.br/ccivil_03/_ato2007-2010/2009/lei/l11947.htm

Brazil. Bill 5487/2009 aims to implement the national policy on payment for environmental services. 24 June 2009. Retrieved 10 Feb 2015. http://www.camara.gov.br/proposicoesWeb/fichadetramitacao?idProposicao=439941

Brazil. Education Ministry, Deliberative Council of the National Education Development Fund Resolution No. 38/2009 creates the National School Meal Program. 16 July 2009. Retrieved 10 Feb 2015. http://www.google.fr/url?sa=t&rct=j&q=&esrc=s&source=web&cd=1&ved=0CCEQFjAA&url=http%3A%2F%2Fwww.fnde.gov.br%2Farquivos%2Fcategory%2F60-2012%3Fdownload%3D57%3Ares038-16072009&ei=wUlCVZ3RHsXLaMSpgdAH&usg=AFQjCNGu_Mkfw1xRAc-cTEcdjfgp7aQp0A&sig2=fVoKPhTfJACD2GP95pFl6-A&bvm=bv.92189499,d.d2s

Brazil. Resolution of the National Environmental Council—CONAMA 429/2011 establishes the methodology of Permanent Protection Areas recovery. 28 Feb 2011. Retrieved 10 Feb 2015. http://www.mma.gov.br/port/conama/legiabre.cfm?codlegi=644

Brazil. Forest Code. Federal Act 12.651/2012. 25 May 2012. Retrieved 10 Feb 2015. http://www.planalto.gov.br/ccivil_03/_ato2011-2014/2012/lei/l12651.htm

Brazil. Presidency of the Republic. Decree 7.794/2012 establishes the National Policy for Agroecology and Organic Production. August, 20, 2012. Retrieved 10 Feb 2015. http://www.planalto.gov.br/ccivil_03/_ato2011-2014/2012/decreto/d7794.htm

Brazil. Presidency of the Republic. Decree 7.830/2012 establishes guidelines regarding the Rural Environmental Registry. 17 Oct 2012. Retrieved 10 Feb 2015. http://www.planalto.gov.br/ccivil_03/_Ato2011-2014/2012/Decreto/D7830.htm

Brazil. ES State Governor. Decree 3182-R/2012 regulates Act 9.864/2012. 21 December 2012. Retrieved 10 Feb 2015. http://www.legisweb.com.br/legislacao/?id=248833

Brazil. ES State Act 9.864/2012 creates the PES Reforestation Program. 27 Dec 2012. Retrieved 10 Feb 2015. https://www.legisweb.com.br/legislacao/?id=242674

Brazil. Federal Act 12.854/2013 provides that the Federal Government should encourage, through public policies, the implementation of agroforestry systems in expropriated or degraded rural areas. 26 Aug 2013. Retrieved 10 Feb 2015. http://www.planalto.gov.br/ccivil_03/_Ato2011-2014/2013/Lei/L12854.htm

Caron P, Cairol D (2008) Multifunctionality: what do we know of it? In: Brouwer F, van Rheenen T, Dhillion SS, Elgersma AM (eds) Sustainable land management: strategies to cope with the marginalisation of agriculture. Edward Elgar Publishing Limited, Massachusetts and Glos, pp 41–60

Clarke WC, Thaman RR (1993) Agroforestry in the Pacific Islands: Systems for Sustainability. United Nations University Press, Tokyo

Convention on Biological Diversity. Retrieved 13 Dec 2014. http://www.cbd.int/convention/

Dal Soglio F, Kubo RR (Organisers) (2009) Agricultura e Sustentabilidade. Porto Alegre: Editora UFRGS

Dutu M (2013) Sustainable development in a globalized world—the rural and agricultural dimension. In: Richli P (ed) Agriculture and the requirements of a sustainable development. XXVI European Congress and Colloquium of Rural Law. L'Harmattan, Paris

Espirito Santo BR (2001) Os caminhos da agricultura brasileira. Evoluir, São Paulo

Ewert M (2014) Incentivos e limites da legislação ambiental brasileira para os sistemas agroflorestais: o caso cooperafloresta. Dissertation. Federal University of Santa Catarina

FAOSTAT (2012) Food and agricultural commodities production Brazil. Retrieved 15 Apr 2015. http://faostat.fao.org/DesktopDefault.aspx?PageID=339&lang=en&country=21

Finatto RA, Salamoni G (2008) Agricultura familiar e agroecologia: perfilda produção de base agroecológica no Município de Pelotas/RS. Uberlândia: Sociedade & Natureza

Flobert P (2000) Dictionnaire Latin-Français. Hachette Livre, Paris

Forestry policies, institutions and programs, Brazil (2010) Retrieved 20 Jan 2015. http://www.fao.org/forestry/country/57479/en/fra/

France. Agricultural Orientation Act (1999) Retrieved 10 Feb 2015. http://www.legifrance.gouv.fr/affichTexte.do?cidTexte=JORFTEXT000000395813

Camara dos Deputados. Frente Parlamentar Agricola (n.d.) Retrieved 15 Feb 2015. http://www.camara.gov.br/internet/deputado/Frente_Parlamentar/356.asp
G1 Journal (2014) Embraba oferece vagas para curso de sistema de tecnologia agroflorestal. Retrieved 27 Jan 2015. http://g1.globo.com/ap/amapa/noticia/2014/06/embrapa-oferece-vagas-para-curso-de-sistema-de-tecnologia-agroflorestal.html
Gitau T, Gitau MW, Valtner-Toews D (2008) Integrated assessment of health and sustainability of agroecosystems. CRC Press, London
IUCN (2012) Ecological restoration for protected areas. Switzerland, Global Protected Areas Program
Kaspary AJ (2004) Redação Oficial: Normas e Modelos. Edita, Rio de Janeiro
King KF, Chandler NT (1978) The wasted lands: the program of work of the International Council for Research in Agroforestry (ICRAF), Nairobi, Kenya
Krieger MG, Maciel AMB, Noble GD, Rocha JCC, Finatto MJB (2008) Dicionário de Direito Ambiental: Terminologia das leis do meio ambiente, 2nd edn. Editora Lexikon
Leff H (2001) Saber ambiental, sustentabilidade, racionalidade, complexidade e poder. Editora Vozes, Petrópolis
Lima A (2008) Ordenamento territorial In Almanaque Brasil Socioambiental. ISA, São Paulo, pp 324–326
Lopes SB, Almeida J (2002) Metodologia para Análise Comparativa de Sustentabilidade em Sistemas Agroflorestais. Revista da Sociedade Brasileira de Economia e Sociologia Rural. Retrieved 15 Apr 2015. http://www.ufrgs.br/pgdr/arquivos/436.pdf
Machado PAL (2012) Legislação Florestal (Lei 12.651/2012) e Competência e Licenciamento Ambiental (Lei Complementar 140/2011). São Paulo: Malheiros Editores
Martins TP, Ranieri VEL (2014) Sistemas agroflorestais como alternativa para as reservas legais. Ambiente e Sociedade, São Paulo. vol XVII(3), pp 79–96
Moreira RM, Carmo MS (2004) Agroecologia na construção do desenvolvimento rural sustentável. Agriculturas São Paulo, vol 51(2), pp 37–56
MPA (2010) Censo Agropecuário confirma: agricultura camponesa é a principal produtora de alimentos do país. Retrieved 9 Apr 2015. https://mpabrasiles.wordpress.com/2010/02/18/censo-agropecuario-confirma-agricultura-camponesa-e-a-principal-produtora-de-alimentos-do-pais/
Müller MW (2006) Importância dos Sistemas Agroflorestais para a sustentabilidade dos biomas tropicais. 28 Semana do Fazendeiro, Caderno I. Uruçuca: Ministério da Agricultura, pp 64–73
Neumayer E (2003) Weak versus strong sustainability: exploring the limits of two opposing paradigms, 2nd edn. Edward Elgar, Cheltenham
OECD—Organisation for Economic Co-operation and Development. Environmental Indicators for Agriculture (2001) Methods and results, 3rd edn. OECD Publications Service, Paris
Pascual U, Perrings C (2009) Developing mechanisms for in situ biodiversity conservation in agricultural landscapes. In: Ninan KN (ed) Conserving and valuing ecosystem services and biodiversity: economic, institutional and social challenges. Earthscan, London, pp 151–174
Raman S (2006) Agricultural sustainability: principles, processes, and prospects. Food Products Press, New York
Reifschneider FJB, Henz GP, Ragassi CF, Anjos UG, Ferraz RM (2010) Novos ângulos da história da agricultura no Brasil. Embrapa Informação Tecnológica, Brasília, DF
Report of the Conference of the Parties to the Convention on Biological Diversity at its fourth meeting. Retrieved 20 Nov 2014. http://www.mma.gov.br/estruturas/sbf_chm_rbbio/_arquivos/cop4ing_72.pdf
WCED (1997) World commission on environment and development report "our common future" also named "Brundtland Report". Retrieved 5 Dec 2015. http://www.un-documents.net/wced-ocf.htm
Rorro R, Miccolis A (eds) (2011) Polticas Públicas para o Desenvolvimento Agroflorestal no Brasil. ICRAF, Belém

Santos AC (2007) A agroforesta Agroecologica: Um momento de sintese da Agroecologia, uma agricultura que cuida do meio ambiente. Departamento de Estudos Socio-Economicos Rurais, no. 156

Seoane CE, Froufe LC, Amaral-Silva J, Arantes ACV, Nogueira R, Steenbock W (2014) Conservação Ambiental Forte Alcançada Através de Sistemas Agroflorestais Multiestratificados. 1—Agroflorestas e a Restauração Ecológica de Florestas. Paper presented at Agroecol—First Seminary of Agroecology of South America. Dourados

Schroth G, da Fonseca GAB, Harvey CA, Gascon C, Vasconcelos HL, Izac A-MN (eds) (2004) Agroforestry and biodiversity conservation in tropical landscapes. Island Press, Washington

Silva VJM (2008) Uso de sistemas agroflorestais como viabilizadores de corredores ecológicos. Dissertation. Federal University of Lavras

Taýra F (2006) Capital natural e graus de sustentabilidade: visões só mundo e objetivos conflitantes. Pensamento e Realidade 19:100–118

Tejeiro G, Stanton M (eds) (2014) Sistemas Estaduais de Pagamento por Serviços Ambientais: Diagnóstico, lições aprendidas e desafios para a futura legislação. Instituto O Direito por um Planeta Verde, São Paulo

Teixeira JC (2005). Modernização da agricultura no Brasil: impactos econômicos, ambientais e sociais. Revista Eletrônica da Associação dos Geógrafos Brasileiros—Seção Três Lagoas Três Lagoas-MS, vol 2(2)

UNEP (2008) Payments for ecosystem services: getting started. Nairobi: UNON. Retrieved 5 Apr 2015. http://www.unep.org/pdf/PaymentsForEcosystemServices_en.pdf

Vivan JL (2001) Diversificação e manejo em sistemas agroflorestais. III Congresso Brasileiro de Sistemas Agroflorestais. Embrapa, Manaus

Voigt C (2013) The principle of sustainable development: integration and ecological integrity. In: Voigt C (ed) Rule of law for nature: new dimensions and ideas in environmental law. Cambridge University Press, Cambridge

Next Generation Rural Natural Resource Governance: A Careful Diagnosis

Paul Martin and Jacqueline Williams

Abstract Achieving "sustainable development" requires the exploitation without diminution of financial capital, manufactured capital, intellectual capital, human capital, social and relationship capital, and (particularly) natural capital. This requires effective natural resource governance, to guide human uses of the earth into sustainable patterns. Agriculture and rural communities are central to sustainable development because among other reasons: agricultural activities typically require natural environments; agriculture is resource dependent and fundamental to society; and rural people are frequently relatively poor. Good rural natural resource governance is thus a prerequisite for sustainable development.

Keywords Human behaviour · Efficiency · Institutions · Justice

1 Introduction

Rural natural resource governance shapes the human behaviour that affects the economic, social or ecological sustainability of natural capital, the sharing of resource benefits, the costs of use, and the costs of conservation (including of the governance system). Particular behaviours drive exploitation, contamination, conservation, and restoration, market transactions and other aspects of resource exploitation or protection. Whilst 'governance' conjures up government and the law, resource use and conservation are driven by people and organisations serving

Paul Martin—Program Leader Invasive Animals CRC 'Effective Community led Action'

P. Martin (✉) · J. Williams
Australian Centre for Agriculture and Law, University of New England, Armidale, Australia
e-mail: pmartin9@une.edu.au

J. Williams
e-mail: Jacqueline.Williams@une.edu.au

V. Mauerhofer (ed.), *Legal Aspects of Sustainable Development*,
DOI 10.1007/978-3-319-26021-1_30

diverse roles using many different types of instrument. These include: government as lawmaker and regulator, land manager, creator and defender of private rights, and instigator and implementer of public policy; primary industries such as mining and farming, and the industries that add value to or deal in rural products; and civil society as purchasers and consumers, political activists, volunteers, taxpayers and voters. The many behaviours that create governance outcomes are shaped by the political system, social dynamics, the economy, culture, technology, and industrial systems. Governance involves responses to complex and dynamic drivers, and requires multi-faceted interventions across society (Martin 2008a; Scarlett et al. 2013).

Governance success depends upon whether the interventions ‘fit’ with the context and can be resourced. The hypothesis this chapter develops is that some preferred rural governance approaches fit poorly with the rural context, and that effectiveness is constrained by insufficient implementation capacity. Whilst our focus is Australian natural resource governance, our diagnosis (and prescriptions) apply more broadly to rural natural resource governance.

Economic, social, and environmental governance outcomes reflect a coupling between dynamic human and biophysical systems, through transactions between people or people and nature. Good rural natural resource governance is thus a process of managing complex socio-ecological systems, adjusting transactions to ideally create patterns of resource use and conservation that deliver human services, justly, whilst preserving natural and other capitals. One governance tool is the law, relying on the powers of the state. The law serves many functions in natural resource governance, enabling environmental and social harm, as well as protection, restoration and social benefits. Legal arrangements include property rights, administrative law, contract, trade practices, intellectual property, indigenous rights, taxation, criminal law, and the laws that govern public and private organisational arrangements.

Institutions are shaping forces, and legal arrangements are part of the mix. Beyond the law and governments, markets dictate many outcomes, as do investment arrangements, cultural and sometimes religious or social norms. Powerful forces like taxation, consumer preferences, and competitive positioning of manufacturers or retailers cannot be ignored. The law sits alongside culture (e.g. indigenous or commercial culture), social norms, voluntary collaboration, industry requirements, and other normative mechanisms. These constitute a complex and often inconsistent ‘regulatory’ mosaic.

These considerations indicate that governance involves strategic management of complex social and economic systems. Effectiveness requires reshaping patterns of transactions that couple human systems to ecological systems (Martin and Gunningham 2011). Fundamental to strategic effectiveness are the “fit” between the strategy and the context, securing sufficient of the right resources for implementation, and the quality of implementation (Martin and Verbeek 2006; Burgman et al. 2009). The alignment of context, strategy (including legal instruments), resources and effective implementation will determine how effective rural natural resource governance will be.

2 New Rural Natural Resource Governance Problem Archetypes

Overconsumption and contamination of natural resources such as minerals, water, forests, and biodiversity are archetypical environmental challenges to which governance systems respond using regulation, economic incentives or social mechanisms to control accumulating harms. Managing overconsumption and contamination problems requires the exercise of restraint. Instruments promoting restraint such as regulation, licensing, and economic incentives to conserve are appropriate tactics. Investment in recovery and restoration may also be required, involving different types of instrument. Governing these accumulative harms is less a 'collective action" problem (Adger 2010) than it is an individual action problem. Whilst conventional accumulative challenges will always be with us, contemporary rural natural resource governance involves other archetypes that require institutional strategies that challenge traditional concepts of legal governance.

Established weeds, human or animal disease, soil erosion and salinity, many social problems, and pest animal problems will increase "under their own steam" once the processes of harm are established. They are 'autopoietic': they adapt to circumstances, and increase without external impetus (Seidl 2004). Responding to an established disease, weed, or a pest animal or insect invasion requires investment of funds and labour by those affected (not just restraint by those who have caused the harm). The investment generally has to be coordinated, comprehensive and sustained, a difficult collective action challenge. Intervention can involve complex behaviours, substantial investment and complex coordination. Depending on the issue, coordinated collective action may need to cover large areas. Such coordinated action is more likely when landholders have the same enterprise type. Cattle farmers have different concerns to lifestyle landholders or managers of public conservation estates, whose concerns differ from those of sheep producers, miners, wheat producers or peri-urban communities. Land use and social fragmentation makes autopoietic problems difficult to manage—we detail this further below. The unique character of this type of problem has not been well recognised in rural natural resource governance. Academic understanding of collective action changing nature of risk is generally based on relatively simple problems where individuals know each other and have reasonably symmetrical abilities and interests (Golembiewski and Olson 1966; Lubell 2007; Ostrom 2010). We do not have a lot of knowledge or a body of successful experience to help govern these archetypes. Traditional legal instruments are ill-suited to forcing positive cooperation and investment across tenures where other forms of persuasion or voluntary cooperation fail.

Rural natural resource governance is affected by the changing nature of risk and risk-management. Traditional problem archetypes involve observable actions of people and observable "linear" responses of people and ecosystems. Causes, effects and actors exist within the same jurisdiction, and issues are easy to define. Emerging social and ecological risks exhibit far more complex nonlinear dynamics and span jurisdictional boundaries, frustrating traditional governance (Dietz et al. 2003; Lim 2014).[1] Many causes interact in un-anticipated ways. Sometimes understanding the issue requires intelligence that is not available to authorities, or challenges science. Often the causes, effects, and actors do not respect jurisdictional boundaries. Those who drive the system that causes the problem may be multi-national corporations or networks of corporations who may be more economically powerful than governments, and operate across many jurisdictions (Vorley 2001).

Increasingly rural natural resource governance is concerned with non-linear cause/effect phenomena (Martin and Gunningham 2011; Mayne and Stern 2013, p. 12). Biodiversity loss may be caused by interacting factors including over-harvesting, indirect poisoning, habitat loss, and interspecies competition, coupling complex biophysical, economic, social or other factors. A particularly challenging problem arises when cause and effect transcend institutional structures (particularly jurisdictions) and test the knowledge or power of governments. Even within state boundaries, depending on local laws, governments often have few practical legal options to force people to take coordinated action on private land, or to sustain the investment of labour or funds for the public good. Even where legal authority exists, political impediments to "excessive" government action are substantial in rural areas, where notions of private property rights are powerful norms. Because an owner (or a state) has the ability to exclude interventions touching their estate, this limits the ability to require coordinated action. Solutions to these problems typically require institutional arrangements, though voluntary coordination will reduce formality (and transaction costs)).

There are limits to rural citizens' decision-making capacity and their willingness to take action. The first consideration is the "tragedy of the commons" (Hardin 1968). The benefits of protecting the resource are distributed across society but only some of the benefit (and most of the cost) of rural resource conservation accrues to those whose action or restraint is needed. Across large areas, the ability of society (largely through government) to supervise those who impact upon the resource is limited by transaction costs and jurisdictional limits. Experiments by Moxnes (2000) demonstrate how difficult it is for farmers to understand the consequences of their own actions. In addition the economic imperatives for harm-doing even when it is clear that resource use is unsustainable are very strong (Moxnes 2000). When these factors combine with political resistance to forced conservation, the hurdles to effective rural natural resource governance are substantial.

The management of trans-boundary groundwater illustrates this type of problem: the physical state and dynamics of the aquifer cannot be readily observed, so there is ambiguity about the physics of the resource; extraction and contamination activities occur in different jurisdictions, frustrating supervision and control; and authority and responsibility are fragmented and limited by sovereignty. The 'tragedy of the

[1]Jurisdictional boundaries are defined by nation or state, and organisational authority and power. Constitutions, the limits to the power of the state over private property or the interests of the citizen, and national sovereignty, limit government action.

commons' is also evident (Martin and Becker 2011; Tan et al. 2012; Arnold and Gunderson 2013; Waslekar and Futehally 2013).

The causes of harms may be episodic or chronic, and the resulting governance strategies for each type are different. Episodic harms are generally managed as "risks": interventions are contingent upon phenomena such as the leak of a pollutant or a damaging incident like a storm. Risk instruments like insurance or disaster management, and managing risk perception and responses, are likely to be part of environmental risk governance. Examples include bio-security or natural disaster risks. Chronic harms however typically call for sustained investment of funds or manpower, requiring sustained resourcing and motivational instruments. The requirement for sustained investment is likely to be particularly draining on less wealthy communities. Biodiversity loss is typically a chronic problem, though episodic factors may play a part.

3 Social and Economic Capacity

We have briefly considered the problem archetypes that challenge conventionally used governance strategies. Identifying more effective rural governance strategies should begin with understanding the challenge, the context, and the resources that may be available. Whilst we focus in detail on Australia, the challenges in this wealthy and ecologically unique country mirror the situation around the world.

We open our exploration of the strategic issues with a comparison of countries that are more and those that are less economically dependent upon agriculture, and Australia. The less agricultural comparators are Switzerland, UK, USA, Canada, and France. The more agriculture dependent jurisdictions are Iceland, Argentina, China, Thailand and Indonesia.[2] The comparisons are indicative rather than definitive, because the total number of countries in the world is relatively small (thus limiting the sample) and given the large number of variables involved (Fig. 1).

Authors' diagram using data from Yale Environmental Performance Index (2014), C.I.A (2014).

All data converted to a common 100 point scale for ease of representation.

Countries that are agriculture dependent are likely to be relatively poor, and perform less well on social welfare and on environmental governance measures. Sustainable development (and thus resource governance) outcomes are worse for rural people than for others, and in many jurisdictions environmental capital continues to depreciate.

Whilst national poverty and institutional deficiencies, failings of stewardship, and incapacity of farmers are partial explanations for this, the effects of geography,

[2]Selection of countries was based upon criteria to ensure a diverse sample, including developed and less developed countries, diverse climate and ecological conditions, and degrees of agricultural activity. The findings are considered to be consistent with other intelligence on agriculture and development.

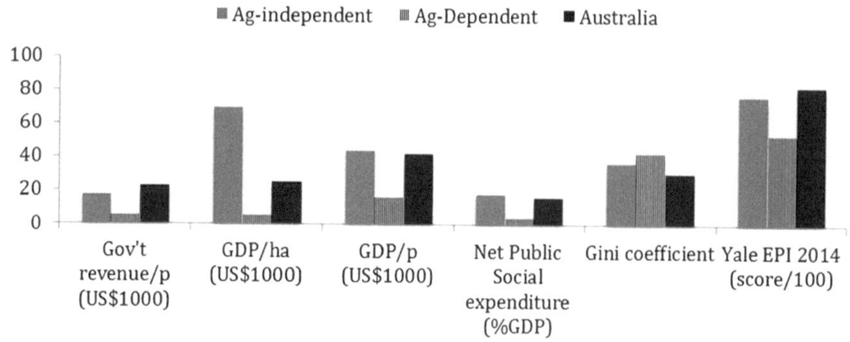

Fig. 1 International comparisons

demographics, ecological and social system dynamics, and the dynamics of markets, are very important. This is in addition to political and cultural considerations including corruption and transparency, education and health, and services. In all of these agriculture dependent jurisdictions the statistics are far worse. This suggests that the drivers of disadvantage and environmental degradation are deeply embedded. It calls into question rural governance strategies that do not address structural causes of governance underperformance. Traditional legal governance relies largely on laws affecting the land manager, rather than those targeting change to the rural system that shapes the transactions of the land manager.

Australia is an informative case study in part because it is rich (comparable to the industrialised USA or the United Kingdom on a per capita basis) unlike many 'developing', agriculturally dependent countries, and it has strong governance institutions. It has reasonable income equality (Whiteford 2014), transparency, and performance across many variables. It has a high standard of literacy, and around 9 % of GDP is spent on a high standard health system. Internationally its public and environmental governance score well. Many of the conventional explanations of poor social or sustainable poor social or sustainable development outcomes in developing countries are thus absent: it is indeed, "The Lucky Country" (Horne 2008, p. 228).[3] However even in Australia patterns of rural social disadvantage and environmental harm mirror developing countries. This has significant implications.

Turning firstly to environmental outcomes a recent paper (Ritchie et al. 2013) by prominent ecological scientists decried the failure of (mainly) rural natural resource governance.

> Australia's highly diverse and predominantly endemic biodiversity is seriously imperilled. In the past two centuries, at least 27 mammals, 23 birds (including island species and subspecies), 4 frogs and over 60 plant species have vanished (Department of Sustainability 2009). In addition, over 1500 mammals, birds, reptiles, amphibians, and plants are currently

[3]The 'lucky country' descriptor was used satirically in the opening lines of the final chapter.

threatened with extinction, along with over 3000 ecosystem types (Keith et al. 2013). In Victoria, for instance, ~30 % of the original native vegetation remains, and some vegetation types, such as grasslands and open woodlands, have been reduced by more than 99 % since European settlement (Bradshaw 2012). The situation for marine systems is far more uncertain owing to data limitations even for economically important species (Beeton et al. 2012; FRDC 2012). In addition, Australia has the world's most recent mammal extinction, the Christmas Island pipistrelle bat (Pipistrellus murrayi) in 2009 (Martin et al. 2012). If current trends continue, many other species such as the Leadbeater's possum (Gymnobelideus leadbeateri) will suffer the same fate. Indeed, Lindenmayer and Possingham (2013) suggested that the Victorian government is knowingly condoning activities that will reduce the viability of this IUCN-listed endangered species (Ritchie et al. 2013, pp. 1133–1134).

There has indeed been a significant decline in biodiversity in Australia over the last 200 years, greater than any other continent, with the extinction of 50 animal and 48 plant species. Threats to biodiversity include land clearing and fragmentation of habitats, over consumption of resources, pollution and invasive species (ABS 2013). Ritchie et al. (2013) draw the conclusion that the cause is the failure to legislate and enforce traditional forms of environmental protection. Australian governments are guilty of passing laws that are not adequately implemented, and have been guilty of 'winding back' legal protections. However political, legal and institutional failures are far from complete diagnosis of the underlying causes.

The first consideration is history. Australia was colonised by Europeans in 1788. Prior to that human occupation was by Aboriginal people who were non-industrial. Colonising countries had already diminished their biodiversity, and therefore had (residual) ecologies that were more resilient to these pressures. Biodiversity decline in Australia was inevitable, though the degree of inevitable loss and the quality of the strategies to stem the losses are important matters for debate.

The second consideration is the macroeconomics of conservation. Australia covers 7.688 million square kilometres (ABS 2014a), and contains only 23.5 million people (ABS 2014b). Around 81 % of the population lives within 50 km (Department of the Environment 2011a) of the coast and 66 % in capital cities (ABS 2014c). The effectiveness of any strategy depends substantially upon the sufficiency and appropriateness of the resources for its implementation. The Australian paradigm for rural natural resource governance relies upon landholder action and investment and government regulation and limited subsidisation. Volunteer labour and philanthropic investment contribute. This resourcing paradigm is not only historically insufficient, it is likely to be increasingly so.

Our estimate (subject to caveats) is that landscape protection and restoration in Australia is likely to require around 2 % of GDP (Martin and Werren 2009a, b), much of this investment being required in non-urban areas. Interventions such as regulation, education or extension, or stewardship incentives can do little without manpower or funds. Australia's GDP of US$46.6 K per person is comparable with France (US$40.4 K), United Kingdom (US$37.7 K), Canada ($44.5 K),

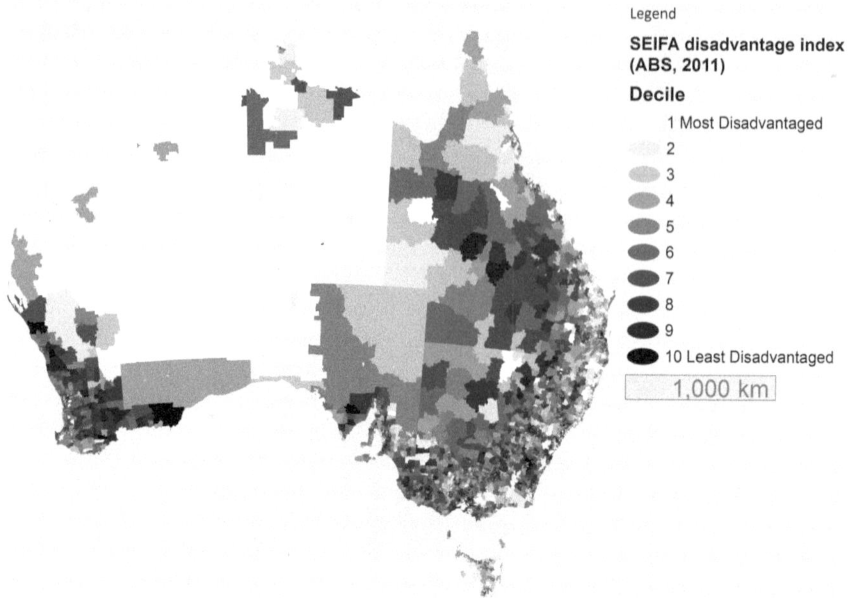

Fig. 2 Distribution of socio-economic disadvantage. Prepared by R Blackley, Queensland Murray-Darling Committee Inc. Unpublished. *Socio-Economic Indexes for Areas* (SEIFA) is a product developed by the Australian Bureau of Statistics, 2011

Switzerland (US$55.2 K), and the United States (US$54.8 K) (C.I.A. 2014).[4] Our GDP per hectare including all economic activities (US$1995) is similar to Argentina (US$1713) and Canada (US$1976) and less than Brazil (US$2628) but our agricultural and mining GDP combined is only 20 % of our national GDP (ABS 2012b; National Farmers Federation 2012). Australia's population density is roughly equivalent to Mongolia or Iceland (around one hundredth of the density of the UK, one tenth of the USA, or one fiftieth of the density of China) (The World Bank 2015). Adjusted for urban concentration the manpower (and funds) for stewardship are minute compared to the challenge. There is limited manpower and money available 'in the bush' for stewardship of a vast land.

The second constraint is the limited capacity of rural people. In developing countries 55 % of the population lives in rural areas. More than 70 % were living in extreme poverty in 2011, compared to 55 % of the total population. Rural poverty is probably under-represented in these statistics because of migration of the rural poor to cities (International Fund for Agricultural Development 2010). Even in wealthy Australia a pattern of relative socio-economic disadvantage persists (Fig. 2).

[4]Australia enjoys a lower level of family income inequity than these countries, measured by the Gini coefficient at 30.3.

The correlation between remoteness from urban centres and disadvantage reflects:

1. relatively low rural incomes;
2. limited availability of professional services
3. a resulting lower average standard of education, health, and other measures of welfare
4. a higher proportion of residents of aboriginal descent (with rural Aboriginal people being the most disadvantaged sub-group of the overall disadvantaged Aboriginal population) (Steering Committee for the Review of Government Service Provision 2014).[5]

Even within a rich and well-governed country, social justice and environmental outcomes from natural resources are unsatisfactory, reflecting the common characteristics of rural areas in many countries.

4 Exporting the Limited Rural Wealth

Location, resource dependence and socio-economic characteristics have implications for what is feasible, effective and fair. A worldwide trend is the declining share of the price of agricultural produce accruing to the producer, achieving a redistribution of wealth away from rural areas (Vorley 2001; Murphy 2006). The low return on capital experienced by many primary producers, the low incomes of rural people, and the poor socio-economic status of rural communities all reflect this. Commodity markets, supplier/buyer chain dynamics and government policies create pressure on farm-gate gross margins. Coupled with other issues of farm economics, this undermines farmer welfare and investment capacity. These pressures contribute to over-exploitation of resources, poor welfare outcomes, and a reduced capacity to protect the environment (Vorley 2001).

Australia's agricultural share of GDP fluctuates around 3–4 % and its percentage of labour force employed in agriculture fluctuates around the same level. Off-farm processing and marketing increases the average contribution of agriculture to around 12 % of GDP, a leverage of at least 3 to 4 (National Farmers Federation 2012). Mining in 2009–10 contributed 8 % to national GDP (ABS 2012b). Transformation of minerals to finished product adds most of the value within the end product. The conclusion: most of the economic value of agricultural production is secured outside rural areas.

[5]Whilst most Aboriginal people now live in urban and inner regional areas, Aboriginal people are a higher proportion of the population of rural and remote areas. For most indicators, outcomes for Aboriginal and Torres Strait Islander Australians worsen as remoteness increases.

The percentage of the wealth which does accrue to the farmer varies with the commodity: for drinking milk the percentage is 25–44 (Australian Competition and Consumer Commission 2008), for poultry it is less than 8 %[6] (Australian Competition and Consumer Commission 2008), for beef the 10 year average is 33 % (Meat and Livestock Australia 2015a), for lamb the 10 year average is 48 % (Meat and Livestock Australia 2015b) and for vegetables 40–50 % (Victorian Farmers Federation 2008). One result is that despite working long hours 'the average weekly disposable income of farmers in 2009–10 ($568) was considerably lower than that of people working in other occupations ($921)' (ABS 2012a).[7] There are many reasons why the economic value of primary production is not retained, but regardless of cause, the consequences include poor rural social and environmental outcomes.

The 'missing markets' problem for environmental values in primary production contributes to unsatisfactory economic and environmental performance. This is beyond the control of rural people. One dimension is the lack of market payment for good stewardship (or market penalty for bad practice). This reduces the incentive for stewardship and the economic capacity to exercise it. Another missing market is for non-production values, which are mainly pursued through regulation or land acquisition rather than market payments. A third missing market is consumer payment for the natural resources embodied in what they consume. This 'caps' the recovery of the full value of natural resources, a key to equitable and efficient conservation.

The degree of embodied environmental values is illustrated by water and greenhouse gas emissions. These indicate the natural resources implicitly exported within primary products (Tables 1 and 2).

Rural producers can do little to create a market for environmental or social values if consumers and distributors, who have market power, are not prepared to pay for this. In a high margin industry it may be feasible for producers to invest to protect unpriced values, but in commodity industries this is less likely.

It is not only through agricultural markets that distributional problems arise between rural communities and the balance of society, relevant to natural resource governance. Society relies on rural landscapes for conservation because urbanisation and industrialisation destroy natural values: continuation of cultural practices (particularly of Indigenous people) is often possible only in rural areas; enjoyment of nature and outdoor recreation often requires rural or wilderness areas. Rural

---

[6]Based on the price the chicken grower receives compared to the retail price of a dressed frozen chicken. This value is much less when compared to other commodities as the chicken grower is essentially a contractor. The processor provides the chicken grower with the majority of the inputs required for the growing process.

[7]Low income does not mean a lack of wealth. Farming is capital intensive. The average wealth of farming households in 2009–10 was $1.3 million, compared to other households ($393,000). Income is the more relevant measure of welfare and of capacity to invest in stewardship.

Table 1 Embodied water for selected agricultural commodities (Extract from p. 40 Hoekstra and Chapagain 2007)

Agricultural commodity	Average embodied water content Australia (m^3/t)	Average embodied water content world (m^3/t)
Rice (husked)	1327	2975
Wheat	1588	1334
Maize	744	909
Soybeans	2106	1789
Beef	17,112	15,497
Eggs	1844	3340
Milk	915	990

Table 2 Embodied greenhouse gas emissions for selected agricultural commodities (Extract from Table 1 Eady et al. 2011)

Agricultural product	Greenhouse gas emissions (kg CO_2 equivalent)
1 kg live weight—40 kg prime lamb	7.9
1 kg live weight—833 kg cull bull	13.7
1 kg live weight—158 kg cull sows	2.2
1 kg live weight—2.5 kg meat chicken	1.7
1 kg sunflower	0.34
1 kg field peas	0.40

conservation is delivered through public and private protected areas and regulation, access controls and initiatives such as regional natural resource management. Rural people typically receive little economic return from resource conservation and restoration and non-production uses, but bear costs of withdrawal of land and other resources from production, fragmentation (Brown et al. 2005; Bryan et al. 2013) (causing management complexities) and restrictions. Conservation for the public benefit, no matter how laudable, does have distributional impacts (Bergstrom et al. 1985; Merenlender et al. 2004) which are systemic, and which affect wealth and welfare, and the transaction costs of governance.

This analysis suggests that there is a problem with the economic model for rural landscape governance. To achieve desired outcomes will require more funds and human capacity than is likely to be available from rural communities, or even from government. This is particularly if the landscape is sparsely populated, and there is little economic activity. Effective rural natural resource governance will require a powerful fiscal model to share the load more equitably across society, engage those with significant economic resources, and deploy funds and manpower in a systematic and coordinated manner (Martin and Werren 2009a, b).

5 Fragmentation Transaction Costs Are a Governance Issue

Resource competition and land use fragmentation increase management complexity, frustrating coordinated action, generating conflict, and impacting the feasibility of some enterprises, particularly in peri-urban areas (Martin et al. 2010). Delivery of conservation and cultural values from rural lands reduces land and resources for primary production, intensifying competition and adding to fragmentation of land and other resource use (Shepheard and Martin 2009). This creates problems for sustained, coordinated collective action. Rural communities incur costs of providing environmental or cultural benefits to society through reduction in access to resources, restrictions upon their use of the resources to which they do have access, and the transaction costs of coordinated action.

Within the broad land use categories shown in Table 3. There is increasing diversification in land uses (for example the 'production' categories encompass traditional grazing and cropping, new species grazing and cropping, 'lifestyle' farming, private conservation) which create problems for coordinated management. However we do not have data on land use change within the broad categories, for example from cropping to animal production or between types of primary product, or conversion of production to lifestyle farming (Department of the Environment 2011b). Land use control relies on legal restrictions over things like biodiversity harm, clearing of native vegetation, of water pollution; and project environmental impact controls (such as for mining, quarrying or wind-farm development). There is also fragmentation of legal rights and institutional arrangements: within private tenures, property rights might include tradeable water rights, biodiversity banking interests, mining interests, and contingent rights such as under family law and government rights of access (Martin et al. 2013). The intersection of many laws administered by many agencies, the three levels of Australian government, proliferating market and other rights, results in institutional complexity (Gibbs et al. 2013). Spatial and economic considerations restrict the ability of government to

Table 3 Change in land use from 1996–97 to 2005–06 for residential, mining and production areas (Australian Collaborative Land Use and Management Program n.d.)

Land use	1996–97 (km^2)	2005–06 (km^2)	Change (km^2)
Urban residential	6102	10,343	4241
Rural residential	9442	9491	49
Mining	1366	1482	116
Production from relatively natural environments	4,455,238	3,673,099	−782,139
Production from dryland agriculture and plantations	423,441	1,000,708	577,267
Production from irrigated agriculture and plantations	24,413	26,847	2434

effectively supervise land use in many parts of Australia, notwithstanding legal responsibilities.

The change in land use in the period 2001 to 2006 in Australia indicates longer-term shifts. There was a reduction of 654,686 km^2 in land used for 'production from relatively natural environments' (Table 3). There was an increase of 517,374 km^2 of 'land used for dryland agriculture and plantation production' and an increase of 142,437 km^2 of 'conservation and natural environment'. Some 'productive natural environment' lands may have been converted to conservation or to agriculture and plantations. There was also an increase in 2006 of land used for intensive purposes of 13,968 km^2. Intensive uses include residential, mining, transport and communications land use (Australian Collaborative Land Use and Management Program n.d) (Table 4).

Many effects of fragmentation on coordinated landscape management arise from a shift from homogenous land use (such as widespread grazing or cropping producing similar products, thus sharing land and environmental management issues) to more diverse land use (a more complex mix of grazing, cropping, urbanisation, lifestyle and conservation uses, and more diverse production increasing heterogeneity among stewards).

Mining occupies 0.2 % of Australian land and 2 % of the country is subject to mining leases (Minerals Council of Australia n.d). A greater percentage is subject to exploration licenses. There is conflict over the expansion of mining and coal-seam gas extraction, with a concern that prime agricultural land is being lost to mining (Geoffrey et al. 2013). According to Millar and Roots (2012) no data on the amount of agricultural land being converted to mining is available.

Increasing land use heterogeneity has governance impacts. We illustrate this with the autopoietic [self-generating, see Maturana and Varela (1980), and for its relevance to law Teubner (1987)] invasive species problem, using wild pigs as an example. Wild pigs cause damage to crops, lambs and other young animals, native vegetation and small native species like frogs or turtles, and can spread disease (Martin et al. 2014). They are adaptive, can travel great distances. If controls are unsuccessful the animals learn to avoid similar control, and pass this knowledge on to other animals (Whitney and Gabler 2008). The effectiveness of control degrades with landholder nonparticipation in coordinated action (Martin 2008b). Unfortunately personal incentives to participate in control differ with the type of enterprise. Other than for social licence reasons (Williams and Martin 2011; Owen and Kemp 2013), a mining company may have no economic reason to invest in

Table 4 Change in land use from 1996–97 to 2005–06 for traditional indigenous uses, nature conservation and native vegetation (Consolidation of data from Australian Collaborative Land Use and Management Program n.d.)

Land use	1996–97 (km^2)	2005–06 (km^2)	Change (km^2)
Traditional indigenous uses	897,311	876,578	−20,733
Nature conservation	446,915	569,240	122,325
Remnant native vegetation	239,059	223,944	−15,115

coordinated control. A grain grower, sugar cane producer or a lamb producer will have different preferences as to when and how control should be implemented. For some landholders the effects of wild pigs upon their enterprise will be significant, but for others it will only be an episodic risk concern. There will be some landholders for whom pig control is not attractive, such as for those who use pigs as a hunting resource, or who are paid for hunting over their lands. For landholders who use dogs for farm management, poisons carry a particular risk. If all landholders have the same type of enterprise many complexities disappear though there will still be differences in attitude and capacity. The increasingly complex mosaic of production, conservation, and indigenous peoples land uses thus increases the costs and reduces the effectiveness of control. It can also result in social conflict (Shepheard and Martin 2009).

For problems with complex causes and effects that span jurisdictions and involve unrelated actors, management requires complex institutional coordination (e.g. collaborative agreements or trans-boundary conventions). Coordination and complexity generate transaction costs: the time (and frustration) cost to the landholder in dealing with agencies; the administration of frontline coordination and coordination across agencies; monitoring widely-spread and diverse activities with a small number of people; and costs of conflict and distrust (Martin and Gunningham 2011). Insufficient attention is paid to the impact of landholder and volunteer transaction costs and their effect in disenabling community action, but this is an important concern (Martin and Gunningham 2011).

Fragmentation, community expectations driving political demands, insufficient funding and weak economic incentives to voluntarily achieve desired standards of conservation, lead to regulation being relied upon in circumstances where other approaches are preferable (Martin and Gunningham 2011). Intense conflicts have emerged over regulatory issues like the protection of native vegetation, mining and gas extraction, animal welfare, water, biodiversity and land development, and many other issues (Paavola 2007; Martin and Becker 2011). Political opposition, the high transaction costs of monitoring and enforcement, and arguably unachievable expectations, have resulted in implementation difficulties, disputes and political challenges. One result is a political 'see-saw' over regulation, in the political arena and within agencies, over enforcement (Martin and Verbeek 2000; House of Representatives Standing Committee on the Environment 2014; Beetles 2015).

When regulations prove to be ineffective new regulations may be created, compounding the structural problem (but at least temporarily minimising the political one). Australia's history of rural regulation demonstrates that whilst effective regulation may be necessary, the creation of a regulatory instrument is far from sufficient to ensure outcomes. The processes through which rural regulation is created and implemented do not adequately address the many variables that need to be taken into account to ensure effective, efficient, and fair laws (Martin et al. 2007; Tan et al. 2012).

6 Prognosis

Governance strategy must, like all forms of strategy, be forward looking. Current or historical conditions are relevant only to the extent that they indicate the challenges and opportunities that the strategy must deal with. We have touched issues where history and trends suggest that many of the known challenges will continue into the future, and that many will follow the existing trajectory which suggests that socio-ecological outcomes seem unlikely to improve markedly unless 'game-changing' contingencies arise.

The regulatory, extension and investment capacity of government is increasingly strained. Australian governments of all persuasions have continued to reduce their staffing of rural agencies including those involved in frontline environmental and production support and the management of protected areas such as national parks (though the area of rural protected areas increases). The shrinking contribution of government in rural natural resource governance is the result of two factors: the overall pressure on government resources and declining political attention to rural issues. The most recent Intergenerational Report indicates that Australia's underlying cash deficit under current legislated arrangements will increase from 2014–15 to around 6 % of GDP by 2054–55 with net debt increasing to almost 6 % of GDP (Anon 2015). Whilst many debates are possible about the reliability of projections and about how resources are allocated, the overwhelming sense is that governments will be under further pressure to cut expenditures, and rural natural resource governance investment is more likely to suffer than not.

The second important contingency at the intersection between ecology, economics and institutional arrangements is climate change. The following chart highlights some alternative futures for regions of Australia, under a variety of scenarios that reflect the climate modelling for the 5th Assessment Report (AR5) of the Intergovernmental Panel on Climate Change adjusted for Australian conditions. What is significant is the enormous variation in possible outcomes in Australian landscape and social consequences, across a variety of issues.

The data in Fig. 3 was interpreted and qualitatively downscaled by an expert ACEAS working group in the context of the RCP descriptions. The + is a positive change in the sustainability metric relative to 2011 levels and—is a decline. Background information to aid interpretation can be found in van Vuuren et al. (2011).

The basic assumptions that underpin these scenarios are biophysical climate change, and the human response to climate factors. The future outcomes of resource governance are highly variable, but that governance factors will be a key determinant of what these outcomes are.

Fig. 3 Australian sustainability indicators: future scenarios 2011–2100. *Source* Dr Brett Bryan, CSIRO (unpublished)

7 Prescription

It seems unlikely that the Australian community could be satisfied with the outcomes that are being achieved from rural natural resource governance, if they were aware of the facts. If past trends continue, particularly with climate change, rural demographics, and declining public investment, sustainable development outcomes will probably be even less gratifying than those currently being achieved. Thankfully, projections are not destiny. Innovation in rural natural resource governance (and in other areas of rural governance) could lead to better outcomes.

In 2013 a diverse group of leading researchers turned their minds to the future of rural Australia and the governance challenges that would have to be met. Their deliberations were inputs into the model cited above. They concluded that

> Biophysical and economic drivers are converging that may induce a second industrial transformation of Australian landscapes. The combination of increased societal expectations of environmental stewardship and the spcctre of climate change require transformational adaptation. We need to adapt to our environment as it changes, and as we change, and we need to do it fast. The six emerging economies for ecosystem services—carbon, water, food, energy, amenity and mining—provide opportunities for the kind of transformation and adaptation that we need and want. These will shape how we use and conserve our unique natural capital but we need to manage the risks of trade-offs and unintended consequences. To achieve positive transformational adaptation of Australian land- scapes and to provide an opportunity to thrive within environmental limits, new partnerships

> between government, science, the private sector and communities are required. This is up to all Australians. Science needs to inform critical environmental limits and develop new ways of understanding the social processes underpinning transformational adaptation. Governments need to regulate environmental limits, streamline existing laws, institutions and governance, and establish and support innovative local adaptation. The private sector needs to better engage with communities, participate in new markets for ecosystem services, and reduce the reliance on government for funding environmental management. Despite being subject to unprecedented pressure already, individual landholders and communities have the responsibility of innovating and implementing evidence-based, collaborative change. These are the first steps needed on an adaptive path to sustainability in the second industrial transformation of Australian landscapes (Bryan et al. 2013, p. 6).

In practice true 'sustainable development' in rural Australia will depend upon fundamental innovation providing solutions that are not currently available to meet major challenges. It will require significant improvement in the profitability of rural land-use, to afford both good incomes and environmental protection. We also have to achieve far greater effectiveness and efficiency in the conservation of natural systems and assets, and find more equitable ways to allocate the benefits and costs of land-use and conservation. Far more funds for environmental stewardship will have to be found, through markets which better reflect environmental inputs, or conservation funding. Given the constraints on government it would be risky to assume that these resources can come from the public sector. We must overcome the conflicts and transaction costs of fragmentation which prevent sustained, well-coordinated collective action over large areas. "Traditional" public and private governance strategies will have to become more effective, less costly for those being governed and for those who govern, and they will have to distribute the benefits and burdens more fairly. Without radical innovation it is hard to be sanguine about sustainable development outcomes in Australian rural communities. Throughout this chapter we have referenced studies that propose reforms that might individually or in combination significantly improve rural natural resource governance outcomes. There are many others. We finish this chapter by highlighting some other possible catalysts for significant change.

One is greater involvement of industry in natural resource governance. Corporations and industries are changing their stance on natural resource governance driven by four strategic considerations: brand value; resource availability and costs; risk-management and corporate citizenship. This involvement comes in many forms, ranging from voluntary reporting, through environmental philanthropy, reducing the product or service footprint, 'greening' corporate strategy, and whole of industry initiatives. There are many industry codes and standards, reporting schemes, environmental stewardship and 'chain of responsibility' initiatives including as those based on the ISO 1400 international standards, or ISO 14040 life-cycle reporting. This is statistically indicated by the growth of environmental labels, with 458 'ecolabels' reported for 197 countries, spanning 25 industry sectors (Ecolabel Index 2015). Increasingly the private sector is central to setting and enforcing environmental and other standards, through initiatives variously termed "hybrid governance", "co-regulation" or other terms (Sorsa 2009; Armitage et al. 2012). Industry has regulatory power because it can control access to the market, vitally important to primary industries. Harnessing this power to address the missing markets problem and the limits of traditional regulation is an important opportunity. There is a contest underway over "who will govern" for the environment: governments, industry, NGOs, or combinations of these roles. It is possible to contemplate a more integrated and coordinated approach that could help transform rural natural resource governance. These strategies will require different types of law, and legally supported integrity mechanisms if the community is to trust in them.

A second opportunity is to reassert (and support) the role of citizens in leading coordinated action. Over recent decades Australia constructed a regional natural resource management programme that involved (with varying degrees of success) citizens and three levels of government. Whilst this led to many successes, dependency on government institutional arrangements was entrenched, sometimes driving out citizen led initiatives (Williams et al. 2008; Curtis et al. 2014). As governments are forced by economic circumstances to reduce their rural support and leadership roles such as extension and project work, a vacuum is emerging. The opportunity is for new models of community led coordinated action, drawing upon developments in community engagement theory and practice, and upon the use of behavioural science to improve the effectiveness of engagement and communications (Whelan and Oliver 2004; Williams et al. 2008; Martin et al. 2012). The law has a significant role to play in ensuring that governance arrangements empower and enable, rather than marginalize, community leadership and coordination.

The next generation of rural natural resource governance must respond effectively to new classes of challenge, whilst dealing more effectively with issues we have always had to deal with since colonisation. Greater effectiveness will require high levels of coordinated collective action, and integrated responses on a "whole of system" management basis. This will undoubtedly be very challenging, and if we do not succeed any aspirations of rural sustainable development in Australia will not be met. There are, thankfully, many possibilities to deal more effectively with these challenges. As always the limiting factor is political will. It is leadership by citizens that will motivate politicians to take the necessary steps. As always, the quality of social capital will determine the economic and ecological outcomes of rural natural resource governance. The law can be a support or an inhibitor to the transformation of rural natural resource governance.

Acknowledgements The research assistance of Dr. Amy Cosby and Roxanne Blackley is gratefully acknowledged. This paper is partly based on research with the support of the Australian Research Council (Project No LP110100659) and project partners: Australian Cotton Research and Development Corporation; Pennsylvania State University; Soil Conservation Service of Iceland; Australian Government Department of Sustainability, Environment, Water, Population and Communities; Tamar NRM, RDA Tasmania, the Tamar and Namoi region communities and the Invasive Animals CRC research program 4 "Facilitating Effective Community Action".

References

ABS (2012a) Australian social trends, Dec 2012. Retrieved 31 Mar 2015, from http://www.abs.gov.au/AUSSTATS/abs@.nsf/allprimarymainfeatures/EED42391CD269634CA257B4800143EAC?opendocument

ABS (2012b) Year book Australia 2012 (1301.0) mining industry. Retrieved 31 Mar 2015, from http://www.abs.gov.au/ausstats/abs@.nsf/Lookup/1301.0Main+Features292012

ABS (2013) Australia's biodiversity. Retrieved 1 Feb 2015, from http://www.abs.gov.au/ausstats/abs@.nsf/Previousproducts/1301.0Feature%20Article12009%E2%80%9310?opendocument&tabname=Summary&prodno=1301.0&issue=2009%9610&num=&view=

ABS (2014a) Australia (S/T). Retrieved 18 May 2015, from http://stat.abs.gov.au/itt/r.jsp?RegionSummary&region=0&dataset=ABS_NRP9_ASGS&geoconcept=REGION&measure=MEASURE&datasetASGS=ABS_NRP9_ASGS&datasetLGA=ABS_NRP9_LGA&regionLGA=REGION&regionASGS=REGION

ABS (2014b) Australian demographic statistics, Sep 2014. Retrieved 18 May 2015, from http://www.abs.gov.au/ausstats/abs@.nsf/Latestproducts/3101.0Main%20Features2Sep%202014?opendocument&tabname=Summary&prodno=3101.0&issue=Sep%202014&num=&view=

ABS (2014c) Feature article: capital cities: past present and future. Retrieved 31 Mar 2015, from http://www.abs.gov.au/AUSSTATS/abs@.nsf/Previousproducts/3218.0Feature%20Article22012-13?opendocument&tabname=Summary&prodno=3218.0&issue=2012-13&num=&view=

Adger NW (2010) Social capital, collective action, and adaptation to climate change. In: Voss M (ed) Der Klimawandel: Sozialwissenschaftliche Perspectiven. VS Verlag für Sozialwissen schaften, Wiesbaden, pp 327–345

Anon (2015) Intergenerational report Australia in 2055, Canberra, Australia

Armitage D, De Loë R, Plummer R (2012) Environmental governance and its implications for conservation practice. Conserv Lett 5(4):245–255

Arnold C, Gunderson L (2013) Adaptive law and resilience. Environ Law Reporter 43:10426–10443

Australian Collaborative Land Use and Management Program (n.d.) National scale land use data. Retrieved 11 Jan 2015, from http://www.agriculture.gov.au/abares/aclump/land-use/data-download

Australian Competition and Consumer Commission (2008) Report of the ACCC inquiry into the competitiveness of retail prices for standard groceries. Retrieved 31 Mar 2015, from https://www.accc.gov.au/system/files/Grocery%20inquiry%20report%20-%20July%202008.pdf

Beetles C (2015) Ag red tape slashed $25 m. The Australian Dairyfarmer. Retrieved 31 Mar 2015, from http://adf.farmonline.com.au/news/nationalrural/agribusiness/general-news/ag-red-tape-slashed-by-25m/2727451.aspx?storypage=0

Bergstrom JC, Dillman BL, Stoll JR (1985) Public environmental amenity benefits of private land; the case of prime agricultural land. South J Agric Econ 17(1):139–150

Brown DG, Johnson KM, Loveland TR, Theobald DM (2005) Rural land-use trends in the conterminous United States, 1950–2000. Ecol Appl 15(6):1851–1863

Bryan B, Meyer W, Campbell A, Harris G, Lefroy T, Lyle G, Martin P, McLean J, Montagu K, Rickards L, Summers D, Thackway R, Wells S, Young M (2013) The second industrial transformation of Australian landscapes. Curr Opin Environ Sustain 5(3):278–287

Burgman M, Walshe T, Godden L, Martin P (2009) Designing regulation for conservation and biosecurity. Australas J Nat Res Law Policy 13(1):93–112

C.I.A (2014) The world factbook. Retrieved 15 Apr 2015, from https://www.cia.gov/library/publications/the-world-factbook/fields/2004.html#85

Curtis A, Ross H, Marshall GR, Baldwin C, Cavaye J, Freeman C, Syme GJ (2014) The great experiment with devolved NRM governance: lessons from community engagement in Australia and New Zealand since the 1980s. Australas J Environ Manage 21(2):175–199

Department of the Environment (2011a) Coasts—major issues for coastal environments: population growth and urban development. Retrieved 31 Mar 2015, from http://www.environment.gov.au/science/soe/2011-report/11-coasts/2-major-issues/2-7-population

Department of the Environment (2011b) State of the environment report 2011—urban environmental efficiency. Retrieved 31 Mar 2015, from http://www.environment.gov.au/science/soe/2011-report/10-built-environment/2-state-and-trends/2-2-urban

Dietz T, Ostrom E, Stern PC (2003) The struggle to govern the commons. Science 302 (5652):1907–1912

Eady SJ, Sanguansri P, Bektash R, Ridoutt B, Simons L, Swiergon P (2011) Carbon footprint for Australian agricultural products and downstream food products in the supermarket. Paper presented at the 7th Australian conference on life cycle assessment, Melbourne

Ecolabel Index (2015) Ecolabel index. Retrieved 15 Apr 2015, from http://www.ecolabelindex.com

Geoffrey L, Richards C, Lyons K (2013) Food security in Australia in an era of neoliberalism, productivism and climate change. J Rural Stud 29:30–39

Gibbs C, Harris-Adams K, Davidson A (2013) Review of selected regulatory burdens on agriculture and forestry businesses. Australian Bureau of Agricultural and Resource Economics and Sciences, Canberra

Golembiewski RT, Olson M (1966) The logic of collective action. Am Sociol Rev

Hardin G (1968) The tragedy of commons. Science 162(3859):1243–1248

Hoekstra AY, Chapagain AK (2007) Water footprint of nations: water use by people as a function of their consumption pattern. Water Resour Manage 21(1):35–48

Horne D (2008) The lucky country, 6th edn. Penguin Modern Classics, Melbourne

House of Representatives Standing Committee on the Environment (2014) Report—streamlining environmental legislation: inquiry into streamlining environmental regulation, 'green tape', and one stop shops. Retrieved 18 May 2015, from http://www.aph.gov.au/Parliamentary_Business/Committees/House/Environment/Green_Tape/Report

International Fund for Agricultural Development (2010) New realities, new challenges: new opportunities for tomorrow's generation. Retrieved 31 Mar 2015, from http://www.ifad.org/rpr2011/report/e/rpr2011.pdf

Lim M (2014) Is water different from biodiversity? Governance criteria for the effective management of transboundary resources. Rev Eur Comp Int Environ Law 23(1):96–110

Lubell M (2007) Familiarity breeds trust: collective action in a policy domain. J Polit 69:237–250

Martin P (2008a) The changing role of law in the pursuit of sustainability. In: Jeffery M, Firestone J, Bubna-Litic K (eds) Biodiversity conservation, law and livelihoods: bridging the North-South divide. Cambridge University Press, New York, pp 49–65

Martin P (2008b) Cross pollination or cross-contamination? Directions for informing the management of invasives with market-economy concepts. Paper presented at 16th Australian weed conference, Cairns

Martin P, Bartel R, Sinden J, Gunningham N, Hannam I (2007) Developing a good regulatory practice model for environmental regulations impacting on farmers, vol 6. Australian Farm Institute, Canberra

Martin P, Becker J (2011) A tale of two systems: conflict, law and the development of water allocation in two common law jurisdictions. Int J Rural Law Policy 1:1–18

Martin P, Gunningham N (2011) Leading reform of natural resource management law: core principles. Environ Plan Law J 28(3):137–158

Martin P, Kennedy A, Page J, Williams J (2013) Environmental property rights in Australia: constructing a new Tower of Babel. Environ Plan Law J 30:531–552

Martin P, Le Gal E, Low Choy D, Marshall G, Dickson K (2014) Improving invasive animal institutions: a citizen-focused approach. A citizen-focused review of institutional arrangements for invasive animal management. Version 1.2. Consultation report invasive animals CRC. Retrieved 18 May 2015, from http://invasives.org.au/files/2014/12/REPORT-Improving-Australia's-invasives-management-institutions-V1.2.pdf

Martin P, Verbeek M (2000) The cartography of law: finding new paths to effective resource use regulation. Land and Water Australia, Canberra
Martin P, Verbeek V (2006) Sustainability strategies. Federation Press, Sydney
Martin P, Verbeek M, Rile S, Bartel R, Le Gal E (2012) Innovations in institutions to improve weed funding, strategy and outcomes. Rural Industries Research and Development Corporation, Canberra
Martin P, Werren K (2009a) The use of taxation incentives to create new eco-service markets critical issues. In: Lye L, Milne J, Ashiabor H, Deketelaere K, Kreiser L (eds) Environmental taxation, vol VII. Oxford University Press, Oxford, pp 511–528
Martin P, Werren K (2009b) Discussion paper: an industry plan for the Victorian environment? Department of Sustainability and Environment, Victoria
Martin P, Williams J, Stone C, Alter T (2010) Researcher lessons from community partnership and trans-disciplinary research in a peri-urban setting: the WISER experience in Western Sydney. CRC for Irrigation Futures
Maturana HR, Varela FJ (1980) Autopoiesis and cognition: the realization of the living. D. Reidel Publishing Company, Dordrecht
Mayne J, Stern E (2013) Impact evaluation of natural resource management resource programs: a broader view. Retrieved 19 May 2015, from http://aciar.gov.au/files/ias84.pdf
Meat and Livestock Australia (2015a) Producer shares climb in February but ease in March. Retrieved 31 Mar 2015, from http://www.mla.com.au/Prices-and-markets/Market-news/Producer-shares-climb-in-February-but-ease-in-March-01042015
Meat and Livestock Australia (2015b) Producers carve out share of retail dollar. Retrieved 31 Mar 2015, from http://www.mla.com.au/Prices-and-markets/Market-news/Producers-carve-out-share-of-retail-dollar
Merenlender AM, Huntsinger L, Guthey G, Fairax SK (2004) Land trusts and conservation easements: who is conserving what for whom? Conserv Biol 18(1):65–76
Millar J, Roots J (2012) Changes in Australian agriculture and land use: implications for future food security. Int J Agric Sustain 10(1):25–39
Minerals Council of Australia (n.d.) The minerals industry. Retrieved 8 Jan 2015, from http://www.minerals.org.au/corporate/about_the_minerals_industry
Moxnes E (2000) Not only the tragedy of the commons: misperceptions of feedback and policies for sustainable development. Syst Dyn Rev 16(Winter):325–348
Murphy S (2006) Concentrated market power and agricultural trade. EcoFair Trade Dialogue Discuss Pap 1:1–41
National Farmers Federation Farm Facts (2012) Farm facts. Retrieved 13 Apr 2015, from http://www.nff.org.au/farm-facts.html
Ostrom E (2010) Analyzing collective action. Agric Econ 41:155–166
Owen JR, Kemp D (2013) Social licence and mining: a critical perspective. Res Policy 38 (1):29–35
Paavola J (2007) Institutions and environmental governance: a reconceptualization. Ecol Econ 63 (1):93–103
Ritchie EG, Bradshaw CJA, Dickman CR, Hobbs R, Christopher N, Johnston EL, Woinarski J (2013) Continental-scale governance failure will hasten loss of Australia's biodiversity. Conserv Biol 27(6):1133–1135
Scarlett PL, Boyd JW, Brittain A, Shabman LA, Brennan TJ (2013) Catalysts for conservation: exploring behavioral science insights for natural resource investments. Retrieved 20 May 2015, from http://www.rff.org/RFF/Documents/RFF-Rpt-BehavioralScienceEconomicInsights.pdf
Seidl D (2004) Luhmann's theory of autopoietic social systems. Munich Business Res Pap 2:1–28
Shepheard M, Martin P (2009) The multiple meanings and practical problems with making a duty of care work for stewardship in agriculture. Macquarie J Int Comp Environ Law 6:191–215
Sorsa K (2009) Self-regulation and co-regulation in value chain—reviewing retail, food and chemical industry examples. Research communications No. 100, National Research Institute of Legal Policy, Helsinki Finland

Steering Committee for the Review of Government Service Provision (2014) Overcoming indigenous disadvantage: key indicators 2014, Productivity Commission, Canberra. Retrieved 18 May 2015, from http://www.pc.gov.au/research/recurring/overcoming-indigenous-disadvantage/key-indicators-2014/key-indicators-2014-overviewbooklet.pdf

Tan PL, Bowmer KH, Baldwin C (2012) Continued challenges in the policy and legal framework for collaborative water planning. J Hydrol 474:84–91

Teubner G (1987) Autopoietic law—a new approach to law and society. Walter De Gruyter & Co, Berlin

The World Bank (2015) Population density (people per sq. km of land area). Retrieved 19 May 2015, from http://data.worldbank.org/indicator/EN.POP.DNST

van Vuuren D, Edmonds J, Kainuma M, Riahi K, Thomson A, Hibbard K, Hurtt G, Kram T, Krey V, Lamarque JF, Masui T, Meinshausen M, Nakicenovic N, Smith S, Rose S (2011) The representative concentration pathways: an overview. Clim Change 109:5–31

Victorian Farmers Federation (2008) Submission to ACCC inquiry into the competitiveness of retail prices for standard groceries. Retrieved 31 Mar 2015, from https://www.accc.gov.au/system/files/126%20(late%2013%20Mar)%20-%20Victorian%20Farmers%20Federation%20(11%20pages).pdf

Vorley B (2001) The chains of agriculture: sustainability and the restructuring of Agri-food markets, International Institute for Environment and Development, London. Retrieved 18 May 2015, from http://pubs.iied.org/pdfs/11009IIED.pdf

Waslekar S, Futehally L (2013) Water cooperation for a secure world. Strategic Foresight Group, Mumbai

Whelan J, Oliver P (2004) Regional community-based planning: the challenge of participatory environmental governance (Technical report 44). In: Cooperative research centre for coastal zone, estuary and waterway management. Retrieved 18 May 2015, from http://www.ozcoasts.gov.au/pdf/CRC/44-regional_community_based_planning.pdf

Whiteford P (2014) Income and wealth inequality: how is Australia faring? The conversation. Retrieved 31 Mar 2015, from http://theconversation.com/income-and-wealth-inequality-how-is-australia-faring-23483

Whitney KD, Gabler CA (2008) Rapid evolution in introduced species, 'invasive traits' and recipient communities: challenges for predicting invasive potential. Divers Distrib 14(4):569–580

Williams J, Beeton RJS, McDonald GT (2008) Success attributes of regional natural resource management. Int J Sustain Dev Plann 3(3):203–222

Williams J, Martin P (2011) Defending the social licence of farming: issues, challenges and new directions for agriculture. CSIRO Publishing, Collingwood

Yale Environmental Performance Index (2014) Results at-a-glance. Retrieved 13 Apr 2015, from http://epi.yale.edu/epi

Part XI
Conclusions

Conclusions

Volker Mauerhofer

Abstract This chapter summarizes the main findings of the previous chapters.

Keywords Findings · Results · Summaries

Legal aspects of Sustainable Development are multi-faceted and pluri-dimensional. This is valid for the issues assessed in this edited volume as well as for the conclusions drawn by the several authors and to be drawn by careful readers.

Thus, such as already mentioned at the end of the introduction, the fourth and final part about conclusions of this edited volume does not strive to show up interrelations among all of the chapters but instead makes short summaries of the findings of each of them, shows certain interlinkages among some of them, provides—where appropriate—short comparisons within the sub-parts and draws a few own conclusions.

1 Horizontal Policy Issues

Ivano Alogna's (Chapter "The Circulation of the Model of Sustainable Development: Tracing the Path in a Comparative Law Perspective") additional "oblique" circulation of legal models allows a more differentiated picture in comparison to the two common vertical and horizontal circulation models and also adds to the discussion a new perspective regarding potential overlaps between these three models when it comes to regional integration procedures.

From the way of how sustainable development law should be transferred, the direction is made towards, what should be transferred.

Thus, Volker Mauerhofer (Chapter "3-D Sustainability and Its Contribution to Governance Assessment in Legal Terms: Examples and Perspectives") innovatively offers—based on ongoing research—several objective-driven solution proposals for

V. Mauerhofer (✉)
University of Vienna, Vienna, Austria
e-mail: volker.mauerhofer@univie.ac.at

V. Mauerhofer (ed.), *Legal Aspects of Sustainable Development*,
DOI 10.1007/978-3-319-26021-1_31

addressing in a sustainable manner geopolitical and organizational scales as well as trade-offs when it comes to re-writing existing environmental legal institutions (*de lege ferenda*). He further provides proposals for the innovative implementation of existing regimes without modifying legislative text (*de lege lata*).

Useful and innovative proposals including the ones based on regional integration explicitly mentioned by Mauerhofer (Chapter "3-D Sustainability and Its Contribution to Governance Assessment in Legal Terms: Examples and Perspectives") as *de lege ferenda* as well as *de lege lata* means are just as good as their factual implementation, supranationally, nationally and subnationally.

Hence, the contribution of Ferdinand Kerschner and Erika Wagner (Chapter "Sustainability—A Long, Hard Road") shows on the example of sustainable development law in connection with regional integration that indeed considerable implementation efforts exist. While their scepticism can be shared whether the concept of sustainability legally works in terms of the desired success of measures taken, inter alia, in the fields of transport, energy, climate change and waste management or whether is it instead a program without (sufficient) implementation or even just a tokenism.

Human rights can be seen in the legal sustainability discussion as another useful tool for better implementation, with regard to social as well as environmental sustainability, and are therefore discussed as a more individualistic approach as well as a more public good oriented approach (see e.g., Francioni 2010).

Rasyikah Md Khalid, Faridah Jalil and Mazlin Bin Mokhtar (Chapter "Environmental Sustainability as a Human Right") acknowledge the trade-offs between the social and environmental dimension as development can require some deforestation and land clearance and that may affect the flora and fauna that depended on environmental sustainability. Their conclusion that 'sustainability' can be used and argued upon in courts in terms of a human right through judicial activism and by relaxing the doctrine of legal standing should be able to obtain considerable attention and support.

Here, the chapter shows clear links, on the one hand to Mauerhofer's (Chapter "3-D Sustainability and Its Contribution to Governance Assessment in Legal Terms: Examples and Perspectives") *de lege lata* approach of increased capacity building about direct effect of Multilateral Environmental Agreements (MEAs). On the other hand the doctrine of legal standing is also intensively discussed in the three chapters related to the horizontal issue 'Public Participation' (Okubo, Chapter "Judicial Control Over Acts of Administrative Omission: Environmental Rule of Law and Recent Case Law in Japan"; Gera, Chapter "Examining the Resilience of Public Participation Structures for Sustainable Mining in the Philippines"; Odumosu, "Public Participation and Constitutional Impediments to Sustainable Development in Nigeria").

Vasilka Sancin and Maša Kovič Dine (Chapter "Ensuring Access to Safe Drinking Water as an Imperative of Sustainable Development") with regard to water similarly to Khalid et al. (Chapter "Environmental Sustainability as a Human Right") conclude that the obligation of ensuring access to safe drinking water is an obligation of conduct and stems from the two intertwined pillars of sustainable development: social development and environmental protection. Furthermore, they point out that such an imperative due diligence obligations derived from sustainable development opens up a possibility of additional legal reasoning, alongside the human rights arguments.

Natalie P. Stoianoff (Chapter "Ensuring a Sustainable Future through Recognizing and Protecting Indigenous Ecological Knowledge") shows how the implementation of particular human rights, namely the rights of indigenous peoples, works in practice, based on an example associated with natural resource management focussed on the Aboriginal Communities of the state of New South Wales in Australia. The different working steps carefully implemented (initial comparative analysis, special Working Party, Aboriginal Community consultation) and the resulting White Paper could be considered as a blue print for similar attempts worldwide.

Within the sub-part III (Communication and Social Enterprise Governance), Jordi Prades and Aitana de la Varga (Chapter "Framing New Environmental Cultures for Sustainability. Communication and Sensemaking in Three Intractable Multiparty Conflicts in the EbreBiosfera, Spain") show impressively how an alternative communication frame can positively influence environmental conflicts. They indicate that in all three conflicts assessed, the latter UNESCO recognition of the Terres de l'Ebre as a Biosphere Reserve (EbreBiosfera) in the sense of communicative legislation or "soft law" , understood as a horizontal interactive two-way dialogue, is more effective and offers more satisfactory long-term results than a traditional top-down approach.

Aikaterini Argyrou, Tineke Elisabeth Lambooy, Robert Jan Blomme, Henk Kievit, Guus Nieuwenhuijzen Kruseman and Duco Hora Siccama authored the second Chapter "An empirical investigation of supportive legal frameworks for social enterprises in Belgium: A cross-sectoral comparison of case studies for social enterprises from the social housing, finance and energy sector Perspective" within this sub-part. They show that even though the Belgian tailor-made legal framework regarding social enterprises i.e. VSO is conducive to employee participation in the decision-making processes, the concept of employee participation differs in the various cooperatives with social purpose from the different sectors. The range found by them spans from processes of direct and formal participation via indirect participation through informal participation settings to prohibition by special legislation.

From stakeholder participation within social enterprises it is just a small step towards Public Participation in Environmental Matters which increasingly becomes a worldwide topic with regard to sustainability (Mauerhofer 2015).

First, Okubo (Chapter "Judicial Control Over Acts of Administrative Omission: Environmental Rule of Law and Recent Case Law in Japan") finds that recently, State liability lawsuits, such as implemented with the Minamata and asbestos cases, have proven to be effective measures to challenge the non-use of the regulatory power. She further points out that although such kind of lawsuit is a direct measure to compensate the victims, it also plays an important role as an indirect measure to push the government to exercise its regulatory power while the (in the future perhaps more effective) mandamus action is more a direct measure to control illegal failures.

Weena Gera (Chapter "Examining the Resilience of Public Participation Structures for Sustainable Mining in the Philippines") then illustrates—with focus on the legislative side—how intersecting forces of predominant political economic interests and emergent civil society networks navigate through legal frameworks to influence dependencies and the shifting boundaries of public engagement in mining governance. She concludes that this results to arbitrary policy compromises that exacerbate prevailing tensions amid power imbalances in the sector, leaving civil society in perpetual square-off against mining corporations.

Finally in this sub-part about Public Participation, Taiwo Odumosu (Chapter "Public Participation and Constitutional Impediments to Sustainable Development in Nigeria") suggests with a view on the 2014 amendments proposed to the Nigerian constitutions that the provisions of the constitution must be made justiciable for sustainable society and development to be achieved. Otherwise, according to his findings, the constitution will be burdened for non-functionality and lack of societal sustainability, especially in terms of Public Participation.

In conclusion, while the outlook for the Japanese judiciary system (Okubo, Chapter "Judicial Control Over Acts of Administrative Omission: Environmental Rule of Law and Recent Case Law in Japan") appears to be rather positive, the case of the Philippine legislation related to public participation in mining (Gera, Chapter "Examining the Resilience of Public Participation Structures for Sustainable Mining in the Philippines") as well as the case of the constitutional amendments in terms of Public Participation for Nigeria (Odumosu, Chapter "Public Participation and Constitutional Impediments to Sustainable Development in Nigeria") both do not seem to provide such an optimistic outlook.

Public Participation is often closely related to the last policy issue "Assessment tools" (sub-part V) on horizontal policy of the first main part of this edited volume.

Volker Mauerhofer (Chapter "A Global Conceptual Framework for Categorizing Environmental Change Based on Property Rights and Compensation") shows this connection clearly in an overall way. When testing his newly proposed framework for the assessment of participation of public and private stakeholders in situations of environmental change by means of a semi-random sample from the academic literature, in general its global applicability and usefulness is found. The framework therein also proofed to be appropriate for formally (rule of law based) and informally (customary law based) institutionalized situations.

Sophie Riley (Chapter "Prioritising the Environment in Sustainable Development: Lessons from Australian Environmental Impact Assessment") finds in her analysis of the more concrete assessment tool of the Environmental Impact Assessment (EIA) of the Australian state of New South Wales that for the assessed projects potentially harmful to the environment the legislation provides little guidance on how to prioritize these criteria and still achieve SD. She further advocates the application of means to curtail the wide discretion available to decision-makers using civic science, which to be effective, needs to be legislatively-based.

At the end of the first part of this edited volume and in its last sub-part, Yao-Ming Hsu (Chapter "Reframing Sustainability in Taiwan: Legal Challenges and Opportunities") finds that Taiwan has shown its capacity and ambition to cope with the need of sustainability by releasing national programs, strategies as well as related laws and regulations, Furthermore, in the approaching future, a new Ministry of Environment and Resources will be installed. But Hsu also points out that those tools still lack the *de jure* international monitoring due to the international *sui generis* nature of Taiwan.

This last sub-part of the horizontal policy issues provides just a glimpse of the wide range of topics related to "Assessment Tools". It spans from tools of assessing categorisations of environmental change situations, via concrete procedures such as EIA to decide whether to proceed with a project proposal potentially harmful to the environment, towards general strategies and organizations for environmental monitoring. Nevertheless, the sub-part provides concrete insights into the different interpretations of assessment tools.

Each of the sub-parts of this first main part of this edited volume deals with horizontal policy issues that can be (mutually) related to each of the five sub-parts of the now following next main part dealing with sectorial policy issues.

2 Sectorial Policy Issues

In the first sub-part on Forest and Water Management of the sectorial policy issues, Pablo Peña (Chapter "A Legal Butterfly Effect: Unexpected Twists and Turns of the Law in Costa Rica's Payment for Ecosystem Services Program") finds that numerous laws and policies (including foreign policies) outside the boundaries of the PES' regulations influence the PES system in its evolvement and functioning. He also highlights that most informants on the ground pointed at a pervasive divide between the technicians and the lawyers wherein the 'lawyers' seem to have displaced the 'technicians' and thus 'rendering legal' nature, which has conflicting implications in the effectiveness of the PES.

Henrik Josefsson (Chapter "From River Basins to Landscapes—Holistic Legal Constructs and Their Differentiation") finds that-by combining the spaces of regulation in the Water Framework Directive (WFD) and the Landscape Convention-a more sustainable space of regulation can be established which both reflect Ostrom's variables and the multiplicity of the social and ecological dimensions. He also proposes that, similar to the Landscape Convention, the general community in the WFD, as the public concerned, should be eligible to participate in the differentiation of spaces of regulation.

Jonida Abazaj (Chapter "Coherence Issues Between Climate, Energy and Water in the Case of European Hydropower: Can We Have It All?") shows pictorially the link between energy-climate-water-sustainability and explains trade-offs, synergies and opportunities for policy integration and coherence. Her analysis reveals that while synergies between climate and energy are straightforward and clearly stated at the EU level, and while it is possible to track this co-ordination between water and biodiversity, challenges persist in relation to the link between the climate and renewable energy package and the Water Framework Directive.

The three papers appear to be quite different. Despite, they can be comparatively assessed based on the legal level of the conflicts identified and show interesting insights. Peña (Chapter "A Legal Butterfly Effect: Unexpected Twists and Turns of the Law in Costa Rica's Payment for Ecosystem Services Program") describes a conflicting situation on the level of implementation while Josefsson (Chapter "From River Basins to Landscapes—Holistic Legal Constructs and Their Differentiation") identifies an issue of lack of public participation already on one legislative level. Latter issue is not that explicit addressed within the three pillars of Public Participation mentioned above (see also sub-part IV). Moreover, Abazaj (Chapter "Coherence Issues Between Climate, Energy and Water in the Case of European Hydropower: Can We Have It All?") points out conflicts also on the legislative level, but between different legislative acts and not necessarily related to Public Participation.

The next sub-part on Renewable Energy shows with regard to the sources of renewable energy some close connections to the previous one.

Marco Citelli (Chapter "Generating Renewable Energy for the Material Realization of Sustainable Development: What Do We Need from Multilateral Cooperation, the Climate Change and the International Trade Regimes?") indicates therein that states have historically been slow in embracing *consensus* on renewable energy options as shown by many soft law developments in the framework of the UN. He also points out some signals for a reversal of this trend that are being provided by the epistemic activities undertaken by the International Renewable Energy Agency encouraging the dissemination of information and knowledge and, to a certain extent, by a few proposals on renewables put forward during the last round of climate negotiations. Furthermore, Citelli calls for the inception of international legal instruments in this field, also due to ongoing legal conflicts.

One of such legal conflicts is addressed by the next contribution. Thomas Dromgool and Daniel Ybarra Enguix (Chapter "The Fair and Equitable Treatment Standard and the Revocation of Feed in Tariffs—Foreign Renewable Energy Investments in Crisis-struck Spain") find in their assessment of the Spanish case of disruptive cuts in tariff regulation supporting photovoltaic energy that the investors may rely on the stability of the Spanish *régimen especial*, despite their own conduct and the State's right to regulate. Even in times of crisis such measures could constitute a breach of legitimate expectations as the very raison d'être of the initial investment is stripped away.

David Grinlinton (Chapter "Horizontal and Vertical Integration of Sustainability into Policymaking, Planning and Implementation of Renewable Energy Projects—The New Zealand Model") finds that the New Zealand approach of incorporating the principle of sustainability influences policy-making and "macro-planning" at the national and regional levels, which in turn influences lower level planning and operational decision-making. According to his findings, the system is integrated both vertically between different levels of government (central, regional and municipal), and horizontally between central and local government and resource management agencies, corporations, public interest groups and individuals.

Christopher Frey contributes the final Chapter "Tackling Climate Change Through the Elimination of Trade Barriers for Low-Carbon Goods: Multilateral, Plurilateral and Regional Approaches" to the long sectorial sub-part on Renewable Energy. He provides an overview of the existing trade barriers and the efforts to eliminate these barriers for renewable energy goods and energy efficient products. Furthermore, Frey identifies the liberalization potentials and main challenges in the multilateral, pluri-lateral and regional contexts. Perhaps surprisingly, according to his findings Mega-Regional Trade Agreements are identified as the most promising venue therefore.

While the first two chapters of this sub-part focus on sources of renewable energy, the third part concentrates on policy making and macroplanning in spatial terms and the fourth one envisages renewable energy goods and energy efficient products. All in common have conflicts to face either based on a mainly soft law basis internationally (Citelli, Chapter "Generating Renewable Energy for the Material Realization of Sustainable Development: What Do We Need from Multilateral Cooperation, the Climate Change and the International Trade Regimes?") combined with contradictory national law-making (Dromgool and Ybarra Enguix, Chapter "The Fair and Equitable Treatment Standard and the Revocation of Feed in Tariffs—Foreign Renewable Energy Investments in Crisis-struck Spain"), conflicting interests of other policy sectors or of individuals (Grinlington, Chapter "Horizontal and Vertical Integration of Sustainability into Policymaking, Planning and Implementation of Renewable Energy Projects—The New Zealand Model") or conflicts with withstanding international "hard law" (Frey, Chapter "Tackling Climate Change Through the Elimination of Trade Barriers for Low-Carbon Goods: Multilateral, Plurilateral and Regional Approaches").

The conclusions of the next sectorial sub-part on Cities, Waste and Material Management differ considerably among each other but can be again compared in a meaningful sense based on effectiveness and efficiency aspects (Mauerhofer 2008).

Ying Yin (Chapter "Environmental Integration in China's Eco-City Development—From an Institutional Perspective") concludes after her legal assessment of the environmental integration in China's Eco-City development that despite some progress, the absence of a holistic perspective and effective guidance and constraints in China's environmental and planning formal rules at different levels appeared as an institutional weakness. Whether environmental requirements and their status were clearly stated in formal rules also considerably affected the enforcement. She also found an important role of actors, such as local leaders and governments.

Rob Hoogmartens, Maarten Dubois and Steven Van Passel, the authors of the next Chapter "Identifying the Interaction Between Landfill Taxes and NIMBY. A Simulation for Flanders (Belgium) Using a Dynamic Optimization Model", find in their simulation of two different scenarios for landfill paths and price paths in Flanders (Belgium) that, when landfill taxes are legally levied, it takes 42 years for landfill exhaustion to occur. Without taxes, this period would be shortened to only 20 years. In addition, they found that the use of taxes increases discounted total welfare significantly. Thus, they conclude that the added value of a tax is considerable in terms of welfare gains.

Thomas J. de Römph (Chapter "Pressing Forward—Developments in the Transition Towards Sustainable Materials Management in EU Environmental Law") finds in his final chapter of this sectorial sub-part that despite of a variety of constructive policies during the past fifteen and fruitful attempts to adjust particular laws, the legal framework on EU's Sustainable Use of Materials as a whole is still lagging behind policy. Nevertheless, he identifies several developments that enhance the legal transition which just started according to his findings (such as a new strategic approach, the Better Regulation and the Circular Economy Packages and an increasing legal emphasis on life cycle of material).

The first and the third chapter of this past sectorial sub-part on Cities, Waste and Material Management look stronger on the effectiveness side of environmental law towards absolute reduction of environmental pressures (Ying, Chapter "Environmental Integration in China's Eco-City Development—From an Institutional Perspective"; de Römph, Chapter "Developments in the Transition Towards Sustainable Materials Management in EU Environmental Law"). While the second part (Hoogmartens, Dubois and Chapter "An Empirical Investigation of Supportive Legal Frameworks for Social Enterprises in Belgium: A Cross-Sectoral Comparison of Case Studies for Social Enterprises from the Social Housing, Finance and Energy Sector Perspective") address the topic more efficiency-based in the sense of an efficient use of sink resources. However the role of a possible rebound effect based on potential additional economic growth in terms of a constantly increased throughput of resources and use of sinks is yet to be assessed (see also Mauerhofer, Chapter "3-D Sustainability and its Contribution to Governance Assessment in Legal Terms: Examples and Perspectives").

The conclusions now again return stronger to the natural capital basis of sustainable development (Mauerhofer 2008) with the next sub-part IX covering policy issues related to Biodiversity, Nature conservation, Oceans and Spatial Planning within this second main part on sectorial policy issues.

Frederik H. Kistenkas (Chapter "Sustainable Development: New Thoughts, New Policy, New Law?") concludes that much of EU's nature conservation and planning legislation predates the common understandings of sustainability and might be able to obstruct sustainable development. He advocates that sustainable growth, usually seen in terms of development for which assets and impacts for ecology, economy and society are brought in balance, should be facilitated by future law rather than being obstructed. He also presents some new improvement directions based on the concept of ecosystem services.

Glen Wright, Julien Rochette and Thomas Greiber (Chapter "Sustainable Development of the Oceans: Closing the Gaps in the International Legal Framework") find that, despite a wide range of international legal instruments related to oceans, in particular the areas beyond national jurisdiction (ABNJ) lack an overarching regulatory framework, with no provisions for marine protected areas, environmental impact assessment, or access and benefit sharing in relation to marine genetic resources. They also identify gaps and weaknesses in the international framework for the exploitation of offshore oil and gas resources as well as outline relevant ongoing processes to fill them, and propose ways forward.

Finally in this sub-part, Judith Preston (Chapter "Knowledge in Sustainable Resource Management in Australia") finds that Indigenous Knowledge (IK) is not sufficiently acknowledged and protected in Australian sustainability legislation. While she shows based on her three protected areas cases that IK is proving to be beneficial in the long term for achieving positive environmental and cultural outcomes. Thus she proposes for Australia to develop a sui generis legal framework to acknowledge, respect and give effect to use of IK for environmental objectives and the thriving of indigenous cultural heritage.

While the first chapter of this sub-part (Kistenkas, Chapter "Sustainable Development: New Thoughts, New Policy, New Law?") indicates insufficient current legal and judicial approaches to address some new sustainability related concepts, such as sustainable growth and ecosystem services, the other two contributions (Wright, Rochette and Greiber, Chapter "Sustainable Development of the Oceans: Closing the Gaps in the International Legal Framework" and Preston, Chapter "Knowledge in Sustainable Resource Management in Australia") find the absence of a (sufficient) basic legal framework to protect marine environment and IK in connection with protected areas respectively.

For Agriculture and Rural Policy (part X), the last sub-part of the main part on sectorial policy issues two chapters present their findings.

Marcia Fajardo Cavalcanti de Albuquerque (Chapter "The Sustainable Use of Biodiversity and Its Implications in Agriculture: The Agroforestry Case in the Brazilian Legal Framework") shows that according to the green economy's theory of strong sustainability, the agroforestry way of production represents a sustainable agricultural system because the quality and the quantity of each component of the natural stock is maintained in its original state. Nevertheless, she concludes that, regarding the Brazilian context, the potential of the agroforestry systems is constraint by the lack of supportive regulatory framework.

Paul Martin and Jacqueline Williams (Chapter "Next Generation Rural natural Resource Governance: A Careful Diagnosis") provide a next generation governance model by mainly offering different opportunities of improved actions. They highlight that innovation in rural natural resource governance (and in other areas of rural governance) is fundamental and could lead to better outcomes. Further possible catalysts for significant change pointed out by them are greater involvement of industry in natural resource governance and stronger involvement of rural communities therein (which relates to the horizontal parts III on communication and IV on public participation).

In comparison to the other sub-part IX particularly dealing with biodiversity, Fajardo Cavalcanti de Albuquerque (Chapter "The Sustainable Use of Biodiversity and its Implications in Agriculture: The Agroforestry Case in the Brazilian Legal Framework") as well as Martin and Williams (Chapter "Next Generation Rural natural Resource Governance: A Careful Diagnosis") mainly cover ecological systems that are defined by and often dependent on a certain type of human use (similar to the one covered by Preston's Chapter "Knowledge in Sustainable Resource Management in Australia"). While the chapters of Kistenkas (Chapter "Sustainable Development: New Thoughts, New Policy, New Law?") and—in particular—of Wright, Rochette and Greiber (Chapter "Sustainable Development of the Oceans: Closing the Gaps in the International Legal Framework)" deal with ecological systems that do not (necessarily) have a certain extent of human use as a precondition for their existence.

These final thoughts brings us back to a core issue of the sustainable development which is the differentiation between an anthropocentric vision wherein humans are central versus an ecocentric vision that sees humans as a part of the environmental system, embedded in and dependent on environmental assets (Mauerhofer 2008).

The recently adopted 17 Sustainable Development Goals and 169 targets (UN 2015) provide substantial guidance for both visions and law is meant to ongoing provide substantial support to achieve the appropriate trade-offs and prioritisations of interests among these two visions in connection with the environmental, social and economic dimensions of sustainable development.

References

Francioni F (2010) International human rights in an environmental horizon. Eur J Int Law 21 (1):41–55

Mauerhofer V (2008) 3-D sustainability: an approach for priority setting in situation of conflicting interests towards a sustainable development. Ecol Econ 66:496–506

Mauerhofer V (2015) Public participation in environmental matters: compendium, challenges and chances globally, land use policy (in press)

UN (2015) Transforming our world: the 2030 agenda for sustainable development. Finalised text for adoption (1 August). Retrieved 14 Aug 2015. https://sustainabledevelopment.un.org/content/documents/7891TRANSFORMING%20OUR%20WORLD.pdf

Index

V. Mauerhofer (ed.), *Legal Aspects of Sustainable Development*,
DOI 10.1007/978-3-319-26021-1

H

I

J

K

L

FSC
www.fsc.org
MIX
Papier | Fördert
gute Waldnutzung
FSC® C083411

Kolibri 360 GmbH
Ferdinand-Jühlke-Straße 7
99095 Erfurt, Deutschland
produktsicherheit@kolibri360.de